# The Design of Active Crossovers

The use of active crossovers is increasing. They are used by almost every sound reinforcement system, and by almost every recording studio monitoring set-up. There is also a big usage of active crossovers in car audio, with the emphasis on routing the bass to enormous low-frequency loudspeakers. Active crossovers are used to a small but rapidly growing extent in domestic hifi, and I argue that their widespread introduction may be the next big step in this field.

*The Design of Active Crossovers* has now been updated and extended for the Second Edition, taking in developments in loudspeaker technology and crossover design. Many more pre-designed filters are included so that crossover development can be faster and more certain, and the result will have a high performance. The Second Edition continues the tradition of the first in avoiding complicated algebra and complex numbers, with the mathematics reduced to the bare minimum; there is nothing more complicated to grapple with than a square root.

New features of the Second Edition include:

*   More on loudspeaker configurations and their crossover requirements:
    MTM Mid-Tweeter-Mid configurations (The d'Appolito arrangement)
    Line arrays (J arrays) for sound reinforcement
       Frequency tapering
       Band zoning
       Power tapering
    Constant-Beamwidth Transducer (CBT) loudspeaker arrays

*   More on specific sound-reinforcement issues like the loss of high frequencies due to the absorption of sound in air and how it varies.
*   Lowpass filters now have their own separate chapter.

    Much more on third, fourth, fifth, and sixth-order lowpass filters.
    Many more examples are given with component values ready-calculated

*   Highpass filters now have their own separate chapter, complementary to the chapter on lowpass filters.
    Much more on third, fourth, fifth, and sixth-order highpass filters.
    Many more examples are given with component values ready-calculated

- A new chapter dealing with filters other than the famous Sallen & Key type. New filter types are introduced such as the third-order multiple feedback filter.
  There is new information on controlling the Q and gain of state-variable filters.

- More on the performance of crossover filters, covering noise, distortion, and the internal overload problems of filters.

- The chapter on bandpass and notch filters is much extended, with in-depth coverage of the Bainter filter, which can produce beautifully deep notches without precision components or adjustment.

- Much more information on the best ways to combine standard components to get very accurate non-standard values. Not only can you get a very accurate nominal value, but also the effective tolerance of the combination can be significantly better than that of the individual components used. There is no need to keep huge numbers of resistor and capacitor values in stock.

- More on low-noise high-performance balanced line inputs for active crossovers, including versions that give extraordinarily high common-mode rejection. (noise rejection)

- Two new appendices giving extensive lists of crossover patents, and crossover-based articles in journals.

This book is packed full of valuable information, with virtually every page revealing nuggets of specialized knowledge never before published. Essential points of theory bearing on practical performance are lucidly and thoroughly explained, with the mathematics kept to an essential minimum. Douglas' background in design for manufacture ensures he keeps a very close eye on the cost of things.

**Douglas Self** studied engineering at Cambridge University, then psychoacoustics at Sussex University. He has spent many years working at the top level of design in both the professional audio and hi-fi industries and has taken out a number of patents in the field of audio technology. He currently acts as a consultant engineer in the field of audio design.

Douglas Self maintains a website at douglas-self.com

# The Design of Active Crossovers

## Second Edition

Douglas Self

Routledge
Taylor & Francis Group

NEW YORK AND LONDON

Second edition published 2018
by Routledge
711 Third Avenue, New York, NY 10017

and by Routledge
2 Park Square, Milton Park, Abingdon, Oxon, OX14 4RN

*Routledge is an imprint of the Taylor & Francis Group, an informa business*

First edition published by Focal Press 2011

*Library of Congress Cataloging-in-Publication Data*
Names: Self, Douglas, author.
Title: The design of active crossovers / Douglas Self.
Description: Second edition. | New York, NY : Routledge, 2018.
Identifiers: LCCN 2017048835 | ISBN 9781138733022
    (hardback : alk. paper) | ISBN 9781138733039 (pbk. : alk. paper) |
    ISBN 9781315187891 (ebook)
Subjects: LCSH: Electric filters, Active—Design and construction. |
    Bridge circuits—Design and construction.
Classification: LCC TK7872.F5 S447 2018 | DDC 621.3815/324—dc23
LC record available at https://lccn.loc.gov/2017048835

ISBN: 978-1-138-73302-2 (hbk)
ISBN: 978-1-138-73303-9 (pbk)
ISBN: 978-1-315-18789-1 (ebk)

Typeset in Times New Roman
by Apex CoVantage, LLC

*To Julie, with all my love*

# Contents

Acknowledgments ........................................................................................ xxi

Preface ...................................................................................................xxiii

**Chapter 1: Crossover Basics**............................................................................ 1

What a Crossover Does ...................................................................... 1

Why a Crossover Is Necessary .............................................................. 1

Beaming and Lobing............................................................................ 2

Passive Crossovers .............................................................................. 4

Active Crossover Applications .............................................................. 5

Bi-Amping and Bi-Wiring .................................................................... 6

Loudspeaker Cables............................................................................ 8

The Advantages and Disadvantages of Active Crossovers............................ 9

    The Advantages of Active Crossovers .............................................9

    Some Illusory Advantages of Active Crossovers ...........................13

    The Disadvantages of Active Crossovers........................................14

The Next Step in Hi-Fi ...................................................................... 16

Active Crossover Systems .................................................................. 16

Matching Crossovers and Loudspeakers ............................................... 20

A Modest Proposal: Popularising Active Crossovers ............................... 21

Multi-Way Connectors........................................................................ 22

Subjectivism .................................................................................... 23

**Chapter 2: How Loudspeakers Work** .......................................................... **25**

Sealed-Box Loudspeakers .................................................................. 26

Reflex (Ported) Loudspeakers............................................................. 27

Auxiliary Bass Radiator (ABR) Loudspeakers......................................... 28

Transmission Line Loudspeakers ......................................................... 28

Horn Loudspeakers............................................................................ 29

Electrostatic Loudspeakers ................................................................. 29

Ribbon Loudspeakers ........................................................................ 30

Electromagnetic Planar Loudspeakers................................................... 30

Air-Motion Transformers.................................................................... 31

Plasma Arc Loudspeakers................................................................... 31

The Rotary Woofer .......................................................................... 32

MTM Tweeter-Mid Configurations (d'Appolito) ..................................... 33

# Contents

Vertical Line Arrays................................................................. 34
    Line Array Amplitude Tapering .......................................37
    Line Array Frequency Tapering .......................................37
CBT Line Arrays.................................................................... 39
Diffraction............................................................................. 39
Sound Absorption in Air ...................................................... 44
Modulation Distortion .......................................................... 46
Drive Unit Distortion............................................................ 47
Doppler Distortion................................................................ 48
Further Reading on Loudspeaker Design .............................. 49

## Chapter 3: Crossover Requirements ................................. 51
General Crossover Requirements ........................................... 51
    1 Adequate Flatness of Summed Amplitude/Frequency Response On-Axis ...........51
    2 Sufficiently Steep Roll-Off Slopes Between the Filter Outputs...........................51
    3 Acceptable Polar Response .........................................52
    4 Acceptable Phase Response ........................................53
    5 Acceptable Group Delay Behaviour.............................53
Further Requirements for Active Crossovers ......................... 54
    1 Negligible Extra Noise .............................................54
    2 Negligible Impairment of System Headroom ...............55
    3 Negligible Extra Distortion ........................................55
    4 Negligible Impairment of Frequency Response ...........56
    5 Negligible Impairment of Reliability ..........................56
Linear Phase.......................................................................... 56
Minimum Phase..................................................................... 57
Absolute Phase ..................................................................... 57
Phase Perception................................................................... 58
Target Functions ................................................................... 59

## Chapter 4: Crossover Types ........................................... 61
All-Pole and Non-All-Pole Crossovers ................................. 61
Symmetric and Asymmetric Crossovers................................. 62
Allpass and Constant-Power Crossovers ............................... 62
Constant-Voltage Crossovers................................................. 63
First-Order Crossovers.......................................................... 63
First-Order Solen Split Crossover ........................................ 69
First-Order Crossovers: 3-Way ............................................. 70
Second-Order Crossovers ..................................................... 70
    Second-Order Butterworth Crossover...........................71
    Second-Order Linkwitz-Riley Crossover .......................78
    Second-Order Bessel Crossover ...................................79
    Second-Order 1.0 dB-Chebyshev Crossover .................80

Third-Order Crossovers ......................................................................... 83
    Third-Order Butterworth Crossover ................................................84
    Third-Order Linkwitz-Riley Crossover ..........................................86
    Third-Order Bessel Crossover .........................................................89
    Third-Order 1.0 dB-Chebyshev Crossover ....................................89
Fourth-Order Crossovers ..................................................................... 92
    Fourth-Order Butterworth Crossover ............................................93
    Fourth-Order Linkwitz-Riley Crossover ........................................95
    Fourth-Order Bessel Crossover ......................................................99
    Fourth-Order 1.0 dB-Chebyshev Crossover ..................................99
    Fourth-Order Linear-Phase Crossover ........................................101
    Fourth-Order Gaussian Crossover ...............................................103
    Fourth-Order Legendre Crossover ...............................................106
Higher-Order Crossovers ................................................................... 108
Determining Frequency Offsets ....................................................... 109
Filler-Driver Crossovers ......................................................................111
The Duelund Crossover ..................................................................... 113
Crossover Topology ........................................................................... 113
Crossover Conclusions ...................................................................... 118

**Chapter 5: Notch Crossovers ........................................................... 121**
Elliptical Filter Crossovers ................................................................ 121
Neville Thiele Method™ (NTM) Crossovers ................................... 125

**Chapter 6: Subtractive Crossovers .................................................. 131**
Subtractive Crossovers ...................................................................... 131
First-Order Subtractive Crossovers .................................................. 132
Second-Order Butterworth Subtractive Crossovers ........................ 133
Third-Order Butterworth Subtractive Crossovers .......................... 135
Fourth-Order Butterworth Subtractive Crossovers ......................... 135
Subtractive Crossovers With Time Delays ....................................... 137
Performing the Subtraction ............................................................... 141

**Chapter 7: Lowpass and Highpass Filter Characteristics .............. 145**
Active Filters ...................................................................................... 145
Lowpass Filters ................................................................................... 146
Highpass Filters .................................................................................. 146
Bandpass Filters ................................................................................. 146
Notch Filters ....................................................................................... 146
Allpass Filters ..................................................................................... 147
All-Stop Filters ................................................................................... 147
Brickwall Filters ................................................................................. 147
The Order of a Filter .......................................................................... 147

Filter Cutoff Frequencies and Characteristic Frequencies ........................ 148
First-Order Filters ............................................................................. 148
Second-Order and Higher-Order Filters ................................................ 149
Filter Characteristics ......................................................................... 149
   Amplitude Peaking and $Q$ ............................................................ 150
   Butterworth Filters .................................................................. 151
   Linkwitz-Riley Filters .............................................................. 153
   Bessel Filters ........................................................................ 155
   Chebyshev Filters .................................................................. 160
   1 dB-Chebyshev Lowpass Filter ............................................... 162
   3 dB-Chebyshev Lowpass Filter ............................................... 162
Higher-Order Filters .......................................................................... 167
   Butterworth Filters up to 8th-Order .......................................... 168
   Linkwitz-Riley Filters up to 8th-Order ....................................... 172
   Bessel Filters up to 8th-Order ................................................. 173
   Chebyshev Filters up to 8th-Order ........................................... 175
More Complex Filters—Adding Zeros .................................................. 176
   Inverse Chebyshev Filters (Chebyshev Type II) ......................... 177
   Elliptical Filters (Cauer Filters) ................................................ 179
Some Lesser-Known Filter Characteristics ........................................... 182
   Transitional Filters ................................................................ 182
   Linear-Phase Filters ............................................................... 182
   Gaussian Filters .................................................................... 184
   Legendre-Papoulis Filters ....................................................... 187
   Laguerre Filters .................................................................... 190
   Synchronous Filters ............................................................... 190
Other Filter Characteristics ............................................................... 193

**Chapter 8: Designing Lowpass Filters: Sallen & Key ........................... 195**
Designing Real Filters ....................................................................... 195
Component Sensitivity ....................................................................... 195
First-Order Lowpass Filters ................................................................ 197
Second-Order Filters ......................................................................... 198
Sallen & Key 2nd-Order Lowpass Filters ............................................. 198
Sallen & Key Lowpass Filter Components ............................................. 199
Sallen & Key 2nd-Order Lowpass: Unity Gain ...................................... 199
Sallen & Key 2nd-Order Lowpass Unity Gain: Component Sensitivity ...... 202
Filter Frequency Scaling .................................................................... 202
Sallen & Key 2nd-Order Lowpass: Equal Capacitor ............................... 204
Sallen & Key 2nd-Order Lowpass Equal-C: Component Sensitivity .......... 206
Sallen & Key 2nd-Order Butterworth Lowpass: Defined Gains ................ 207
Sallen & Key 2nd-Order Lowpass: Non-Equal Resistors ......................... 207
Sallen & Key 2nd-Order Lowpass: Optimisation .................................... 209

Sallen & Key 3rd-Order Lowpass: Two Stages.................................................. 209
Sallen & Key 3rd-Order Lowpass: Single Stage .............................................. 210
Sallen & Key 3rd-Order Lowpass in a Single Stage: Non-Equal Resistors............ 214
Sallen & Key 4th-Order Lowpass: Two Stages .................................................. 215
Sallen & Key 4th-Order Lowpass: Single-Stage Butterworth............................ 217
Sallen & Key 4th-Order Lowpass: Single-Stage Linkwitz-Riley........................ 221
Sallen & Key 4th-Order Lowpass: Single Stage With
    Non-Equal Resistors ................................................................................ 224
Sallen & Key 4th-Order Lowpass: Single Stage With Other
    Filter Characteristics ............................................................................... 224
Sallen & Key 5th-Order Lowpass: Three Stages ............................................... 225
Sallen & Key 5th-Order Lowpass: Two Stages ................................................. 228
Sallen & Key 5th-Order Lowpass: Single Stage ............................................... 229
Sallen & Key 6th-Order Lowpass: Three Stages ............................................... 230
Sallen & Key 6th-Order Lowpass: Single Stage ............................................... 232
Sallen & Key Lowpass: Input Impedance ........................................................ 234
Linkwitz-Riley Lowpass With Sallen & Key Filters: Loading Effects .................. 234
Lowpass Filters With Attenuation ................................................................. 236
Bandwidth Definition Filters ......................................................................... 237
Bandwidth Definition: Butterworth Versus Bessel .......................................... 237
Variable-Frequency Lowpass Filters: Sallen & Key ......................................... 239

**Chapter 9: Designing Highpass Filters ......................................................... 241**
First-Order Highpass Filters .......................................................................... 241
Sallen & Key 2nd-Order Filters...................................................................... 242
Sallen & Key 2nd-Order Highpass Filters....................................................... 242
Sallen & Key Highpass Filter Components...................................................... 243
Sallen & Key 2nd-Order Highpass: Unity Gain .............................................. 243
Sallen & Key 2nd-Order Highpass: Equal Resistors........................................ 244
Sallen & Key 2nd-Order Butterworth Highpass: Defined Gains...................... 245
Sallen & Key 2nd-Order Highpass: Non-Equal Capacitors ............................. 247
Sallen & Key 3rd-Order Highpass: Two Stages ............................................... 248
Sallen & Key 3rd-Order Highpass in a Single Stage........................................ 249
Sallen & Key 4th-Order Highpass: Two Stages ............................................... 251
Sallen & Key 4th-Order Highpass: Butterworth in a Single Stage .................... 252
Sallen & Key 4th-Order Highpass: Linkwitz-Riley in a Single Stage ................ 254
Sallen & Key 4th-Order Highpass: Single-Stage With Other
    Filter Characteristics ............................................................................... 256
Sallen & Key 5th-Order Highpass: Three Stages ............................................. 257
Sallen & Key 5th-Order Butterworth Filter: Two Stages ................................. 259
Sallen & Key 5th-Order Highpass: Single Stage.............................................. 260
Sallen & Key 6th-Order Highpass: Three Stages ............................................. 260
Sallen & Key 6th-Order Highpass: Single Stage.............................................. 262

Sallen & Key Highpass: Input Impedance................................................262
Bandwidth Definition Filters ...............................................................262
Bandwidth Definition: Subsonic Filters ................................................263
Bandwidth Definition: Combined Ultrasonic and Subsonic Filters ........264
Variable-Frequency Highpass Filters: Sallen & Key...............................267

*Chapter 10: Other Lowpass and Highpass Filters............................ 271*
Designing Filters..................................................................................271
Multiple-Feedback Filters.....................................................................272
Multiple-Feedback 2nd-Order Lowpass Filters.......................................273
Multiple-Feedback 2nd-Order Highpass Filters .....................................274
Multiple-Feedback 3rd-Order Filters.....................................................274
    Multiple-Feedback 3rd-Order Lowpass Filters ....................................275
    Multiple-Feedback 3rd-Order Highpass Filters ...................................275
Biquad Filters .....................................................................................276
Akerberg-Mossberg Lowpass Filter ......................................................276
Akerberg-Mossberg Highpass Filters ....................................................279
Tow-Thomas Biquad Lowpass and Bandpass Filter ...............................280
Tow-Thomas Biquad Notch and Allpass Responses ..............................285
Tow-Thomas Biquad Highpass Filter ....................................................286
State-Variable Filters ...........................................................................288
Variable-Frequency Filters: State-Variable 2nd Order ...........................292
Variable-Frequency Filters: State-Variable 4th-Order ...........................293
Variable-Frequency Filters: Other Orders of State-Variable ...................295
Other Filters........................................................................................296

*Chapter 11: Lowpass and Highpass Filter Performance........................ 297*
Aspects of Filter Performance: Noise and Distortion...............................297
Distortion in Active Filters ...................................................................297
Distortion in Sallen & Key Filters: The Distortion Aggravation Factor .................298
Distortion in Sallen & Key Filters: Looking for DAF................................303
Distortion in Sallen & Key Filters: 2nd-Order Lowpass..........................305
Distortion in Sallen & Key Filters: 2nd-Order Highpass .........................308
Mixed Capacitors in Low-Distortion 2nd-Order Sallen & Key Filters .................310
Distortion in Sallen & Key Filters: 3rd-Order Lowpass Single Stage..................311
Distortion in Sallen & Key Filters: 3rd-Order Highpass Single Stage.................312
Distortion in Sallen & Key Filters: 4th-Order Lowpass Single Stage..................314
Distortion in Sallen & Key Filters: 4th-Order Highpass Single Stage.................317
Distortion in Sallen & Key Filters: Simulations......................................318
Distortion in Sallen & Key Filters: Capacitor Conclusions......................320
Distortion in Multiple-Feedback Filters: The Distortion Aggravation Factor.........321
Distortion in Multiple-Feedback Filters: 2nd-Order Lowpass ..................322
Distortion in Multiple-Feedback Filters: 2nd-Order Highpass.................323

Distortion in Tow-Thomas Filters: 2nd-Order Lowpass.............................................. 324
Distortion in Tow-Thomas Filters: 2nd-Order Highpass............................................ 326
Noise in Active Filters .............................................................................................. 327
Noise and Bandwidth................................................................................................ 328
Noise in Sallen & Key Filters: 2nd-Order Lowpass ................................................. 329
Noise in Sallen & Key Filters: 2nd-Order Highpass ................................................ 330
Noise in Sallen & Key Filters: 3rd-Order Lowpass Single Stage ............................ 330
Noise in Sallen & Key Filters: 3rd-Order Highpass Single Stage............................ 330
Noise in Sallen & Key Filters: 4th-Order Lowpass Single Stage ............................ 331
Noise in Sallen & Key Filters: 4th-Order Highpass Single Stage............................ 331
Noise in Multiple-Feedback Filters: 2nd-Order Lowpass ........................................ 331
Noise in Multiple-Feedback Filters: 2nd-Order Highpass........................................ 332
Noise in Tow-Thomas Filters .................................................................................... 332

**Chapter 12: Bandpass and Notch Filters .................................................. 333**
Multiple-Feedback Bandpass Filters ........................................................................ 333
High-$Q$ Bandpass Filters ........................................................................................ 334
Notch Filters .............................................................................................................. 335
The Twin-T Notch Filter............................................................................................ 336
The 1-Bandpass Notch Filter ..................................................................................... 337
The Bainter Notch Filter ............................................................................................ 337
    Bainter Notch Filter Design ............................................................................339
    Bainter Notch Filter Example .........................................................................341
    An Elliptical Filter Using a Bainter Highpass Notch .....................................342
The Bridged-Differentiator Notch Filter ................................................................... 342
Boctor Notch Filters .................................................................................................. 343
Other Notch Filters .................................................................................................... 345
Simulating Notch Filters ............................................................................................ 345

**Chapter 13: Time-Delay Filters ............................................................... 347**
The Requirement for Delay Compensation ............................................................... 347
Calculating the Required Delays ............................................................................... 349
Signal Summation...................................................................................................... 352
Physical Methods of Delay Compensation................................................................ 353
Delay Filter Technology ............................................................................................ 355
Sample Crossover and Delay Filter Specification ..................................................... 355
Allpass Filters in General .......................................................................................... 355
First-Order Allpass Filters ......................................................................................... 356
Distortion and Noise in 1st-Order Allpass Filters..................................................... 361
Cascaded 1st-Order Allpass Filters............................................................................ 363
Second-Order Allpass Filters..................................................................................... 364
Distortion and Noise in 2nd-Order Allpass Filters ................................................... 368
Third-Order Allpass Filters........................................................................................ 369

Distortion and Noise in 3rd-Order Allpass Filters.................................370
Higher-Order Allpass Filters........................................................373
Delay Lines for Subtractive Crossovers ............................................379
Variable Allpass Time Delays.......................................................381
Lowpass Filters for Time Delays....................................................382

**Chapter 14: Equalisation.............................................................385**

The Need for Equalisation..........................................................385
What Equalisation Can and Can't Do.................................................386
Loudspeaker Equalisation...........................................................387
1 Drive Unit Equalisation.........................................................387
2 6 dB/octave Dipole Equalisation.................................................388
3 Bass Response Extension ........................................................388
4 Diffraction Compensation Equalisation ..........................................389
5 Room Interaction Correction ....................................................390
Equalisation Circuits..............................................................393
HF-Boost and LF-Cut Equaliser.....................................................393
HF-Cut and LF-Boost Equaliser.....................................................395
Combined HF-Boost and HF-Cut Equaliser............................................398
Adjustable Peak/Dip Equalisers: Fixed Frequency and Low $Q$ ......................398
Adjustable Peak/Dip Equalisers: Variable Centre Frequency and Low $Q$ ..............400
Adjustable Peak/Dip Equalisers With High $Q$......................................402
Parametric Equalisers .............................................................405
The Bridged-T Equaliser ...........................................................406
The Biquad Equaliser ..............................................................407
Capacitance Multiplication for the Biquad Equaliser...............................414
Equalisers With Non-Standard Slopes................................................415
    Equalisers With −3 dB/Octave Slopes ..........................................415
    Equalisers With −3 dB/Octave Slopes Over Limited Range........................419
    Equalisers With −4.5 dB/Octave Slopes ........................................420
    Equalisers With Other Slopes..................................................420
Equalisation by Filter Frequency Offset...........................................421
Equalisation by Adjusting All Filter Parameters ..................................422

**Chapter 15: Passive Components for Active Crossovers...........................425**

Component Values...................................................................425
Resistors..........................................................................425
    Through-Hole Resistors .......................................................426
    Surface-Mount Resistors ......................................................427
Resistors: Values and Tolerances .................................................428
Improving Accuracy With Multiple Components: Gaussian Distribution ...............431
Resistor Value Distributions .....................................................435

Improving Accuracy With Multiple Components: Uniform Distribution ................. 437
Obtaining Arbitrary Resistance Values ..................................................................... 438
Other Resistor Combinations ................................................................................... 439
Resistor Noise: Johnson and Excess Noise ............................................................. 441
Resistor Non-Linearity ............................................................................................. 443
Capacitors: Values and Tolerances .......................................................................... 445
Obtaining Arbitrary Capacitance Values ................................................................. 446
Capacitor Shortcomings ........................................................................................... 447
Non-Electrolytic Capacitor Non-Linearity ............................................................. 449
Electrolytic Capacitor Non-Linearity ...................................................................... 454

**Chapter 16: Opamps for Active Crossovers ..................................................... 459**
Active Devices for Active Crossovers ...................................................................... 459
Opamp Types ............................................................................................................ 460
Opamp Properties: Noise .......................................................................................... 460
Opamp Properties: Slew Rate ................................................................................... 462
Opamp Properties: Common-Mode Range .............................................................. 463
Opamp Properties: Input Offset Voltage ................................................................. 463
Opamp Properties: Bias Current ............................................................................... 463
Opamp Properties: Cost ............................................................................................ 464
Opamp Properties: Internal Distortion ..................................................................... 465
Opamp Properties: Slew Rate Limiting Distortion .................................................. 466
Opamp Properties: Distortion Due to Loading ........................................................ 466
Opamp Properties: Common-Mode Distortion ........................................................ 467
Opamps Surveyed ..................................................................................................... 468
The TL072 Opamp .................................................................................................... 469
The NE5532 and 5534 Opamps ................................................................................ 471
    The 5532 With Shunt Feedback .......................................................................... 471
    5532 Output Loading in Shunt-Feedback Mode ................................................ 472
    The 5532 With Series Feedback .......................................................................... 474
    Common-Mode Distortion in the 5532 .............................................................. 474
    Reducing 5532 Distortion by Output Stage Biasing .......................................... 477
    Which 5532? ........................................................................................................ 482
    The 5534 Opamp ................................................................................................. 483
The LM4562 Opamp ................................................................................................. 485
    Common-Mode Distortion in the LM4562 ........................................................ 486
The LME49990 Opamp ............................................................................................. 489
    Common-Mode Distortion in the LME49990 .................................................... 491
The AD797 Opamp .................................................................................................... 492
    Common-Mode Distortion in the AD797 ........................................................... 494
The OP27 Opamp ...................................................................................................... 494
Opamp Selection ....................................................................................................... 497

Contents

***Chapter 17: Active Crossover System Design*** ..................................................... **499**

   Crossover Features ............................................................................................... 499
       Input Level Controls .................................................................................... 499
       Subsonic Filters ........................................................................................... 499
       Ultrasonic Filters ......................................................................................... 500
       Output Level Trims ..................................................................................... 500
       Output Mute Switches, Output Phase-Reverse Switches .............................. 500
       Control Protection ....................................................................................... 500
   Features Usually Absent ...................................................................................... 501
       Metering ...................................................................................................... 501
       Relay Output Muting ................................................................................... 501
   Switchable Crossover Modes .............................................................................. 501
   Noise, Headroom, and Internal Levels ............................................................... 503
   Circuit Noise and Low-Impedance Design .......................................................... 504
   Using Raised Internal Levels .............................................................................. 504
   Placing the Output Attenuator ........................................................................... 506
   The Amplitude/Frequency Distribution of Musical Signals and
       Internal Levels ............................................................................................ 507
   Gain Structures .................................................................................................. 510
   Noise Gain ......................................................................................................... 513
   Active Gain Controls .......................................................................................... 514
   Filter Order in the Signal Path ........................................................................... 516
   Output Level Controls ........................................................................................ 518
   Mute Switches .................................................................................................... 519
   Phase-Invert Switches ......................................................................................... 520
   Distributed Peak Detection ................................................................................. 520
   Power Amplifier Considerations ......................................................................... 522

***Chapter 18: Subwoofer Crossovers*** ....................................................................... **525**

   Subwoofer Applications ...................................................................................... 525
   Subwoofer Technologies ..................................................................................... 525
       Sealed-Box (Infinite Baffle) Subwoofers ..................................................... 526
       Reflex (Ported) Subwoofers ........................................................................ 527
       Auxiliary Bass Radiator (ABR) Subwoofers ................................................ 527
       Transmission Line Subwoofers .................................................................... 528
       Bandpass Subwoofers .................................................................................. 528
       Isobaric Subwoofers .................................................................................... 529
       Dipole Subwoofers ...................................................................................... 529
       Horn-Loaded Subwoofers ........................................................................... 530
   Subwoofer Drive Units ....................................................................................... 530
   Hi-Fi Subwoofers ............................................................................................... 530
   Home Entertainment Subwoofers ....................................................................... 531
       Low-Level Inputs (Unbalanced) .................................................................. 532

Low-Level Inputs (Balanced)........................................................................532
High-Level Inputs .........................................................................................532
High-Level Outputs ......................................................................................533
Mono Summing.............................................................................................533
LFE Input .....................................................................................................533
Level Control................................................................................................533
Crossover In/Out Switch ..............................................................................534
Crossover Frequency Control (Lowpass Filter)...........................................534
Highpass Subsonic Filter..............................................................................534
Phase Switch (Normal/Inverted) ..................................................................534
Variable Phase Control .................................................................................535
Signal Activation Out of Standby .................................................................535
Home Entertainment Crossovers ......................................................................535
Fixed Frequency ...........................................................................................536
Variable Frequency .......................................................................................536
Multiple Variable ..........................................................................................536
Power Amplifiers for Home Entertainment Subwoofers ...................................536
Subwoofer Integration ......................................................................................537
Sound-Reinforcement Subwoofers....................................................................538
Line or Area Arrays ......................................................................................539
Cardioid Subwoofer Arrays ..........................................................................539
Aux-Fed Subwoofers .........................................................................................539
Automotive Audio Subwoofers .........................................................................540

**Chapter 19: Motional Feedback Loudspeakers .......................................... 543**
Motional Feedback Loudspeakers .....................................................................543
History ...............................................................................................................544
Feedback of Position..........................................................................................544
Feedback of Velocity .........................................................................................545
Feedback of Acceleration ..................................................................................548
Other MFB Speakers ..........................................................................................551
Published Projects..............................................................................................552
Conclusions .......................................................................................................552

**Chapter 20: Line Inputs .............................................................................. 555**
External Signal Levels........................................................................................555
Internal Signal Levels.........................................................................................555
Input Amplifier Functions..................................................................................556
Unbalanced Inputs .............................................................................................556
Balanced Interconnections .................................................................................559
The Advantages of Balanced Interconnections...................................................560
The Disadvantages of Balanced Interconnections..............................................560
Balanced Cables and Interference ......................................................................560
Balanced Connectors ..........................................................................................562

Balanced Signal Levels.................................................................................... 563
Electronic vs Transformer Balanced Inputs................................................... 563
Common-Mode Rejection Ratio (CMRR).................................................... 563
The Basic Electronic Balanced Input ........................................................... 565
Common-Mode Rejection Ratio: Opamp Gain.............................................. 568
Common-Mode Rejection Ratio: Opamp Frequency Response...................... 569
Common-Mode Rejection Ratio: Opamp CMRR .......................................... 570
Common-Mode Rejection Ratio: Amplifier Component Mismatches .............. 571
A Practical Balanced Input ............................................................................ 573
Variations on the Balanced Input Stage ......................................................... 576
Combined Unbalanced and Balanced Inputs................................................... 576
The Superbal Input ........................................................................................ 577
Switched-Gain Balanced Inputs .................................................................... 578
Variable-Gain Balanced Inputs ..................................................................... 580
The Self Variable-Gain Balanced Input......................................................... 581
High Input Impedance Balanced Inputs ......................................................... 582
The Instrumentation Amplifier ...................................................................... 583
Instrumentation Amplifier Applications ........................................................ 584
The Instrumentation Amplifier With 4x Gain................................................ 585
The Instrumentation Amplifier at Unity Gain................................................ 588
Transformer Balanced Inputs......................................................................... 590
Input Overvoltage Protection......................................................................... 592
Noise and Balanced Inputs ............................................................................ 593
Low-Noise Balanced Inputs .......................................................................... 593
Low-Noise Balanced Inputs in Real Life ....................................................... 598
Ultra-Low-Noise Balanced Inputs ................................................................ 598

**Chapter 21: Line Outputs ......................................................................... 603**
Unbalanced Outputs....................................................................................... 603
Zero-Impedance Outputs ............................................................................... 604
Ground-Cancelling Outputs............................................................................ 605
Balanced Outputs........................................................................................... 606
Transformer Balanced Outputs ...................................................................... 607
Output Transformer Frequency Response ...................................................... 608
Transformer Distortion .................................................................................. 609
Reducing Transformer Distortion .................................................................. 611

**Chapter 22: Power Supply Design .......................................................... 615**
Opamp Supply Rail Voltages......................................................................... 615
Designing a ±15 V Supply.............................................................................. 616
Designing a ±17 V Supply.............................................................................. 619
Using Variable-Voltage Regulators................................................................ 620
Improving Ripple Performance ...................................................................... 621

Dual Supplies From a Single Winding ........................................................... 622
Mutual Shutdown Circuitry ......................................................................... 623
Power Supplies for Discrete Circuitry .......................................................... 624

*Chapter 23: An Active Crossover Design* ......................................... **625**
Design Principles .......................................................................................... 625
Example Crossover Specification ................................................................. 625
The Gain Structure ...................................................................................... 626
Resistor Selection ........................................................................................ 627
Capacitor Selection ..................................................................................... 627
The Balanced Line Input Stage .................................................................... 627
The Bandwidth Definition Filter .................................................................. 628
The HF Path: 3 kHz Linkwitz-Riley Highpass Filter .................................... 628
The HF Path: Time-Delay Compensation .................................................... 631
The MID Path: Topology .............................................................................. 632
The MID Path: 400 Hz Linkwitz-Riley Highpass Filter ............................... 633
The MID Path: 3 kHz Linkwitz-Riley Lowpass Filter ................................... 634
The MID Path: Time-Delay Compensation .................................................. 634
The LF Path: 400 Hz Linkwitz-Riley Lowpass Filter .................................... 635
The LF Path: No Time-Delay Compensation ................................................ 636
Output Attenuators and Level Trim Controls ............................................... 636
Balanced Outputs ........................................................................................ 638
Crossover Programming ............................................................................... 638
Noise Analysis: Input Circuitry ................................................................... 639
Noise Analysis: HF Path .............................................................................. 640
Noise Analysis: MID Path ............................................................................ 641
Noise Analysis: LF Path ............................................................................... 642
Improving the Noise Performance: The MID Path ........................................ 642
Improving the Noise Performance: The Input Circuitry ............................... 643
The Noise Performance: Comparisons With Power Amplifier Noise .............. 646
Conclusion ................................................................................................... 647

*Appendix 1  Crossover Design References* .................................... **649**
*Appendix 2  US Crossover Patents* ................................................ **653**
*Appendix 3  Crossover and Loudspeaker Articles in* Wireless World/
                Electronics World ...................................................... **655**
*Appendix 4  Loudspeaker Design References* ................................ **657**
*Appendix 5  Component Series E3 to E96* ...................................... **659**
*Index* ............................................................................................... **661**

# *Acknowledgments*

My heartfelt thanks to:

Gareth Connor of The Signal Transfer Company for unfailing encouragement, providing the facilities with which some of the experiments in this book were done, and with much appreciation of our long collaboration in the field of audio.

# *Preface*

When the first edition of this book was published, I imagined that its rather specialised appeal might limit its sales compared with my other books on preamplifiers and power amplifiers. Rather remarkably, that has not proved to be the case, and so I am very happy to be able to offer a new and improved second edition. The book has been considerably expanded, growing from 19 to 23 chapters, and increasing in length by 30%.

The new material includes

- A much, much, broader palette of ready-made filter designs for crossover construction, including single-stage 4th-order Linkwitz-Riley filters, which makes it possible to build a 2-way 4th-order active crossover with one dual opamp (that's just one stereo channel, obviously). Other additions are many new ways to make 4th- and 5th-order filters, new 5th- and 6th-order highpass filters, and more advanced filters such as the Tow-Thomas biquad, which allow greater precision in defining the response.
- More on notch filters, especially full design details and procedures for the extremely flexible and useful Bainter notch filter.
- Economical ways of making high-order bandwidth definition filters.
- A big new chapter on distortion and noise in active filters, covering Sallen & Key, multiple-feedback, and Tow-Thomas filters.
- More on equalisation circuitry, including filter slopes of −3 dB and −4.5 dB/octave.
- More on different loudspeaker types and their specific crossover requirements. New stuff on speaker non-linearity and Doppler distortion, and sound absorption in air.
- A new section on MTM (d'Appolito) Tweeter-Mid configurations.
- A full explanation of vertical line arrays of loudspeakers, including J-arrays and amplitude and frequency tapering.
- A whole new chapter on motional feedback loudspeakers and interfacing them with active crossovers.
- Instrumentation amplifiers as low-noise balanced inputs. They have long been praised for having superb common-mode rejection, but this has been hard to exploit in audio applications.

I demonstrate how to do it and get lower noise at the same time. Also, what I modestly call the Self variable-gain balanced input, which gives excellent channel balance on stereo inputs without using expensive parts.

- Much more on the ingenious but little-known technology of ground-cancelling outputs, showing how they can give a noise advantage over conventional balanced interconnections. Cunning ways of substantially reducing line output transformer distortion at near-zero cost are described.
- New material on achieving awkward resistor and capacitor values.

However, what you most emphatically will *not* find here is any truck with the religious dogma of audio Subjectivism; the directional cables, the oxygen-free copper, the World War One vintage triodes still spattered with the mud of the Somme, and all the other depressing paraphernalia of pseudo-science and anti-science. I have spent more time than I care to contemplate in double-blind listening tests—properly conducted ones, with rigorous statistical analysis—and every time the answer was that if you couldn't measure it you couldn't hear it. Very often if you could measure it you still couldn't hear it. However, faith-based audio is not going away any time soon because few people (apart of course from the unfortunate customers) have any interest in it doing so. If you want to know more about my experiences and reasoning in this area, there is a full discussion in my book *Audio Power Amplifier Design* (Sixth edition).

As in the first edition of the current volume, the final implementation of each crossover design is analogue, using opamps. For the digital domain you just split off at the point where the frequencies and $Q$'s of the various filter stages are fixed. You then set your IIR filter coefficients and so on. Purely digital issues like word length and noise shaping are left for you to grapple with.

Also as in the first edition, you will find here the barest minimum of mathematics required to get a design job done. There is no complex algebra, no Laplace variables, and the square root of minus one is conspicuous by its absence. This approach is only possible because SPICE simulators are downloadable for free (e.g. LTspice) and can give you the frequency response, phase, and group delay of a filter before you could write down the first line of complex algebra. The design and analysis equations for 1st- and 2nd-order filters are given because they are simple, but beyond that the design procedures are long and complicated and often involve solving systems of equations rather than just plugging in numbers. Explaining them would take up far too much space, and also, some of them are part of my personal stock-in-trade as a consultant in the audio business. For all those reasons, those procedures are not described here.

Each filter is given as an example, usually with a cutoff frequency of 1 kHz. Full instructions are given in Chapter 8 for scaling the component values to give the desired frequency. Nothing more complicated than multiplication is required.

There is unquestionably a need for high-quality analogue circuitry. For example, a good microphone preamplifier needs a gain range from 0 to +80 dB if it is to get any signal it is likely to encounter up to a workable nominal value; there is little prospect of ever being able to connect an A-to-D converter directly to a microphone. If you are starting at line level and have an analogue power amplifier, and all you need is a relatively simple but high-quality active crossover, there is little incentive to convert to digital via an expensive ADC, perform the very straightforward arithmetic manipulations in the digital domain, then go back to analogue via an expensive DAC. Certainly you can do just about anything in the digital domain just by writing code, but you first have to get the signal safely *into* the digital domain. A little while ago I was involved in a telemetry project where sophisticated analogue processing, including allpass filters for time compensation, was required before the signal could be accurately digitised.

Therefore analogue circuitry is often the way to go. This book describes how to achieve high performance without spending a lot of money. As was remarked in a review of the first edition of *Active Crossover Design*, duplicating the performance attained here in the digital domain is not at all a trivial business.

In the pursuit of high quality at low cost, there are certain principles that pervade this book:

- Low-impedance design. This reduces the effects of Johnson noise and current noise without making voltage noise worse; the only downside is that a low impedance requires an opamp capable of driving it effectively, and sometimes more than one. This is a small price to pay; since an active crossover is an extra piece of equipment inserted into the signal path, it needs to be as transparent as possible, and low noise is an important aspect of this.
- A sophisticated approach to implementing awkward component values. Active crossover design— and filter design in general—is a prime example of a requirement for non-standard component values (RIAA networks for phono preamplifiers, as well as variable equalisers, are some other important applications in the audio business). Very often in filter design, only one component can be chosen to be a preferred value from the widely used E24 resistor series. While resistors are freely available in the E96 series, if you are faced with a required value that is effectively random, you will do much better by using two E24 values in series, or, preferably, in parallel. The nominal value will on average be three times more accurate than for E96. If you go a little further and use three E24 values combined, the accuracy of the nominal value is further improved by ten times. This may appear a bit profligate with components, but resistors are cheap. There is another advantage to the multiple component approach that is less obvious: the effective tolerance is reduced. By how much depends on how much the values vary; the maximum advantage with 2xE24 is $\sqrt{2}$ times and with 3xE24 is $\sqrt{3}$ times. The problem is more acute with capacitors as they are not often available in a closely spaced series like E24. They are also more expensive.
- The principle of optimisation. This may sound grand, but what it comes down to is closely scrutinising each circuit block is to see if it is possible to improve it by doing a bit more thinking rather than a bit more spending.

And now a little about the nuts and bolts of this book:

Phrases describing filter order such as "second-order" and "third-order" recur over and over again in this book. To save space and make them stand out better, they have been rendered as "2nd-order" and "3rd-order". All other ordinal numbers are spelt out fully. The active elements are called "amplifiers" though most of the time they will be opamps; however the Sallen & Key filter configuration in particular can be easily implemented with discrete emitter-followers if desired, so I have used the more general term "amplifiers".

All measurements were performed with an Audio Precision SYS-2702.

So far as I am aware, no supernatural assistance was received in the making of this book. All suggestions for its improvement that do not involve combustion or a baying crowd will be gratefully received. You can find my email address on the front page of my website at douglas-self.com, where I will be adding supplementary material on active crossover design as it occurs.

Further information, and PCBs, kits and built circuit boards of some of my designs, such as phono input stages and complete preamplifiers, can be found at: www.signaltransfer.freeuk.com.

A good deal of thought and experiment has gone into this book, and I dare to hope that I have moved analogue audio design a bit further forward. I hope you find it useful, and I hope you enjoy it too.

Douglas Self

London, September 2017

# CHAPTER 1

# *Crossover Basics*

I do hope you've read the Preface. It is not mere amiable meanderings, but gives an oversight as to how this book is constructed and what it is intended to do.

## What a Crossover Does

The basic function of any crossover, be it passive or active, analogue or digital, is to take the audio spectrum that stretches roughly from 20 Hz to 20 kHz and split it into two, three, or more bands so they can be applied to loudspeaker drive units adapted for those frequencies. In hi-fi use the crossover frequencies are usually fixed and intended for work with one particular loudspeaker design, but for sound-reinforcement applications the crossover frequencies are normally variable by front panel controls.

There are also other functions that are sometimes but not always performed by crossovers, and we may list them all as follows, roughly in order of importance:

1.  Equalisation to correct drive unit frequency responses
2.  Correction for unmatched drive unit sensitivities
3.  The introduction of time delays into the crossover outputs to correct for the physical alignment of drive units
4.  Equalisation to correct for interactions between drive units and the enclosure, e.g. diffraction compensation
5.  Equalisation to correct for loudspeaker-room interactions, such as operating in half-space as opposed to quarter-space
6.  Enhancement of the natural LF response of the bass drive unit and enclosure by applying controlled bass boost

Most of these functions are equally applicable to both passive and active crossovers, but the time-delay function is rarely implemented in passive crossovers because it requires a lot of expensive components and involves significant power losses.

Some of these functions will probably appear wholly opaque if you are just starting this book, but stay with me. All will be revealed.

## Why a Crossover Is Necessary

The need for any crossover at all is rooted in the impracticability of making a drive unit that can handle the whole ten-octave audio spectrum satisfactorily. This is not merely because the technology of loudspeaker construction is inadequate but is also rooted in some basic physics. Ideally the acoustic output of a loudspeaker would come from a single point; with such a source the sound field is uniform, because there can be no interference effects that result from multiple sources or from a source of finite size.

A tweeter has a small physical size, with a dome usually around an inch (2 or 3 cm) in diameter, and approximates fairly well to a point source. This technology works very well for high frequencies, say down to 1 kHz, but is hopelessly inadequate for bass reproduction because such a small area cannot move much air, and to reproduce bass frequencies you need to move a lot of it. Low-frequency drive units are therefore of much greater diameter, up to 12 inches for domestic hi-fi and up to 18 inches or more for sound-reinforcement applications. As cone area is proportional to the square of diameter, a 12-inch drive unit has 144 times the area of a typical tweeter, and an 18-inch unit has 324 times the area.

It is not at present technically possible to make a big low-frequency drive unit that works accurately up to 20 kHz, because as the frequency increases the cone ceases to move as a unit—it is not one of those most desirable "infinitely rigid pistons" that are always cropping up in loudspeaker theory but never in manufacturer's catalogues. This effect is often called "cone break-up" not because the cone physically falls apart but because, due to its finite stiffness, with rising frequency its surface divides up into separate areas of vibration. This unhappy state of affairs is put to advantage in so-called full-range loudspeakers which have a "parasitic tweeter" in the form of a small cone attached to the voice coil. The idea is that at higher frequencies the main cone does its own thing and is effectively decoupled from the voice coil and the tweeter cone, allowing the latter to radiate high frequencies without being restrained by the much greater mass of the main cone—what you might call a mechanical crossover. As you might imagine, there are many compromises involved in such a simple arrangement, and the response is generally much inferior to a good 2-way loudspeaker with separate bass unit and tweeter.

Nonetheless, the field of audio being what it is, there are a certain number of hi-fi enthusiasts who advocate full-range speakers for various reasons. Eliminating a passive crossover naturally increases power efficiency (as none is lost in the crossover components) and reduces cost.

## Beaming and Lobing

Even if it was possible to make a large-diameter drive unit which covered the whole audio spectrum perfectly, there is a powerful reason why it is far from certain that this would be a good idea. If a radiating surface is of finite size, then if you stand to the side of the central axis, sound from one area of the drive unit will arrive at your ear at a slightly different time compared with another area because of the differing path lengths through the air. This will cause interference between the two signals and, given the right amount of path difference, complete cancellation. There will therefore be major irregularities in the frequency response anywhere but on the central axis. This is called "beaming" or "lobing", and it occurs when the diameter of the radiating object is comparable with the wavelength of the sound. It is obviously to be avoided as much as possible; the variation in response as the angle between the listener and the centre axis changes is called the polar response, and a "uniform polar response" is much sought after in loudspeaker design.

**Table 1.1: Onset of beaming versus drive unit diameter.**

| Driver Diameter (inches) | Beaming Onset Frequency (Hz) |
|---|---|
| 1 | 13,680 |
| 2 | 6840 |
| 5 | 3316 |
| 6.5 | 2672 |
| 8 | 2105 |
| 10 | 1658 |
| 12 | 1335 |
| 15 | 1052 |
| 18 | 903 |

The beaming phenomenon is why a tweeter has to be of small diameter if it is to approach having a uniform polar response. Deciding when beaming becomes significant depends on the application, but the figures in Table 1.1 [1] have been put forward as shown above.

These frequencies are approximately those at which the wavelength in air equals the driver diameter. The whole business of beaming, lobing, and polar response generally is obviously much too complex to be summed up in a single table, but it does give some indication of when you need to start worrying about it.

There is of course much more to a crossover than simply splitting the audio signal into separate frequency bands. The vital point to understand is that the splitting has then to be followed by summation. The frequency bands have to be joined together again seamlessly. This requires the acoustic signals be summed to be correct not only in amplitude but in phase. The crossover and speaker system can only create the exactly correct signal at one point in space, which is unfortunate, as we have two ears and each listener therefore needs the signal to be correct at *two* points in space. Crossover design is always a matter of compromise to some degree.

It is not sufficient to get a perfect response on-axis, even if one interprets this as being capable of summing correctly at both ears. The off-axis output from the loudspeaker will not only be heard by those in the room unfortunate enough to not get the best seat on the sofa, but it also creates the ambient sound environment through room reflections and reverberation. If it has serious response irregularities, then these will detract from the listening experience, even if the direct on-axis sound is beyond reproach.

The term "lobing" is also used to describe the reinforcements and cancellations that occur when two separate drive units are radiating; in this case their size is relatively unimportant because interference would still occur even if both were point sources. When the radiation pattern is shifted at the crossover frequency because the signals to the two drive units are not in phase, this is called "lobing error". There is much more on this in Chapters 3 and 4.

## Passive Crossovers

Passive crossovers use only passive components, mostly capacitors and inductors. Resistors are also sometimes used, and the inductors are sometimes elaborated into auto-transformers, often with multiple taps for setting LF/HF balance. Figure 1.1 shows a typical 2-way crossover with a couple of common elaborations. C1, L1 make up a 2nd-order highpass filter, and C2, L2 make up a 2nd-order lowpass filter. The crossover slopes are therefore 12 dB/octave. The elaborations here consist of the addition of R1, to reduce the sensitivity of the tweeter to match that of the LF unit, and the addition of the Zobel network R2, C3 across the LF unit to reduce its impedance peak at resonance, and so make the crossover design easier. No component values are given as they depend very much on the desired crossover frequency and the properties of the drive units used.

It needs to be stated firmly here that using a simple crossover such as that in Figure 1.1, and then assuming the drive unit impedances to have a constant (i.e. resistive) impedance will not give anything like the desired result. Loudspeaker impedance is a complicated business and absolutely must be taken into account in passive crossover design; in active crossover design it can be blissfully ignored. Loudspeaker impedance and its effects on amplifiers are dealt with in detail in [2].

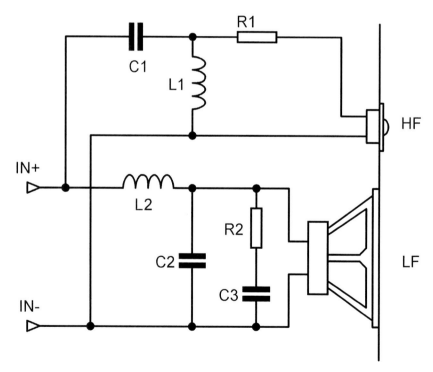

**Figure 1.1: A typical 2-way passive crossover, including HF unit sensitivity adjustment by R1 and a Zobel network R2, C3 across the LF unit.**

# Active Crossover Applications

The main fields of application for active crossovers in association with multi-way loudspeakers are high-end hi-fi, sound reinforcement, automotive audio, sound recording studios, cinema theatres, and film studios. In hi-fi, active crossover technology offers better and more consistent quality than passive crossovers. We shall look closely at why this is so later in this chapter.

In the area of sound reinforcement the use of active crossovers is virtually mandated by the need to use banks of loudspeakers with different characteristics, especially subwoofers. The size and number of the loudspeaker cabinets used means that it is physically impossible to put them close together, and hence sophisticated control of time delays is essential to obtain the desired coverage and polar responses. The large amount of power used in a typical sound-reinforcement system means that the losses inherent in the use of passive crossovers cannot be tolerated. The high power requirement also means that multiple power amplifiers are always used, and the extra cost of an active crossover system is very small by comparison.

Automotive audio marches to its own drummer, so to speak, the priority of most of its exponents being the maximum possible level of bass at all costs. This is perhaps not the place to speculate on whether this is driven by an appreciation of musical aesthetics or macho territorial aggression, but the result is that subwoofer systems are very popular, and so naturally some sort of crossover system is required. This is usually an active crossover, because the high power levels once again make the losses in a passive crossover unacceptable. This is particularly true because 4 $\Omega$ loudspeakers are normally used, so the current levels in inductors are doubled and $I^2R$ power losses are quadrupled, compared with the 8 $\Omega$ situation.

Active crossovers do have other important applications besides driving multi-way loudspeakers. They are also used in multi-band signal processing, of which the most common example is multi-band compression. A multi-band compressor uses a set of filters, working on exactly the same principle as a loudspeaker crossover, to split the audio signal into two, three, four or even more frequency bands; three- or four-band compressors are the most popular. On emerging from the crossover, each band is fed to a separate compressor, after which the signals are recombined, usually by simple summation. This is delightfully simple compared with the acoustic summation that recombines the outputs of the different drive units of a multi-way loudspeaker, because there are no problems with polar response or time delays, but it still needs attention to the same considerations of correct phase and frequency-offsetting.

The great advantage of multi-band compression is that a peak in level in one frequency band will not cause any gain reduction in the other bands; a high-level transient from bass guitar or kick drum will not depress the level of the whole mix. Another feature is the ability to use different attack/decay times for different frequency bands. You may be thinking at this point that it would have been more sensible to compress the kick drum *before* you mixed it in with everything else, and you are of course quite right. However, in many situations you are not doing the mixing but dealing with fully mixed audio as it comes along. Radio stations (not excluding the BBC) make considerable use of multi-band compression and limiting on existing stereo material to maximise the impact of their transmissions.

Other applications for multi-band processing include multi-band distortion, where splitting the distortion operation into separate bands prevents intermodulation turning the sound into an unpleasant muddle. A simple example of this is the "high-frequency exciter" or "psychoacoustic enhancer" technology, where a filter selects some part of the upper reaches of the audio spectrum and applies distortion to it in order to replace missing or under-strength harmonics. Multi-band techniques have also been used in noise-reduction techniques, notably in the Dolby A and SR noise-reduction systems. [3]

A new addition to non-loudspeaker applications is the so-called DJ isolator used in disco work. Typically the audio spectrum is divided into LF, MF, and HF bands, with the level of each set by a rotary control having 0 dB or +6 dB when fully clockwise and "off" when fully counter-clockwise; the bands are then recombined. The controls are twiddled up and down to "improve" the music by isolating one or two bands. [4] Whether you think that is a good idea or not, it is clearly an active crossover in all but name, and was indeed directly derived from active crossovers used in DJ venues. Research shows that 18 dB/octave filter slopes (3rd order) seem to be standard for high-quality isolators, but one isolator, the Formula Sound FSM-3, has 48 dB/octave slopes, i.e. 8th-order filtering; [5] this is accomplished in the analogue domain. Isolator crossover frequencies are usually fixed because of the large number of component values that have to be altered simultaneously to change them.

The active crossovers used for these multi-band signal processing applications do essentially the same job as those for loudspeakers, and this book will be most useful in their design.

## Bi-Amping and Bi-Wiring

The use of two separate amplifiers driven by an active crossover is sometimes called bi-amping. This is nothing to do with the use of two amplifier channels for stereo. If there are three separate amplifiers powering three drive units, this is called tri-amping or simply multi-amping. Bi-amping should not be confused with bi-wiring, which is a completely different idea. It is worth taking a quick look at it because the alleged benefits of bi-wiring are very relevant to crossover operation.

The most common method of wiring between an amplifier and a multi-way loudspeaker is to use a single cable, as shown in Figure 1.2a. The passive crossover networks shown are the simplest possible, and real passive crossovers will very likely have more components.

In bi-wiring (or tri-wiring, or "multi-cabling") only one amplifier is used, but separate cables run from it to separate sections of a passive crossover that is mounted as usual inside the loudspeaker enclosure; this is illustrated in Figure 1.2b. Bi-wiring requires individual access to the passive crossover sections for the LF unit and tweeter, which is arranged for by having two sets of terminals on the rear of the loudspeaker connected together with shorting links that can be easily removed.

Arranging things for tri-wiring is a little more expensive and difficult, as it requires six terminals and four shorting links, plus the facility of connecting two shorting links to one terminal in a tidy manner. This is probably why loudspeakers with three drive units are more often connected for bi-wiring rather than tri-wiring. Normally the tweeter and mid drive units are connected to one pair of terminals and

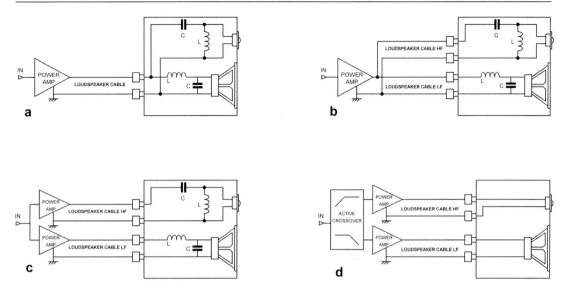

Figure 1.2: (a) Normal loudspeaker wiring, (b) bi-wiring, (c) full-band multi-amplification, and (d) active crossover multi-amplification.

the bass drive unit to the other. There is also the point that if your taste runs to expensive loudspeaker cables of undefined superiority, having three sets of them is significantly more costly than two.

The question is, just what is gained by bi-wiring? The hard fact is: not much. Its enthusiasts usually claim that separating the high and low frequencies into different cables prevents "intermodulation" of some sort in the cabling. I must point out at once that even the cheapest copper cable is absolutely linear and can generate no distortion whatsoever, even when passing heavy currents. The same absolutely linearity is shown at the smallest signal levels that can be measured. [6] From time to time some of our less bright audio commentators speculate in the vaguest of terms that a metallic conductor is actually some sort of swamp of "micro-diodes", and that non-linearity might just conceivably be found if the test signal levels were made low enough. This is categorically untrue and quite impossible with it. The physics of conduction in metals has been completely understood for a long time, and there simply is no threshold effect for metallic conduction. The only way that a cabling system can introduce non-linear distortion is if the connectors are defective, which in this context pretty much means either that the contacts are corroded or the connector is about to fall apart altogether and is barely making physical contact.

Actually, a lot of audiophiles are much less definite about the advantages of bi-wiring. They tend to say something pitifully vaporous like: "If only one pair of conductors delivers the whole signal, there is a danger that the bass frequencies can affect the transmission of the more delicate and subtle treble frequencies." Affect the transmission? Exactly how, pray? And how much danger is there? Not much? Immediate extreme peril? Danger, Will Robinson!

Where bi-wiring can be of use is in minimising the interaction of loudspeaker impedance variations with cable resistances. Let's suppose that the combination of LF crossover and bass drive unit has a dip in its impedance (the drive unit alone cannot show an impedance lower than its voice coil resistance): this dip will cause a greater voltage drop across the cable resistance, giving a dip in driver output that would not have existed if the cable (and the amplifier) had zero resistance. If we suppose again, and this is a bit more of a stretch, that impedance dips in the HF and LF impedance might overlap, then applying them to separate cables will reduce the level variations. However, the same effect could be got by doubling up a non-bi-wired loudspeaker cable, as this would halve its resistance.

Bi-wiring has a certain enduring appeal because it enables the experimenter to feel he has done something sophisticated to his system but requires little effort and relatively little cost (though this of course depends on how much you think you need to pay for acceptable loudspeaker cables), and the chances of anything going wrong are minimal. This is not of course the case if you start delving about inside a power amplifier.

Bi-amping means a separate amplifier for each crossover section. This almost always implies the use of an active crossover to split the audio spectrum and feed different sections to different amplifiers. Occasionally, however, you hear advocated what might be called full-range multi-amplification. This is shown in Figure 1.2c, where two power amplifiers are fed with identical signals and drive the passive crossover sections separately. This is an expensive way to gain relatively small advantages. The loading on each amplifier will be less, and you might get some of the advantages of bi-amping, such as less stress on each amplifier and reduced intermodulation of LF and HF signals in the amplifiers. The advantage of the latter should be negligible with well-designed solid-state amplifiers, but I suppose there are valve amplifiers to consider. There is also the possibility of using a smaller power amplifier for the HF path. In general, however, this approach combines most of the disadvantages of passive crossovers with the extra cost of active crossovers.

Figure 1.2d shows standard bi-amping, with two power amplifiers fed with the appropriate frequency bands by an active crossover, the passive crossover no longer being present. This is the approach dealt with in this book.

## Loudspeaker Cables

It would be somewhat off-topic to go into a lot more detail about loudspeaker cable design, but this might be a good place to point out that there is absolutely nothing magical or mysterious about it. My findings are that, looking at the amplifier-cable-loudspeaker system as a whole, the amplifier and cable impedances have the following effects:

- Frequency response variation due to the cable resistance forming the upper arm of a potential divider with the loudspeaker load as the lower arm. The effect of the resistive component from the amplifier output impedance is usually negligible with a solid-state design, but this is unlikely to be the case for valve amplifiers. This is at least potentially audible; thick cables will minimise the total resistance. It is not a good reason for using anything other than ordinary copper cable.
- A high-frequency roll-off due to the cable inductance forming an LR lowpass filter with the loudspeaker load. The amplifier's series output inductor (almost always present to give stability with capacitive loads) adds directly to this to make up the total series inductance. The shunt-capacitance of any normal speaker cable is trivially small and can have no significant effect on frequency response or anything else.

The main factors in speaker cable selection are therefore series resistance and inductance. If these parameters are less than 100 mΩ for the round-trip resistance and less than 3 µH for the total inductance, any effects will be imperceptible. [7] These conditions can be met by standard 13-Amp mains cable (I'm not quite sure how the equivalent cable is labelled in the USA). This cable has three conductors (live, neutral, and earth) each of 1.25 sq mm cross-section, made up of 40 x 0.2 mm strands. Using just two of the three conductors, a 100 mΩ round-trip resistance allows 3.7 metres of cable. The lowest cable resistance is obtained if all three conductors are used, normally by paralleling the neutral and earth conductor on the cold (grounded) side of the cable; the maximum length for 100 mΩ is now 5.0 metres, which should do for most of us. This three-conductor method does give what I suppose you might call an "asymmetric" cable, which could offend some delicate sensibilities, but I can assure you that it works very nicely.

The loudspeaker cables that I have in daily use are indeed made of such 13-Amp mains cable, bought from an ordinary hardware shop nearly 40 years ago. Should a passing audiophile query the propriety of using such humble cabling, I usually tell them that with so much passage of time in regular use, the electrons have been thoroughly shaken loose and move about with the greatest possible freedom. I do hope nobody reading this book is going to take that seriously.

## The Advantages and Disadvantages of Active Crossovers

Here I have tried to put down all the advantages and disadvantages of the active crossover approach. Some of them may not be very comprehensible until you have read the relevant chapter of this book. My initial plan was to attempt to put them in order of importance, but this is not an easy thing to do. The order here is therefore to some extent subjective, if I may use the term . . .

### *The Advantages of Active Crossovers*

The advantages of active crossovers are:

1. The over-riding advantage of an active crossover is it offers ultimate freedom of design, as virtually any frequency or phase response that can be imagined can be used. The filter slopes of the crossover can be made as steep as required without using large numbers of big and relatively expensive components. Any increase in passive crossover complexity means a significant increase in cost.

2. The design of passive crossovers is restricted by the need to keep the loading on the power amplifier within reasonable limits. With an active crossover, correction of the response for each driver is much simpler, as it can be undertaken without having to worry about the combined load becoming too low in impedance for the average amplifier.

3. The design of passive crossovers is further complicated by the need to keep the power losses in the crossover within reasonable limits. The losses in the resistors and in the inductors (because of their inevitable series resistance) of a passive crossover, especially a complex design employing high-order filters or time-delay compensation, can be very serious. In a big sound-reinforcement system the losses would be measured in tens of kilowatts. Not only does this seriously degrade the power efficiency of the overall system by wasting power that could be better applied to the drive units,

but it also means that the crossover components have to be able to dissipate a significant amount of heat and are correspondingly big, heavy, and expensive. It is far better to do the processing at the small-signal level; the power used by even a sophisticated active crossover is trivial.

4. If one of the power amplifiers is driven into clipping, that clipping is confined to its own band. Clipping is usually less audible in the bass, so long as there is no intermodulation with high frequencies. It has been stated that an active crossover system can be run 4 dB louder for the same subjective impairment. This is equivalent to more than twice the power but less than twice the perceived volume, which would require a 10 dB increase in sound pressure level SPL.

5. Delays can be added to compensate for differing acoustic centres for the drive units quite easily. Passive delay lines can be built but are prodigal in their use of expensive, lossy, and potentially non-linear inductors, and as a result have high overall losses.

6. Tweeters and mid drive units can have resonances outside their normal operating range, which are not well suppressed by a passive crossover because it does not put a very low source impedance across the voice coil. The presence of a series capacitor can greatly reduce the damping of the main resonance, [8] and it is also possible for a series capacitor to resonate with the tweeter voice coil inductance, [9] causing an unwanted rise in level above 10 kHz or thereabouts.

7. Drivers of very different sensitivities can be used, if they happen to have the best characteristics for the job, without the need for large power-wasting resistances or expensive and potentially non-linear transformers or auto-transformers. If level controls for the drive units are required, these are very straightforward to implement in an infinitely variable fashion with variable resistors. When passive crossovers are fitted with level controls (typically for the mid unit or tweeter, or both) these have to use tapped auto-transformers or resistor chains, because the power levels are too high for variable resistors, and so control is only possible in discrete steps.

8. The distortion of the drive unit itself may be reduced by direct connection to a low-impedance amplifier output. [10] It is generally agreed that the current drawn by a moving-coil drive unit may be significantly non-linear, so if it is taken from a non-zero impedance, the voltage applied to the drive unit will also be distorted. This may be related to out-of-band tweeter resonance; Jean-Claude Gaertner states that tweeters can have increased distortion below 1 kHz. [11] I do not know for sure, but I very strongly suspect that when drive units are being developed they are driven from amplifiers with effectively zero output impedance and that linearity is optimised under this condition. Any other approach would mean guessing at the source impedance, which, given the number of ways in which it could vary, would be a quite hopeless exercise.

9. With modern opamps and suitable design techniques, an active crossover can be essentially distortion free, though care must be taken with the selection of capacitors in the filters. It will not however be noise-free, though the noise levels can be made very low indeed by the use of appropriate techniques; these are described later in this book.

A passive crossover contains inductors, which if ferrite or iron-cored will introduce distortion. It also contains capacitors, often in the form of non-polar electrolytics, which are not noted for their linearity or the stability of their value over time. I haven't been able to find any published data on either of these problems. Capacitor linearity is very definitely an issue because they are being used in filters and therefore have significant voltage across them. It is possible for capacitor distortion to occur in active crossovers too, but the signal voltages are much lower and one can expect the amount of distortion generated to be much less. See Chapter 15, where using the worst sort of capacitor increases the distortion from 0.0005% to 0.005%, with a signal level of 10 Vrms. In contrast, the distortion from a passive crossover can easily exceed 1%.

10.  With the rise of AV there is more experience in making multi-channel amplifiers economically. The separate-module-for-each-channel approach, where each module has a small toroid mounted right up at one end, while the input circuitry is at the other, is more expensive to manufacture but can give an excellent hum and crosstalk performance. The main alternative is the huddle-around-the-big-central-toroid approach, which has some serious and intractable hum issues.

11.  If a protection system is fitted that is intended to guard the drive units against excessive levels, then it can be closely tailored for each drive unit.

12.  Voice coil heating will increase the resistance of the wire in its windings, reducing the output. This is known as thermal compression. It also increases the impedance of a drive unit, and if it is part of a complex passive crossover, the interaction can be such that there are much greater effects on the response than that of thermal compression alone. In one set of tests conducted by Phil Ward, [12] the voice coil temperatures of four different loudspeakers showed a maximum of 195°C and a rise in resistance of 176%. That sort of variation has got to cause interaction with almost any sort of passive crossover.

13.  It has been proposed that active crossovers can allow the modelling of voice coil heating by calculations based on signal level, frequency, and known thermal time-constants. Thus the effects of thermal compression (the reduction in output as the voice coil resistance rises with temperature) could be compensated for. It does however imply relatively complex computation that would be better carried out by digital processing rather than in analogue circuitry. There would have to be A to D conversion of the signal and perhaps D to A conversion of the control parameters, even if the actual crossover function was kept in the analogue domain. Controlling the active crossover parameters with analogue switches or VCAs without compromising signal quality is going to be hard to do. Modern volume control chips have excellent linearity, but they are not really adapted to general control, and using a lot of them would be rather expensive. If you are undertaking this sort of complex stuff then it's probably going to be best to do all the processing in the digital domain.

Clearly this plan can only work if the crossover is programmed with the thermal parameters for a given loudspeaker and its set of drive units; this information would have to be provided by the loudspeaker manufacturers, and once again we see the need for the active crossover to be matched to the loudspeaker.

14.  Drive unit production tolerances can be trimmed out. It has been said that changes in driver characteristics due to aging can also be trimmed out, but since aging is not likely to be an absolutely predictable process, this would ideally require some sort of periodic acoustic testing. For a reference loudspeaker in a laboratory or monitors in a recording studio this is entirely practical, but it is less so in the home environment because of the need for an accurate measuring microphone, or, more likely, one whose response deviations are sufficiently predictable for them to be allowed for. Extra electronics are of course required to implement the testing procedure.

15.  The active filter crossover components will have stable values. The inductors in a passive crossover should be stable (though I suppose turns could shift under heavy vibration), but the non-polar electrolytic capacitors that are often used have a bad reputation for shifting value over time. The stability of these components has improved in recent years, but it is still a cause for concern. It has been stated that electrolytics in high-end passive crossovers should be regarded as having a lifetime

of no more than ten years. [13] Plastic film crossover capacitors such as polypropylene show better stability but are very expensive. A fashion has grown up recently for bypassing big passive crossover capacitors with smaller ones—whether this has any beneficial effects is very questionable indeed.

16. The active filter crossover components will not change in the short-term due to internal heating. In a passive crossover the capacitors will have large voltages across them and large currents through them; dielectric losses and ohmic losses in the ESR (equivalent series resistance) may cause these capacitors to heat up with sustained high power and change in value. Non-polar electrolytic capacitors (basically two ordinary electrolytics back-to-back) are considered particularly susceptible to this effect because their relatively small size for a given capacitance-voltage product means they have less area to dissipate heat, and so the temperature rise will be greater.

17. The relatively small capacitors used in active crossover filters can be economically chosen to be types that do not exhibit non-linear distortion—polystyrene and polypropylene capacitors have this useful property. Non-polar electrolytic capacitors when used in passive crossovers are known to generate relatively large amounts of distortion.

18. No inductors are required in active crossover circuitry (apart perhaps for a few small ones at inputs and outputs for EMC filtering). Inductors are notorious for being awkward and expensive. If they have ferromagnetic cores they are heavy and generate large amounts of non-linear distortion. If they are air-cored distortion is not a problem, but many more turns of copper wire are needed to get the same inductance, and the result is a bulky and expensive component. Martin Colloms has stated [14] that if an inadequate ferromagnetic core is pushed into saturation by a large transient, the resulting sudden drop in inductance can cause a drastic drop in the impedance seen at the loudspeaker terminals, and this sort of thing does not make life easier for power amplifiers.

19. Passive crossovers typically use a number of inductors, and it may be difficult to mount these so there is no magnetic coupling between them; unwanted coupling is likely to lead to frequency response irregularities. (It should be said that some types of passive crossover use transformers or auto-transformers, where the coupling is of course entirely deliberate.)

20. When a passive crossover is designed, it is absolutely not permissible to treat a drive unit as if its impedance was simply that of an 8 Ω resistor. The peaky impedance rise at resonance and the gentler rise at HF due to the voice coil inductance have to be taken into account to get even halfway acceptable results. This naturally complicates the filter design process considerably, and one way of dealing with this is to attempt to compensate for these impedance variations by placing across the drive unit terminals a series-resonant LCR circuit (to cancel the resonance peak) and an RC Zobel network (to cancel the voice coil inductance rise). [10] This is often called "conjugate impedance compensation", and while it may make the crossover design easier, it means there are at least five more components associated with just one drive unit, and they all have to be big enough to cope with large signals. There may also be changes in impedance due to changes in acoustic loading across the drive unit's passband. In an active crossover system the drive unit is simply connected directly to its power amplifier, and assuming that amplifier has an adequate ability to drive reactive loads, the details of the drive unit impedance curve can be ignored. Determined use of conjugate impedance compensation (also called "conjugate loading") can turn a peaky but essentially 8 Ω impedance plot into a more or less flat 4 Ω plot at the terminals. In my view, artificially reducing a loudspeaker's impedance to make its plot flatter is a poor idea, as all amplifiers, so far as I know without exception, give more distortion with heavier loading.

21.  The presence of an active crossover in the system makes it easy to add bass-end equalisation of the loudspeaker; this is distinct from equalising individual drive units. An equaliser circuit is added which provides a carefully controlled amount of bass boost to counteract the natural roll-off of the bass drive unit in conjunction with its enclosure, thus extending the level bass response to lower frequencies. This sort of thing always has to be approached with care, as too much equalisation will lead to excessive drive unit displacements and permanent damage. This is a particular problem with ported enclosures that put no loading on the drive unit cone at low frequencies.

22.  An active crossover system also opens up the possibility of motional feedback, where the position, velocity, etc of the bass drive unit cone are monitored by an accelerometer or another method, and the information is used to correct frequency response irregularities and non-linear distortion. This is obviously more practical where the crossover and power amplifiers are built into the loudspeaker enclosure. Motional feedback is dealt with in more detail in Chapter 19, but it may be remarked in passing that while it sounds like a first-class idea, it has never achieved much success in the marketplace.

That is a pretty formidable list of real advantages for active crossovers. There are also some claimed advantages that are pretty much bogus, and to be fair we had better have a quick look at these before we move on:

## Some Illusory Advantages of Active Crossovers

1.  A commonly quoted "advantage" is that there is less intermodulation distortion in the power amplifiers because each one handles a restricted frequency range. This may once have been a significant issue, but nowadays designing highly linear power amplifiers is not hard when you know what pitfalls to avoid. I have demonstrated this many times. [15] I suppose this might well be a valid advantage if you insist on using valve amplifiers, which have generally poor linearity, at its worst at low frequencies due to the well-known limitations of output transformers.

2.  Similarly, it has been claimed that the danger of slew rate limiting is much reduced in the more powerful LF and MF power amplifiers, as they do not handle high frequencies. This is a specious argument, as there is no difficulty whatsoever in designing a power amplifier that has a faster slew rate than could ever conceivably be needed for an audio signal. [16]

3.  It is sometimes stated that an active crossover system results in simpler loading for each power amplifier, as it is connected to only one drive unit. This is true, as regards human understanding of what the impedance curves look like—it is often difficult to see what causes the impedance ups and downs of a complex passive crossover, whereas for a single moving-coil drive unit the effects of the resonance, the voice coil resistance, and the voice coil inductance are all readily identifiable. However, amplifiers are not sentient, and no amount of anthropomorphic thinking will make them so. An impedance curve that is hard to understand, with a cross-section like the Pyrenees, is not necessarily hard to drive. What puts most stress on a power amplifier are impedance dips and high levels of reactance; these increase the peak and average dissipation in the amplifier output devices. It is not a question of the amplifier having to think harder about it, but the amplifier designer may have to.

4.  Following from the previous point, an amplifier connected to a single moving-coil drive unit sees basically a mixture of resistance and inductance. Purely capacitive loading is not possible, and it has been said that amplifier stability is therefore easier to obtain. In reality, ensuring amplifier

stability into capacitive loads is not a significant a problem—you just put the traditional inductor in series with the amplifier—so this is not a big deal. The inductors used have only a few turns and are not expensive. If you are constructing a system where the amplifiers are *permanently* connected to the drive units—for example, if the amplifiers are built into the loudspeaker enclosure—then there is the possibility of omitting the power amplifier output inductors, as a single drive unit is essentially resistive and inductive, but I'll say right now that I've never tried this.

5.  It has sometimes been claimed that it is easier or cheaper to design a power amplifier that only has to handle a limited bandwidth. This is not true. It is absolutely straightforward to design a good power amplifier that not only covers the whole audio spectrum but several octaves on either side of it. Likewise a limited bandwidth does not reduce costs much. You could make the capacitor in the power amplifier negative feedback smaller in the MID and HF channels, but that is a matter of a penny or two; if the amplifiers have separate power supplies, then those for the MID and HF could have smaller reservoir capacitors, which might save tens of pennies. It is possible to save a significant amount of money by designing the HF power amplifier for a lower power output, but that is a different matter from the bandwidth issue and is dealt with in Chapter 17 on crossover system design.

6.  If we are dealing with an integrated crossover/power amplifier/loudspeaker system with everything installed in one enclosure, it is often claimed there is no need to fit short-circuit protection to the amplifiers, as their outputs are not externally accessible, and so money can be saved. This is a dubious approach, because it assumes that voice coils will never fail short-circuit; if one does there could be serious safety issues. It also assumes that there will be no accidents when testing the amplifiers. The cost saving is very small and is not recommended.

It is still of course very necessary to provide all the power amplifiers with effective DC-offset protection, for safety reasons if for no other. This is a special challenge when driving tweeters, as described later. This is a good deal more expensive than short-circuit protection, as it usually requires an output relay for each power amplifier.

## The Disadvantages of Active Crossovers

The disadvantages of active crossovers, once again in what I think to be their order of importance, are:

1.  Much greater electronic complexity. The number of power amplifiers is doubled, or more likely tripled, and quite possibly quadrupled. This could potentially lead to lower reliability, as the failure of any one of the power amplifiers or of the active crossover itself renders the whole system unusable. However, with proper design techniques and the use of adequate safety margins, the failure-rate of a modern solid-state power amplifier or active crossover should be so low that reliability is not an issue.

2.  As a direct result of the first statement, there is much more hardware to fit into the living room. Not just more electronic boxes, but all the cabling between them. Ways of tackling this issue are considered later in this chapter.

3.  Also flowing from (1) is the issue of greater cost. The high-frequency amplifiers can be of lower power output than the low-frequency amplifiers, but this definitely does not reduce their cost proportionally. There may be economies of scale to be made by making all the power amplifiers identical, but with the high-frequency ones fed from lower supply voltages. This is perfectly feasible without any compromise on amplifier performance.

Against this must be set the fact that precise, stable, and generally high-quality components for passive crossovers are not cheap, and a top-end passive crossover can easily end up costing more than an active crossover; this does not however take into account the extra power amplifiers.

4. The vast majority of active crossover loudspeakers have been built as one unit. A mono active crossover and the two or three power amplifiers required are put inside into the loudspeaker enclosure. While this makes a very convenient package, which requires only a mains lead to each loudspeaker as extra wiring, it has a deadly disadvantage. The simple truth is that people want to be able to choose their own power amplifiers. Not everybody is prepared to believe that a company that specialises in loudspeakers would be able to come up with a good power amplifier, and it is entirely understandable that people prefer to buy each audio component from specialists in the relevant field.

5. The active crossover must be matched to the loudspeaker and cannot be bought off the shelf. Many speakers allow for bi-wiring, but this still leaves the passive crossover components in place. Few, if any, loudspeakers allow direct access to the drive units without some serious dismantling, and so any domestic system must be either home-built, or custom-built, with a commensurately high cost.

6. Tweeters, and to a lesser extent, midrange drive units are much more exposed to amplifier DC-offset faults. When they are connected to an amplifier via a passive crossover, there will be at least one capacitor between the amplifier and the voice coil. This will be of modest size, as it is not intended to pass low frequencies, and so provides very good protection for tweeters and midrange drive units which do not have the displacement capability or thermal inertia of bass drivers. Tweeters in particular can be destroyed in an instant by a large DC offset. This seems to me a very good, if not irrebuttable, argument for using a low-power amplifier to drive the tweeter in an active crossover system.

   Any DC-coupled amplifier connected to a loudspeaker should of course have DC-offset protection, but since such a fault is usually detected by putting the output signal through a lowpass filter with a very low cutoff frequency to remove all the audio and then applying it to a comparator, there is inevitably some delay in its operation; an output muting relay also takes a significant amount of time to open its contacts and break the circuit. If the amplifier driving the tweeter is specialised for the task then it can be given DC protection that reacts more quickly, as its lowpass filter will not need to reject high amplitude bass signals; it can therefore have a higher cutoff frequency and will respond faster.

7. There will be a greater power consumption due to quiescent current flowing in more amplifiers. In a solid-state Class-B power amplifier, how much power is involved depends on the details of the output stage design. Class-A amplifiers, built with any technology, uselessly dissipate almost all the power so trustingly fed into them, showing efficiencies of around 1% with musical signals, as opposed to sine wave testing. [17] There is also the power consumption of the active crossover itself, but this is not likely to be significant, even for a very complex analogue active crossover.

8. Passive crossovers cannot be radically maladjusted by the user (though some allow vernier adjustments, typically of HF output level). In the sound-reinforcement business there is always the possibility that some passing "expert" may decide to try and use the crossover controls as a sort of equaliser. This is unlikely to be effective and can result in severe damage to loudspeakers. Lockable security covers solve this problem.

9. Active crossovers add extra electronics to the signal path which would otherwise not be there and so must be beyond reproach, or at any rate not cause any significant signal degradation. With the intelligent use of either opamp or discrete circuitry this should not be a serious problem.

## The Next Step in Hi-Fi

As we have just seen, active crossovers have a long and convincing list of technical advantages. The score is 22 very real advantages and five not-too-convincing advantages, as opposed to nine advantages for passive crossovers. It is generally accepted that active crossover hi-fi systems sound obviously better than their passive crossover counterparts. Any sort of consensus is rare in the wide field of audio, so this is highly significant. I strongly suspect that the widespread adoption of active crossovers, suitably matched to their loudspeakers, would be The Next Big Step in Hi-Fi, and possibly even The Last Big Step possible with current technology.

Nonetheless, it is undeniable that active crossovers, despite their compelling advantages, have made very little headway in the domestic market so far, though they are used in all but the smallest sound-reinforcement systems, and extensively in automotive audio. The big question is how to make active crossover technology more acceptable in the marketplace. The first thing we shall do is look at ways of solving the "too many boxes and wires" problem.

## Active Crossover Systems

We will consider the various ways in which an active crossover system can be configured, with an especially hard look at making it acceptable in a domestic environment. Sound-reinforcement systems are a separate issue. One of the significant disincentives to the active crossover approach is the sheer amount of hardware required, in terms of electronic boxes and cables. This can be hard to fit into a minimalist decoration scheme—and indeed often hard to fit into any sort of decoration scheme at all. It is desirable to package the technical functionality as neatly as possible. I am assuming here that a high-quality system is intended, and so 3-way active crossovers will be used.

In terms of cabling and equipment the tidiest setup is undoubtedly achieved with a mono active crossover and its three power amplifiers built into each loudspeaker enclosure, but it is vital to realise that this is not an acceptable approach for people who take their power amplifiers at all seriously.

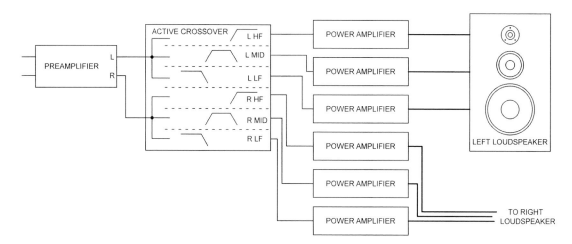

**Figure 1.3: Active crossover system using six monobloc power amplifiers.**

Figure 1.3 shows an active crossover system using six separate monobloc power amplifiers. This could be configured with all six amplifiers in one location, in which case putting them all near one of the loudspeakers reduces the length of at least one set of loudspeaker cables to a minimum. The amplifiers may be all of the same type, but all that is really required of them is that they have the same gain. Even if we have six nominally identical amplifiers made at the same time by the same manufacturer, somewhere in each amplifier will be at least two gain-setting resistors, each with a tolerance; nonetheless, in a competent design the variation should be comfortably less than variations in the drive units.

Alternatively, three of the power amplifiers could be placed adjacent to each loudspeaker, considerably shortening the total length of what might be expensive loudspeaker cable, as three of the connections are now very short. Three of the line-level cables to the power amplifiers naturally become correspondingly longer, but since their resistance is of much less importance than that of the loudspeaker cables, this is overall a good thing. The increased resistance of long line cables makes the link more susceptible to voltages induced by currents flowing through the ground connection, but I think it is fair to assume that a hi-fi system with active crossovers would use balanced connections to cancel such noise. Ultimately the placing of the various parts of the system is going to be influenced by furniture arrangement and the availability of handy (and hopefully well-ventilated) cupboards in which to stash the boxes.

Adding it up, a system configured in this way consists of seven electronic boxes, (not including the preamplifier) eight line-level cables and six loudspeaker cables. If the connection from the preamp to the crossover is a 2-way cable (i.e. two parallel cables joined together along their length), that is reduced to seven boxes, seven line cables, and six loudspeaker cables, which is hardly a great improvement.

However, it must be said that there are excellent technical reasons for using 2-way cables when you can. Their construction keeps the grounds for the two links physically close together and prevents them forming a loop that could pick up magnetic fields which would induce current flow; this current would cause voltage drops in the ground resistance and degrade the signal. The use of balanced connections greatly reduces the effects of ground currents, but it is of course much sounder to prevent the currents arising in the first place.

In Figure 1.4, the system is configured with three stereo power amplifiers. This has the advantage that stereo amplifiers are the most common sort and give the greatest choice. The ones used here

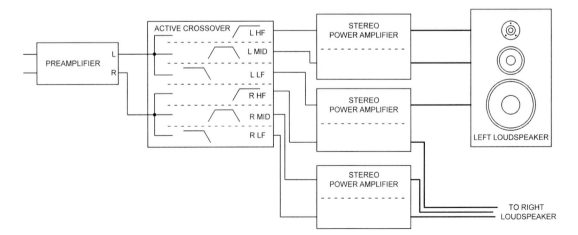

**Figure 1.4: Active crossover system using three stereo power amplifiers.**

are assumed to be identical; if a lower-power stereo amplifier is chosen to drive the tweeters, then Figure 1.4 would need to have its connections rearranged. Figure 1.4 uses five electronic boxes, eight line cables, and six loudspeaker cables. Using 2-way cables between the preamplifier and crossover and also to the power amplifiers simplifies this to five boxes, four line cables, and six loudspeaker cables. There are always going to be six loudspeaker cables.

Figure 1.5 shows a variation on this approach which puts one of the stereo amplifiers adjacent to the right loudspeaker instead of piling them all up on the left side. This cuts down the total length of loudspeaker cable required, but there is still a long run from one amplifier on the left to the right loudspeaker on the other side of the room.

The opinion is held in some quarters that very high degrees of isolation between left and right channels is essential to obtain an optimal stereo image. This is wholly untrue, but audio is not a field in which rational argument can be relied upon to convince everybody. The configuration of Figure 1.5 could be criticised on the grounds that left and right channels pass through one stereo power amplifier, and this might compromise the crosstalk figures. It is not actually harder to get a good crosstalk performance from a stereo power amplifier than from a stereo preamplifier; in fact it is usually easier because the preamplifier has more complex signal routing for source selection and the like. Nonetheless it is only fair to point out that there might be objections to the 3x stereo amplifier arrangement because however it is configured, at least one amplifier will have to handle both right and left signals.

We will now take a radical step and assume the ready availability of three-channel power amplifiers.

Multi-channel power amplifiers at a reasonable cost have been available for surround-sound systems for many years, to deal with 5:1 formats and so on. The last multi-channel power amplifier I designed (The TAG 100x5R:10) could be configured for ten channels of 80 W/8 Ω each. A three-channel power amplifier of high quality presents absolutely no new technical challenges at all.

As you can see from Figure 1.6, using two three-channel power amplifiers simplifies things considerably. There are now four electronic boxes, eight line cables, and six loudspeaker cables. Using a 2-way cable from preamp to crossover and 3-way cables between the crossover and amplifiers is well worthwhile and reduces the parts count to four boxes, three line cables, and six loudspeaker

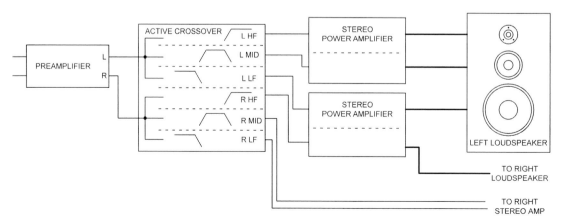

**Figure 1.5: Alternative setup of three stereo power amplifiers, with one placed on right speaker side.**

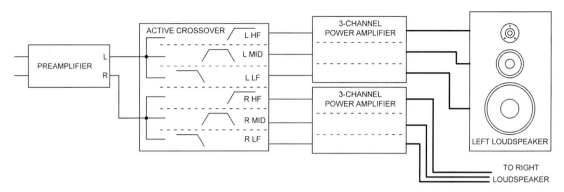

**Figure 1.6: Active crossover system using two three-channel power amplifiers.**

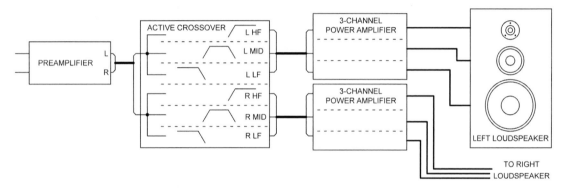

**Figure 1.7: Active crossover system using two three-channel power amplifiers and multi-way cables.**

cables. This important configuration is shown in Figure 1.7. One of the three-channel power amplifiers could be sited over by the right loudspeaker, and this is much to be preferred as it minimises total loudspeaker cable length. I think it is fairly clear that this is the best way to configure things, with the least number of separate parts and the possibility of keeping the loudspeaker cables very short indeed if each power amplifier is sited right behind its loudspeaker.

The only real difficulty is those three-channel power amplifiers. Some do exist, evidently intended for multi-channel AV use rather than in active crossover systems; two current examples are the Classé CA-3200 three-channel power amplifier [18] and the Teac A-L700P 3-Channel Amplifier. There appear, however to be no three-channel amplifiers specifically designed for our application here. Such an amplifier would be able to economise on its total power output by having a big output for the LF driver, a medium output for the mid drive unit, and a smaller output again for the tweeter. The downside to that plan is that it would be less versatile than a three-channel amplifier with equal outputs, which could be pressed into stereo or multi-channel service if required.

3-way cables should present no problems; in the UK, the widespread Maplin chain sells four-way audio line cables with individually lap-screened cores at a very reasonable price. Individual screened cores are of course highly desirable to prevent capacitive crosstalk.

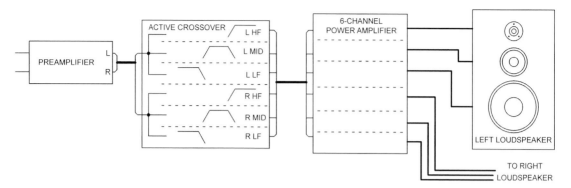

**Figure 1.8: Active crossover system using a single six-channel power amplifier and multi-way cables.**

To take things to their logical conclusion, we can use a six-channel power amplifier. These are used extensively in AV applications so are not hard to obtain. If we assume the use of multi-way cables from the start, we get Figure 1.8. This cuts down the separate pieces of hardware to the minimum, giving three electronic boxes, two line cables, and the usual six loudspeaker cables. Unfortunately this configuration brings back the need for long and possibly expensive loudspeaker cables to drive the right loudspeaker, and it also raises again the question of left-right crosstalk in the power amplifier.

In my view the configuration using two three-channel amplifiers is clearly the best approach. There is one more box but the loudspeaker cables can be of minimal length.

## Matching Crossovers and Loudspeakers

Having looked at the best way to package and configure an active crossover system, there are some other issues to deal with. In the sound-reinforcement business, active crossovers are sold with wide-ranging control over crossover frequencies, time-delay compensation, and so on. This is feasible in a situation where skilled operators set up the system. In a hi-fi situation it is more common to have fixed crossover parameters that are carefully matched to the loudspeaker characteristics. These parameters include not only crossover frequencies and delays but also things like drive unit frequency response equalisation, diffraction compensation equalisation, and so on. The crossover has to be set up for one model of loudspeaker only. This is fine if the crossover and amplifiers are built into the loudspeaker, but as we have seen, this is not going to work for potential customers who want to choose their own power amplifiers, and that is most of them.

The solution to this problem is to sell active crossovers and loudspeakers as a matched package, but leave the power amplifiers to be bought separately. All that would be required of the amplifiers is that they should all have the same gain. This is nothing new; THX-spec amplifiers all have a gain of +29 dB for an unbalanced input. For the higher reaches of the high end, each active crossover unit could be individually calibrated by acoustic testing to match each loudspeaker with its drive units at the factory, authoritatively solving the problem of variation in drive unit parameters. The loudspeakers and crossover would then be tied together by their serial numbers. Obviously it would be essential to not swap over the Left and Right channels at any point. The calibration could be done by plugging in resistors until the optimal result is obtained and then soldering them in permanently. This kind of hand tuning would not of course be cheap, but with the right measuring gear and an intelligent algorithm for adjusting the various crossover parameters, I am sure it could be done at a price some folk would be prepared to pay.

What it would not do is compensate for drift in drive unit parameters over time. This could only be addressed by regular recalibration, which would be unpopular with most people if it involved sending the crossover and loudspeakers back to the factory. And handing over some more money, no doubt.

For this scheme it would be necessary to construct loudspeakers so that direct access could be gained to the drive units, optionally bypassing the internal passive crossover, if one is fitted at all. Providing facilities for bi-wiring on the back of a loudspeaker is straightforward; you just put four terminals on the back and add a pair of shorting links that can be easily removed. Tri-wiring is a little more difficult but still thoroughly doable. Bypassing a passive crossover is however more complex, as it is necessary to disconnect it not only from the input terminals but also from two, three, or possibly four drive units.

## A Modest Proposal: Popularising Active Crossovers

Let us suppose that it becomes accepted practice to sell crossovers and speakers as a package, but the amplifiers are bought separately. As we have just seen the best and neatest way to configure the complexities of a 3-way active crossover system is to use three- or six-channel power amplifiers, and one might hope that a market for these would develop. One wonders if there might be some sort of psychological resistance to buying two parts of the audio chain with a gap (the power amplifier) between them. We can, I think, confidently predict that some enthusiasts would favour different makes of power amplifier for the LF, MID, and HF channels. There would be much room for entertaining debate if you like that sort of thing. We might even speculate that third-party active crossovers might be marketed to replace those from the speaker manufacturer.

We also saw that it is highly desirable to use multi-way cables between the active crossover and the power amplifiers, for tidiness as well as the best technical performance. A market could emerge for 3-way or six-way leads. Ideally each audio line should be individually screened to prevent capacitive crosstalk between them, especially on long cable runs, but this puts the cost of the cable up substantially. Conventional line outputs have a series resistor in the output to ensure stability when faced with cable capacitance from signal to ground, resulting in a typical output impedance of 50–100 $\Omega$. This is enough to allow the capacitance between adjacent and unshielded signal conductors to significantly degrade the crosstalk performance. A most effective solution to this problem is the use of so-called Zero-impedance line outputs on the crossover. This technology typically achieves an output impedance of a fraction of an ohm and can reduce crosstalk by 40 dB or more; it is equally stable into long cable runs, and the extra cost is trivial. Zero-impedance outputs are fully described in Chapter 21 on line output stages.

An unbalanced six-way interconnection uses six signal feeds and at least one ground wire, and so requires at least seven pins in the connector used.

Since an active crossover system is expected to reach high levels of quality, the use of balanced interconnections needs examination. The hot and cold signals from each channel can be wrapped by a single grounded screen, but ideally there would be separate assemblies of this sort for each channel, which makes for rather non-standard cable. Conventional balanced operation naturally increases the number of contacts on the connectors used to at least $(6 \times 2) + 1 = 13$, though multiple ground connections would be good, not only to reduce voltage drops due to ground currents but also to avoid alarming the superstitious end of the audio market.

This assumes that the outputs are truly balanced, in that there are two outputs in anti-phase. This gives a handy 6 dB increase in level over the link, which is very useful in coping not only with noise on the link itself, but also relatively noisy balanced input amplifiers; see Chapter 20. If however we drop the requirement for anti-phase outputs, then we could have a quasi-balanced cable with six signal feeds, one cold line to sense the source-end ground, and one ground wire. The single cold line would be distributed to the cold inputs of six balanced input amplifiers at the receiving (power amplifier) end. The danger is, of course, that people would condemn it as "not a real balanced connection", though it would almost certainly give an equally good performance as regards common-mode rejection. This plan would require a minimum of eight pins in the connector.

## Multi-Way Connectors

A multi-way cable requires multi-way connectors. It would be nice if active crossovers became so popular that a committee designed a special connector for us (like the HDMI connector, for example), but realistically that isn't likely to happen soon. It is therefore worth looking at what existing connectors are capable of meeting our needs. What we must strive to avoid is a connector configuration already in common use, because in the real world people are prone to plug stuff into any socket that will physically accept it.

XLR connectors have the benefit of being fairly familiar, but they only go up to seven pins. [19] This is fine for 3-way balanced use, but only allows unbalanced six-way operation (six audio feeds plus a common ground). Since we are dealing with higher-end systems, it seems inappropriate to rule out balanced operation.

The familiar standard DIN connectors have a 13.2 mm diameter metal body and go up to eight pins, [20] so we can have fully balanced 3-way operation or six-way quasi-balanced operation, but not fully balanced six-way usage. These connectors have some unfortunate associations with the low-quality DINs in the past, but today reliable high-quality versions are freely available and are used for MIDI links and stage lighting control. They are not popular with those who have to solder cables into them, but then this is probably true of any small multi-way connector.

The smaller Mini-DIN connectors [21] are 9.5 mm in diameter and officially come in seven patterns, with the number of pins from three to nine, though there are at least two non-standard ten-pin versions which are not approved by the Deutsches Institut für Normung (DIN), the German standards body. Once more we can have fully balanced 3-way operation, or six-way quasi-balanced operation, but not a fully balanced six-way mode. Mini-DINs are sometimes called "video camera connectors" though they have in fact been used for a very wide variety of uses, including computer power supplies, so there is a definite element of risk there.

The nine-way D-type connector [22] is inexpensive and offers screw retention, but it is hard to argue that it has anything of a high-end audio air about it. It is also likely to get plugged into the wrong place, whereas standard-format connectors like XLRs or DINs with an unusually large number of pins are much safer. Once more we can have fully balanced 3-way operation, or six-way quasi-balanced operation, but not a fully balanced six-way mode.

The 15-way D-type connector has the same problem of lack of glamour as the nine-way, but it is the only easily-sourced connector which will allow fully balanced six-way operation. The two spare ways can be usefully pressed into service as extra ground connections, or used for control functions such as 12 V trigger to activate amplifiers.

There are various proprietary connector systems; for example the eight-way Neutricon by Neutrik is a robust circular metal connector. [23] It is stocked by Farnell so is easy to source. Neutrik also make the 12-way miniCON. [24]

## Subjectivism

As I warned you in the preface, this book has no truck with faith-based audio. There is no discussion of oxygen-free copper, signal cables that only work one way, magic capacitors hand-rolled on the thighs of Burmese virgins under a full moon, or loudspeaker cables that cost more than a decent car. Valve technology is ignored because it is inefficient and obsolete and, despite much ill-informed special pleading, absolutely has no magical redeeming features. You will find here no gas-fired pentodes, nor superheated triodes fed with the best Welsh steam coal and still bespattered with the mud of the Somme. It would of course be possible to design a complex active crossover using valves, but even if you accepted mediocre performance as regards noise and distortion, the result would be very expensive, very hot, and very heavy.

You will, I think, find enough real and intriguing intellectual challenges in crossover design to make it unnecessary to seek out non-existent ones.

## References

[1] www.talkbass.com/wiki/index.php/Driver_diameter_&_pistonic_beaming_frequencies
[2] Self, Douglas "Audio Power Amplifier Design" Sixth Edn, Newnes, 2013, pp. 370–371, ISBN: 978-0-240-52613-3, Chapter 14
[3] https://en.wikipedia.org/wiki/Dolby_noise-reduction_system; Accessed December 2016
[4] http://djtechtools.com/2011/12/11/an-introduction-to-mixing-with-dj-isolator-mixers/; Accessed December 2016
[5] http://formula-sound.co.uk/1ru-solutions/fsm-3-1ru-isolator/; Accessed December 2016
[6] Self, Douglas "Ultra-Low-Noise Amplifiers & Granularity Distortion" JAES, November 1987, pp. 907–915
[7] Self, Douglas "Audio Power Amplifier Design" Sixth Edn, Newnes, 2013, pp. 370–371, ISBN: 978-0-240-52613-3 (cable effects)
[8] Colloms, Martin, "High Performance Loudspeakers" Third Edn, Pentech Press, 1985, p. 174, ISBN 0-7273-0806-8
[9] Colloms, Martin, "High Performance Loudspeakers" Third Edn, Pentech Press, 1985, p. 178, ISBN 0-7273-0806-8
[10] Borwick, John (ed.)"Loudspeaker & Headphone Handbook" Second Edn, Focal Press, 1994, p. 234, ISBN 0-240-51371-1
[11] Gaertner, Jean-Claude "Project 21 Part 1" Linear Audio Volume 0, pub J Didden 2010, p. 132 ISBN: 9–789490–919015

[12] Ward, Phil  www.soundonsound.com/sos/jul02/articles/monitors2.asp (thermal compression)

[13] Newell, Phillip "Recording Studio Design" Second Edn, Focal Press, p. 511, ISBN: 978-0-240-52086-5

[14] Colloms, Martin "High Performance Loudspeakers" Third Edn, Pentech Press, 1985, ISBN 0-7273-0806-8

[15] Self, Douglas "Audio Power Amplifier Design" Sixth Edn, Newnes, 2013, Chapter 12, ISBN: 978-0-240-52613-3(power amp distortion)

[16] Self, Douglas "Audio Power Amplifier Design" Sixth Edn, Newnes, 2013, Chapter 15, ISBN: 978-0-240-52613-3 (slew-rate limiting)

[17] Self, Douglas "Self On Audio (Audio Power Analysis)"Third Edn, Newnes, 2016, Chapter 41, ISBN: 978-1-138-85445-1 hbk(Class-A efficiency)

[18] Rubinson, Kalman  www.stereophile.com/solidpoweramps/207classe/ (review)

[19] https://en.wikipedia.org/wiki/XLR_connector; Accessed December 2016

[20] https://en.wikipedia.org/wiki/DIN_connector; Accessed December 2016

[21] https://en.wikipedia.org/wiki/Mini-DIN_connector; Accessed December 2016

[22] https://en.wikipedia.org/wiki/D-subminiature; Accessed December 2016

[23] www.neutrik.com/en/industrial/circular-connectors/neutricon/neutricon-connectors/; Accessed December 2016

[24] www.neutrik.com/en/industrial/circular-connectors/minicon/minicon-connectors/; Accessed December 2016

# CHAPTER 2

# How Loudspeakers Work

This chapter is not in any way a guide to how to design loudspeakers, nor is it a complete explanation of how they work; what it does do is look at their functioning from the point of view of the crossover designer. The specific topic of time-aligning the drive units is dealt with in Chapter 13 on delay compensation.

There are many ways to make a loudspeaker. Some of the less common types are electrostatics, ribbon loudspeakers, electromagnetic planar loudspeakers, Heil air-motion transformers, and the rather worrying Ionophone. These are dealt with only briefly in this chapter, but plenty of references are given for further reading. Here I must concentrate on the most popular concepts: sealed boxes, ported reflex boxes, ABR systems, transmission line loudspeakers, and horns (see Figure 2.1).

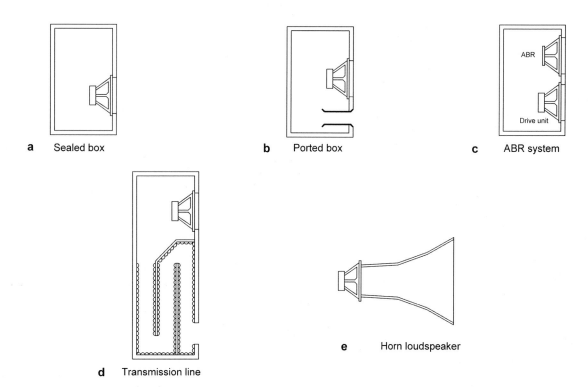

**a**    Sealed box           **b**    Ported box           **c**    ABR system

**d**    Transmission line           **e**    Horn loudspeaker

**Figure 2.1: The basic loudspeaker arrangements: sealed box, ported reflex, ABR, transmission line, and horn.**

## Sealed-Box Loudspeakers

A sealed enclosure, as shown in Figure 2.1a, is the simplest. It gives a good transient response, assuming a correct choice of Q; good low-frequency power handling because the drive unit cone is always loaded; and has less sensitivity to misaligned parameters than other types. On the other hand, sealed enclosures have higher low-frequency cutoff points and lower sensitivity than the other loudspeaker types for the same box volume.

A loudspeaker in a sealed box has two important factors working on it—the mass of the moving parts of the driver, and the compliance ("springiness") of the air in the box. These two factors make up a 2nd-order system, like a weight bouncing on a spring, and the effect is that the loudspeaker response is that of a 2nd-order lowpass filter with a 12 dB/octave roll-off as frequency falls. Driver cone excursion increases at 12 dB/octave, with decreasing frequency for constant SPL, so cone excursion tends to be constant below the cutoff frequency.

Such a filter can have different values of Q; the higher the $Q$ the more peaked the response (as shown in Figure 2.2). The $Q$ is an important choice in the design, and is often referred to as the "alignment" by reference to a filter type. Thus a Butterworth alignment uses a $Q$ of 0.707 (1/√2) and gives a maximally flat response, just like a 2nd-order Butterworth filter. This is not necessarily the ideal Q; higher $Q$'s give significantly more LF output below the cutoff frequency, but at the expense of response peaking, and too high a $Q$ gives a boom-box or "one-note bass" effect, which is not usually what is wanted. A $Q$ of 0.5 gives a good transient response ($Q$ = 0.707 gives overshoot on square waves) but is often criticised for being "too taut" or overdamped, and the LF output clearly suffers.

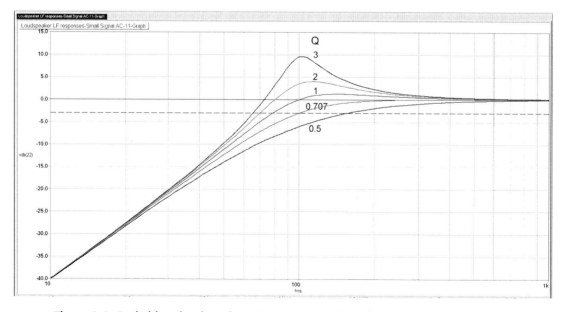

**Figure 2.2: Sealed-box loudspeaker LF responses with a cutoff frequency of 100 Hz and various Qs. Dotted line is at −3 dB**

For a given drive unit, the system *Q* is determined by the box volume. The bigger the box, the lower the Q; the responses in Figure 2.2 correspond to a range of box volumes of about 50 times.

The box may be left completely empty, or it may be lined or wholly filled with acoustic damping material, such as bonded acetate fibre (BAF) or long fibre wool. This not only absorbs internal reflections but, for the "fill-'er-up" approach, alters the thermodynamic properties of the enclosed air so that the enclosure behaves as though slightly larger than its physical size.

Sealed-box loudspeakers are normally classified into two kinds: the infinite baffle type and the air suspension type. If you have a large enclosure where the compliance of the air inside is greater than the compliance of the drive unit suspension, so that most of the restoring force comes from the latter, it is regarded as an infinite baffle type. A small enclosure where the compliance of the air inside is less than the compliance of the drive unit suspension by a factor of three or more is regarded as an air suspension type.

The efficiency of a drive unit in a sealed box is proportional to the cube of the drive unit resonance frequency (the Thiele-Small parameter $F_s$), [1] so attempts to get a lot of low-frequency extension with a given drive unit and a small box volume inevitably result in low efficiency. Nowadays amplifier power is relatively cheap, but there may be problems with large cone excursions and voice coil heating. Other sorts of loudspeaker enclosures that reinforce the LF output with the rear cone radiation are often used to increase the efficiency, as we shall now see.

## Reflex (Ported) Loudspeakers

The alternative loudspeaker types all, in different ways, return low-frequency energy from the rear of the drive unit in the correct phase to reinforce that radiated from the front.

Reflex or ported loudspeakers (sometimes called vented boxes), as shown in Figure 2.1b, have a pipe-like port which allows the passage of air in and out of the box. The mass of air in the port resonates with the compliance of the air in the box, causing the port output to be in phase with the forward radiation of the drive unit at low frequencies, and allows the bass response to be extended without the lower efficiency that results when that is done with a sealed box. At frequencies below the port resonance the port output is in anti-phase, causing the output to fall at 24 dB/octave rather than 12 dB/octave, as for the sealed box. The response looks like Figure 2.2 but with steeper roll-offs. This does not result in a notch in the response, as occurs in the ABR system, but the increased slope makes things more difficult if you are planning low-frequency response extension by equalisation in the crossover, as the matching of equaliser and driver/box responses is more critical.

While a ported loudspeaker can have lower drive unit distortion, greater power handling, and a lower cutoff frequency than a sealed-box system using the same drive unit, there are snags. Below the cutoff frequency the drive unit quickly becomes unloaded, and its excursion increases dramatically, worsening non-linear distortions and threatening mechanical damage. To guard against this a subsonic filter should always be used when the source is vinyl. It has been persuasively argued that the arrival of CDs, with extended bass but no subsonic disturbances, caused a major increase in the popularity of ported loudspeakers. The transient response of a ported loudspeaker is considered to be not so good as for a sealed-box system with the same drive unit. Ported-box systems are significantly more sensitive to misaligned parameters than are sealed-box systems and require more careful design.

At low frequencies large amounts of air are moving through the port; to reduce chuffing noises from turbulent air flow it should have the largest diameter possible consistent with the resonance required and be flared at both ends.

## Auxiliary Bass Radiator (ABR) Loudspeakers

Auxiliary bass radiator loudspeakers (also called passive radiator loudspeakers) work in a similar way to ported loudspeakers. The mass of air in the port is replaced by the mass of the auxiliary bass radiator, which is essentially an LF drive unit with the usual cone and suspension but no voice coil or magnet, as shown in Figure 2.1c. The response of an auxiliary bass radiator loudspeaker is therefore very similar to that of a ported loudspeaker using the same driver. However, the LF cutoff frequency will be somewhat higher, and the roll-off slope will be steeper, due to the presence of a notch in the frequency response which corresponds to the free air resonant frequency of the ABR. This causes a steeper roll-off below the system's tuned frequency Fb (the resonant frequency of LF driver and box together) and a poorer transient response. Normally the notch is well below the loudspeaker passband (say below 20 Hz), where the response has already dropped considerably, so despite the rapid phase changes associated with the notch, it is normally considered to be of little or no audible significance; it is not practical to equalise it away. The larger the ABR, the more mass its cone will have, and the lower its resonance frequency will be for the same target Fb, pushing the notch further away from the bottom of the loudspeaker passband. There is much more flexibility in design, as the ABR cone mass can be altered. The ABR unit may be mounted above or below the main LF drive unit. Two ABR units are sometimes combined with a single active LF driver unit; if so, they are mounted on either side so that the inertial forces cancel out and put less excitation into the front baffle.

ABRs eliminate port turbulence noises and port resonances and give good LF extension for a given box volume. They are however more complex to design than ported types and are more costly due to the extra auxiliary bass radiator.

## Transmission Line Loudspeakers

A true transmission line or "acoustic labyrinth" loudspeaker sends the energy from the rear of the drive unit down an infinite pipe, where it is gradually absorbed and never heard of again. Since infinite pipes rarely fit in well with domestic surroundings, practical "transmission line" loudspeakers are a compromise. The rear radiation passes down a length of duct chosen so that it is in the correct phase to reinforce the forward radiation at low frequencies. To get worthwhile reinforcement the duct needs to be a quarter of a wavelength long at the frequency of interest, so it shifts the phase of the rear output by 90°. At 40 Hz the wavelength of sound is 8.58 metres, and a quarter of that is 2.145 metres, which is an impractically long duct unless it is folded at least twice, as shown in Figure 2.1d, where the bobbles represent absorbent material lining the duct. Transmission line loudspeakers are often described as non-resonant, but without the quarter-wave resonance (which could be suppressed by using enough absorbent stuffing in the duct) there will be no enhancement of the bass response.

The low-frequency reinforcement works in the same way as a ported box, and the response falls at 24 dB/octave, once again making low-frequency response extension by equalisation in the crossover more difficult. Transmission line loudspeakers are not used in sound reinforcement because the duct greatly increases the volume and weight of the box without offering any advantages over a ported system; they have the same disadvantage in the domestic environment. Compared with sealed-box, ported, and ABR designs, they are not popular.

## Horn Loudspeakers

An acoustic horn is used to improve the coupling efficiency between a drive unit and the air. It acts as an acoustic transformer that provides mechanical impedance matching, converting large pressure variations over a small area into low pressure variations over a large area, giving a greater acoustic output from a given driver. Horns have a low-frequency limit set by the flare rate and the mouth size; the slower the flare rate, the lower the frequencies a horn can reproduce for a given length. A horn with an area flare rate of 30% per foot is considered to be effective down to about 30 Hz. The flare may have an exponential or tractrix flare; conical flares, as used on old phonographs, have a poorer low-frequency response; the flare in Figure 2.1e is illustrative only. For these reasons horns with an extended low-frequency response are physically large, and only practical if folded in some way. This is not quite as clever as it appears because bouncing the sound around corners can cause frequency response anomalies at the upper end of the working range, due to reflections and resonances.

Horn loudspeakers are widely used in sound reinforcement because of their high efficiency, which makes a significant savings in the number of power amplifiers and loudspeakers you need.

Constant directivity horns are a relatively recent development where an initial exponential section is combined with a final conical flare, dispersing the shorter high-frequency wavelengths more effectively. They require special equalisation to allow for this greater spread, and this is covered in Chapter 14.

## Electrostatic Loudspeakers

Electrostatic loudspeakers consist of large flat diaphragms moved by electrostatic forces. [2] The most famous examples of these have been made by Quad, beginning with the ESL-57 in 1957. [3] Its 1981 successor, the ESL-63, [4] was a more complex design having eight separate panels driven from a delay line, intended to improve both the bass response and the dispersion; the latter was very poor at high frequencies in the ESL-57. The ESL-63 was also provided with protection against over-driving.

To obtain tolerable distortion characteristics electrostatics have two diaphragms driven in push-pull, on either side of a fixed electrode carrying a high DC polarising voltage. Another requirement for acceptable distortion is the use of constant-charge mode; the moving film is coated with a high-resistance layer that prevents charge moving in or out over a cycle of the lowest audio frequency of interest. Electrostatic loudspeakers usually have no rear enclosure and so operate as dipoles, meaning that low frequencies tend to sidle around the edge of the panel and cancel. Thus there is a −3 dB bass roll-off at the frequency where the narrowest dimension of the panel equals a quarter-wavelength. A Quad ESL-63 is 0.66 metres wide, and the −3 dB point occurs at around 130 Hz. This is a high bass roll-off frequency compared with moving-coil loudspeakers, and so upward-shelving equalisation is sometimes used. This sort of bass extension has limited effectiveness because it has to be done with very great care indeed; over-driving an electrostatic speaker can cause the film to touch the fixed electrode, burning a hole in the film. For real bass extension, crossing over to one or two subwoofers is recommended. See Chapter 18 on the crossover requirements for subwoofers.

Electrostatics require high drive voltages, usually provided by a step-up transformer driven by a solid-state power amplifier, though direct drive from suitable valve amplifiers is possible and can in theory at least side-step the imperfections of a transformer. An electrostatic appears pretty much as a purely capacitive load, which when multiplied by the square of the step-up ratio can cause problems for some amplifiers. The almost totally reactive load greatly increases the peak power dissipation in the output devices and can also cause HF instability. The use of an appropriate air-cored series output inductor in a solid-state power amplifier eliminates the instability problem.

Much information on electrostatic loudspeaker design issues can be found in [5].

## Ribbon Loudspeakers

Ribbon loudspeakers consist of a very thin metal film ribbon suspended in a strong magnetic field. Aluminium is normally used for its good combination of low density and high electrical conductivity. The lightness of the ribbon means it can accelerate quickly and give an excellent high-frequency response, but this also makes it very fragile both physically and electrically. The small radiating area means that ribbons are only suitable for use as tweeters. Most ribbon tweeters emit sound in a dipole pattern, but some have the back closed. They are expensive to build as the ribbon is a single conductor (as opposed to the multiple conductors of a voice coil), and so very strong magnetic fields are required to get an acceptable sensitivity. The impedance of ribbon speakers is inevitably very low, ranging from 1 $\Omega$ to 0.1 $\Omega$, and they cannot be driven with conventional solid-state amplifiers. Traditionally they were driven through step-down transformers, with conversion from 0.2 $\Omega$ to 8 $\Omega$ requiring a step-down ratio of 6.3:1. This is difficult because of the need to keep the secondary winding resistance very low indeed to minimise power losses. Connections from the transformer to the ribbon also need to be of very low resistance and hence very short; in some ribbon drive units, such as those by Raal, the transformer is built into the ribbon drive unit. Raal have also developed 175:1 ratio transformers that allow valve amplifiers to drive their ribbons; the impedance conversion is therefore 30,600:1, bringing a 0.1 ohm ribbon up to 3 k$\Omega$, high enough for the anode loading of many power triodes.

A few specialised solid-state amplifiers have been designed to drive ribbons directly. A design published in Elektor in 1992 [6] used multiple output bipolar output devices to drive 140 Wrms into 0.4 $\Omega$.

If the ribbon drive unit is used as a dipole source, then at a sufficiently low frequency there will be cancellation between the front and back radiation, and upward-shelving equalisation may be desirable; however, it must be used with great caution, as ribbons have negligible thermal inertia and can be burnt out very easily.

## Electromagnetic Planar Loudspeakers

Electromagnetic planar loudspeakers consist of a flexible membrane, often rectangular, with flat conductors making up a voice coil printed onto it. The conductor currents interact with the field of magnets on either side of the diaphragm, giving a substantially uniform force over its area.

The membrane therefore tends to move all as one piece without flexing, reducing resonances and improving linearity. Planar loudspeakers are normally used with closed or ported enclosures, much as for cone drivers.

Magnetic-planar loudspeakers tend to have low efficiency. They have a lower BL product than conventional cone drive units because the pole-to-pole magnet spacing is much larger than in the conventional coil and magnet assembly. Also, there is usually a shorter length of conductor in the magnetic field.

There is no rigid demarcation between magnetic-planar and ribbon drive units. A magnetic-planar unit with a very light and flexible diaphragm, with relatively few conductors attached, is sometimes described as a ribbon unit.

## Air-Motion Transformers

Air-motion transformers [7] use mechanical means to make the air move faster than the drive unit element. A typical version has an accordion-pleated element carrying flexible metal conductors. As this opens and closes under the force of the conductors in a magnetic field, the air is squeezed in and out of the folds. More air is thus set in motion faster than would be the case for a cone or electrostatic driver of the same visible surface area. A 1-inch-wide air-motion transformer (AMT) strip has an effective driver area that is comparable to an 8-inch diameter cone drive unit. The compact format means that it acts more like a point source than does the much larger equivalent cone drive unit. The small range of motion should much reduce Doppler distortion (see end of chapter).

The classic example of this technology was the Heil air-motion transformer. I first experienced these in the mid-'70s. A subjective assessment from so long ago is of little value, but I will record that the treble was beautifully clear and clean.

## Plasma Arc Loudspeakers

Plasma arc loudspeakers [8] are unique in that they have no mechanical moving part that acts on the air. They use ionised gas (electrical plasma) created by an electric discharge as a radiating element. Plasma has almost no mass, but it is charged and therefore can be manipulated by an electric field. The ionisation is usually induced by a high-voltage RF field, onto which the audio is amplitude modulated.

The idea goes back to the singing arc of William Duddell. In 1899 carbon arc lamps were coming into use for street lighting in England, and while they gave a brilliant light a downside was the humming and shrieking noise emitted by the arc. While trying to suppress this Duddell discovered that, with appropriate circuitry, tunes could be played on the arc. It was not until 1946 that Siegfried Klein confined an arc in a small-diameter quartz tube that was joined at one end to an exponential horn to improve coupling efficiency to the air. Klein worked with DuKane to produce the Ionovac from 1956, and in 1958 it was marketed through Electro-Voice and cost a hefty $147. The discharge was produced by a high RF voltage, initially at 100 kHz, but later raised to 3 MHz to suppress hissing noises. This caused severe EMC emission problems; later versions raised this to 27 MHz. In England at the time this was the radio-control model band, and presumably there was a hope that no one would notice interference there, at least not until they got hit by a runaway model plane. Unfortunately the

DuKane electrode assembly lasted only 200 to 300 hours; much work increased this to 1200 hours. In 1965, Fane Acoustics in England produced the Ionofane 601 loudspeaker, which appears to have been essentially a copy of the DuKane Ionovac; it had little success in the marketplace.

A serious disadvantage of plasma arc technology is the generation of ozone and nitrogen oxides (NOX) from the air; you don't want to breathe either. In 1978 Alan Hill of the Air Force Weapons Laboratory in Albuquerque, NM, designed the Plasmatronics Hill Type I, whose plasma was generated from helium gas, which solved that problem but replaced it with a need for gas cylinders that had to be changed every month. It is worth pondering that helium is the ultimate non-renewable resource.

Paul Klipsch measured a plasma speaker that was presumably an Ionophone [9] and concluded that its Doppler distortion was severe because of the small size of the effective piston area, which meant it had to move very fast at normal volumes. [10] Most plasma arc references are some way back in the past; these are worth checking out if you want to know more. [11][12] There is a fascinating 1954 report by the BBC Research Department that can be downloaded for free. [13] Despite the difficulties, interest in these devices continues; Deraedt investigated a corona-discharge plasma transducer in 1991. There has been a recent article in *HiFi News* [14] where it emerges that ozone poisoning is a real risk. There are at least two plasma tweeters currently on the market, both coming from Germany; the Acapella ION TW 1S, [15] which claims a frequency response reaching to 50 kHz, and the Lansche Corona, [16] which claims a frequency response to 150 kHz. Neither company mentions ozone or NOX in its advertising material.

A variation on this is the flame loudspeaker, which creates plasma thermally but is even less practical. Hydrogen and oxygen are recommended as fuel and oxidiser to avoid toxic combustion products, but you may still get some NOX. This is not my idea of a practical technology.

## The Rotary Woofer

This is a special form of loudspeaker that few of us are likely to encounter, but it has a certain steampunk allure. It was introduced by Bruce Thigpen. It is essentially a rotary fan with variable-pitch blades, [17] their pitch being controlled by the audio signal. Blade pitch can move both positive and negative with respect to zero pitch, so the fan can either suck or blow, and this can be used to create very low-frequency sound below 20 Hz when the fan is installed in a wall to act as a baffle. If the fan is set in the wall of a sealed room, it can generate subsonic frequencies right down to DC; in other words it can maintain a static pressure differential simply by compressing the air in the sealed space. This is not of great value unless you are trying to reproduce the barometric pressure prevailing when a recording was made. Since the audio signal only changes the fan blade pitch and does not create acoustic power directly, only relatively small power amplifiers are required; most of the power comes from the fan motor. The load on the amplifier is essentially resistive and thus easy to drive. There is little musical content that goes below 20 Hz, and rotary woofers are normally restricted to VLF sound effects and special applications such as the emulation of very long organ pipes.

One example of this was Trinity Church in Wall Street, New York, where two rotary woofers were temporarily installed in a specially sealed-off side-entry porch (I think in 2008); this was effectively an infinite baffle and suppressed the fan motor noise. The rotary woofers emulate the sound of a 32-foot organ stop, which has a bottom note of "double pedal C0", which is 16.35 Hz. Obviously a couple of fans are going to take up much less space than a set of 32-foot organ pipes; the pipes also need a powerful air compressor to function. This is by no means as low as you can go; there are a few organs in the world that have 64-foot stops, the lowest note being 8 Hz. Examples are in the Sydney Town Hall, Australia, and the Atlantic City Convention Hall, USA. Be aware there can be dirty work at the crossroads in the pipe organ business; some stops attempt to emulate a very cumbersome 64-foot stop by combining a 32-foot stop with a Fifth extension 21–1/3-foot stop; mixing these gives a beat note at 8 Hz that gives something of an impression of a real 64-foot stop. If you are intrigued by extreme pipe organs consult [18].

The only commercial rotary woofer on the market at present is the TRW-17, manufactured by Eminent Technology. [19] They claim a frequency response of 1–30 Hz ±4 dB in a suitable installation and require 150–200 W of audio drive power. The crossover requirements are essentially those for a subwoofer, but at lower frequencies (see Chapter 18). The manufacturers recommend a crossover frequency of 20 Hz with an 18 dB/octave slope. At these frequencies it seems doubtful if there's much point in worrying about the phase, but a phase-inversion switch might well be useful in some circumstances. Join the fan club. . .

## MTM Tweeter-Mid Configurations (d'Appolito)

The MTM or D'Appolito configuration is designed to greatly reduce lobing errors. Joseph D'Appolito introduced it in 1983. [20] As described in Chapters 1, 3, and 4, when two drive units are radiating simultaneously—in other words at or near their crossover frequency—the driver outputs cancel and reinforce to give lobes in the polar response. When the drivers are not time-aligned, the main lobe is tilted unhelpfully away from the main axis. Lobing error can also be caused by crossovers that have phase differences between the output signals; if so the main lobe is tilted towards the drive unit that is phase lagging, which is usually the LF drive unit. This is shown for the conventional tweeter/mid (TM) configuration in Figure 2.3.

If two LF/MID drive units (sometimes called midwoofers) are placed on the vertical axis, equally spaced from the tweeter, then the symmetry of the MTM situation causes the main lobe to always be

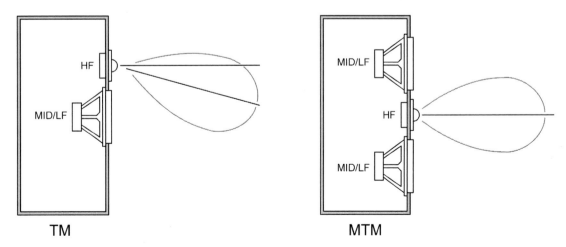

**TM**

**MTM**

**Figure 2.3: TM and MTM (D'Appolito) drive unit configurations.**

directed forward. MTM designs are commonly 2-way, using LF/MID drive units crossing over to the tweeter at a relatively high frequency. This is also known as a (3,2) configuration—i.e. three drive units but a 2-way crossover; the TM case is described as (2,2), having two drive units and a 2-way crossover.

There are disadvantages:

1.  The main forward lobe is rather narrower than in the TM case because of cancellation and reinforcement between the two LF/MID units.
2.  It was pointed out by Mithat Konar [21] that interaction between the two LF/MID drive units below the crossover frequency causes irregularities in the frequency response well off the central axis. These can be minimised by mounting the three drive units as close together as is mechanically possible; this is rarely done in commercial loudspeaker designs, apparently because there is a consensus that it looks better to have some wood between the drive units. It is also helpful to place the crossover frequency low enough so that there is enough output from the tweeter to mask nulls caused by the two LF/MID units cancelling.
3.  The need for two LF/MID drive units means that it is harder to design an MTM loudspeaker for an 8 Ω nominal impedance. Most versions have a 4 Ω nominal impedance.
4.  A 2-way design needs a relatively high crossover frequency, typically around 1 kHz, and this puts considerable demands on the LF/MID drive unit capabilities.

The narrow forward lobe can be addressed by ensuring that the tweeter lags the midwoofer by 90° in phase at the crossover frequency, as in the 3rd-order Butterworth crossover originally used by D'Appolito. A more recent approach by Hughes [22] is the suppression of the cancellation nulls by adding allpass and lowpass filters to the binary crossover structure put forward by D'Appolito. [23]

## Vertical Line Arrays

Vertical line arrays of loudspeakers have become very popular for PA use; at most big gigs nowadays there will be two or more vertical columns of identical loudspeakers (boxes) suspended, almost always with the lower speakers pointed somewhat downward; this is often called a J-array. If subwoofers are used they are likely to be too big and heavy for suspension and will be placed on the stage below the vertical array.

The major advantage is that the vertical stacking of drive units causes the acoustic output to be mostly forward, and so directed at the audience. In the horizontal plane the array will have the same polar response as a single drive unit; the horizontal coverage is typically around 90° wide. Relatively little acoustic energy goes up or down. In particular, side lobes in the vertical polar response are minimised, so little energy is directed upwards. This has several advantages:

1.  More of the energy is directed forward, so the acoustic efficiency is higher.
2.  If little energy is directed upwards, even less will be reflected from the roof and back down to the audience. This reflected sound suffers an extra delay due to the longer path and will seriously degrade the overall sound quality.
3.  Reverberation in the upper part of the auditorium will not be excited, and this will give a cleaner sound in the audience positions.

Straight and J-array vertical columns are shown in Figure 2.4. The number of boxes and drive units will vary in different line array systems.

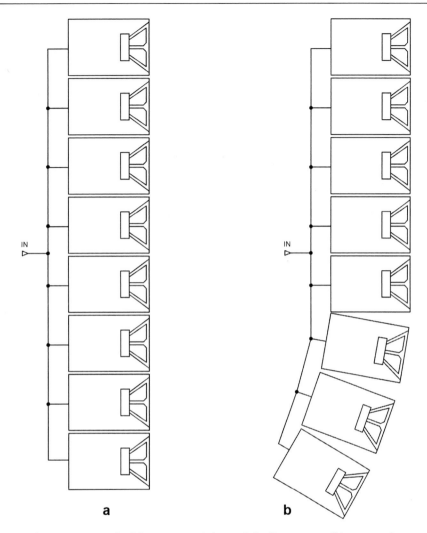

**Figure 2.4: Vertical line arrays: (a) Straight-line array; (b) J-array for better coverage of front areas.**

Medium-sized line arrays may be either two- or 3-way; the LF drive units are typically 10 to 15 inches in diameter. Large line array loudspeakers (boxes) designed for big venues or outdoor festivals are usually 3-way and typically use, in each box, an HF compression drive unit with multiple MID and LF cone drive units placed symmetrically around the HF driver, as in the MTM (D'Appolito) configuration. This prevents the main forward lobes from being deflected sideways. The LF drive units are typically 15 or 18 inches in diameter for adequate power handling.

Vertical loudspeaker arrays are not a new technology. They were described by Olson in his classic 1957 text, *Acoustical Engineering*, [24] but he described the basic operation of a line source a long time before that, in 1940. [25] An early use was for sermon reinforcement in big churches, where reducing the excitation of the roof space, which would have an enormous reverberation time, improved intelligibility considerably. Interestingly, a comprehensive account of the application of line arrays in St Paul's cathedral, London, appeared in *Wireless World* in 1952, [26] long before Olson's 1957 book. The implementation included amplitude tapering to broaden the main forward lobe and reduce the side lobes, both in amplitude and number; this is shown in Figure 2.5, which is after the original in the *WW* article; it seems to have been the cutting-edge of the technology of the time. Amplitude tapering is explained later.

Another major advantage of the vertical line array is that the sound directed forward falls off more slowly with distance than for other loudspeaker configurations. A point source propagates sound equally in all directions as a sphere. The surface area of this sphere quadruples as the distance from the loudspeaker doubles, so the acoustic power is spread out over four times the area. This translates into a −6 dB loss in sound pressure level (SPL) as distance doubles. In contrast, a line array is ideally a one-dimensional line source; sound from it propagates in the shape of a cylinder, and now every doubling of distance only doubles the surface area; this translates into a −3 dB loss of SPL as distance doubles. This area is often called the near-field. However there are limits to this; with increasing distance the line source starts to look more like a point source, and the loss changes from −3 dB per doubling of distance to −6 dB per doubling; after this you are in the far-field. The distance at which this effect becomes significant is sometimes known as the critical distance.

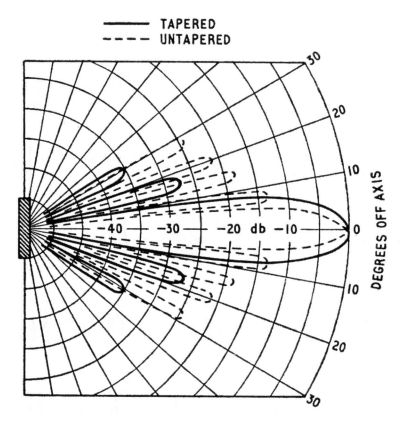

**Figure 2.5: Line array with and without amplitude tapering. From *Wireless World*, 1952.**

Vertical line arrays maintain this −3 dB loss for the doubling of distance for much further at high frequencies than at low frequencies, and this can be useful for counteracting the preferential absorption of high frequencies by the atmosphere—there is more on this at the end of the chapter.

Straight-line arrays are not much used today, as the forward lobe becomes too narrow at high frequencies. The J-array (sometimes called an articulated array) much reduces this problem, as the downward-pointing drive units at the bottom fill in the front of the audience area which would otherwise miss out. A development of this is the so-called spiral array, in which the angle between the enclosures increases steadily from the top to the bottom of the array. This gives better lobe control over the audio bandwidth.

A greater length of array gives a narrower vertical dispersion, and the longer the array, the lower the frequency down to which the polar pattern control will be effective.

## Line Array Amplitude Tapering

It was mentioned earlier that if the power to each loudspeaker is reduced steadily from the middle to each end of the array, the polar response can be improved; this amplitude tapering technique was in use as early as 1952, in the St Paul's cathedral installation. The method is demonstrated in Figure 2.6.

This technique is also called power tapering or amplitude shading. It is widely used in current line array systems to give better front fill coverage, where the lower curve of a J-array is supplying the very near-field listeners in the front rows. St Paul's cathedral used resistive attenuators for the tapering, but this would be hopelessly inefficient at modern power levels. The tapering could be done by using several small power amplifiers with separate level-controlled feeds to each from the active crossover; this would give some redundancy against failure but is rather complicated. Modern practice is to use a relatively small number of high-power amplifiers, with amplitude tapering implemented by connecting differing numbers of drive units in series. A typical arrangement is shown in Figure 2.6; this is called the 2/4/6 combination because in the centre of the array (boxes 6 and 7) two boxes are in series. Further out in each direction there are four boxes in series; boxes 4, 5, 8, and 9 are in series together. At the ends of the array there are six boxes in series; 1, 2, 3, 10, 11, and 12 are all in series together. Each series section is connected in parallel to the power amplifier output.

Figure 2.6 also shows an alternative 2/3/3/4 combination where boxes 6 and 7 are in series, 3, 4, and 5 are in series, 8, 9, and 10 are in series, and 1, 2, 11, and 12 are in series. In contrast a uniform (non-tapered) array might be set up as 4/4/4, in other words with four boxes in series at top, middle, and bottom.

When selecting the combinations for series connection it must be remembered that they are in this case all paralleled at the amplifier output, and the total impedance seen by the amplifier must not fall too low. If 8 Ω boxes are used the 2/4/6 combination gives a total load of 8.73 Ω on the amplifier, while 2/3/3/4 gives a rather heavier load of 5.65 Ω. If the loading becomes too heavy, it is of course possible to drive the array with two or more amplifiers with the series-connected sections split between them.

## Line Array Frequency Tapering

The forward coverage of a vertical line array narrows as frequency increases, leading to an uneven sound balance in some areas. If a line array is driven through a bank of lowpass filters at successively

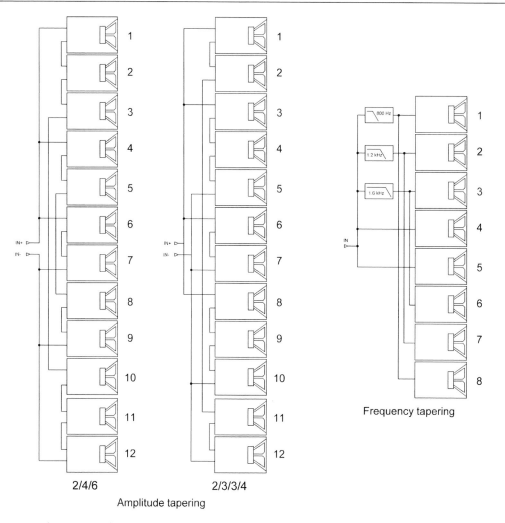

2/4/6     2/3/3/4

Amplitude tapering

Frequency tapering

**Figure 2.6: Line arrays with two forms of amplitude tapering, and line array
with frequency tapering.**

lower frequencies for drive units placed farther out towards the upper and lower ends of the vertical array, then we effectively have a long column at long wavelengths and a shorter column at shorter wavelengths, producing better matched polar patterns and critical distances over a wider range of frequencies, which in turn gives a more balanced frequency response at varying listening distances and angles. The basic method is demonstrated in Figure 2.6; it can be implemented by active crossover techniques at low level, but this may lead to an excessive number of power amplifiers (the amplifiers are not shown in Figure 2.6 for clarity). Alternatively the filters may be implemented with passive crossover techniques at high level, and despite the well-known limitations of passive crossovers this is the simplest and most popular approach.

The article on St Paul's cathedral described frequency tapering but concluded it was over-complex for the application. Even more remarkably, what were effectively delay towers were used. The line arrays away from the pulpit were delayed by 5, 10, and 15 ms, explicitly making use of the Haas effect. The delays were implemented by tape heads contacting a rotating magnetic disc. For the time it was a remarkable system.

## CBT Line Arrays

CBT stands for "constant beamwidth transducer". It is a technique developed by Don Keene [27] [28] and is proprietary to JBL. CBT combines amplitude tapering with steadily increasing time delays to the outer loudspeakers in the array, giving a much wider and smoother coverage off-axis than amplitude tapering alone. The technology originated in beam-forming systems for submarine sonar. The tapering and delays can be conveniently implemented as a passive lossy delay line, which means much less hardware than differing digital delays feeding multiple power amplifiers. JBL have produced an excellent introductory guide to this technology. [29] One of the patents that cover it is EP2247120A2, [30] which is assigned to Harman International.

Figure 2.7 shows an eight-box CBT array, similar to the JBLCBT-40 array. The undelayed signal directly feeds boxes 4 and 5, which are connected in series. Boxes 3 and 6 are fed in series with a delayed signal after L1 and C1, boxes 2 and 7 are in series after the second delay stage L2-C2, and the outer boxes 1 and 8 are fed in series after the final delay section L3-C3. The European patent gives a full description of the delay process, but is rather vague about how the amplitude tapering is done; it seems likely that in part at least it relies on losses in the series resistance of the inductors. The inductance and capacitor values increase steadily from the input to the end of the delay line.

## Diffraction

Diffraction affects loudspeaker operation for the simple reason that high frequencies have shorter wavelengths than long frequencies. At high frequencies, a loudspeaker cone radiates sound mostly in a forward direction, into what is called "half-space": a hemisphere facing forwards. At low frequencies, the longer wavelengths of the sound bend, or diffract, around the sides and rear of the enclosure, so radiation occurs into "full-space", which is a sphere with the loudspeaker at its centre. In other words the loudspeaker becomes more or less omnidirectional. The difference between these two modes appears as a lower output on the forward axis at low frequencies. In the quite unrealistic conditions of an anechoic chamber the low-frequency loss is 6 dB, halving the SPL. In the special case where the loudspeaker enclosure is spherical, the transition between the two modes is smooth and can be approximated by a 1st-order shelving response, as shown in Figure 2.8, which is derived from the classic paper on the subject by Olson in 1969. [31]

This effect is sometimes called the "6 dB baffle step" or the "diffraction loss" of the enclosure ("step" is not a very good description—it is actually a very gentle rise). The frequency at which this occurs is proportional to the enclosure dimensions, and the effect can be compensated for by simple shelving equalisation.

While spherical loudspeaker enclosures are perfectly practical, a notable example being the Cabasse La Sphère, [32] which claims to be "A true 4-way co-axial point source", they are never likely to be very popular. Apart from anything else, they need special stands to stop them rolling around the room.

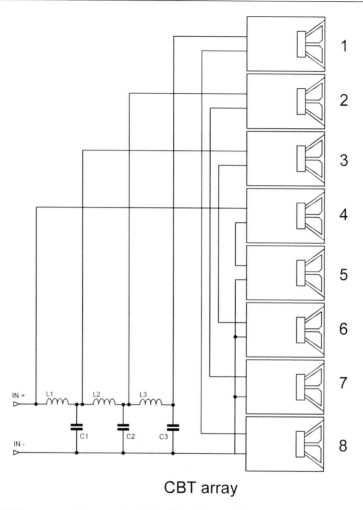

CBT array

**Figure 2.7: CBT line array with passive delay line gives wider and smoother coverage off-axis. Based on European Patent EP2247120A2.**

There are also problems because the internal reflection paths are all the same length. I once knew a chap who built a pair of spherical loudspeakers out of GRP, using full-range drive units. He was unable to resist the temptation to paint them like a giant pair of eye-balls, which did not create a restful effect.

For any shape other than spherical, or perhaps ellipsoidal, the diffraction business becomes more complicated. Any other shape has some sort of edge or edges, and when the sound waves from the drive unit strike them, a negative pressure wave is formed which radiates forward and sums with the direct output from the drive unit. This creates an interference effect which adds wobbles to the basic frequency response with its 6 dB rise; these wobbles can be as large as the step itself. This edge effect is illustrated in Figure 2.9.

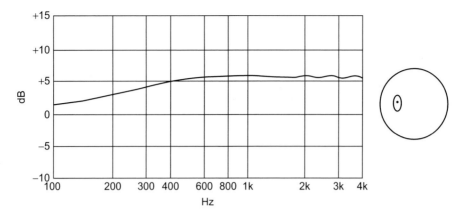

Figure 2.8: HF rise of 6 dB due to diffraction around a spherical loudspeaker enclosure 24 inches in diameter (after Olson, 1969).

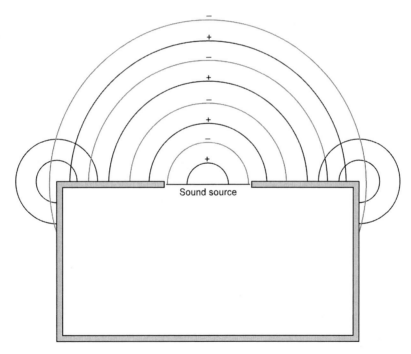

Figure 2.9: Extra sound waves generated by diffraction at the corners of a loudspeaker enclosure.

About the worst thing you can do when choosing an enclosure for a loudspeaker is to mount it in the end of a cylinder, as in Figure 2.10. This has an equal distance from the drive unit to every part of the edge at the front, and so produces the most accentuated response deviations, exceeding 10 dB. You will note that the horizontal distance between peaks or between dips on the graph decreases

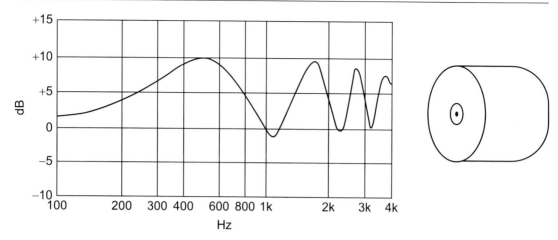

Figure 2.10: Serious response disturbances due to corners, superimposed on 6 dB rise; cylindrical loudspeaker enclosure 24 inches in diameter and 24 inches long (after Olson, 1969).

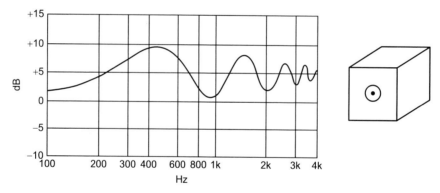

Figure 2.11: Reduced response disturbances due to corners, superimposed on 6 dB rise; cubic loudspeaker enclosure 24 inches on a side (after Olson, 1969).

as the frequency increases; this is simply because the graph has a logarithmic frequency scale, but the diffraction effect, being based on multiples of wavelength, is linear. Although the response to be corrected should be constant, because it depends on the mechanical dimensions of the enclosure, clearly any attempt to equalise away these variations would take quite a bit of doing. The obvious conclusion is that this shape of enclosure is of no use to man or beast, and should be shunned.

Rectangular (or to be strictly accurate, rectangular parallelepiped, i.e. with all edges parallel) boxes are of course much easier to fabricate than spheres or cylinders, and are correspondingly more popular. The simplest shape is the cube, which has the advantage over the cylinder that the path length from the drive unit to the front edges varies. This helps to smooth out the response wobbles, but as Figure 2.11 shows, they are still substantial, the difference between the first peak and the first dip being about 8 dB.

Figure 2.12 shows the result of converting the flat face of the forward baffle into a truncated pyramid, to make the edges less sharp. The frequency response deviations are now much less pronounced, not exceeding 2 dB, apart from the inherent 6 dB rise. Olson gives the following information about its dimensions:

> The length of the edges of the truncated surface is 1 ft. The height of the truncated pyramid is 6 in. The lengths of the edges of the rectangular parallelepiped are 1 ft. and 2 ft. The loudspeaker mechanism is mounted in the centre of the truncated surface. The lengths of the edges of the rectangular parallelepiped are 2 ft. and 3 ft. The loudspeaker mechanism is mounted midway between two long edges and 1 ft. from one short edge.

The vast majority of loudspeakers are of course rectangular boxes, as shown in Figure 2.13. These give a much worse response than the previous example because of the sharp edges of the front baffle, with deviations of up to 6 dB. The lengths of the edges of the rectangular box were

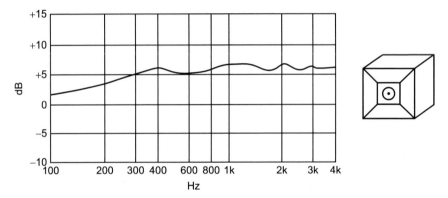

**Figure 2.12: Much-reduced response disturbances superimposed on 6 dB rise; truncated pyramid and rectangular parallelepiped loudspeaker enclosure; dimensions given in text (after Olson, 1969).**

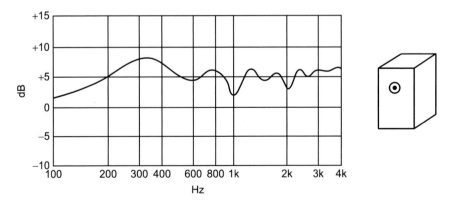

**Figure 2.13: Response disturbances due to sharp corners, superimposed on 6 dB rise; rectangular loudspeaker enclosure; dimensions given in text (after Olson, 1969).**

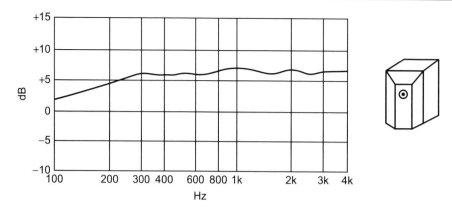

**Figure 2.14: Much-reduced response disturbances due to corners, added to 6 dB rise; Olson optimal loudspeaker enclosure; dimensions given in text (after Olson, 1969).**

2 feet and 3 feet. The drive unit was mounted midway between two long edges and 1 foot from the top short edge.

Using the information derived from these enclosure shapes, Olson suggested an enclosure that would give good results while still fitting into a domestic setting rather better than a sphere. The results are given in Figure 2.14, and the response deviations are about 2 dB; not as good as the sphere, but far better than the simple rectangular box. Olson says about its dimensions:

> A rectangular truncated pyramid is mounted upon a rectangular parallelepiped. The lengths of the edges of the rectangular parallelepiped are 1, 2, and 3 ft. The lengths of the edges of the truncated surface are 1 ft. and 2½ ft. The height of the truncated pyramid is 6 in. One surface of the pyramid and one surface of the parallelepiped lie in the same plane.

There are a few interesting points about the Olson tests which are not normally quoted. First, the drive unit he used was specially designed and had a cone diameter of only 7/8 of an inch, so it was small compared with the wavelengths in question and could be treated as a perfect piston; that is why it looks so tiny on the enclosure sketches. All the measurements were done on-axis in an anechoic chamber. Second, while the reference that is always given is the 1969 *JAES* paper, because it is accessible, the original paper was presented at the Second Annual Convention of the AES in 1950. This accounts for the somewhat steampunk look of the measuring equipment pictured in the paper.

Other workers in this field have reported that further reduction in the response ripples can be obtained by optimising the driver position and rounding the enclosure edges.

## Sound Absorption in Air

The further you are from a source of sound, the lower the SPL you experience, as the energy has spread out more; this is sometimes called "geometric spreading". However, there is another form of attenuation that increases with distance—part of the sound is absorbed as it is conducted through the air. Some of the acoustic energy is converted to heat by viscosity effects in the air itself and by vibration of water molecules mixed with the air; the attenuation therefore increases with rate of movement and so with frequency. It is also affected by the air temperature and pressure, but primarily by frequency and the relative humidity. Figure 2.15 shows the effects; virtually nothing happens below 2 kHz, but the attenuation increases quickly up to 10 kHz. The exact attenuation can be calculated, but it is a formidable equation; [33] it is said to be valid for an air temperature of less than 57°C, a pressure under 2 atmospheres, and altitudes up to 3 km. I hope that will cover most open-air concerts.

Since the amount of attenuation depends on the distance the sound travels and is negligible for short distances, this is nothing to do with hi-fi, and in PA work is normally only a significant issue for large outdoor performances. A rule of thumb I have heard more than once is: "If at 100 feet the effects are audible, at 500 feet they are likely to be serious."

From Figure 2.15, if the humidity is 40%, 2kHz will only be attenuated by 0.3 dB at 100 feet, but 10kHz is down by 4.5 dB. The decibel losses are linear with distance, so at 200 feet 2kHz suffers 0.6 dB loss, but 10 kHz is now down by a rather serious 9.0 dB. Moving out to 500 feet gives us 1.5 dB loss at 2 kHz and a 22.5 dB loss at 10 kHz, which I would agree is "serious". Generally you are unlikely to encounter humidity outside of the range 30–80%.

Figure 2.16 shows the data replotted to give a better idea of how the high frequencies are attenuated; up to 10 kHz and down to 30% humidity the slope is no greater than 6 dB/octave, and the roll-off can be compensated with simple 1st-order equalisers.

Because the attenuation increases with distance, there are limits to what can be done with overall HF equalisation. Applying 22 dB of boost at 10 kHz is not practical—it might sound good at 500 feet but would give horrible results at 100 feet. There are at least two answers to this problem—the first is the

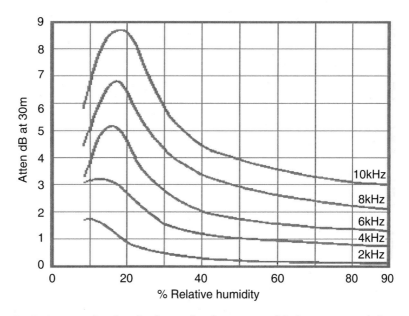

**Figure 2.15: Attenuation by air absorption increases with frequency and decreases with humidity. Results for 30m (approximately 100 feet).**

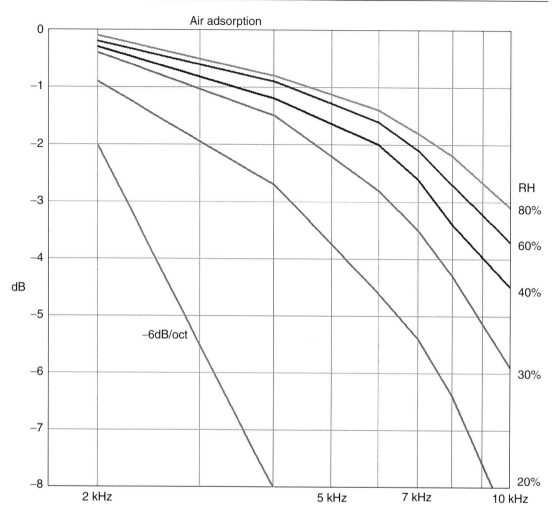

**Figure 2.16: Attenuation by air absorption plotted to show the roll-off with frequency. Results for 30 m (approximately 100 feet). The straight line is at −6 dB/octave.**

reintroduction of delay towers to shorten the air path to the listener, but nobody really wants to do that because of the extra hardware, cabling, security, and so on. A much better solution is to exploit the useful property of vertical line arrays so that they maintain a −3 dB loss per doubling of distance for much further at high frequencies than at low frequencies.

## Modulation Distortion

With any crossover the drive units have to handle a range of frequencies, and this leads to the problem known as modulation distortion. What has been called "total modulation distortion" [34] is made up of two components: amplitude modulation distortion and frequency modulation distortion. Both modulation distortion components are reduced by limiting the range of frequencies that the drive units handle and by using steep crossover slopes.

The first component is intermodulation distortion caused by non-linearities in the drive unit. This depends on the details of drive unit construction and the amount of cone movement that a given loudspeaker design requires to give the desired sound pressure level. There is more on this in the next section.

The second is a result of the Doppler effect, a low-frequency movement of the cone causing frequency modulation of the higher frequencies present.

## Drive Unit Distortion

This occurs in the drive unit itself, due to the relationship between applied voltage and cone movement not being completely linear. It is normally at its greatest at the bottom of the frequency range of the LF drive unit, as this is where the cone movements are greatest and put the greatest demands on the suspension linearity. Figures of 10% THD or more at 20 Hz are common, and the contrast with an attached power amplifier that can easily give less than 0.0004% at this frequency and full power gives one pause for thought.

There are many sources of non-linearity in the conventional moving-coil drive unit. Wolfgang Klippel has done a good deal of work in this area, [35] and he identifies significant non-linearities in the following parameters:

1.  Suspension stiffness. Due to the effect of the spider and cone surround, the force for a given displacement increases faster than linearly, often looking very like a square law.
2.  Force-factor $Bl_x$ of the coil/magnet system. This depends on the geometry of voice coil and magnet; a typical "overhung" coil is considerably longer than the magnetic gap. There is thus always the same length of coil in the gap, and so the force-factor is constant with displacement until it falls off rapidly at the extremes. This distortion mechanism is not frequency dependent.
3.  Back-EMF from (2) causing non-linear damping of voice coil movement
4.  Coil inductance changing with displacement due to variation in relationship of the voice coil and the magnetic material.
5.  Inductance variation as a result of permeability changes in the AC magnetic field. This sometimes is called "flux modulation".
6.  Young's modulus of the cone material varying with stress and strain.
7.  Non-linear air flow in reflex ports. The port must be large enough to keep the air velocity down and avoid chuffing noises while still performing its acoustic function. The port should have symmetric in-flow and out-flow characteristics to prevent a steady pressure building up in the enclosure, which will cause an offset of the voice coil position and increase distortions (1) and (2) above.

Nonlinearities (1) to (6) are clearly all matters for the drive unit designer rather than the active crossover designer, and likewise (7) depends on loudspeaker cabinet construction.

Motional feedback of the cone behaviour can be very effective in reducing drive unit distortion, and also in improving the overall frequency response; see Chapter 19 for more details.

Another approach that avoids the problems of attaching sensors to drive units is the insertion of an open-loop distortion correction system in the signal feed to the power amplifier. Since the amount of distortion that needs to be cancelled varies strongly with both level and frequency, designing such a system is no trivial matter, and I regret there is no room to go deeply into it here. Obviously a distortion correction system has to be matched to a particular model of drive unit, as does an active crossover, but it also needs to be matched to a particular example of that drive unit model; there are further difficulties with quantities like suspension compliance that are liable to drift with time. Nevertheless, if open-loop distortion correction is to be used, it makes sense to integrate it with the active crossover rather than have a separate box. Manipulating a frequency response with active filters is reasonably straightforward in the analogue domain. However, setting up very precise non-linearities that vary accurately with frequency and level is almost certainly best done with DSP.

Klippel has done a good deal of work in this area, introducing the idea of a mirror filter that could be configured to cancel out drive unit inadequacies. [36] Schurer, Slump, and Herrmann put forward an intriguing approach to cancellation-unit design in 1996. [37]

Electrostatic loudspeakers are generally considered to have much lower distortion than moving-coil designs, as they have none of the distortion mechanisms listed earlier, apart perhaps from (6) as relating to the film material; factors of one or two orders of magnitude better have been quoted. This naturally assumes proper constant-charge push-pull operation.

## Doppler Distortion

Doppler distortion is the frequency modulation of a high frequency by a lower frequency. If a cone is reproducing both high and low frequencies, when the cone is moving forward at the low frequency, the high-frequency signal is effectively increased in frequency, and vice versa. The amount of Doppler distortion generated depends on the speed at which the drive unit cone moves. Small diameter drive units have to move their cones faster to move the same volume of air in a given time, so larger drive units would be preferred. Also, splitting up the audio spectrum in a 2-way or 3-way loudspeaker, whether by an active or a passive crossover, will much reduce the possibility of Doppler effects. Because it depends on cone velocity, Doppler distortion naturally increases with level.

Obviously the frequency modulation is in theory undesirable. However, before worrying too much about this issue, you should consult the paper by Allison and Villchur, [38] who after much ingenious experimentation (for example, how do you get a Doppler-free signal from a loudspeaker? You use a microphone 90° off the main axis) and listening tests concluded that for practical cone velocities in a multi-way speaker, Doppler distortion was inaudible by a large margin. However, not all investigators have reached the same conclusion; those thinking otherwise include Paul Klipsch [39] and James Moir. [40]

It would be in principle possible to compensate for Doppler distortion. If the crossover/correction unit held an internal model of the drive unit cone velocity, it could cancel the frequency modulation by varying a suitable time-delay. Obviously this would be best done by DSP. It is only going to work below the cone-breakup frequency. Doppler compensation is important in radio communication with fast moving objects like satellites, and mobile phones on the GSM system do it to allow for high-speed trains, etc. Bats have been doing it with their ultrasonic sonar long before us. [43] I have however been unable to find anything currently going on that attempts to cancel Doppler distortion in loudspeakers. It seems to be an unfashionable subject.

## Further Reading on Loudspeaker Design

If you want to get deeper into loudspeaker design issues, two excellent books are *Loudspeakers* by Newell and Holland, [41] and *The Loudspeaker Design Cookbook* by Vance Dickason. [42] There are also a great number of useful loudspeaker design references in Appendices 3 and 4.

## References

[1] https://en.wikipedia.org/wiki/Thiele/Small; Accessed January 2017

[2] https://en.wikipedia.org/wiki/Electrostatic_loudspeaker; Accessed February 2017

[3] https://en.wikipedia.org/wiki/Quad_Electrostatic_loudspeaker; Accessed February 2017

[4] Holt, J. G.  www.stereophile.com/floorloudspeakers/416/#XpdkFtGLdWflIGB4.97; Accessed February 2017

[5] Borwick, John (ed.)"Loudspeaker and Headphone Handbook" Third Edn, Focal Press, 2001, ISBN-10: 0240515781, ISBN-13: 978–0240515786

[6] Giesberts, Ton "Output Amplifier for Ribbon Loudspeakers" Elektor, November–December 1992

[7] https://en.wikipedia.org/wiki/Air_Motion_Transformer; Accessed February 2017

[8] https://en.wikipedia.org/wiki/Plasma_speaker; Accessed February 2017

[9] Klipsch, Paul "Modulation Distortion in Loudspeakers" JAES 17(2), April 1969, pp. 194–206

[10] Deraedt, Alain "Electrostatic Transducer Using Corona Effect" AES 90th Convention, Paris, February 1991 (Preprint 3037)

[11] Anon "The Ionophone" Radio-Electronics, November 1951

[12] Anon "Loudspeaker Without Diaphragm" Wireless World, January 1952

[13] Chew, J. R.  www.bbc.co.uk/rd/publications/rdreport_1954_13; Accessed February 2017

[14] Howard, Keith "Back to the Ion Age" HiFi News, August 2017, pp. 24–29

[15] www.acapella.de/en/ion-tw-1s; Accessed July 2017

[16] www.lansche-audio.com/products/plasmatweeter/; Accessed July 2017

[17] https://en.wikipedia.org/wiki/Rotary_woofer; Accessed February 2017

[18] www.die-orgelseite.de/kurioses_e.htm#register; Accessed February 2017

[19] www.rotarywoofer.com; Accessed February 2017

[20] D'Appolito, Joseph A."A Geometric Approach to Eliminating Lobing Error in Multiway Loudspeakers" AES 74th Convention, New York, September 1983

[21] Konar, M. "Vertically Symmetric 2-way Loudspeaker Arrays Reconsidered" AES 100th Convention, Copenhagen, May 1996 (Preprint 4186)

[22] Hughes, Charles "Loudspeaker Directivity Improvement Using Low Pass and All Pass Filters" AES 125th Convention October 2008, San Francisco (Convention Paper 7536)

[23] D'Appolito, Joseph A. "Active Realization of Multi-Way Allpass Crossover Systems" AES 76th Convention October 1984, New York (Convention Paper 2125)

[24] Olson, Harry F. "Acoustical Engineering" Van Nostrand, 1957

[25] Olson, Harry F. "Elements of Acoustical Engineering" Van Nostrand, 1940, pp. 24–25

[26] Parkin, P. H. and R. H. Taylor Wireless World, February–March 1952

[27] www.xlrtechs.com/dbkeele.com/CBT.php; Accessed February 2017

[28] Keele, D. B. Jr., "The Application of Broadband Constant Beamwidth Transducer (CBT) Theory to Loudspeaker Arrays" AES 109th Convention, Los Angeles, September 2000

[29] www.jblpro.com/ProductAttachments/CBT_Tech_Note_Vol1No35_091007.pdf; Accessed 30 February 2017 https://docs.google.com/viewer?url=patentimages.storage.googleapis.com/pdfs/dbc4f6e813a18db71b87/EP2247120A2.pdf

[30] Olson, H. F. "Direct Radiator Loudspeaker Enclosures" JAES 17(1), January 1969, pp. 22–29

[31] Cabasse, www.cabasse.com/us/; Accessed Feb 2017

[32] https://en.wikibooks.org/wiki/Engineering_Acoustics/Outdoor_Sound_Propagation#Attenuation_by_atmospheric_absorption.5B5.5D_.5B6.5D_.5B7.5D; Accessed July 2016

[33] Klipsch, Paul W. "Modulation Distortion in Loudspeakers" AES Convention #34, April 1968, Paper #562

[34] Klippel, Wolfgang "Loudspeaker Nonlinearities—Causes, Parameters, Symptoms" AES 119th Convention October 2005, New York (Convention Paper 6584)

[35] Klippel, Wolfgang "The Mirror Filter: A New Basis for Reducing Nonlinear Distortion & Equalizing Response in Woofer Systems" JAES 40(9), September 1992, pp. 675–691

[36] Schurer, H., Cornelis H. Slump, and O.E. Herrmann "Exact Input-Output Linearization of an Electrodynamical Loudspeaker" AES 101st Convention, Los Angeles, November 1996 (Preprint 4334)

[37] Allison, Roy and Edgar Villchur "On the Magnitude and Audibility of FM Distortion in Loudspeakers" JAES 30(10), October 1982, pp. 694–700

[38] Klipsch, Paul "Modulation Distortion in Loudspeakers" JAES 17(2), April 1969, pp. 194–206

[39] Moir, James "Doppler Distortion in Loudspeakers" AES 46th Convention, September 1973 (Preprint 925)

[40] Newell, Philip and Keith Holland "Loudspeakers for Music Recording and Reproduction" Focal Press, 2007, ISBN-13: 978-0-240-52014-8ISBN-10: 0-2405-2014-9

[41] Dickason, Vance "The Loudspeaker Design Cookbook" Seventh Edn, Audio Amateur Press, January 2006, ISBN: 1882580478

[42] https://en.wikipedia.org/wiki/Doppler_Shift_Compensation; Accessed July 2017

# CHAPTER 3

# *Crossover Requirements*

## General Crossover Requirements

The desirable characteristics of a crossover system are easy to state but not so easy to achieve in practice. There is a general consensus that there are five principal requirements that apply to all crossovers, be they active or passive, and that they should be ranked in order of importance thus:

1. Adequate flatness of summed amplitude/frequency response on-axis
2. Sufficiently steep roll-off slopes between the output bands
3. Acceptable polar response
4. Acceptable phase response
5. Acceptable group delay behaviour

To some extent the amount of space devoted in this book to each requirement is dependent on their relative importance as set out in this list.

### 1 Adequate Flatness of Summed Amplitude/Frequency Response On-Axis

This requires that the output of each filter is appropriate in both amplitude and phase over a sufficient frequency range so that when summation occurs the overall amplitude/frequency response is flat. Note that this requirement does *not* place restrictions on the phase of the summed result; in many cases this will show considerable phase-shift that varies across the audio band. This issue is addressed in requirement 4.

Crossovers that sum to a completely flat amplitude response include the 1st-order crossover, the 2nd-order Linkwitz-Riley crossover, the 3rd-order Butterworth crossover and the 4th-order Linkwitz-Riley crossover. A new addition to this list is the Neville Thiele Method notch crossover.

There are many crossover types that can be made to sum very nearly flat by tweaking the filter cutoff frequencies. For example, the 2nd-order Bessel crossover can be made flat to within ±0.07 dB by using a frequency-offset ratio of 1.45 times, and the 3rd-order Linkwitz-Riley crossover can be made flat to within ±0.33 dB by applying a frequency-offset ratio of 0.872 times. There is much more on this in Chapter 4.

### 2 Sufficiently Steep Roll-Off Slopes Between the Filter Outputs

The filter roll-off slopes must be fast enough to prevent driver damage. Even a small amount of LF energy can rapidly wreck a tweeter. The slopes must be steep enough to not excite areas of poor drive unit frequency response, such as resonances outside the normal band of usage; this applies only to midrange drive units and tweeters, as the LF drive unit resonance will always be used. In addition, the linearity of drive units is very often worse outside their intended frequency range, so steeper slopes will give less non-linear distortion, and that has got to be a good thing.

Restricting the frequency range sent to each drive unit will also minimise frequency modulation distortion caused by the Doppler effect (see Chapter 2). It is also desirable to make the frequency range over which crossover occurs as narrow as possible to minimise the band over which lobing occurs due to two drive units radiating simultaneously.

Unless specially designed drive units are used, the minimum practical slope is usually considered to be 12 dB/octave, which requires a 2nd-order crossover. Steeper slopes such as 18 dB/octave (3rd-order) and 24 dB/octave (4th-order) are generally considered to be very desirable. Eighth-order (48 dB/octave) slopes are sometimes used in sound reinforcement.

## 3  Acceptable Polar Response

An even and well-spread polar response is desirable because it increases the amount of space in which a good sound is obtained. It is also desirable to avoid a large amount of radiated energy from being directed at the floor in front of the loudspeaker, from which it will reflect and cause unwanted comb-filtering effects by interference. Loudspeakers normally give a good polar response in the horizontal on-axis plane, assuming the drive units are mounted in a vertical line as usual. This however causes problems in the vertical plane, for in the crossover region two drive units separated in position are radiating simultaneously, and their outputs will interfere, giving reinforcements and cancellations in the radiation pattern at different angles, known as lobing. This a result of having two drive units separated in space, and there is nothing the crossover designer can do about this except make the crossover frequency range as small as possible.

However, it gets worse. If the crossover outputs to each drive unit are in phase, then the main lobe points forward on the horizontal axis, and stays there. If, however, there is a constant phase-shift between the outputs, as for 1st order crossovers (90° phase-shift) or 3rd-order crossovers (270° phase-shift), then the main lobe is tilted towards the drive unit that is phase lagging. This is usually the LF unit, so the main energy is being unhelpfully directed towards the floor. This is a frequency-dependent effect, as it only occurs in the crossover region and is at its greatest at the crossover frequency itself. It is sometimes simply called "lobing error".

It is therefore clear that, if we are going to keep the main lobe of the summed acoustic output on the axis, it is highly desirable that the lowpass and highpass outputs are in phase in the crossover region. [1] This property is possessed only by 2nd-order crossovers (assuming one output is inverted to get a flat response, otherwise the phase-shift is 180°) and 4th-order crossovers with no inversion. This is one reason for the popularity of the 4th-order Linkwitz-Riley crossover.

All of this depends on the drive unit time-delay compensation being correct; the drive units must be either physically mounted or electrically compensated so that the direct sound from each one arrives at the listener's ear at the same time over the whole of the crossover frequency range. Otherwise the main lobe will have a frequency-dependent tilt towards the driver with the longest air path to the ear.

A good polar response therefore requires that the crossover outputs be in phase and that the time-delay compensation be correct. Another approach to reducing or eliminating the lobing error is the MTM(D'Appolito) configuration, where two MID/LF drive units are placed one above and one below the tweeter. See Chapter 2 for more details on this.

## 4 Acceptable Phase Response

An acceptable phase response for the *combined* output is also required. Most crossovers are not linear phase or minimum phase but have the phase response of a 1st-order allpass filter, with the phase changing by 180° over the audio band. The best known of these are the 1st-order (inverted), the 2nd-order Linkwitz-Riley, and 3rd-order Butterworth crossovers. The 4th-order Linkwitz-Riley crossover has the phase response of a 2nd-order allpass filter, with the phase changing by 360° over the audio band. These phase responses are generally agreed to be inaudible on music signals so the fourth requirement is not too onerous. There is a separate section on the issue of phase perception later in this chapter.

## 5 Acceptable Group Delay Behaviour

Group delay is simply a measure of how much a signal is delayed. This is directly connected with an acceptable phase response for the combined output, and in fact the group delay is completely determined by the phase-shift. Group delay is mathematically the rate of change of the total phase-shift with respect to angular frequency (i.e. frequency measured in radians per second rather than Hertz).

Group delay would be of little interest if it was constant, but as the rate of change of phase varies across the audio band, with the phase response of an allpass filter, the group delay also varies. The change is sometimes smooth but may show a pronounced peak near the crossover frequency. This variation would, if it was sufficiently severe, cause a time-smearing of acoustical events and would sound truly dreadful, but you must not mistake this with the use of the word "smearing" in hi-fi reviews, where it is purely imaginary. The thresholds for the perception of group delay variation are well known because of their historical importance on long telephone lines. The most accepted thresholds were given by Blauer and Laws in 1978 [2] and are shown in Table 3.1.

**Table 3.1: Variation of group delay threshold with frequency.**

| Frequency | Group delay |
|---|---|
| 500 Hz | 3.2 ms |
| 1 kHz | 2.0 ms |
| 2 kHz | 1.0 ms |
| 4 kHz | 1.5 ms |
| 8 kHz | 2.0 ms |

These times are given in milliseconds, and a typical group delay for a 1 kHz crossover would be something like ten times less. The section on phase perception later in this chapter is concerned with the audibility of allpass filters, and the conclusion is firm that neither their phase-shift nor their group delay can be heard on normal musical signals.

The word "group" is derived from "group velocity" in wave-propagation theory, but for our purposes it is simply the amount by which a signal at a given frequency is delayed.

## Further Requirements for Active Crossovers

In addition to the general requirements for all crossovers given earlier, there are further special requirements for active crossovers. As explained in Chapter 1, if you are adding an extra unit into the signal path it must be as transparent as possible if overall quality is not to take a step backwards. The ultimate goal is total transparency, so that the introduction of the crossover causes no degradation at all.

Some further requirements for active crossovers, in no particular order, are:

1.  Negligible extra noise
2.  Negligible impairment of system headroom
3.  Negligible extra distortion
4.  Negligible impairment of frequency response
5.  Negligible impairment of reliability

It would be easy to add further requirements to this list, such as no degradation in EMC immunity, or of safety, or a modest power consumption, but these apply to any piece of electrical equipment. Those listed above are pretty clearly the most important.

### 1 Negligible Extra Noise

While passive crossovers have many limitations, as described in Chapter 1, they do not add noise to the signal. I am sure someone will now point out that crossover inductors do have some resistance and thus must generate Johnson noise, but I am quite certain that −152 dBu of noise (which is what you would get from a 1 Ω resistance) added to a loudspeaker-level signal is one of the lesser problems facing the audio business.

In contrast, the average active crossover performs its processing at line level, or possibly below, and the relative complex structure of a high-quality crossover may allow lots of opportunities for the signal-to-noise ratio to be degraded. Since active crossovers are placed after the main system volume control, turning down the volume will not turn down its noise contribution, and with poor design the result could easily be unwelcome levels of hiss from the loudspeakers.

For this reason I have deemed it important to consider noise performance at every step while describing the varied internal circuit blocks of an active crossover. Chapter 17 describes how to minimise noise by using low-impedance design, by using active gain controls, and by adopting the best order for the stages in the crossover. It also deals with the very important possibility of running the crossover at higher nominal internal levels than the input and output signals, providing the placing of gain controls in the whole sound system makes this feasible, so the circuit noise is relatively lower and we get better signal/noise ratios. The intriguing possibility of running the HF crossover path at a still higher level than the others because of the relatively small amount of energy at the top end of the audio spectrum is looked at in detail. Still another technique is optimising the order of stages in each crossover path; making sure the lowpass filters come last, after the highpass and allpass delay-compensation filters in the path, means that the noise from upstream is lowpass filtered and will be reduced by several dB. Chapter 20 on line inputs demonstrates that the conventional balanced line input stage is a rather noisy beast and shows a number of ways in which to improve it.

I have attempted to bring all these techniques together in a demonstration crossover design in Chapter 23 at the end of this book. The measured results for the all-5532 version show a signal/noise ratio of 117. 5 dB for the HF path output, 122.2 dB for the MID output, and127.4 dB for the LF output, which I think proves that using all the aforementioned noise-reduction techniques together can give some pretty stunning results.

Note however, that this crossover will still be a few dB noisier than a naked power amplifier where the input goes straight to the input transistor pair; such power amplifiers have an equivalent input noise (EIN) of about −120 dBu if well designed. [3] However, almost any sort of balanced input in the power amplifier will reverse this situation, and the crossover will produce less noise than the amplifier. This point is examined in more depth at the end of Chapter 23.

## 2 Negligible Impairment of System Headroom

As described in the previous section, it can be very advantageous to the noise performance to run an active crossover with elevated internal levels, so long as the placement of the gain controls in the whole system permits this. The scenario you are trying to avoid is having a level control between the crossover and the power amplifier that somehow (not me, guv!) gets turned down so it attenuates excessively. Then somebody turns up the volume control on the preamp or pushes up the mixing desk output faders to compensate. This means there is an excessive level inside the crossover, an unexpected signal peak comes along, and . . . crunch.

This is not a hard situation to avoid. If the active crossover has output level controls that are essentially gain trims with a limited range then it should not be possible to introduce excessive attenuation. In other cases headroom problems are avoided simply by having one competent person with control over the whole system. In hi-fi applications the maximum input levels are fairly well defined by the FSD of the digital source. In other cases the mechanical limits of the wax cylinder or the vinyl disc impose a less well-defined but still very real limit. In sound-reinforcement applications the input levels are much less predictable, but a combination of control from the mixing desk and the use of compressor/limiters should prevent excessive levels from getting as far as the active crossover.

## 3 Negligible Extra Distortion

The most significant source of distortion in the average sound system is either the loudspeakers or, if the signal source is vinyl disc, the process of cutting grooves and then wiggling needles about in them. This has never been and never will be accepted as justification for giving up on the design of very linear preamplifiers and power amplifiers. While progress has been made towards making power amplifiers as distortion free as small-signal circuitry, there are still major technical challenges to be overcome, and at present the most significant source of distortion in the electronic domain is almost always the power amplifier. For this reason it may not be too hard to design a signal path which is significantly more free of distortion, especially the crossover distortion that we all abhor, than the average power amplifier. Nevertheless, as I have said before, we are inserting an extra signal processing box into the signal path, and it behoves us to make the degradation it introduces as small as economically possible. This is not too hard from the cost point of view, as the active crossover, even if built to the highest standards, is likely to be much cheaper than the extra power amplifiers required for multi-way operation.

### 4 Negligible Impairment of Frequency Response

This may seem like a strange requirement in a piece of equipment whose whole *raison d'être* is radical modification of the spectrum of the signals it handles, but here I want to distinguish between the frequency response modifications you want and those you don't. Even if we assume that an active crossover has a well-conceived filter structure, correct filter characteristics, and suitably low component sensitivities and is built with components of accurate value, it would still be possible to degrade the frequency response by using an over-aggressive ultrasonic filter or an inappropriate subsonic filter, and these stages should get their full share of attention.

### 5 Negligible Impairment of Reliability

There is nothing that upsets the paying customer as much as equipment that stops working (OK, if it sets fire to the house that would probably annoy them rather more), so I make no apology for putting this rather general requirement in. We are dealing with opamp circuitry using modest supply rail voltages, the general levels of voltage and current are low, and the opamps even have internal overload protection. So as long as the designer knows what they're about there is no reason why any components should be much stressed. There are a couple of not-quite-obvious things that could go wrong in the power supply, such as regulator heatsinks that are normally OK but prove to be too small when the mains is 10% high in a hot country, or ill-conceived decoupling capacitors that cause the supply to fail to start on an unpredictable basis. These design landmines are dealt with in Chapter 22 on power supplies.

## Linear Phase

A linear-phase crossover has a combined output phase-shifted by an amount proportional to frequency; in other words it introduces a pure time-delay only, and the group delay is constant. Non-linear-phase crossovers have phase-shifts that change non-linearly with frequency and act like allpass filters; for this reason they are often called allpass crossovers. Now while linear phase is clearly desirable on a purely theoretical basis, in that it makes the crossover more transparent and closer to perfection, it is difficult to achieve. There is certainly no settled consensus that it is necessary for a good acoustic performance, and the bulk of evidence is that it is simply not necessary for satisfactory results on normal music. There is more on this issue in the upcoming section on phase perception.

The best-known crossover types with linear phase are 1st-order non-inverted crossovers, filler-driver crossovers, and subtractive crossovers with time delay, such as those put forward by Lipshitz and Vanderkooy in 1983. [1]

## Minimum Phase

Minimum phase, or minimal phase, is a term that is sometimes confused with linear phase. In fact they are not only not the same thing, but almost opposites. A minimum-phase system, in our case a filter, has the minimum phase-shift possible to get the amplitude/frequency response it shows. A minimum-phase filter is also one where the phase/frequency response can be mathematically derived from the amplitude/frequency response, and vice versa. Furthermore, the effect of a minimum-phase filter can be completely cancelled out in both phase and amplitude by using a reciprocal filter that has the opposite effect. A good example of this is given in Figure 14.1 in Chapter 14 on equalisation, where it is demonstrated that a peaking equaliser can be exactly cancelled out by a dip equaliser, and a square wave put through them both is reconstructed.

There are, as you might expect, several much more precise mathematical ways of defining the minimum-phase condition, [4] but they are not helpful here. In general you cannot say that a minimum-phase filter is better than a non-minimum-phase filter, as it depends what you are trying to do with said filter.

Most crossover filters, such as highpass and lowpass types in their various kinds, are inherently minimum phase. The classic exception to this is the allpass filter used for time-delay correction. Since an accurate allpass filter has a completely flat amplitude/frequency response, you cannot deduce anything at all from it about the phase/frequency response. The phase-shift of an allpass filter in fact varies strongly with frequency, as described in Chapter 13 on time-delay compensation, but the amplitude/frequency response gives you no clue to that. It is also impossible to undo the effect of an allpass filter, because its particular phase-shift characteristics give a constant delay at suitably low frequencies, and you cannot make a filter that has a negative delay. That would mean foretelling the future and on the whole would probably not be a good thing.

The 1st-order crossover is minimum phase when its outputs are summed normally; it has a flat phase plot at 0°. If one output is inverted, however, while the SPL and power responses are still flat, the summed output has a 1st-order allpass phase response, the phase swinging from 0° to −180° over the frequency range. It is therefore no longer minimum phase.

As we just noted, a linear-phase crossover acts as a pure time delay and so cannot be minimum phase. You cannot, however, say that a crossover which includes allpass filters for time-delay correction can never be minimum phase, because they are correcting for physical misalignments, and what counts is the summed signal at the ear of the listener.

## Absolute Phase

Another phase issue is the perception of absolute phase. In other words, if the polarity of a signal is inverted (it is not relevant whether it is heard via a single- or multi-way loudspeaker), does it sound different? The answer is yes, providing you use a single tone with a markedly asymmetrical waveform, such as a half-wave rectified sine wave or a single unaccompanied human voice. Otherwise, with more complex signals such as music, no difference is heard.

Obviously an active crossover must have all its outputs in the correct phase with each other (in some cases correct means phase inverted), or dire response errors will result, but it is also necessary to make sure that the outputs are in the correct phase relationship to the crossover input signal.

Almost all hi-fi equipment such as preamplifiers and power amplifiers are now designed to preserve absolute phase. Mixing consoles have always been so designed to prevent unwanted cancellation effects on mixing signals.

## Phase Perception

Some of the crossovers described in this book have quite dramatic phase changes in the summed output around the crossover points, so the sensitivity of human hearing to phase-shift is an important consideration. If the phase-shift is proportional to frequency, then the group delay is constant with frequency and this is a linear-phase system, as described earlier; we just get a pure time delay with no audible consequences. However, in most cases the phase-shift is not remotely proportional to frequency, and so the group delay varies with frequency. This is sometimes called group delay distortion, which is perhaps not ideal, as "distortion" implies non-linearity to most people, while here we are talking about a linear process.

Most of the components in the microphone-recording-loudspeaker chain are minimum phase; they impose only the phase-shift that would be expected and can be predicted from their amplitude/frequency response. The great exception to this is . . . the multi-way loudspeaker. The other great exception was the analogue magnetic tape-recorder, which showed rapid phase changes at the bottom of the audio spectrum, usually going several times round the clock. [5] In the first edition of this book in 2011, I wrote "Fortunately we don't need to worry about *that* anymore," but it appears I spoke too soon. There were several vintage multi-track tape recorders displayed at the Paris AES Convention in June 2016.

Whatever happens with tape recorders, we are certainly going to have multi-way loudspeaker systems around for the foreseeable future, and most of them have allpass crossovers. Clearly an understanding of what degradation—if any—this allpass behaviour causes is vital. Much experimentation has been done, and there is only space for a summary here.

One of the earliest findings on phase perception was Ohm's law. No, not that one, but Ohm's *other* law, which is usually called Ohm's acoustic law, and was proposed in 1843. [6] In its original form it simply said that a musical sound is perceived by the ear as a set of sinusoidal harmonics. The great researcher Hermann von Helmholtz extended it in the 1860s into what today is known as Ohm's acoustic law by stating that the timbre of musical tone depends solely on the number and relative level of its harmonics, and *not* on their relative phases. This is a good start, but does not ensure the inaudibility of an allpass response.

An important paper on the audibility of midrange phase distortion was published by Lipshitz, Pocock, and Vanderkooy in 1982, [7] and they summarised their conclusions as follows:

1. Quite small phase non-linearities can be audible using suitable test signals.
2. Phase audibility is far more pronounced when using headphones instead of loudspeakers.

3. Simple acoustic signals generated in an anechoic environment show clear phase audibility when headphones are used.
4. On normal music or speech signals, phase distortion is not generally audible.

At the end of the abstract of their paper the authors say: "It is stressed that none of these experiments thus far has indicated a present requirement for phase linearity in loudspeakers for the reproduction of music and speech." James Moir also reached the same conclusion. [8]

An interesting paper on the audibility of 2nd-order allpass filters was published in 2007, [9] which describes a perception of "ringing" due to the exponentially decaying sine wave in the impulse response of high $Q$ allpass filters (for example $Q = 10$). It was found that isolated clicks show this effect best, while it was much more difficult to detect, if audible at all, with test signals such as speech, music, or random noise. That is the usual finding in this sort of experiment—that only isolated clicks show any audible difference. While we learn that high-$Q$ allpass filters should be avoided in crossover design, I think most people would have thought that was the case anyway.

Siegfried Linkwitz has done listening tests where either a 1st-order allpass filter, a 2nd-order allpass filter (both at 100 Hz), or a direct connection could be switched into the audio path. [10] These filters have similar phase characteristics to allpass crossovers and cause gross visible distortions of a square waveform but are in practice inaudible. He reports "I have not found a signal for which I can hear a difference. This seems to confirm Ohm's Acoustic Law that we do not hear waveform distortion."

If we now consider the findings of neurophysiologists, we note that the auditory nerves do not fire in synchrony with the sound waveform above 2 kHz, so unless some truly subtle encoding is going on (and there is no reason to suppose that there is), then perception of phase above this frequency would appear to be inherently impossible.

Having said this, it should not be supposed that the ear operates simply as a spectrum analyser. This is known not to be the case. A classic demonstration of this is the phenomenon of "beats". If a 1000 Hz tone and a 1005 Hz tone are applied to the ear together, it is common knowledge that a pulsation at 5 Hz is heard. There is no actual physical component at 5 Hz, as summing the two tones is a linear process. (If instead the two tones were multiplied, as in a radio mixer stage, there *would* be new components generated.) Likewise, non-linearity in the ear itself can be ruled out if appropriate levels are used.

What the brain is actually responding to is the envelope or peak amplitude of the combined tones, which does indeed go up and down at 5 Hz as the phase relationship between the two waveforms continuously changes. Thus the ear is in this case acting more like an oscilloscope than a spectrum analyser. It does not however seem to work as any sort of phase-sensitive detector.

The conclusion we can draw is that a crossover whose summed phase response is that of a 1st-order or 2nd-order allpass filter is wholly acceptable. This obviously implies that a group delay characteristic that emulates a first or 2nd-order allpass filter is also completely acceptable.

## Target Functions

A target function for a loudspeaker system is the combined crossover and loudspeaker response that you are aiming for. Drive units are hopefully fairly flat over the frequency range that we hope to use them, but if this is not the case then their response obviously has to be taken into account. Response irregularities may be corrected by equalisation (see Chapter 14), performed either by adding dedicated equalisation stages to the relevant crossover path or by modifying the characteristics of the crossover filters. The latter uses less hardware but is much more difficult to understand unless properly documented.

In some cases the inherent properties of the drive unit and the enclosure may form part of the target function. For example, a suitably damped LF unit and enclosure will have a 2nd-order Butterworth-type maximally flat roll-off at 12 dB/octave. If this is combined with a 2nd-order Butterworth highpass filter in the crossover, then this makes up a Linkwitz-Riley 4th-order alignment, which can be used to crossover to a separate subwoofer.

## References

[1] Lipshitz, Stanley P. and John Vanderkooy "A Family of Linear-Phase Crossover Networks of High Slope Derived by Time Delay" JAES 31, January/February 1983, p. 1

[2] Blauert, Jens and P. Laws "Group Delay Distortions in Electroacoustical Systems", Journal of the Acoustical Society of America (JASA) 63(5), pp. 1478–1483, May 1978

[3] Self, Douglas "Audio Power Amplifier Design" Sixth Edn, Newnes, 2013, pp. 150–157, ISBN: 978-0-240-52613-3 (power amp input noise)

[4] Anon http://en.wikipedia.org/wiki/Minimum_phase

[5] Fincham, Laurie "The Subjective Importance of Uniform Group Delay at Low Frequencies" JAES33(6), June 1985, pp. 436–439

[6] https://en.wikipedia.org/wiki/Ohm%27s_acoustic_law

[7] Lipshitz, Stanley P., Mark Pocock and John Vanderkooy "On The Audibility of Midrange Phase Distortion in Audio Systems" JAES, September 1982, pp. 580–595

[8] James Moir Comment on "Audibility of Midrange Phase Distortion in Audio Systems" JAES, December 1983, p. 939

[9] Møller, Henrik, Pauli Minnaar, Soren Krarup Olesen, Flemming Christensen and Jan Plogsties "On the Audibility of All-Pass Phase in Electroacoustical Transfer Functions" JAES, 55(3), March 2007, pp. 113–134

[10] www.linkwitzlab.com/phs-dist.htm (allpass audibility)

# CHAPTER 4

# *Crossover Types*

There are many types of crossover, classified in various ways. The categories listed here define what the crossover does, not how it does it. It is essentially a catalogue of target functions. It does not matter how the filter responses are obtained; that may be by filtering alone or a combination of filtering and drive unit responses.

## All-Pole and Non-All-Pole Crossovers

In discussions of crossover design, and in filter design generally, the terms "pole" and "zero" tend to be freely scattered through the text. These terms relate to the complex mathematical equation that describes the response of a filter or other frequency-dependent system such as a servomechanism. A complex equation is one involving j, the square root of −1, and not just a *complicated* equation. This equation is often called the transfer function; it usually comes in the form of a fraction, and the poles derive from the bottom half (the denominator) while the zeros derive from the top half (the numerator). Both are distinguished by their frequencies, so you might say: "There is a pole at 1 kHz."

If that is as clear as mud, don't panic. As I said in the preface (and I hope you have read it), this book attempts to avoid getting involved in the whole complex algebra business. For our purposes, a "pole" causes the frequency response to turn downwards by 6 dB/octave, as in a simple 1st-order RC lowpass filter. A "zero" causes the response at that frequency to be zero; in other words there is a notch, theoretically infinitely deep.

Poles sometimes occur in pairs (called conjugate pairs or complex pairs), and such a pair acts a resonator, creating a response peak whose height depends upon the $Q$ of the circuit. Zeros can also occur in conjugate pairs; a notch filter has a conjugate pair of poles combined with a conjugate pair of zeros.

An all-pole crossover is composed entirely of all-pole filters. These are filters such as Butterworth, Bessel, and Linkwitz-Riley types which have a monotonic roll-off; in other words, once the response starts going down it does not come back up again.

Non-all-pole crossovers contain filters such as the Inverse-Chebyshev and elliptical (Cauer) types have zeros as well as poles in their responses, and their roll-offs are not monotonic; for example an Inverse-Chebyshev filter has notches in the stopband. Notch-type crossovers contain either inverse-Chebyshev filters or notch filters as such, like the Bainter filter. These filters incorporate zeros as well as poles, so they are also non-all-pole crossovers. Notch crossovers are dealt with in Chapter 5.

## Symmetric and Asymmetric Crossovers

Symmetric crossovers have filters with the same slope at the crossover point. For example, a 2-way 2nd-order crossover has 12 dB/octave slopes for both HF and LF filters.

Asymmetric crossovers have differing slopes at the crossover point; the HF filter might have a 18 dB/octave slope while the LF filter has a 12 dB/octave slopes. Asymmetric crossovers do not in general sum to flat unless a roll-off in the response of one of the drivers is being used to make the slopes equal in terms of acoustic output.

Subtractive crossovers in their straightforward (non-delayed) form always give asymmetric slopes for orders greater than one because the output derived by subtraction always has a 6 dB/octave slope, whatever the slope of the filter. A slope of 6 dB/octave is usually inadequate unless especially capable wide-range drive units are employed. Adding a time delay in the path going to the subtraction stage can give equal slopes, but the delay needs to be very precise for this to extend over an adequate frequency range, and this appears to be a major problem with the concept. This problem is explored in Chapter 6 on subtractive crossovers.

## Allpass and Constant-Power Crossovers

Crossovers are classified as either "allpass crossovers" (APC) or "constant-power crossovers" (CPC) in accordance with the way that the outputs recombine.

An allpass crossover has filter outputs that sum in the air in front of the loudspeaker to create a sound pressure level (SPL) with a flat amplitude/frequency response. The human ear is only sensitive to the pressure changes at the ear hole and has no way to integrate the acoustic power bouncing around a room. If the filter outputs are summed electrically instead of acoustically, as a test of proper recombination, the filter outputs should sum to a flat voltage response, voltage being equivalent to SPL; this is sometimes called the amplitude response. Summation is by vector addition (i.e. phase must be taken into account) of the highpass output $V_{HP}$ with the lowpass output $V_{LP}$:

$$V_{SUM} = V_{HP} + V_{LP} \qquad\qquad 4.1$$

APC crossovers are the usual choice because they give the loudspeaker a flat on-axis frequency response, and this is usually where the person who pays for the audio system sits; this certainly does not guarantee a flat off-axis response. The phrase "allpass" rather than something like "flat response" is used to emphasise that while the summed amplitude/frequency response may be flat, the summed phase response is that of an allpass filter. The phase response of some crossovers is that of a 1st-order allpass filter, with the phase changing by 180° over the audio band, the best known of these being the 1st-order (inverted), the 2nd-order Linkwitz-Riley and 3rd-order Butterworth types. The 4th-order Linkwitz-Riley crossover has the phase response of a 2nd-order allpass filter, with the phase changing by 360°. There is more information on allpass filters in Chapter 13.

A constant-power crossover has filter outputs that sum in the air in front of the loudspeaker to create a flat frequency response in terms of power rather than sound pressure level. The power response of a loudspeaker is the sum of all its off-axis and on-axis amplitude/frequency responses; it is the frequency response of the total acoustical power radiated into a given listening space. Because the signals are not phase-correlated, phase is ignored, and summation is by RMS addition of the highpass output $V_{HP}$ with the lowpass output $V_{LP}$:

$$V_{SUM} = \sqrt{\left(V_{HP}\right)^2 + \left(V_{LP}\right)^2}$$

$$4.2$$

CPC crossovers are not popular because they do not in general give a flat on-axis frequency response, but they are sometimes stated to be beneficial in reverberant environments where the off-axis output makes a significant contribution to the sound arriving at the listener. This is usually an undesirable (but sometimes unavoidable) situation, as the characteristics of the listening space then have a greater effect on sound quality. For this reason the principles of CPC crossover design constitute a relatively small part of this book.

## Constant-Voltage Crossovers

It sounds as though a constant-voltage crossover would be the same thing as an allpass crossover, but in fact they are quite different. Constant-voltage crossovers operate by subtracting the output of a filter (which is one output) from the unfiltered input to generate the other output; in this book they are called "subtractive crossovers", as I think that term is a bit clearer. If the output from a lowpass filter is subtracted from the unfiltered input you get a highpass output. Constant-voltage crossovers have the property, rare amongst crossover types, of being linear phase and so able to "reconstruct the waveform"; in other words when you sum the two outputs you get back the waveform you put in. This process is described in detail here and also in Chapter 6 on subtractive crossovers. The big snag is that the output obtained by subtraction has a slope of only 6 dB/octave, whatever the slope of the filter, and this is usually inadequate.

The constant-voltage crossover was first properly described by Dick Small, one of the great pioneers of scientific loudspeaker design, in 1971. [1]

## First-Order Crossovers

A 1st-order crossover has the great advantage that it is the only type that can produce linear-phase and minimum-phase response and a flat amplitude response. As a consequence, it is the only type that allows a waveform to be reconstructed when an HF and LF path are added together. It is potentially attractive for passive crossovers, as it is the simplest crossover and so requires the smallest number of expensive crossover components and has the lowest power losses.

There is only one type of 1st-order crossover, just as there is only one sort of 1st-order filter. It is only when you go to 2nd-order filters that you get a choice of characteristics, such Butterworth, Bessel, and so on. It is however possible to use different cutoff frequencies for the highpass and lowpass filters to manipulate the crossover behaviour—see the Solen split 1st-order crossover described next.

A 1st-order crossover has some serious disadvantages. The slopes of a 1st-order crossover are a gentle 6 dB/octave, and this does not normally provide enough separation of the frequencies sent to the drive units. Tweeters are likely to be damaged by excessive coil excursions when fed with inappropriately low frequencies. LF drive units are unlikely to be damaged by high frequencies, but if operated outside their frequency design range they are likely to show an irregular response and poor linearity. Likewise, the 6 dB/octave slope is usually too shallow to prevent activation of a tweeter's resonance frequency (which is usually below the crossover frequency), giving rise to an unwanted peak in the response. It is generally considered that if a 1st-order crossover is used, the drive units must be well-behaved for at least two and preferably three octaves either side of the crossover frequency.

Actually achieving the maximally flat amplitude response and minimum-phase operation requires very careful driver alignment and also requires the listener to be exactly the same distance from each drive unit. These problems are consequences of the large overlap in operation of the drive units.

A basic 1st-order crossover is shown in Figure 4.1, where the 1st-order crossover filters are shown as simple RC networks. The crossover frequency, which is the −3 dB cutoff frequency for each filter, is 1 kHz; this can be obtained by using 220 nF capacitors for $C$ and 723.4 $\Omega$ resistors for $R$. This crossover frequency is used for all the examples in this chapter, not because it is a good crossover frequency—very often it isn't—but because it's a convenient way of displaying the results, putting the crossover point nicely in the middle of the plot. Using other crossover frequencies gives exactly the same behaviour, but the plots are shifted sideways. The circular summing element to the right represents how the two acoustic outputs from the HF and LF drive units add linearly in the air in front of the loudspeaker. It represents pure mathematical addition and does not load or affect the two filters in any way.

Figure 4.2 shows the amplitude response of the two outputs and their sum. The sum is completely flat with frequency, lying on top of the 0 dB line in the plot. Neither output has been phase inverted before summation. If one output is phase inverted, the summed response is still dead flat—this is a property that no higher-order crossover possesses.

The power response is shown in Figure 4.3. Since the two contributions to the total power are uncorrelated (because of multiple room reflections and so on), they add in an RMS fashion; in other words you take the square root of the sum of the squares of the two levels. For the 1st-order crossover, the power response is also absolutely flat. Figure 4.3 therefore looks identical to Figure 4.2, though it was obtained by RMS summation rather than simple addition.

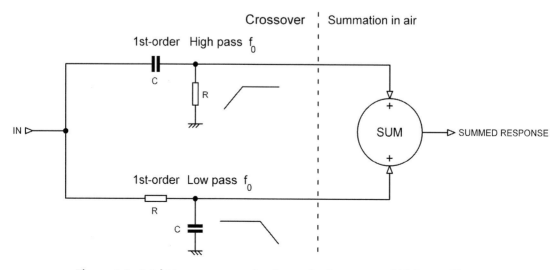

**Figure 4.1: A 1 kHz crossover using 1st-order lowpass and highpass filters.**

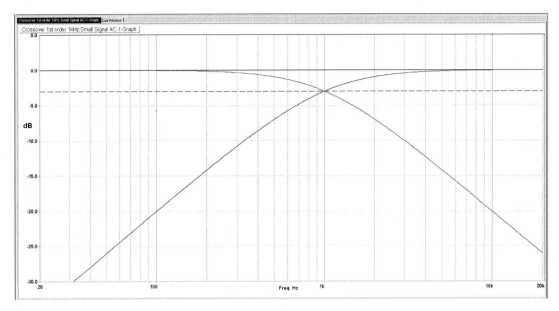

**Figure 4.2: Frequency response of 1st-order 1 kHz crossover; both filter outputs plus their sum (straight line at 0 dB).**

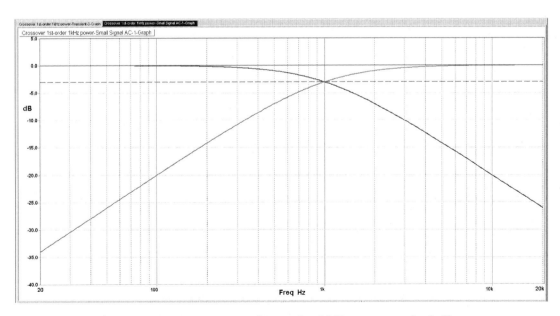

**Figure 4.3: Power response of 1st-order 1 kHz crossover; both filter outputs plus their sum (straight line at 0 dB).**

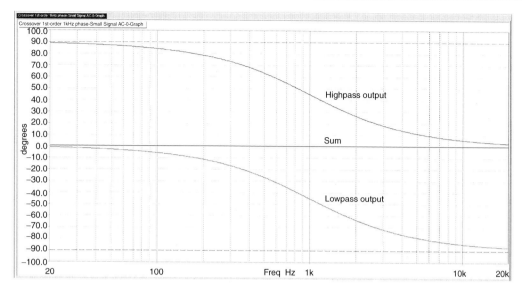

**Figure 4.4: Phase response of 1st-order 1 kHz crossover; both filter outputs plus their sum (straight line at 0°).**

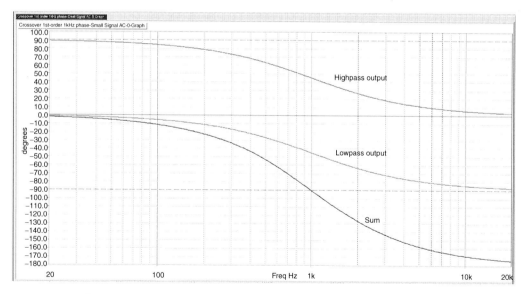

**Figure 4.5: Phase response of 1st-order 1 kHz crossover; both filter outputs plus allpass sum when highpass output is inverted.**

The two outputs have a constant 90° phase-shift between them, as shown in Figure 4.4. The phase of the summed outputs is always zero, as shown by the horizontal line at 0°. The crossover is minimum phase.

If however the highpass output has been phase inverted before summation, the summed response has a phase that swings from 0° to −180°, as shown in Figure 4.5. This is precisely the phase response of a 1st-order allpass filter, as described in Chapter 13, and demonstrates why the phrase "allpass crossover" crops up so often in this subject. The crossover is no longer minimum phase. Note that in Figure 4.5 the phase of the highpass output is shown *before* it is inverted. Please note that for this and all other phase plots in this chapter the 0° reference is the lowpass output at 20 Hz.

While a phase response that moves through 180° over the audio band may appear to be highly questionable, the consensus is that it is completely inaudible with normal music, and can only be detected by the use of special test signals such as isolated clicks. This is discussed in detail in Chapter 3.

With one output phase inverted, the two outputs always have a phase difference of 90°, and this affects the radiation pattern of the two drive units in the frequency range where they are both delivering a significant output. Because of the gentle filter slopes, this range is wider than in higher-order crossovers. With the normal (non-inverted) version, the result is "lobing error", which is a tilting of the vertical coverage pattern, pointing it downwards by 15° from the horizontal; when the crossover has one output inverted, the tilt is 15° upwards. This assumes that the time alignment of the tweeter and woofer are correct at the crossover frequency to prevent differing time delays to the listener's ear; 1st-order crossovers are particularly sensitive to this because of the broad driver overlap. The tilt will increase and lobing will become more severe if the drivers are unduly separated on the baffle face. For a 1st-order crossover this effect is considered to be significant over at least two octaves.

The normal (non-inverted) connection shows a flat group delay because the phase response is flat at zero. The inverted connection, however, has the group delay versus frequency response shown in Figure 4.6, with a flat section at 318 usec, rolling off slowly to zero over the crossover region.

The wide frequency overlap of a 1st-order crossover means that quite small time-alignment errors can cause large anomalies in the amplitude/frequency response. Vance Dickason, in his excellent

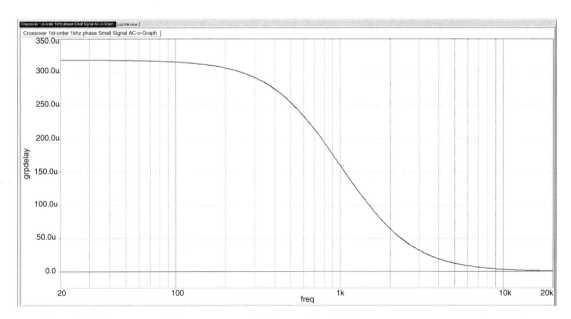

**Figure 4.6: The group delay response of a 1st-order 1 kHz crossover, one output inverted.**

*Loudspeaker Design Cookbook*, [2] shows how a mere 0.5 inch misalignment can cause a peak or dip of 2.5 dB in a broad band centred on 3 kHz, the peak or dip effect depending on whether one output is inverted or not. A 1 inch misalignment causes amplitude ripples of up to ±4 dB, while a 2 inch misalignment causes errors of up to 10 dB. This is a serious argument against 1st-order crossovers.

A 1st-order crossover has the unique property that it can reconstruct the waveform put into it. If you put a square wave into the crossover, and then sum the LF and HF outputs, you get a square wave back; see Figure 4.7. The input waveform has levels of 0 V and +1.0 V; this is purely to make the middle plot clearer, and input levels of −0.5 V and +0.5 V, symmetrical about zero, give exactly the same reconstruction of the waveform. No other type of all-pole crossover has this ability.

As you read through this chapter, you will see that in some ways 1st-order crossovers have some uniquely desirable features, such as the ability to reconstruct the waveform, but also some serious

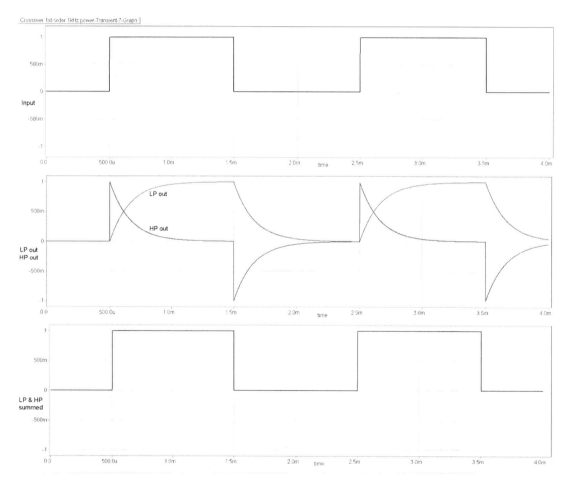

**Figure 4.7: Reconstruction of a square wave by summing the highpass and lowpass output of a 1st-order crossover. Non-inverted connection.**

disadvantages. The6 dB/octave slopes are too gentle for use with normal drive units, and time alignment is frighteningly critical. Despite these disadvantages the positive features of a 1st-order crossover are so powerful that there does seem to be a distinct trend towards investing in driver technology in order to create units that are suitable for 1st-order operation. A recent example is the KEF *Q*-series of loudspeakers. [3]

## First-Order Solen Split Crossover

The so-called Solen split 1st-order crossover is a variation on the standard 1st-order crossover (Solen is a company that supplies passive crossover parts). This scheme attempts to improve the poor slow crossover from one drive unit to the other by putting the crossover point for each filter at −6 dB rather than the usual −3 dB, as in Figure 4.2. This applies a frequency offset to the cutoff (−3 dB) frequencies of each filter, so that the lowpass filter now has a cutoff of 579 Hz and the highpass filter now has a cutoff of 1.726 kHz, pulling apart them apart by a factor of 2.98 times or 1.68 octaves; the pulling-apart process is presumably where the term "split" comes from. It may somewhat ease the demands on the drive units, but what of the amplitude response?

Figure 4.8 shows that, as we might expect, pulling apart the two cutoff frequencies has caused the summed response to sag in the middle, by 6 dB in fact. This is obviously going to sound like rubbish, but we get a better result if we reverse the phase to one of the drive units—a common manoeuvre in crossover design.

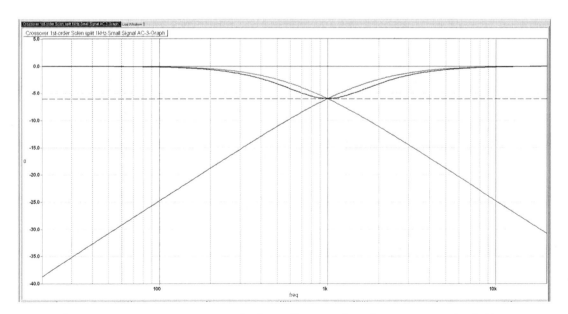

**Figure 4.8: Frequency response of Solen split 1st-order 1 kHz crossover; both filter outputs plus their in-phase sum.**

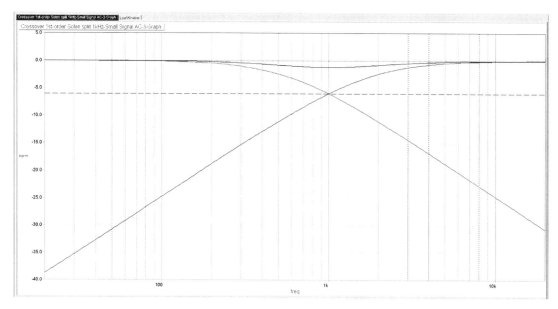

**Figure 4.9: Frequency response of Solen split 1st-order 1 kHz crossover; both filter outputs plus the sum, with one output reversed (phase inverted).**

In passive crossover design, phase-inverting one of the outputs is extremely simple; just swap over the two wires to the drive unit in question; everything is done in the box, and no-one is any the wiser. Active crossovers, however, will need to use some kind of phase-inverting stage.

Figure 4.9 demonstrates that with one of the phases reversed we still get a dip, but it is now only 1.2 dB deep; information on this crossover scheme is scanty, but that is presumably how it is supposed to work. It might be possible to reduce the deviation from perfect flatness by partial cancellation of the dip with a response irregularity in one of the drivers. It is however still difficult to get enthusiastic about a crossover with 6 dB/octave slopes.

## First-Order Crossovers: 3-Way

It is difficult to make a 1st-order 3-way crossover because the slow 6 dB/octave slopes do not provide adequate separation into three bands across the audio spectrum. There is in any case little point because drive units capable of handling the wide frequency ranges inherent in such a crossover would probably be equally suitable for a 1st-order 2-way crossover setup.

## Second-Order Crossovers

The use of a 2nd-order crossover promises relief from the lobing, tilting, and time-alignment criticality of 1st-order crossovers, because the filter slopes are now twice as steep at 12 dB/octave. A very large number of passive crossovers are 2nd-order, because they are still relatively simple, and this simplicity is very welcome, as it reduces power losses and cuts the total cost of the large crossover components required. Neither of these factors applies to active crossovers; the extra power consumption and the extra cost of making a 4th-order crossover rather than a 2nd-order crossover are very small.

All 2nd-order filters have a 180° phase-shift between the two outputs, which causes a deep cancellation notch in the response at the crossover frequency when the HF and LF outputs are summed. Such a response is of no use whatever, and the standard cure is to invert the polarity of one of the outputs. With a 2nd-order crossover using Butterworth filters, this gives not a flat response but a +3 dB hump at the crossover frequency. As we shall see, the size of the hump can be much reduced by using a frequency offset; in other words the highpass and lowpass filters are given different cutoff frequencies. An offset factor of 1.30 turns the +3 dB hump into symmetrical amplitude ripples of ±0.45 dB, which is the flattest response that can be achieved for the Butterworth crossover by this method.

Second-order crossovers have much less sensitivity to driver time-misalignments because of their 12 dB/octave slopes. Vance Dickason [2] has shown that for a 2nd-order Butterworth crossover, a 1 inch time-misalignment gives errors of only fractions of a dB, while a 2 inch misalignment gives maximal errors of 2 dB. The corresponding figures for a 1st-order crossover are 4 dB and 10 dB respectively. The frequency-offset technique can also be used to reduce the effect of time-alignment errors on the amplitude/frequency response.

A 2nd-order crossover gives better, though by no means stunning, protection of the drive units against inappropriate frequencies, less excitement of unwanted behaviour outside their intended frequency range, and less modulation distortion. Since one output has to be inverted to get a usable amplitude response, the outputs are in phase instead of 180° phase-shifted, and so there should be no lobing error, i.e. tilt in the vertical coverage pattern.

### *Second-Order Butterworth Crossover*

The 2nd-order Butterworth crossover is perhaps the best-known type, despite the fact that it is far from satisfactory. A classic bit of crossover misdesign that has been published in circuit ideas columns and the like a thousand times is shown in Figure 4.10. You take two 2nd-order Butterworth filters with the same cutoff frequency, one highpass and one lowpass, and there you have your two outputs. As before, the summing device represents how the two outputs add linearly in the air in front of the loudspeaker.

As Figure 4.11 shows all too clearly, this does not work well; in fact "catastrophic" would be a more accurate description. Each filter gives a 90° phase-shift at the crossover frequency, one leading and one lagging. The signals being summed are therefore a total of 180° out of phase and cancel out completely. This causes the deep notch at the crossover frequency seen in Figure 4.11.

Since a 180° phase-shift is the root of the problem, that at least can be eliminated by the simple expedient of reversing the connections to one of the drivers, normally the high-frequency one of the pair. Figure 4.12 shows the result—the yawning gulf is transformed into a much less frightening +3 dB hump centred at the crossover frequency.

In a passive crossover this reversed connection can be hidden inside the speaker enclosure along with the crossover components, but in an active crossover the issue is more exposed. You will need to either

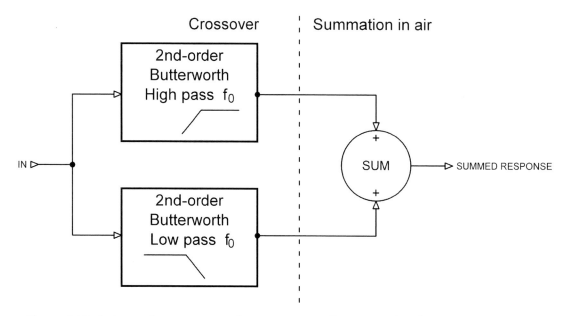

Figure 4.10: A doomed attempt to make a crossover using two 2nd-order Butterworth filters.

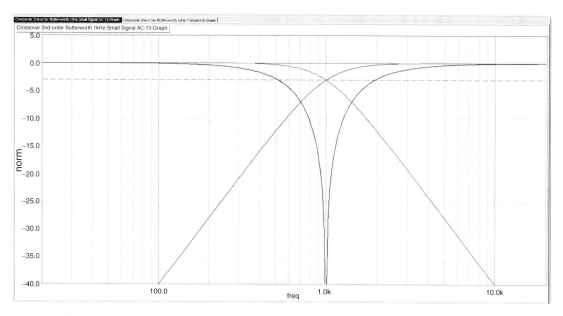

Figure 4.11: The frequency response resulting from the in-phase summation of two 2nd-order Butterworth filters: a disconcerting crevasse in the combined response. The dashed line is at −3 dB.

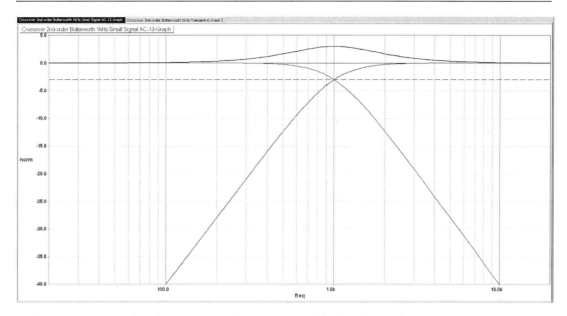

**Figure 4.12: Second-order Butterworth crossover, with the phase of one output reversed; the crevasse has become a more usable +3 dB hump. The dashed line is at −3 dB.**

build a phase inversion into the active crossover, which again effectively hides the phase reversal from the user, or specify that one of the power amplifier-speaker cables be reversed. A lot of users are going to feel that there is something not right about such an instruction, and building the inversion into the crossover is strongly recommended.

Clearly our +3 dB hump is much better than an audio grand canyon, but accepting that much deviation from a flat response is clearly not a good foundation for a crossover design. That hump is going to be very audible. It might be cancelled out by an equalisation circuit with a corresponding dip in its response, but there is a simpler approach. Looking at Figure 4.12, it may well occur to you, as it has to many others, that something might be done by pulling apart the two filter cutoff frequencies so the response sags a bit in the middle, as it were. There is no rule in crossover design, be it active or passive, that requires the two halves of the filtering to have the same cutoff frequency.

Figure 4.13 shows the result of offsetting each of the filter cutoff frequencies by a factor of 1.30 times. This means that the highpass filter cutoff frequency is changed from 1.00 kHz to 1.30 kHz, while the lowpass cutoff becomes 1.00/1.30 = 0.769 kHz. Crossover now occurs at −6 dB. With the phase inversion, the offset factor of 1.30 turns the hump into symmetrical amplitude ripples of ±0.45 dB above and below the 0 dB line; this represents the minimum possible response deviation obtainable in this way. Now the amplitude response is looking a good deal more respectable, if not exactly mathematically perfect, and it is very questionable whether response ripples of this size could ever be audible. You may wonder if the frequency-offset process has rescued the response with outputs in-phase; the answer is that it is not much better; there is no longer a notch with theoretically

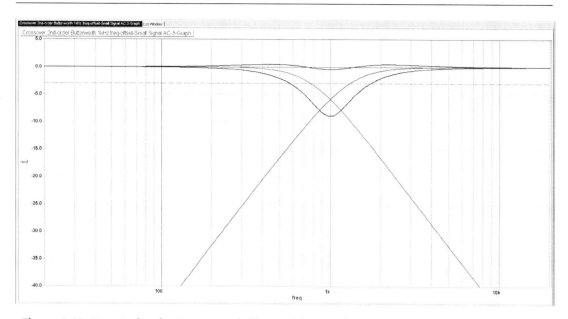

**Figure 4.13: Two 2nd-order Butterworth filters with 1.30 times frequency offset, in normal and reversed connection. With the phase of one output reversed, the +3 dB hump is smoothed out to a mere ± 0.45 dB ripple. The normal-phase connection still has a serious dip 9 dB deep. The dashed line is at −3 dB; the two filter cutoff (−3 dB) frequencies are now different.**

infinite depth, but the there is a great big dip 9 dB deep at the bottom, and such a response is still of no use at all.

The frequency offsets required for maximal flatness with various types of crossover are summarised in Table 4.1 at the end of this chapter.

The power response for the 2nd-order Butterworth crossover with no frequency offset is shown in Figure 4.14; it is a perfect straight line. This qualifies it as a constant-power crossover (CPC); it is a pity that the power response is of much less importance than the pressure response. It is important to realise that the power response is the same whether or not one of the outputs is phase reversed. When the two outputs are squared and added to get the total power, the negative sign of the reversed output disappears in the squaring process.

When looking at this power response plot, it is important to appreciate that each crossover output is still shown at −3 dB at the crossover frequency, meaning the power output from it is halved. The two lots of half-power sum to unity, in other words 0 dB.

Earlier we saw that a frequency offset of 1.30 times was required to get near-flat amplitude response. How is that going to affect the perfectly flat power response of Figure 4.14? The answer, predictably, is that any change is going to be for the worse, and Figure 4.15 shows that there is now an 8 dB dip in the power response.

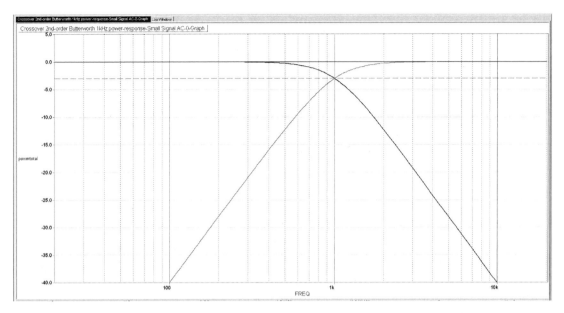

Figure 4.14: Power response of 2nd-order Butterworth crossover with no frequency offset—the sum is perfectly flat at 0 dB. Each crossover output is still at −3 dB at the crossover frequency, meaning the power is halved. The two half-powers sum to 0 dB.

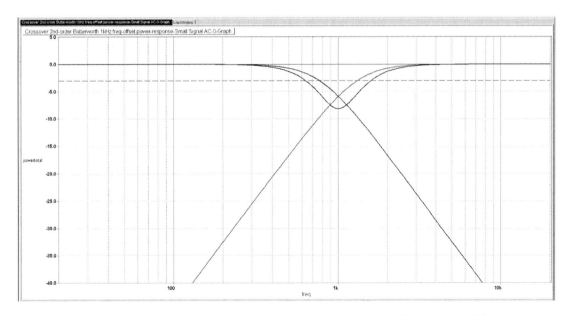

Figure 4.15: Power amplitude response of 2nd-order Butterworth crossover with 1.30x frequency offset—rather less than perfect, showing an−8 dB dip at the crossover frequency.

As mentioned earlier, 2nd order filters have a 180 degree phase-shift between the two outputs. This is shown in Figure 4.16, where the phase of the sum lies exactly on top of the trace for the lowpass output. The phase of the sum is that of a 1st-order allpass filter, the same phase characteristic as that of the 1st-order crossover. This is inaudible with normal music signals. For reasons of space, phase and group delay plots are from here on only given for the more interesting crossover types.

The summed group delay for the inverted connection is shown in Figure 4.17. It has a level section at 226 usec and a gentle peak of 275 usec just below the crossover frequency. Note that the level section shows less group delay than the 1st-order crossover.

When we were looking at 1st-order crossovers, you will recall that it was said that no other crossover could solve the waveform reconstruction problem, i.e. to get out the waveform that we put in after summing the outputs. How does a 2nd-order Butterworth cope with the square wave reconstruction problem?

The answer from Figure 4.18 is that it fails completely, and inverting one of the filter outputs makes things even worse. The phase-shifts introduced by the 2nd-order filters make it impossible for reconstruction to occur. While it is not very obvious from Figure 4.18, the first and second cycles of the simulation are not quite identical; since we have a circuit with energy-storage elements (capacitors) and we are starting from scratch, it takes a little time for things to settle down so you obtain the result for continuous operation. In some cases it is not uncommon for 20 cycles to be required to reach equilibrium. Not every writer on the subject of audio has appreciated this fact, and major embarrassment has resulted.

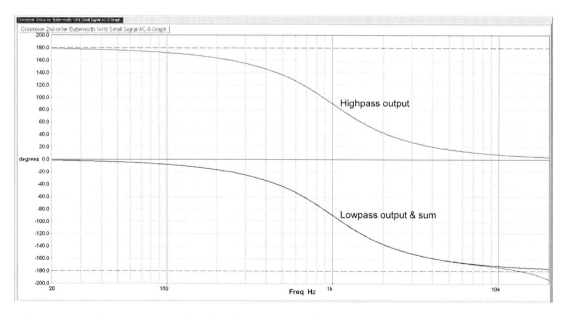

**Figure 4.16: Phase response of 2nd-order Butterworth crossover with one output inverted; the outputs are always 180° out of phase. The phase of the sum lies exactly on top of that of the lowpass output. Highpass output shown *before* inversion.**

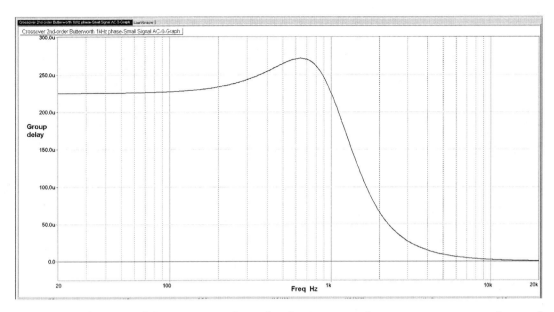

Figure 4.17: The group delay response of a 2nd-order Butterworth crossover, one output inverted.

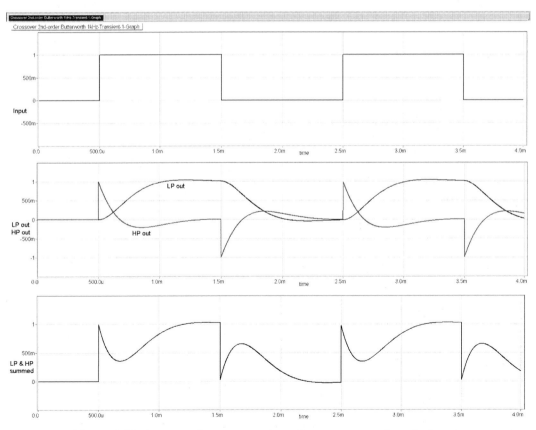

Figure 4.18: Attempted reconstruction of a square wave by a 2nd-order
Butterworth crossover without offset. Total failure!

The 2nd-order Butterworth crossover does not sum to flat even with one output inverted, though frequency offsetting can give a big improvement. The 12 dB/octave slopes are usually considered inadequate.

## Second-Order Linkwitz-Riley Crossover

The Butterworth crossover filter can be made very nearly flat by tweaking the cutoff frequencies of the two filters. An alternative and much better approach in the 2nd-order case is to alter the $Q$'s of the filters. Setting the $Q$ of each filter to 0.5, with identical cutoff frequencies, turns the 2nd-order Butterworth crossover into a 2nd-order Linkwitz-Riley crossover with each output −6 dB at the crossover point, as seen in Figure 4.19. The flat response qualifies it as an allpass crossover network.

The two outputs are still 180° out of phase for the normal connection and give the same yawning gulf in the response as the Butterworth. With the reversed connection the signals add as for the Butterworth case, but they are now 3 dB lower, so there is no hump. The summation gives a completely flat response, without the ripple you get with a frequency-offset 2nd-order Butterworth crossover.

The reversed connection has a −3 dB dip in the power response at the crossover frequency. In this respect it is not as good as the 1st-order crossover, which has a flat power response as well as a flat voltage (or SPL) response. However, as we have seen, the power response of a crossover is usually a minor consideration.

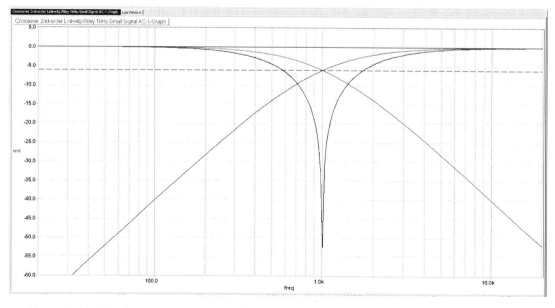

**Figure 4.19: The frequency response of a 2nd-order Linkwitz-Riley crossover. The in-phase summation of two filters still has a cancellation crevasse, but with one output reversed the sum is exactly flat at 0 dB. The dashed line is at the −6 dB crossover level.**

The 2nd-order Linkwitz-Riley crossover is neither linear-phase nor minimum phase. The phase response plot looks indistinguishable from that of the 2nd-order Butterworth crossover, though there are minor differences. The summed group delay does not peak but rolls off slowly around the crossover frequency.

### Second-Order Bessel Crossover

The Bessel filter has a much slower roll-off than the Butterworth but also has a maximally flat group delay; in other words it stays flat as long as possible before it rolls off, while the Butterworth group delay has a peak in it. This makes the Bessel filter an interesting possibility for crossovers with flat group delay characteristics.

The 2nd-order Bessel crossover without any frequency offset gives a −8 dB dip for the in-phase connection, as in Figure 4.20, but a more promising broad +2.7 dB hump with one output reversed, as in Figure 4.21. This looks very like the Butterworth hump in Figure 4.12, so it seems very likely we can also reduce this one by applying frequency offset to the filter cutoffs.

The first attempt; a frequency offset of 1.30 times, as used in the 2nd-order Butterworth case, reduces the size of the hump to +1.1 dB at its centre, but this time no dips below the 0 dB line have appeared; this differing behaviour looks ominous and suggests that the hump may get flatter and flatter with increasing offset but never actually reach a definite maximally flat condition.

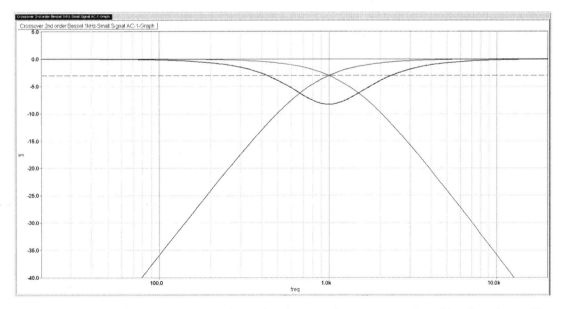

**Figure 4.20: The frequency response of a 2nd-order Bessel crossover summed in-phase has a dip going down to −8 dB. The dashed line is at −3 dB.**

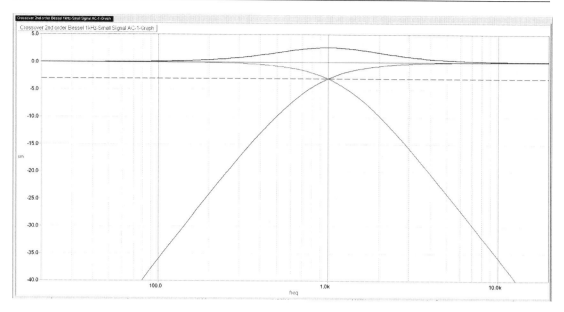

**Figure 4.21: The frequency response of a 2nd-order Bessel crossover summed with one output phase reversed has a +2.7 dB hump. The dashed line is at −3 dB.**

However, this is a great example of a situation where you should not give up too soon. If we keep increasing the offset, then the shape of the summed response changes, until at an offset ratio of 1.45 we obtain a dip and two flanking peaks, as in Figure 4.22. The deviation from 0 dB is less than ±0.07 dB, so this result is actually much better than the 2nd-order Butterworth crossover, which at maximal flatness had deviations of ±0.45 dB. Such small deviations as ±0.07 dB will be utterly lost in drive unit tolerances. This offset ratio gives crossover at −6.0 dB.

The phase response plot looks very similar to that of the 2nd-order Butterworth crossover, but the rate of change around the crossover region is slightly slower due to the lower $Q$ of the filters. The summed group delay does not peak but rolls off slightly more slowly than the Butterworth around the crossover frequency.

The 2nd-order Bessel is not linear-phase, though it deviates from it less than do the 2nd-order Butterworth or Linkwitz-Riley types. It is not minimum phase.

A very good discussion of Bessel crossovers is given in [4].

## Second-Order 1.0 dB-Chebyshev Crossover

All Chebyshev filters have ripples in their passband response, and given the problems we have had achieving a near-flat response when we were using filters without such ripples, things don't look too hopeful. Using 1.0 dB-Chebyshev filters, which in 2nd-order form peak by 1 dB just before roll-off,

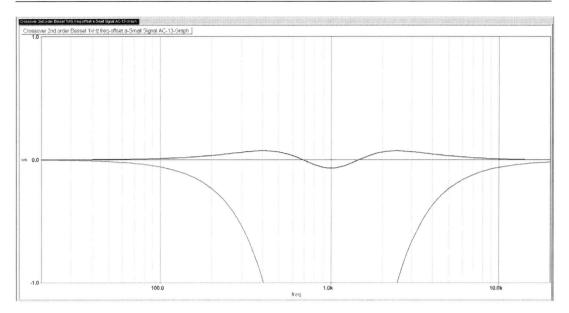

**Figure 4.22: The frequency response of a 2nd-order Bessel crossover summed with one output phase reversed and a frequency offset of 1.449 times has deviations from the 0 dB line of less than 0.07 dB. Note much enlarged vertical scale covering only ±1 dB.**

we get Figure 4.23, which shows the in-phase result. There is a deep central dip rather like that of the 2nd-order Bessel crossover, except that this one is even deeper, at −14 dB.

Reversing the phase of one of the outputs gives us a 6 dB hump, substantially higher than that of the reversed-phase 2nd-order Bessel crossover; see Figure 4.24. Flattening this by using frequency offset is a tall order, but we will have a go.

As the frequency-offset ratio is increased, a dip develops in the centre of the hump and moves below the 0 dB line, until we reach the optimally flat condition, which unfortunately is not that flat. The deviations are ±1.6 dB at an offset ratio of 1.53 times and look too big to be a basis for sound crossover design; see Figure 4.25. The large 1.53 times offset ratio causes the crossover point to be at −5.6 dB.

The power response has significantly more ripple than the optimally flat amplitude response.

The phase response plot looks very similar to that of the 2nd-order Butterworth crossover but changes faster around the crossover frequency. The summed group delay has a very big peak just below the crossover frequency; while it is probably not audible, it is certainly not desirable.

It would be possible to try other types of Chebyshev filters as 2nd-order crossovers; Chapter 7 gives details on how to design Chebyshev filters with 0.5 dB, 1 dB, 2 dB, and 3 dB of passband ripple. There seems to be no reason to think that the versions with greater passband ripple would be any better than the 1.0 dB version and every reason to think that they would be worse. The high filter $Q$'s required to realise the 2 dB and 3 dB filters imply very poor group delay characteristics with serious peaking.

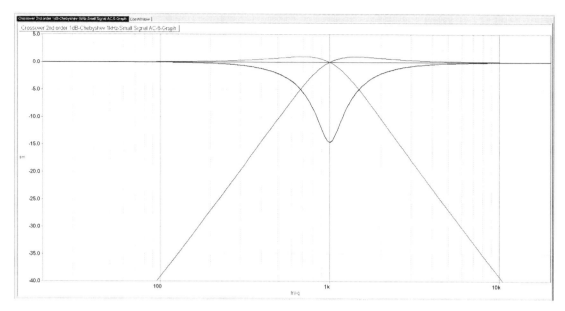

Figure 4.23: The frequency response of a 2nd-order 1.0 dB-Chebyshev crossover. In-phase summation gives a deep dip of −14 dB. The crossover is at 0 dB, because the cutoff frequency of this filter is defined as the point in the roll-off where the response returns to 0 dB after the 1.0 dB peak.

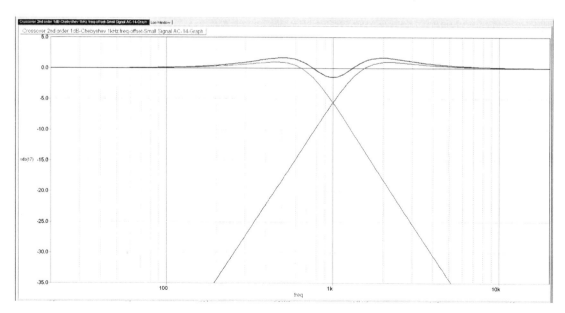

Figure 4.24: The frequency response of a 2nd-order 1.0 dB-Chebyshev crossover. Summation with one output phase reversed gives a peak of +6 dB. Vertical scale has been moved up by 5 dB to accommodate the peak.

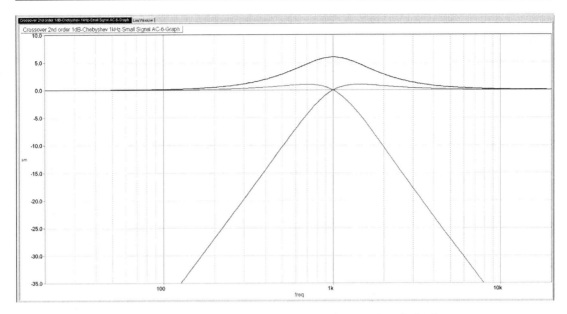

**Figure 4.25: The frequency response of a 2nd-order 1.0 dB-Chebyshev crossover
with one output phase reversed and a frequency offset of 1.53 times.
The deviation is ±1.6 dB about the 0 dB line.**

## Third-Order Crossovers

Third-order crossovers have the advantage of greater separation between the drive units than either
1st- or 2nd-order designs can give. The steeper 18 dB/octave slopes make it less likely that drive
unit irregularities such as the tweeter resonance will be excited and allow modulation distortion to be
reduced. Third-order crossovers are also less sensitive to driver time-delay misalignments because of
there is less frequency overlap in the filter outputs. The 3rd-order Butterworth crossover gives a flat
amplitude response *and* a flat power response.

On the other side of the ledger, the filter outputs are always 270° apart in phase, which can result
in lobing and tilting of the coverage pattern in the range where the drivers overlap, but this range is
narrower than for 2nd-order crossovers. Inverting one output reduces this to 90°, and so as with the
1st-order crossover, there is a −15 degree downward tilt in the crossover region.

Third-order crossovers have further reduced sensitivity to driver time-misalignments because of their
steeper 18 dB/octave slopes. According to Vance Dickason [2] a 3rd-order Butterworth crossover
and a 2 inch time-misalignment gives a maximal error of 2.5 dB, while the corresponding error for a
1st-order crossover is 10 dB. As before, the frequency-offset technique can also be used to reduce the
effect of time-alignment errors on the amplitude response.

Third-order crossovers are generally the most complicated passive types that are popular, though 4th-
order and 5th-order passive crossovers have been used.

### Third-Order Butterworth Crossover

Figure 4.26 shows the amplitude response of a 3rd-order Butterworth crossover with outputs in-phase and with no frequency offset. The response is ruler flat, and it remains ruler flat when one of the outputs is reversed. The power response is also flat, as it was for the 1st-order crossover; all odd-order Butterworth crossovers have a flat power response. It is thus both an APC and a CPC crossover; it is not linear phase or minimum phase. The crossover point is at −3 dB.

The flat power response is illustrated in Figure 4.27.

The summed phase response of the normal (non-inverted) connection is that of a 2nd-order allpass filter, with the phase changing by 360° over the audio band. If the polarity of one output is inverted, the amplitude response remains perfectly flat, but the summed phase response is improved to that of a 1st-order allpass filter, with the phase changing by only 180° over the audio band. This is shown in Figure 4.28.

With outputs normal (not inverted) their phase difference is a constant 270°, but with one output inverted this is reduced to 90°, the same as the 1st-order crossover. It therefore tilts the polar pattern in the same way, but the frequency range over which this is significant is much reduced because of the steeper slopes of a 3rd-order crossover. The lobe tilt can be eliminated by the using the d'Appolito MTM drive unit configuration, which places a LF/MID drive unit on each side of the tweeter. The main forward lobe is however narrower; 3rd-order crossovers are considered good at minimising this effect. Interaction between the two LF/MID drive units can cause response irregularities off axis and below the crossover frequency. There is more on d'Appolito configurations in Chapter 2.

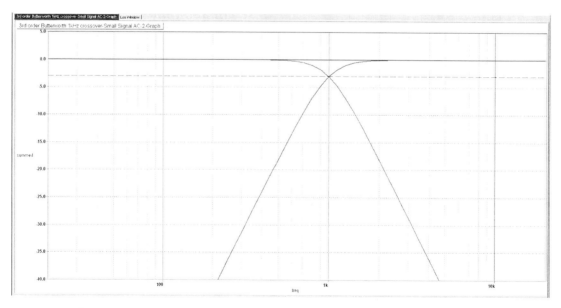

**Figure 4.26: The frequency response of a 3rd-order Butterworth crossover.**
**In-phase or phase-reversed summation of the two outputs gives a flat line at 0 dB.**
**The dashed line is at the −3 dB crossover level.**

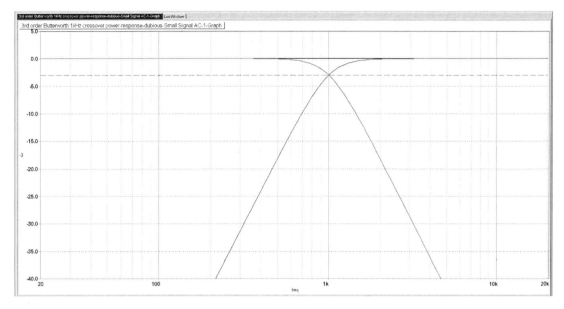

Figure 4.27: The power response of a 3rd-order Butterworth crossover. The RMS summation of the two filter outputs gives a straight line at 0 dB. The dashed line is at the −3 dB level.

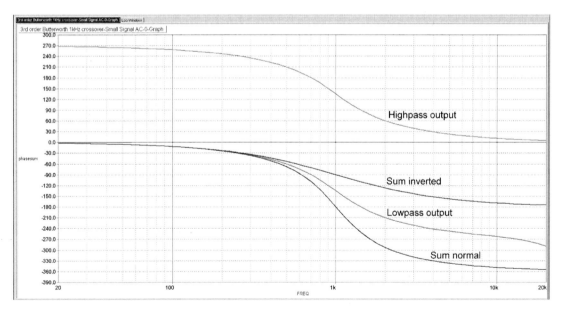

Figure 4.28: The phase response of a 3rd-order Butterworth crossover, for normal and inverted connections.

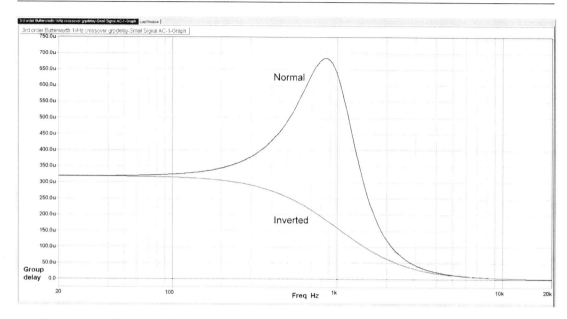

**Figure 4.29: The group delay response of a 3rd-order Butterworth crossover. The normal connection has a big peak in delay, but with one output inverted there is just a roll-off.**

Figure 4.29 shows how the faster phase changes for the normal connection cause a big peak in the group delay. The inverted connection has no peak and looks much more satisfactory. Note that the group delay at low frequencies is 318 usec, exactly the same as for the 1st-order crossover.

The 3rd-order Butterworth crossover is one of the better ones. It has flat amplitude and power responses, and a 1st-order phase response when one output is inverted. On the downside, it has the lobe-tilting problem of all odd-order crossovers (though this can be addressed by the d'Appolito configuration), and many people feel that the 18 dB/octave slopes are not really steep enough.

### Third-Order Linkwitz-Riley Crossover

The response of the 3rd-order Linkwitz-Riley crossover alignment is shown in Figure 4.30. The summed amplitude response has a dip of exactly −3 dB at the crossover frequency, and each filter output is at −6 dB. In this case, reversing the phase of one output does not convert it into a hump but gives exactly the same response with the same dip, because of the differing phase relationships. As we have seen, a hump in the response can be flattened by using a frequency offset that lowers the cutoff frequency of the lowpass filter and raises the cutoff frequency of the highpass filter. Clearly a dip cannot be dealt with like that, as pulling apart the crossover frequencies, as we have done before, is just going to make the dip deeper.

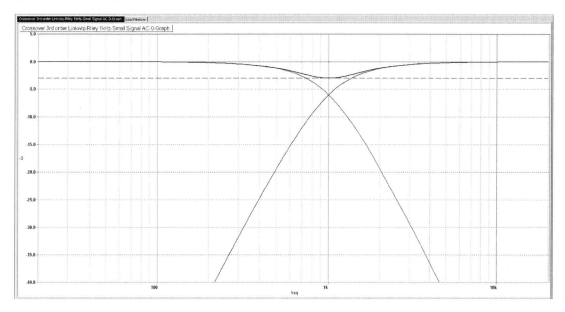

**Figure 4.30: The frequency response of a 3rd-order Linkwitz-Riley crossover. The in-phase summation of the two filter outputs gives a −3 dB dip. The dashed line is at−3 dB.**

Instead, if we take the phase-reversed case and push the two cutoff frequencies together by using an offset ratio of 0.872 times, so the HP cutoff is now 0.872 kHz and the LP cutoff is 1.15 kHz, we get the maximally flat result with only±0.33 dB of ripple. See Figures 4.31 and 4.32. The crossover point is now at −4.5 dB. This procedure only works with one phase reversed—if applied to the in-phase case we just dig a deeper hole for ourselves, the dip becoming −4.4 dB deep at the bottom.

There is no point in inverting one output as regards the amplitude response, because it is unchanged. We do, however, get a much better phase response, as for the 3rd-order Butterworth. A 2nd-order allpass phase response is converted to a 1st-order allpass phase response. The latter only is shown in Figure 4.33.

You will notice that the phase response shown in Figure 4.33 looks very much the same as for the Butterworth version in Figure 4.28. There seems little point in using valuable space by repeating diagrams that are very similar, and so phase responses are only shown for selected crossovers.

The 3rd-order Linkwitz-Riley (without frequency offset) has a power response with a −3 dB dip at crossover; it is not a CPC type.

The 3rd-order Linkwitz-Riley crossover is neither linear phase nor minimum phase. The group delay has a peak just below the crossover frequency.

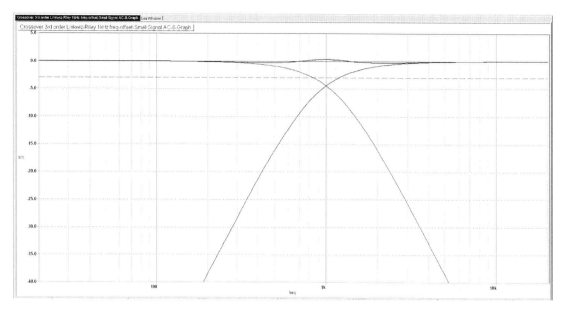

Figure 4.31: The frequency response of a 3rd-order Linkwitz-Riley crossover, with frequency offset of 0.872 times. The reversed-phase summation of the two filter outputs now gives a ripple of only ±0.33 dB.

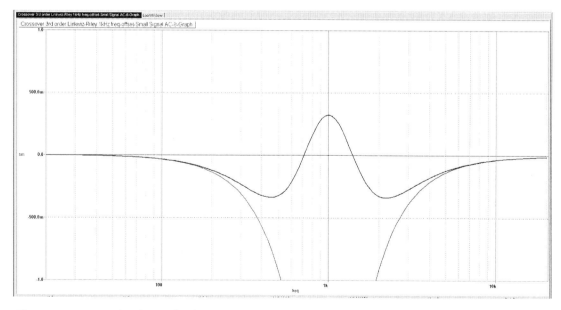

Figure 4.32: Zooming in on the frequency response of a 3rd-order Linkwitz-Riley crossover, with frequency offset of 0.872 times and reversed-phase summation. Ripple is only ±0.33 dB.

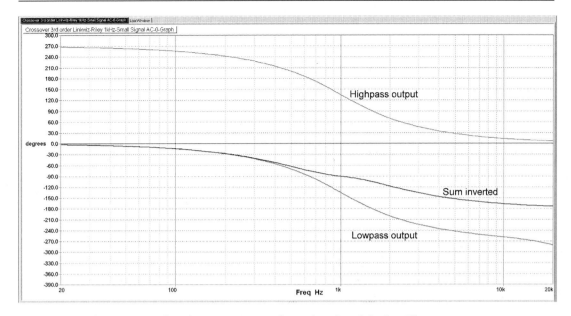

**Figure 4.33: The phase response of a 3rd-order Linkwitz-Riley crossover, for inverted connection only.**

## Third-Order Bessel Crossover

If a Bessel filter characteristic is used for a 3rd-order crossover, the in-phase summation, as in Figure 4.34, shows a deep dip that bottoms out at −13.5 dB. This looks distinctly unpromising.

Phase-inverting one of the outputs gives a +3.0 dB hump, as in Figure 4.35, which is very reminiscent of what we got from the 2nd-order Butterworth crossover with one output phase reversed, though in this case the hump is rather narrower because we are using 3rd-order filters. This immediately suggests that the hump could be dealt with as before, by frequency-offsetting the filters and so spreading their cutoff frequencies apart.

In this case offsetting each filter cutoff frequency by a ratio of 1.22, so that the highpass cutoff is at 1.22 kHz while the lowpass cutoff is at $1/1.22 = 0.820$ kHz, gives the maximally flat sum, with a peak and two dips all 0.35 dB away from the 0 dB line; see Figure 4.36. The crossover point is now at −4.6 dB instead of −3 dB.

## Third-Order 1.0 dB-Chebyshev Crossover

Using 1.0 dB-Chebyshev filters gives us the unpleasant peak at +4 dB seen in Figure 4.37, with 1 dB dips on either side. This does not look like a good starting point for a flat crossover. You can see that the crossover point is now at −1 dB, because the cutoff frequency of a 3rd-order Chebyshev filter is defined as the point where the roll-off response has fallen again to the level of the 1 dB dip after rising briefly back to 0 dB.

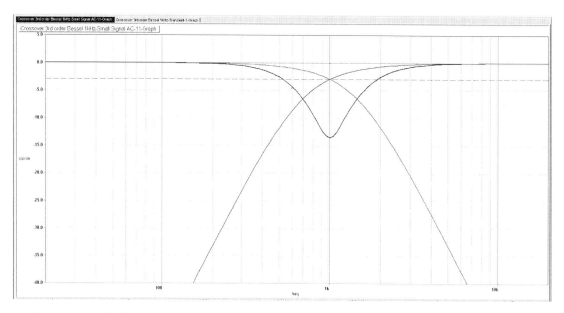

Figure 4.34: The frequency response of a 3rd-order Bessel crossover resulting from in-phase summation and giving a nasty dip. The dashed line is at −3 dB.

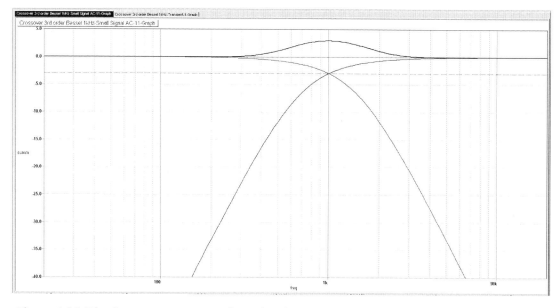

Figure 4.35: The frequency response of a 3rd-order Bessel crossover, resulting from summation with one of the filter outputs phase reversed. The dashed line is at −3 dB.

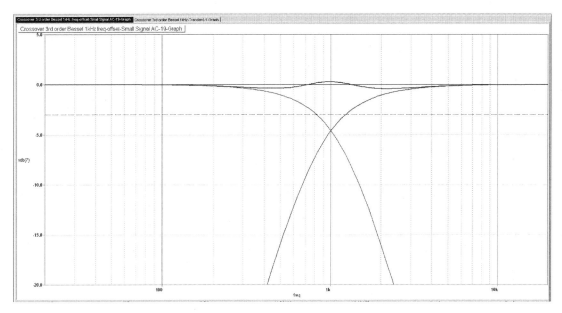

Figure 4.36: The frequency response of 3rd-order Bessel crossover, summation with one output phase reversed and both cutoff frequencies offset by 1.22x for maximal flatness. Ripple is only ±0.35 dB. Note vertical scale now has −20 dB at bottom.

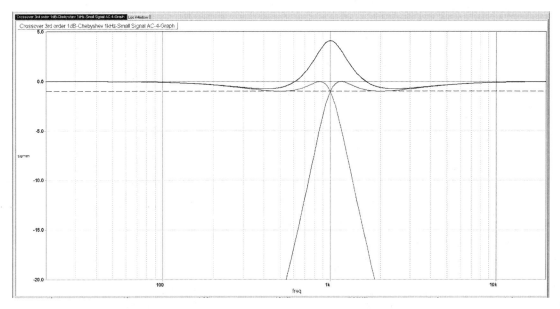

Figure 4.37: The frequency response of a 3rd-order 1.0 dB-Chebyshev crossover, with in-phase summation; a +4 dB peak with dips either side. The dashed line is now at −1 dB.

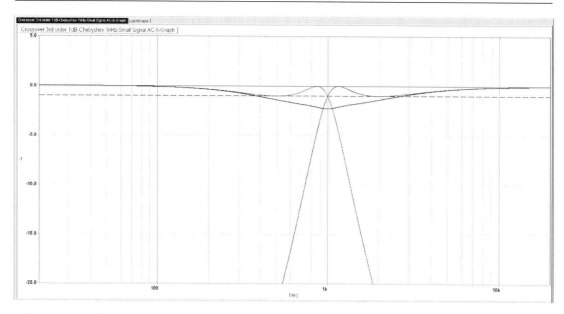

**Figure 4.38: The frequency response of a 3rd-order 1.0 dB-Chebyshev crossover, resulting from summation with one of the filter outputs phase reversed. The dashed line is at −1 dB.**

If one output is inverted we get instead a gentle dip, as in Figure 4.38, which looks more promising as a subject for frequency-offsetting. As we saw with the 3rd-order Linkwitz-Riley crossover, to tackle a dip in the combined response, it is necessary to push the curves together rather than pull them apart, and so the offset ratio will be less than 1.0.

An offset ratio of 0.946 (making the highpass cutoff 0.946 kHz and the lowpass cutoff 1/0.946 = 1.057 kHz, gives the maximally flat response in Figure 4.39. The response deviations are ±1.6 dB, and once more the Chebyshev filter does not look like a promising start for a crossover design.

## Fourth-Order Crossovers

The 4th-order crossovers give still greater separation between the drive units than the 1st-, 2nd-, or 3rd-order types. The 24 dB/octave slopes minimise the chances that out-of-band drive unit irregularities or tweeter and midrange resonances will be excited. Modulation distortion will be further reduced, though probably not to a great extent. The sensitivity to driver time-delay misalignments is further lessened due to the narrower crossover region. The 4th-order Butterworth crossover does not gives a flat summed amplitude response, but the 4th-order Linkwitz-Riley famously does. For the Butterworth and Linkwitz-Riley versions, inverting one output gives a useless response with a deep notch at the crossover frequency.

Since 4th-order filters are used, their extra phase-shift gives outputs that are 360° apart, which is the same as being in-phase and so eliminates lobing errors and tilting of the vertical coverage pattern in the crossover region; this is a major advantage.

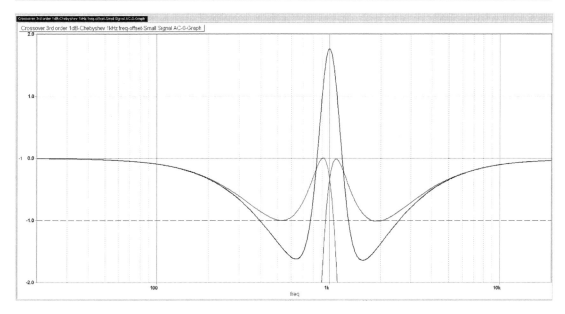

Crossover 3rd order 1dB-Chebyshev 1kHz freq-offset-Small Signal AC-0-Graph

**Figure 4.39: Zooming in on the frequency response of a 3rd-order 1.0 dB-Chebyshev crossover, with frequency offset of 0.946 times and reversed-phase summation. Ripple is ±1.6 dB, vertical scale ±2 dB.**

Fourth-order crossovers have further reduced sensitivity to driver time-misalignments because of their steeper 24 dB/octave slopes. Vance Dickason records [2] that a 4th-order Butterworth crossover and a 2 inch time-misalignment gives an amplitude response error of about 1 dB, smaller than that of any other crossover examined so far.

Despite their advantages, 4th-order filters are rarely used in passive crossovers because the greater number of expensive inductors increases the losses due to their resistance, and increasing the inductor wire gauge to reduce these losses puts the cost up even more. Fourth-order crossovers also require a greater number of big capacitors.

I was inspired by Vance Dickason [2] to investigate some more exotic filters as possible candidates for crossovers; the linear-phase filter, the Gaussian filter, and the Legendre filter. It has to be said that on examination none look very promising. Other unusual filters such as transitional and synchronous types are looked at in Chapter 7.

### Fourth-Order Butterworth Crossover

The 4th-order Butterworth sums to give a +3.0 dB hump at the crossover frequency, as in Figure 4.40, while the summation with one phase reversed gives a deep crevasse, as in Figure 4.41; this is the exact opposite of the behaviour of the 2nd-order Butterworth. A frequency offset of 1.128 times reduces the hump to a maximally flat ripple of ±0.47 dB, as seen in Figure 4.42.

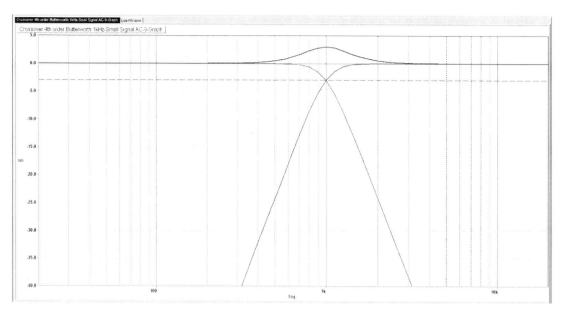

Figure 4.40: The frequency response of a 4th-order Butterworth crossover, resulting from in-phase summation. The dashed line is at −3 dB.

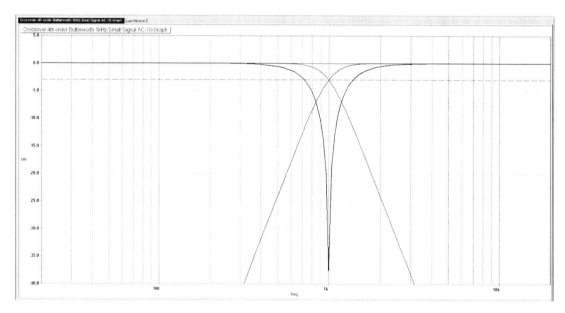

Figure 4.41: The frequency response of a 4th-order Butterworth crossover with reverse-phase summation. The dashed line is at −3 dB.

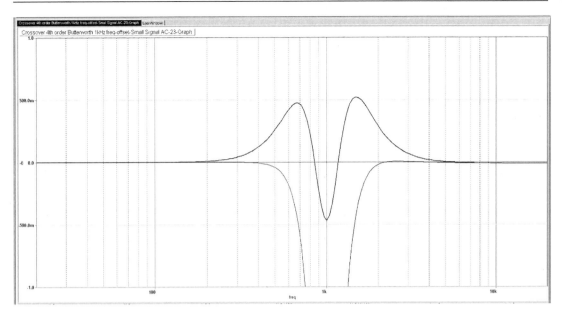

**Figure 4.42: Zooming in on the frequency response of a 4th-order Butterworth crossover with in-phase summation, and an optimal frequency-offset ratio of 1.128 times. The error is ±0.47 dB.**

The 4th-order Butterworth crossover produces a flat power response, and so it is a CPC crossover. The outputs are 360° apart in phase at all times. This is equivalent to being in-phase, and so there is no tilting of the vertical coverage pattern in the crossover region. The summed group delay has a significant peak just below the crossover frequency.

Given that the best possible amplitude response flatness obtainable by frequency offset is ±0.47 dB, there seems no reason to use the 4th-order Butterworth in preference to the 4th-order Linkwitz-Riley crossover. The phase and group delay plots are therefore not shown.

### Fourth-Order Linkwitz-Riley Crossover

The 4th-order Linkwitz-Riley is considered by many the best crossover alignment of the lot. The in-phase response sums to completely flat, making it an APC type; see Figure 4.43. (The reversed-phase response in Figure 4.44 has a deep crevasse at the crossover frequency and is of no value.) The outputs are 360° apart in phase at all times, so there is no lobe tilting in the crossover region. The crossover point is at −6 dB. There is a −3 dB dip in the power response, so it is not a CPC crossover.

One of the beauties of this type of crossover is the ease of its design. It is normally made by cascading two 2nd-order Butterworth filters with identical cutoff frequencies and identical $Q$'s of 0.7071 ($1/\sqrt{2}$). Thus it is sometimes called a "squared Butterworth" filter, or, less logically, a "−6 dB Butterworth" filter.

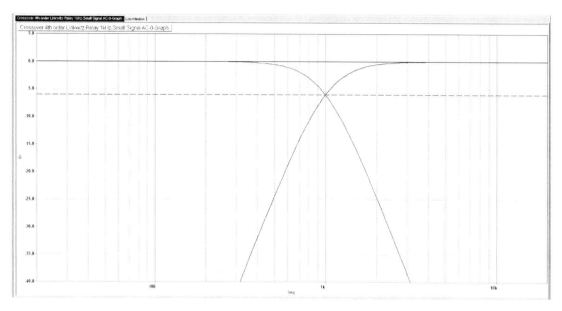

Figure 4.43: The frequency response of a 4th-order Linkwitz-Riley crossover, resulting from in-phase summation. The dashed line is at −6 dB.

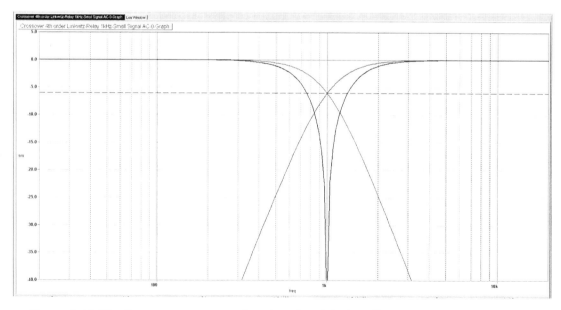

Figure 4.44: The frequency response of a 4th-order Linkwitz-Riley crossover, resulting from phase-inverted summation. The dashed line is at −6 dB.

Figure 4.45 shows the −3 dB dip in the power response. The two outputs are at −6 dB at the crossover point and are regarded as uncorrelated, so they RMS-sum to give −3 dB. While this is not ideal, it is a modest dip, and the use of 4th-order filters means it is not wide, so the effect on the reverberant energy in a listening space will be correspondingly small.

Figure 4.46 illustrates how the outputs are 360° apart in phase at all times. This is equivalent to being in-phase and so eliminates tilting of the vertical coverage lobes in the crossover region. The summed phase swings through 360° across the audio band and thus emulates a 2nd-order allpass filter.

The summed group delay in Figure 4.47 has a flat LF region at 450 usec, which is longer than any of the lower-order crossovers we have looked at; however, that is much less than the 3.2 ms audibility threshold at 500 Hz, as quoted in Chapter 3. The group delay shows a moderate peak of 540 usec just below the crossover frequency.

Passive versions of the 4th-order Linkwitz-Riley are relatively uncommon because of the power losses and the number of components in a passive 4th-order crossover, but one example of a loudspeaker using the technology was the KEF Model 105, [5] of which the first version was released in 1977.

The 4th-order Linkwitz-Riley crossover is widely considered to be the best. It sums to a flat amplitude response, and its power response has a dip of limited width which is only −3 dB deep. There is no lobe tilting, and the 24 dB/octave slopes are considered adequate for the vast majority of drive units. On the debit side, the phase response is that of a 2nd-order allpass filter rather than a 1st-order, and the group delay is relatively long, at 450 usec, and has a peak.

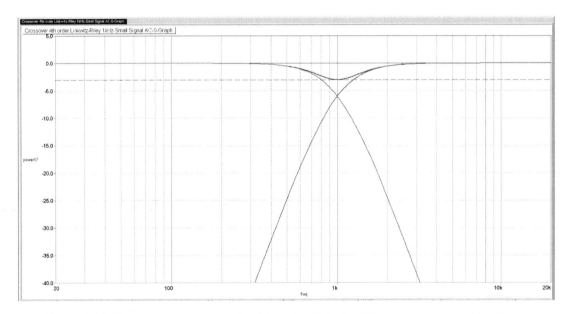

**Figure 4.45: The power response of a 4th-order Linkwitz-Riley crossover, resulting from in-phase summation. The dashed line is at −3 dB.**

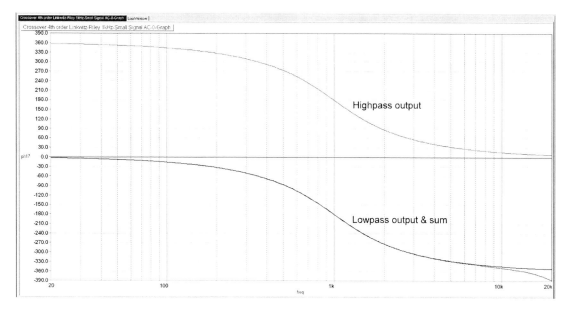

Figure 4.46: The phase response of a 4th-order Linkwitz-Riley crossover, in-phase connection.
There is a constant 360° phase difference between the two outputs (the summed phase trace is
on top of the lowpass output trace).

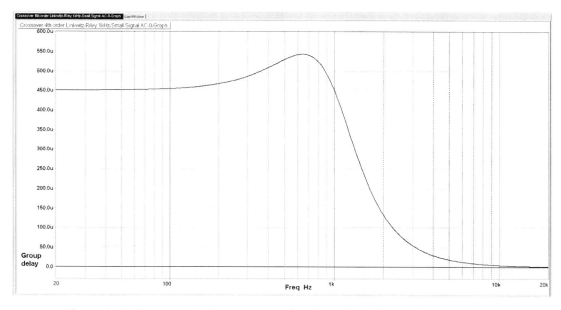

Figure 4.47: The group delay response of a 4th-order Linkwitz-Riley crossover,
in-phase connection.

### Fourth-Order Bessel Crossover

The 4th-order Bessel has its crossover point at −3 dB. It sums to give a −2.6 dB dip at the crossover frequency, as in Figure 4.48, while the summation with one phase reversed gives a +2 dB hump and two −1 dB dips as in Figure 4.49; the phase-reversed case does not look like a suitable case for frequency-offset treatment, as pulling the cutoff frequencies apart will pull down the hump but deepen the dips, while pushing them together will pull up the dips but make the hump worse.

It looks more promising for the in-phase case, but pushing the cutoffs together for this crossover actually deepens the dip due to the phase-shifts involved. Pulling them apart by an increasing amount makes the dip more shallow to begin with, but before the response begins to straddle the 0 dB line, it undergoes a fairly complicated set of changes, with two new dips appearing either side of the central one, and they deepen as the offset ratio increases.

In this situation it is hard to say what constitutes the maximally flat solution, but a promising candidate is shown in Figure 4.50, where a frequency offset of ratio of 1.229 times gives only −0.18 dB at the crossover frequency, with dips either side that are −0.3 dB deep.

The summed group delay response has a moderate peak just below the crossover frequency.

### Fourth-Order 1.0 dB-Chebyshev Crossover

Using the 1.0 dB-Chebyshev characteristic for each filter gives us Figure 4.51. The in-phase summed response has central peak at the crossover frequency of +2.22 dB, with the dips either side going down

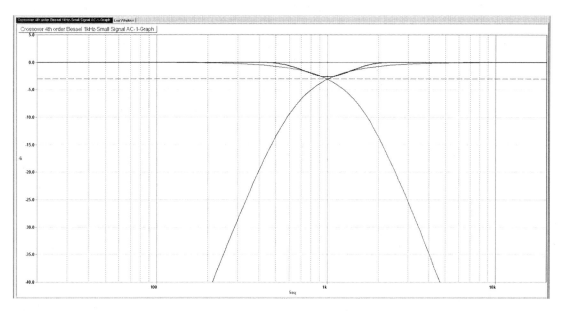

**Figure 4.48: The frequency response of a 4th-order Bessel crossover, resulting from in-phase summation. The dashed line is at −3 dB.**

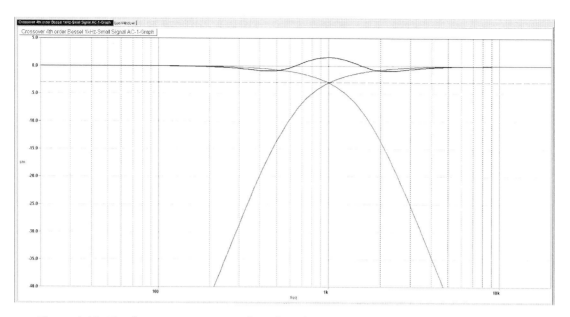

Figure 4.49: The frequency response of a 4th-order Bessel crossover, with reversed-phase summation. The dashed line is at −3 dB.

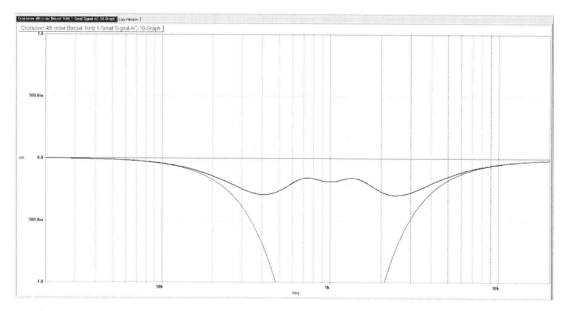

Figure 4.50: Zooming in on the frequency response of a 4th-order Bessel crossover, with an offset ratio of 1.229 times and in-phase summation. Vertical scale is ±1 dB.

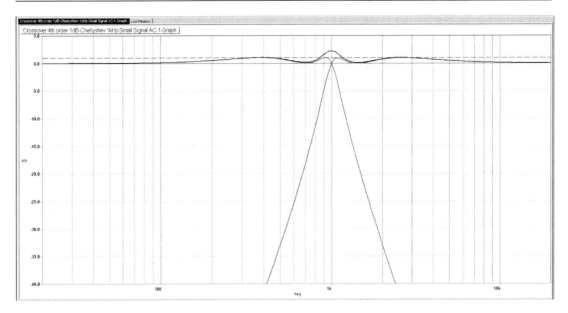

**Figure 4.51: The frequency response of a 4th-order Chebyshev crossover, resulting from in-phase summation. The dashed line is at −3 dB.**

to +0.15 dB, as in Figure 4.51. These dips are above0 dB because a 4th-order Chebyshev filter has its passband ripples above the 0 dB line. All even-order Chebyshev filters have response peaks above 0 dB, while all odd-order Chebyshev filters have dips below it.

The summed response with one output phase inverted seen in Figure 4.52 looks very much the same, except that the central peak is taller at +3.65 dB. Neither summed response looks as though it could be significantly flattened by the use of a frequency offset, and I have not attempted it.

### Fourth-Order Linear-Phase Crossover

There is not a great deal of information out there about the design of linear-phase filters, and some of what there is appears to be contradictory. The crossover here uses linear-phase filters as defined by Linear-X Systems Filtershop, [6] which for a 4th-order linear-phase filter consists of two cascaded 2nd-order stages; the first has a cutoff frequency of 1.334 and a $Q$ of 1.316, while the 2nd has a cutoff frequency of 0.7496 and a $Q$ of 0.607. With this filter structure the crossover point is at −4.5 dB. The summed response with the outputs in phase has a gentle +1.2 dB hump at crossover, dipping very slightly below the 0 dB line on either side; see Figure 4.53.

The summation with one output phase reversed has a deep hole in it, but is notable because it has an unusual flat portion at −8.7 dB around the crossover frequency, as in Figure 4.54. It is in fact very flat indeed, to within 0.01 dB across the visibly flat part. While this is strange and rather interesting, the deep dip does not look like a good starting point for the frequency-offset process.

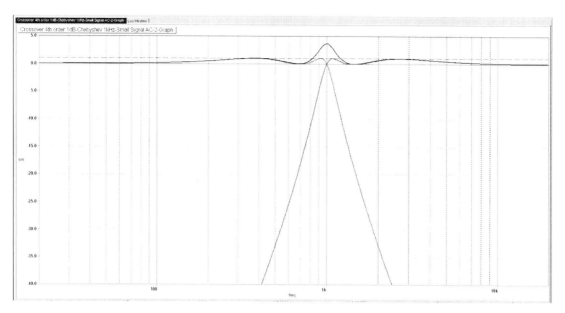

Figure 4.52: The frequency response of a 4th-order Chebyshev crossover, resulting from reversed-phase summation. The dashed line is at −3 dB.

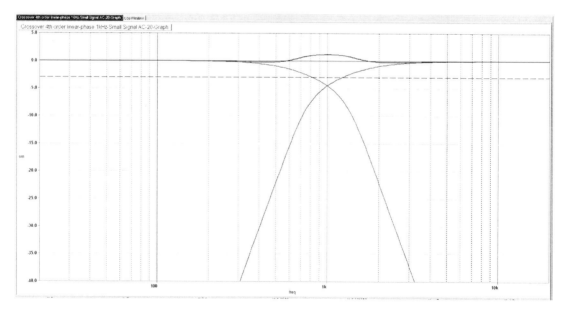

Figure 4.53: The frequency response of a 4th-order linear-phase crossover, resulting from in-phase summation. The dashed line is at −3 dB.

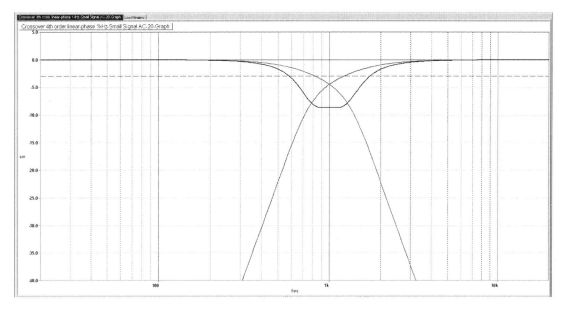

**Figure 4.54: The frequency response of a 4th-order linear-phase crossover, resulting from reversed-phase summation. The dashed line is at −3 dB.**

Applying frequency offset to the in-phase case of Figure 4.53, we find that an offset ratio of 1.0845 times gives a maximally flat response, with deviation of ±0.66 dB around the 0 dB line; see Figure 4.55 for a close-up. This is poor compared with other crossovers.

Despite its promising name, the linear-phase filter crossover cannot be made to sum flatter than ± 0.66 dB, which makes it a very doubtful choice.

### Fourth-Order Gaussian Crossover

As explained in Chapter 7, the Gaussian filter characteristic is based on time-domain considerations, being designed for no overshoot on a step function input while keeping rise and fall times as fast as possible. This response is closely connected to the fact that the Gaussian filter has the minimum possible group delay. Gaussian filters come in various kinds identified by a dB suffix, such as "Gaussian-6 dB" and "Gaussian-12 dB", though the differences in amplitude response are very small. The amplitude response is very similar indeed to that of a Bessel filter.

Designing the lowpass and highpass filters for a cutoff of 1 kHz gives a crossover point at −2.8 dB, as in Figure 4.56. The reversed-phase connection gives the crevasse in Figure 4.57, which will not be considered further.

Looking at Figure 4.56, it is clear that the filter cutoff frequencies need to be pushed together to get a flatter summed response. The depth of the dip is minimised at −1.05 dB by a frequency-offset ratio of 0.8045 times; thus the highpass filter cutoff becomes 0.8045 kHz and the lowpass filter cutoff becomes 1/0.8045 = 1.243 kHz, giving the response in Figure 4.58. Altering the ratio either up or down gives a deeper dip.

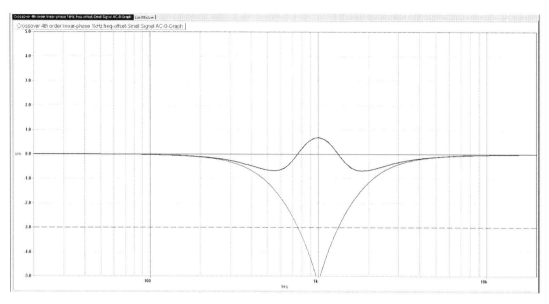

Figure 4.55: The frequency response of a 4th-order linear-phase crossover with a frequency offset of 1.0845 times, and in-phase summation, showing errors of ± 0.66 dB. The dashed line is at −3 dB.

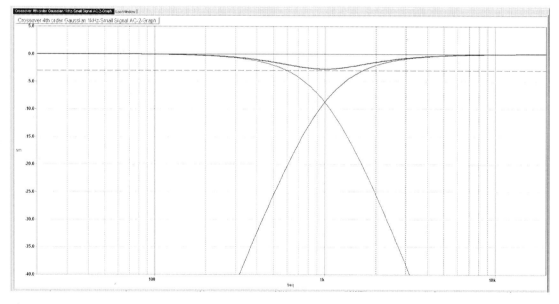

Figure 4.56: The frequency response of a 4th-order Gaussian crossover, resulting from in-phase summation. The dashed line is at −3 dB.

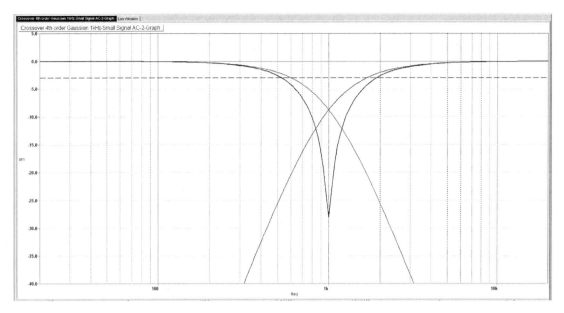

Figure 4.57: The frequency response of a 4th-order Gaussian crossover, resulting from reversed-phase summation. The dashed line is at −3 dB.

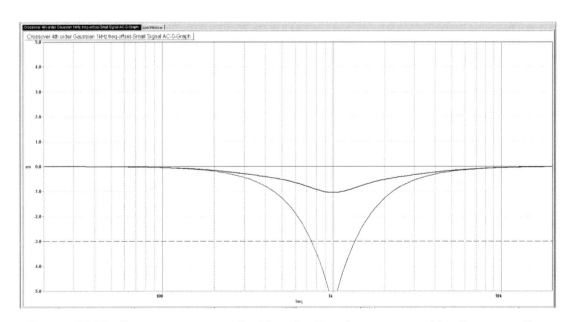

Figure 4.58: The frequency response of a 4th-order Gaussian crossover with a frequency offset of 0.8045 times to give minimal depth of dip. In-phase summation. The dashed line is at −3 dB.

Once again, there is not a great deal of information about on Gaussian filters, and some of what there is appears to be contradictory. The crossover here uses Gaussian filters as defined by Linear-X Systems Filtershop, [6] which for the 4th-order consist of two cascaded 2nd-order stages; the first has a cutoff frequency of 0.9930 and a $Q$ of 0.636, while the second has a cutoff frequency of 1.0594 and a $Q$ of 0.548. It is very possible that if you use a different design methodology from me—see Chapter 7 for more details—you will get different cutoff frequencies. This is irritating but not a cause for despair, as adjusting the frequency-offset ratio to suit is straightforward.

However, the Gaussian alignment appears to have no great advantages. Its roll-offs are slow; if that suits your design intentions, the Bessel crossover has a better maximal flatness (the Bessel has two dips −0.3 dB deep, as opposed to the Gaussian −1.05 dB single dip).

The summed group delay response has a moderate peak just below the crossover frequency.

### Fourth-Order Legendre Crossover

The Legendre filter (sometimes called the Legendre-Papoulis filter) is optimised for the greatest possible slope at the passband edge without passband ripples—in other words it is a monotonic filter, with a response that always goes downwards. It gives a faster roll-off than the Butterworth characteristic, but the drawback is that the passband is not maximally flat, as is the Butterworth; instead it begins to slope gently down until the rapid roll-off begins. This can be seen in Figure 4.59, which also shows that in-phase summation gives a narrow hump reaching +1.4 dB at the crossover

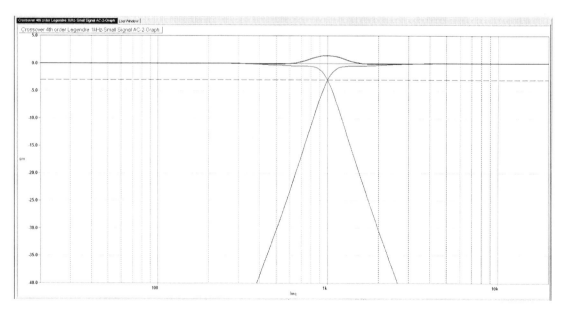

**Figure 4.59: The frequency response of a 4th-order Legendre crossover, resulting from in-phase summation. The dashed line is at −3 dB.**

point. It may not be visible in the plot, but the summed response actually dips below the 0 dB line by 0.1 dB on either side of the hump.

A 4th-order Legendre filter is normally constructed by cascading a 2nd-order stage with a $Q$ of 2.10 with another 2nd-order stage of $Q = 0.597$. A $Q$ of 2.10 is quite high for a standard Sallen & Key filter and requires a capacitor ratio of 17.6 to achieve it; this indicates that component sensitivity will be higher than usual and care will be needed to get an accurate response.

The summation with one phase reversed has a shallow dip of $-2.0$ dB around the crossover point, as shown in Figure 4.60. It is not too clear which of these can be best flattened by the use of a frequency offset, so I tried both.

The rapid roll-off of the Legendre filters means that quite small amounts of frequency offset have large effects. An offset ratio of only 1.029 times gives the maximally flat response for the in-phase condition, illustrated in Figure 4.61. The twin peaks are $+0.3$ dB high and the central dip is $-0.39$ dB deep.

Applying a frequency offset to the reversed-phase case gives an intriguing shape to the curve but larger errors, the $-2$ dB dip reaching its minimum depth of $-1.05$ dB with an offset ratio of 1.041 times, as shown in Figure 4.62. Clearly the in-phase option is better, unless you are trying to keep abreast of the times, and perhaps compensating for a 1 dB peak in a drive unit response.

The group delay of the summed output has a significant peak just below the crossover frequency.

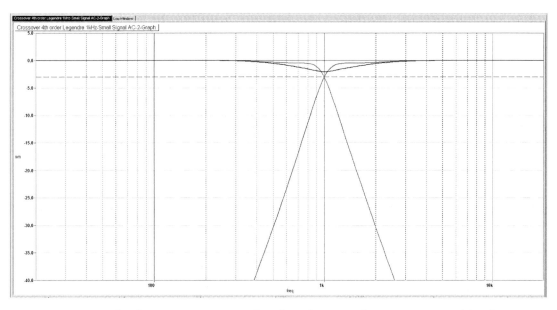

**Figure 4.60: The frequency response of a 4th-order Legendre crossover, resulting from reversed-phase summation. The dashed line is at −3 dB.**

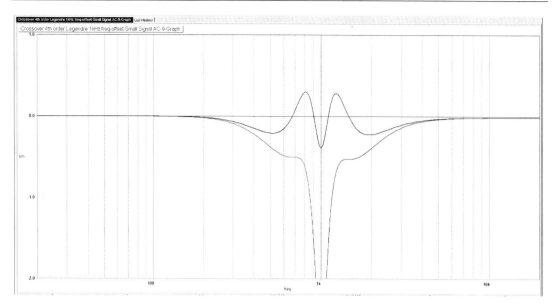

**Figure 4.61: The frequency response of a 4th-order Legendre crossover with a frequency offset of 1.029 times, and in-phase summation. Zoomed-in.**

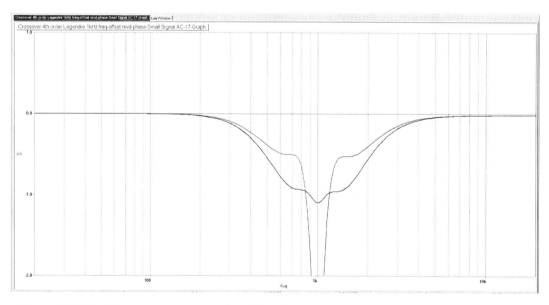

**Figure 4.62: The frequency response of a 4th-order Legendre crossover with a frequency offset of 1.041 times, and reversed-phase summation. Zoomed-in.**

## Higher-Order Crossovers

As we have seen, 4th-order crossovers provide a wide range of possible alignments and good slopes of 24 dB/octave, which are generally considered to be more than adequate for maintaining separation between drive units. The 4th-order Linkwitz-Riley, with the absolute flatness of its summed response and its delightfully straightforward filter design, is considered by many the best crossover alignment known at present. The question nonetheless arises if crossovers of 5th-, 6th- or higher order could give any further advantages. There seems to be no consensus as to whether the extra steepness of the slopes gives significant benefits, and there is an issue about the perceptibility of the group delay. According to Siegfried Linkwitz, "Crossover filters of higher order than LR4 are probably not useful, because of an increasing peak in group delay around the crossover frequency."

Nonetheless, crossovers of up to the 8th-order (with 48 dB/octave slopes) are sometimes used, mainly for sound-reinforcement applications. These are usually of the Linkwitz-Riley type. According to Dennis Bohn of the Rane Corporation, [7] going from a 4th-order Linkwitz-Riley to an 8th-order Linkwitz-Riley halves the effective width of the crossover region from 1.5 octaves to 0.75 octaves, and gives more linear drive unit operation and greater driver protection, and hence better power handling. He makes the point that the system designer must have a very good knowledge of the drive unit characteristics to use such steep slopes effectively.

Eighth-order crossover filters raise questions about component sensitivity. Expensive precision capacitors (and possibly resistors) are likely to be required. Eighth-order crossovers are probably best realised by DSP techniques.

At least one company makes 5th-order passive crossovers. CDT Audio [8] manufacture crossovers for automotive audio which are stated to effectively give 5th-order slopes by the use of elliptical filters.

## Determining Frequency Offsets

The frequency offsets required for maximal flatness are given earlier for the common (and some not-so-common) crossover alignments. However, if you are bravely striking out on your own with an unconventional filter not covered here, you may find it useful to be able to determine your own frequency offsets for maximal response flatness.

There are several ways of doing this. It would be possible to do it purely mathematically, working from the complex filter responses and manipulating the cutoff frequencies by some sort of optimisation technique. However, the rather heavyweight mathematics involved in that approach does not appeal to all of us. An alternative simulation method which requires a bit of work but no turn-the-paper-sideways algebraic manipulation is as follows:

1.  Simulate the lowpass and highpass filters, designed for the same cutoff frequency (typically 1.0 kHz) and driven from the same signal source. Check that they are giving the response shapes and cutoff frequencies that you expect.
2.  Sum the two filter outputs. In most simulators this can be done simply by plotting the arithmetical sum, so there is no need to add an electrical summing stage to the simulation. The reverse-phase connection can be checked by plotting the difference of the filter outputs instead of the sum.
3.  If the summed response is not ruler flat, determine how it needs to be altered. Very roughly, if there is a peak around the crossover frequency you will need to move the filter cutoff frequencies apart (offset ratio greater than 1), and if there is a dip you need to move them together (offset ratio less than 1).
4.  Now there is a bit of work; move one of the cutoff frequencies by changing the component values in the relevant filter and see how the response changes. Normally five or six attempts will get

the response as flat as you are going to get it; in many cases this means equal deviations above and below the 0 dB line. Changing the cutoff frequencies is much easier if you have designed the filters as Sallen & Key types and scaled the components so that they are equal. For example, a 4th-order highpass Sallen & Key filter commonly consists of two cascaded 2nd-order stages. These will have two equal capacitors in each stage, and by manipulating the resistor values it can be arranged that all four capacitors have the same value. The cutoff frequency can then be changed by at worst typing in four equal values; if your simulator has parameter facilities you should be able to set things up so typing in a single value changes all four capacitors.

5.   You will note that in the previous step we performed the cut-and-try on one filter only, to save effort. This means that the crossover frequency will move away from the original design value, but ignore this as you concentrate on the shape of the summed response. When you have the best response you can get, one cutoff frequency will be unchanged at 1.0 kHz, but the other is altered to, say, 1.4 kHz. To get the crossover point back where it should be, the 1.4 times ratio needs to be equally distributed between the two filter cutoff frequencies. This is done by taking the square root of the ratio, so the 1.4 kHz cutoff becomes 1.183 kHz, and the 1 kHz cutoff becomes 1/1.183 = 0.845 kHz. If you have got the calculations right, the crossover frequency will now be back at 1 kHz. Several examples of this are given in the descriptions of the crossover alignments earlier in the chapter.

Table 4.1 includes a summary of the frequency offsets required for maximal flatness with the crossover alignments dealt with in this chapter.

**Table 4.1: The properties of the crossovers examined so far, including the frequency offsets required for maximal flatness (n/d = not determined).**

| Order | Crossover type | Phase | APC? | CPC? | Lobe error? | Freq-offset x | Deviation ±dB |
|:---:|:---:|:---:|:---:|:---:|:---:|:---:|:---:|
| 1 | 1st-order | In-phase | Linear | YES | YES | None | Flat |
| 1 | 1st-order | Reversed | YES | YES | YES | None | Flat |
| 2 | Butterworth | Reversed | NO | YES | NO | 1.30 | 0.45 |
| 2 | Linkwitz-Riley | In-phase | YES | NO | NO | None | Flat |
| 2 | Bessel | Reversed | NO | NO | NO | 1.45 | 0.07 |
| 2 | 1.0 dB-Chebyshev | Reversed | NO | NO | NO | 1.53 | 1.60 |
| 3 | Butterworth | Either | YES | YES | YES | None | Flat |
| 3 | Linkwitz-Riley | Reversed | NO | NO | YES | 0.872 | 0.33 |
| 3 | Bessel | Reversed | NO | NO | YES | 1.22 | 0.35 |
| 3 | 1.0 dB-Chebyshev | Reversed | NO | NO | n/d | 0.946 | 1.60 |
| 4 | Butterworth | In-phase | NO | YES | NO | 1.128 | 0.47 |
| 4 | Linkwitz-Riley | In-phase | YES | NO | NO | None | Flat |
| 4 | Bessel | In-phase | NO | NO | n/d | 1.229 | 0.30 |
| 4 | 1.0 dB-Chebyshev | In-phase | NO | NO | n/d | n/d | n/d |
| 4 | Linear phase | In-phase | NO | NO | n/d | 1.084 | 0.66 |
| 4 | Gaussian | In-phase | NO | NO | n/d | 0.804 | 1.05 |
| 4 | Legendre | In-phase | NO | NO | n/d | 1.029 | 0.39 |

# Filler-Driver Crossovers

The filler-driver crossover concept was introduced by Erik Baekgaard of Bang & Olufsen in 1977. [9] It uses an extra drive unit in order to obtain a linear-phase response. This is one of those not-too-common situations where wading through the complex algebra that describes the crossover response is instructive. If you take the equation for the response of a 2nd-order Butterworth crossover, and then compare it with the corresponding mathematical description of a linear-phase crossover, there is a missing term in the former. The idea is to add an extra drive unit, called a "filler-driver", that will supply that missing term, which is equivalent to a bandpass filter of low $Q$ centred on the crossover frequency.

The 2nd-order Butterworth alignment is normally used with one output inverted to get somewhere near a flat response (without frequency offset it gives a broad +3 dB hump at crossover). If the outputs are in-phase, then there is a deep notch at the crossover frequency. This notch is filled in precisely by the filler drive unit, as shown in Figure 4.63, where the summed response is exactly flat, and additionally we get a flat phase response at 0° and a flat group delay response. If we assume the filler drive unit itself has a flat response either side of the crossover frequency, then it must be fed via a bandpass filter that is −3 dB down an octave away from the centre frequency on each side; this corresponds to a $Q$ of 0.6667. Ways of realizing the required filter appear to be strangely absent from most discussions of the filler-driver technique, so I present here a straightforward method.

Since the required $Q$ of 0.6667 is less than 0.707, it cannot be obtained with a standard MFB (Rauch) filter. The Deliyannis version is used instead (see Chapter 10 for the design procedure); the passband gain cannot be set independently, and we find we get a loss of 1.125 times (−1.03 dB), which must

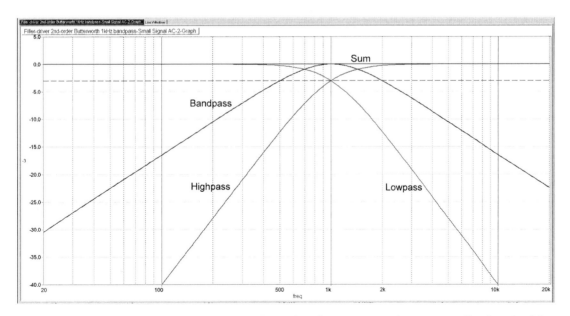

**Figure 4.63: The flat amplitude response of a 2nd-order Butterworth crossover (in-phase) with an added bandpass filler-driver.**

be made up. MFB filters also give a phase inversion that must be undone, as the filler contribution must be in phase; if its phase is reversed, the summed response is still dead flat, but it is allpass rather than linear phase, with a peak in the group delay. The A3 stage in Figure 4.64 neatly performs both corrections. Note that the value of R6 eerily echoes the resistor values in the other filters. Component tolerance errors will cause ripples in the amplitude, phase, and group delay responses.

Since the bandpass filter is a simple 2nd-order affair, the slopes at either side of the crossover frequency are only 6 dB/octave, which places quite severe demands on the filler drive unit.

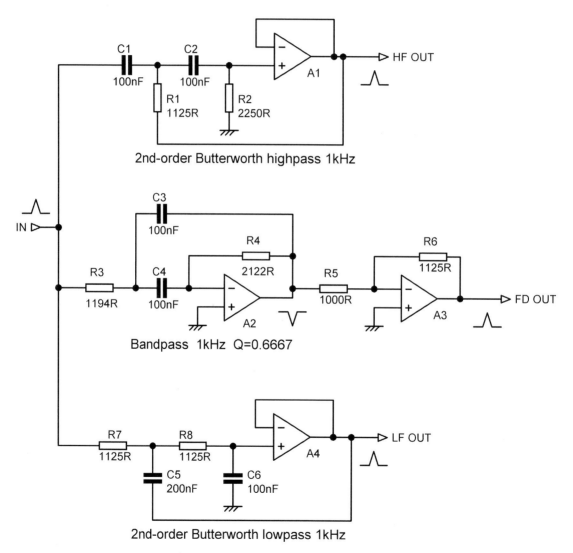

**Figure 4.64: The schematic of a 2nd-order Butterworth crossover (in-phase) with added bandpass filler-driver.**

The filler-driver concept can also be used with 4th-order Linkwitz-Riley crossovers, [10] but the filler slopes remain at 6 dB/octave.

While the filler-driver concept is unquestionably ingenious, it has not caught on. The filler-driver takes up room on the baffle and makes it harder to get the two main drivers close together, worsening time-alignment problems. It significantly complicates an active crossover audio system, as we need an extra set of filters and an extra pair of power amplifiers, not to mention the two extra drive units (assuming stereo). These last will not be cheap because of the wide bandwidth of operation required, and cost is added to the loudspeaker enclosures because of the extra cut-outs, terminals, etc that are required.

## The Duelund Crossover

This idea, originated by Steen Duelund, [11] is also usually regarded as a filler-driver crossover concept. It is even harder to explain without recourse to complex algebra, but the basic plan is to use two cascaded bandpass filters to generate a MID output, rather than simply filling in a notch. This output has 12 dB/octave slopes on each side rather than the 6 dB/octave slopes of the Baekgaard filler-driver. The demands on the driver concerned, however, are still quite severe. There is much more information on the Duelund crossover in [12] and [13].

## Crossover Topology

For the sake of simplicity, all the crossovers we have looked at in this chapter are 2-way; they divide the audio spectrum into two bands only. For the purposes of illustration a 1 kHz crossover frequency was used throughout. However, most active crossover systems will use three- or four-way crossovers to split the audio spectrum into three or four bands, to reduce the demands on the drive units and generally get the full benefits of active crossover technology. This introduces some extra complications.

It would appear to be very simple to make a 3-way crossover by combining two 2-way crossovers. We will use our usual example crossover specification with an LF/MID crossover frequency of 400 Hz and a MID/HF crossover frequency of 3 kHz. There are two highpass filter and two lowpass filters, and the standard way of connecting them is shown in Figure 4.65.

It has however been pointed out by several people, amongst them Siegfried Linkwitz in [13] and [14], that this crossover topology is defective. If all three outputs are summed, they do not sum to exactly flat. This is best illustrated when two crossover frequencies are close together. Consider the 4th-order Linkwitz-Riley crossover in Figure 4.66a, where the MID crosses over to the LF at 200 Hz, but this crosses over again to a subwoofer at 50 Hz; the rest of the crossover is left out to keep things simple. This is what we might call an in-line crossover, because the filters follow one after the other; the outputs from it are shown in Figure 4.67, where you will note that the LF output never gets up to 0 dB, as the crossover frequencies are only two octaves apart. As a check we sum the subwoofer output SUB with the LF output of the 50 Hz highpass filter (before it has been through the 200 Hz lowpass filter) and we get the dead flat line labelled "50 Hz sum", which proves that the 50 Hz Linkwitz-Riley filters are doing their stuff.

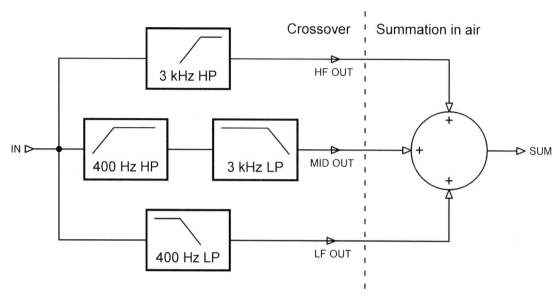

**Figure 4.65: A 3-way crossover made by combining a 400 Hz 2-way crossover with a 3 kHz 2-way crossover. In-line topology.**

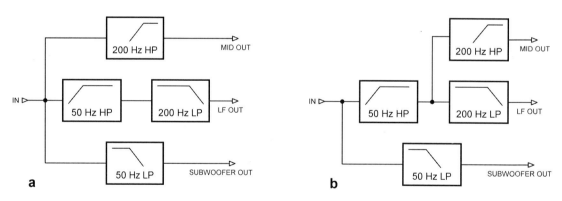

**Figure 4.66: 3-way crossovers with frequencies at 50 Hz and 200 Hz: (a) in-line topology; (b) branching topology.**

Next we sum the MID and LF outputs; we expect an output that will roll-off around 50 Hz as we have not included the subwoofer output. What we get is labelled "200 Hz sum" in Figure 4.67, and I think you can see at once that it has a warped look which indicates something is wrong. The problem is that the LF signal has passed through the 50 Hz highpass filter, while the MID signal has not. Fifty Hz is only two octaves away from 200 Hz, and while the amplitude response of the 50 Hz highpass filter is only down by 0.53 dB at 100 Hz and 0.03 dB at 200 Hz, the phase-shift is still 40° short of its ultimate value of 360°, and this spoils the MID-LF summation.

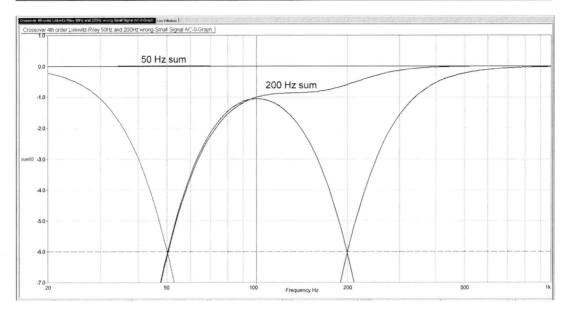

Crossover 4th order Linkwitz-Riley 50Hz and 200Hz wrong-Small Signal AC-0-Graph

Crossover 4th order Linkwitz-Riley 50Hz and 200Hz wrong-Small Signal AC-0-Graph

**Figure 4.67: Summed LF and MID outputs with in-line filter topology as in Figure 4.66a. The 200Hz sum response is bent out of shape.**

The cure for this problem is to rearrange our in-line crossover into a branching topology, as in Figure 4.66b. Now both signals going to the 200 Hz filters have passed through the 50 Hz highpass filter, and there is no extra relative phase-shift to mess things up. The result is seen in Figure 4.68, where the "200 Hz sum" now has the response we expect. The error with the in-line topology reaches a maximum of −0.74 dB at 150 Hz.

That demonstrates the phase-shift effect. But what happens when we sum all three outputs, as we will in real life? Figure 4.69 gives the unwelcome answer that something is amiss with the summation of the SUB and LF outputs, the summed level being 0.8 dB low around 65 Hz. The problem is similar to the one we have just solved; the LF output has been phase-shifted by going through the 200 Hz lowpass filter, but the SUB output has not. Going back to the in-line topology makes it worse; the sag in the summed response is now slightly deeper and much wider, extending from 40 Hz to 200 Hz.

Before we panic, we must recognise that bringing in a subwoofer output as demonstration deliberately put the crossover frequencies as close together as is plausible, and most crossovers will have a much greater frequency spacing. If we go back to our example crossover frequencies of 400 Hz and 3 kHz, using the branching topology as in Figure 4.70a, and sum all three outputs, we get the much smaller error of a −0.20 dB dip around 530 Hz. If we try the alternative branching topology in Figure 4.70b, there is still a −0.20 dB dip, but around 2.2 kHz, as in Figure 4.71.The in-line topology gives the same depth of dip but it is once more much wider, extending between the crossover frequencies.

Clearly the branching topology is superior, but we do seem to have uncovered an inherent problem with 3-way crossovers. The MID output must go through two filters, and one of the other outputs can be put through two filters by branching; but no matter which of the two ways in Figure 4.70 you

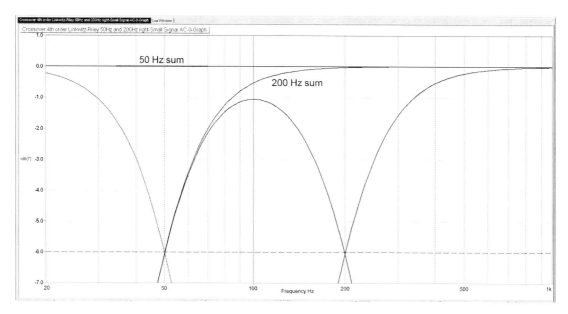

Figure 4.68: Summed LF and MID outputs with branching filter topology as in Figure 4.66b. The 200Hz sum response is now as expected.

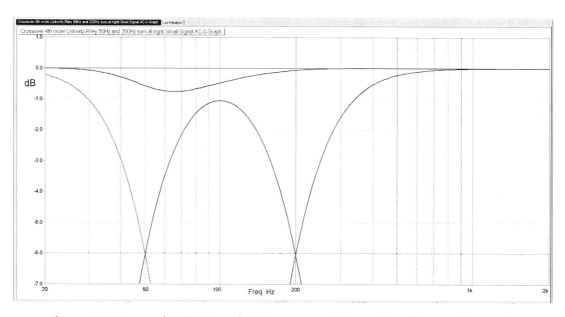

Figure 4.69: Summed SUB, LF, and MID outputs with branching filter topology as in Figure 4.66b.

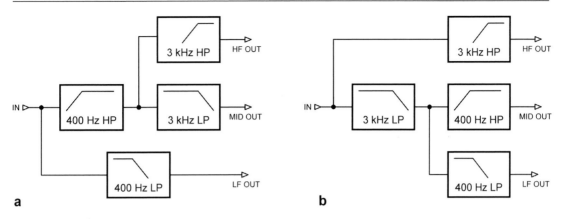

a                                          b

**Figure 4.70: The 3-way crossover of Figure 4.65 converted to the branching filter topology in two different ways.**

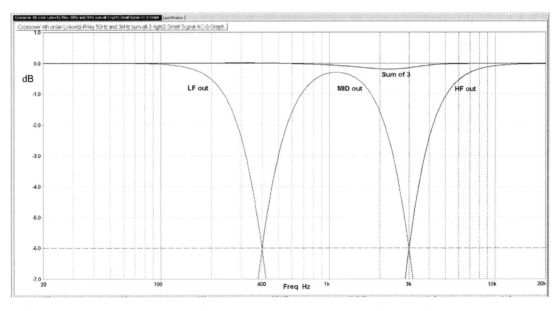

**Figure 4.71: The summed LF, MID, and HF outputs, with branching crossover topology as in Figure 4.70b, show a −0.2 dB dip around 2.2 kHz.**

choose, the remaining output of the three can only go through one filter. There will always be a phase-shift problem, though fortunately an error of 0.20 dB is going to be negligible compared with driver irregularities.

Earlier in this chapter we made extensive use of offsetting the filter frequencies to minimise response deviations. We can do that here. Applying an offset of 0.961 times to the 3 kHz highpass filter, moving

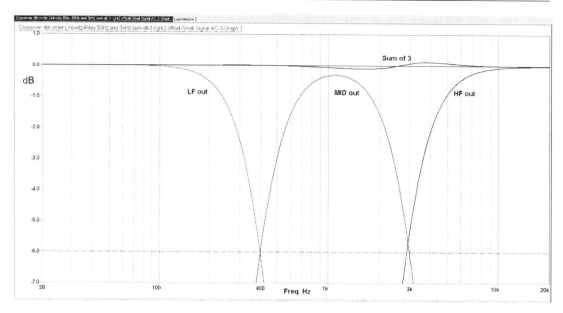

**Figure 4.72: The summed LF, MID, and HF outputs as in Figure 4.71, but with the HF 3 kHz highpass filter cutoff frequency offset by 0.961 times to 2.88 kHz.**

its cutoff down to 2.88 kHz, converts the −0.20 dB dip into an innocuous ±0.11 dB ripple in the summed response, as in Figure 4.72. I think we can live with that.

## Crossover Conclusions

This has been a long chapter, and we have looked at a lot of different crossover types, with yet more to be found in Chapter 5 on notch crossovers and Chapter 6 on subtractive crossovers. To briefly summarise the best ones:

First-order crossovers have tempting properties but put great demands on the drive units.
The 2nd-order Linkwitz-Riley crossover sums flat but does not have adequate slopes.
The 3rd-order Butterworth crossover sums flat but shows lobing errors.
The 3rd-order Linkwitz-Riley crossover does not sum flat.
The 4th-order Linkwitz-Riley crossover sums flat, with no lobing errors and good slopes.

Of the crossover types not yet examined, I draw your attention to the very clever NTM crossover in Chapter 5.

## References

[1] Small, Richard H. "Constant Voltage Crossover Network Design" JAES, 19(1), January 1971, pp. 12–19
[2] Dickason, Vance "The Loudspeaker Design Cookbook" Seventh Edn, Audio Amateur Press, January 2006 ISBN: 1882580478

[3] KEF  www.kef.com/GB/Loudspeakers/QSeries2010/Floorstanding

[4] Miller, Ray "A Bessel Filter Crossover, and Its Relation to Others" Rane Note 147, Rane Corporation, 2006

[5] KEF "KEF 105 Description"  www.kef.com/history/en/1970/model105.asp

[6] Strahm, Chris "Linear-X Systems Filtershop"  www.linearx.com/products/software/FilterShop/Fshop_01.htm

[7] Bohm, Dennis "Linkwitz-Riley Crossovers: A Primer" Rane Note 160, October 2005, Rane Corporation

[8] cdtAudio  www.cdtaudio.com/crossovers_08.htm

[9] Baekgaard, E. "A Novel Approach to Linear Phase Loudspeakers Using Passive Crossover Networks" JAES 25(5), 1977, p. 284 (filler-driver)

[10] Megann, Alex  www.soton.ac.uk/~apm3/diyaudio/Filler_drivers.html December 1999 (Second and 4th-order crossovers with filler-drivers)

[11] Duelund, Steen  www.duelundaudio.com/Articles_by_Steen_Duelund.asp

[12] Kreskovsky, John  www.musicanddesign.com/Duelund_and_Beyond.html

[13] Linkwitz, Siegfried  www.linkwitzlab.com/crossovers.htm (Duelund crossover)

[14] Linkwitz, Siegfried, "STEREO-From Live to Recorded and Reproduced." Linear Audio Volume 0, p. 109 published by Jan Didden 2010. www.linearaudio.nl/ (crossover topology)

# CHAPTER 5

# Notch Crossovers

The crossovers examined in Chapter 4 are all-pole crossovers, which means that the filters used are relatively simple lowpass and highpass types, though of varying order and filter characteristics (Butterworth, Bessel, etc). It is also possible to contrive crossovers that have notches (or to get mathematical, zeros) built into the roll-off, typically giving a much steeper filter slope, to begin with at least, than an all-pole crossover of a practical order. This can be very useful when drive units that are otherwise acceptable misbehave badly when taken just outside their intended operating range. Neville Thiele [1] gives the example of a horn loudspeaker being used near its cutoff frequency. He also cites the case of a mono subwoofer, where its contribution must be rolled-off as quickly as possible out of band to prevent it contaminating the stereo localisation cues from the main speakers.

## Elliptical Filter Crossovers

A 4th-order crossover design based on the use of elliptical filters was published by Bill Hardman in *Electronics World* in 1999, [2] and this has been the basis for much discussion on the subject. The amplitude response of the published design is shown in Figure 5.1.The original version had a crossover point at 1.5 kHz, but I have modified the filter frequencies for 1 kHz instead, to match all the other crossover response examples in this book. Nothing else that might affect the response has been changed.

The two deep notches symmetrically placed on either side of the crossover point are at 529 Hz and 1.84 kHz, less than an octave away from the crossover point. It is obvious that their presence greatly increases the initial roll-off slope, making it more effective than a 4th-order Linkwitz-Riley alignment from this point of view.

It is, however, vital that the outputs of a crossover should sum to as nearly flat as possible, and looking at the spiky nature of Figure 5.1, you are probably thinking that this is going to be very difficult to arrange. In fact it can be done relatively easily. Figure 5.2 shows the in-phase summation; the central hump is only −0.13 dB deep, while the dips on either side are about −0.9 dB deep. The phase-reversed summation has a deep notch at the crossover frequency and is of no value.

Looking at Figure 5.2, one cannot help feeling it might be possible to improve the flatness of the summation by tweaking the filter frequencies. However, applying a frequency offset is in fact not very helpful, because the centre rises much more than the dips do; for example, an offset ratio of 1.045 gives a central peak of +0.4 dB with dips of −0.7 dB, which is not much improvement, if any, on the original Hardman alignment.

The elliptical filters used in this crossover are the usual combinations of notch filters and all-pole filters. Figure 5.3 shows that the lowpass path is composed of a lowpass notch and a 2nd-order lowpass Sallen &

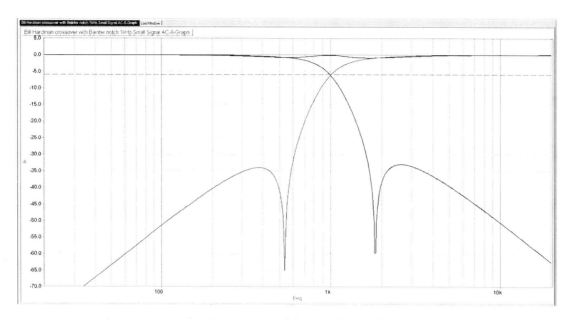

Figure 5.1: Amplitude response of the Hardman elliptical crossover.

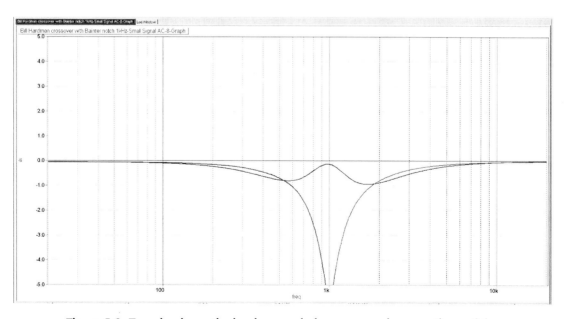

Figure 5.2: Zooming in on the in-phase and phase-reversed summations of the Hardman elliptical crossover.

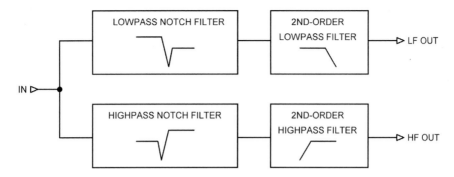

**Figure 5.3: Block diagram of the Hardman elliptical crossover architecture.**

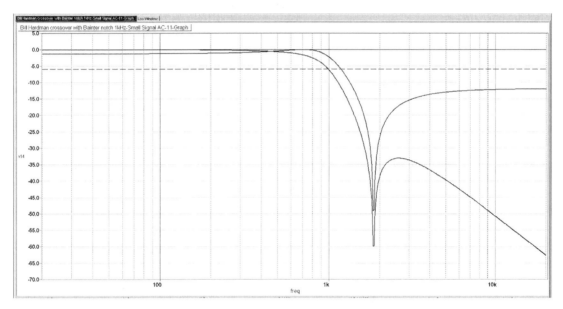

**Figure 5.4: How the Hardman elliptical lowpass filter works. The upper plot is the output of the lowpass notch filter only; the second plot adds in the effect of the lowpass filter.**

Key filter; the latter has equal component values but a non-standard Q, which places it somewhere between a Linkwitz-Riley and a Bessel filter characteristic. The highpass path is composed of a highpass notch and a 2nd-order highpass Sallen & Key filter with the same non-standard value of $Q$.

Figure 5.3 underlines the difference between a lowpass notch and a highpass notch. These are not your more familiar symmetrical notches that go up to 0 dB on either side of the central crevasse. A lowpass notch response starts out at 0 dB at low frequencies, drops down into the crevasse, then comes back to level out at a lower gain, say −10 dB; the lowpass notch can be seen in both Figures 5.1 and 5.4. Conversely, a highpass notch has a response that is 0 dB at high frequencies, drops down, then comes back up to, say,−10 dB at low frequencies.

Figure 5.4 shows how it works for the lowpass path. The notch filter provides a lowpass notch; as frequency increases there is some peaking just before the roll-off into the crevasse. This is cancelled out by the low-Q, relatively slow roll-off from the all-pole lowpass filter, giving a fast roll-off around the crossover frequency, and also giving a notch deepened by the increasing attenuation of the lowpass filter. The ultimate roll-off slope, above the notch frequency, is only 12 dB/octave because it comes entirely from the 2nd-order lowpass filter.

The original Hardman crossover used Bainter filters to create the notches followed by 2nd-order Sallen & Key filters. Bainter filters are popular for making elliptical filters of the type shown here because they give good deep notches where the depth does not depend on the matching of passive components, but only on the open-loop gain of the opamps. You can see lovely deep notches in Figures 5.1 and 5.4; it is in a sense a pity that the notch depth is not in any way critical to get a good summed response. So long as a distinct notch is visible the effect on the summed response is negligible.

A design for a Hardman crossover is shown in Figure 5.5. While the essential filter characteristics are closely based on those in the original article, [2] the crossover frequency, as noted, has been altered from 1.5 kHz to 1.0 kHz, and the general impedance level of the circuitry reduced by a factor of approximately 20, to reduce noise. Any further impedance reduction would risk overloading the opamps (assumed to be 5532s) with a consequent increase in distortion. As with many other circuits in this book, it has been designed to use standard capacitor values, with the resistor values coming out as whatever they do. It is much cheaper to get an exact value by combining resistors rather than by combining capacitors. The simplest way to scale the circuit to implement other crossover frequencies is to change all the capacitors in the same ratio.

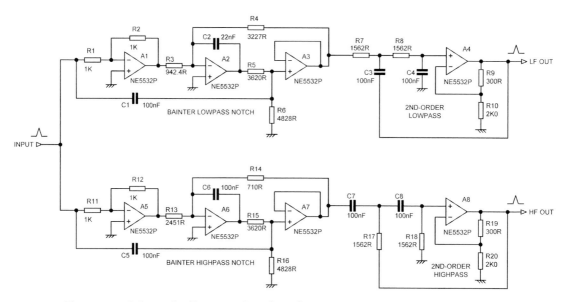

**Figure 5.5: Schematic diagram of a 4th-order Hardman elliptical crossover with a crossover frequency of 1 kHz. The outputs are at different levels—see text for details. Note C2 is meant to be 22 nF.**

Figure 5.5 shows that the two Bainter notch filters for lowpass and highpass are very similar, with only three components differing in value. This sort of convenient behaviour is what makes the Bainter filter so popular. The relationship between R3 and R4 in the lowpass filter, and between R13 and R14 in the highpass filter, determine the kind of notch produced. If R4 is greater than R3, you get a lowpass notch. If R3 = R4, you get the standard symmetrical notch. If R3 is greater than R4, you get a highpass notch. The value of R6 (or R16) sets the notch $Q$.

There are however some aspects of this circuit that are less convenient. The lowpass notch response does not actually start out at 0 dB at low frequencies; instead it has a gain of +10.6 there. The response then drops into the crevasse and comes back up to 0 dB. There is then the gain of +1.3 dB from the lowpass filter, giving a total of 11.9 dB of passband gain, which may not fit well into the gain/headroom scheme planned for the crossover.

The highpass notch response has a passband gain of 0 dB at high frequencies, and at frequencies below the notch comes back up to −10.8 dB. With the final highpass filter added the passband gain is +1.2 dB. Therefore we have a level difference of $11.9 - 1.2 = 10.7$ dB between the two outputs, and this will have to be accommodated somewhere in the crossover system design. There is more information on the Bainter filter and other notch filters in Chapter 12.

A most interesting paper on the use of Chebyshev filters (ripples in the passband), inverse-Chebyshev (notches in the stopband), and elliptical filters (ripples in the passband *and* notches in the stopband), with the added feature of a variable crossover frequency, was published in the *JAES* by Regalia, Fujii, Mitra, Sanjit, and Neuvo in 1987. [3] It is well worth studying.

## Neville Thiele Method™ (NTM) Crossovers

One of the better-known notch crossovers is that known as the Neville Thiele Method™ crossover, introduced by Neville Thiele in an *AES* paper in 2000. [1] This does not appear to consist of elliptical filters as such (as far as my knowledge of elliptical filters goes, anyway) but a rather more subtle arrangement that sums to unity much more accurately than the Hardman crossover we have just looked at. One of the few examinations of this technique that has been published is that by Rod Elliot. [4]

I should say at once that the Neville Thiele Method™ or NTM is a proprietary technology addressed by US Patent 6,854,005 and assigned to Techstream Pty Ltd, Victoria, AU, and that if you plan to use it for anything other than a private project you might want to talk to them about licensing issues. The information given here is published by permission and is derived solely from the public-domain References [1] and [5], which I have to say are not an easy read. I have never seen a schematic of a manufactured crossover using this technology, nor have I ever deconstructed any related hardware.

Using References [1] and [4], it appears that the lowpass path of a 6th-order NTM crossover filter (8th-order versions are also possible) consists of a bridged-T lowpass notch filter, followed by a 2nd-order lowpass filter with a $Q$ of about 1.6; this is followed in turn by two 1st-order filters. This structure, together with the matching highpass path, is shown in Figure 5.6.

The result of this rather complicated-looking block diagram is shown in Figure 5.7. Each filter output has a fast roll-off after the crossover point, terminating in a shallow notch; the response then comes

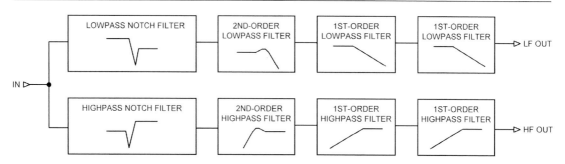

**Figure 5.6: Block diagram of a 6th-order NTM crossover.**

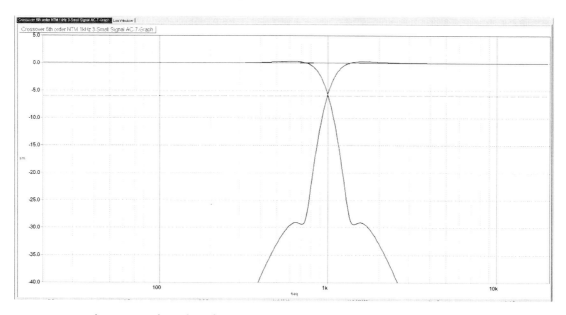

**Figure 5.7: The 6th-order NTM crossover. The dashed line is at −6 dB.**

back up a bit but then settles down to an ultimate 24 dB/octave roll-off. The filter responses may not look very promising for summation, but in fact they do add up to an almost perfectly flat response when the phase of one output is reversed. If summed in-phase there is a central crevasse about 12 dB deep, of no use to anyone.

Figure 5.8 shows a much closer view of the summed response. The bump below the crossover frequency peaks at +0.056 dB, while the flat error in the level above the crossover frequency is at −0.031 dB. Since these very small errors are asymmetrical about the crossover point, it seems more likely that they are due to opamp limitations or similar causes, rather than anything inherent in the crossover. They are negligible compared with transducer tolerances, or with the summation errors of crossovers that approximate flatness by using frequency offsetting; the performance is much better than that of the Hardman elliptical crossover, which has flatness errors of 0.9 dB.

Figure 5.9 attempts to show how the NTM crossover works; the response of each filter stage in the lowpass path is shown separately. (The responses of the two 1st-order filters have been combined

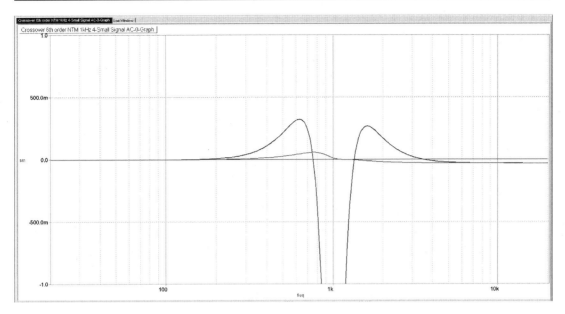

**Figure 5.8: Zooming in on the summation of the 6th-order NTM crossover, showing very small errors. The vertical scale is ±1 dB.**

into a single response for the two when cascaded.) The bridged-T filter creates a lowpass response that comes back up to −5 dB after it has been down in the notch. You will note that the notch is much shallower than that of the more complex Bainter filter used for the Hardman crossover, but this has only a very small effect on the summed response, and the simplicity of the bridged-T circuit is welcome.

The next stage in the crossover is a 2nd-order lowpass filter with a cutoff frequency of 1.0 kHz and a $Q$ of approximately 1.6. This is quite a high $Q$ for a 2nd-order filter and gives a +4.5 dB peaking of the response before the ultimate 12 dB/octave roll-off. This will need to be taken into account when planning the gain structure; rearranging the order so the 1st-order filters come before the 2nd-order should ease the situation considerably.

The two 1st-order filters both have a cutoff frequency of 1.0 kHz, and so their combined response is down −6 dB at 1.0 kHz, with an ultimate slope of 12 dB/octave. The action of these two stages together is the same as that of a 2nd-order synchronous filter, as described in Chapter 7. If implemented as simple RC networks, they must be separated by a suitable unity-gain buffer to give the correct response.

Figure 5.10 shows the schematic of my version of an NTM crossover based wholly on the information given in [1] and [4].

An interesting point is that in each path the two 1st-order filters can be combined into a single Sallen & Key 2nd-order stage with a $Q$ of 0.5, thus saving an opamp in each path. As Figure 5.11 shows, the component values used are exactly the same. This is my modest contribution to the NTM.

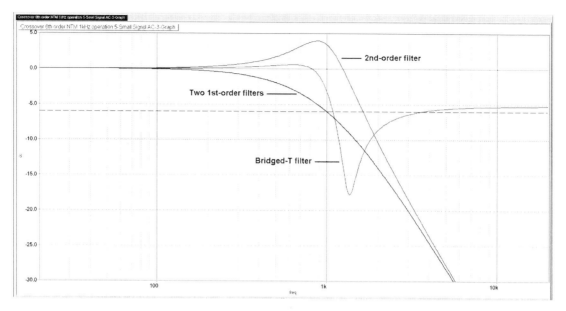

Figure 5.9: The operation of the LF path of a 6th-order NTM crossover filter, which combines the lowpass notch, a peaking 2nd-order lowpass filter, and two cascaded 1st-order lowpass filters (the last are shown as one trace here).

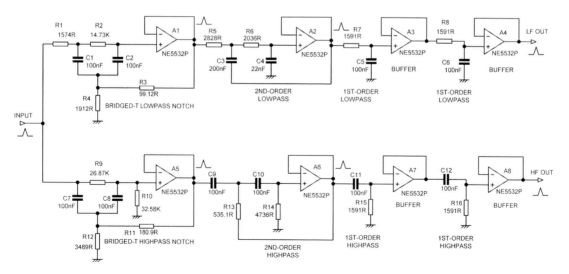

Figure 5.10: Schematic of my NTM implementation with a crossover frequency of 1 kHz.

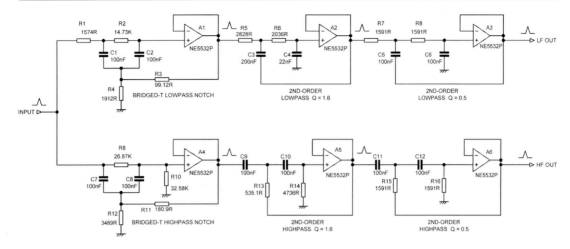

**Figure 5.11: Replacing the two 1st-order filters with an equivalent 2nd-order Sallen & Key stage. This saves two opamps, as no output buffers are required.**

# References

[1]  Thiele, Neville "Loudspeaker Crossovers with Notched Responses" JAES, 48(9), September 2000, p. 784

[2]  Hardman, Bill "Precise Active Crossover" Electronics World, August 1999, p. 652

[3]  Regalia, Philip A., Nobuo Fujii, Sanjit K. Mitra and Yrjo Neuvo "Active RC Crossover Networks With Adjustable Characteristics" JAES 35 (1/2), January/February 1987, pp. 24–30

[4]  Elliot, Rod http://sound.westhost.com/articles/ntm-xover.htm

[5]  Thiele, Neville "Crossover Filter System and Method" US Patent 6,854,005 Feb 2005 (Assigned to Techstream Pty Ltd, Victoria, AU)

# CHAPTER 6

# *Subtractive Crossovers*

## Subtractive Crossovers

What you might call the standard crossover architecture has a lowpass filter to generate the LF signal for the bass drive unit, and a corresponding highpass filter to create the HF signal for the tweeter/midrange driver. (Obviously this only describes a 2-way crossover, and a 3-way crossover has more filters.) That is not, however, the only way to do it, and one of the alternatives is the subtractive crossover. In this you just have one filter, say the lowpass, which gives the LF signal, and you make the HF signal by subtracting the LF signal from the original input, so that HF = 1 − LF. This process is illustrated in Figure 6.1, which shows a 1st-order subtractive crossover. Higher-order subtractive crossovers can be implemented by replacing the 1st-order filter with a higher-order version; Figure 6.3 shows a 2nd-order subtractive crossover. There is no necessity for the one filter used to be the lowpass type. It is equally possible to use a highpass filter and generate the LF signal by subtracting the highpass filter output from the original input. Subtractive crossovers are also called derived-crossovers, as one output is derived from the other instead of being filtered independently, and constant-voltage crossovers because of the way that their outputs sum to reconstruct the original waveform. The constant-voltage crossover, first properly described by Dick Small, [1] one of the great pioneers of scientific loudspeaker design, is a subtractive crossover.

The attraction of the subtractive crossover is that it promises a perfect response—surely the subtraction process, followed by the summation that takes place in the air in front of the speaker, must give a ruler-flat combined response? Well, the answer is that it does—but not in a way that is generally useful. Exactly why this is so will be described shortly.

Subtractive crossovers also promise—and indeed deliver—perfect waveform reconstruction. It is described in Chapter 4 how a 1st-order conventional crossover can do this reconstruction, but 2nd-order and higher conventional crossovers cannot, because of their phase-shift behaviour. Nonetheless, this ability is not as advantageous as it appears, because it is of little value when compared with the serious compromises inherent in the simple subtractive crossover.

The subtractive process also offers perfect matching between the crossover characteristics of the HF and LF paths. Since there is only one filter, there cannot be a mismatch between filters because of component tolerances. Instead, we have the requirement that the subtraction must be done accurately, so that the HF signal really is 1 −HF rather than 1 −0.99HF. This is straightforward to arrange because the subtractor will be subject only to resistor tolerances, and accurate resistors are much cheaper than accurate capacitors.

An intriguing aspect of the subtractive crossover is the prospect of saving some serious money on filter capacitors. Since only one filter is used, the number of expensive precision capacitors is halved. Certainly we have to pay for the subtractor circuit, but that is cheap by comparison.

Relatively few designs for subtractive crossovers have been published; one example that deserves examination was put forward by Christhof Heinzerling in *Electronics World* in 2000. [2] This design includes what looks to me like rather ambitious biquad equalisation to extend the bass response of the LF unit and a boost/cut control for very low frequencies based on the Baxandall circuit. While I don't agree with every statement made in the article, it is well worth tracking down and perusing.

Probably the best-known subtractive crossover is a design incorporating a time delay, based on the work of Lipshitz and Vanderkooy [3] and published by Harry Baggen in *Elektor* in 1987. [4] The purpose of adding a time delay in one of the paths of a subtractive crossover is to make the filter slopes equal (without it the output derived by subtraction always has an inadequate 6 dB/octave slope) and obtain a linear-phase crossover. A linear-phase crossover has a combined output phase-shifted by an amount proportional to frequency; in other words it only delays the signal, while other crossovers have phase-shifts that change non-linearly with frequency and act like allpass filters. The snag is that the time delay has to be very accurately matched to the filter to get the steeper crossover slopes.

All these issues are looked at in detail here.

## First-Order Subtractive Crossovers

A 1st-order subtractive crossover is shown in Figure 6.1. The circular summing element to the right is not part of the crossover but simply shows how the acoustic outputs from the HF and LF drive units sum together as air pressure in front of the loudspeaker box. It signifies the purely mathematical

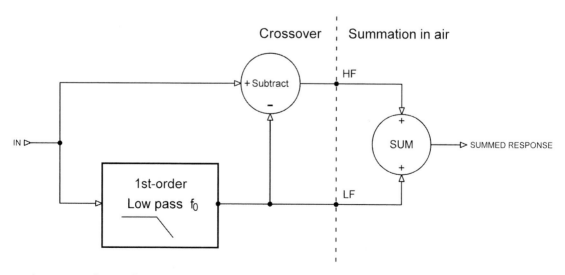

**Figure 6.1: First-order subtractive crossover. The lowpass filter output is subtracted from the incoming signal to get the HF output.**

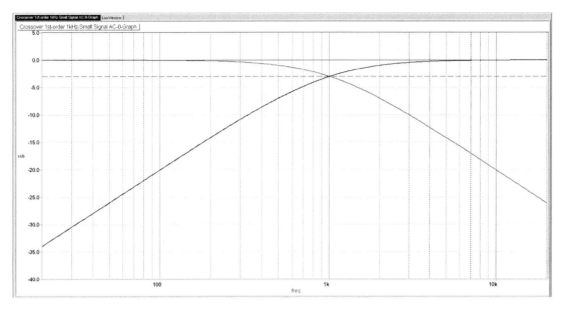

**Figure 6.2: Frequency response of 1st-order crossover; both filter outputs plus their sum (straight line at 0 dB). Dashed line is at −3 dB.**

process of addition, and it does not affect the operation of the crossover itself in any way. The circular subtracting element is of course part of the crossover, but likewise it performs a pure subtraction and nothing else.

Figure 6.2 shows the response. The LF lowpass filter output is as expected, with its −3 dB point at 1 kHz. The HF output obtained by subtraction is exactly the same as would be obtained from a 1st-order highpass filter with a 1 kHz −3 dB cutoff frequency. The summed response is a flat line at 0 dB, exactly as for the conventional 1st-order crossover. The phase behaviour is also identical, and it reconstructs waveforms in the same way.

The main advantage is that there is only one filter, so there is no problem with filter matching. The economic advantages are small in this case because we have saved only one capacitor, and the subtractor circuit may cost more than that. As always, the unsolvable problem with 1st-order crossovers is that the 6 dB/octave slopes are just not adequate for directing frequencies to the appropriate drive unit.

## Second-Order Butterworth Subtractive Crossovers

We have just seen how the 1st-order subtractive crossover gives some interesting results, though these are in fact the same as you get with a conventional two-filter crossover, and the advantages of the subtractive method are very limited. A 2nd-order crossover that can reconstruct a waveform (as a subtractive crossover can) sounds however like it might be more interesting. Figure 6.3 shows

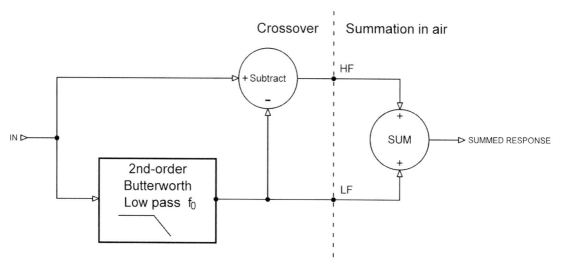

**Figure 6.3: Second-order subtractive crossover. The Butterworth lowpass filter output is subtracted from the incoming signal to get the HF output.**

a 2nd-order subtractive crossover, which you will note is exactly the same as the 1st-order version, except that the 1st-order filter has been replaced by a 2nd-order filter.

If we choose a 2nd-order Butterworth for our filter, we get the response of Figure 6.4, which was probably not what you were expecting. The lowpass output is a maximally flat Butterworth response, as it must be, but the derived HF output is very different. First, its slope is only 6 dB/octave, while the LF output has the expected 12 dB/octave slope. Worse still, the HF output has a peak +2.4 dB high just above the crossover frequency; while the final summed response may be flat, that peak represents a lot of extra energy being delivered to the midrange/tweeter at a frequency below the crossover point, which is not a good place for it, probably leading to excessive coil excursions and increased distortion. It is the phase-shift of the lowpass filter that causes the 6 dB/octave slope and the response peaking.

What we have built here is an asymmetrical crossover—one with unequal slopes. The 6 dB/octave slope of the HF output is not adequate for frequency separation of normal drive units. You will note that the method I have used here derives the HF output from the LF output, and I did that simply to draw attention to the fact that it is not a good idea. The derived output always has a slope of only 6 dB/octave, as you will shortly see, and deriving the HF output makes it very clear that its feeble slope extends well into the bass end of the plot, so an excessive amount of LF energy is going to get into the midrange/tweeter, with a high risk of damage. On the other hand, if you really wanted to use a subtractive crossover of the simple sort described here, you would derive the LF output from the highpass filter output. That would give a decent slope to the highpass output, and the 6 dB/octave lowpass slope would then be failing to keep HF out of the LF drive unit. This will probably lead to some unfortunate response irregularities, as the LF drive unit will be working outside its intended frequency range, but it is not likely to be damaged.

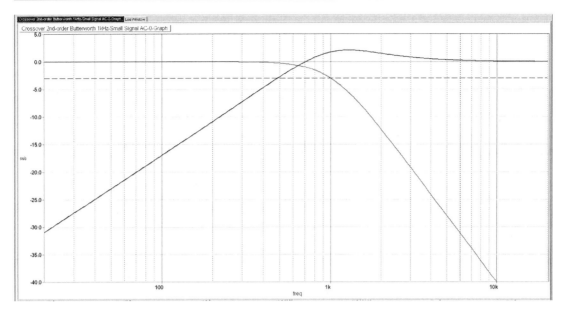

**Figure 6.4: Frequency responses of 2nd-order Butterworth subtractive crossover; both filter outputs plus their sum (straight line at 0 dB). Dashed line is at −3 dB.**

## Third-Order Butterworth Subtractive Crossovers

A 3rd-order subtractive crossover can be made in just the same way by replacing the 2nd-order lowpass filter with a 3rd-order one and carrying out the same subtraction. If we plug in a 3rd-order Butterworth, we find that the results are no better—in fact they are rather worse. The crossover is still asymmetrical, for despite the use of a 3rd-order filter instead of a 2nd-order one, the HF output still only has a slope of 6 dB/octave. The unwelcome peak in the response is still there; now it is at slightly below the crossover frequency and it has grown to +4.0 dB in height. Third-order filters are clearly not the answer.

## Fourth-Order Butterworth Subtractive Crossovers

If we try a 4th-order Butterworth as the lowpass filter, the results are much the same. The LF output is the direct output of the lowpass filter and thus is what we expect, rolling off at a satisfactory 24 dB/octave. The HF output slope stays stubbornly at 6 dB/octave, and the peak moves down a little in frequency and grows in height to +5.2 dB. The crossover is still asymmetric—in fact it is more asymmetric, with the LF slope now being four times that of the HF slope.

You may be thinking at this point that we are making a crass mistake by using Butterworth filters, and some other filter characteristic like Bessel or Chebyshev would give better results. The most popular 4th-order crossover is the Linkwitz-Riley alignment (equivalent to two cascaded 2nd-order Butterworth filters), so let's see if using that for the lowpass filter makes a revolutionary difference.

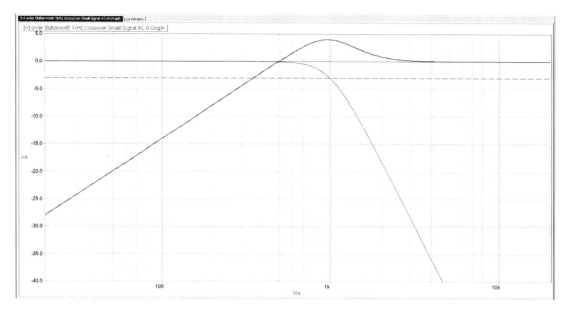

Figure 6.5: Frequency response of 3rd-order Butterworth subtractive crossover; both filter outputs plus their sum (straight line at 0 dB). Dashed line is at −3 dB.

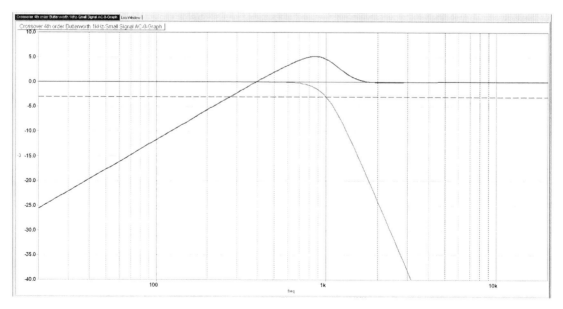

Figure 6.6: Frequency response of 4th-order Butterworth subtractive crossover; both filter outputs plus their sum (straight line at 0 dB). Dashed line is at −3 dB.

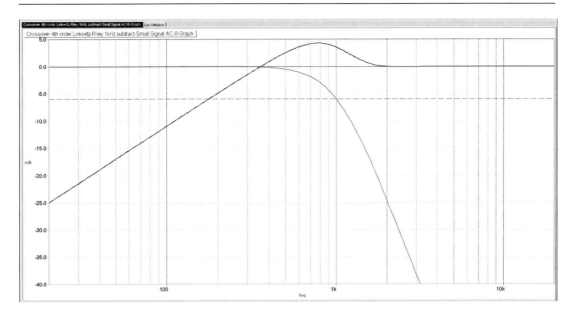

**Figure 6.7: Frequency response of 4th-order Linkwitz-Riley subtractive crossover; both filter outputs plus their sum (straight line at 0 dB). Dashed line is at −6 dB.**

Not at all. As Figure 6.7 shows, the crossover is still highly asymmetrical because the HF output still has that useless 6 dB/octave slope. The height of the peak is slightly less at +4.3 dB, but that's precious little help.

## Subtractive Crossovers With Time Delays

In 1983 Lipshitz & Vanderkooy [3] proposed that linear-phase crossover networks could be produced by a subtractive method, the key idea being that a time delay inserted in the unfiltered path would compensate for the phase-shift in the lowpass filter and allow crossovers to be designed with symmetrical slopes of useful steepness. The basic arrangement is shown in Figure 6.8b.

To the best of my knowledge, the only practical design of this sort of crossover that has been published was by Harry Baggen, in a famous article in *Elektor* in 1987. [4] It was a 3-way crossover based on 4th-order Linkwitz-Riley filters. Since the highpass outputs were derived by subtraction, using the time-delay concept, only two Linkwitz-Riley filters were required, to some extent making up for the extra cost of the subtractors and the 2nd-order allpass filters used to create the delays. This crossover may be over 20 years old, but its conceptual significance is such that it is still being actively discussed today.

The block diagram of this crossover is shown in Figure 6.9. The crossover frequencies were nominally 500 Hz and 5 kHz, but the actual frequencies calculated from the original component values are 512 Hz and 5.12 kHz. The 512 Hz 4th-order Linkwitz-Riley lowpass filter gives the LF output, while its phase-shift is compensated for in the lower path by the delay filter t1. A highpass signal is derived

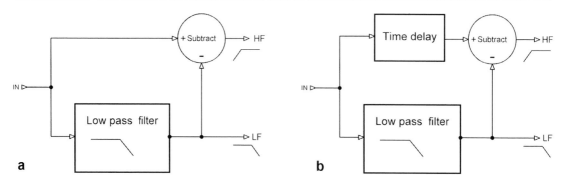

Figure 6.8: Basic subtractive crossover is at (a). Adding a time delay in the unfiltered path (b) allows symmetrical-slope crossover outputs to be derived.

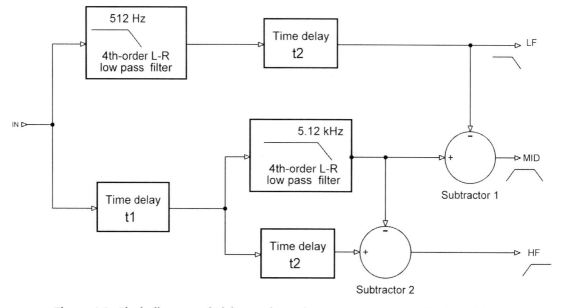

Figure 6.9: Block diagram of Elektor subtractive 3-way crossover with time delays.

from it by Subtractor 1. The circuit section, including the 5.12 kHz lowpass filter, a delay block t2, and Subtractor 2, is as shown in Figure 6.8b and derives the HF output. The signal from the 5.12 kHz lowpass filter then has the signal from the 512 Hz lowpass filter subtracted from it to create the MID output; note that another t2 delay block is inserted into this path to allow for the phase-shift in the 5.12 kHz lowpass filter.

Since the 3-way nature of the crossover makes it quite complex, I thought it best to examine the time-delay principle by looking at only one section of it. The MID/HF subtractive crossover circuitry is shown in Figure 6.10, with the original component values. The 4th-order Linkwitz-Riley lowpass filter is a standard configuration made up of two cascaded Butterworth 2nd-order filters A1, A2. The

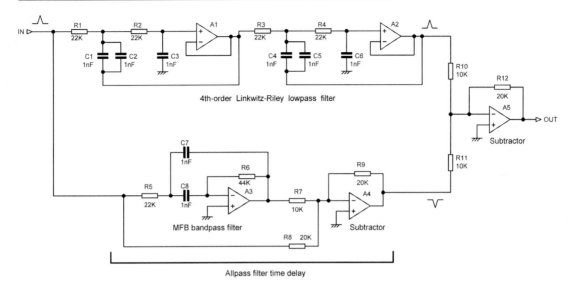

4th-order Linkwitz-Riley lowpass filter

MFB bandpass filter

Subtractor

Allpass filter time delay

**Figure 6.10: Schematic of the MID/HF section of the Elektor subtractive 3-way crossover with the original component values. R6 is two 22k resistors in series.**

time delay t2 in the other path is realised by a 2nd-order allpass filter, made up of a multiple-feedback bandpass filter A3 and the shunt-feedback stage A4. This implements the 1-2BP 2nd-order allpass configuration, where the signal is fed to a 2nd-order bandpass filter, multiplied by two, and then subtracted from the original signal. It is not what you might call intuitively obvious, but this process gives a flat amplitude response and a 2nd-order allpass phase response. Since the MFB bandpass filter phase-inverts, the subtraction can be performed by simple summation using A4. The MFB bandpass filter has unity gain at its resonance peak, so R7 needs to be half the value of R8 to implement the scaling by two. The operation and characteristics of this configuration is much more fully described in Chapter 13 on time-domain filtering.

We now have two signals, one lowpass filtered and one time delayed, and the former must be subtracted from the latter to derive the highpass output. This can again be done by a simple summing stage, in this case A5, because the delayed signal has been phase inverted by A4, so summing is equivalent to subtraction. The alert reader—and I trust there is no other sort here—will have noticed that the phase of the signals going to the subtractor A5 in Figure 6.10 is the opposite of those shown in Figure 6.9; this is because in the complete crossover the signal entering the MID/HF crossover circuitry has already been phase inverted by the delay circuitry t1.

You are possibly thinking that the impedance levels at which this circuitry operates are rather higher than recommended in this book, and you are quite right. When the Elektor crossover was published in 1987, the 5532 opamp was still expensive, and so the crossover used TL072s. These opamps have a much inferior load-driving capability, with even light loading degrading their distortion performance, so low-impedance design was not practicable.

Figure 6.11 shows the two outputs, with nice symmetrical 24 dB/octave slopes, crossing over at −6 dB very close to 5 kHz. However, for this plot the vertical scale has been extended down to −80 dB,

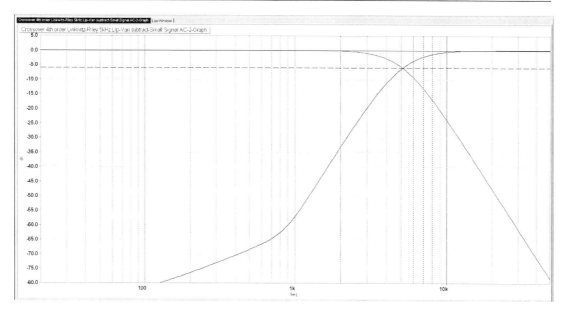

**Figure 6.11: The MID/HF crossover is only symmetrical down to −60 dB. Dashed line is at −6 dB.**

and you can see that something goes wrong at about −60 dB, with the derived HF output 24 dB/octave slope quite suddenly reverting to a shallow 6 dB/octave. It is highly unlikely that a shallower slope at such a low level could cause any drive unit problems, but alarm bells ring in the distance because this is a simulation, and one of the most dangerous traps in simulation is that it enables you to come up with an apparently sound circuit that actually depends critically on component values being exactly correct. Further investigation is therefore called for . . .

I suspected that the abrupt shallowing of the slope was due to the delay not being exactly matched to the lowpass filter characteristics, and to test this hypothesis I increased the allpass delay by about 2% by changing R5 to 22.5 kΩ and R6 to 45 kΩ. This raised the level at which the derived highpass output slope became shallower quite dramatically to −25 dB, as shown in Figure 6.12.

In reality you would probably find that the delay errors were larger, as they also depend on C7 and C8 in the allpass filter, and these may not be more accurate than ±5%. If all other components are completely accurate, setting C7 and C8 so they are both 5% high causes the 24 dB/octave slope of the derived signal to become 6 dB/octave at only −15 dB, which is certainly going to interfere with proper crossover operation. Setting both 5% low gives the same result. Various other twiddlings and tweakings of C7 and C8 have similar effects on the slope, which always reverts to −6 dB/octave. This is obviously not a statistically rigorous analysis of the likely errors in the delay filter, but things are not looking promising.

I then turned to the lowpass filter, introducing assorted 5% errors into the four capacitors there. It did not come as a total shock to find once more that the derived signal slope was severely compromised.

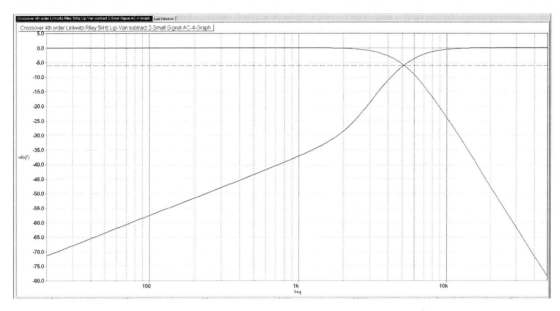

**Figure 6.12: With a 2% time error the subtractive crossover is only symmetrical down to −25 dB. Dashed line is at −6 dB.**

Since the whole process depends on subtraction, it seemed very likely that the components defining the accuracy of the subtraction would be critical, and so it proved. For accurate subtraction R10 has to be equal to R11; errors in R12 can only result in a gain error. Increasing R11 by just 1% to 22.2 kΩ gives the disconcerting result seen in Figure 6.13, where a notch has appeared at 1.6 kHz; above the notch the slope is increased from 3 kHz on down, while at lower frequencies the response has become flat at just below −40 dB. Changing R11 by 1% in the other direction, to 21.8 kΩ, gives a similar levelling-out just below −40 dB, but the notch is absent. This is a really discouraging result, stemming from a limit-of-tolerance (1%) change to one single component. Any process depending on subtraction will have problems when the output is required to be low, because of the way that errors are magnified when the difference between two large quantities is taken. This is bad enough when just two components affect the result, but here we have one relatively complex circuit trying to cancel out a property of another relatively complex circuit. There is a lot that can go wrong.

These results strongly suggest that while a linear-phase crossover may be a theoretically desirable goal, (though not everyone would agree with on that—see Chapter 3) this sort of subtractive crossover needs to be constructed with an impractical degree of precision to work properly. Rod Elliot has come to the same conclusion. [5]

## Performing the Subtraction

It is not hard to come up with a circuit that performs a subtraction. Any balanced line input stage does this; it takes the in-phase (hot) input and subtracts the out-of-phase (cold) input from it to remove

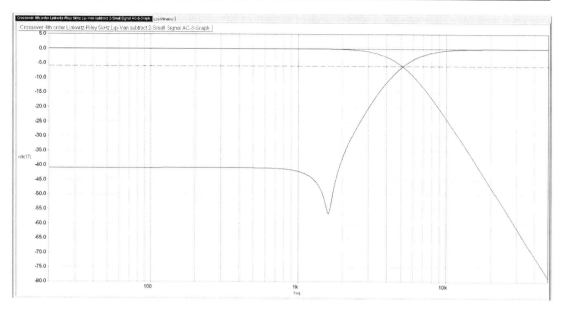

**Figure 6.13: Altering the subtraction resistor R11 by 1% gives the derived highpass signal a notch and a flat section in its response. Dashed line is at −6 dB.**

common-mode signals occurring in the input cable or elsewhere. This process is described in detail in Chapter 20, where a variety of balanced input stages are described. Here we only need a simple subtraction without variable gain or any other complicating features. We do however need an accurate subtraction—in the world of balanced line inputs, this is described as having a high common-mode rejection ratio.

The standard balanced input/subtractor is shown in Figure 6.14a. It is very often built with four 10 kΩ resistors as a compromise between noise performance and achieving reasonably high input impedances, and the measured noise output using a NE5532 opamp is −105.1 dBu. However, in an active crossover application we can assume that both inputs will be driven from opamp outputs (almost certainly from more NE5532's), and so the resistor values can be reduced drastically. This lowers the Johnson noise from the resistors and also means that the opamp current noise is flowing through less resistance and so creating less voltage noise. The effect of the opamp voltage noise is unchanged.

Reducing the four resistors from 10 kΩ to 820 Ω, as in Figure 6.14b, reduces the noise output from −105.1 dBu to −111.7 dBu, a helpful improvement of 6.6 dB in performance for no cost at all. The value of 820 Ω is chosen as it can be driven by another NE5532 without significantly impairing its distortion performance. This is a good example of low-impedance design.

Since R3 and R4 are in series, the load at the crossover input is 2 × 820 Ω = 1640 Ω. These two resistors could therefore be reduced to, say, 470 Ω without overloading the opamp driving them, and this would give a further small reduction in noise.

If one of the inputs to be subtracted is phase inverted, then simply summing the two inputs together is equivalent to a subtraction. This is the method used in the Elektor crossover described earlier in this chapter.

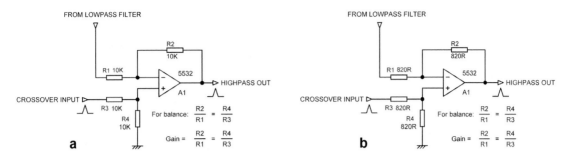

**Figure 6.14: Subtractor circuits: the version in (a) is 6.6 dB noisier than that at (b), which has lower resistor values.**

# References

[1]  Small, Richard H. "Constant Voltage Crossover Network Design" JAES 19(1), January 1971, pp. 12–19
[2]  Heinzerling, Christhof "Adaptable Active Speaker System" (subtractive) Electronics World, February 2000, p. 105
[3]  Lipshitz, Stanley P. and John Vanderkooy "A Family of Linear-Phase Crossover Networks of High Slope Derived by Time Delay" JAES 31, January/February 1983, p. 2
[4]  Baggen, Harry "Active Phase-Linear Crossover Network" Elektor, September 1987, p. 61
[5]  Elliot, Rod Private Communication, October 2010

# CHAPTER 7

# *Lowpass and Highpass Filter Characteristics*

Active filters are the building blocks of active crossovers. There are many excellent and comprehensive textbooks on active filters, [1][2][3] but most of them assume familiarity with Laplace *s*-variables, complex algebra, and so on. This chapter however contains nothing more fearsome than a square root, and I hope shows in a straightforward way how to put together the blocks that make up a filter. The design of each block to component value level is covered in Chapters 8 and 9.

Filter design is at the root highly mathematical, and it is no accident that popular filter characteristics such as Bessel and Chebyshev are named after mathematicians, who usually did not live long enough to see their mathematics applied to practical filters. A notable exception is the Butterworth characteristic, probably the most popular and useful characteristic of all; Stephen Butterworth was a British engineer.

Here however I am going to avoid the complexities of pole and zero placement etc and concentrate on practical filter designs that can be adapted for use at different frequencies by a simple process of scaling component values. Most filter textbooks give complicated equations for calculating the amplitude and phase response at any desired frequency. This gets very cumbersome if you are dealing with a lot of different filters. Here I have assumed access to a simulator, which gives all the information you could possibly want much more quickly and efficiently than wrestling with calculations manually or in a spreadsheet. Free simulator packages can be downloaded.

Filters are either passive or active. Passive or LCR filters use resistors, inductors, and capacitors only. Active filters use resistors, capacitors, and gain elements such as opamps; active filter technology is usually adopted with the specific intent of avoiding inductors and their well-known limitations. Nevertheless, there are some applications where LCR filters are essential, such as the removal of high frequencies from the output of Class-D power amplifiers, which would otherwise upset audio testgear. The answer, as described by Bruce Hofer, [4] is a passive LCR roofing filter.

There is no reason why you could not make a line-level passive crossover using only inductors, capacitors, and resistors, but it is difficult to think of any reason why you'd want to. Likewise, you could design a crossover made up of active filters using only inductors and resistors (i.e. no capacitors), but this would also come under the heading of "perverse electronics".

## Active Filters

Active filters do not normally use inductors as such, though configurations such as gyrators that explicitly model the action of an inductor are sometimes used. The active element need not be an opamp; the Sallen & Key configuration requires only a voltage-follower, which in some cases can be a simple BJT emitter-follower. Opamps are usual nowadays, however. This chapter deals only with lowpass and highpass filters; bandpass, notch, and allpass filters are dealt with in later chapters.

## Lowpass Filters

Lowpass filters have obvious uses in crossover design, keeping the MID and HF material out of the LF path and keeping HF material out of the MID path. They are also used to explicitly define the upper limit of the audio bandwidth in a system, say at 50 kHz. This is much better than letting it happen by the casual accumulation of a lot of 1st-order roll-offs in succeeding stages. This bandwidth definition is not a duplication of input RF filtering, which, as described in Chapter 20, must be passive and positioned before the incoming signals encounter any electronics at all, as active circuitry can demodulate RF. Lowpass filters are used in sound-reinforcement systems to protect power amplifiers and loudspeakers against ultrasonic oscillation from outside sources or in the system itself. Lowpass filters are defined by their order, Q, and cutoff frequency.

## Highpass Filters

Highpass filters are used in active crossovers for keeping the LF out of the MID and HF paths and keeping LF and MID material out of the HF path. They are also commonly used to explicitly define the lower limit of the audio bandwidth in a system, say at 20 Hz. In this case relying on the happenstance accumulation of 1st-order LF roll-offs is distinctly dangerous because loudspeakers are vulnerable to damage from unduly low frequencies. Highpass filters are defined by their order, Q, and cutoff frequency.

Lowpass and highpass filters are the types most used in active crossover design, but there are others that sometimes see use:

## Bandpass Filters

Bandpass filters as such are not much used in crossover design, except perhaps occasionally as part of an equalisation scheme, but they are of great utility as a basis from which to make notch filters and allpass filters by subtractive methods.

The $Q$ required rarely exceeds a value of 5, which can be implemented with relatively simple active filters, such as the multiple-feedback type. Higher $Q$'s or independent control of all the resonance parameters require the use of the more complex biquad or state-variable filters. Bandpass and notch filters are said to be "tuneable" if their centre frequency can be altered relatively easily, say by changing only one component value. Bandpass filters are dealt with in detail in Chapter 12. They are defined by their $Q$ and centre frequency.

## Notch Filters

Notch filters are used in some active crossovers; for example, see [5]. They are an integral part of elliptic filters, inserting zeros (notches or nulls) in the stopband. When someone says "notch filter" we naturally think of the most common version—the symmetrical notch which goes back up to 0 dB on either side of the central crevasse. However, there are also lowpass notch and highpass notch filters, and these are highly relevant to crossover design. Notch filters are dealt with in detail in Chapter 12. They are defined by their $Q$ and centre frequency.

## Allpass Filters

Allpass filters are so-called because they have a flat frequency response, and so pass all frequencies equally. Their point is that they have a phase-shift that *does* vary with frequency, and this is often used for delay correction in active crossovers. Allpass filters are dealt with in detail in Chapter 13.

## All-Stop Filters

The all-stop filter passes no signal at any frequency, and therefore is not hard to design; you just omit to design or build it at all. It is of course a filter-designer's joke, and is indeed probably the only joke associated with the dourly serious business of filter design. For an example found in the wild, see [6].

## Brickwall Filters

Brickwall filters may be lowpass or highpass; their distinguishing characteristic is a very fast roll-off that looks almost vertical on the response plot. Their main use was in anti-aliasing and reconstruction filters in the early years of digital audio. They were typically 9th-order elliptical filters with significant passband ripple; they were hard to design, and also hard to construct, because of high component sensitivities. Mercifully they are no longer required now we have oversampling ADCs and DACs. A brickwall filter is not intended to keep unwanted masonry out of your electronics; that is a matter for your builder.

## The Order of a Filter

Strictly speaking, the order of a filter is the highest power of frequency that occurs in the complex algebraic equation that describes its behaviour. Since that is the sort of complication we are trying to avoid, we have to here rely on somewhat less formal definitions. For example, the ultimate roll-off slope of a lowpass or highpass filter is 6 dB/octave times the order of the filter. A 1st-order filter (which is just a single RC time-constant) rolls off at 6 dB/octave, a 2nd-order filter at 12 dB/octave, and so on. For almost all kinds of active filter the number of capacitors is equal to the order of the filter; the 2nd-order highpass MFB filter is a notable exception. See Chapter 10.

A bandpass filter must be at least 2nd-order; it can also be 4th-order, which represents two bandpass filters cascaded, perhaps with their centre frequencies offset to give a flattish section rather than a sharp peak at the frequencies of interest; in radio receivers this is referred to as staggered tuning. It is impossible to have a 1st-order bandpass or notch (band-reject) response.

## Filter Cutoff Frequencies and Characteristic Frequencies

The cutoff frequency of a lowpass or highpass filter is a measure of where in the spectrum its roll-off starts. In many cases it is defined as the frequency at which the gain is down by $-3$ dB ($= 1/\sqrt{2}$). The word "cutoff" is perhaps unfortunate because it seems to imply a frequency response that drops suddenly, like falling off a cliff. So-called brickwall filters with very fast roll-offs do exist, but the filters used in active crossover design are rarely higher than 4th-order, and the start of the roll-off is actually quite gentle. However, nobody's going to change the word now.

Filter types like the Butterworth and Bessel characteristics have their cutoff frequencies defined at the $-3$ dB point, whereas the Linkwitz-Riley filter may be defined at the $-3$ dB or the $-6$ dB point. To an extent the choice of attenuation is arbitrary; it would be quite possible to design these filters with cutoff defined at $-4.5$ dB or whatever, but there would be no point at all in doing so. For the Butterworth filter in particular, using $-3$ dB makes the mathematics very simple.

Other filters, like the Chebyshev type, which has amplitude ripples in the passband, have their cutoff defined as the point where the gain passes through 0 dB for the last time as the frequency increases (in the lowpass case) and the steady roll-off begins.

The characteristic frequency of a lowpass or highpass filter is not at all the same thing as its cutoff frequency, and the two terms should not be confused. Taking lowpass filters as an example, the amplitude response of all 2nd-order filters will eventually become a 12 dB/octave straight line heading downwards. If you extend that line upwards and to the left it will eventually cut the 0 dB gain line, and the point where it cuts it is the characteristic frequency. You can design a Bessel filter and a Chebyshev filter so that their responses converge on the same 12 dB/octave line and so have the same characteristic frequency, but they will then have different cutoff frequencies.

Characteristic frequencies are not widely used in crossover design, but they are useful when dealing with state-variable filters; this is discussed in an excellent article by Ramkumar Ramaswamy. [7] This book uses cutoff frequencies exclusively.

## First-Order Filters

A 1st-order filter is just a single RC time-constant, with an ultimate roll-off of 6 dB/octave. Some writers regard 1st-order filters as having a fixed $Q$ of 0.5; this is not helpful when dealing with the amplitude response but is useful when working in the time domain. This is absolutely NOT the same thing as a 2nd-order filter with a $Q$ of 0.5, which is equivalent to *two* 1st-order filters. There is no such thing as a Butterworth 1st-order filter or a Bessel 1st-order filter; just a 1st-order filter. Most filter design software does not issue a warning if you ask for a "Butterworth 1st-order filter"—it simply serves up a 1st-order filter with no further comment. The cutoff frequency for a 1st-order filter is always that frequency at which the output has fallen to $-3$ dB ($1/\sqrt{2}$) of the passband gain.

A filter made up only of cascaded 1st-order stages with the same cutoff frequency is called a synchronous filter; it has a very slow roll-offend is therefore in general not very useful. It is however described later in this chapter.

# Second-Order and Higher-Order Filters

Second-order bandpass responses are basically all the same, being completely defined by centre frequency, Q, and gain. Second-order and higher highpass and lowpass filter characteristics are also defined by cutoff frequency, Q, and gain but come in many different types, one of which is selected as a compromise between the need for a rapid roll-off, flatness in the passband, a clean transient response, and desirable group delay characteristics. The Butterworth (maximally flat) characteristic is the most popular for many applications, not least because it is refreshingly simple to design. Filters with passband ripples in their amplitude response, such as the Chebyshev or the elliptical types, have not found favour for in-band filtering, such as in active crossovers, but were once widely used for applications like 9th-order anti-aliasing filters in front of ADCs and similar reconstruction filters after DACs, the passband ripples being accepted as inescapable if the filters were to be kept to an acceptable level of complexity. Such filters have mercifully been made obsolete by oversampling, pushing the unwanted frequencies a long way up the spectrum, so post-DAC filters are nowadays usually simple 2nd- or 3rd-order Butterworth types.

The Bessel filter characteristic gives a maximally flat group delay (maximally linear-phase response) across the passband and so better preserves the waveform of filtered signals, but it has a much slower roll-off in amplitude response than the Butterworth.

The cutoff frequency for 2nd and higher-order filters is to a certain extent a matter of definition. The cutoff frequency of a Butterworth filter is always taken as the −3 dB frequency because this fits in very neatly with the delightfully straightforward design equations. Bessel filters normally use the same definition, but a Linkwitz-Riley filter might be specified at either −3 dB or −6 dB, the latter being useful because when a lowpass-highpass pair of these filters is used to make a Linkwitz-Riley crossover, the filters are −6 dB at the crossover frequency for proper summation. The cutoff frequency definitions for Chebyshev filters are more complicated, depending on whether the filter is odd-order or even-order; this is dealt with later in this chapter.

# Filter Characteristics

There are many different types of recognised filter response, each with their own special properties. Their study is complicated by the fact that some of them go by several different names, and you need to keep a close eye on the terminology that an author is using. The well-known filter types are given in Table 7.1 and some of the less common types in Table 7.2.

We will look at some of these filter types, in each case first examining the simplest possible instances, the 2nd-order versions; there are no 1st-order versions because a 1st-order filter is just a 1st-order filter, in other words a simple single time-constant. All the filters in this section are lowpass filters with a cutoff frequency of 1 kHz. In every case there is an equivalent highpass filter.

For the simpler filters, the Butterworth, Linkwitz-Riley, and the Bessel, we will look at the amplitude response, the phase-shift response, the response to a step input and the group delay curve; we then go on to compare the response of the higher-order versions. Chebyshev filters have a more complicated

**Table 7.1: The popular filter types**

| Common name | Other names used | Main features |
|---|---|---|
| Butterworth | Maximally flat | Maximally flat amplitude, easy design |
| Linkwitz-Riley | Butterworth-squared, Butterworth-6 dB | Two cascaded make 4th-order filter summing to allpass |
| Bessel | Thomson, Bessel-Thomson | Maximally flat group delay, slower roll-off than Butterworth |
| Chebyshev | Chebyshev Type-I | Passband ripple, faster roll-off than Butterworth |
| Inverse Chebyshev | Chebyshev Type II | Stopband has notches |
| Elliptical | Cauer, Zolotarev, complete-Chebyshev | Passband ripple and stopband notches, fastest roll-off |

**Table 7.2: Lesser-known filter types**

| Common name | Other names used | Main features |
|---|---|---|
| Transitional | | Compromise between any two filter characteristics |
| Linear Phase | Butterworth-Thomson | Compromise between Butterworth and Bessel |
| Gaussian | | Optimised for step risetime without overshoot |
| Legendre | Legendre-Papoulis | Passband monotonic but not maximally flat |
| Synchronous | | Very slow roll-off |
| Ultraspherical | | Generalisation of other filter types |
| Halpern | | Similar to Legendre-Papoulis |

response, and some explanation of this is required before we get into the same details as for the earlier filters.

The filters used to generate the information are shown in Figure 7.1. They are all unity-gain Sallen & Key lowpass configurations, differing only in $Q$. Full details of how to design lowpass Sallen & Key stages for a desired cutoff frequency and $Q$ are given in Chapter 8. For highpass S&K filters see Chapter 9.

## Amplitude Peaking and Q

All 2nd-order lowpass and highpass filters show peaking in the frequency response when the $Q$ is greater than 0.7071 ($1/\sqrt{2}$ or −3 dB), and a clear appreciation of this is important, as 2nd-order filters are commonly used as the building blocks for higher-order and more complex filters. The height of the peak ($A_{max}$) and the frequency of its maximum ($f_p$) relative to the filter cutoff frequency ($f_0$) are completely determined by the $Q$ value chosen and nothing else, and are related by these equations:

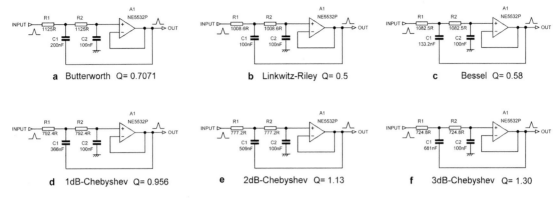

**a** Butterworth  Q= 0.7071        **b** Linkwitz-Riley  Q= 0.5        **c** Bessel  Q= 0.58

**d** 1dB-Chebyshev  Q= 0.956    **e** 2dB-Chebyshev  Q= 1.13    **f** 3dB-Chebyshev  Q= 1.30

**Figure 7.1: Second-order Butterworth, Linkwitz-Riley, Bessel and Chebyshev lowpass filters. Cutoff frequency is 1.00 kHz.**

$$\text{Lowpass and highpass: } A_{max} = \frac{Q}{\sqrt{1 - \dfrac{1}{4Q^2}}} \qquad\qquad 7.1$$

$$\text{Lowpass only: } f_p = f_0 \sqrt{1 - \frac{1}{2Q^2}} \qquad\qquad 7.2$$

$$\text{Highpass only: } f_p = f_0 \frac{1}{\sqrt{1 - \dfrac{1}{2Q^2}}} \qquad\qquad 7.3$$

The peaking caused by different $Q$'s in lowpass and highpass filters are shown in Figures 7.2 and 7.3. Given the frequency response of a 2nd-order filter, it is straightforward to work out its $Q$. Note that the peak amplitude can only be found for $Q$ greater than or equal to $1/\sqrt{2}$; for lower $Q$ values the results are meaningless, as there is no peak to describe. Likewise the equations for $f_p$ will only work for $Q$ greater than $1/\sqrt{2}$, as the number in the big square root must be positive to give a meaningful answer.

## Butterworth Filters

The Butterworth filter characteristic is one of the most popular because it has a maximally flat frequency response.

The cutoff frequency is defined as the −3 dB point. Butterworth filter design is relatively easy in its high-order versions (3rd and above) because every stage making up the complete filter has the same cutoff frequency—only the $Q$'s vary. The 2nd-order Butterworth filter has a $Q$ of $1\sqrt{2}$ (= 0.7071). Some books use the term "damping factor" $a$ or $\alpha$ (= $1/Q$) rather than $Q$, but here we will stick with $Q$. Pay attention, 007.

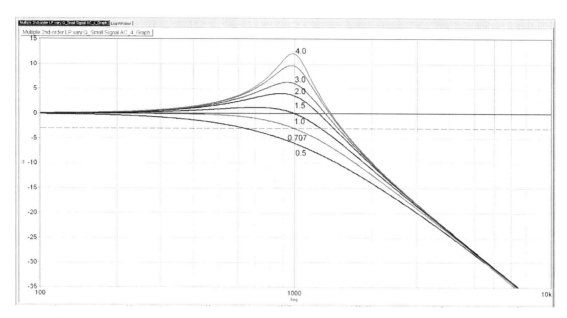

Figure 7.2: Lowpass 2nd-order amplitude response peaking for different values of $Q$.

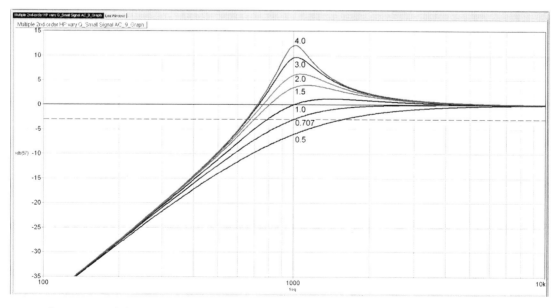

Figure 7.3: Highpass 2nd-order amplitude response peaking for different values of $Q$.

The Butterworth response was introduced by Stephen Butterworth (1885–1958) in 1930. [8] Compared with other filters, it is remarkable that not only was Butterworth a physicist /engineer rather than a mathematician, but in this case the filter is actually named after the person that put it into use.

Figure 7.4 shows the amplitude response, with a sharper "knee" than other filters such as the Bessel. The ultimate slope is 12 dB/octave, determined by the fact that this is a 2nd-order filter. The phase-shift is shown in Figure 7.5; it is 90° at the cutoff frequency.

As Figure 7.6 demonstrates, the Butterworth response gives a modest degree of overshoot when it is faced with a step input. Maximal flatness in the frequency response does *not* mean no overshoot and does *not* mean a flat group delay (Figure 7.7). For that you need a Bessel filter.

### Linkwitz-Riley Filters

The lowest $Q$ normally encountered in a 2nd-order filter is 0.50, which is used for 2nd-order Linkwitz-Riley (LR-2) crossovers. This is less than the $Q$ of 0.58 used for 2nd-order Bessel filters. Note that 0.5 is the square of the Butterworth $Q$ which is0.7071 (1/√2). They are sometimes called Butterworth-squared filters because the 4th-order version is often implemented by cascading two identical 2nd-order Butterworth stages. They are also called Butterworth-6 dB filters because there is a −6 dB level at the cutoff frequency of the two cascaded filters.

Designing for 1.00 kHz using the equations given later, you will actually get −6.0 dB at 1.00 kHz; the −3 dB point is at 633.8 Hz. The phase-shift is 90° at 1.00 kHz. Use a frequency scaling factor (FSF) of

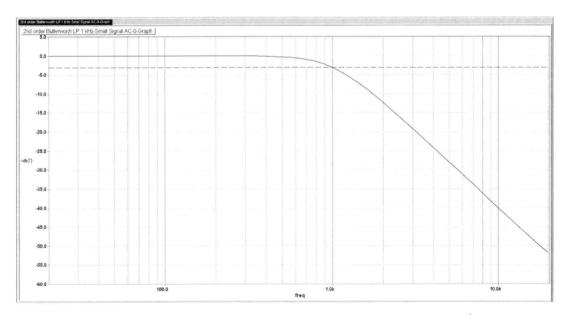

**Figure 7.4: Amplitude response of 2nd-order Butterworth lowpass filter ($Q$ = 1√2 = 0.7071). Cutoff frequency is 1.00 kHz. Attenuation at 10 kHz is −40 dB.**

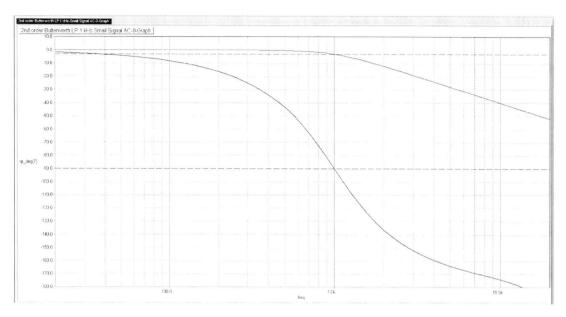

Figure 7.5: Phase response of 2nd-order Butterworth lowpass filter. Cutoff frequency is 1.00 kHz. The phase-shift is 90° at 1.00 kHz. Upper trace is amplitude response.

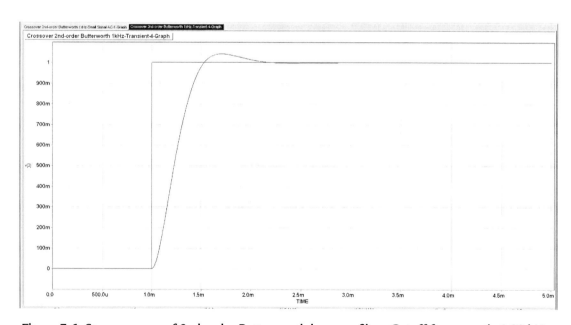

Figure 7.6: Step response of 2nd-order Butterworth lowpass filter. Cutoff frequency is 1.00 kHz. There is significant overshoot of about 4%.

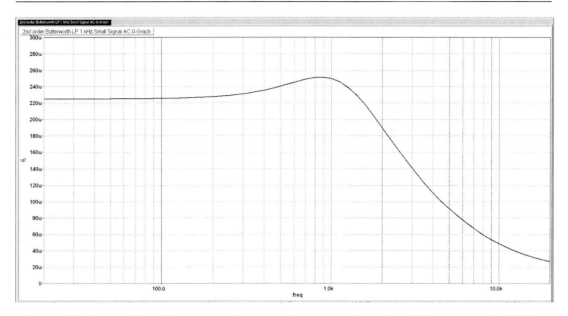

**Figure 7.7: Group delay of 2nd-order Butterworth lowpass filter. Cutoff frequency is 1.00 kHz.**
**The delay peaks by 15% near the cutoff frequency.**

1.578, i.e. design for 1.578 kHz, and you will then get −3 dB at 1.00 kHz. The phase-shift is then 90° at 1.578 kHz.

The cutoff frequency is defined as the −3-dB point.

Figure 7.8 demonstrates the slower roll-off of the Linkwitz-Riley compared with the Butterworth filter. The attenuation at 10 kHz is −32.5 dB instead of −40 dB for the Butterworth. The ultimate slope is still 12 dB/octave, as this is determined by the fact that this is a 2nd-order filter and is not affected by the $Q$ chosen. Figure 7.9 shows the phase response. There is no step-response overshoot (Figure 7.10) and no group delay peak (Figure 7.11).

## Bessel Filters

As is common in the world of filters, Bessel filters as such were not invented by Bessel at all. Friedrich Wilhelm Bessel (1784–1846) was a German mathematician and astronomer; he was long dead by the time that anyone thought of applying his mathematics to electrical filtering. He systematised the Bessel functions, which, to keep the level of confusion up, were actually discovered by Daniel Bernoulli (1700–1782).

The first man to put Bessel functions to work in filters was W. E. Thomson, [9] and that is why Bessel filters are sometimes called Thomson filters or Bessel-Thomson filters. According to Ray Miller, [10] Thomson was actually anticipated by Kiyasu [11] working in Japan in 1943, but given the date

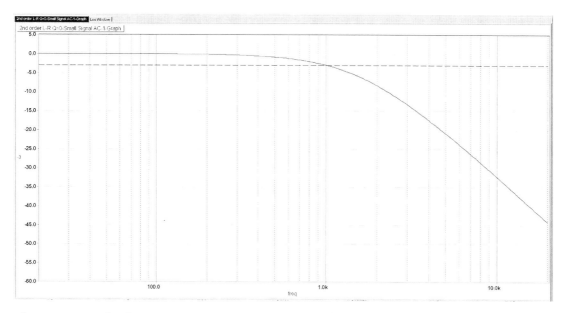

Figure 7.8: Amplitude response of 2nd-order Linkwitz-Riley lowpass filter ($Q$ = 0.50); −3 dB at 1 kHz. Attenuation at 10 kHz is −32.5 dB.

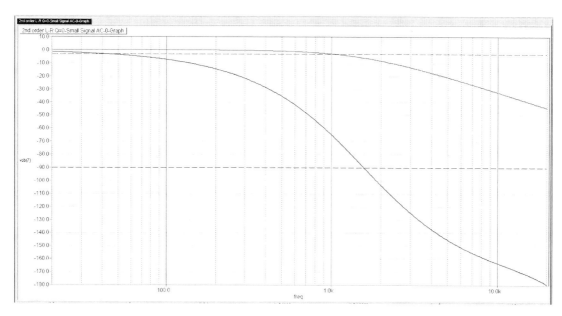

Figure 7.9: Phase response of 2nd-order Linkwitz-Riley lowpass filter;−3 dB at 1.00 kHz. The phase-shift is 90° at 1.578 kHz. Upper trace is amplitude response.

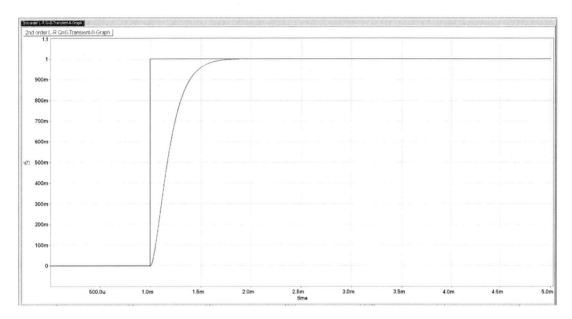

Figure 7.10: Step response of 2nd-order Linkwitz-Riley lowpass filter; −3 dB at 1.00 kHz. There is no overshoot.

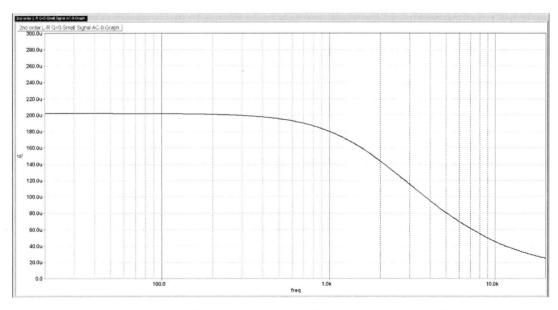

Figure 7.11: Group delay of 2nd-order Linkwitz-Riley lowpass filter; −3 dB at1.00 kHz. There is no peaking in the delay.

it is not surprising that communications with the West were somewhat compromised. It is important to remember that the term Bessel-Thomson does *not* refer to a hybrid between Bessel and Thomson filters, because they are the same thing. This is in contrast to Butterworth-Thomson transitional filters, which *are* hybrids between Butterworth and Thomson (i.e. Bessel) filters.

A $Q$ of $1/\sqrt{3} = 0.5773$ is used for 2nd-order Bessel filters. The Bessel filter gives the closest approach to constant group delay (Figure 7.15); in other words the group delay curve is maximally flat. As a result the time response is very good (Figure 7.14), and the overshoot of a step function is only 0.43% of the input amplitude. This is often not visible on plots and has led some people to think that there is no overshoot at all; it is certainly very small and unlikely to cause trouble in most applications, but it does exist.

The downside of the Bessel filter is that the amplitude response roll-off is slow—actually very slow compared with a Butterworth filter; see Figure 7.12. The cutoff frequency is defined as the −3-dB point. Figure 7.13 gives the phase response.

If you design for 1.00 kHz using the equations, you will actually get −4.9 dB at 1.00 kHz; the −3 dB point is at 777 Hz. There is however a 90° phase-shift at 1.00 kHz. Use a frequency scaling factor (FSF) of 1.2736, i.e. design for 1.2736 kHz, and you will get −3 dB at 1.00 kHz.

The attenuation at 10 kHz is −36 dB, which is worse than the Butterworth (−40 dB) but better than the Linkwitz-Riley (−32.5 dB).

Note that the flat part of the group delay curve extends further up in frequency than for the Linkwitz-Riley filter, and yet the group delay does not peak like the Butterworth filter. If you want to combine

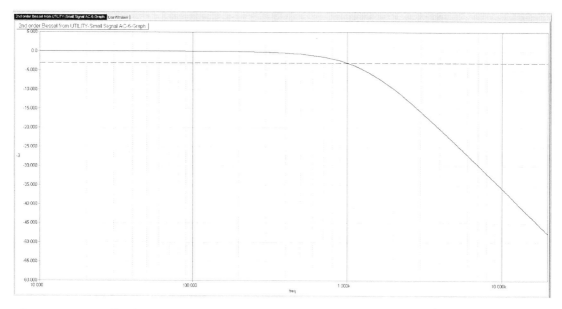

**Figure 7.12: Amplitude response of 2nd-order Bessel lowpass filter ($Q$ = $1/\sqrt{3}$ = 0.5773). Cutoff frequency is 1 kHz. Attenuation at 10 kHz is −36 dB.**

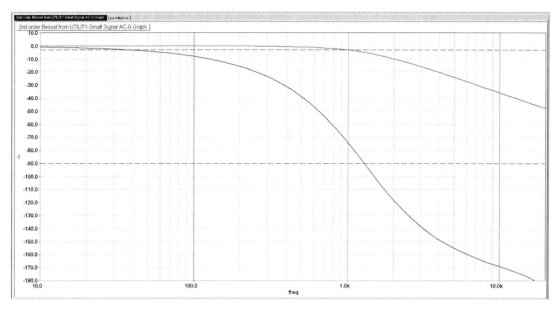

Figure 7.13: Phase response of 2nd-order Bessel lowpass filter; −3 dB at 1.00 kHz. The phase-shift is 90° at 1.3 kHz. Upper trace is amplitude response.

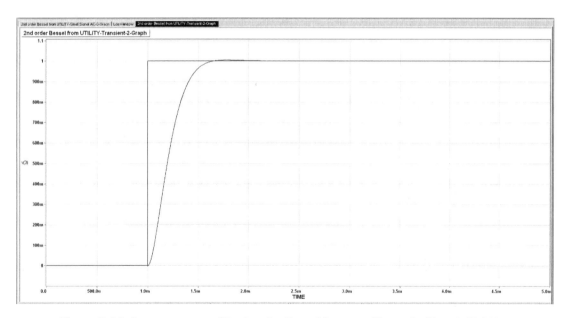

Figure 7.14: Step response of 2nd-order Bessel lowpass filter; −3 dB at 1.00 kHz. The 0.43% overshoot is just visible.

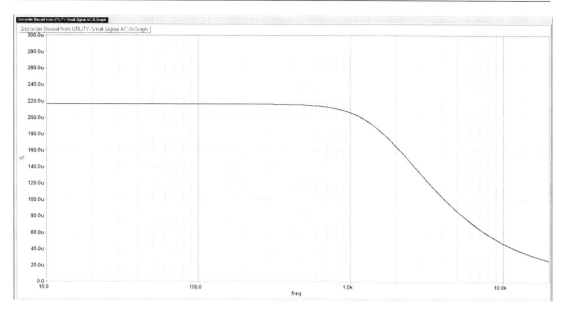

**Figure 7.15: Group delay of 2nd-order Bessel lowpass filter; −3 dB at 1.00 kHz.
There is no peaking, and it is maximally flat.**

maximally flat group delay with filtering, the Bessel filter is mathematically the best answer possible. If you want to create group delay *without* filtering, then you need an allpass filter which gives delay with a flat frequency response. See Chapter 13 on time-domain filtering.

## Chebyshev Filters

Pafnuty Lvovich Chebyshev (1821–1894) was a Russian mathematician. His name can be alternatively transliterated as Chebychev, Chebyshov, Tchebycheff, or Tschebyscheff.

There is only one kind of Bessel response, and only one kind of Butterworth response, but there are an infinite number of Chebyshev responses, depending on the amount of passband ripple you are prepared to put up with in order to get a faster roll-off. As the ripple increases, the roll-off becomes sharper. The passband ripple might range from 0.5 dB to 3 dB in practical use, with all values between possible. There is however nothing whatever to stop you designing a Chebyshev filter with 0.1, 0.01, or even 0.0001 dB passband ripple, though there is very little point in doing it, as the result is in practice indistinguishable from a Butterworth characteristic; in fact with zero ripple it *is* a Butterworth characteristic. Filter design programs like Filtershop will not let you attempt Chebyshev filters with more than 3 dB of passband ripple, and very few references to greater ripple can be found anywhere. This is not (to my knowledge) an inherent limitation of Chebyshev filters but a reflection of the fact that a filter with big ripples in the passband is of little use to anybody.

The Chebyshev filter accepts a non-flat passband as the price for a faster roll-off. The transient response of a Chebyshev filter to a pulse input shows more overshoot and ringing than a Butterworth filter.

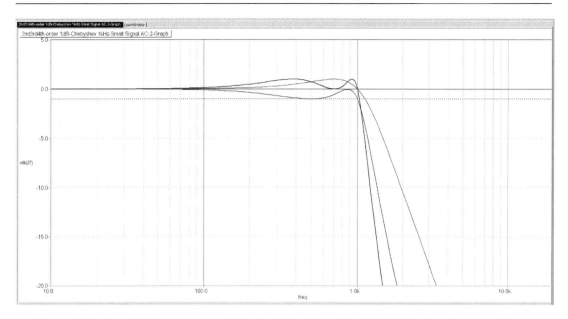

**Figure 7.16: Amplitude responses of 2nd-, 3rd-, 4th-order 1 dB-Chebyshev lowpass filters, showing the 1 dB ripples in the passband. Cutoff frequency is 1 kHz.**

It should be said that the famous "passband ripple" does not look much like ripple for lower-order Chebyshev filters; in the 2nd-order case it simply means that the response peaks gently above the 0 dB line by the amount selected and then falls away as usual; see the 1 dB-Chebyshev response in Figure 7.16, which peaks by the desired +1 dB at 750 Hz, and then passes through 0 dB at the cutoff frequency.

The 4th-order 1 dB-Chebyshev filter has two +1 dB peaks before roll-off, and this is starting to look more like a real "ripple"; the response also passes through 0 dB at the cutoff frequency. For all Chebyshev filters, even-order versions have peaks *above* the 0 dB line and a response going through 0 dB at cutoff. Odd-order versions have dips *below* it and instead pass through the ripple value (here −1 dB) at cutoff. For higher-order filters, each deviation from and return to the 0 dB line requires a 2nd-order stage in the complete filter.

Figure 7.17 shows how the response of an odd-order (3rd, in this case) 1 dB-Chebyshev filter dips 1 dB below the 0 dB line, then goes back up to exactly 0 dB again before falling to pass through the −1 dB line at the cutoff frequency. In the same way a 2 dB-Chebyshev filter dips 2 dB below the 0 dB line, then goes back up to 0 dB before falling to go through the −2 dB line at the cutoff frequency, and the same applies to a 3 dB-Chebyshev filter.

It is important to realise that in a Chebyshev filter the frequencies at which the passband ripples occur cannot be controlled independently. Once you have specified the passband ripple and the cutoff frequency of the filter, that's it—you have no more control over the response. This means that attempts to use the ripples of a Chebyshev crossover filter to equalise humps or dips in a drive unit response are unlikely to be successful. It is far better to keep the crossover and equalisation functions in separate circuit blocks.

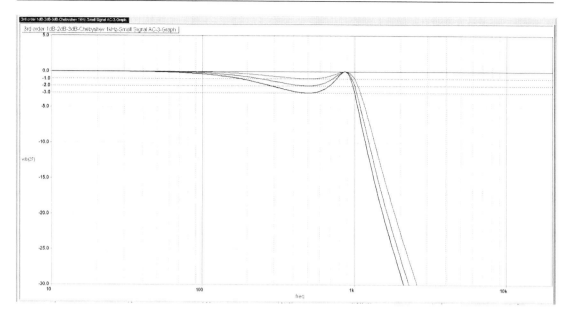

**Figure 7.17: Amplitude responses of 3rd-order Chebyshev lowpass filter designed for 1 dB, 2 dB, and 3 dB of passband ripple. The curves go through the ripple value of −1 dB, −2 dB, and −3 dB at the cuto ff frequency of 1 kHz.**

For a given filter order, a steeper cutoff can be achieved by allowing more passband ripple, as shown by the 4th-order Chebyshev response in Figure 7.18, which has been designed for 1 dB, 2 dB, and 3 dB of passband ripple. The improvement is, however, not exactly stunning. Looking at the response at 3 kHz, the 1 dB case is down −33 dB, the 2 dB case is down −35 dB, and the 3 dB case is down −37 dB. In general, if you want a steeper roll-off, you go for a higher-order filter.

## 1 dB-Chebyshev Lowpass Filter

The 2nd-order 1 dB-Chebyshev lowpass filter has a 1 dB peak in the passband at 750 Hz. The amplitude response is shown in Figure 7.19; note that the attenuation at 10 kHz is −39 dB, which is actually slightly worse than the −40 dB of the Butterworth filter. The $Q$ is 0.9565. Phase response, step response, and group delay are shown in Figures 7.20, 7.21, and 7.22.

## 3 dB-Chebyshev Lowpass Filter

The 2nd-order 3 dB-Chebyshev lowpass filter has a 3 dB peak in the passband at 710 Hz, slightly lower than for the 1 dB-Chebyshev. The amplitude response is shown in Figure 7.23; note that the attenuation at 10 kHz is −43 dB, which is somewhat better than the −40 dB of the Butterworth filter. The $Q$ for a 2nd-order 3 dB-Chebyshev is 1.305. Phase response, step response, and group delay are shown in Figures 7.24, 7.25, and 7.26.

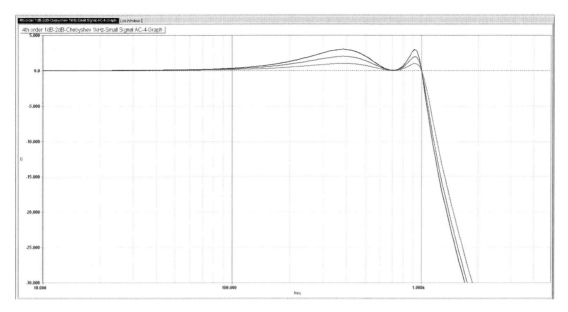

Figure 7.18: Amplitude responses of 4th-order Chebyshev lowpass filter designed for 1 dB, 2 dB, and 3 dB of passband ripple. All the curves go through 0 dB at the cutoff frequency of 1 kHz.

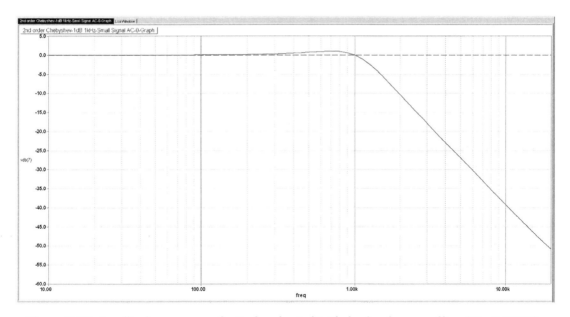

Figure 7.19: Amplitude response of a 2nd-order 1 dB-Chebyshev lowpass filter ($Q$ = 0.9565). Cutoff frequency is 1 kHz. Attenuation at 10 kHz is −39 dB.

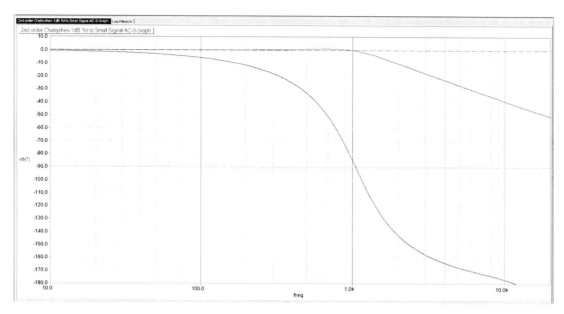

Figure 7.20: Phase response of a 2nd-order 1 dB-Chebyshev lowpass filter; 0 dB at 1.00 kHz. The phase-shift is 90° at 1.1 kHz. Upper trace is amplitude response.

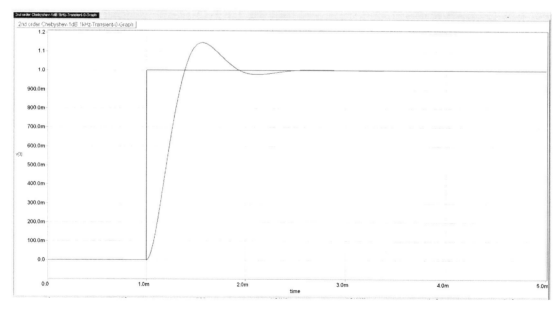

Figure 7.21: Step response of a 2nd-order 1 dB-Chebyshev lowpass filter; 0 dB at 1.00 kHz. There is significant overshoot of 14%, followed by undershoot of 2%.

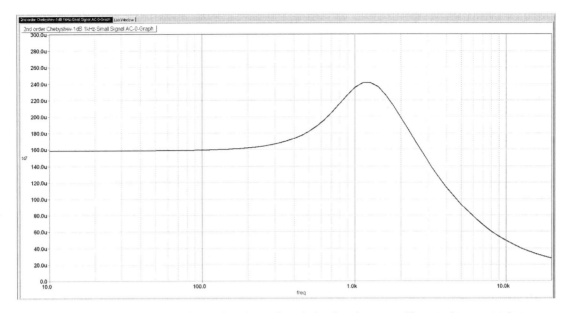

Figure 7.22: Group delay of a 2nd-order 1 dB-Chebyshev lowpass filter; 0 dB at 1.00 kHz. There is now considerable peaking in the delay, reaching 50%.

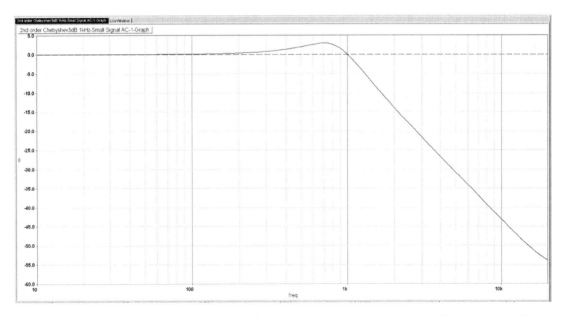

Figure 7.23: Amplitude response of a 2nd-order 3 dB-Chebyshev lowpass filter ($Q$ = 1.305). Cutoff frequency is 1 kHz. Attenuation at 10 kHz is −43 dB.

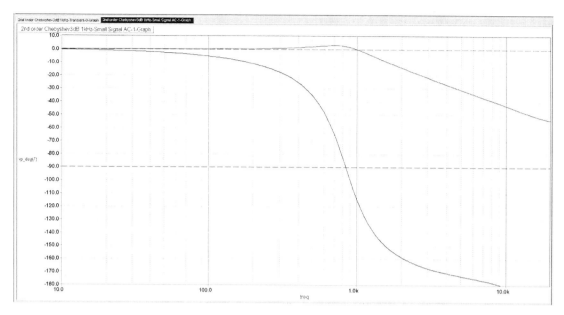

Figure 7.24: Phase response of a 2nd-order 3 dB-Chebyshev lowpass filter; 0 dB at 1.00 kHz. The phase-shift is 90° at 0.84 kHz. Upper trace is amplitude response.

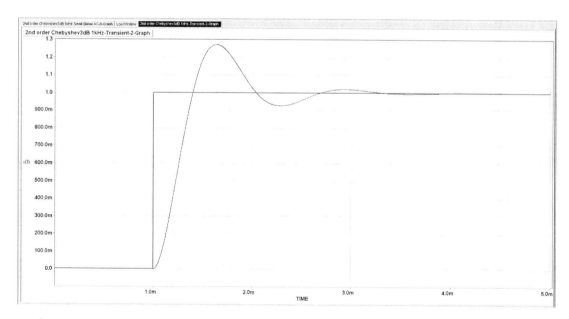

Figure 7.25: Step response of a 2nd-order 3 dB-Chebyshev lowpass filter; 0 dB at 1.00 kHz. There is a serious overshoot of 28%, followed by an undershoot of 8% and continuing damped oscillation.

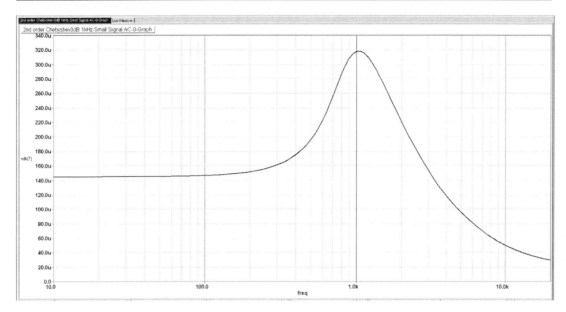

**Figure 7.26: Group delay of a 2nd-order 3 dB-Chebyshev lowpass filter; 0 dB at 1.00 kHz.
The peaking is now huge at 120%; note scale change to fit it in.**

## Higher-Order Filters

In many cases 2nd-order filters do not give a fast enough roll-off. As we have seen, going from a
2nd-order Butterworth filter to a 2nd-order 3 dB-Chebyshev filter does not help much; in a 1 kHz filter
we gain only 3 dB more attenuation at the expense of a 3 dB peak in the passband, which is not very
helpful in most circumstances.

If you want a faster roll-off as well as a maximally flat passband, then the answer is to go to a higher
order of Butterworth filter. Likewise, if you have a 1 dB-Chebyshev filter that gives a passband
ripple you either want or can put up with, but need a steeper roll-off, then a higher-order filter is the
way to go.

The mathematical equations describing the response of a 2nd-order filter contain the square of
frequency and cannot be simplified, but it is a mathematical fact (which I devoutly hope you will
take on trust, because this is not the place for me to prove it) that any 3rd-order equation of this sort
(technically, a polynomial) can be broken down into a 2nd-order equation multiplied by a 1st-order
equation, which represents a 2nd-order filter cascaded with a 1st-order filter. Furthermore, it is equally
a fact that any complex equation describing a high-order filter can be factorised into a combination of
2nd order equations multiplied together for even-order filters and a combination of 2nd order equations
with one 1st-order equation multiplied together for odd-ordered filters. This leads directly to making
a high-order filter by cascading 1st- and 2nd-order filters. Thus a 4th-order filter can be made by
cascading two 2nd-order filters, and a 5th-order filter can be made by cascading two 2nd-order filters
and a 1st-order filter, and so on.

It is not compulsory to break things down as much as possible, into 2nd- and 1st-order stages; it is quite possible to make a 5th-order filter by cascading a 3rd-order stage with a 2nd-order stage, as described in Chapters 8 and 9. However, using only 2nd- and 1st-order stages is almost always the method adopted because, although it may use more amplifiers, it is the best way to achieve minimum component sensitivities; in other words the circuits are more tolerant of component value tolerances. There is much more on component sensitivity in the next chapter.

One thing you cannot do is make a 5th-order filter by cascading five 1st-order stages; nor in fact can you do it by cascading one 2nd-order stage with three 1st-order stages. The rules are that an odd-order filter will be composed of a number of 2nd-order stages and one 1st-order stage, while an even-order filter will be made up solely of 2nd-order stages.

Putting together higher-order filters of the all-pole type (that term will be explained shortly) is actually quite straightforward if you have the information you need. It is provided here.

The order of filter stages is important. Filter textbooks usually put the stages in order of increasing $Q$, so that large signals will be attenuated by earlier stages and so are less likely to cause clipping in the high-$Q$ stages. This policy probably derives from telecommunications practice, where clipping is much more of a problem than a small increase in the noise floor. However, for high-order lowpass filters at least, if noise performance is more important, then the order should be reversed, so that the low-gain low-$Q$ stages attenuate the higher noise from the earlier high-$Q$ stages.

### Butterworth Filters up to 8th-Order

Figure 7.27 shows how to make Butterworth filters up to the 8th-order. All you need to know is the cutoff frequency and the $Q$ of each 2nd-order stage and the cutoff frequency of the 1st-order stages.

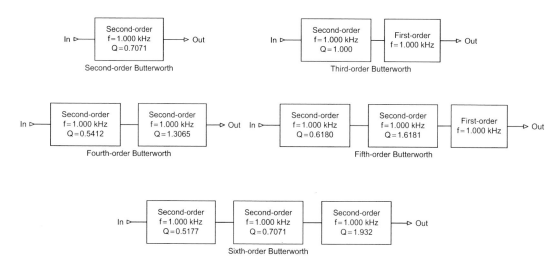

Figure 7.27: Butterworth lowpass filters up to 6th-order made from 2nd- and 1st-order stages cascaded; all stage have the same cutoff frequency of 1.00 kHz.

**Table 7.3: Frequencies and $Q$'s for Butterworth filters up to 8th-order. Stages are arranged in order of $Q$, with the 1st-order section at the end.**

| Order | Freq 1 | $Q$ 1 | Freq 2 | $Q$ 2 | Freq 3 | $Q$ 3 | Freq 4 | $Q$ 4 |
|-------|--------|-------|--------|-------|--------|-------|--------|-------|
| 2 | 1.0000 | 0.7071 | | | | | | |
| 3 | 1.0000 | 1.0000 | 1.0000 | n/a | | | | |
| 4 | 1.0000 | 0.5412 | 1.0000 | 1.3065 | | | | |
| 5 | 1.0000 | 0.6180 | 1.0000 | 1.6181 | 1.0000 | n/a | | |
| 6 | 1.0000 | 0.5177 | 1.0000 | 0.7071 | 1.0000 | 1.9320 | | |
| 7 | 1.0000 | 0.5549 | 1.0000 | 0.8019 | 1.0000 | 2.2472 | 1.0000 | n/a |
| 8 | 1.0000 | 0.5098 | 1.0000 | 0.6013 | 1.0000 | 0.8999 | 1.0000 | 2.5628 |

This information is given in Table 7.3. You are not very likely to need 7th- or 8th-order filters in a crossover design, but if you do the cutoff frequencies and $Q$s are given at the end of Table 7.3; both 7th and 8th-order filters require four stages. The stages are arranged in the conventional way, in order of increasing $Q$, though as explained in the previous section this may not always be the best order for crossover work. The parameters are given to four decimal places, though not all of these will be needed for practical circuit design.

Butterworth filter design is relatively easy because every stage has the same cutoff frequency—only the $Q$'s vary. To design a 2nd-order Butterworth filter that is −3 dB at 1 kHz, you use the standard 2nd-order filter design equations (see Chapters 8 and 9) to make a single stage with a cutoff frequency of 1 kHz and a $Q$ of 0.7071.

For a 3rd-order Butterworth filter you consult Table 7.3, which tells you need a 2nd-order stage with a cutoff frequency of 1 kHz and a $Q$ of 1.000, followed by a 1st-order stage that also has a cutoff frequency of 1 kHz.

A 4th-order Butterworth filter has a 2nd-order stage with a cutoff frequency of 1 kHz and a $Q$ of 0.5412, followed by another 2nd-order stage with a cutoff frequency of 1 kHz and a $Q$ of 1.3065. The $Q$'s are always different. The process is the same for higher-order filters.

Table 7.3 shows that the higher the order of the filter, the wider the ranges of $Q$'s used in it. It is a handy property of Butterworth filters that the maximum $Q$'s needed are relatively low and therefore do not require very precise components to achieve acceptable accuracy.

Figure 7.28 compares the amplitude responses of 2nd-, 3rd-, and 4th-order Butterworth lowpass filters.

The next few diagrams (Figures 7.29 to 7.31) demonstrate just how these high-order filters work. In Figure 7.29 the upper trace is the output of the 2nd-order stage; because its $Q$ is 1.0 rather than 0.7071 it peaks by +1 dB just before the roll-off. When this is combined with the slow roll-off of the 1st-order stage, the combined response is held up by the peak to give a maximally flat response until it falls rapidly around the cutoff frequency. This naturally only works if the 1st-order cutoff frequency exactly matches the 2nd-order cutoff frequency; mismatching these will give either an unduly slow roll-off or

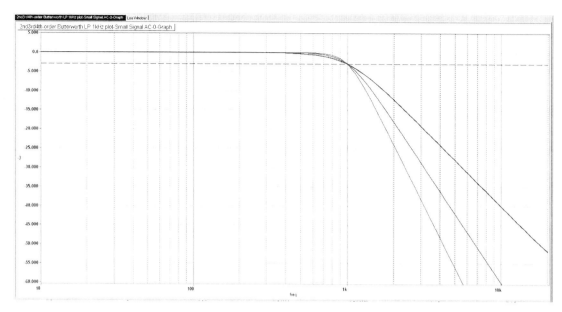

Figure 7.28: Amplitude responses of 2nd-, 3rd-, and 4th-order Butterworth lowpass filters. Cutoff frequency 1.00 kHz, dashed line at −3 dB.

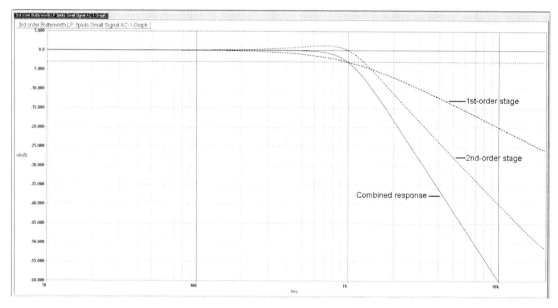

Figure 7.29: Amplitude response of 2nd-order stage and the final output for a 3rd-order Butterworth lowpass filter. Cutoff frequency 1 kHz, dashed line at −3 dB.

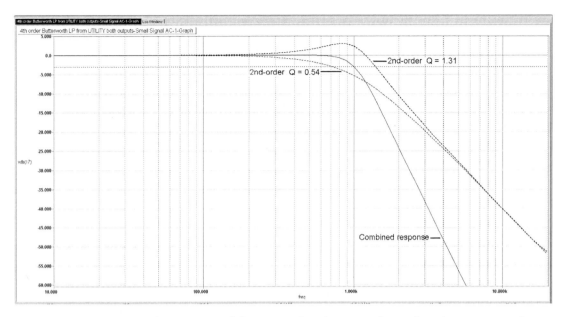

Figure 7.30: Amplitude response of the two 2nd-order stages in a 4th-order Butterworth lowpass filter. Cutoff frequency 1 kHz, dashed line at −3 dB.

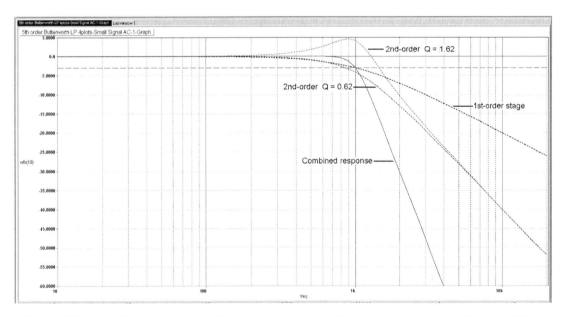

Figure 7.31: Amplitude response of the three stages in a 5th-order Butterworth lowpass filter. Cutoff frequency 1 kHz, dashed line at −3 dB.

unwanted peaking in the passband. The $Q$ of the 2nd-order stages also needs to be just right to get the requisite amount of peaking. The higher the filter order, the more precise this matching needs to be, and the greater the demands on component accuracy.

Because we have a 2nd-order stage cascaded with a 1st-order stage, the ultimate roll-off rate is 18 dB/octave.

The 4th-order filter is made up of two 2nd-order stages cascaded. One of these has a relatively low $Q$ of 0.54, while the other has a relatively high $Q$ of 1.31. Figure 7.30 shows that the latter gives a sharper peak in the passband than the 2nd-order stage did in the 3rd-order filter, and this sharper peak is cancelled out by the other 2nd-order stage, which rolls off faster than a 1st-order stage. Because we have two 2nd-order stages cascaded, the ultimate roll-off rate is 24 dB/octave.

The 5th order Butterworth filter of Figure 7.31 is a bit more complicated, being made up of three stages. As with the 4th-order filter, there are two 2nd-order stages, one with high $Q$ and the other with low $Q$. The peaking is higher but interacts with the low $Q$ stage and the 1st-order stage to once more give a maximally flat passband.

### Linkwitz-Riley Filters up to 8th-Order

Linkwitz-Riley filters, with Butterworth filters, are the only ones that have all stages set to the same cutoff frequency, as shown in Table 7.4. Unlike Butterworth filters, only a few different values of $Q$ are used. Using the table with the stage cutoff frequencies shown will give an amplitude response −6 dB down at the cutoff frequency; this is the "natural" cutoff attenuation for a Linkwitz-Riley filter. If you want to use the more familiar −3 dB criterion for cutoff, then you will need to change all the stage frequencies in Table 7.4 to 1.543.

A comparison of 2nd-, 3rd-, and 4th-order Linkwitz-Riley filters is shown in Figure 7.32.

Table 7.4: Frequencies and $Q$'s for Linkwitz-Riley filters up to 8th-order. Stages are arranged in order of increasing $Q$; odd-order filters have the 1st-order section at the end with no $Q$ shown.

| Order | Freq 1 | $Q$ 1 | Freq 2 | $Q$ 2 | Freq 3 | $Q$ 3 | Freq 4 | $Q$ 4 |
|-------|--------|-------|--------|-------|--------|-------|--------|-------|
| 2 | 1.0000 | 0.5000 | | | | | | |
| 3 | 1.0000 | 0.7071 | 1.0000 | n/a | | | | |
| 4 | 1.0000 | 0.7071 | 1.0000 | 0.7071 | | | | |
| 5 | 1.0000 | 0.7071 | 1.0000 | 1.0000 | 1.0000 | n/a | | |
| 6 | 1.0000 | 0.5000 | 1.0000 | 1.0000 | 1.0000 | 1.0000 | | |
| 7 | 1.0000 | 0.5412 | 1.0000 | 1.0000 | 1.0000 | 1.3066 | 1.0000 | n/a |
| 8 | 1.0000 | 0.5412 | 1.0000 | 0.5412 | 1.0000 | 1.3066 | 1.0000 | 1.3066 |

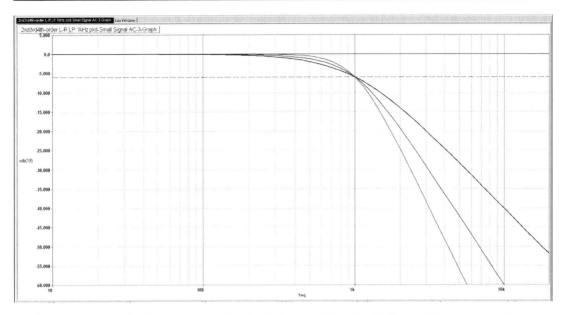

**Figure 7.32: Amplitude response of 2nd-, 3rd-, and 4th-order Linkwitz-Riley lowpass filters. Note that 1 kHz cutoff is at −6 dB.**

You will note that the 4th-order filter is made up of two cascaded Butterworth filters with $Q = 1/\sqrt{2}$ (= 0.7071), and this is the most common arrangement for 4th-order Linkwitz-Riley crossovers. For Linkwitz-Riley filters above 4th-order, higher $Q$ values must be used.

### Bessel Filters up to 8th-Order

Bessel filters have different cutoff frequencies as well as different $Q$'s for each stage in the higher-order filters. The cutoff frequencies here are based on the amplitude response being 3 dB down at the required frequency. Thus, to design a 2nd-order lowpass Bessel filter that is −3 dB at 1 kHz, you use the standard 2nd-order filter design equations in Chapter 8 to make a stage with a cutoff frequency of 1.27 kHz and a $Q$ of 0.5773.

A comparison of 2nd-, 3rd-, and 4th-order Bessel filters is shown in Figure 7.33.

Table 7.5 shows the cutoff frequencies and $Q$'s for each stage. These are no longer all the same, as they were for Butterworth and Linkwitz-Riley filters, so a bit more care is required. The frequencies are normalised on 1.000, so if you want a 3rd-order Bessel lowpass filter with a cutoff of 1.00 kHz, you design a 2nd-order stage with a cutoff frequency of 1.452 kHz and a $Q$ of 0.691, and cascade it with a 1st-order section having a cutoff frequency of 1.327 kHz. A vital point is that if you want

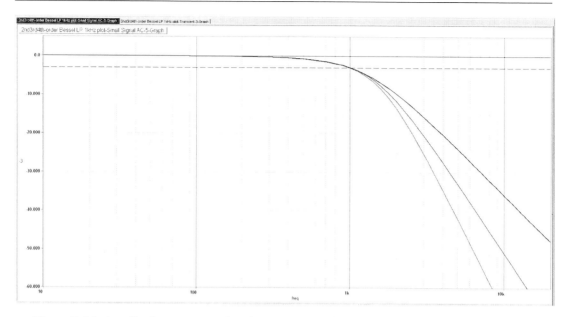

**Figure 7.33: Amplitude response of 2nd-, 3rd- and 4th-order Bessel lowpass filters. Cutoff frequency 1 kHz. Compare the Butterworth filters in Figure 7.28.**

**Table 7.5: Frequencies and Q's for Bessel lowpass filters up to 8th-order. For highpass filters use the reciprocal of the frequency. Stages are arranged in order of increasing Q; odd-order filters have the 1st-order section at the end with no Q shown.**

| Order | Freq 1 | Q 1 | Freq 2 | Q 2 | Freq 3 | Q 3 | Freq 4 | Q 4 |
|-------|--------|--------|--------|--------|--------|--------|--------|--------|
| 2 | 1.2736 | 0.5773 | | | | | | |
| 3 | 1.4524 | 0.6910 | 1.3270 | n/a | | | | |
| 4 | 1.4192 | 0.5219 | 1.5912 | 0.8055 | | | | |
| 5 | 1.5611 | 0.5635 | 1.7607 | 0.9165 | 1.5069 | n/a | | |
| 6 | 1.6060 | 0.5103 | 1.6913 | 0.6112 | 1.9071 | 1.0234 | | |
| 7 | 1.7174 | 0.5324 | 1.8235 | 0.6608 | 2.0507 | 1.1262 | 1.6853 | n/a |
| 8 | 1.7837 | 0.5060 | 1.8376 | 0.5596 | 1.9591 | 0.7109 | 2.1953 | 1.2258 |

a highpass Bessel filter with a cutoff of 1.00 kHz for the other half of a crossover, you must use the reciprocal of the frequency in each case, because a highpass filter is a lowpass filter mirrored along the frequency axis, if you see what I mean. Thus a 3rd-order highpass Bessel filter a is made up of a 2nd-order stage with a cutoff frequency of 1/1.452 kHz = 688.5 Hz and a Q of 0.691, cascaded with a 1st-order section having a cutoff frequency of 1/1.327 kHz = 753.6 Hz. These two 3rd-order filters are both −3 dB at 1 kHz.

## Chebyshev Filters up to 8th-Order

Like Bessel filters, Chebyshev filters have different cutoff frequencies as well as different $Q$'s for each stage in the higher-order filters. Once again, if you want a highpass Chebyshev filter, you must use the reciprocal of the frequency in each case.

The stage frequencies and $Q$'s for Chebyshev filters with passband ripple of 0.5, 1, 2, and 3 dB are shown in Tables 7.6 to 7.9. You can see that as the passband ripple increases (and the steepness of roll-off also increases), higher $Q$'s are required. For the high-order filters, these high $Q$'s are a serious problem, as they require great component precision to implement them with the required accuracy. The usual Sallen & Key and multiple-feedback filter configurations are not up to the job, and more sophisticated circuitry is necessary. It is fair to say that filters like these nowadays are avoided like the plague; if an 8th-order 3 dB-Chebyshev filter is the answer, then you might want to look at changing

**Table 7.6: Frequencies and $Q$'s for 0.5 dB-Chebyshev lowpass filters up to 8th-order. For highpass filters use the reciprocal of the frequency. Stages are arranged in order of increasing $Q$; odd-order filters have the 1st-order section at the end with no $Q$ shown.**

| Order | Freq 1 | $Q$ 1 | Freq 2 | $Q$ 2 | Freq 3 | $Q$ 3 | Freq 4 | $Q$ 4 |
|---|---|---|---|---|---|---|---|---|
| 2 | 1.2313 | 0.8637 | | | | | | |
| 3 | 1.0689 | 1.7062 | 0.6265 | n/a | | | | |
| 4 | 0.5970 | 0.7051 | 1.0313 | 2.9406 | | | | |
| 5 | 0.6905 | 1.1778 | 1.0177 | 4.5450 | 0.3623 | n/a | | |
| 6 | 0.3962 | 0.6836 | 0.7681 | 1.8104 | 1.0114 | 6.5128 | | |
| 7 | 0.5039 | 1.0916 | 0.8227 | 2.5755 | 1.0080 | 8.8418 | 0.2562 | n/a |
| 8 | 0.2967 | 0.6766 | 0.5989 | 1.6107 | 0.8610 | 3.4657 | 1.0059 | 11.5308 |

**Table 7.7: Frequencies and $Q$'s for 1 dB-Chebyshev lowpass filters up to 8th-order. For highpass filters use the reciprocal of the frequency. Stages are arranged in order of increasing $Q$; odd-order filters have the 1st-order section at the end with no $Q$ shown.**

| Order | Freq 1 | $Q$ 1 | Freq 2 | $Q$ 2 | Freq 3 | $Q$ 3 | Freq 4 | $Q$ 4 |
|---|---|---|---|---|---|---|---|---|
| 2 | 1.0500 | 0.9565 | | | | | | |
| 3 | 0.9971 | 2.0176 | 0.4942 | n/a | | | | |
| 4 | 0.5286 | 0.7845 | 0.9932 | 3.5600 | | | | |
| 5 | 0.6552 | 1.3988 | 0.9941 | 5.5538 | 0.2895 | n/a | | |
| 6 | 0.3532 | 0.7608 | 0.7468 | 2.1977 | 0.9953 | 8.0012 | | |
| 7 | 0.4800 | 1.2967 | 0.8084 | 3.1554 | 0.9963 | 10.9010 | 0.2054 | n/a |
| 8 | 0.2651 | 0.7530 | 0.5838 | 1.9564 | 0.8506 | 4.2661 | 0.9971 | 14.2445 |

Table 7.8: Frequencies and $Q$'s for 2 dB-Chebyshev lowpass filters up to 8th order. For highpass filters use the reciprocal of the frequency. Stages are arranged in order of increasing $Q$; odd-order filters have the 1st-order section at the end with no $Q$ shown.

| Order | Freq 1 | $Q$ 1 | Freq 2 | $Q$ 2 | Freq 3 | $Q$ 3 | Freq 4 | $Q$ 4 |
|---|---|---|---|---|---|---|---|---|
| 2 | 0.9072 | 1.1286 | | | | | | |
| 3 | 0.9413 | 2.5516 | 0.3689 | n/a | | | | |
| 4 | 0.4707 | 0.9294 | 0.9637 | 4.5939 | | | | |
| 5 | 0.6270 | 1.7751 | 0.9758 | 7.2323 | 0.2183 | n/a | | |
| 6 | 0.3161 | 0.9016 | 0.7300 | 2.8443 | 0.9828 | 10.4616 | | |
| 7 | 0.4609 | 1.6464 | 0.7971 | 4.1151 | 0.9872 | 14.2802 | 0.1553 | n/a |
| 8 | 0.2377 | 0.8924 | 0.5719 | 2.5327 | 0.8425 | 5.5835 | 0.9901 | 18.6873 |

Table 7.9: Frequencies and $Q$'s for 3 dB-Chebyshev lowpass filters up to 8th-order. For highpass filters use the reciprocal of the frequency. Stages are arranged in order of increasing $Q$; odd-order filters have the 1st-order section at the end with no $Q$ shown.

| Order | Freq 1 | $Q$ 1 | Freq 2 | $Q$ 2 | Freq 3 | $Q$ 3 | Freq 4 | $Q$ 4 |
|---|---|---|---|---|---|---|---|---|
| 2 | 0.8414 | 1.3049 | | | | | | |
| 3 | 0.9160 | 3.0678 | 0.2986 | n/a | | | | |
| 4 | 0.4426 | 1.0765 | 0.9503 | 5.5770 | | | | |
| 5 | 0.6140 | 2.1380 | 0.9675 | 8.8111 | 0.1775 | n/a | | |
| 6 | 0.2980 | 1.0441 | 0.7224 | 3.4597 | 0.9771 | 12.7899 | | |
| 7 | 0.4519 | 1.9821 | 0.7920 | 5.0193 | 0.9831 | 17.4929 | 0.1265 | n/a |
| 8 | 0.2243 | 1.0337 | 0.5665 | 3.0789 | 0.8388 | 6.8251 | 0.9870 | 22.8704 |

the question. If such sharp filtering is really required (and that situation is not very likely in crossover design), then it will probably be cheaper to convert to the digital domain and use DSP techniques, which can provide stable filtering of pretty much any kind you can imagine.

Comparisons of 2nd-, 3rd-, and 4th-order order Chebyshev filters of various kinds can be seen in Figures 7.16 to 7.18.

# More Complex Filters—Adding Zeros

All the filter types we have looked at so far can be made by plugging together 2nd-order and 1st-order lowpass or highpass stages in cascade. This is true no matter how high the filter order. They are technically known as "all-pole filters", which basically means that they combine different sorts of roll-off, but always the response *stays* rolled-off; it does not come back up again. There are no deep notches in the amplitude response; once the roll-off is established, it just keeps on going down. However, filters with notches in their stopband that plunge to the infinite depths (in theory, at least) have their uses, and their study makes up a large proportion of filter theory.

When a faster roll-off than Butterworth is required, without the passband amplitude ripples of the Chebyshev, one possibility is the Inverse Chebyshev, which adds a notch in the response just outside the passband. Notches have slopes that get steeper and steeper as you approach the actual notch frequency, so the roll-off is much accelerated.

Elliptical (Cauer) filters permit ripple in the passband *and* have notches in the response just outside the passband, and offer even steeper roll-off slopes. They are also very economical on hardware. Suppose you need a serious lowpass filter that must roll-off from −0.5 dB to −66 dB in a single octave (not very likely in crossover design, but stay with me). This would need a 13th-order Butterworth filter or an 8th-order Chebyshev, but a 5th-order Cauer filter can do the job with much greater economy in components and also in power, because fewer amplifiers are required.

In filter design the notch frequencies are known as "zeros" because they are the frequencies at which the complex equations describing the filter response give a value of zero—in other words, infinite attenuation. Real filters do not have infinitely deep notches, as the depth usually depends on component tolerances and amplifier gain-bandwidths. An exception is the Bainter filter, which gives beautifully deep notches without tweaking; more on the Bainter later.

The design procedures for these filters are not at all straightforward, and I am simply going to show some design examples. These can have their component values scaled in the usual away to obtain different cutoff frequencies.

It is of course always possible to add notches to a filter response by, for example, cascading a Butterworth filter with a notch filter placed suitably in the stopband. This is however not as efficient as Inverse Chebyshev or Cauer filters (though it is conceptually much simpler), because in the latter the notch is properly integrated with the filter response, and so better passband flatness and sharper roll-offs are obtained.

### Inverse Chebyshev Filters (Chebyshev Type II)

The Inverse Chebyshev filter, also known as the Chebyshev Type II filter, does not have amplitude ripples in the passband; instead it has notches (zeros) in the stopband. Like the Chebyshev filter, it is directed towards getting a faster roll-off than a Butterworth filter while meeting other conditions; it offers a maximally flat passband, a moderate group delay, and an equi-ripple stopband. The cutoff frequency is usually defined to be at the −3 dB level, though other definitions can be used. The Inverse Chebyshev filter uses zeros, so it is not an all-pole filter. These filters are complicated to design, and it would not be a good use of space to try and plod through the procedures here. Instead I am presenting a finished design which can be easily scaled for different frequencies.

Figure 7.34 shows a simple 5th-order Inverse Chebyshev filter with a cutoff frequency of 1 kHz. It is made up of two lowpass notch filters followed by a 1st-order lowpass filter. The most familiar

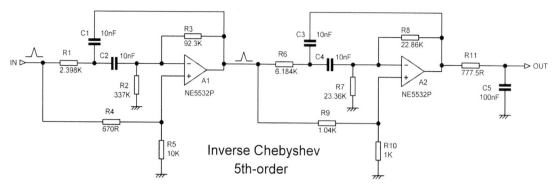

**Figure 7.34: A 5th-order Inverse Chebyshev lowpass filter made up of two lowpass notch filters followed by a 1st-order lowpass. Cutoff frequency 1 kHz.**

notch filter is the symmetrical sort, where the gains on either side of the notch are the same; there are however also lowpass notch filters where, as the frequency increases, the response dives down into the notch but comes up again to level out at a lower gain, and highpass notch filters where the gain is lower on the low-frequency side of the notch. There is more on this in Chapter 12 on bandpass and notch filters.

Here the two lowpass notch filters are of the Deliyannis-Friend type, [12][13] which consists basically of a multiple-feedback (MFB) lowpass filter with an extra signal path via the non-inverting input that generates the notch by cancellation. It is not exactly obvious, but these filter stages are non-inverting in the passband. The notch depth is critically dependent on the accuracy of the ratio set by R4, R5. The complete filter has an overall gain in the passband of +1.3 dB. The filter structure is based on an example given by Van Valkenburg. [14]

The amplitude response in Figure 7.35 shows how the first notch in the stopband, at 1.22 kHz, is very narrow and makes the roll-off very steep indeed. However, the response naturally also starts to come back up rapidly, but this is suppressed by the second notch (at 1.88 kHz), which has a lower $Q$ and is therefore broader and buys time, so to speak, for the final 1st-order filter to start bringing in a useful amount of attenuation. Each time the response comes back up it reaches −17 dB; this is what is meant by an equi-ripple stopband. Variations on this type of filter with non-equal stopband ripples may not be officially Inverse Chebyshev filters, but they can be useful in specific cases.

To scale this circuit for different frequencies you can alter the capacitors, keeping C1 = C2 and C3 = C4 and the ratios of C1, C3 and C5 the same, or you can change resistors R1, R2, R3 and R6, R7, R8 and R11, keeping all their ratios the same. You can of course change both capacitors and resistors to get appropriate circuit impedances. R4 and R5 can be altered, but their ratio must remain the same, and likewise with R9 and R10; remember that the accuracy of this ratio controls notch depth. R11 and C5 can be altered as you wish, quite separately from any other alterations to the circuit, so long as its time-constant is unchanged.

More details on the design of Inverse Chebyshev filters can be found in Van Valkenburg. [15]

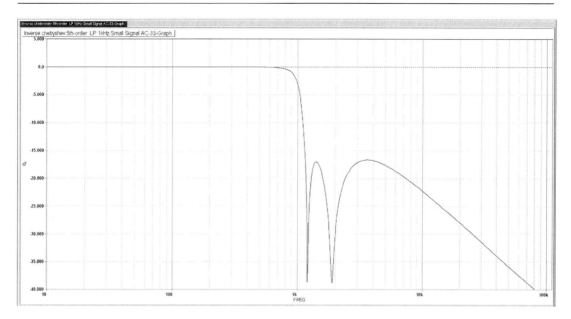

Figure 7.35: The amplitude response of a 5th-order Inverse Chebyshev lowpass filter. Note how the passband is maximally flat (no ripples) before the very steep roll-off. Cutoff frequency 1 kHz.

## Elliptical Filters (Cauer Filters)

Elliptical filters (also called Cauer filters) are basically a combination of Chebyshev and Inverse Chebyshev filters; amplitude ripples in the passband are accepted as the price of a faster roll-off, and there are also one or notches (zeros) in the stopband to steepen the roll-off rate. They have a sharp cutoff, high group delay, and greatest stopband attenuation. They are sometimes called complete-Chebyshev filters or Zolotarev filters.

As for the Chebyshev filter, the definition of an elliptic filter cutoff frequency depends on the passband ripple amplitude. In most filter design software any value of attenuation can be defined as the cutoff point.

The design of elliptical filters is not simple, and even the authors of filter textbooks that are a morass of foot-long complex equations are inclined to say things like "it is rather involved . . ." and recommend you use published tables to derive component values. Regrettably, the use of these tables is in itself rather hard going, so here I am just going to give one example of how these filters are put together. This can be easily scaled for different frequencies. There is another example in Chapter 12 on notch crossovers.

The amount of passband ripple in an elliptic filter is sometimes quoted as a "reflection coefficient" percentage $\rho$, which as you might imagine is a hangover from transmission line theory, and not in my opinion a very helpful way of putting it.

Wilhelm Cauer (June 24, 1900–April 22, 1945) was a German mathematician and scientist. He is most noted for his work on the analysis and synthesis of electrical filters, [16] and his work marked the beginning of the field of network synthesis. He was shot dead in his garden in Berlin-Marienfelde in Berlin by Soviet soldiers during the capture of the city in 1945.

Elliptical filters are commonly implemented by combining a notch filter with an all-pole filter such as a Butterworth type. The notches used are not in general symmetrical notches that go up to 0 dB either side of the central crevasse—they are usually lowpass or highpass notch filters. A lowpass notch response starts out at 0 dB at low frequencies, plunges into the crevasse, and then comes up again to flatten out at a lower level, often −10 dB. When combined with an all-pole lowpass filter this gives much improved high-frequency attenuation rather than a symmetrical notch. Conversely, a highpass notch has a response that is 0 dB at high frequencies but comes back up to, say, −10 dB at low frequencies. There is more on lowpass and highpass notches in Chapter 12 on bandpass and notch filters.

Figure 7.36 shows a simple 3rd-order elliptical filter with a cutoff frequency of 1 kHz and a reflection coefficient of 20%. The passband ripple is therefore very small at 0.2 dB, and there is only one notch in the stopband. The filter structure is based on an example given by Williams and Taylor [17]. The first stage around A1 is a lowpass notch filter, made up of a Twin-T notch filter with its Q (notch sharpness) enhanced by positive feedback through C1, the amount being fixed by R5, R6, which set the closed-loop gain of A1. C4 is added to make a lowpass notch rather than a symmetrical notch. The output of this 2nd-order stage is the upper trace in Figure 7.37, and you can see it has been arrange to peak gently just before the roll-off. When this is combined with the 1st-order lowpass filter R7,

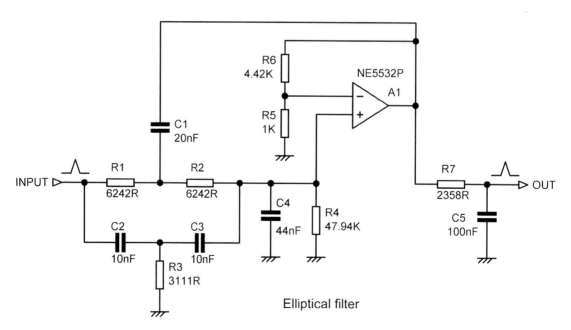

**Figure 7.36: A 3rd-order elliptical lowpass filter made up of a lowpass notch filter followed by a 1st-order lowpass. Cutoff frequency 1 kHz.**

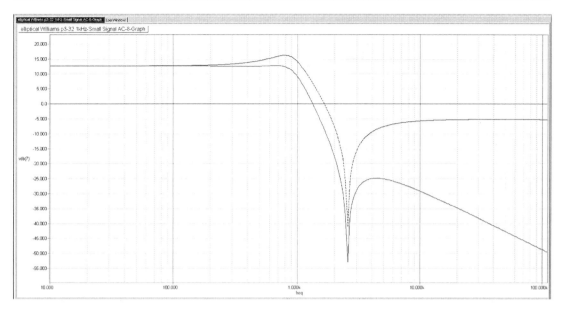

**Figure 7.37: The amplitude response of the 3rd-order elliptical lowpass filter. The lower trace is the final output, and the upper trace is the signal from the first stage. Cutoff frequency 1 kHz.**

C5, the final response is the 3rd-order lower trace in Figure 7.37. The 0.2 dB ripple in the passband is just visible. You will observe that as the frequency increases, once the drama of the notch is over the ultimate roll-off slope is only 6 dB/octave because the lowpass notch response is now flat, and so only the final 1st-order filter is contributing to the roll-off. This is the price you pay for implementing a fast roll-off with a filter that is only 3rd-order. Fourth-order filters that have an ultimate roll-off slope of 12 dB/octave are described in Chapter 5 on notch crossovers.

To scale this circuit for different frequencies you can alter all the capacitors, keeping their ratios to each other the same, or the resistors R1–4 and R7. You can of course change both to get appropriate circuit impedances. R5 and R6 can be altered, but their ratio must remain the same.

We have compared the roll-off of the previous filters by looking at the attenuation at 10 kHz, a decade above the 1 kHz cutoff frequency. That is less helpful here because of the way the amplitude response comes back up at high frequencies; the response at 10 kHz is −42 dB, but in my simulation the attenuation at the bottom of the notch (at 2.6 kHz) was about−65 dB. The passband gain is +12.7 dB because of the positive feedback applied to the notch network, and in many cases this will be less than convenient. Elliptical filters can sometimes be very useful for crossover use, for if we have an otherwise good drive unit with some nasty behaviour just outside its intended frequency range, the notch can be dropped right on top of it.

This elliptical filter and the Inverse Chebyshev filter are the only ones in this chapter that are not all-pole filters. More details on the design of elliptical filters can be found in Williams and Taylor [18] and Van Valkenburg. [19]

# Some Lesser-Known Filter Characteristics

The filter types described earlier are the classic types. Their characteristics depend on particular mathematical relationships such as Bessel functions or Chebyshev polynomials. There is however no rule that you must restrict yourself to using the "official" filters. There are in fact an infinite number of filter characteristics that can be used if necessary. As an example, look at a simple 2nd-order filter. It only has two parameters that define its behaviour—the cutoff frequency and the $Q$. It is the $Q$ that determines the filter type; we have seen that a $Q$ of 0.50 gives a Linkwitz-Riley filter, a $Q$ of 0.578 gives a Bessel filter, a $Q$ of 0.707 gives a Butterworth filter, a $Q$ of 0.956 gives a 1 dB-Chebyshev filter, and so on; as we have seen, the cutoff frequencies must be scaled appropriately for most filter types. There is however nothing whatever to stop you using a $Q$ of 0.55, 0.666, or whatever you feel you need.

The recognised kinds of 2nd-order filter are summarised in Table 7.10 Things become more complicated with filters of 3rd-order and above, where more parameters are needed to define the filter.

## Transitional Filters

When people talk about "transitional filters", they are most often referring to filters that are a compromise between Bessel and Butterworth characteristics. [20] Filtershop offers transitional filters, but it describes them as a Gaussian passband spliced together with a Chebyshev stopband. There are an infinite number of versions of such a filter, depending on the attenuation level at which the splicing occurs; Filtershop offers 3 dB, 6 dB, and 12 dB options. All the filters described here are all-pole filters.

## Linear-Phase Filters

As we have seen, Bessel filters have a maximally flat group delay, avoiding the delay peak that you get with a Butterworth filter. Discussions on filters always remark that the Bessel alignment has a

**Table 7.10: Frequencies and $Q$'s for recognised 2nd-order filter types.**

| Filter type | Freq | $Q$ |
|---|---|---|
| Linkwitz-Riley | 1.578 | 0.5000 |
| Bessel | 1.2736 | 0.5773 |
| Butterworth | 1.0000 | 0.7071 |
| Chebyshev 0.5 dB | 1.2313 | 0.8637 |
| Chebyshev 1.0 dB | 1.0500 | 0.9565 |
| Chebyshev 2.0 dB | 0.9072 | 1.1286 |
| Chebyshev 3.0 dB | 0.8414 | 1.3049 |

slower roll-off, but they often fail to emphasise that it is a *much* slower roll-off. If the requirement for a maximally flat delay is relaxed to allow equi-ripple group delay of a specified amount, then the amplitude roll-off can be much faster. The equi-ripple group delay characteristic is more efficient in that the group delay remains flat (within the set limits) further into the stopband.

This is very much the same sort of compromise as in the Chebyshev filter, where the rate of amplitude roll-off is increased by tolerating a certain amount of ripple in the passband amplitude response. Linear-phase filters are all-pole filters.

Filters with of this kind are frequently referred to as "linear phase" filters but are sometimes (and more accurately) called Butterworth-Thomson filters. For some reason they never seem to be called Butterworth-Bessel filters, though this means exactly the same thing. The parameter $m$ describes the move from Thomson to Butterworth, with $m = 0$ meaning pure Butterworth and $m = 1$ meaning pure Thomson. Any intermediate value of $m$ yields a valid transitional filter. Linear-phase filters can alternatively be characterised by an angle parameter, so you can have a linear-phase 0.05 degree filter or a linear-phase 0.5 degree filter. This refers to the amount of deviation in the filter phase characteristics caused by allowing ripples in the group delay curve.

Figure 7.38 shows the amplitude responses of Bessel, linear-phase 0.05 degree, linear-phase 0.5 degree, and Butterworth lowpass filters. This makes it very clear how linear-phase filters provide intermediate solutions between the Bessel and Butterworth characteristics. Stage frequencies and Q's are given in Table 7.11.

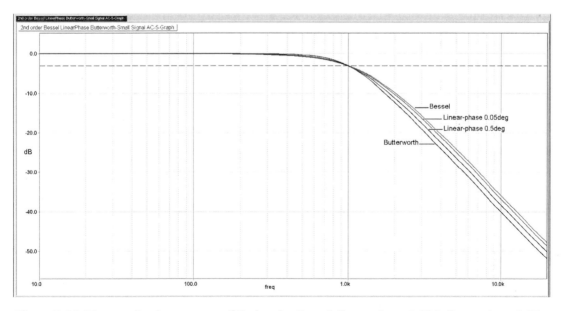

**Figure 7.38: The amplitude response of 2nd-order Bessel, linear-phase 0.05°, linear-phase 0.5°, Butterworth lowpass filters. Cutoff frequency is 1 kHz in all cases.**

**Table 7.11:** Frequencies and *Q*'s for linear-phase filters up to 8th-order. Stages are arranged in order of increasing *Q*; odd-order filters have the 1st-order section at the end with no *Q* shown.

| Order | Freq 1 | Q 1 | Freq 2 | Q 2 | Freq 3 | Q 3 | Freq 4 | Q 4 |
|-------|--------|--------|--------|--------|--------|--------|--------|--------|
| 2 | 1.0000 | 0.6304 | | | | | | |
| 3 | 1.2622 | 0.9370 | 0.7923 | n/a | | | | |
| 4 | 1.3340 | 1.3161 | 0.7496 | 0.6074 | | | | |
| 5 | 1.6566 | 1.7545 | 1.0067 | 0.8679 | 0.5997 | n/a | | |
| 6 | 1.6091 | 2.1870 | 1.0741 | 1.1804 | 0.5786 | 0.6077 | | |
| 7 | 1.9162 | 2.6679 | 1.3704 | 1.5426 | 0.8066 | 0.8639 | 0.4721 | n/a |
| 8 | 1.7962 | 3.1146 | 1.3538 | 1.8914 | 0.8801 | 1.1660 | 0.4673 | 0.6088 |

## Gaussian Filters

A Gaussian filter is essentially a time-domain filter, optimised give no overshoot to a step function input while minimising the rise and fall time; this response is closely connected to the fact that the Gaussian filter has the minimum possible group delay.

A great deal has been published about Gaussian filters, but much of it is based on digital implementations, particularly for image-processing. I can vouch for the fact that "Gaussian blur" is a very useful function when you are cleaning up old images. However, much less has been published on analogue Gaussian filters.

The first problem is that a Gaussian filter is impossible to build. The slow roll-off that gives the good time response is obtained by cascading a series of 1st-order stages of differing and carefully chosen frequencies. Unfortunately the Gaussian roll-off slope steadily increases with frequency and is infinite at infinite frequency, and you therefore need an infinite number of 1st-order stages, which makes construction rather difficult. The frequency response of a true Gaussian filter is shown in Figure 7.39. This plot was produced by calculation rather than simulation, for it is no more practical to simulate a true Gaussian filter than it is to build it. The equation for the Gaussian response is given by Hamill [21] and is shown here as Equation 7.1.

$$A = e^{\left(\left(\frac{freq}{freq_{-3}}\right)\frac{\log_e 2}{2}\right)} \qquad\qquad 7.1$$

Where *A* is the gain, *freq* is the frequency of interest, and *freq*$_{-3}$ is the cutoff frequency.

The natural log term is just a constant.

The Gaussian response is plotted in Figure 7.39 down to −50 dB, where you can see that the −3 dB point is at 1 kHz, and the increasing slope is very obvious. However, if we look at the curve between −30 and −50 dB, we might be convinced that it is straightening out into a fixed slope. It is not. Figure 7.40 extends the plot down to the subterranean depths of −200 dB, and makes it clear that the slope really does continue to increase indefinitely.

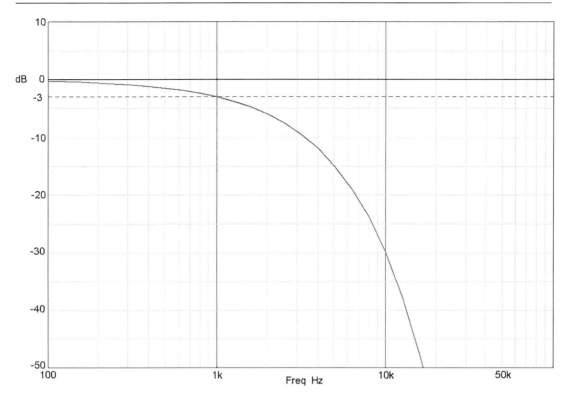

**Figure 7.39: The calculated amplitude response of a true Gaussian filter down to −50 dB. Dashed line is at −3 dB.**

While you cannot build a true Gaussian filter, it is however entirely practical to build an approximation to a Gaussian filter by keeping the same shape for the initial roll-off but then smoothly splicing this to a constant, and therefore more practical, filter slope such as the 18 dB/octave of a 3rd-order Butterworth or Bessel filter. All real Gaussian filters are therefore both approximations and transitional filters. As its order increases, a Bessel filter approaches a Gaussian characteristic but never reaches it unless it is of infinite order: impossible again.

Gaussian filters generated by software come in various kinds identified by a dB suffix, such as "Gaussian-6 dB" and "Gaussian-12 dB", which indicate the attenuation level at which the slow Gaussian roll-off steepens into the constant-slope section, though the amplitude response differences between them are very small. For 3rd-order filters the amplitude response is also very similar indeed to that of a Bessel filter, as demonstrated in Figure 7.41.

Gaussian filters are named after Johann Carl Friedrich Gauss (1777–1855), German mathematician and scientist, because their mathematical derivation stems from the same basic equations used to derive the Gaussian distribution in statistics.

Figure 7.42 compares the step response of 3rd-order Gaussian-12 dB and Butterworth lowpass filters. The Gaussian output rises faster than the Butterworth and has no overshoot at all. The Bessel step response is similar to that of the Gaussian. Gaussian filters are all-pole filters.

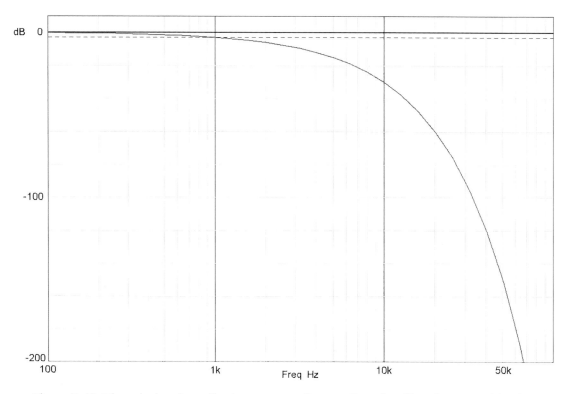

Figure 7.40: The calculated amplitude response of a true Gaussian filter down to −200 dB. The slope continues to steepen forever.

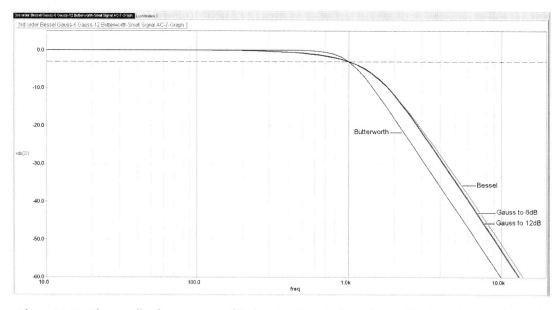

Figure 7.41: The amplitude response of 3rd-order Bessel, Gaussian-6 dB, Gaussian-12 dB, and Butterworth lowpass filters. Cutoff frequency is 1 kHz in all cases.

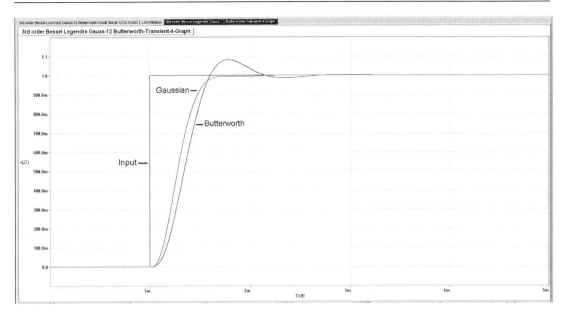

**Figure 7.42: The step response of 3rd-order Gaussian-12 dB and Butterworth lowpass filters. Cutoff frequency is 1 kHz.**

**Table 7.12: Frequencies and $Q$'s for Gaussian filters up to 8th-order. Stages are arranged in order of increasing $Q$; odd-order filters have the 1st-order section at the end with no $Q$ shown.**

| Order | Freq 1 | $Q$ 1 | Freq 2 | $Q$ 2 | Freq 3 | $Q$ 3 | Freq 4 | $Q$ 4 |
|-------|--------|-------|--------|-------|--------|-------|--------|-------|
| 2 | 0.9170 | 0.6013 | | | | | | |
| 3 | 0.9923 | 0.5653 | 0.9452 | n/a | | | | |
| 4 | 0.9930 | 0.6362 | 1.0594 | 0.5475 | | | | |
| 5 | 1.0427 | 0.6000 | 1.1192 | 0.5370 | 1.0218 | n/a | | |
| 6 | 1.0580 | 0.6538 | 1.0906 | 0.5783 | 1.1728 | 0.5302 | | |
| 7 | 1.0958 | 0.6212 | 1.1358 | 0.5639 | 1.2215 | 0.5254 | 1.0838 | n/a |
| 8 | 1.1134 | 0.6644 | 1.1333 | 0.5994 | 1.1782 | 0.5537 | 1.2662 | 0.5219 |

## Legendre-Papoulis Filters

The Legendre-Papoulis filter is a monotonic all-pole filter; in other words the response is always downwards. It is optimised for the greatest slope at the passband edge, given this condition. It gives faster attenuation than the Butterworth characteristic, but the snag is that the passband is not maximally flat, as is the Butterworth; instead it slopes gently until the rapid roll-off begins. This is the crucial difference. Legendre-Papoulis filters can be useful in applications which need a steep cutoff at the passband edge but cannot tolerate passband ripples, or in cases where a Chebyshev I filter produces too much group delay at the passband edge. It is essentially a compromise between the

Butterworth and the Chebyshev filters; it has the maximum possible roll-off rate for a given filter order while maintaining a monotonic amplitude response. Legendre-Papoulis filters are often simply called Legendre filters.

A 2nd-order Legendre-Papoulis response is exactly the same as the Butterworth response. With a 2nd-order filter the only parameter you have to play with is $Q$, and the highest $Q$ that can be used without causing peaking (in which case the response would no longer be monotonic) is the familiar $1/\sqrt{2}$ which gives the Butterworth response.

Legendre filters are only different from Butterworth filters for 3rd-order and above.

The amplitude responses of 3rd-order Legendre-Papoulis and Butterworth filters are compared in Figure 7.43. You will note that the differences are not very great. Compared with the Butterworth, the Legendre-Papoulis sags by 0.5 dB around 500 Hz, which may or may not be negligible depending on your application. Once the roll-off has begun, the Legendre-Papoulis is usefully if not dramatically superior; at 2 kHz it gives 3.7 dB more attenuation, while at 3 kHz it gives 4.3 dB more. After that the difference is effectively constant at 4.4 dB, as the two curves must ultimately run parallel at −18 dB/octave, both filters being 3rd-order. Legendre-Papoulis filters are all-pole filters.

The Legendre-Papoulis filter was proposed by Athanasios Papoulis in 1958. [22] It is also sometimes known simply as a Legendre filter, an "Optimum L" filter or just an "Optimum" filter. The filter design is based on Legendre polynomials; their French inventor, Adrien-Marie Legendre (1752–1833), was yet another mathematician who did not live to see his mathematics applied to electrical filters.

The fact that the response is monotonic but not maximally flat means that Legendre-Papoulis filters are of little use in many hi-fi applications. For example, a subsonic filter for a phono input needs to

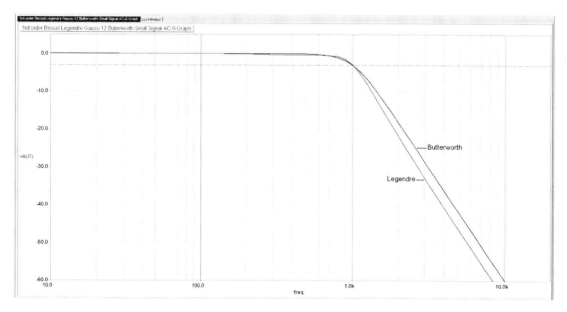

**Figure 7.43: Amplitude response of 3rd-order Legendre and Butterworth filters.**
**Cutoff frequencies are 1 kHz.**

be maximally flat, as a slow early roll-off will cause increased RIAA errors in the low-frequency part of the audio range. However, it is possible that Legendre-Papoulis filters may be of use in crossover design.

The step response of the Legendre-Papoulis filter is similar to that of the Butterworth. Figure 7.44 shows that the response is slightly slower, with about the same amount of overshoot, but after that the undershoot is significantly greater. The Legendre-Papoulis is essentially a frequency-domain filter.

The Legendre filter should not be confused with the Laguerre filter. *C'est magnifique, mais ce n'est pas Laguerre.* [23]

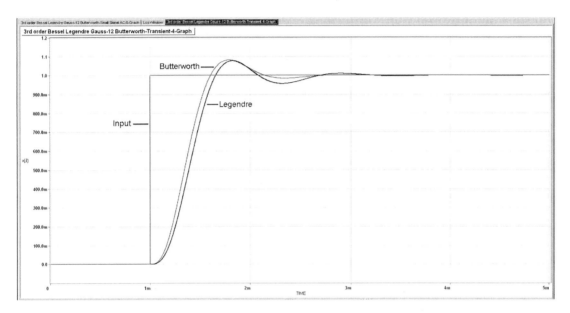

**Figure 7.44: Step response of 3rd-order Legendre and Butterworth filters.
Cutoff frequencies are 1 kHz.**

**Table 7.13: Frequencies and Q's for Legendre-Papoulis filters up to 8th-order. Stages are arranged in order of increasing Q; odd-order filters have the 1st-order section at the end with no Q shown.**

| Order | Freq 1 | Q 1 | Freq 2 | Q 2 | Freq 3 | Q 3 | Freq 4 | Q 4 |
|---|---|---|---|---|---|---|---|---|
| 2 | 1.000 | 0.707 | | | | | | |
| 3 | 0.9647 | 1.3974 | 0.6200 | n/a | | | | |
| 4 | 0.9734 | 2.1008 | 0.6563 | 0.5969 | | | | |
| 5 | 0.9802 | 3.1912 | 0.7050 | 0.9082 | 0.4680 | n/a | | |
| 6 | 0.9846 | 4.2740 | 0.7634 | 1.2355 | 0.5002 | 0.570 | | |
| 7 | 0.9881 | 5.7310 | 0.8137 | 1.7135 | 0.5531 | 0.7919 | 0.3821 | n/a |
| 8 | 0.9903 | 7.1826 | 0.8473 | 2.1807 | 0.6187 | 1.0303 | 0.4093 | 0.5573 |

## Laguerre Filters

The Laguerre filter is a generalisation of a transversal filter. These filters use only delay stages or a tapped delay line, and the weighted sum of the taps gives the required frequency response. The Laguerre filter is implemented in the analogue domain by replacing each true delay stage of the transversal filter by a 1st-order allpass section and by applying a 1st-order lowpass filter with the same cutoff frequency as the allpass sections to the filter input signal. The Laguerre filter is normally implemented in the DSP domain as a FIR filter. In the analogue domain the amount of hardware to realise the stages is likely to be excessive.

There are several filters of this sort; if the delay line is composed of 1st-order lowpass sections it is called a Gamma filter. [24] Such a filter may also be made of 1st-order allpass filters [25] or a combination of 1st-order lowpass and allpass filters, in which case it is a Laguerre filter, [26] which as you may have seen coming, is based on the Laguerre polynomials, [27] introduced by Edmond Laguerre (1834–1886). I am not aware that these kinds of filter have been used in crossovers, but it is as well to know their names in case you do come across them.

It is unsettling to find that Laguerre filters are very popular for analysing financial data in the hope of making a quick buck on stock price movements.

## Synchronous Filters

Synchronous filters consist of a number of identical 1st-order stages in cascade. Since there are no 2nd-order stages, the response is inevitably monotonic. The term "synchronous" refers to the fact that all the stages have an identical cutoff frequencies and therefore identical time-responses; it has nothing to do with system clocks or digital data transmission. You cannot produce a synchronous filter simply stringing together a chain of R's and C's; each 1st-order RC stage must be isolated from the next by a buffer stage to prevent loading and interaction effects. Figure 7.45 shows 2nd-, 3rd-, and 4th-order synchronous filters made in this way. All are designed to be −3 dB at 1 kHz overall, so the stage frequencies for the higher-order filters are increased so that the overall response passes through this point. For example, the 4th-order filter has four stages with cutoff (−3 dB) frequencies of 2.303 kHz, each of which give −0.75 dB at 1 kHz; when the four stages are used together the attenuation is therefore −3 dB at 1 kHz. The cutoff frequencies required for 2nd-, 3rd-, and 4th-order synchronous filters are given in Table 7.14.

Figure 7.46 shows the amplitude response, which has a slower roll-off than the Bessel filter because of a complete lack of internal peaking. The response differences below the −3 dB point are very small. It does not at present appear very likely that synchronous filters will be useful in crossover filters as such because of their very slow rate of roll-off, but it is a branch of filter technology to be aware of because they may be appropriate for specific equalisation of drive unit roll-offs.

Looking at Figure 7.45, you can see that making high-order synchronous filters by stringing together 1st-order stages uses a lot of amplifiers for buffering. However, a 2nd-order stage with a $Q$ of 0.5 is equivalent to two cascaded 1st-order stages, and the circuit shown in Figure 7.47a saves one amplifier section for a 2nd-order filter. A 2nd-order stage with a $Q$ of 0.5 is also equivalent to a 2nd-order Linkwitz-Riley filter, but this relationship does not hold for 3rd-order and higher filters. It is

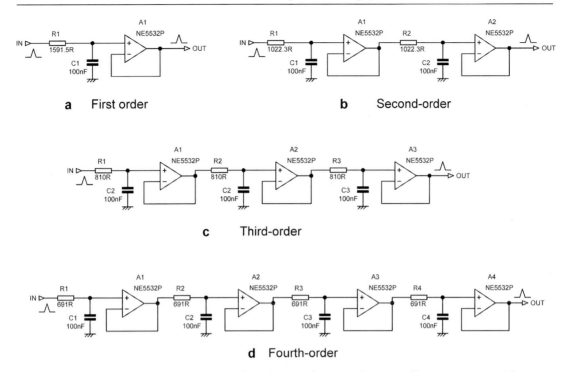

a   First order

b   Second-order

c   Third-order

d   Fourth-order

**Figure 7.45: First-, 2nd-, 3rd-, and 4th-order synchronous lowpass filters constructed from repeated 1st-order sections. All four filters are −3 dB at 1 kHz.**

**Table 7.14: Cutoff frequencies for synchronous filters that give cutoff at 1.000 when stages are cascaded.**

| Order | Cutoff frequency |
|---|---|
| 1 | 1.0000 |
| 2 | 1.5568 |
| 3 | 1.9649 |
| 4 | 2.3033 |

not possible to replace three cascaded 1st-order stages with a 2nd-order stage of any $Q$ because the ultimate slope is 12 dB/octave rather than 18 dB/octave.

A 3rd-order synchronous filter can be built in two stages, as in Figure 7.47b, by simply putting another synchronous 1st-order stage after the 2nd-order stage, but only one amplifier is saved because of the need to buffer the final pole R3–C3 with A2. A better solution is the one-stage Geffe filter at Figure 7.47c, [28] which saves two amplifiers compared with Figure 7.45c; the component ratios are R2 = 2R1, R3 = R1, C2 = C1, C3 = C1/2. A 4th-order synchronous filter can be economically built

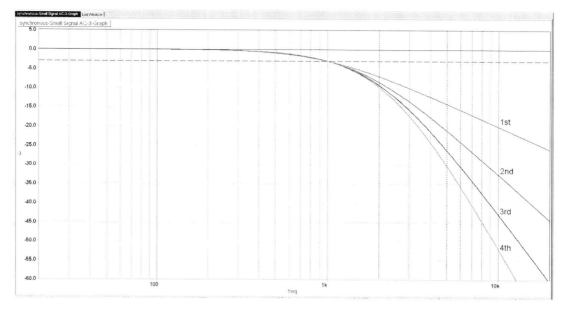

Figure 7.46: Amplitude response of 2nd-, 3rd-, and 4th-order synchronous lowpass filters, all designed to be −3 dB at 1 kHz.

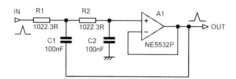

**a    2nd-order synchronous**

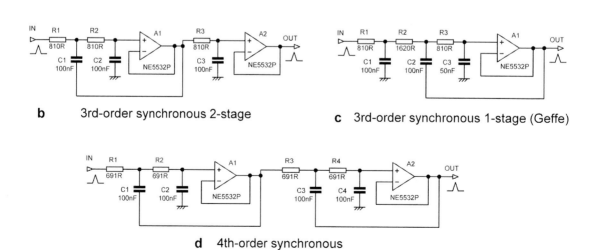

**b    3rd-order synchronous 2-stage**

**c    3rd-order synchronous 1-stage (Geffe)**

**d    4th-order synchronous**

Figure 7.47: More efficient ways to make higher-order synchronous lowpass filters using 2nd-order filters with $Q = 0.5$. All filters are −3 dB at 1 kHz.

from two 2nd-order stages, each with a $Q$ of 0.5, as in Figure 7.47d, saving two amplifiers compared with Figure 7.45d.

These stages can be plugged together in various ways to make higher-order synchronous filters. Cascading a 2nd-order stage with a 3rd-order stage gives a 5th-order filter, and cascading two 3rd-order stages gives a 6th-order synchronous filter.

It is possible to make 4th-, 5th-, and 6th-order highpass and lowpass filters in one stage; see Chapter 8 and 9. It may therefore be possible to make high-order synchronous filters in the same way; if anybody knows the answer I would be glad to hear it.

## Other Filter Characteristics

We have just looked at a number of obscure filter types; this however by no means exhausts the recognised filter characteristics that exist.

Ultraspherical filters [29] are based on ultraspherical polynomials. How, you may inquire, can something be more spherical than spherical? To answer that in a suitably short space is quite beyond my powers, but rest assured that the whole business is solidly rooted in higher mathematics. Ultraspherical polynomials are sometimes called Gegenbauer polynomials, after their inventor, and inhabit a fairly elevated area of mathematics. They are generalisations of Legendre polynomials and Chebyshev polynomials and are special cases of Jacobi polynomials. The response of an ultraspherical filter is set by four parameters; the filter order and the cutoff frequency as usual, plus $\varepsilon$, which controls the attenuation at the cutoff frequency, and $\alpha$, which controls the shape of the amplitude response. Ultraspherical filters are a class of filters rather than a distinct type, and they include Butterworth, Legendre, and Chebyshev filters as special cases; for example $\alpha = \infty$ gives a Butterworth filter, while $\alpha = -0.5$ gives a Chebyshev.

There are also inverse ultraspherical filters, in the same way that there are inverse-Chebyshev filters. [30]

Another rare bird is the Halpern filter. [31] This is related to the Legendre-Papoulis filter in that it rolls off faster than a Butterworth but slower than a Chebyshev. Likewise, the Halpern filter has a monotonic passband without being maximally flat. The peak group delay near the cutoff is significantly lower than that of a Chebyshev filter but not as good as Butterworth or Legendre-Papoulis.

I don't pretend that the information in this final section is going to be digestible to anyone without a quite advanced knowledge of mathematics, but do not fret. There is at present nothing to suggest that these filter types have anything to offer crossover design that is not already covered by the better-known filter types dealt with in detail earlier, so there seems no point in going further into the rather heavy mathematics required.

## References

[1] Williams and Taylor, "Electronic Filter Design Handbook" Fourth Edn, McGraw-Hill, 2006, ISBN: 0-07-147171-5
[2] Valkenburg, Van "Analog Filter Design" Holt-Saunders International Editions, 1982, ISBN: 4-8338-0091-3

[3] Berlin, H. M. "Design of Active Filters With Experiments" Blacksburg, 1978, p. 85

[4] Hofer, Bruce "Switch-Mode Amplifier Performance Measurements" Electronics World September 2005, p. 30

[5] Hardman, Bill "Precise Active Crossover" Electronics World, August 2009, p. 652

[6] Hickman, Ian "Top Notches" (notch filter design) Electronics World, February 2000, p. 120

[7] Ramaswamy, Ramkumar "A Universal Continuous-Time Active Filter" Linear Audio Volume 1 Publ. Jan Didden, 2011

[8] Butterworth, S. "On the Theory of Filter Amplifiers," Experimental Wireless and the Wireless Engineer, 7, 1930, pp. 536–541.

[9] Thomson, W.E. "Delay Networks Having Maximally Flat Frequency Characteristics" Proc IEEE, part 3, 96, November 1949, pp. 487–490

[10] Miller, Ray "A Bessel Filter Crossover, and Its Relation to Others" RaneNote 147, Rane Corporation, 2006

[11] Kiyasu, Z. "On a Design Method of Delay Networks" Journal of The Institution of Engineers, Japan, 26, pp. 598–610, August 1943

[12] Delyiannis, T. "High-$Q$ Factor Circuit with Reduced Sensitivity" Electronics Letters, December 1968, p. 577

[13] Friend, J. "A Single Operational-Amplifier Biquadratic Filter Section" IEEE ISCT Digest Technical Papers 1970, p. 189

[14] Valkenburg, Van "Analog Filter Design" Holt-Saunders International Editions, 1982, p. 375, ISBN: 4-8338-0091-3

[15] Valkenburg, Van "Analog Filter Design" As [14], p. 365

[16] Cauer, W. "Theorie der linearen Wechselstromschaltungen", Vol. I, Akad. Verlags-Gesellschaft Becker und Erler, Leipzig, 1941. (in German) (Theory of Linear AC Circuits, Vol. 1)

[17] Williams and Taylor, "Electronic Filter Design Handbook" Fourth Edn, 2006, p 2.76 ISBN: 0-07-147171-5

[18] Williams and Taylor "Electronic Filter Design Handbook" As [17]

[19] Valkenburg, Van "Analog Filter Design" As [12], p. 385

[20] Paarmann, L. "Design & Analysis of Analog Filters: A Signal Processing Perspective" Springer, 2001, pp. 234,235, ISBN: 0792373731 (Transitional filters)

[21] Hamill, D. "Transient Response of Audio Filters" Wireless World, August 1981, p. 59

[22] Papoulis, A. "Optimum Filters With Monotonic Response" Proc IRE 46, 1958, pp. 606–609

[23] Bousquet, P. He actually said "C'est magnifique, mais ce n'est pas la guerre: c'est de la folie." on observing the Charge of The Light Brigade.

[24] Juan, Harris, and Principe "Analog Hard-Ware Implementation of Adaptive Filter Structures," Proceedings of the International Joint Conference on Neural Networks

[25] Bult, Klaas and Hans Wallinga "A CMOS Analog Continuous-Time Delay Line With Adaptive Delay-Time Control," IEEE Journal of Solid-State Circuits, 23 (3), pp. 759–766, 1988

[26] Silva, Tomas "Laguerre Filters: An Introduction" Revista do Detua 1(3), January 1995, http://sweet.ua.pt/tos/bib/5.2pdf; Accessed January 2016

[27] https://en.wikipedia.org/wiki/Laguerre_polynomials; Accessed July 2013

[28] Anon Personal Communication, postmarked 'San Francisco'

[29] Paarmann, L. "Design & Analysis of Analog Filters: A Signal Processing Perspective" As [20] p. 245

[30] Corral, C. A. "Inverse Ultraspherical Filters" Journal of Analog Integrated Circuits and Signal Processing, 42(2), pp. 147–159, January 2005, Springer Netherlands ISSN 0925–1030

[31] Paarmann, L. "Design & Analysis of Analog Filters: A Signal Processing Perspective" As [20] p. 255

# CHAPTER 8

# Designing Lowpass Filters
## *Sallen & Key*

## Designing Real Filters

In Chapter 7 we saw how just about any filter you can dream up can be implemented by cascading 1st- and 2nd-order filter blocks. In this chapter we look at the best ways to turn those theoretical blocks into reality. There are many ways to do this, but for the moderate $Q$ values normally required, one of the simplest and the best methods is the well-known Sallen & Key configuration. This chapter and Chapter 9 on highpass filters focus entirely on various versions of the Sallen & Key filter. Other configurations, such as the multiple-feedback filter (MFB), are described in Chapter 10. To accurately implement high $Q$'s, more complicated multi-amplifier configurations such as the Tow-Thomas filter are preferred, and these can also be found in Chapter 10.

All the filters in this chapter and the next have been designed with a cutoff frequency of 1 kHz, which allows quick comparisons and quick scaling of the examples to get the filter frequency you want. The use of the term "cutoff" might be taken to imply that the filter has a response that drops steeply; this is in general not the case, and even the most gently sloping filter has a "cutoff" frequency. For the Butterworth filter the cutoff frequency is the point at which the response has fallen by 3 dB, i.e. to $1/\sqrt{2}$ of its passband value. Other filters such as the Chebyshev have different definitions of cutoff. The term "transition frequency" means the same thing but is rarely encountered.

In the examples shown here, the capacitor values have been chosen to put the resistor values mostly in the range 700 Ω–1 kΩ in order to minimise Johnson noise and the effects of device current noise without putting excessive loading on opamps that will increase distortion. Most examples are of the lowpass version.

## Component Sensitivity

The accuracy of the response of a filter obviously depends on the precision of the components with which it is made. What is less obvious is that in some cases the response is affected very greatly by small changes in certain components. This sort of "sensitivity" has nothing whatever to do with the sensitivity of an equipment input, as in "Sensitivity: 500 mVrms for full output". It is unfortunate that two different concepts have ended up with the same name, but no one is going to change it now.

As described in Chapter 7, high-order filters are usually made up from cascaded 2nd-order sections, plus a 1st-order section for the odd-order versions. These sections usually have different $Q$s, and the higher the $Q$, the greater the component sensitivity and the more the stage amplifies component errors.

Precision components are expensive, and so there is a strong incentive to use the lowest $Q$s possible for a filter design.

While cascaded 2nd-order sections are the norm, identical filters can be made up from higher-order sections, so that a 5th-order filter could be redesigned as one 3rd-order section and one 2nd-order section instead of two 2nd-order sections plus a 1st-order section. Combinations that appear to be impossible are a 4th-order filter made from three sections and a 6th-order filter made from two cascaded 3rd-order sections instead of the usual three 2nd-order sections; do not regard that as my final conclusion. The situation is summarised in Table 8.1, where "YES" means that filter configuration is possible and is included in this chapter and the next. Note the combinations marked "not applicable" because there is no point to them; while it is hard to see how you could make a 1st-order filter in two or three stages, it is even harder to see why you would want to. Likewise there is no way to make a 2nd-order filter in two or three stages, and no point in it if you could.

Actually, 3rd-order sections are not often used even though they save an amplifier. A 3rd-order section has greater component sensitivity, and any saving made in the number of opamps may well be less than the extra cost of more precise components to get a filter of the same accuracy as the original; they are also much more difficult to design. It is also possible to make a 4th-order filter in a single stage, and even 5th- and 6th-order, as revealed in this chapter, but the component sensitivity is predictably even worse, and so is the complexity of the design procedure.

Component sensitivity may appear to have a somewhat academic flavour about it, but in fact keeping it as low as possible is crucial to making an economic design, as the cost of components, especially capacitors, increases very steeply with increased accuracy. The alternative is some sort of cut-and-try in production, which is going to be costly and time-consuming and will require accurate measuring equipment.

Component sensitivity is expressed as a ratio. There is a different figure for every component and how it affects every parameter of the stage. Thus, every component in a 2nd-order filter has two sensitivities: one for its effect on cutoff frequency and one for its effect on $Q$. Thus if a certain resistor R in a filter has a sensitivity with respect to cutoff frequency of $-0.5$, that means that a 1% increase in the component will cause a $-0.5$% reduction in the cutoff frequency. This is the usual frequency sensitivity in 2nd-order Sallen & Key filters.

Component sensitivities are dealt with in the relevant section for each type of filter described.

**Table 8.1: Possible Sallen & Key filter configurations**

| Filter order | One-stage | Two-stage | Three-stage |
|---|---|---|---|
| 1st | YES | n/a | n/a |
| 2nd | YES | n/a | n/a |
| 3rd | YES | YES | n/a |
| 4th | YES | YES | NO |
| 5th | YES | YES | YES |
| 6th | YES | NO | YES |

# First-Order Lowpass Filters

First-order filters are the simplest possible; they are completely defined by one parameter: the cutoff frequency.

The usual first-order filter is just a resistor and capacitor. Figure 8.1a shows the normal (non-inverting) lowpass version in all its beautiful simplicity. To give the calculated first-order response it must be driven from a very low impedance and see a very high impedance looking in to the next stage. It is frequently driven from an opamp output, which provides the low driving impedance, but often requires a unity-gain stage to buffer its output, as shown. In general, all filters in this book assume that low-impedance drive is available, which is typically the case, but explicitly show output buffers when they are required.

Sometimes it is necessary to invert the phase of the audio to correct an inversion in a previous stage. This can be handily combined with a shunt-feedback 1st-order lowpass filter, as shown in Figure 8.1b. This stage inherently provides a low output impedance, so there are no loading effects from circuitry downstream. The signal level can be adjusted either up or down by giving R1 and R2 the required ratio.

Here are the design and analysis equations for the 1st-order filters. The design equations give the component values required for given cutoff frequency; the analysis equations give the cutoff frequency when the existing component values are plugged in. Analysis equations are useful for diagnosing why a filter is not doing what you planned.

For both versions choose a suitable preferred value for $C$:

The design equation is:                     and the analysis equation is:

$$R = \frac{1}{2\pi f_0 C} \qquad\qquad\qquad f_0 = \frac{1}{2\pi RC} \qquad\qquad 8.1, 8.2$$

For the inverting lowpass version you must use R2 and not R1 in the equations; its passband gain A = R2/R1.

There are no issues with component sensitivities in a 1st-order filter; the cutoff frequency is inversely proportional to the product of $R$ and $C$, so the sensitivity for either component is always 1.0. In other words a change of 1% in either component will give a 1% change in the cutoff frequency.

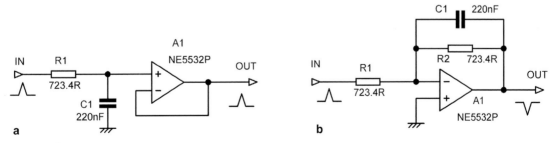

**Figure 8.1: First-order filters: (a) non-inverting lowpass; (b) inverting lowpass; cutoff frequency (−3 dB) is 1 kHz in all cases.**

## Second-Order Filters

Second-order filters are much more versatile. They are defined by two parameters; the cutoff frequency and the $Q$, and as described in Chapter 7, higher-order filters are usually made by cascading 2nd-order stages with carefully chosen cutoff frequencies and $Q$'s. Odd-order filters require a 1st-order stage as well.

There are many ways to make a 2nd-order filter. The simplest and most popular are the well-known Sallen & Key configuration and the multiple-feedback (MFB) filters. The latter are better-known in their bandpass form but can be configured for either lowpass or highpass operation. MFB filters give a phase inversion so can only be used in pairs (e.g. in a Linkwitz-Riley 4th-order configuration) or in conjunction with another stage that re-inverts the signal to get it back in phase.

These two filter types are not suitable for producing high $Q$ characteristics as for $Q$s above about 3, such as might be needed in a high-order Chebyshev filter, as they begin to show excessive component sensitivities and also may start to be affected by the finite gain and bandwidth of the opamps involved. For higher $Q$s more complex multi-opamp circuit configurations are used that do not suffer from these problems. Having said that, very high $Q$ filters are not normally required in crossover design.

## Sallen & Key 2nd-Order Lowpass Filters

The Sallen & Key filter configuration was introduced by R.P. Sallen and E. L. Key of the MIT Lincoln Laboratory in 1955. [1] It became popular as it is relatively easy to design and the only active element required is a unity-gain buffer, so in pre-opamp days it could be effectively implemented with a simple emitter-follower, and before that with a cathode-follower.

An example of a 2nd-order lowpass Sallen & Key filter is shown in Figure 8.2. Its operation is very simple; at low frequencies the capacitors are effectively open circuit, so it acts as a simple voltage-follower. At higher frequencies C2 begins to shunt the signal to ground, and this reduces the output of the follower, causing the "bootstrapping" of C1 to be less effective, and so causing it to also shunt signal away. Hence there are two roll-off mechanisms working at once, and we get the familiar 12 dB/octave filter slope.

There are two basic ways to control the $Q$ of a 2nd-order lowpass Sallen & Key stage. In the first, the two capacitors are made unequal, and the greater the ratio between them the higher the $Q$; a unity-gain buffer is always used. In the second method, the two capacitors are made equal (which will usually be cheaper as well as more convenient), and $Q$ is fixed by setting the gain of the amplifier to a value greater than one. The first method usually gives superior performance, as filter gain is often not wanted in a crossover. The $Q$ of a Sallen & Key filter can also be controlled by using non-equal resistors in the lowpass case or non-equal capacitors in the highpass case, but this is unusual.

There are likewise two methods of $Q$-control for a 2nd-order highpass Sallen & Key stage, but in this case it is the two resistors that are being set in a ratio, or held equal while the amplifier gain is altered.

While the basic Sallen & Key configuration is a simple and easy to design, it is actually extremely versatile, as you will appreciate from the many different versions in this chapter. It is possible to design bandpass Sallen & Key filters, (see Chapter 12) but this chapter is confined to lowpass versions and the next chapter to highpass versions.

## Sallen & Key Lowpass Filter Components

A fundamental difficulty with lowpass Sallen & Key filters is that the resistor values are usually all the same, but in general the capacitors are awkward values. This is the wrong way round from the designer's point of view; using two resistors in parallel or series to get the exact required value—or at any rate close enough to it—is cheap. Capacitors however are relatively expensive, and paralleling them to get a given value is therefore a relatively costly process. Putting capacitors in series to get the right value is not sensible, because you will have to use bigger and more expensive capacitors. A 110 nF capacitance could be obtained by putting two 220 nF parts in series, but the alternative of 100 nF plus 10 nF in parallel is going to be half the cost or less and also occupy less PCB area. The relatively precise capacitors required for accurate crossover filters do not come cheap, especially if the polypropylene type is chosen to prevent capacitor distortion; they are almost certainly the most expensive components on the PCB, and in a sophisticated crossover there may be a lot of them. For this reason I have gone to some trouble to present lowpass filter configurations that keep as many capacitors as possible the same value; the more of one value you buy, the cheaper they are.

Highpass filters do not have this problem. The capacitors are usually all the same value, and it is the resistors that come in awkward values; paralleling them is cheap and takes up little PCB space.

## Sallen & Key 2nd-Order Lowpass: Unity Gain

Figure 8.2 shows a very familiar circuit—a 2nd-order lowpass Sallen & Key filter with a cutoff (−3 dB) frequency of 1 kHz and a $Q$ of 0.707, giving a maximally flat Butterworth response. The pleasingly simple design equations for cutoff (−3 dB) frequency $f_0$ and $Q$ are included. Other recognised 2nd-order responses can be designed by taking the $Q$ value from Table 8.2.

In this chapter the component values are exact, just as they came out of the calculations, with no consideration given to preferred values or other component availability factors; these are dealt with later in Chapter 15. However it is worth pointing out now that in Figure 8.2, C1 would have to be made up of two 100 nF capacitors in parallel, or the errors resulting from using a 220 nF part accepted.

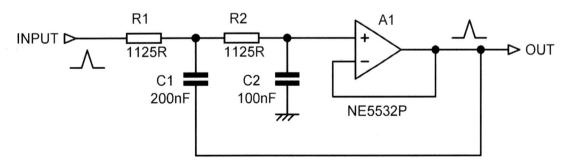

**Figure 8.2: The classic 2nd-order lowpass unity-gain Sallen & Key filter. Cutoff frequency is 1 kHz, and $Q$ = 0.7071 for a Butterworth response.**

The main difference in the circuit that you will notice from textbook filters is that the resistor values are rather low and the capacitor values correspondingly high. This is an example of low-impedance design, where low resistor values minimise Johnson noise and reduce the effect of the opamp current noise and common-mode distortion. The measured noise output is−117.4 dBu if a 5532 section is used. This is after correction by subtracting the testgear noise floor.

In this case the component choice for the capacitors is simple. C2 is a preferred value and C1 can be made up of two 100 nF in parallel. The resistance required is not a preferred value; R1 = R2 are shown to four significant figures in Figure 8.2, but a more exact value is 1125.41 Ω. By pure chance this is very close to the E24 value of 1100 Ω, and using such resistors will give a nominal cutoff frequency of 1023.1 Hz which is only 2.3% in error. If this is not acceptable, there are three useful options for more accuracy. These options are described in more detail in Chapter 15:

1.   Use a single E96 value. The nearest to 1125.39 Ω is 1130 Ω, which gives a nominal cutoff frequency of 995.93 Hz, 0.41 % low.
2.   Use two E24 values combined (usually in parallel) with the values as near equal as possible to maximise the improvement in effective tolerance (see Chapter 15). The best answer depends on how you prioritise an accurate nominal value versus improvement in effective tolerance, but here a convincing result is 1800 Ω paralleled with 3000 Ω, giving a combined value of 1125.0 Ω, as in Figure 8.2. This gives a nominal cutoff frequency of 1000.35 Hz, only 0.035 % high. To get the benefit of this it will be necessary to use 0.1% tolerance resistors; the effective tolerance is improved by the use of two resistors to 0.073%.
3.   Use three E24 values combined (usually in parallel) with the values as near equal as possible to maximise the improvement in effective tolerance. For maximum accuracy of the nominal value, 1500 Ω,8200 Ω, and 10000 Ω in parallel gives 1125.343 Ω and a cutoff frequency of 1000.05 Hz, a tiny 0.005 % high. This is far more accurate than is likely to ever be necessary but shows little reduction in effective tolerance (1% resistors will give 0.77% overall), as the resistors are far from equal. Somewhat better in this respect is 2000 Ω, 2700 Ω, and 56000 Ω in parallel, giving 1125.84 Ω and a nominal cutoff frequency of 999.61 Hz, which is only 0.039 % low. Using 1% resistors will give an effective tolerance of 0.70%.

This gives you something of the flavour of choosing between various ways of implementing awkward component values. In general C1 will not be an integer multiple of C2, and so a similar process will have to be applied, though the practice is rather different because capacitors come in much sparser series than resistors, usually E3, E6 or E12. This matter is dealt with in Chapter 15.

Here are the design and analysis equations for the general unity-gain Sallen & Key lowpass filter; in other words they cover all characteristics, Butterworth, Bessel, etc. The design equations give the component values required for given cutoff frequency and $Q$; the analysis equations give the cutoff frequency and $Q$ when the existing component values are fed in.

Design equations:

Begin by choosing a suitable value for C2

$$R1 = R2 = R = \frac{1}{2Q(2\pi f_0)C_2} \qquad C1 = 4Q^2 C_2 \qquad\qquad 8.3, 8.4$$

Analysis equations:

$$f_0 = \frac{1}{2\pi R\sqrt{C_1 C_2}} \qquad Q = \frac{1}{2}\sqrt{\frac{C_1}{C_2}} \qquad\qquad\qquad 8.5,\ 8.6$$

Table 8.2 gives the $Q$s and capacitor ratios required for the well-known 2nd-order filter characteristics.

Note that these FSFs apply *only* to 2nd-order filters. Higher-order filters in general have different design frequencies for each stage. These are given in the tables in Chapter 7.

It is important to remember that a $Q$ of 1 does *not* give the maximally flat Butterworth response; the value required is $1/\sqrt{2}$, i.e. 0.7071. The lowest $Q$'s you are likely to encounter are Sallen & Key filters, with a $Q$ of 0.5 sometimes used in 2nd-order Linkwitz-Riley crossovers, but these are not favoured because the 12 dB/octave roll-off of the highpass filter is not steep enough to reduce the excursion of a driver when a flat frequency response is obtained, nor to attenuate drive unit response irregularities. [2]

The input impedance of this circuit is of importance, because if it loads the previous stage excessively, then the distortion of that stage will suffer. The input impedance is high at low frequencies, where the series impedance of the shunt capacitors is high. It then falls with increasing frequency, reaching 2.26 k$\Omega$ at the 1 kHz cutoff frequency. Above this frequency it falls further, finally levelling out around 4 kHz at 1.125 k$\Omega$, the value of R1. This is because at high frequencies there is no significant signal at the opamp non-inverting input, so C1 is effectively connected to ground at one end; its impedance is now very low, so the input impedance of the filter looks as if R1 was connected directly to ground at its inner terminal.

Another consideration when contemplating opamp loading and distortion is the amount of current that the filter opamp must deliver into the capacitor C1. This is significant. If the circuit of Figure 8.2 is driven with 10 Vrms, the current through C1 is very low at LF, and then peaks gently at 1.9 mA rms at 650 Hz. The C1 current then reverses direction as frequency increases, and C1 is required to absorb

**Table 8.2: Second-order Sallen & Key unity-gain lowpass $Q$s and capacitor ratios for various filter types.**

| Type | FSF | $Q$ | C1/C2 ratio |
|---|---|---|---|
| Linkwitz-Riley | 1.578 | 0.500 | 1.000 |
| Bessel | 1.2736 | 0.5773 | 1.336 |
| Linear-Phase 0.05deg | 1.210 | 0.600 | 1.440 |
| Linear-Phase 0.5deg | 1.107 | 0.640 | 1.638 |
| Butterworth | 1.000 | 0.707 | 2.000 |
| 0.5 dB-Chebyshev | 1.2313 | 0.8637 | 2.986 |
| 1.0 dB-Chebyshev | 1.0500 | 0.9565 | 3.663 |
| 2.0 dB-Chebyshev | 0.9072 | 1.1286 | 5.098 |
| 3.0 dB-Chebyshev | 0.8414 | 1.3049 | 6.812 |

most of the input current, reaching −8.9 mA at 20 kHz. This sounds alarming, but at these frequencies the filter output is very low (about 32 dB down), so the effect on the opamp is not as bad as it sounds.

Sallen & Key lowpass filters have a lurking problem. When implemented with some opamps, the response does not carry on falling for ever at the filter slope—instead it reaches a minimum and starts to come back up at 6 dB/octave. This is because the filter action relies on C1 seeing a low impedance to ground, and the impedance of the opamp output rises with frequency due to falling open-loop gain and hence falling negative feedback. When the circuit of Figure 8.2 is built using a TL072, the maximum attenuation is −57 dB at 21 kHz, rising again and flattening out at −15 dB at 5 MHz. More capable opamps such as the 5532 give much better results, though the effect is still present. While the rising response can be countered by adding a suitable 1st-order RC filter (preferably in front of the Sallen & Key filter to reduce possible intermodulation), this type of filter should not be used to reject frequencies well above the audio band; a lowpass version of the multiple-feedback filter is preferred.

## Sallen & Key 2nd-Order Lowpass Unity Gain: Component Sensitivity

The unity-gain Sallen & Key filter is noted for having low component sensitivities, as shown in Table 8.3. The sensitivities with respect to cutoff frequency are all −0.5, which makes excellent sense, as if we wanted to deliberately reduce the cutoff frequency by 1%, we would, from the design equations 8.3 to 8.6 given earlier, increase either both resistors or both capacitors by 1%.

Note that the resistor values have sensitivities of zero with respect to $Q$. As the design equations show, $Q$ depends only on the capacitor ratio.

## Filter Frequency Scaling

This seems as good a place as any to demonstrate the scaling of an existing filter design to operate at a different cutoff frequency. This idea is frequently referred to in this book; it avoids having to create tables of component values for different frequencies, which would be quick to use but very unwieldy, even with a limited frequency resolution.

As an example we will take the classic 1 kHz cutoff 2nd-order lowpass unity-gain Sallen & Key filter in Figure 8.2 and scale it for a −3 dB cutoff frequency of 1.4 kHz. It is a fact of electronic life that resistors

### Table 8.3: Second-order Sallen & Key unity-gain lowpass component sensitivities

| Component | Cutoff frequency | $Q$ |
|:---:|:---:|:---:|
| R1 | −0.5 | 0 |
| R2 | −0.5 | 0 |
| C1 | −0.5 | 0.5 |
| C2 | −0.5 | −0.5 |

come in many more values than capacitors; any value of resistance can be achieved by combining two or more of them, so the best approach is to scale the resistors and leave the capacitors alone. If you are scaling by a large factor you will probably have to scale the capacitors as well to prevent the resistors from becoming too large or too small. Since the ratio of the capacitances is exactly two, if three capacitors are used in the filter, with C1 made up of two 100 nF in parallel, the capacitor values are at least *nominally* correct. Obviously, the capacitors will have tolerances about the nominal value.

From earlier in the chapter we know that the cutoff frequency is inversely proportional to the value of the two resistors. To obtain a cutoff frequency of 1.4 kHz the value of 1125.41 Ω is divided by a factor of 1.4, giving 803.86 Ω. Before going any further we must check that this lower resistance will not overload the previous stage.

If all is well, we now choose a strategy for obtaining that awkward nominal resistance. Remember, we are just setting the nominal value, and resistor tolerances are something else.

The nearest E24 resistor value is 820 Ω, which is a 2.0% error, giving a cutoff of 1372.43 kHz, 2.0% away from the 1.4 kHz desired. It is not desirable to build this sort of error into a crossover design.

The nearest E96 resistor value is 806 Ω, which is a 0.27 % error, giving a cutoff of 1396.27 kHz. This sort of error will be acceptable in many cases.

The best answer using two E24 resistors in parallel is 1100 Ω and 3000 Ω, which gives a 0.13 % error and a cutoff of 1398.22 kHz. This will be accurate enough for even precision crossover work.

The best answer using three E24 resistors in parallel is 2200 Ω, 2200 Ω, and 3000 Ω, which to five significant figures has zero error. This demonstrates the power of three-resistor combinations. The near-equal values will give a very good reduction in the effective tolerance.

The example here describes a relatively minor change in filter cutoff frequency, where it's fairly obvious that the capacitors are fine as they are and can be left alone. However, supposing we want to change the cutoff frequency to, say, 4 kHz? The resistor values now come out as 281.35 Ω, which is going to load the previous stage excessively. (Just how the filter loads the previous stage is described at the end of this chapter.) Clearly we are going to have to also scale the capacitor values to get reasonable resistor values. This ratio is just to get convenient capacitor values and is independent of the resistor scaling, though the latter is of course affected by the new capacitor values. Scaling the capacitors by a factor of 10 is very simple; C1 becomes 2 × 10 nF and C2 is 10 nF. On re-doing the calculations this increases the resistor value to 2813.5 Ω, which is not going to overload anything upstream. Scaling capacitors by a factor other than 10 is more difficult because of the sparse number of values available.

However, if you are practising low-impedance design to keep noise down, you may feel that 2813.5 Ω is a bit high, and you would prefer something more in the 1000 Ω region. The capacitors will then need to be scaled by a less friendly factor than 10 times, and there will be two scaling factors to deal with. If we make C2 to be 47 nF and C1 = 2 × 47 nF, then the resistor value comes out as 598.62 Ω, which in most cases will be too low. With C2 = 22 nF and C1 = 2 × 22 nF the resistor value is 1278.87 Ω, which should be very satisfactory.

The frequency scaling of other filters proceeds in the same way, though there may be more components to scale. In the case of a 3rd-order Sallen & Key lowpass filter, there are three resistors to be scaled by the same ratio, and also three capacitors to possibly be altered if a large frequency change is made. For a 4th-order Sallen & Key lowpass filter there are four resistors to be scaled and four capacitors to possibly change, and so on. With Sallen & Key filters it is usually obvious which components have to be scaled; if the stage has gain, the gain is *not* altered. With some filter types it is less clear, and so specific guidance on frequency scaling is given when they are described.

## Sallen & Key 2nd-Order Lowpass: Equal Capacitor

In this version of the Sallen & Key configuration, the two capacitors are made equal and the $Q$ is controlled by increasing the gain of the amplifier above unity; this avoids difficulties with awkward capacitor ratios. The gain must not exceed 3 or the filter becomes unstable and oscillates; this is shown in the analysis equation for $Q$ (Equation 8.11), where the bottom of the fraction becomes 0 when $A = 3$, implying infinite $Q$.

The circuit is shown in Figure 8.3, with $Q$ set to 0.7071 to give a 2nd-order Butterworth response. The unity-gain buffer is now replaced with a voltage gain-stage, and if the gain set by R3 and R4 is 1.586 times (+4.00 dB) for a $Q$ of 0.707, then this allows C1 and C2 to be the same value. The gain introduced here is often an embarrassment in crossover designs, but there is another and more subtle snag to this neat-looking circuit; the component sensitivity is worse; this issue is dealt with in the next section. An equal-resistor-value highpass filter can be made in exactly the same way; this is also described in Chapter 9.

An interesting feature of this circuit is that its filter characteristic can be smoothly altered from Linkwitz-Riley through to 3 dB-Chebyshev, passing through Bessel, linear phase, Butterworth, and all the lower-ripple Chebyshevs on the way, by making R3, R4 a variable potentiometer. Unfortunately the passband gain varies at the same time, though this could be solved by taking the output from the junction of R3 and R4 via a unity-gain buffer; there will however still be a headroom issue due to the higher signal level at the opamp output.

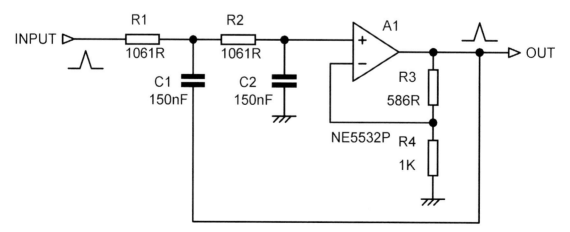

**Figure 8.3: Equal-C 2nd-order lowpass Butterworth filter with a cutoff frequency of 1061 Hz. Gain must be 1.586 times for Butterworth maximally flat response ($Q$ = 0.7071).**

Here are the design and analysis equations for the equal-C Sallen & Key filter. The design equations give the component values required for given cutoff frequency and $Q$; the analysis equations give the cutoff frequency and $Q$ when component values are entered. The $Q$s and gains for various filter types are summarised in Table 8.4.

Design equations:

Choose a value for $C$ $(= C1 = C2)$

$$R_1 = R_2 = \frac{1}{(2\pi f_0)C} \qquad \text{Passband gain } A = 3 - \frac{1}{Q} \qquad \text{8.7, 8.8}$$

Then choose R3, R4 for required passband gain for the chosen $Q$

Analysis equations:

$$f_0 = \frac{1}{2\pi RC} \qquad \text{(R's and C's are both equal)} \qquad \text{8.9}$$

$$\text{Passband gain} \quad A = \frac{R3 + R4}{R4} \qquad\qquad Q = \frac{1}{3-A} \qquad \text{8.10, 8.11}$$

Note that for the 3.0 dB-Chebyshev case the high $Q$ required means we are getting uncomfortably close to the gain limit of three times.

The Chebyshev filters have a peak in their response which it may be possible to utilise for equalisation purposes. The height and frequency of the peak is, however, fixed for a given $Q$, making it a very inflexible method. This sort of approach can make modifications to deal with driver changes, etc, problematic, compared with using a dedicated equalisation stage; it is very helpful if you can say *"this*

### Table 8.4: Second-order Sallen & Key equal-C lowpass: $Q$s and gains for various filter types.

| Type | FSF | $Q$ | Gain A |
|---|---|---|---|
| Linkwitz-Riley | 1.578 | 0.500 | 1.000 |
| Bessel | 1.274 | 0.578 | 1.270 |
| Linear-Phase 0.05deg | 1.210 | 0.600 | 1.333 |
| Linear-Phase 0.5deg | 1.107 | 0.640 | 1.437 |
| Butterworth | 1.000 | 0.707 | 1.586 |
| 0.5 dB-Chebyshev | 1.231 | 0.864 | 1.842 |
| 1.0 dB-Chebyshev | 1.050 | 0.956 | 1.954 |
| 2.0 dB-Chebyshev | 0.907 | 1.129 | 2.114 |
| 3.0 dB-Chebyshev | 0.841 | 1.305 | 2.234 |

modifies *this*". It is however economical, and since fewer stages are required, noise and distortion may be somewhat improved.

## Sallen & Key 2nd-Order Lowpass Equal-C: Component Sensitivity

The equal-C Sallen & Key configuration has sensitivities for cutoff frequency that are the same as for the unity-gain version, but it has a more complicated set of sensitivities for $Q$, with their actual value depending on the $Q$ value chosen, as shown in Table 8.5.

Table 8.6 shows how the sensitivity for $Q$ varies with the $Q$ value chosen. For $Q = 0.7071$ ($1/\sqrt{2}$) which is the most popular value in crossover design, the capacitor sensitivities for $Q$ are almost twice the 0.5 of the unity-gain version (see Table 8.3). It may be a tricky decision as to whether the convenience of having equal capacitors makes up for the fact that they need to be twice as accurate for the same precision of response.

For $Q = 2$, the sensitivities are significantly greater. If you could get away with 5% capacitors for $Q = 0.7071$, now you will need 1% to get the same precision, and they will be significantly more expensive. The Linkwitz-Riley filter has the lowest $Q$ and so the lowest sensitivities.

Table 8.5: Second-order Sallen & Key equal-C lowpass cutoff frequency sensitivities (after Van Valkenburg).

| Component | Cutoff frequency sensitivity | Q sensitivity |
|---|---|---|
| R1 | −0.5 | $-0.5 + Q$ |
| R2 | −0.5 | $0.5 - Q$ |
| C1 | −0.5 | $-0.5 + Q$ |
| C2 | −0.5 | $0.5 - 2Q$ |
| R3 | 0 | $2Q - 1$ |
| R4 | 0 | $-(2Q - 1)$ |

Table 8.6: Second-order Sallen & Key equal-C lowpass Q sensitivities (after Van Valkenburg).

| Component | Q sensitivity (Q = 0.7071) | Q sensitivity (Q = 2) | Q sensitivity (Q = 8) |
|---|---|---|---|
| R1 | 0.207 | 1.5 | 7.5 |
| R2 | −0.207 | −1.5 | −7.5 |
| C1 | 0.914 | 2.5 | 15.5 |
| C2 | −0.914 | −2.5 | −15.5 |
| R3 | −1.00 | −5.00 | −15.00 |
| R4 | 1.00 | 5.00 | 15.00 |

If you tried to generate a $Q$ of 8, which is the sort of value that can turn up in high-order elliptical filters, things are dire indeed, with sensitivities of 15 or more. If you had 5% capacitors for $Q = 0.7071$, you now will need 0.3% to get the same precision, and they will be very expensive, if they can be obtained at all.

If you do need this sort of $Q$, there are multiple opamp configurations which work much more dependably.

## Sallen & Key 2nd-Order Butterworth Lowpass: Defined Gains

We saw in the previous section that we could select convenient capacitor values for the lowpass filter by using a particular gain in the filter. Looking at it another way, we can have whatever gain we want, and whatever $Q$ we want, by altering the capacitor values. As I have said several times, gain in a filter is often unwanted, but there are occasions when a defined amount of gain *is* wanted, as in the HF path of the crossover design example in Chapter 23, and building it into a filter will save an amplifier stage and reduce cost while possibly also reducing noise and distortion, as the signal has gone through one less stage. On the other hand, distortion may increase somewhat because the filter stage giving the gain has less negative feedback; the outcome depends on the opamp types used and the exact circuit conditions.

It would be take up an enormous amount of space to tabulate all the combinations of $Q$ and gain, so I will concentrate on the ever-useful Butterworth filter. The exact resistor values required for gains from 0 to +8 dB are shown in Figure 8.4 and summarised in Table 8.7, which also includes the standard cases of unity gain and of equal capacitors. In the latter case the gain required is extremely close to +4 dB. All the filters have a 1 kHz cutoff frequency and resistors set to 1 k$\Omega$; other frequencies can be obtained by scaling the component values. It is assumed that R4 in the gain network is fixed at 1 k$\Omega$, though in the higher-gain cases it would be advisable to reduce this, to lower the impedance of the gain network and so minimise noise.

There is no obvious reason why higher gains than this could not be obtained, but they are unlikely to be useful, as the further the filter gain gets away from unity, the greater the compromises on noise and headroom will be. The gain limit of 3 times given in the earlier section on equal-capacitor lowpass filters does not apply here because we are adjusting the capacitor values to keep the $Q$ constant at 0.707 instead of letting it shoot off to infinity. There is however the consideration that reducing the negative feedback factor of the opamp will worsen the distortion performance.

Similar design data for 2nd-order Butterworth Sallen & Key highpass filters is given in Chapter 9 on highpass filters. Not surprisingly, the ratios between the capacitors are exactly the same as the ratios of the resistors in the highpass case.

## Sallen & Key 2nd-Order Lowpass: Non-Equal Resistors

In lowpass Sallen & Key filters the resistors are almost always made equal. There is no absolute requirement that this be so—perfectly respectable filters can be made with non-equal resistors, but as you might imagine this somewhat complicates the calculations, and it seems as if there is little to be gained by doing it.

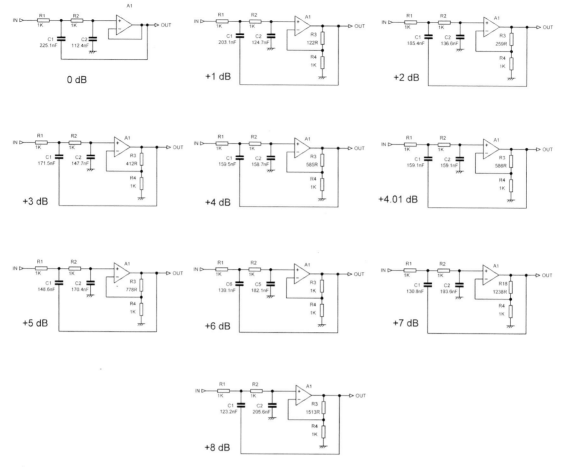

Figure 8.4: Lowpass Butterworth 1 kHz S&K filters with defined passband gains from 0 to +8 dB.

Table 8.7: Component values for defined gains in a Butterworth 2nd-order S&K lowpass filter.

| Gain dB | Gain Times | C1 nF | C2 nF | C1/C2 ratio Times | R3 Ohms | R4 Ohms |
|---|---|---|---|---|---|---|
| 0.00 | 1.000 | 225.1 | 112.4 | 2.000 | n/a | n/a |
| +1.00 | 1.122 | 203.1 | 124.7 | 1.628 | 122 | 1000 |
| +2.00 | 1.259 | 185.4 | 136.6 | 1.357 | 259 | 1000 |
| +3.00 | 1.412 | 171.5 | 147.7 | 1.161 | 412 | 1000 |
| +4.00 | 1.585 | 159.5 | 158.7 | 1.004 | 585 | 1000 |
| +4.01 | 1.586 | 159.1 | 159.1 | 1.000 | 586 | 1000 |
| +5.00 | 1.778 | 148.6 | 170.4 | 0.871 | 778 | 1000 |
| +6.00 | 2.000 | 139.1 | 182.1 | 0.764 | 1000 | 1000 |
| +7.00 | 2.238 | 130.8 | 193.6 | 0.675 | 1238 | 1000 |
| +8.00 | 2.512 | 123.2 | 205.6 | 0.599 | 1513 | 1000 |

I did think at one point it might be possible to make useful 2nd-order filters such as Butterworths with equal capacitor values combined with unity gain by choosing non-equal resistors, but this is regrettably not true. With equal capacitors and a unity-gain amplifier the maximum $Q$ obtainable is 0.5, when the resistors are equal; this is useful for 2nd-order Linkwitz-Riley filters but for no other type. As soon as the resistors are made non-equal in either direction the $Q$ falls below 0.5.

Non-equal resistors are however useful for third and higher-order filters constructed as a single stage, because they allow all the capacitors to be made the same value. There is more on this later.

## Sallen & Key 2nd-Order Lowpass: Optimisation

In Chapter 11 on filter performance, dealing with distortion and noise, the concept of the distortion aggravation factor (DAF) is introduced. In this case it is the ratio by which the amplifier distortion is increased by the positive feedback operation of a Sallen & Key filter. This is discussed in detail in Chapter 11, which covers optimisation of the filter to minimise the DAF by using both non-equal resistors and non-standard capacitor values. Since the DAF of a standard S&K Butterworth filter is only 2, peaking around the cutoff frequency, this effect is not normally measurable so long as low-distortion opamps such as the 5532 are used. Chapter 11 shows how to reduce this DAF from 2 to 1.7, but this is unlikely to be useful, so this form of filter optimisation is not considered further here.

## Sallen & Key 3rd-Order Lowpass: Two Stages

Figure 8.5 shows the classical method of making a 3rd-order lowpass filter by cascading a 2nd-order stage and a 1st-order stage. This is the usual sequence of stages found in the filter textbooks. For the Butterworth version both cutoff frequencies are the same as the cutoff frequency of the complete filter, while the $Q$ of the 2nd-order stage is exactly 1. Output buffer A2 ensures that R3–C3 are not affected by external loading; if the following stage has a high impedance, then A2 can be omitted. If R1, R2 are set to 1 k$\Omega$, then C1 and C2 are non-preferred values; they are in a ratio of 4:1, giving the $Q$ of 1. Alternatively, C2 can be set to the preferred value of 100 nF, which gives C1 = 400 nF, perhaps made up of 4 × 100 nF in parallel, and then R1 = R2 = 795.77 $\Omega$. In both cases R3 is accepted as an awkward value, so C3 can be the preferred value of 100 nF.

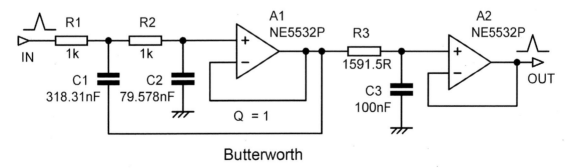

**Figure 8.5: Third-order Sallen & Key 1 kHz lowpass Butterworth filter implemented in two stages. Cutoff frequency 1 kHz.**

**Table 8.8: Third-order two-stage Sallen & Key lowpass: component values for various filter types.**

| Type | R1 = R2 Ω | C1 nF | C2 nF | R3 Ω | C3 nF | Cutoff dB |
|---|---|---|---|---|---|---|
| Linkwitz-Riley | 1125.49 | 200 | 100 | 1591.5 | 100 | −6 |
| Bessel | 792.91 | 190.99 | 100 | 1199.36 | 100 | −3 |
| Linear-Phase 5% delay ripple | 1431.61 | 165.06 | 47 | 2008.90 | 100 | −3 |
| Butterworth | 795.77 | 400 | 100 | 1591.5 | 100 | −3 |
| 1 dB-Chebyshev | 841.62 | 765.29 | 47 | 3220.5 | 100 | −1 |
| 2 dB-Chebyshev | 704.92 | 1224.0 | 47 | 4314.2 | 100 | −2 |
| 3 dB-Chebyshev | 602.52 | 1769.34 | 47 | 5330.0 | 100 | −3 |

The order of the two stages can be reversed without affecting the frequency response, but it does affect the clipping behaviour. Since the A1 stage has a $Q$ of 1, greater than 0.7071, it has a gain that peaks at greater than unity. With $Q = 1$ there is +1.2 dB of peaking around the roll-off, which commensurately reduces the headroom. If the order of the stages is reversed, putting the 1st-order stage first, its roll-off reduces the signal amplitudes before they reach the 2nd-order stage, and there is no restriction on headroom. I call this the New Order, and it will crop up several times as we examine filter technology.

Component values for the other types of two-stage 3rd-order lowpass filter are given in Table 8.8. C2 and C3 have been selected as preferred values, giving non-preferred values for C1 and all the resistors. The Butterworth version shown in the table will be cheaper to implement than that in Figure 8.5, which uses two non-preferred capacitors rather than one. C2 has been adjusted as required to prevent C1 from becoming too large. The amount of gain peaking is different for each type, increasing as the filter moves from Linkwitz-Riley to a 3 dB-Chebyshev characteristic.

The component values were calculated using the tables of stage frequency and $Q$ given in Chapter 7, and checked by simulation. The design process depends on what definition of cutoff attenuation is used; throughout this book I have employed the most common definition, set out in the rightmost column of the table. The linear-phase filter has a 5% group delay ripple. The maximum $Q$ of 3.07 occurs in the 2nd-order stage of the 3 dB-Chebyshev filter; note the very large ratio between C1 and C2.

The cutoff frequency component sensitivities for the Butterworth version are unity in the 1st-order stage and very close to unity in the 2nd-order stage, with the exception of C2, which has a sensitivity of 1.33.

## Sallen & Key 3rd-Order Lowpass: Single Stage

So far we have created first and 2nd-order responses by using a single stage containing a single opamp. It is also possible to create 3rd- and higher-order responses in a single stage, though I should warn at once that the component sensitivities are worse. Figure 8.6a shows a single-stage 3rd-order Butterworth lowpass Sallen & Key filter with unity gain. This is known as the Geffe configuration, first presented in [3]. There are no simple design equations for this configuration; setting the pole positions to create a complex pole pair plus a single pole requires solving three simultaneous equations; this is best done by computer. For more information on the rather complicated process see [4]. Single-stage filters have the advantage that there can be no internal gain peaking to compromise headroom. Also, because in a single-stage filter all of the filtering takes place before the amplifier, this allows large input voltages at frequencies where the filtering action reduces the output amplitude.

Table 8.9 gives exact component values for 3rd-order unity-gain single-stage filters of various types. Note that for the Chebyshev filters, as the passband ripple increases, the capacitor ratios get very large; it is probably not a good idea to make Chebyshev filters as a single stage. The Linkwitz-Riley values look out of line with the others because the cutoff frequency is specified at −6 dB rather than −3 dB as for Bessel, linear phase, and Butterworth.

The practical way to use these single-stage filters, and also the more complex ones that follow, is to take one of the examples given in this section and scale the resistor and capacitor values to get the cutoff frequency you want, bearing in mind that the resistor values must not be too high (or there will be excess noise) or too low (because there will be excess distortion due to unduly heavy opamp loading). Filter frequency scaling is described earlier in this chapter, and there is more guidance on this issue in Chapter 16 on the use of opamps.

As for the 2nd-order stages, the filter design can be done in two different ways; with a unity-gain buffer or by using voltage amplification. By altering the capacitor ratios one can obtain any 3rd-order characteristic with any gain. Figure 8.6a shows a unity-gain lowpass filter; starting with R = 1 kΩ, we find that there are no simple capacitor ratios, and C2 is inconveniently big at 562 nF, more than twice the size of C1. Figure 8.6b shows an alternative circuit with a gain of 1.5 times, which gives the same Butterworth response as Figure 8.6a. As usual, this gain may be an embarrassment rather than a help in a crossover filter.

The interesting thing about Figure 8.6b is that by using a gain of 1.5 times, we find that C2 is now smaller than C1, and a much more manageable 188.4 nF. This led me to ponder that there must be an intermediate value of gain which would give equal values for C1 and C2; this would be very convenient. There is indeed, and a good deal more mathematics leads to Figure 8.6c, where a gain of 1.26 times yields values for C1 and C2 that are very close to equality. Figure 8.6d shows the circuit

**Table 8.9: Component values for 3rd-order unity-gain single-stage S&K lowpass filters with 1 kHz cutoff.**

| Type | R1 = R2 = R3 Ω | C1 | C2 | C3 |
|---|---|---|---|---|
| Linkwitz-Riley | 1 k | 215 nF | 382 nF | 54.90 nF |
| Bessel | 1 k | 155.4 nF | 224.0 nF | 39.95 nF |
| Linear-Phase 5% delay ripple | 1 k | 237 nF | 402 nF | 33.51 nF |
| Butterworth | 1 k | 221 nF | 564 nF | 32.22 nF |
| 0.5 dB-Chebyshev | 1 k | 304.9 nF | 1523 nF | 12.13 nF |
| 1.0 dB-Chebyshev | 1 k | 373 nF | 2353 nF | 9.346 nF |
| 2.0 dB-Chebyshev | 1 k | 479.7 nF | 4287 nF | 5.997 nF |
| 3.0 dB-Chebyshev | 1 k | 577.5 nF | 6910 nF | 4.031 nF |

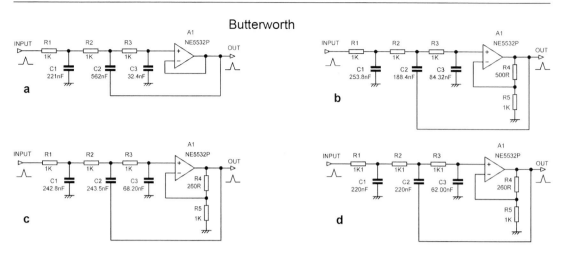

**Figure 8.6: Third-order Sallen & Key 1 kHz lowpass Butterworth filters implemented as a single stage (a) unity gain; (b) with gain = 1.5; (c) equal-C with gain = 1.26; (d) equal-C with preferred values, gain = 1.26. Cutoff frequency 1 kHz for all.**

with R and C values scaled to give C1 = C2 = 220 nF. By happy coincidence R1 = R2 comes out at the E24 value of 1.1k and gives a very accurate Butterworth response. To the best of my knowledge this way of using identical capacitors in a 3rd-order single-stage filter is a new idea.

Since there are no more degrees of freedom to explore, we are in general stuck with an awkward value for C3; here, by happy coincidence again, it comes out exactly as an E24 value. Capacitor series with E24 values are however not common, so we would use two E6 capacitors in parallel thus: 47 nF + 15 nF = 62 nF; yet another happy coincidence. I'm starting to think there is something spooky about this circuit.

Note that it is the two largest capacitors we have made equal and preferred values, leaving the much smaller C3 to be made up of a parallel combination; the smaller capacitors required will be significantly cheaper.

I warned you that the component sensitivities would be worse, and the living proof is in Table 8.10 and Table 8.11. I have only been able to calculate the cutoff frequency sensitivity. A difficulty is that while a 2nd-order filter is completely described by cutoff frequency and Q, a 3rd-order filter is going to have a third sensitivity parameter (Thirdiness ??).

The same process can be used to generate any 3rd-order filter characteristic with equal-C values by choosing the correct gain. Figure 8.7 shows two examples. Figure 8.7a shows a unity-gain 3rd-order lowpass Bessel filter, and Figure 8.7b a 3rd-order Bessel filter with two equal capacitors. Figures 8.7c and Figure 8.7d show the same two variations for a 1 dB-Chebyshev lowpass filter.

Table 8.12 shows the ratio between the two equal capacitors C1, C2 and the remaining capacitor C3, and the gain A required for a wider range of filter types. The linear-phase filter has a 5% group delay ripple.

**Table 8.10:** Third-order Butterworth Sallen & Key single-stage (unity-gain) lowpass component sensitivities.

| Component | Cutoff frequency |
|-----------|------------------|
| R1 | 0.9974 |
| R2 | 0.9994 |
| R3 | 1.2180 |
| C1 | 0.9943 |
| C2 | 1.1098 |
| C3 | 1.3445 |

**Table 8.11:** Third-order Butterworth Sallen & Key equal-C single-stage (gain = 1.26x) lowpass component sensitivities.

| Component | Cutoff frequency |
|-----------|------------------|
| R1 | 0.9979 |
| R2 | 0.9984 |
| R3 | 1.3357 |
| C1 | 0.9946 |
| C2 | 1.0050 |
| C3 | 1.7646 |
| R4 | 0.9200 |
| R5 | 0.9186 |

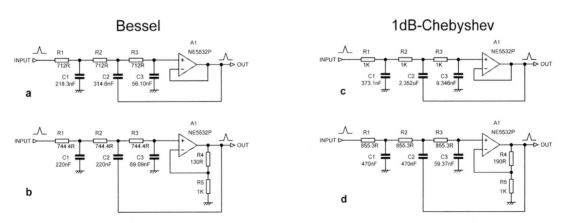

**Figure 8.7:** Third-order Sallen & Key 1 kHz lowpass filters implemented as a single stage: (a) unity-gain Bessel; (b) equal-C Bessel with gain = 1.13 and preferred C values; (c) unity-gain 1 dB-Chebyshev; (d) equal-C 1 dB-Chebyshev with gain = 1.19 and preferred C values. Cutoff frequency 1 kHz for all.

**Table 8.12: Capacitor ratios and gains for equal-C 3rd-order single-stage filters, and cutoff-frequency sensitivity.**

| Type | C1/C3 ratio | Gain A | C3 sensitivity for cutoff freq |
|---|---|---|---|
| Linkwitz-Riley | 0.326 | 1.16 | 1.377 |
| Bessel | 0.314 | 1.13 | 1.339 |
| Linear-Phase 5% delay ripple | 0.210 | 1.14 | 1.544 |
| Butterworth | 0.282 | 126 | 1.765 |
| 0.5 dB-Chebyshev | 0.157 | 1.22 | 2.453 |
| 1.0 dB-Chebyshev | 0.126 | 1.19 | 2.682 |
| 2.0 dB-Chebyshev | 0.0949 | 1.15 | 3.072 |
| 3.0 dB-Chebyshev | 0.0633 | 1.123 | 3.331 |

In every case in Table 8.12 the worst sensitivity for frequency is shown by C3. The values are tabulated in the last column, where it can be seen that as expected, the response demanding the highest $Q$ has the worst sensitivity, and that sensitivity is noticeably worse than for the normal method of making a 3rd-order filter by cascading suitable 2nd-order and 1st-order filters, as indicated earlier. For the Butterworth response, the worst capacitor sensitivity for frequency is 1.765 rather than the 1.33 we saw for the 3rd-order two-stage filter; nothing to panic about but something to keep an eye on. In a given application you might have to use a 1% capacitor instead of a 2% part to achieve a given accuracy, and that may undo the cost saving of using only one amplifier. This is the sort of situation you must expect when higher-order filters are implemented in a single-stage.

This means that the single-stage approach is not very suitable for really precise crossover filters, but it is not without value. There are some applications, such as subsonic or ultrasonic filtering, where a very accurate amplitude response is not essential. A 3rd-order filter ensures there is very little roll-off in the passband, so minor variations in it are of negligible importance. A good example of the method is its use in the combined subsonic-ultrasonic filter described later in this chapter.

## Sallen & Key 3rd-Order Lowpass in a Single Stage: Non-Equal Resistors

Earlier in this chapter we noted that there was little to be gained by using non-equal resistor values in a 2nd-order Sallen & Key stage. However, it can be useful in a 3rd-order stage. As we just saw, two of the capacitors in such a stage can be made equal by suitable choice of the amplifier gain. If we allow non-equal resistors, the extra degrees of freedom allow us to design 3rd-order stages with *three* equal capacitors. Figure 8.8 shows Butterworth and Bessel versions; note that the required amplifier gain required is relatively high at 2.2 times (+ 6.8 dB), which is rarely likely to be convenient in crossover design. Chebyshev filters of this type can also be designed.

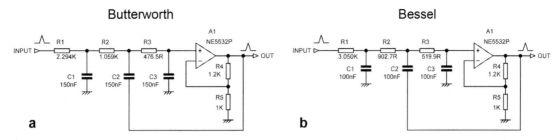

**Figure 8.8: Third-order Sallen & Key 1 kHz lowpass filters implemented as a single stage, with three equal capacitors: (a) Butterworth with gain = 2.2 and exact values; (b) Bessel with gain = 2.2 and exact values.**

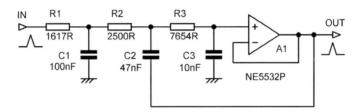

**Figure 8.9: Third-order Sallen & Key Bessel 1 kHz lowpass filter implemented in a single stage.**

Once again there are no manageable design equations, and you should scale the component values given, which are as always for a 1 kHz cutoff, to get the filter you want.

Figure 8.8 also shows that the capacitor values are in general lower than for the previous filters—150 and 100 nF instead of 220 nF. This could give a significant saving when buying close-tolerance capacitors; another cost benefit is that three, and three only, capacitors are required, as it is never necessary to parallel components to achieve awkward capacitance values. Note that the resistor values range over a ratio of 4.81.

It is also possible to design 3rd-order single-stage filters with non-equal resistors to have unity gain, which is usually preferred for crossover work. The design in Figure 8.9 not only achieves this but uses the extra degrees of freedom in the non-equal resistors to make all three capacitors preferred values.

This very clever design is derived by impedance and frequency scaling from a filter published by Tietze and Schenk. [5] To them goes the credit.

## Sallen & Key 4th-Order Lowpass: Two Stages

A 4th-order filter is usually built by cascading two 2nd-order stages, as shown in Figure 8.10 designed for 1 kΩ resistors. For the Butterworth version the two cutoff frequencies are the same, but one stage

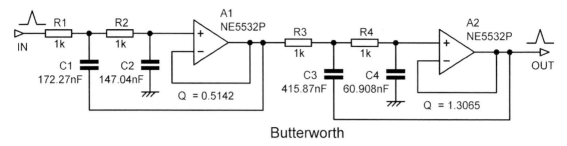

Butterworth

**Figure 8.10: Fourth-order Sallen & Key Butterworth 1 kHz lowpass filter implemented in two stages.**

**Table 8.13: Fourth-order two-stage Sallen & Key lowpass: component values for various filter types. Cutoff 1 kHz.**

| Type | R1 = R2 Ω | C1 nF | C2 nF | R3 = R4 Ω | C3 nF | C4 nF | Cutoff dB |
|---|---|---|---|---|---|---|---|
| Linkwitz-Riley | 1125.41 | 200 | 100 | 1125.41 | 200 | 100 | −6 |
| Bessel | 1074.2 | 108.99 | 100 | 620.48 | 259.85 | 100 | −3 |
| Linear-Phase 5% delay ripple | 964.38 | 325.64 | 47 | 1747.73 | 147.57 | 100 | −3 |
| Butterworth | 1470.39 | 117.16 | 100 | 609.09 | 682.78 | 100 | −3 |
| 1 dB-Chebyshev | 1918.98 | 246.18 | 100 | 2250.63 | 506.94 | 10 | 0 |
| 2 dB-Chebyshev | 1819.00 | 345.51 | 100 | 1797.53 | 844.16 | 10 | 0 |
| 3 dB-Chebyshev | 1670.19 | 463.54 | 100 | 1501.51 | 1244.1 | 10 | 0 |

has a $Q$ of 0.5142 and the other a $Q$ of 1.3065. The latter is quite a high $Q$ for a Sallen & Key filter; since the $Q$ is set by the ratio of the two capacitors, C3 is large, being 6.8 times bigger than C4. Alternatively C2 and C4 can be made preferred values, giving non-preferred values for the resistors.

The first stage has a $Q$ less than 0.7071 (1/√2) and so shows no gain peaking. The second stage has a peak of +3.0 dB at 847 Hz; with the stage order shown this is not an issue, as the first stage attenuates the signal before it reaches the second stage. If the stage order were reversed there would be a serious loss of headroom in the region of 500 Hz–1 kHz.

Component values for the common types of two-stage 4th-order lowpass filter are given in Table 8.13. C2 and C4 have been selected as preferred values, giving non-preferred values for the resistors, C1, and C3. C4 has been adjusted as required to prevent C3 from becoming too large. The amount of gain peaking is different for each type, increasing as the filter moves from Linkwitz-Riley to a 3 dB-Chebyshev characteristic. The linear-phase filter has a 5% group delay ripple.

Note that even-order Chebyshev filters always pass through 0 dB at the cutoff frequency.

The component values were calculated using the tables of stage frequency and $Q$ given in Chapter 7, and checked by simulation. The design process depends on what definition of cutoff attenuation is

**Table 8.14: Fourth-order two-stage Sallen & Key Butterworth lowpass cutoff frequency sensitivities.**

| Component | Cutoff frequency sensitivity |
|:---:|:---:|
| R1 | 0.9979 |
| C1 | 0.9924 |
| R2 | 0.9965 |
| C2 | 1.1408 |
| R3 | 1.1764 |
| C3 | 1.0582 |
| R4 | 1.2400 |
| C4 | 1.4720 |

used. Throughout this book I have employed the most common definition, set out in the rightmost column of Table 8.13.

The maximum $Q$ of 5.58 occurs in the second stage of the 3 dB-Chebyshev filter. This is normally considered too high for practical use, as the component sensitivity of the Sallen & Key configuration becomes high at high Qs; the component values are given here for comparison purposes. When high-$Q$ stages are required, other filter configurations are used which require more amplifiers but have lower component sensitivities, such as the Tow-Thomas three-amplifier configuration described in Chapter 10.

The cutoff frequency sensitivities for the Butterworth version (Figure 8.10) are given in Table 8.14. The worst sensitivity of 1.47 occurs for the second capacitor in the second (higher Q) stage. Note it is somewhat worse than the worst sensitivity of 1.33 shown by the 3rd-order two-stage Butterworth, and this is to be expected as filter order increases.

## Sallen & Key 4th-Order Lowpass: Single-Stage Butterworth

Having seen how it is possible—if not always advisable—to make a 3rd-order filter block using a single opamp, one's mind naturally turns to pondering if a 4th-order filter block can be constructed in the same kind of way. The answer is yes, with some reservations about the gain of the stage. There are no simple design equations for this configuration; calculating the component values involves solving simultaneous complex equations, and this really has to be done by computer. To the best of my knowledge there is no practical way of doing it by hand.

As before, you can make any filter characteristic, but not with unity gain, so far as I can determine. Attempts to do it founder because as the gain approaches unity, C3 becomes infinitely larger than C4. With a gain of 1.01 times, C3 is 244 times larger than C4. This is a great pity because very often unity gain is what you want; any significant gain can cause headroom problems. It is however possible to design practical filters with a relatively low gain of only 1.09 times (+0.75 dB). which it should be possible to fit into any crossover gain structure without significant compromise on either noise or headroom. Figure 8.11 shows a 1 kHz lowpass version.

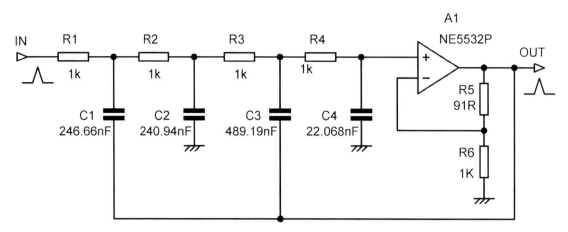

**Figure 8.11: Fourth-order Sallen & Key Butterworth 1 kHz lowpass filters
implemented as a single stage. Gain = 1.091 times.**

Two further examples with non-equal resistors are shown in Figure 8.12. In Figure 8.12a the gain
has been chosen to make C1 and C2 exactly 220 nF for convenience; C3 and C4 just have to fall
as they may, and C3 in particular will have to be made up of two or more capacitors. Clearly
220 nF +100 nF +100 nF is extremely close. However, since the cost of capacitors seems to go
up very roughly as a square root law (see Chapter 15), the best economy results if the biggest
capacitance is a single preferred-value component. We can set C3 to 470 nF by scaling the
resistors R1 to R4 so that C3 becomes 470 nF. The gain is unchanged so C1 still equals C2, but
their value is less convenient; see Figure 8.12b. To make this sort of design decision properly you
need to be working with actual prices from your intended supplier; there is more on capacitor
costs in Chapter 15.

If more gain is acceptable, then it is possible to come up with designs that keep R1 to R4 the same
value (you'll note they don't differ by very much in Figure 8.12), though awkward resistor values are
in general much less of a problem than awkward capacitors. You can at the same time make C1 equal
to C3 for convenience. Figure 8.13a shows the exact values for R = 1 kΩ and the usual 1 kHz cutoff
frequency. The obvious step is to make C1 and C3 the nearest value E3 of 220 nF and scale the other
components accordingly to keep the cutoff frequency the same. This gives us Figure 8.13b, where the
capacitors are kept in the same ratio and the resistor values adjusted. Note that more gain is required
(1.43x = +3.1 dB) and this is unlikely to be convenient.

The component sensitivities for the filter in Figure 8.13b are shown in Table 8.15; for example a 1%
change in R1 causes a 1.14% change in the cutoff frequency. They are not as bad as might be feared,
because of the relatively low $Q$'s involved in a 4th-order Butterworth characteristic. Even so, the worst
is 3.05 for C4, which is six times worse than a 2nd-order Butterworth stage, and C3 is not much better
at 2.83. If you have the option of using some precision capacitors, this is the place to put them. Note
also R4 with a sensitivity of 2.13.

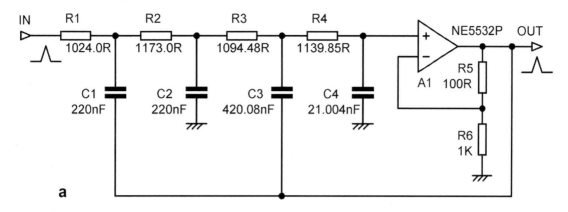

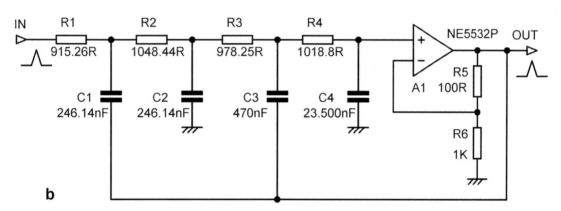

**Figure 8.12: Fourth-order Sallen & Key Butterworth 1 kHz lowpass filters implemented as a single stage (a) with C1 = C2 = 220 nF; (b) with C3 = 470 nF; both with gain = 1.10 times.**

It was earlier mentioned in passing that a disadvantage of the Sallen & Key lowpass configuration is that the frequency response at very high frequencies would start to come back up, because one of the shunt capacitors is terminated at the opamp output, which has a non-zero output impedance that rises with frequency. This effect is likely to be more of a problem with a 4th-order single-stage lowpass filter, because there are now two capacitors terminated at the opamp output.

Take the Butterworth filter in Figure 8.13b; if unwisely implemented with an elderly TL072 it will be found that the response has risen above the theoretical characteristic by 1 dB as low as 4.6 kHz. Admittedly this is at a theoretical attenuation of 52 dB, so will probably be of no consequence in a crossover, but it is something to keep an eye on. This difficulty is eliminated effectively by using opamps like the 5532 that can maintain a low output impedance up to considerably higher frequencies. The TL072 really is obsolete for almost all audio purposes and is only mentioned here to illustrate what can go wrong if you use it in a lowpass filter.

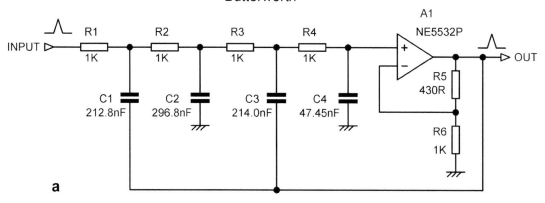

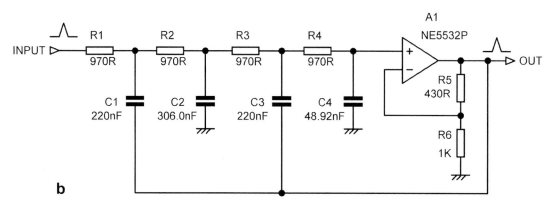

Figure 8.13: Fourth-order Sallen & Key Butterworth 1 kHz lowpass filters implemented as a single stage (a) C1 = C3 (very nearly) with gain = 1.43 and exact values; (b) C1 = C3 (preferred value of 220 nF) with gain = 1.43.

Table 8.15: Fourth-order single-stage Sallen & Key equal-C Butterworth lowpass cutoff frequency sensitivities.

| Component | Cutoff frequency sensitivity |
|---|---|
| R1 | 1.14 |
| R2 | 1.25 |
| R3 | 1.54 |
| R4 | 2.13 |
| C1 | 1.04 |
| C2 | 2.29 |
| C3 | 2.83 |
| C4 | 3.05 |
| R5 | 2.99 |
| R6 | 2.98 |

## Sallen & Key 4th-Order Lowpass: Single-Stage Linkwitz-Riley

By far the most popular application of 4th-order filters is in Linkwitz-Riley crossovers, so a 4th-order Linkwitz-Riley filter in a single stage would be a very useful item. It would be possible to make a complete 2-way 4th-order active crossover using just one dual opamp per channel. Figure 8.14 shows an example; the gain is set to 1.13 times (+1.06 dB) so that C1 and C2 conveniently have identical E3 preferred values. The very low gain minimises noise/headroom compromises. Resistors R1 to R4 have the same value, temptingly close to the preferred E24 value of 1.1 kΩ; the error is however 2%, a bit high for precision work. However, for little more cost each resistor can be replaced by a 2xE24 parallel pair; 18 kΩ in parallel with 30 kΩ is only 0.24% high, and the effective tolerance is improved to 0.73% if 1% resistors are used. This approach is demonstrated in the practical example shown in Figure 8.15. The example assumes you want a crossover frequency of exactly 1 kHz; for other frequencies the value of R1 to R4 must be scaled before converting them to the 2xE24 format, as described earlier in this chapter and in Chapter 15.

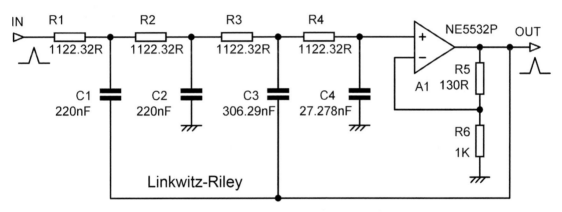

**Figure 8.14: Fourth-order Sallen & Key Linkwitz-Riley 1 kHz lowpass filter. Gain = 1.13x (+1.06 dB).**

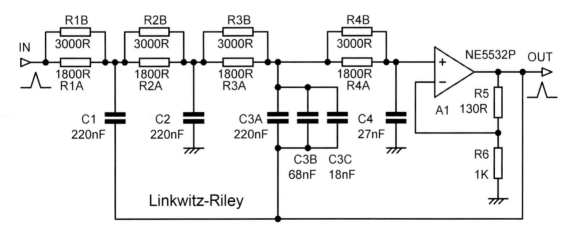

**Figure 8.15: Practical version of Figure 8.14 with 2xE24 resistors and E12 series capacitors.**

In the practical filter of Figure 8.15, C3 is made up of 220 nF + 68 nF + 18 nF = 306 nF, which assumes your capacitors are available in the E12 series. By pure luck this comes out very well, being only 0.093% low. C4 is simply made to be 27 nF; this is 1.03% low but, this makes a negligible difference to the response when it is compared with a conventional two-stage 4th-order Linkwitz-Riley filter as described in Table 8.13, despite the fact that C4 shows the highest component sensitivity.

The component sensitivities for Figure 8.14 and Figure 8.15 are shown in Table 8.16; you will note that they are rather more favourable than for those for the single-stage Butterworth filter in Table 8.14 because of the lower $Q$'s in a Linkwitz-Riley filter. Things look pretty good, with only C2, C4, and R4 needing a mild eye to be kept on them.

Alternatively, it may be more cost-effective to make C1 = C3. Figure 8.16a shows the precise values that emerge when R1–R4 is 1 kΩ and the gain is adjusted to give the desired filter characteristic, while this time making C1 and C3 very nearly equal. Figure 8.16b shows the result of scaling the circuit values of Figure 8.16a to make C1 = C3 = 220 nF, while keeping the same 1 kHz cutoff frequency. Resistors R1–R4 now have the awkward value of 1.016 kΩ. This can be obtained using the 2xE24 combination of 1300 Ω and 4700 Ω in parallel, which has a nominal value only 0.23% low. A gain of 1.33x (+2.5 dB) is now required, which is likely to be less convenient. The economics of capacitor choice in the 4th-order single-stage circuit are explored in more detail in the previous section on the Butterworth version of this filter.

The overall response of the circuit is exactly the same as two cascaded 2nd-order Butterworth filters with the same cutoff frequency, which is the usual way of making a 4th-order Linkwitz-Riley filter. We have saved the cost of an opamp and probably reduced noise and distortion somewhat due to its absence, but the extra series resistances (four instead of two) may cause trouble with increased common-mode distortion. See Chapter 11 on filter performance.

**Table 8.16: Fourth-order single-stage Sallen & Key Linkwitz-Riley lowpass cutoff frequency sensitivities. Gain = 1.13x.**

| Component | Cutoff frequency sensitivity |
|---|---|
| R1 | 1.0153 |
| R2 | 0.9929 |
| R3 | 1.0246 |
| R4 | 1.4632 |
| C1 | 0.8850 |
| C2 | 1.5944 |
| C3 | 1.1321 |
| C4 | 1.7355 |
| R5 | 1.0105 |
| R6 | 1.0087 |

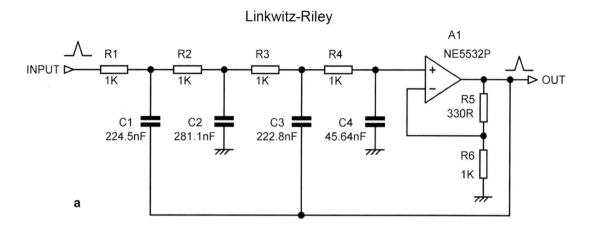

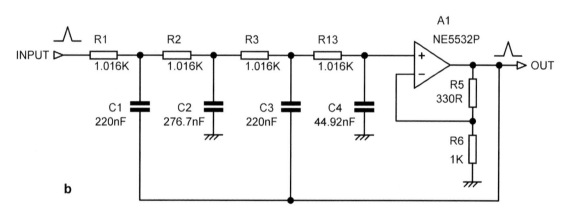

**Figure 8.16: Fourth-order Sallen & Key Linkwitz-Riley 1 kHz lowpass filters implemented as a single stage: (a) equal-C with gain = 1.33 and exact capacitor values; (b) equal-C with gain = 1.33 and preferred values for C1 and C3.**

The component sensitivities for the filter in Figure 8.16b are shown in Table 8.17; they are slightly better than for the 4th-order Butterworth because of the lower $Q$'s involved in a Linkwitz-Riley characteristic but worse than the low-gain Linkwitz-Riley of Figure 8.14 and Figure 8.15. The worst, for C4 again, is 2.15, which is now only about four times worse than a 2nd-order Butterworth stage, and with appropriate capacitor sourcing may make this approach a viable alternative. However, overall it may not be the optimal strategy; we have saved an opamp but need more precise capacitors to achieve the same accuracy, and the net cost may be greater.

The loading effects of these filters on the preceding stage is much as for the 2nd- and 3rd-order Sallen & Key filters. At low frequencies the input impedance is high because the capacitors are effectively open circuits. As the frequency rises to the cutoff point, the input impedance falls until it is

**Table 8.17: Fourth-order single-stage Sallen & Key equal-C Linkwitz-Riley lowpass cutoff frequency sensitivities.**

| Component | Cutoff frequency |
|:---:|:---:|
| R1 | 1.02 |
| R2 | 1.02 |
| R3 | 1.02 |
| R4 | 1.60 |
| C1 | 0.88 |
| C2 | 1.79 |
| C3 | 1.64 |
| C4 | 2.15 |
| R5 | 1.77 |
| R6 | 1.77 |

only slightly greater than the series input resistor R1. Above that the input impedance rises again by about 10% and then stays flat with increasing frequency.

## Sallen & Key 4th-Order Lowpass: Single Stage With Non-Equal Resistors

Earlier in this chapter we saw that by using non-equal resistors, it was possible to make a filter with three equal capacitors. This principle can be extended to 4th-order filters, as shown in Figure 8.17. The required gain is still 2.2 times, but note that the resistors now range over a ratio of 13.5 instead of 4.81.

These circuits also have potentially worse problems with the response coming back up again in the audio band. If you are living in the past with TL072s, the use of the 5534/2 instead prevents any problems.

## Sallen & Key 4th-Order Lowpass: Single Stage With Other Filter Characteristics

The component values for other filter characteristics such as Bessel and Chebyshev are laid out in Table 8.18, together with a summary of the Butterworth and Linkwitz-Riley options given earlier. Using a gain of 1.090 times (+0.75 dB) gives reasonable capacitor values with only tiny compromises on noise and headroom. For the filters that incorporate higher $Q$'s, such as the Chebyshev versions, the capacitor ratios become unwieldy, and this is probably not a good route to take; a 2-stage implementation may require another amplifier, but this will almost certainly be outweighed by lower capacitor costs.

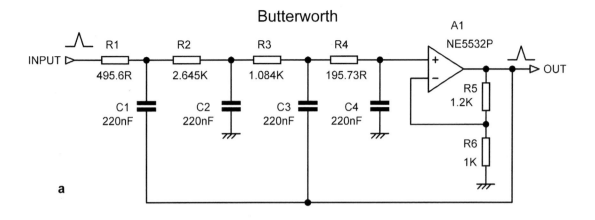

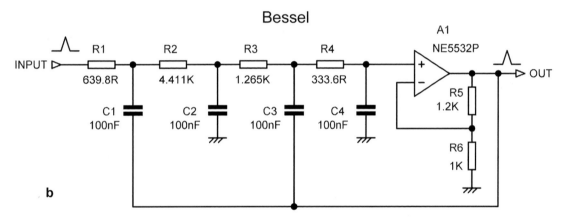

**Figure 8.17: Fourth-order Sallen & Key 1 kHz lowpass filters implemented as a single stage, with four equal capacitors, gain = 2.2 and exact values: (a) Butterworth; (b) Bessel.**

## Sallen & Key 5th-Order Lowpass: Three Stages

A 5th-order filter is usually built by cascading two 2nd-order stages and one 1st-order stage. Output buffer A3 ensures that R5-C5 are not affected by external loading; if the following stage has a suitably high impedance, then A3 can be omitted. For the Butterworth version the three cutoff frequencies are the same, but one 2nd-order stage has a $Q$ of 0.620 and the other a $Q$ of 1.620. The latter is an increase on the highest $Q$ in a 4th-order filter (which is 1.3065), and so the ratio C3/C4 is now even greater at 10.5. If C4 was set to 100 nF, then C3 becomes 1.05 uF, which is inconveniently large.

In the 4th-order two-stage lowpass filter, the resistors were set to be 1 kΩ, and the capacitor values came out as whatever they did. This makes the operation of the filter clearer, but in practice it is easier

**Table 8.18: Component values for 4th-order unity-gain single-stage S&K lowpass filters with 1 kHz cutoff. All cutoffs at −3 dB, except Linkwitz-Riley cutoff is at −6 dB, Chebyshev cutoff is at 0 dB.**

| Type | R1 = R2 = R3 = R4 Ω | C1 | C2 | C3 | C4 | Gain x |
|---|---|---|---|---|---|---|
| Linkwitz-Riley | 1 kΩ | 224.5 nF | 281.1 nF | 222.8 nF | 45.64 nF | 1.330 |
| Linkwitz-Riley | 1.016 kΩ | 220 nF | 276.7 nF | 220 nF | 44.92 nF | 1.330 |
| Linkwitz-Riley | 1.12232 kΩ | 220 nF | 220 nF | 306.29 nF | 27.278 nF | 1.130 |
| Bessel | 1 kΩ | 144.79 nF | 165.76 nF | 243.22 nF | 20.774 nF | 1.090 |
| Linear-Phase 5% delay ripple | 1 kΩ | 240.30 nF | 212.16 nF | 376.85 nF | 15.288 nF | 1.090 |
| Butterworth | 1 kΩ | 212.8 nF | 296.8 nF | 214.0 nF | 47.45 nF | 1.430 |
| Butterworth | 970 Ω | 220 nF | 306.0 nF | 220 nF | 48.92 nF | 1.430 |
| Butterworth | 1 kΩ | 246.66 nF | 240.94 nF | 489.1 nF | 22.068 nF | 1.090 |
| 0.5 dB-Chebyshev | 1 kΩ | 471.95 nF | 307.96 nF | 757.07 nF | 15.383 nF | 1.090 |
| 1.0 dB-Chebyshev | 1 kΩ | 577.68 nF | 323.58 nF | 836.66 nF | 14.884 nF | 1.090 |
| 2.0 dB-Chebyshev | 1 kΩ | 740.13 nF | 329.21 nF | 917.96 nF | 13.941 nF | 1.090 |
| 3.0 dB-Chebyshev | 1 kΩ | 880.58 nF | 325.26 nF | 962.51 nF | 13.150 nF | 1.090 |

Values all checked by SPICE simulation.

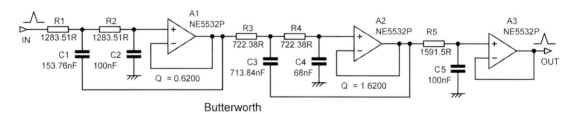

Butterworth

**Figure 8.18: Fifth-order Sallen & Key Butterworth 1 kHz lowpass filter, implemented in three stages.**

and cheaper to use awkward resistor values rather than awkward capacitor values, and so I have demonstrated that approach in this example. By suitable choice of resistor values three out of the five capacitors can be chosen to be preferred values; see Figure 8.18.

C4 was chosen as 68 nF rather than 100 nF to reduce the size of C3, but it is still rather large. The input impedance of a lowpass Sallen & Key filter falls almost to the value of R3 at high frequencies, and care must be taken that it does not excessively load the previous stage. Setting C4 to 68 nF rather than 100 nF reduces C3 from 1.049 uF to 714 nF and also makes R3 and R4 722 Ω rather than 491 Ω, which should reduce the loading to the point where the distortion performance of a good opamp is not compromised. If a lighter loading on A1 is desired, then C4 could be made 47 nF, which gives C3 = 493 nF and R3 = R4 = 1045.15 Ω. Alternatively C3 can be made a preferred value, which is usually more convenient. C4 is small enough to be made up of a number of inexpensive 1%

polystyrene capacitors. Table 8.19 gives some more useful component options for the second stage. With C5 = 100 nF, R5 is large enough to present no loading difficulties for A2.

The first 2nd-order stage has a $Q$ less than 0.7071 (1/√2) and so shows no gain peaking. Clearly the 1st-order third stage cannot show any peaking. The second stage has a substantial gain peak of +4.6 dB at 900 Hz; however, with the stage order shown the headroom loss is only +1.7 dB because of the preliminary attenuation in the first stage. Rearranging the stage order again, so the 1st-order stage comes before the low-$Q$ stage, followed by the high-$Q$ stage, eliminates any headroom loss. I call this the New Order. It may however impact the noise performance, because one advantage of having a 1st-order final stage is that it effectively attenuates the noise from previous stages.

Component values for the common types of two-stage 5th-order lowpass filter are given in Table 8.20. C2, C4, and C5 have been selected as preferred values, giving non-preferred values for all the resistors and for C1 and C3. The value of C4 has been adjusted as required to prevent C3 from becoming too large. The amount of gain peaking is different for each type, increasing as the filter moves from Linkwitz-Riley to a 3 dB-Chebyshev characteristic. The linear-phase filter has a 5% group delay ripple.

The component values were calculated using the tables of stage frequency and $Q$ given in Chapter 7, and checked by simulation. The design process depends on what definition of cutoff attenuation is

**Table 8.19: Options for component values in third stage of a 5th-order Butterworth lowpass filter.**

| C3 nF | C4 nF | R3 = R4 Ω |
|---|---|---|
| 713.84 | 68 | 722.83 |
| 493 | 47 | 1045.15 |
| 440 | 41.914 | 1171.96 |
| 330 | 31.436 | 1562.61 |
| 220 | 20.957 | 2343.92 |

**Table 8.20: Fifth-order three-stage Sallen & Key lowpass: component values for various filter types. Cutoff 1 kHz.**

| Type | R1 = R2 Ω | C1 nF | C2 nF | R3 = R4 Ω | C3 nF | C4 nF | R5 Ω | C5 nF | Cutoff dB |
|---|---|---|---|---|---|---|---|---|---|
| Linkwitz-Riley | 1125.41 | 200 | 100 | 795.77 | 400 | 100 | 1591.55 | 100 | −6 |
| Bessel | 904.62 | 127.01 | 100 | 1049.24 | 157.91 | 47 | 1056.17 | 100 | −3 |
| Linear-Phase 5% delay | 829.67 | 406.33 | 33 | 910.79 | 301.30 | 100 | 1206.3 | 220 | −3 |
| Butterworth | 1283.51 | 153.76 | 100 | 1045.15 | 493.39 | 47 | 1591.55 | 100 | −3 |
| 1 dB-Chebyshev | 868.28 | 782.66 | 100 | 1441.35 | 1233.79 | 10 | 5497.6 | 100 | −1 |
| 2 dB-Chebyshev | 714.97 | 1260.39 | 100 | 1127.61 | 2092.05 | 10 | 7290.4 | 100 | −2 |
| 3 dB-Chebyshev | 606.20 | 1828.42 | 100 | 933.49 | 3105.42 | 10 | 8966.5 | 100 | −3 |

Table 8.21: Fifth-order three-stage Sallen & Key Butterworth lowpass:
cutoff frequency sensitivities.

| Component | Cutoff frequency sensitivity |
|:---:|:---:|
| R1 | 0.9989 |
| C1 | 0.9951 |
| R2 | 0.9984 |
| C2 | 1.1708 |
| R3 | 1.3431 |
| C3 | 1.1713 |
| R4 | 1.3586 |
| C4 | 1.6180 |
| R5 | 1.00 |
| C5 | 1.00 |

used. Throughout this book I have employed the most common definition, set out in the rightmost column of the table.

The maximum $Q$ of 8.81 occurs in the second stage of the 3 dB-Chebyshev filter. This is perhaps a bit high for practical use with the Sallen & Key configuration, as the component sensitivity becomes high at high Qs; on the other hand, in practical measurements I found a Sallen & Key filter with a $Q$ of 10 to be quite tractable and easy to use; see Chapter 11. When really accurate high-$Q$ stages are required, other filter configurations such as the Tow-Thomas can be used, requiring more amplifiers but having lower component sensitivities.

The cutoff frequency sensitivities for the Butterworth version in Figure 8.18 are given in Table 8.21; as before, the worst sensitivity of 1.62 appears for the second capacitor in the 2nd-order stage with the highest $Q$. Comparing this with the worst-case sensitivities of 1.47 for the two-stage 4th-order and 1.33 for the two-stage 3rd-order, it can be seen that the increase is fairly slow.

## Sallen & Key 5th-Order Lowpass: Two Stages

Earlier in this chapter it was shown how to make a 3rd-order filter in one stage, using only one amplifier. The same process can be used to create a two-stage 5th-order filter. The first stage gives a complex pole pair with the low $Q$ of 0.618, plus a single pole; this is equivalent to the combination of the first stage and the last stage in Figure 8.18. The second stage gives another pole pair with the higher $Q$ of 1.618 and does the same job as the second stage in Figure 8.18. The two stages cascaded together give a 5th-order response in the same way as the three-stage 5th-order filter, but do it with two amplifiers rather than three.

In Figure 8.19 the values of R1, R2, R3 have been twiddled to make C3 the preferred value of 68 nF, and R4, R5 have been twiddled to make C5 the preferred value of 47 nF. These values were chosen so

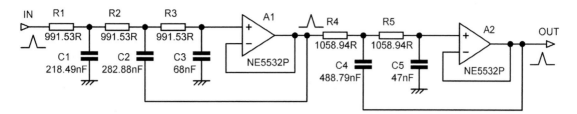

**Figure 8.19: Fifth-order Sallen & Key Butterworth lowpass filter in two stages.
Cutoff −3 dB at 1 kHz.**

that the input impedance of each stage above the cutoff frequency does not fall too low. The resistors and the remaining three capacitors inevitably come out as awkward values.

The 991.53 Ω resistors can be the 2xE24 pair 1800 Ω and 2200 Ω in parallel; the nominal value is only 0.15% low.

The 1058.94 Ω resistors can be the 2xE24 pair 1500 Ω and 3600 Ω in parallel; the nominal value is only 0.01% low.

The response of the first stage is monotonic, i.e. without gain peaking, because of the low $Q$ of its pole pair. The second stage shows a gain peak of +4.9 dB at 884 Hz (for a 1 kHz cutoff). Because of the attenuation in the first stage, the combination rolls off smoothly, and there is no way that the filter can clip internally at A1 before it does so at the output.

I have not been able to calculate component sensitivities for this configuration, but I think it's a fairly safe bet that either C3 or C5 will be the most critical components; probably C5.

## Sallen & Key 5th-Order Lowpass: Single Stage

Until relatively recently I had never seen an attempt to make a 5th-order active filter in one circuit block, and I had a dark suspicion that it might be mathematically as well as physically impossible. However, in the stacks of a proper book-type library I discovered a 1972 paper by Aikens and Kerwin [6] that gives the ratios between the capacitors for Butterworth and Bessel/Thomson filters up to 6th-order. These were derived by solving sets of non-linear equations by computer. In the first edition of this book, I implied that Aitken and Kerwin had got one of the parameters for the 5th-order filter wrong. They hadn't, I had, and I humbly apologise. The 5th-order lowpass is shown in Figure 8.20, which definitely gives a clean roll-off and a 30 dB/octave ultimate slope, and generally behaves as it should.

Attempting to calculate the sensitivities of these single stage circuits is unlikely to be a rewarding experience, and I have not undertaken it. What you can always do, of course, is to simulate the filter and try out the effects of altering one component at a time by a small amount, such as 1%. The effect on the cutoff frequency is then easy to assess, though the effect on other filter parameters such as $Q$ is rather harder to determine; you will have to examine the frequency response and see which of its features are of most interest to you. In many cases 1% is too small to give a change that can be

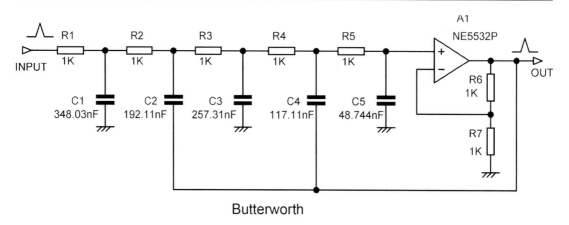

Butterworth

**Figure 8.20: Fifth-order Sallen & Key Butterworth lowpass filter, implemented as a single stage, with gain = 2 and exact values. Cutoff frequency 1 kHz.**

**Table 8.22: Fifth-order single-stage Sallen & Key Butterworth lowpass cutoff frequency sensitivities: see Figure 8.20.**

| Component | Cutoff frequency sensitivity | Component | Cutoff frequency sensitivity |
|---|---|---|---|
| R1 | −0.6 | C1 | −0.19 |
| R2 | −0.21 | C2 | −0.58 |
| R3 | −1.34 | C3 | −3.7 |
| R4 | +1.45 | C4 | +1.4 |
| R5 | −0.73 | C5 | −1.7 |

measured accurately; I used 10% when I applied this method to the 5th-order circuit in Figure 8.20, and the results are shown in Table 8.22; C3 is by far the worst for sensitivity. Please bear in mind that these figures are read from a cursor on a graph, and their accuracy is thereby limited.

## Sallen & Key 6th-Order Lowpass: Three Stages

A 6th-order lowpass filter is usually constructed by cascading three 2nd-order lowpass stages, as in Figure 8.21. For the Butterworth version the three cutoff frequencies are all the same, but the first 2nd-order stage has a $Q$ of 0.517, the second a $Q$ of 0.7071 (1/√2), and the final one a $Q$ of 1.932. The last is an increase on the highest $Q$ in a three-stage 5th-order filter (which is 1.620), and so the ratio C5/C6 is now even larger at 14.9. If C6 was set to 100 nF, then C3 would be1.49 uF, which is inconveniently large; R5 and R6 would also be undesirably small. As it is, 701 nF is still a hefty size. You might want to explore making C6 even smaller and setting C5 to a preferred value or multiple thereof. For example, if C5 is 440 nF (= 2 × 220 nF), then C6 becomes 29.469 nF, and R5 = R6 = 1397.67 Ω. C6 is

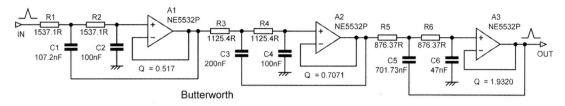

**Figure 8.21: Sixth-order Sallen & Key Butterworth lowpass filter, implemented in three stages. Cutoff frequency 1 kHz.**

**Table 8.23: Options for component values in third stage of 6th-order Butterworth lowpass filter.**

| C5 nF | C6 nF | R5 = R6 $\Omega$ |
|-------|--------|------------------|
| 701.73 | 47 | 876.37 |
| 440 | 29.469 | 1397.67 |
| 330 | 22.102 | 1863.56 |
| 220 | 14.735 | 2795.34 |

small enough to be made up of a number of inexpensive 1% polystyrene capacitors. Table 8.23 gives some additional useful options for the third stage.

There is no gain peaking in either the first or second stage, but significant peaking of +6.02 dB in the third stage. With the stage order shown this is not an issue, as the first low-$Q$ stage attenuates the signal before it reaches the second stage. If the third stage came first there would be a serious loss of headroom in the region of 500–1500 kHz.

Component values for the common types of three-stage 6th-order lowpass filter are given in Table 8.24. C2, C4 and C6 have been selected as preferred values, giving non-preferred values for all the resistors and for C1, C3, and C5. C6 has been adjusted as required to prevent C5 from becoming too large. The amount of gain peaking is different for each type, increasing as the filter moves from Linkwitz-Riley to a 3 dB-Chebyshev characteristic. The linear-phase filter has a 5% group delay ripple.

The component values were calculated using the tables of stage frequency and $Q$ given in Chapter 7, and checked by simulation. The design process depends on what definition of cutoff attenuation is used. Throughout this book I have employed the most common definition, set out in the rightmost column of the table.

The maximum $Q$ of 12.79 occurs in the third stage of the 3 dB-Chebyshev filter. This may be a bit high for practical use, as the component sensitivity of the Sallen & Key configuration becomes high at high Qs; the component values are given here for comparison purposes. When high-$Q$ stages are required, other filter configurations are used which require more amplifiers but have lower component sensitivities; see Chapter 10. Having said that, there is currently no evidence that a 6th-order 3 dB-Chebyshev filter is of any particular use in active crossover design.

**Table 8.24: Sixth-order three-stage Sallen & Key lowpass:
component values for various filter types. Cutoff 1 kHz.**

| Type | R1 = R2 Ω | C1 nF | C2 nF | R3 = R4 Ω | C3 nF | C4 nF | R5 = R6 Ω | C5 nF | C6 nF | Cutoff dB |
|---|---|---|---|---|---|---|---|---|---|---|
| Linkwitz-Riley | 1591.55 | 100 | 100 | 795.77 | 400 | 100 | 795.77 | 400 | 100 | −6 |
| Bessel | 971.00 | 104.16 | 100 | 769.81 | 149.43 | 100 | 867.51 | 196.90 | 47 | −3 |
| Linear-Phase 5% delay ripple | 1028.72 | 147.72 | 100 | 1335.42 | 261.95 | 47 | 1027.86 | 420.90 | 22 | −3 |
| Butterworth | 1537.1 | 107.2 | 100 | 1125.4 | 200 | 100 | 876.37 | 701.73 | 47 | −3 |
| 1 dB-Chebyshev | 2961.41 | 231.53 | 100 | 1031.62 | 908.02 | 47 | 999.27 | 2560.77 | 10 | 0 |
| 2 dB-Chebyshev | 2792.14 | 325.15 | 100 | 1742.02 | 711.92 | 22 | 773.95 | 4377.80 | 10 | 0 |
| 3 dB-Chebyshev | 2557.60 | 436.06 | 100 | 3184.01 | 478.78 | 10 | 636.77 | 636.77 | 10 | 0 |

**Table 8.25: Sixth-order three-stage Sallen & Key Butterworth lowpass:
cutoff frequency sensitivities.**

| Component | Cutoff frequency sensitivity |
|---|---|
| R1 | 0.9976 |
| C1 | 0.9895 |
| R2 | 0.9945 |
| C2 | 1.1316 |
| R3 | 0.9999 |
| C3 | 0.9979 |
| R4 | 1.0008 |
| C4 | 1.2071 |
| R5 | 1.4810 |
| C5 | 1.3037 |
| R6 | 1.5195 |
| C6 | 1.7671 |

The cutoff frequency sensitivities for the Butterworth version in Figure 8.21 are given in Table 8.25; as before the worst sensitivity of 1.77 appears for the second capacitor in the 2nd-order stage with the highest $Q$. This is only a moderate increase on the worst-case sensitivities of 1.62 for the two-stage 5th-order, 1.47 for the two-stage 4th-order, and 1.33 for the two-stage 3rd-order, and suggests this is a practical filter.

## Sallen & Key 6th-Order Lowpass: Single Stage

A 6th-order single-stage Butterworth filter is shown in Figure 8.22. It gives an accurate Butterworth response with a 36 dB/octave ultimate slope. The capacitor ratios were derived from Aikens & Kerwin,

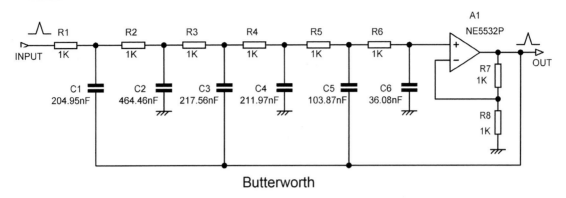

Butterworth

**Figure 8.22: Sixth-order Sallen & Key Butterworth lowpass filter, implemented as single stage, with gain = 2 and exact values. Cutoff frequency 1 kHz.**

**Table 8.26: Sixth-order single-stage Sallen & Key Butterworth lowpass cutoff frequency sensitivities for Figure 8.22.**

| Component | Frequency sensitivity | Component | Frequency sensitivity |
|-----------|-----------------------|-----------|-----------------------|
| R1 | −0.8 | C1 | −0.8 |
| R2 | −0.2 | C2 | +0.4 |
| R3 | −0.6 | C3 | −1.2 |
| R4 | −2.7 | C4 | −4.1 |
| R5 | +2.7 | C5 | +5.2 |
| R6 | −2.2 | C6 | −4.5 |

[6] as for the 5th-order version. The filter has a gain of 2 times, and this may not be convenient. The sensitivity results, obtained in the same way as for the 5th-order single-stage filter, are shown in Table 8.26.

Before I began, I thought that the sensitivities for a single-stage 6th-order would be very bad, and I imagined the value of C6 would be especially critical. In fact C5 has the worst figure, and at 5.2 it is not as awful as I expected. However, compare that with the worst-case sensitivity of only 1.77 in the three-stage 6th-order Butterworth, and note that C4 and C6 also have high sensitivities of more than 4. Also compare the 4th-order Sallen & Key equal-C Butterworth, with a worst-case frequency sensitivity of 3.05.

The worst-case cutoff frequency sensitivities for the various Butterworth filter designs are summarised in Table 8.27.

The 5th- and 6th-order single-stage filters are in some ways clever designs, as no less than two opamp sections have been saved, but the higher sensitivities than multi-stage designs may cause trouble. Attempts to make high-$Q$ single-stage filters could be difficult because some components with high sensitivity would need to be precision parts.

**Table 8.27: Worst-case cutoff frequency sensitivities for the various Butterworth filter designs.**

| Filter order | Number of stages | Worst-case sensitivity |
|---|---|---|
| 1st-order | One | 1.0 |
| 2nd-order | One | −0.5 |
| 3rd-order | Two | 1.33 |
| 3rd-order | One | 1.33 |
| 4th-order | Two | 1.47 |
| 4th-order | One | 3.05 |
| 5th-order | Three | 1.62 |
| 5th-order | Two | n/a |
| 5th-order | One | −3.7 |
| 6th-order | Three | 1.77 |
| 6th-order | One | 5.2 |

I have not attempted to make single-stage filters of higher than 6th-order. I can predict with confidence that the component sensitivities will be yet worse, and so I see no powerful reason to attempt what will almost certainly turn out to be a very complicated procedure.

## Sallen & Key Lowpass: Input Impedance

The input impedance of a Sallen & Key lowpass filter is high at low frequencies, because the shunt capacitor C1 is fully bootstrapped by the unity-gain amplifier, and so it is effectively an open circuit.

At high frequencies, the input impedance of a 2nd-order Sallen & Key lowpass filter approaches the value of R1. This is because, at low frequencies, no signal reaches the opamp because of R2-C2, and so the opamp output is effectively at signal ground, and C1 is connected directly to this output. It therefore loads R1.

## Linkwitz-Riley Lowpass With Sallen & Key Filters: Loading Effects

Here is an example of loading effects and input impedance considerations for Sallen & Key lowpass filters. The most common implementation of the popular 4th-order Linkwitz-Riley filter is made by cascading two 2nd-order Butterworth Sallen & Key filters. This is why the Linkwitz-Riley is sometimes called a "Butterworth-squared" filter. A typical implementation of a 3 kHz Linkwitz-Riley lowpass filter is shown in Figure 8.23, where the resistor values have been reduced as much as appears to be prudent in the quest for low noise. (The resistor values just happen to come out as 797 Ω; this is not some sort of subtle post-modernist reference to the AD797 opamp.) It is therefore necessary to consider carefully the way that the loading from the second filter will affect the first filter that drives it. As noted earlier, the input impedance of a 2nd-order Sallen & Key lowpass filter is high at low

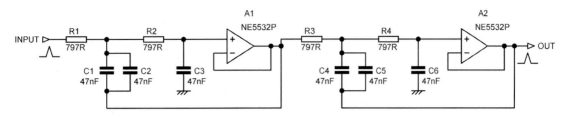

**Figure 8.23: Fourth-order Linkwitz-Riley filter made by cascading two 2nd-order Butterworth Sallen & Key filters. Cutoff is −6 dB at 3 kHz.**

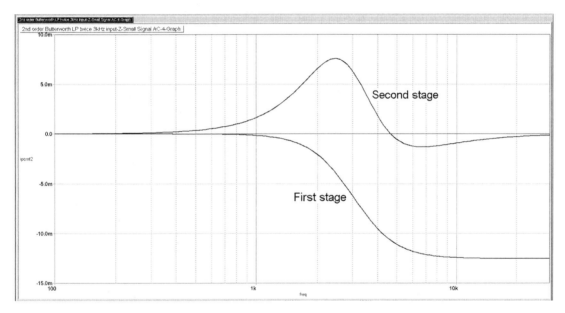

**Figure 8.24: The current drawn at the inputs of the first and second stages of the Linkwitz-Riley filter in Figure 8.23 Input voltage was 10 Vrms.**

frequencies, where the shunt capacitor has a high effective impedance, but with increasing frequency falls to the value of R1, i.e. 797 Ω. A previous 5532 stage should be able to drive this to full level without significant deterioration in the distortion performance.

The second Sallen & Key lowpass filter is identical, so naturally acts in the same way. The current it draws from opamp A1 is, however, less because the amplitude of the signal from A1 is falling as frequency increases. If the input to R1 is 10 Vrms, the maximum current drawn from the previous stage at high frequencies is 12.5 mA rms. The current drawn by the second stage from A1, however, only reaches a maximum of7.6 mA, peaking gently around 2.5 kHz. This is illustrated in Figure 8.24.

The overall distortion of the two stages is therefore somewhat less than might be expected from a casual glance. Similar considerations apply to the equivalent Linkwitz-Riley highpass filter.

## Lowpass Filters With Attenuation

It may occur that you need to attenuate a signal before applying it to a lowpass Sallen & Key filter. Since all the filter designs in this chapter require to be driven from a very low source impedance to get the desired response, you might think that the only solution is to use a potential divider with reasonably high values, so as not to unduly load upstream circuitry, and then put a unity-gain buffer between that and the filter. In fact, with the Sallen & Key lowpass configuration it is usually unnecessary to indulge in an extra amplifier.

If R1 and R3 in Figure 8.25b are chosen so that their ratio gives the required attenuation, but the output impedance of the divider (their value as a parallel combination) is equal to the R1 in Figure 8.25a, no buffer is required, with a saving in power consumption and parts cost and hopefully some reduction in noise and distortion.

The equivalent circuit for the Sallen & Key highpass filter, which has a capacitor at its input, would be a capacitor potential divider with an output reactance equivalent to the original capacitor. This however is likely to lead to high-frequency overload problems in the previous stage, as the impedance to ground

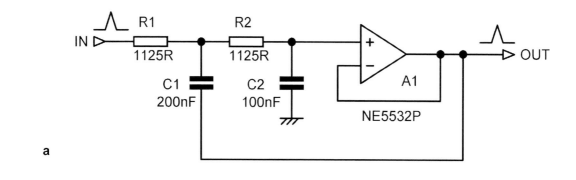

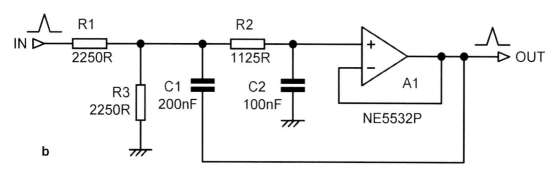

**Figure 8.25: Building attenuation into a filter: (a) the original lowpass filter; (b) with a 6 dB input divider R1, R3 scaled to give the same driving impedance as R1 in the first filter. Cutoff frequency is 1 kHz.**

of the divider will fall with frequency. There will probably also be issues with the stability of the previous stage due to the purely capacitive loading.

## Bandwidth Definition Filters

A very desirable feature of active crossovers for sound reinforcement is bandwidth definition, to keep subsonic and ultrasonic signals out of the amplifier/speaker system. Subsonic signals of sufficient level will cause mechanical and possibly thermal damage to LF drivers, but even smaller amplitudes will erode precious headroom. Ultrasonic signals can burn out tweeters and damage power amplifiers. You might wish to call this "bandwidth limitation", but that somehow sounds a bit less appealing in a marketing sense.

The design of lowpass filters to stop subsonic signals and highpass filters to stop ultrasonic signals is always something of a compromise, because you must steer a course between the most effective filtering and intrusion on the audio bandwidth that you want to keep. A complicating factor is the variation of opinion on the importance of phase-shifts introduced near the edges of the audio band; a hi-fi application would be more likely to set the filter cutoff frequencies further away from the audio band edges, whereas a sound-reinforcement crossover would be more likely to encounter ultrasonic oscillation and will have filters working closer in to maximise the protection and minimise the chance of damage to expensive banks of loudspeakers.

## Bandwidth Definition: Butterworth Versus Bessel

The lowpass filters used to define the upper limit of the audio bandwidth are usually 2nd-order with roll-off rates of 12 dB/octave; 3rd-order 18 dB/octave filters are rather rarer, probably because there seems to be a general feeling that phase changes might be more audible at the top end of the audio spectrum than the bottom. Either the Butterworth (maximally flat frequency response) or Bessel type (maximally flat group delay) can be used, and this gives us an excellent opportunity to make a real-life comparison of the two types. It is unlikely that there is any real audible difference between the two types of filter in this application, as just about everything happens above 20 kHz, but using the Bessel alignment does enforce a compromise in filtering effectiveness because of its slow roll-off. I will demonstrate (see Figure 8.26).

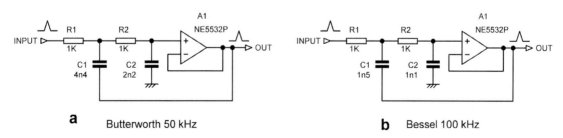

**Figure 8.26: Second-order Sallen & Key lowpass circuits for ultrasonic filtering:**
**(a) Butterworth; (b) Bessel. Both have a loss of less than 0.2 dB at 20 kHz.**

The 2nd-order Butterworth lowpass filter in Figure 8.26a has its −3 dB point set to 50 kHz, and this gives a loss of only 0.08 dB at 20 kHz, so there is minimal intrusion into the audio band; see Figure 8.27. The response is usefully down by −11.6 dB at 100 kHz and by an authoritative −24.9 dB at 200 kHz. C1 is composed of two 2n2 capacitors in parallel.

But let us suppose we are deeply concerned about linear phase at high frequencies, and we decide to use a Bessel filter with the same cutoff frequency instead. The only circuit change is that C1 is now 1.335 times as big as C2 instead of 2 times, but the amplitude response is very different. If we design for −3 dB at 50 kHz once more, we find that the response is −0.47 dB at 20 kHz; a good deal worse than 0.08 dB, and not exactly a stunning figure for your spec sheet. If we decide we can live with−0.2 dB at 20 kHz from the Bessel filter, then it has to be designed for −3 dB at 72 kHz. Due to the inherently slower roll-off of a Bessel filter, the response is now only down −5.6 dB at 100 kHz, and −14.9 dB at 200 kHz, as seen in Figure 8.27; the latter figure is 10 dB worse than for the Butterworth. The measured noise output for both versions is −114.7 dBu (corrected).

If we want to keep the 20 kHz loss to 0.1 dB, the Bessel filter has to be designed for −3 dB at 100 kHz, and the response is now only −10.4 dB down at 200 kHz, more than 14 dB less effective than the Butterworth; this is the design shown in Figure 8.26b. The results are summarised in Table 8.28.

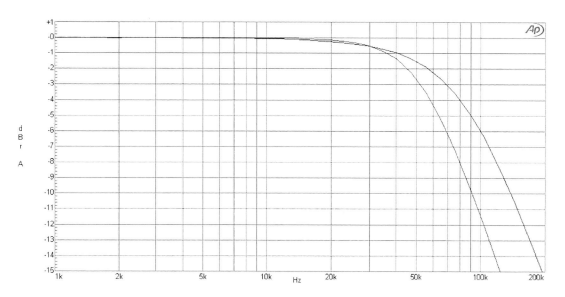

**Figure 8.27: Measured frequency response of 2nd-order 50 kHz
Butterworth and 72 kHz Bessel filters.**

**Table 8.28: The frequency response of various ultrasonic filter options.**

| Frequency | Butterworth 50 kHz | Bessel 50 kHz | Bessel 72 kHz | Bessel 100 kHz |
|---|---|---|---|---|
| 20 kHz | −0.08 dB | −0.47 dB | −0.2 dB | −0.1 dB |
| 100 kHz | −11.6 dB | −10.0 dB | −5.6 dB | −3.0 dB |
| 200 kHz | −24.9 dB | −20.9 dB | −14.9 dB | −10.4 dB |

Discussions on filters always remark that the Bessel alignment has a slower roll-off, but often fail to emphasise that it is a *much* slower roll-off. You should think hard before you decide to go for the Bessel option in this sort of application.

It is always worth checking how the input impedance of a filter loads the previous stage, to make sure it is not loaded to the point where its distortion is significantly increased. The input impedance will vary with frequency. In this case, the input impedance is high in the passband, but above the roll-off point it falls until it reaches the value of R1, which here is 1 k$\Omega$. This is because at high frequencies C1 is not bootstrapped, and the input goes through R1 and C1 to the low-impedance opamp output, which is effectively at ground. Fortunately this low impedance only occurs at high frequencies, where one hopes the level of the signals to be filtered out will be low.

Another important consideration with lowpass filters is the balance between the R and C values in terms of noise performance. R1 and R2 are in series with the input, and their Johnson noise and the voltage result of opamp current noise flowing in them will be added directly to the signal. Here the two 1 k$\Omega$ resistors generate −119.2 dBu of noise (22 kHz bandwidth, 25°C). The obvious conclusion is that R1 and R2 should be made as low in value as possible without causing excess loading (1 k$\Omega$ is a good compromise), with C1, C2 scaled to maintain the desired roll-off frequency.

## Variable-Frequency Lowpass Filters: Sallen & Key

Active crossovers for sound-reinforcement applications commonly have variable crossover frequencies, so they can be used with a wide range of loudspeaker systems. This presents problems when the usual Sallen & Key filters are used. A typical variable Sallen & Key lowpass filter is shown in Figure 8.28, where ganged variable resistors alter the cutoff frequency over a 10:1 range from 100 Hz to 1 kHz, such as might be used for the LF-MID crossover point. R1 and R2 are end-stop resistors to limit the frequency range at the upper end.

The variable resistors are normally of the reverse-log law, so that the frequency calibration is approximately linear in octaves. One problem with this circuit is that it relies on the matching of the resistance values of the variable resistor tracks, and also matching of the value obtained at a given degree of rotation, which is worsened by the need for a log law. This gives rise to errors in the cutoff frequency; the effect on the filter $Q$ is much less. Be aware that in Figure 8.28 no precautions have

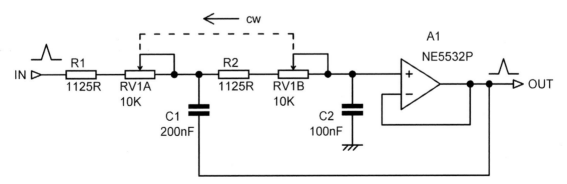

**Figure 8.28: Variable-frequency 2nd-order Sallen & Key lowpass filter, cutoff 100 Hz to 1 kHz.**

been taken to deal with the opamp input bias current flowing through the variable resistors. This could lead to noises as the controls are altered unless suitable DC-blocking capacitors are added.

A more serious difficulty is the very high degree of control-ganging that is required in a practical crossover. If we assume the filter in Figure 8.28 is being used in a 4th-order Linkwitz-Riley crossover, which will very often be the case, then to vary the frequency of the two cascaded Butterworth 2nd-order filters in one crossover path we need a four-gang part, which is obtainable but costly compared with a two-gang. Since we need to vary the frequency of the other path simultaneously (to change the LF-MID crossover point we must alter the frequencies in both the LF and MID paths), we must procure an eight-gang control, which is not only going to be relatively hard to get and expensive but will also probably have a poor control feel due to the increased friction. Fortunately, this is the sort of control that is rarely altered, so it is not the issue that it might be on a hi-fi preamplifier volume control.

We now pause and think nervously of the need to control a stereo crossover. That implies a 16-gang control, which is not really a practical component to specify in a design. Matters are even worse if we try to use multiple-feedback (MFB) filters. The highpass version still requires two resistors to be altered, while the lowpass demands three.

Variable lowpass filters similar to that in Figure 8.28 are most commonly used in mono subwoofer crossovers where only a single 2nd-order filter is required, and so an easily obtained dual control is all that is necessary.

# References

[1]    Sallen & Key "A Practical Method of Designing RC Active Filters" IRE Transactions on Circuit Theory 2(1), 1955–03, pp. 74–85.
[2]    Linkwitz, Siegfried "Active Crossover Networks for Non-Coincident Drivers" Journal of the Audio Engineering Society, January/February 1976
[3]    Geffe, Philip "How to Build High-Quality Filters Out of Low-Quality Parts" Electronics, November 1976, pp. 111–113
[4]    Williams and Taylor "Electronic Filter Design Handbook" Third Edn, McGraw-Hill, 1995, pp. 3.16–3.19
[5]    Tietze and Schenk, "Electronic Circuits" Second Edn, 2002, Springer, p. 815, ISBN: 978-3-540-00429-5
[6]    Aikens and Kerwin, "Single Amplifier, Minimal RC, Butterworth Thomson and Chebyshev Filters to Sixth Order" Proceeds of International Filter Symposium, Santa Monica, April 1972.

# CHAPTER 9

# Designing Highpass Filters

## First-Order Highpass Filters

First-order filters are the simplest possible; they are completely defined by one parameter: the cutoff frequency.

The usual 1st-order highpass filter is just a capacitor and a resistor. Figure 9.1a shows the normal (non-inverting) highpass version in all its beautiful simplicity. To give the calculated 1st-order response it must be driven from a very low impedance and see a very high impedance looking into the next stage. It is frequently driven from an opamp output, which provides the low driving impedance, but often requires a unity-gain stage A1 to buffer its output, as shown here.

Sometimes it is necessary to invert the phase of the audio to correct an inversion in a previous stage. This can be handily combined with a shunt-feedback 1st-order lowpass filter, as shown in Figure 9.1b. This stage inherently provides a low output impedance, so there are no loading effects from circuitry downstream. The signal level can be adjusted either up or down by giving R1 and R2 the required ratio.

Here are the design and analysis equations for the 1st-order filters. The design equations give the component values required for given cutoff frequency; the analysis equations give the cutoff frequency when the existing component values are plugged in. Analysis equations are useful for diagnosing why a filter is not doing what you planned.

For both versions, the design equation is:          and the analysis equation is:

$$R = \frac{1}{2\pi f_0 C} \qquad \text{(choose C)} \qquad\qquad f_0 = \frac{1}{2\pi RC} \qquad\qquad 9.1, 9.2$$

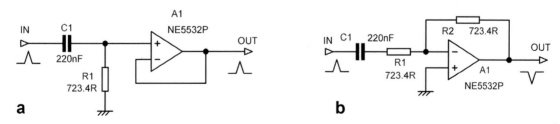

a                                                          b

**Figure 9.1: First-order highpass filters: (a) non-inverting highpass; (b) inverting highpass. Cutoff frequency (−3 dB) is 1 kHz for both.**

241

In both versions you must use R1 and C1 in the equations. R2 only controls the passband gain A = R2/R1. In the inverting version, it may be advisable to put a small (say, 100 pF) capacitor across R2 to ensure HF stability.

There are no issues with component sensitivities in a 1st-order filter; the cutoff frequency is inversely proportional to the product of $R$ and $C$ so the sensitivity for either component is always 1.0. In other words a change of 1% in either component will give a 1% change in the cutoff frequency.

## Sallen & Key 2nd-Order Filters

Second-order filters are much more versatile. They are defined by two parameters, the cutoff frequency and the Q, and as described in Chapter 7, higher-order filters are usually made by cascading 2nd-order stages with carefully chosen cutoff frequencies and $Q$'s. Odd-order filters require a 1st-order stage as well.

There are many ways to make a 2nd-order filter. The simplest and most popular are the well-known Sallen & Key configuration and the multiple-feedback (MFB) filters. The latter are better-known in their bandpass form but can be configured for either lowpass or highpass operation. MFB filters give a phase inversion, so can only be used in pairs (e.g. in a Linkwitz-Riley 4th-order configuration) or in conjunction with another stage that re-inverts the signal to get it back in phase.

These two filter types are not suitable for producing high $Q$ characteristics, as for Qs above about 3, such as might be needed in a high-order Chebyshev filter, as they begin to show excessive component sensitivities and also may start to be affected by the finite gain and bandwidth of the opamps involved. For higher Qs, more complex multi-opamp circuit configurations are used that do not suffer from these problems. Having said that, high $Q$ filters are not normally required in crossover design.

## Sallen & Key 2nd-Order Highpass Filters

An example of a 2nd-order highpass Sallen & Key filter is shown in Figure 9.2. Its operation is simple; at high frequencies the capacitors are effectively short-circuit, so it acts as a simple voltage-follower.

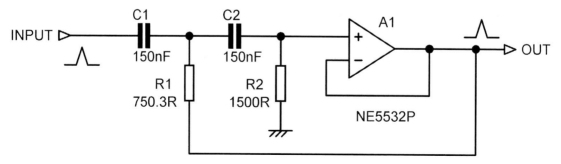

**Figure 9.2: The classic 2nd-order highpass Sallen & Key filter. Cutoff frequency is 1 kHz.**
**$Q$ = 0.7071 for a Butterworth response.**

At lower frequencies C2 and R2 pass less signal, reducing the output of the voltage-follower A1, causing the "bootstrapping" of R1 to be less effective, and so causing it to shunt signal away. There are two roll-off mechanisms working at once, and so we get the familiar 12 dB/octave filter slope.

There are two basic ways to control the $Q$ of a 2nd-order highpass Sallen & Key stage. In the first, the two resistors are made unequal, and the greater the ratio between them the higher the Q; a unity-gain buffer is always used. In the second method, the two resistors are made equal and $Q$ is fixed by setting the gain of the amplifier to a value greater than one. The first method usually gives superior performance, as filter gain is often not wanted in a crossover. The $Q$ of a Sallen & Key highpass filter can also be controlled by using non-equal capacitors, but this is unusual, as it makes the calculations more complicated.

While the basic Sallen & Key configuration is simple and easy to design, it is also extremely versatile, as you will appreciate from the many different versions in this chapter and Chapter 8. It is even possible to design bandpass Sallen & Key filters; see Chapter 12.

## Sallen & Key Highpass Filter Components

Highpass Sallen & Key filters are basically the same as their lowpass brothers, with the R's and C's swapped in their circuit positions. All the considerations described for the lowpass filters, such as the component sensitivities of particular configurations, are applicable so long as this is kept in mind.

As mentioned earlier, highpass filters have the advantage that the capacitors are usually all the same value; it is the resistors that come in awkward values, and paralleling them is cheap and takes up little PCB space. Nonetheless, in an ideal world we would have as many identical resistors as possible to ease sourcing, and some of the highpass filters in this section are configured to achieve this.

## Sallen & Key 2nd-Order Highpass: Unity Gain

Sallen & Key highpass filters are very much the same as the lowpass filters, with the R's and the C's swapped over.

Figure 9.2 shows a 2nd-order highpass Sallen & Key unity-gain filter with a cutoff frequency of 1 kHz and a $Q$ of 0.707 to obtain a Butterworth characteristic. Now the capacitors are equal, while the resistors have a ratio of 2 to define the required $Q$. The measured noise output of this filter is −115.2 dBu (corrected), with its design equations:

**Design equations:**

Choose R2:

$$C = \frac{2Q}{(2\pi f_0)R_2} \qquad\qquad R_1 = \frac{R_2}{4Q^2} \qquad\qquad 9.3, 9.4$$

**Analysis equations:**

$$f_0 = \frac{1}{2\pi C \sqrt{R_1 R_2}} \qquad\qquad Q = \frac{1}{2}\sqrt{\frac{R_2}{R_1}} \qquad\qquad 9.5, 9.6$$

The input impedance of this circuit is relatively high at low frequencies, where the impedance of the series capacitors C1, C2 is high. It then falls with increasing frequency, reaching a minimum just above the −3 dB cutoff frequency. Above this it levels out to the value of R2, as the capacitor impedance is now negligible. R1 does not add to the loading because it is bootstrapped by the voltage-follower A1.

The TL072 is a very poor choice for S&K highpass filters. Its common-mode problems are sharply revealed if it is used in highpass configurations; on clipping, the output hits the top rail and then shoots down to hit the bottom rail as the opamp internals go into phase reversal, which is about the worst sort of clipping you could imagine. In the Bad Old Days when TL072s had to be used in audio paths for cost reasons, if a highpass filter was required, the signal would be attenuated by 6 dB so the common-mode limits could never be reached, and the gain was recovered elsewhere. The TL072 also has a lot of CM distortion and generates a lot of extra distortion when its output is loaded even lightly, and it must be regarded as obsolete for almost all audio applications.

## Sallen & Key 2nd-Order Highpass: Equal Resistors

You can make a lowpass Sallen & Key filter with conveniently equal capacitor values if you configure the amplifier to give voltage gain instead of acting as a unity-gain buffer. A highpass filter can be made in exactly the same way with equal resistor values, though there is less of an advantage, because awkward resistor values are much less of a problem than awkward capacitor values. However, this approach is valuable when you need to make a variable-frequency filter using two-gang pots with equal sections; see the end of this chapter. Figure 9.3 shows a 2nd-order equal-R Butterworth highpass

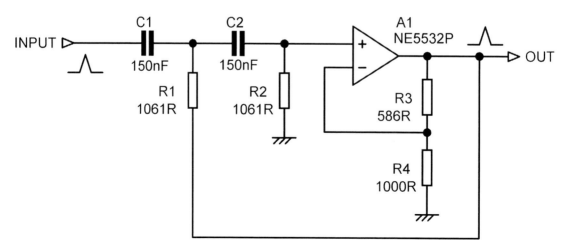

Figure 9.3: Equal-R 2nd-order highpass Butterworth filter with a cutoff frequency of 1 kHz. Gain must be 1.586 times for maximally flat Butterworth response ($Q$ = 0.7071).

S&K filter; you will note that the component values are exactly the same as in the lowpass equal-C filter in Figure 8.3, and the same gain of 1.586 times (+4.0 dB) is required to get a Butterworth $Q$ of 0.7071. The awkward value of 586 $\Omega$ can be obtained in three ways. The nearest E96 value is 590 $\Omega$, with a nominal error of 0.68%. The best 2xE24 parallel pair is 750 $\Omega$ and 2700 $\Omega$, with a nominal error of 0.16%. Using the 3xE24 format by putting 680 $\Omega$, 4700 $\Omega$, and 43 k$\Omega$ in parallel reduces the nominal error to 0.007%, but it is unlikely that this degree of accuracy will be useful, bearing in mind the usual resistor tolerances.

Due to the equal-value R and C components, the design and analysis equations are just the same as for the lowpass equal-C 2nd-order Butterworth described in Chapter 8.

**Design equations:**

Choose a value for $C$ ($= C1 = C2$)

$$R_1 = R_2 = \frac{1}{(2\pi f_0)C} \qquad \text{Passband gain } A = 3 - \frac{1}{Q} \qquad \text{9.7, 9.8}$$

Then choose R3, R4 to get the required passband gain $A$ for the chosen Q

Analysis equations:

$$f_0 = \frac{1}{2\pi RC} \qquad \text{(R's and C's are both equal)} \qquad 9.9$$

$$\text{Passband gain } A = \frac{R3 + R4}{R4} \qquad Q = \frac{1}{3 - A} \qquad \text{9.10, 9.11}$$

# Sallen & Key 2nd-Order Butterworth Highpass: Defined Gains

We saw in the previous section that we could select convenient resistor values for a 2nd-order Sallen & Key highpass filter by selecting the gain in the filter. We can choose both the gain and the $Q$ by altering the resistor values. Filter gain is often unwelcome, but it is sometimes required, as in the HF path of the crossover design example in Chapter 23, and building it into a filter will save an amplifier stage and reduce cost, while possibly also reducing noise and distortion, as the signal has gone through one less stage. On the other hand, distortion may increase somewhat because the filter stage has less negative feedback; this depends on the opamp types used and the exact circuit conditions.

As for the lowpass case, I will focus on the Butterworth filter characteristic. The exact resistor values required for gains from 0 to +8 dB are shown in Figure 9.4, where the cutoff frequency in each case is 1 kHz, and the capacitors have been set to 100 nF; other frequencies can be obtained by scaling the component values. The resistor values summarised in Table 9.1 also include the standard cases of unity gain and of equal resistors. For equal resistors the gain required is extremely close to +4 dB. It is assumed that R4 in the gain network is fixed at 1 k$\Omega$; in the higher-gain cases it would be advisable to reduce its value to lower the impedance of the gain network and thus minimise noise.

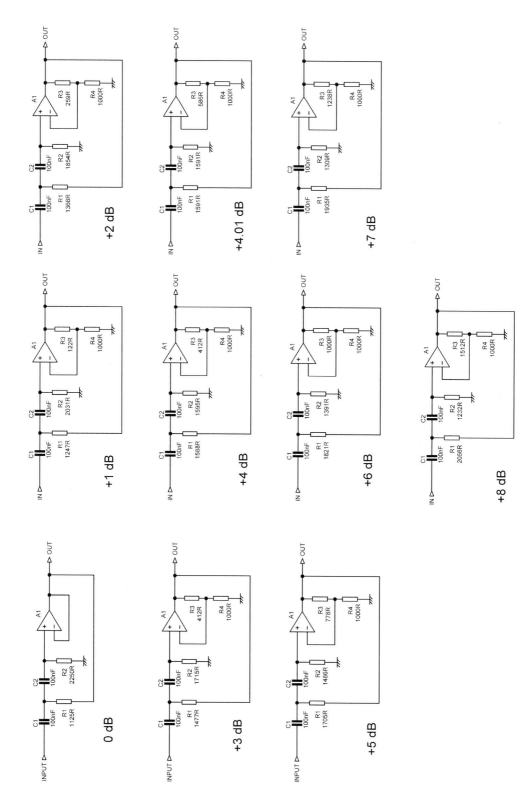

**Figure 9.4: Highpass 1 kHz Butterworth S&K filters with defined passband gains from 0 to +8 dB.**

**Table 9.1: Component values for defined gains in a Butterworth 2nd-order S&K highpass filter.**

| Gain | Gain | R1 | R2 | R2/R1 ratio | R3 | R4 |
|---|---|---|---|---|---|---|
| dB | Times | Ohms | Ohms | times | Ohms | Ohms |
| 0.00 | 1.000 | 1125 | 2250 | 2.000 | n/a | n/a |
| +1.00 | 1.122 | 1247 | 2031 | 1.628 | 122 | 1000 |
| +2.00 | 1.259 | 1366 | 1854 | 1.357 | 259 | 1000 |
| +3.00 | 1.412 | 1477 | 1715 | 1.161 | 412 | 1000 |
| +4.00 | 1.585 | 1588 | 1595 | 1.004 | 585 | 1000 |
| +4.01 | 1.586 | 1591 | 1591 | 1.000 | 586 | 1000 |
| +5.00 | 1.778 | 1705 | 1486 | 0.871 | 778 | 1000 |
| +6.00 | 2.000 | 1821 | 1391 | 0.764 | 1000 | 1000 |
| +7.00 | 2.238 | 1935 | 1309 | 0.637 | 1238 | 1000 |
| +8.00 | 2.512 | 2056 | 1232 | 0.599 | 1512 | 1000 |

Higher gains than this can be obtained. The gain limit of 3 times given in the previous section on equal-resistor highpass filters does not apply here because we are adjusting the resistor values to keep the $Q$ constant at 0.707 instead of letting it shoot off to infinity, or indeed beyond. There is however the consideration that further reducing the negative feedback factor of the opamp will worsen the distortion performance.

Similar design data for 2nd-order Butterworth Sallen & Key lowpass filters is given in the earlier chapter on lowpass filters. Not surprisingly, the ratios between the resistors here are exactly the same as the ratios of the capacitors in the lowpass case.

## Sallen & Key 2nd-Order Highpass: Non-Equal Capacitors

We noted that for lowpass Sallen & Key filters, the resistors are almost always made equal. In the same way, the capacitors in highpass filters are almost always made equal; this is not an essential property of Sallen & Key filters but does make the calculations simpler. There seems to be little point in using non-equal capacitors in any 2nd-order Sallen & Key highpass stage, and even less point in using them in 3rd- and higher-order highpass filters constructed as a single stage; by analogy with the lowpass equivalent, it would allow all the resistors to be made the same value, but because of the sparse nature of the available capacitor values, having equal capacitors is much more useful than having equal resistors; both the total capacitor cost and the PCB area occupied can be minimised.

Equal resistors are however very useful when making a variable-frequency filter, as you can use two-gang pots with equal-value sections. Equal resistors are obtained with a stage gain of 1.586 times (+4.01 dB), as described at the end of this chapter, but this may lead to problems with the gain structure of the crossover. Unfortunately, when the resistors are set equal and the gain to unity, with equal capacitors, we get a $Q$ of only 0.5, and we need 0.707 for a Butterworth filter. The $Q$ can be altered by using unequal capacitors, but only *reduced* from 0.5, and not increased. This is most inconvenient.

# Sallen & Key 3rd-Order Highpass: Two Stages

Figure 9.5 shows the classical method of making a 3rd-order highpass filter by cascading a 2nd-order stage and a 1st-order stage. Output buffer A2 ensures that C3-R3 are not affected by external loading; if the following stage has a high impedance, then A2 can be omitted. The order of the two stages can be reversed without affecting the frequency response, but it does affect the clipping behaviour.

Since the A1 stage has a $Q$ of 1, greater than 0.7071, it has a gain that peaks at more than unity. With $Q = 1$ there is 1.2 dB of peaking around the roll-off, which commensurately reduces the headroom. If the order of the stages is reversed, giving what I call the New Order, the 1st-order roll-off at A2 reduces the signal amplitudes before they reach the 2nd-order stage, and there is no restriction on headroom.

Component values for other types of two-stage 3rd-order highpass filter are given in Table 9.2. All capacitors have been selected as preferred values, giving non-preferred values for the resistors. The

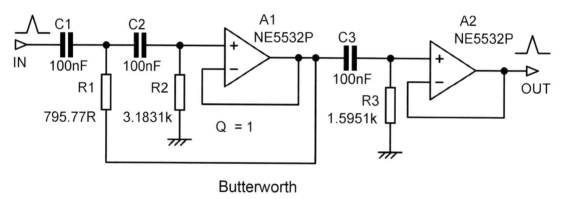

Butterworth

**Figure 9.5: Third-order Sallen & Key highpass Butterworth filter implemented in two stages. Cutoff frequency 1 kHz.**

**Table 9.2: Third-order two-stage Sallen & Key highpass: component values for various filter types. Cutoff 1 kHz.**

| Type | C1 = C2 nF | R1 Ω | R2 Ω | C3 nF | R3 Ω | Cutoff dB |
|---|---|---|---|---|---|---|
| Linkwitz-Riley | 100 | 1125.41 | 2250.77 | 100 | 1591.5 | −6 |
| Bessel | 100 | 1672.62 | 3194.58 | 100 | 2111.99 | −3 |
| Linear-Phase 5% delay ripple | 100 | 1071.96 | 3764.60 | 100 | 1260.9 | −3 |
| Butterworth | 100 | 795.77 | 3183.1 | 100 | 1591.5 | −3 |
| 1 dB-Chebyshev | 47 | 836.74 | 13624.4 | 68 | 1156.7 | −1 |
| 2 dB-Chebyshev | 47 | 624.62 | 16266.6 | 47 | 1249.2 | −2 |
| 3 dB-Chebyshev | 33 | 720.02 | 27105.6 | 47 | 1011.1 | −3 |

capacitors are altered as necessary to prevent R1 and R3 from becoming too small and causing loading problems. The amount of gain peaking is different for each type, increasing as the filter moves from Linkwitz-Riley to a 3 dB-Chebyshev characteristic.

The component values were calculated using the tables of stage frequency and $Q$ given in Chapter 7, and checked by simulation. The design process depends on what definition of cutoff attenuation is used; throughout this book I have employed the most common definition, set out in the rightmost column of the table.

The maximum $Q$ of 3.07 occurs in the 2nd-order stage of the 3 dB-Chebyshev filter. This leads to a very large ratio between R1 and R2; this is typical of Sallen & Key filters.

## Sallen & Key 3rd-Order Highpass in a Single Stage

Third-order highpass filters can also be made in a single stage, using the Geffe configuration, first presented in [1].There are no simple design equations for this configuration; setting the pole positions to create a complex pole pair plus a single pole requires solving three simultaneous equations; this is best done by computer. For more information on the process, see [2]. Single-stage filters have the advantage that there can be no internal gain peaking to compromise headroom. Also, because in a single-stage filter all off the filtering takes place before the amplifier, this allows large input voltages at frequencies where the filtering action reduces the output amplitude.

In Figure 9.6a is shown a Geffe 3rd-order Butterworth highpass using a unity-gain amplifier. As expected this gives us three different resistor values for R1, R2, and R3. As an example, the exact values are converted into 1xE96 and 2xE24 format in Table 9.3.

Figure 9.6b shows an alternative version where the amplifier gain is chosen to at 1.27 times (+2.08 dB) to make R1 and R2 almost identical in value. Note that this method requires larger capacitors if the lowest resistor value is to be kept in the range 700–1000 Ω. R4 very conveniently comes out as the E12 value of 270 Ω.

The alert reader will have spotted that the gain required for the most accurate resistor equality is 1.27 times, as opposed to the 1.26 times in the equivalent lowpass filter in Figure 8.6. This is because the calculation method used only allows gain to be specified to two decimal places and has no other significance. Either value may be used with very small response errors. The exact values are converted into 1xE96 and 2xE24 format in Table 9.4.

Component values for the other filter types are given in Table 9.5.

**Table 9.3: 1xE96 and 2xE24 values for Figure 9.6a, with error in nominal value.**

|  | Exact value Ω | 1xE96 Ω | Error | 2xE24 Ω | | Error |
|---|---|---|---|---|---|---|
| R1 | 2431 | 2430 | −0.0005% | 4300 | 5600 | +0.05% |
| R2 | 954.6 | 953 | +0.1% | 1300 | 3600 | +0.05% |
| R3 | 16.73k | 16.9k | +1.0% | 24k | 56k | +0.42% |

## Butterworth

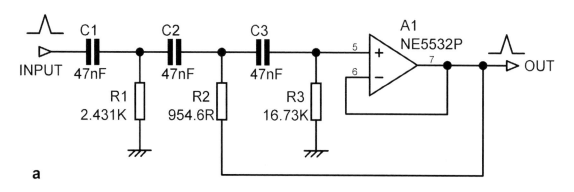

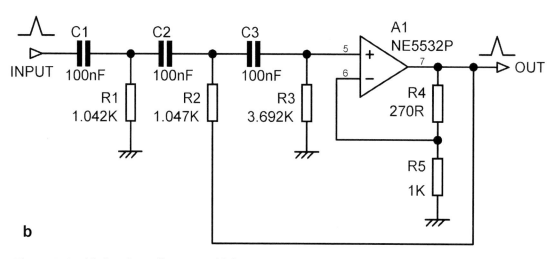

Figure 9.6: Third-order Sallen & Key highpass Butterworth filters implemented as a single stage:
(a) unity gain; (b) equal-R with exact values, gain = 1.27 times. Cutoff frequency 1 kHz.

Table 9.4: 1xE96 and 2xE24 values for Figure 9.6b, with error in nominal value.

|  | Exact value Ω | 1xE96 Ω | Error | 2xE24 Ω | | Error |
|---|---|---|---|---|---|---|
| R1 | 1042 | 1050 | +0.77% | 1600 | 3000 | +0.14% |
| R2 | 1047 | 1050 | +0.29% | 2000 | 2200 | +0.06% |
| R3 | 3692 | 3650 | −1.1% | 6200 | 9100 | −0.12% |

**Table 9.5: Component values for 3rd-order unity-gain single-stage
S&K highpass filters with 1 kHz cutoff.**

| Type | C1 = C2 = C3 | R1 | R2 | R3 |
|---|---|---|---|---|
| Linkwitz-Riley | 68 nF | 1709 Ω | 1116 Ω | 6721 Ω |
| Bessel | 100 nF | 1608 Ω | 1116 Ω | 6258 Ω |
| Linear-Phase 5% delay ripple | 68 nF | 1585 Ω | 918.1 Ω | 11.117 kΩ |
| Butterworth | 47 nF | 2431 Ω | 954.6 Ω | 16.73 kΩ |
| 0.5 dB-Chebyshev | 22 nF | 3775.6 Ω | 756.06 Ω | 94.900 kΩ |
| 1.0 dB-Chebyshev | 10 nF | 6790 Ω | 1048.5 Ω | 271.1 kΩ |
| 2.0 dB-Chebyshev | 10 nF | 5280.5 Ω | 590.84 Ω | 422.38 kΩ |
| 3.0 dB-Chebyshev | 10 nF | 4386.1 Ω | 366.51 Ω | 628.40 kΩ |

## Sallen & Key 4th-Order Highpass: Two Stages

A 4th-order highpass filter is usually built by cascading two 2nd-order highpass stages. For the Butterworth version the two cutoff frequencies are the same, but one stage has a $Q$ of 0.5142 and the other a $Q$ of 1.3065. The latter is quite a high $Q$ for a Sallen & Key filter; since the $Q$ is set by the ratio of the two resistors, the value of R3 is low, being 6.8 times smaller than R4.

In Figure 9.7 the first stage has a $Q$ less than 0.7071 ($1/\sqrt{2}$) and so shows no gain peaking. The second stage has a peak of +2.9 dB at 847 Hz; with the stage order shown this is not an issue, as the first stage attenuates the signal before it reaches the second stage. If the stage order were reversed there would be a serious loss of headroom in the region of 1 kHz–2 kHz.

Component values for the other types of two-stage 4th-order highpass filter are given in Table 9.6. All capacitors have been selected as preferred values, giving non-preferred values for the resistors. The capacitors are chosen to prevent any of the resistors from becoming too small and so presenting excessive loading. The amount of gain peaking is different for each type, increasing as the filter moves from Linkwitz-Riley to a 3 dB-Chebyshev characteristic.

The component values were calculated using the tables of stage frequency and $Q$ given in Chapter 7, and checked by simulation. The design process depends on what definition of cutoff attenuation is used; throughout this book I have employed the most common definition, set out in the rightmost column of the table.

The maximum $Q$ of 5.58 occurs in the second stage of the 3 dB-Chebyshev filter; note the very large ratio between R3 and R4.

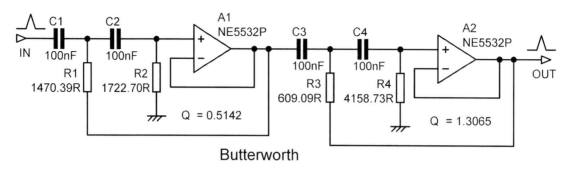

Figure 9.7: Fourth-order Sallen & Key Butterworth highpass filter, implemented in two stages. Cutoff frequency 1 kHz.

Table 9.6: Fourth-order two-stage Sallen & Key highpass:
component values for various filter types. Cutoff 1 kHz.

| Type | C1 = C2 nF | R1 Ω | R2 Ω | C3 = C4 nF | R3 Ω | R4 Ω | Cutoff dB |
|---|---|---|---|---|---|---|---|
| Linkwitz-Riley | 100 | 1125.41 | 2250.77 | 100 | 1125.41 | 2250.77 | −6 |
| Bessel | 100 | 2163.95 | 2357.66 | 100 | 1571.99 | 4079.81 | −3 |
| Linear-Phase 5% delay ripple | 100 | 1193.92 | 1755.7 | 100 | 980.61 | 6794.1 | −3 |
| Butterworth | 100 | 1470.39 | 1722.70 | 100 | 609.09 | 4158.73 | −3 |
| 1 dB-Chebyshev | 47 | 1140.77 | 2808.31 | 22 | 1009.47 | 51.145 k | 0 |
| 2 dB-Chebyshev | 47 | 857.52 | 2962.85 | 22 | 758.78 | 64053.2 | 0 |
| 3 dB-Chebyshev | 47 | 696.13 | 3226.84 | 22 | 616.35 | 76681.1 | 0 |

## Sallen & Key 4th-Order Highpass: Butterworth in a Single Stage

As described for lowpass 4th-order single-stage filters, you can make any highpass filter characteristic, but there are again restrictions on the stage gain. Unity gain is usually desirable to optimise noise and headroom in the signal path, but as the gain gets closer to unity, the ratio between the values of R3 and R4 becomes excessive. As for the single-stage 4th-order filter in Chapter 8, choosing a gain of 1.1 times (+0.83 dB) gives an acceptable R4/R3 ratio of about 20, and only very minor compromises on noise and headroom; see Figure 9.8. This circuit has a relatively high impedance at low frequencies because of C1, but this falls rapidly around the cutoff frequency to a minimum of approximately 1.2 kΩ at 2 kHz. At higher frequencies the stage input impedance plateaus at 1.3 kΩ; note that the low value of R3 does not mean that the input impedance is at any point equally low.

The 1xE96 and 2xE24 values for Figure 9.8 are given in Table 9.7. Note that the 1xE96 values are exact (not an accident); however, not everyone wants to get involved in stocking all the E96 values, so the 2xE24 solutions are also given.

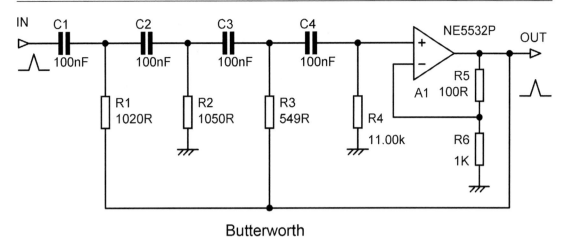

Butterworth

**Figure 9.8: Fourth-order 1 kHz Sallen & Key Butterworth highpass filter in a single stage; gain = 1.1 times (+ 0.82 dB).**

**Table 9.7: 1xE96 and 2xE24 values for Figure 9.8, with error in nominal value.**

|     | Exact value Ω | 1xE96 Ω | Error | 2xE24 Ω | | Error |
|-----|---------------|---------|-------|---------|------|-------|
| R1  | 1020          | 1020    | 0.0%  | 1200    | 6800 | 0.0%  |
| R2  | 1050          | 1050    | 0.0%  | 2000    | 2200 | −0.23% |
| R3  | 549           | 549     | 0.0%  | 1100    | 1100 | +0.18% |
| R4  | 11k           | 11k     | 0.0%  | 22k     | 22k  | 0.0%  |

It is also possible to choose a filter characteristic and then choose a gain which makes two of the resistors equal. There are four resistors, and the other two will in general be non-convenient values, though as explained before this is much less of a problem than awkward capacitor values. A 4th-order Butterworth highpass filter designed on this basis is shown in Figure 9.9a, with the exact component values that emerge from the design process.

You will have spotted that R1 and R3 in Figure 9.9a are both extremely close to the E12 value of 1.2 kΩ. Making them so introduces response errors of less than 0.05 dB from 10 Hz to 2 kHz, and above that the errors are negligible; the resulting circuit is shown in Figure 9.9b. R2 and R4 are much further from E24 preferred values; in 1xE96 format R2 = 845 Ω and R4 = 5.23 kΩ. In 2xE24 format, good parallel combinations are R2 = 1600 Ω in parallel with 1800 Ω (−0.26% error) and R4 = 8.2 kΩ in parallel with 15 kΩ (0.39% low). Note that R5 could be very handily made up of two 220 Ω resistors in series.

The gain required for the most accurate resistor equality is 1.44 times, as opposed to the 1.43 times in the equivalent lowpass filter in 8.13. Once again this is because the gain is only specified to two decimal places. Either value may be used with very small response errors.

## Butterworth

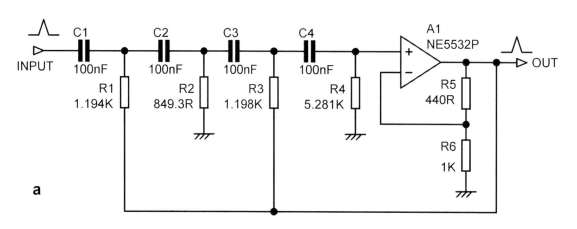

a

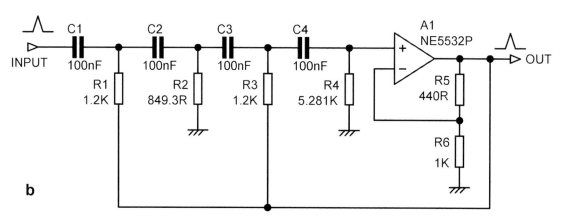

b

**Figure 9.9: Fourth-order 1 kHz Sallen & Key Butterworth highpass filter in a single stage: (a) equal-C with gain = 1.43 and exact values; (b) equal-C with gain = 1.43 and preferred values for R1 and R3.**

## Sallen & Key 4th-Order Highpass: Linkwitz-Riley in a Single Stage

You will find it argued in many places in this book and elsewhere that usually the best crossover is the highly popular 4th-order Linkwitz-Riley configuration. Thus while a 4th-order Butterworth single stage with low gain would undoubtedly be handy for making an 8th-order Linkwitz-Riley, this is inevitably going to be a minority interest, and a 4th-order Linkwitz-Riley single stage with low gain is likely to be much more useful; see Figure 9.10. It makes it possible to build one channel of a 2-way 4th-order active crossover using a single dual opamp.

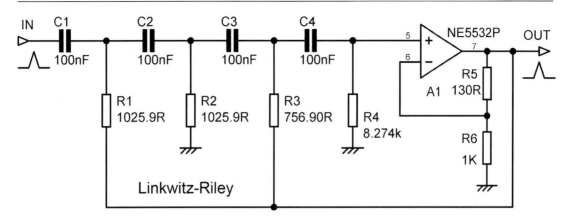

**Figure 9.10: Fourth-order Sallen & Key Linkwitz-Riley highpass filter. Gain =+0 dB.**

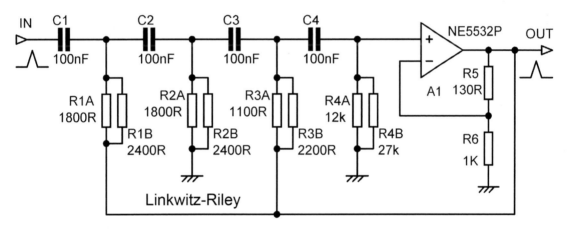

**Figure 9.11: Fourth-order Sallen & Key Linkwitz-Riley highpass filter. Gain =+0 dB.**

Most of the filters described in this book have not been converted to 2xE24 or 3xE24 format, as it is unlikely you will want a crossover frequency of exactly 1 kHz. I have made an exception here, as I think that this filter may prove to be extremely useful, so Figure 9.11 gives a practical 2xE24 implementation. For a 1xE96 version use R1 = 1020 Ω, R2 = 1020 Ω, R3 = 750 Ω, and R4 = 8.25 kΩ.

Table 9.8 gives the cutoff frequency sensitivities for Figures 9.10 and 9.11; they are all fairly low, but you may need to keep an eye on R4 and C4.

As with the Butterworth single-stage described earlier, if you are prepared to let the stage have more gain, then some of the component values can be manipulated to be convenient. There is less incentive to do this in highpass filters, for it is the resistor values rather than the capacitors that are manipulated, and awkward resistor values are much less of a problem. An example is shown in Figure 9.12. The gain for best equality of R1 and R3 is 1.32 times (+2.48 dB) rather than 1.33, for the same reason as before. This time we have been very lucky with the values, as R1 and R3 in Figure 9.12a are both very

Table 9.8: Fourth-order one-stage Sallen & Key Linkwitz-Riley
sensitivities. Gain = 1.13x.

| Component | Cutoff frequency sensitivity |
|-----------|------------------------------|
| R1 | 1.0216 |
| R2 | 1.5949 |
| R3 | 1.1346 |
| R4 | 1.7381 |
| C1 | 1.0152 |
| C2 | 1.0203 |
| C3 | 1.0328 |
| C4 | 1.4651 |
| R5 | 1.0106 |
| R6 | 1.0088 |

close to the E12 value of 1.1 kΩ, R2 is close to 910 Ω and R4 is very close to 5.6 kΩ; R5 has been set to 330 Ω, giving a gain of 1.33 times. A version using these values is shown in Figure 9.12b; the errors are +0.3 dB from 10 Hz to 900 Hz, but there is an error peak of−0.7 dB around 1.4 kHz, falling off to−0.1 dB at 10 kHz.

This example of the use of 1xE24 preferred values shows that it is tricky with 4th-order single-stage filters because of the increased component sensitivity over low-order filters. The greatest error here is introduced by taking R5 to be 330 Ω rather than 320 Ω, so if you do decide to use this kind of filter, you might consider making R5 a 2xE24 parallel pair. The combination of 560 Ω in parallel with 750 Ω is only 0.19% low relative to 320 Ω, and the effective tolerance is improved to 0.71% if 1% components are used. For a 1xE96 version use R1 = 1130 Ω, R2 = 909 Ω, R3 = 1130 Ω, and R4 = 5.62 kΩ.

## Sallen & Key 4th-Order Highpass: Single-Stage With Other Filter Characteristics

Table 9.9 summarises the Butterworth and Linkwitz-Riley designs described earlier, gives a number of extra gain options for Butterworth filters, and also gives the component values for other filter characteristics such as Bessel and Chebyshev.

It is noticeable that as the amplifier gain for the Butterworth case is reduced, the value of R3 steadily decreases, until at a gain of 1.05 times it is down to 374 Ω, which is likely to cause loading problems for the previous stage. It will usually be better to accept an unwanted gain of 1 dB or so rather than have excessively low values for R3. Table 9.9 offers several extra Butterworth and Linkwitz-Riley gain options to demonstrate how this works.

## Linkwitz-Riley

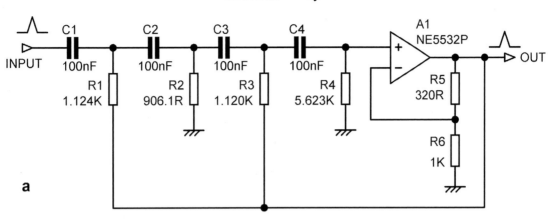

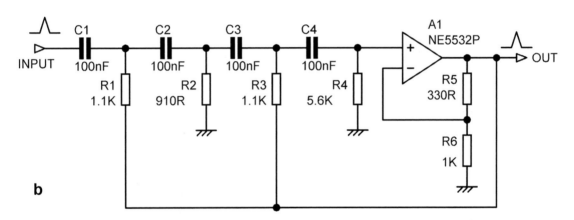

**Figure 9.12: Fourth-order Sallen & Key Linkwitz-Riley highpass filters implemented as a single stages: (a) equal-R with gain = 1.32 and exact values; (b) equal-R with gain = 1.33 and E24 preferred values for all resistors.**

## Sallen & Key 5th-Order Highpass: Three Stages

A 5th-order highpass filter is usually built by cascading two 2nd-order highpass stages and one 1st-order highpass stage. For the Butterworth version the three cutoff frequencies are the same, but one 2nd-order stage has a $Q$ of 0.620 and the other a $Q$ of 1.620. The latter $Q$ is an increase on the highest $Q$ in a 4th-order Butterworth, which is 1.3065.

In Figure 9.13 the first stage has a $Q$ less than 0.7071 ($1/\sqrt{2}$) and so shows no gain peaking. The second stage has a gain peak of +4.63 dB, but with the stage order shown this is not an issue, as the

**Table 9.9: Component values for 4th-order unity-gain single-stage S&K highpass filters with 1 kHz cutoff. All cutoffs at −3 dB, except Linkwitz-Riley cutoff is at −6 dB and Chebyshev cutoff is at 0 dB.**

| Type | C1 = C2 = C3 = C4 Ω | R1 | R2 | R3 | R4 | Gain x |
|---|---|---|---|---|---|---|
| Linkwitz-Riley | 100 nF | 1100 Ω | 910 Ω | 1100 Ω | 5600 Ω | 1.330 |
| Linkwitz-Riley | 100 nF | 1124 Ω | 906.1 Ω | 1120 Ω | 5623 Ω | 1.320 |
| Linkwitz-Riley | 100 nF | 1025.9 Ω | 1025.9 Ω | 756.90 Ω | 8274 Ω | 1.130 |
| Linkwitz-Riley | 100 nF | 1000.3 Ω | 1060.7 Ω | 625.13 Ω | 9673.7 Ω | 1.090 |
| Bessel | 100 nF | 1749.4 Ω | 1528.1 Ω | 1041.5 Ω | 12.193 kΩ | 1.090 |
| Linear-Phase 5% delay ripple | 100 nF | 1054.1 Ω | 1193.9 Ω | 672.16 Ω | 16.569 kΩ | 1.090 |
| Butterworth | 100 nF | 1200 Ω | 849.3 Ω | 1200 Ω | 5281 Ω | 1.440 |
| Butterworth | 100 nF | 1137.7 Ω | 911.98 Ω | 978.08 Ω | 6322.2 Ω | 1.300 |
| Butterworth | 100 nF | 1090.4 Ω | 968.53 Ω | 789.83 Ω | 7692.4 Ω | 1.200 |
| Butterworth | 100 nF | 1020 Ω | 1050 Ω | 549 Ω | 11.0 kΩ | 1.100 |
| Butterworth | 100 nF | 1026.9 Ω | 1051.3 Ω | 517.8 Ω | 11.478 kΩ | 1.090 |
| Butterworth | 100 nF | 1000 Ω | 1100 Ω | 374.0 Ω | 15.4 kΩ | 1.050 |
| 0.5 dB-Chebyshev | 47 nF | 1141.9 Ω | 1750.0 Ω | 711.88 Ω | 35.034 kΩ | 1.090 |
| 1.0 dB-Chebyshev | 47 nF | 932.94 Ω | 1665.5 Ω | 644.14 Ω | 36.209 kΩ | 1.090 |
| 2.0 dB-Chebyshev | 47 nF | 728.17 Ω | 1637.0 Ω | 587.10 Ω | 38.659 kΩ | 1.090 |
| 3.0 dB-Chebyshev | 47 nF | 612.07 Ω | 1657.0 Ω | 559.93 Ω | 40.983 kΩ | 1.090 |

Values all checked by SPICE simulation.

first low-$Q$ stage attenuates the signal before it reaches the second stage. If the 2nd-order stages were interchanged, there would be a serious loss of headroom in the region 1 kHz–2 kHz.

Component values for the other types of three-stage 5th-order highpass filter are given in Table 9.10. All capacitors have been selected as preferred values, giving non-preferred values for the resistors. The capacitors are chosen to prevent any of the resistors from becoming too small and so presenting excessive loading. The amount of gain peaking is different for each type, increasing as the filter moves from Linkwitz-Riley to a 3 dB-Chebyshev characteristic.

The component values were calculated using the tables of stage frequency and $Q$ given in Chapter 7, and checked by simulation. The design process depends on what definition of cutoff attenuation is used; throughout this book I have employed the most common definition, set out in the rightmost column of the table.

The maximum $Q$ of 8.81 occurs in the 2nd-order stage of the 3 dB-Chebyshev filter; note the very large ratio between R3 and R4.

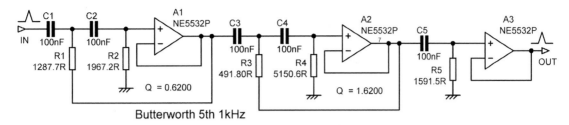

Butterworth 5th 1kHz

**Figure 9.13: Fifth-order Sallen & Key Butterworth highpass filter, implemented in three stages. Cutoff frequency 1 kHz.**

**Table 9.10: Fifth-order three-stage Sallen & Key highpass: component values for various filter types. Cutoff 1 kHz.**

| Type | C1 = C2 nF | R1 Ω | R2 Ω | C3 = C4 nF | R3 Ω | R4 Ω | C5 | R5 | Cutoff dB |
|---|---|---|---|---|---|---|---|---|---|
| Linkwitz-Riley | 100 | 1125.41 | 2250.8 | 100 | 795.77 | 3183.10 | 100 | 1591.5 | −6 |
| Bessel | 100 | 2204.57 | 2800.09 | 100 | 1528.77 | 5136.51 | 100 | 2398.31 | −3 |
| Butterworth | 100 | 1287.7 | 1967.2 | 100 | 491.80 | 5150.6 | 100 | 1591.5 | −3 |
| 1 dB-Chebyshev | 47 | 793.07 | 6206.99 | 22 | 647.45 | 79,881 | 47 | 980.33 | −1 |
| 2 dB-Chebyshev | 33 | 851.79 | 10,735.9 | 10 | 1073.67 | 224,637 | 47 | 739.25 | −2 |
| 3 dB-Chebyshev | 33 | 692.52 | 12,662.2 | 10 | 873.79 | 271,346 | 33 | 856.06 | −3 |

## Sallen & Key 5th-Order Butterworth Filter: Two Stages

A 5th-order filter is commonly built in three stages, as in Figure 9.13. This requires three amplifiers, as the single pole at the end almost always needs buffering. However, with suitable cunning you can combine a pair of complex poles and a single pole in a single stage as in Figure 9.14. This is also done in the 3rd-order Geffe filter of Figure 9.6. The second stage need not have the same capacitor values as the first but must have the correct CR products; this is demonstrated here, where C4 and C5 have been approximately halved in value so the resistors R4, R5 are doubled, preventing excessive loading on the output of A1. There is no gain peaking at the first stage output, and so there are no internal headroom problems.

The second stage is straightforward to design, using a cutoff frequency of 1 kHz and a $Q$ of 1.62, as in Figure 7.27 and Table 7.3. Designing the Geffe first stage is not exactly intuitive; you have to solve three simultaneous equations to create a complex pole pair plus a single pole; but really all you need is the resistor ratios given here; I am afraid designs for other filter types are not available. Table 9.11 gives an example of making up the awkward resistor values from two paralleled E24 parts (2xE24 format).

To my mind this is a very neat bit of circuitry, and it's a pity that 5th-order crossovers are not more useful. The matching two-stage 5th-order lowpass filter is described in Chapter 8.

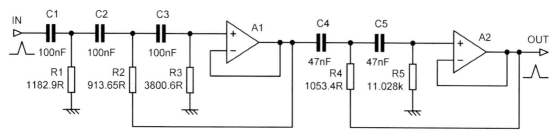

**Figure 9.14: Fifth-order Sallen & Key Butterworth highpass filter in two stages. Cutoff −3 dB at 1 kHz.**

**Table 9.11: Fifth-order two-stage Sallen & Key highpass resistor values in 2xE24 format.**

| Type | R1 Ω | R1a | R1b | Error % | R2 Ω | R2a | R2b | Error % | R3 Ω | R3a | R3b | Error % |
|---|---|---|---|---|---|---|---|---|---|---|---|---|
| Butterworth | 1182.9 | 1500 | 5600 | +0.02 | 913.65 | 1200 | 3900 | +0.44% | 3800.6 | 5100 | 15k | +0.13% |
| Type | R4 Ω | R4a | R4b | Error% | R5 Ω | R5a | R5b | | Error % | | | |
| Butterworth | 1053.4 | 1300 | 5600 | +0.16% | 11.028k | 16k | 36k | | +0.44% | | | |

## Sallen & Key 5th-Order Highpass: Single Stage

A 5th-order highpass filter can be built in a single stage, as in Figure 9.15. The capacitor ratios were derived from Aikens and Kerwin [3] and transformed into resistor values by taking the reciprocal and then appropriately scaling. Aikens and Kerwin only give values for a gain of +6 dB, and this will rarely be convenient in crossover work—an exception being in the HF path, where nominal signal values can be doubled (see Chapters 17 and 23).

The component sensitivities will be inferior to that of a three-stage 5th-order filter.

## Sallen & Key 6th-Order Highpass: Three Stages

A 6th-order highpass filter is usually built by cascading three 2nd-order highpass stages. For the Butterworth version the three cutoff frequencies are the same, but the first 2nd-order stage has a $Q$ of 0.5177, the next a $Q$ of 0.707, and the final a $Q$ of 1.9320, as in Figure 9.16. The latter is a high $Q$ for a Sallen & Key configuration, giving a very large ratio of 14.9 times between R5 and R6. It is therefore desirable to use 47 nF for C5 and C6 to prevent R5 being too small and excessively loading amplifier A2.

There is no gain peaking in either the first or second stage, but significant peaking of +6.02 dB in the third stage. With the stage order shown this is not an issue, as the first low-$Q$ stage attenuates the signal before it reaches the second stage. If the third stage was first in the cascade, there would be a serious loss of headroom in the region of 1 kHz–2 kHz.

Component values for the other types of three-stage 6th-order highpass filter are given in Table 9.12. All capacitors have been selected as preferred values, giving non-preferred values for the resistors. The capacitors are chosen to prevent any of the resistors from becoming too small and so presenting

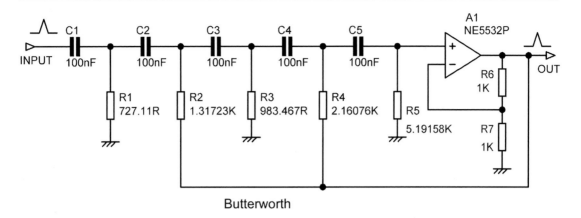

Figure 9.15: Fifth-order Sallen & Key Butterworth highpass filter, implemented as single stage, with gain = 2 and exact values. Cutoff frequencies 1 kHz.

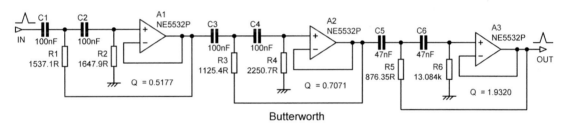

Figure 9.16: Sixth-order Sallen & Key Butterworth highpass filter, implemented in three stages. Cutoff frequency 1 kHz.

Table 9.12: Sixth-order three-stage Sallen & Key highpass: component values for various filter types. Cutoff 1 kHz.

| Type | C1 = C2 nF | R1 Ω | R2 Ω | C3 = C4 nF | R3 Ω | R4 Ω | C5 = C6 nF | R5 | R6 | Cutoff dB |
|---|---|---|---|---|---|---|---|---|---|---|
| Linkwitz-Riley | 100 | 1591.55 | 1591.55 | 100 | 795.77 | 3183.10 | 100 | 795.77 | 3183.10 | −6 |
| Bessel | 100 | 2504.42 | 2608.67 | 100 | 2202.05 | 3290.44 | 100 | 1482.92 | 6212.53 | −3 |
| Butterworth | 100 | 1537.13 | 1647.89 | 100 | 1125.41 | 2250.8 | 47 | 876.351 | 13084.3 | −3 |
| 1 dB-Chebyshev | 22 | 1679.26 | 3887.93 | 22 | 1229.11 | 23,745.9 | 15 | 659.93 | 168,992 | −1 |
| 2 dB-Chebyshev | 22 | 1268.21 | 4123.64 | 22 | 928.39 | 30,042.9 | 10 | 747.60 | 327,284 | −2 |
| 3 dB-Chebyshev | 22 | 1032.38 | 4501.79 | 22 | 755.28 | 36,161.2 | 10 | 607.94 | 397,792 | −3 |

excessive loading. The amount of gain peaking is different for each type, increasing as the filter moves from Linkwitz-Riley to a 3 dB-Chebyshev characteristic.

The component values were calculated using the tables of stage frequency and $Q$ given in Chapter 7, and checked by simulation. The design process depends on what definition of cutoff attenuation is used; throughout this book I have employed the most common definition, set out in the rightmost column of the table.

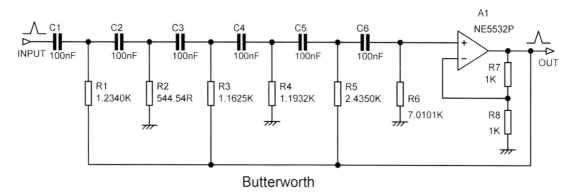

Figure 9.17: Sixth-order Sallen & Key Butterworth highpass filter, implemented as single stage, with gain = 2 and exact values. Cutoff frequencies 1 kHz.

The maximum $Q$ of 12.79 occurs in the 2nd-order stage of the 3 dB-Chebyshev filter; note the very large ratio between R3 and R4.

## Sallen & Key 6th-Order Highpass: Single Stage

A 6th-order highpass filter can be built in a single stage, as in Figure 9.17. The capacitor ratios were derived from Aikens and Kerwin [3] and transformed into resistor values by taking the reciprocal and appropriate scaling. The component sensitivity will be inferior to that of a three-stage 6th-order filter.

There is not, as far as I am aware, any way to make a 6th-order filter using two Geffe 3rd-order filters, because this will give two pole pairs and two 1st-order poles, when what you need is three pole pairs. If I am wrong please let me know.

## Sallen & Key Highpass: Input Impedance

The input impedance of a Sallen & Key highpass filter is high at low frequencies because of the high reactance of the series input capacitor.

At low frequencies, the input impedance of a 2nd-order Sallen & Key highpass filter does not approach the value of R1. At low frequencies no signal reaches the opamp because of C2-R2, and so the opamp output is effectively at signal ground, and R1 is connected directly to this output; however, the input impedance is greater than R1 because of the high impedance of C1.

## Bandwidth Definition Filters

This seems a good point to look at some real filter applications that show how our requirements drive the choice of filter type and its detailed design. An important feature of crossovers for sound reinforcement is the implementation of band definition (or you might wish to call it bandwidth limitation, but that somehow sounds a bit less appealing) to keep subsonic and ultrasonic signals from getting into the amplifier/speaker system. There is more on the design requirements for this in Chapter 8.

## Bandwidth Definition: Subsonic Filters

Highpass filters used for subsonic protection are usually 2nd-order or 3rd-order Butterworth types with roll-offs at 12 dB/octave and 18 dB/octave. Fourth-order filters with 24 dB/octave slopes are less used, no doubt because of fears about the possible audibility of the faster phase changes generated by the steeper filter. If 4th-order filters are used, the cutoff frequency is made lower to space any possible effects further away from the bottom of the audio band. I am not aware of any use of fifth or 6th-order subsonic filters.

There is something of a consensus that either a 3rd-order Butterworth filter with a cutoff frequency of 20 Hz or a 4th-order filter with a cutoff at 15 Hz will give adequate protection. The usual Sallen & Key filters are normally used; one problem is that the low cutoff frequencies mean that large capacitor values are required if the circuit impedances are kept low to minimise noise. A 3rd-order filter will require three of these components, and a 4th-order filter four. While it is desirable to use polypropylene to prevent capacitor distortion, this gets expensive, and polyester types are often used in subsonic filters, in the not unreasonable hope that the levels involved will normally be low and that very low-frequency distortion is not very audible.

Figure 9.18 shows a 3rd-order Butterworth subsonic filter designed as a single stage, as described earlier in this chapter. This requires to be fed from the usual low-impedance source to give an accurate response but has the advantage over the usual 2nd-order + 1st-order configuration that the output impedance is low without the need for buffering by an extra opamp stage. The exact resistor values required are given.

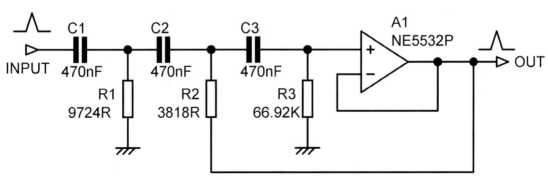

**Figure 9.18: A 3rd-order Butterworth single-stage subsonic filter with a cutoff frequency of 25 Hz.**

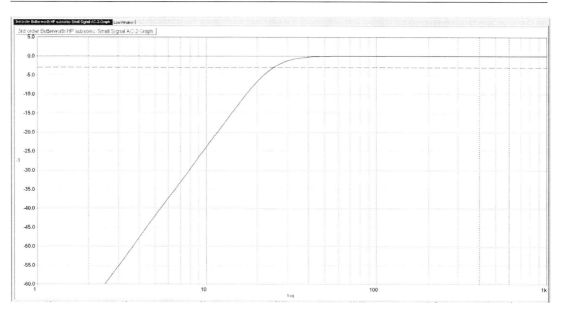

**Figure 9.19: Frequency response of the 3rd-order Butterworth single-stage subsonic filter, −3 dB at 25 Hz.**

Figure 9.19 gives the frequency response of the filter in Figure 9.18; it is 24 dB down at 10 Hz and gives good protection against subsonic disturbances.

## Bandwidth Definition: Combined Ultrasonic and Subsonic Filters

In some cases it is possible to economically combine highpass and lowpass filters into one Sallen & Key stage using only one opamp (ultrasonic filters are dealt with in Chapter 8). I must say at once that this cunning plan is workable only when the highpass and lowpass turnover frequencies are widely different, and its usefulness for crossover filters as such is limited. However, it can be very handy when you wish to explicitly define the bandwidth of an audio signal path by using both subsonic and ultrasonic filters; a sophisticated active crossover will include such bandwidth definition. Combined filters have the advantage that the signal now passes through one opamp rather than two, which may reduce noise and distortion, as well as saving money. This approach can be extremely useful if you only have one opamp section left.

Figure 9.20 shows a 20 Hz 3rd-order Butterworth subsonic filter, much the same as that just described, apart from a slightly lower cutoff frequency; note that this time E12 preferred values have deliberately been used for the resistors, to demonstrate that even so the cutoff frequency comes out as 21 Hz, very close to the desired value, and such an error is not likely to be of much significance in a subsonic filter. This filter is combined with a 2nd-order 50 kHz Butterworth lowpass ultrasonic filter, taken from Chapter 8, and the frequency response of the combination is exactly the same as expected for each separately. The lowpass filter is cautiously designed to prevent significant loss in the audio band and

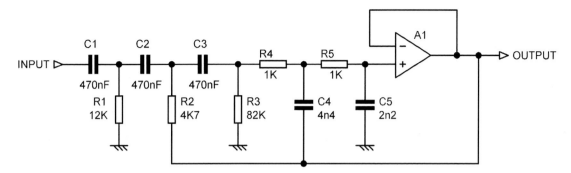

**Figure 9.20: A 3rd-order Butterworth subsonic filter combined with a 2nd-order ultrasonic filter.**

has a −3 dB point at 50 kHz, giving very close to 0.0 dB at 20 kHz; the response is −12.6 dB down at 100 kHz and −24.9 dB at 200 kHz. C4 is made up of two 2n2 capacitors in parallel.

The only compromise is that the midband gain of the combined filter is −0.15 dB rather than exactly unity; this is not exactly a cause for panic. The loss occurs because the series combination of C1, C2, and C3, together with C5, forms a capacitive potential divider attenuating by 0.15 dB, and this is one reason why the turnover frequencies need to be widely separated for filter combining to work well. If they were closer together, then C1, C2, C3 would be smaller, C5 would be bigger, and the capacitive divider loss would be greater. That is why this filter is one of the few in this book that uses 470 nF capacitors rather than 220 nF parts.

While this is an ingenious circuit, if I do say so myself, it occurred to me that it could be improved if designed as two combined 3rd-order filters, which could be implemented by just one amplifier so long as it can be assumed that the loading on the output is sufficiently light for the final pole R6-C6 to not be significantly affected. There is some flexibility here because an ultrasonic filter does not need to be so accurate as a crossover filter or an RIAA network, because almost everything it does is above the range of audibility. The 2nd-order part of the lowpass filter is set by C4 and C5 to a $Q$ of 1.00, as in conventional two-stage 3rd-order filters. This as usual causes a gain peak of +0.87 dB at 35 kHz; it seems unlikely this is going to cause any headroom problems. Note that a 5534A model was used for the amplifier; a TL072 model gave the usual oh-no-it's-coming-back-up-again behaviour above 100 kHz. TL072's are obsolete and deprecated.

The circuit is shown in Figure 9.21. In this case I used 220 nF capacitors for economy of cash and PCB area, and as predicted the passband attenuation was a bit greater, at −0.28 dB. I think this is still small enough to be tolerated and definitely saves significant money on capacitors. The frequency response with its two 18 dB/octave slopes is shown in Figure 9.22.

I was going to leave it there, but the temptation to explore a topic just a little further is irresistible. To me, anyway. The result of a bit more night thought was a 4th-order highpass combined with a 4th-order lowpass, in just two stages, with each stage implementing both a 2nd-order highpass and a 2nd-order lowpass, with different $Q$'s in the two stages, as required for Butterworth or other filter types. All the combined filters in this section are Butterworths; I suppose you might say they were Butterworthy . . .

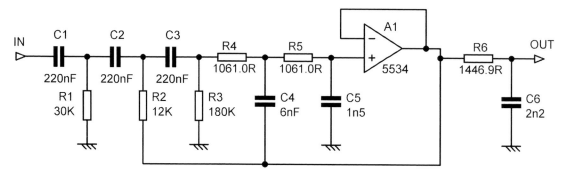

**Figure 9.21: A 3rd-order Butterworth 20 Hz subsonic filter combined with a 3rd-order Butterworth 50 kHz ultrasonic filter.**

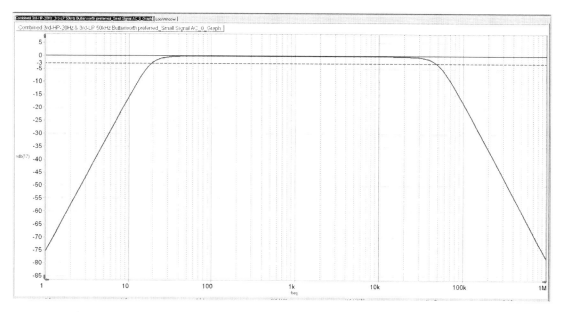

**Figure 9.22: Frequency response of the combined 3rd-order/3rd-order filter.**

The result is shown in Figure 9.23. The passband loss is smaller than that of the previous filter, at only −0.26 dB. Most of this (−0.18 dB) occurs in the first stage due to the loading of C4 on C1 and C2; the loading effect in the second stage is less because C8 is smaller. It would be possible to scale the components R3, R4, C3, C4 to reduce the passband loss, but there seems to be no pressing need to do so. There is no gain peaking in the first stage at either LF or HF, because in both cases it has low Q's, so there will be no headroom problems. There is no final pole that is likely to require buffering.

We therefore have two 4th-order filters implemented with just two amplifiers, which I fondly believe to be a completely new idea. The frequency response with its two 24 dB/octave slopes is shown in Figure 9.24.

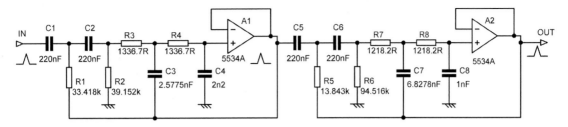

**Figure 9.23: A 4th-order Butterworth 20 Hz subsonic filter combined with a 4th-order Butterworth 50 kHz ultrasonic filter.**

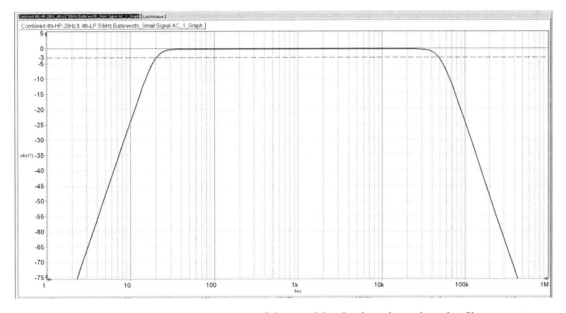

**Figure 9.24: Frequency response of the combined 4th-order/4th-order filter.**

It might be worth repeating at this point that combined filters only work well because the highpass and lowpass cutoff frequencies are well separated; in the case of 20 Hz and 50 kHz, by 11.3 octaves.

## Variable-Frequency Highpass Filters: Sallen & Key

Active crossovers for sound-reinforcement applications commonly have variable crossover frequencies so they can be used with a wide range of loudspeaker systems. This presents problems when the standard Sallen & Key filters are used, because you need ganged pots with different values in the sections. A typical variable-frequency Sallen & Key variable highpass filter that avoids this issue is shown in Figure 9.25, where ganged variable resistors alter the cutoff frequency over a 10:1 range from 100 Hz to 1 kHz, such as might be used for the LF-MID crossover point. R1 and R2 are end-stop resistors to limit the frequency range at the upper end.

Equal resistors, allowing the use of a standard two-gang pot as in Figure 9.25, can be obtained with a stage gain of 1.586 times (+4.00 dB), but this may lead to problems with the gain structure of the crossover. The stage can be made to appear unity-gain by taking the output from the junction of R3 and R4, but +4 dB of gain still exists at the opamp output and will still cause headroom issues. Unity gain would be much preferred.

A unity-gain Butterworth highpass requires R2 to be twice the value of R1. It is possible to obtain to special order pots with different values in their sections, but here lurks a snag. The vast majority of pot track resistances are in the E3 series, running 1, 2.2, 4.7, 10, etc, so you are not going to get closer to the desired 2 times ratio than 2.2 times; see Figure 9.26 (note the capacitor values have been reduced to get the same frequency range). In fact the response errors this creates are not enormous, being about 0.7 dB around the cutoff point with no unwanted gain peaking, but it is not exactly a precision

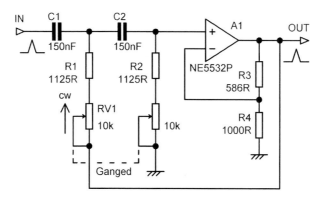

**Figure 9.25: Variable-frequency 2nd-order Sallen & Key highpass Butterworth filter; cutoff 100 Hz to 1 kHz.**

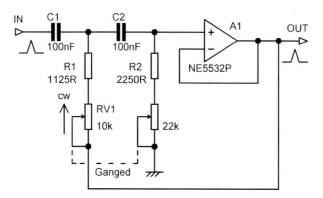

**Figure 9.26: Variable-frequency 2nd-order Sallen & Key highpass Butterworth filter with unity gain; cutoff 100 Hz to 1 kHz.**

solution. For accuracy without headroom issues a state-variable filter would be much superior, but it has a lot of flexibility we don't need and uses three amplifiers instead of one. See Chapter 10 for more details on state-variable filters.

You may think this is not a very satisfactory outcome. We have so far offered an accurate S&K filter with unwanted gain, an inaccurate S&K filter without it, or a much more complicated state-variable filter. And yet . . . in the wonderful world of electronics, there is almost always a way to weasel out of problems. In Figure 9.27, behold the Decoupled Sallen & Key filter, [4] also known as a Bach filter, after its originator. [5] It is so-called because an extra amplifier A2 decouples the R1-C1 time-constant from the R2-C2 time-constant, so the components can be chosen independently. This allows us to make R1 = R2, so long as we double the size of C1 compared with C2, combining a standard pot with overall unity gain. There is however some minor internal peaking of about +2 dB at A2 output just above the cutoff frequency (at around 1.3 kHz with cutoff set to 1 kHz), but I hope you will agree that this is a much better proposition than +4 dB at all passband frequencies.

This configuration also has the advantage of very low cutoff frequency component sensitivities of −0.5, which I think for a 2nd-order filter is as low as they get. There are equivalent Decoupled Sallen & Key lowpass and bandpass filters.

The variable resistors will normally be of the reverse-log law so that the frequency calibration is approximately linear in octaves. This circuit relies on the matching of the resistance values of the variable resistor tracks and also the matching of the value obtained at a given degree of rotation, which is worsened by the need for a log law. This gives rise to errors in the cutoff frequency; the effect on the filter $Q$ is much less. Be aware that in Figures 9.25, 9.26, and 9.27 no precautions have been taken to deal with the opamp input bias current flowing through the variable resistors. This could lead to noises as the controls are altered, unless appropriate DC-blocking capacitors are added.

A serious difficulty is the very high degree of control-ganging that is required in a practical crossover. In a 4th-order Linkwitz-Riley crossover, to vary the frequency of the two cascaded Butterworth 2nd-order filters, we need a four-gang part: obtainable but costly compared with two-gang. We need

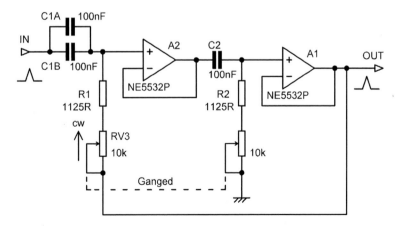

**Figure 9.27: Variable-frequency 2nd-order Decoupled Sallen & Key highpass Butterworth filter; cutoff 100 Hz to 1 kHz.**

to vary the frequency of the another path simultaneously (to change the LF-MID crossover point we must alter the frequencies in both the LF and MID paths), so we must source an eight-gang control; relatively hard to get and expensive. Variable-frequency crossovers of higher order are best implemented by DSP.

# References

[1]  Geffe, Philip "How to Build High-Quality Filters Out of Low-Quality Parts" Electronics, November 1976, pp. 111–113
[2]  Williams and Taylor, "Electronic Filter Design Handbook" Third Edn, McGraw-Hill, 1995, pp. 3.16–3.19
[3]  Aitken and Kerwin," Single Amplifier, Minimal RC, Butterworth Thomson and Chebyshev Filters to Sixth Order" Proceeds of International Filter Symposium, Santa Monica, April 1972.
[4]  Chen, Wai-Kai. "Passive, Active, and Digital Filters" Second Edn, CRC Press, 2009, pp. 13–2 to 13–4, ISBN 978 142 005 8857
[5]  Bach, R. E. "Selecting R-C Values for Active Filters" Electronics, 33, 1960, pp. 82–83

# CHAPTER 10

# *Other Lowpass and Highpass Filters*

## Designing Filters

In Chapters 8 and 9 we saw the stunning versatility of the Sallen & Key configuration for making lowpass and highpass filters, but while it is probably the most useful and popular active filter it is by no means the only type out there. The multiple-feedback or MFB filter configuration is very likely the second best known; it has some potential advantages, but there is an inherent and inconvenient phase inversion. Other filters require more than one amplifier but in return give lower sensitivity to component tolerances or to varying opamp open-loop gain, or give convenient adjustment such as independent control of cutoff frequency and $Q$. Biquad is a term often used in filter design, being short for "biquadratic", and it means that the complex equation that defines the filter response is the ratio of two quadratic equations, giving great flexibility in the filter response. You don't need to get involved in that if you don't wish to; just follow the design examples given here. All examples have a cutoff at 1 kHz; scaling the components to get the cutoff frequency you want is described in Chapter 8.

Table 10.1 gives a very brief summary of the filter configurations you are most likely to come across in textbooks; there are other and more obscure filters which are noted at the end of this chapter. The opamp numbers are for a single 2nd-order stage.

In the early days of active filter design amplifiers were expensive, and much effort was put into minimising their number. The Deliyannis-Friend filter [1] is an example of this. It is essentially a multiple-feedback or MFB filter (sometimes called a Deliyannis filter) with two positive feedback paths added by Friend, which allows zeros to be generated so that notch and allpass responses can be produced. It is therefore a single-amplifier biquad or SAB, a term much used in the filter literature. For simple lowpass and highpass applications the Deliyannis-Friend filter requires more precision passive components than the Sallen & Key and is harder to design. It is not suitable for high $Q$'s. In present day audio the cost of another opamp or two is usually very acceptable if it gives higher performance; SABs are therefore not considered further here. There are other types of SAB based on the Twin-T network described in Chapter 12; more information can be found in filter textbooks.

The Fliege filter presents an interesting intermediate complexity, as it use two opamps rather than one or three. It will not be considered further here because each 2nd-order Fliege stage inherently has an unwanted +6 dB of gain, which will bring some awkward noise/headroom compromises, especially when these stages are cascaded.

**Table 10.1: Summary of active filter types.**

| Filter type | Advantage | Disadvantage | Opamps |
| --- | --- | --- | --- |
| Sallen & Key | Output in phase | Common-mode distortion | 1 |
| Multiple-Feedback (Deliyannis) | No common-mode distortion | Phase inversion | 1 |
| Deliyannis-Friend | Only one amplifier | More passive parts | 1 |
| Akerberg-Mossberg biquad | Insensitive to amplifier gain | Phase inversion Stability issues | 3 |
| Tow-Thomas biquad | No common-mode distortion. Low component sensitivity Orthogonal tuning Output in phase | Uses three amplifiers | 3 |
| Fliege | Lower component sensitivity | Gain is always +6 dB | 2 |

## Multiple-Feedback Filters

The multiple-feedback (MFB) filter is also called the Deliyannis filter; it is most familiar as a bandpass filter working at a modest Q, but the basic configuration can also be used to make lowpass and highpass filters. The variations are shown in Figure 10.1; the bandpass version is dealt with in more detail in Chapter 12.

Multiple-feedback (MFB) filters have the advantage that since they use shunt feedback, with a virtual earth at the inverting input, there is no common-mode voltage to cause distortion in the opamp. This is in contrast with the Sallen & Key filter, where the opamp is working as a voltage-follower, and so the full output signal voltage appears at the opamp inputs; it is the worst case for common-mode distortion. This potential problem is dealt with at length in Chapter 16, but it is worth pointing out here that if you use bipolar opamps such as the 5532 and relatively low source impedances, common-mode distortion is not likely to be a serious difficulty.

One awkward point that stands out is that the highpass MFB uses three capacitors rather than two, which is very unusual in a 2nd-order circuit. Since capacitors are more expensive than resistors, this is a definite drawback. The highpass MFB filter also seems to have some unhelpful performance issues, such as higher noise and distortion than the Sallen & Key equivalent. There is more on the noise

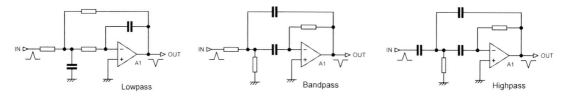

**Figure 10.1: Multiple-feedback (MFB) filters: lowpass, bandpass, and highpass.**

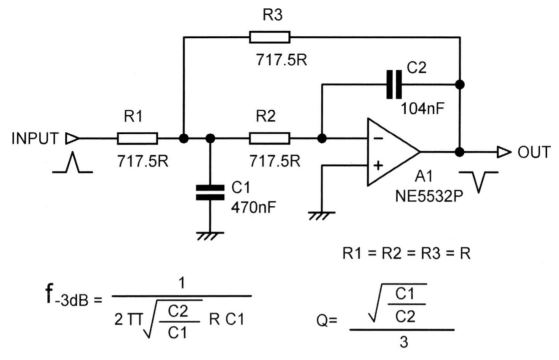

Figure 10.2: A 2nd-order lowpass Butterworth multiple-feedback (MFB) filter with the analysis equations. Cutoff frequency is 1 kHz, $Q$ = 0.707, and passband gain is unity.

and distortion performance of MFB filters in Chapter 11, but I'll tell you now, I would not rush into highpass MFB filters.

## Multiple-Feedback 2nd-Order Lowpass Filters

A practical version of an MFB lowpass filter is shown in Figure 10.2, designed for a Butterworth characteristic ($Q$ = 0.7071) and a cutoff frequency of 1 kHz; the passband gain is unity. Note that if C1 is a preferred value, C2 will in general not be; the value of 104 nF shown here would in practice be approximated by 100 nF in parallel with 4n7. As usual the resistor values do not work out conveniently, but very close approximations to these values can be cheaply made up using 2xE24 parallel pairs of preferred values. MFB bandpass, highpass, and lowpass filters all inherently give a phase inversion. This is no problem if they are used in pairs in a 4th-order Linkwitz-Riley crossover, but otherwise could be inconvenient, in the worst case requiring another stage that does nothing but invert the signal to get it in-phase again.

It is worth pointing out that the MFB lowpass filter does not depend on a low opamp output impedance to maintain stopband attenuation at high frequencies, and so avoids the oh-no-it's-coming-back-up-again behaviour of Sallen & Key lowpass filters. It is doubtful however if this is of much relevance to crossover applications.

## Multiple-Feedback 2nd-Order Highpass Filters

The multiple-feedback highpass filter is the multiple-feedback lowpass filter with the resistors and capacitors interchanged. A practical version of an MFB highpass filter is shown in Figure 10.3, once more designed for a Butterworth characteristic ($Q = 0.7071$), a cutoff frequency of 1 kHz, and unity passband gain.

This time we have three identical capacitors which can be conveniently chosen from the E6 series, dealing with the awkward resistor values in the usual way; as it happens, in this case R1 comes out as the E24 value of 750 Ω. There may be three capacitors, but this is still a 2nd-order circuit. Given the relatively high cost of capacitors compared with other components, this is not a particularly appealing configuration.

## Multiple-Feedback 3rd-Order Filters

Filters of higher order using MFB stages can be made in the same way as for Sallen & Key stages; the appropriate cutoff frequencies and $Q$'s for each stage are taken from Chapter 7 and the stages placed in series, in a suitable order that minimises headroom restrictions due to gain peaking. Buffer amplifiers are used as required to make sure that 1st-order circuits are not loaded by succeeding stages.

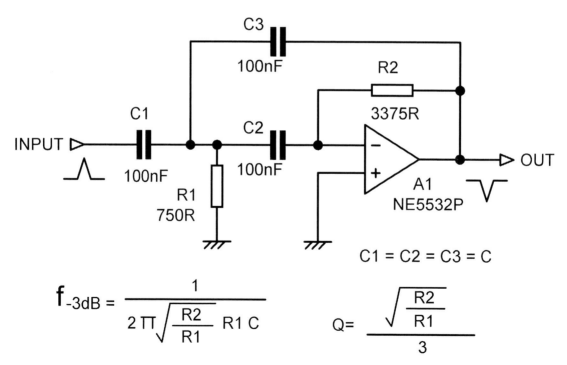

$$f_{-3dB} = \frac{1}{2\pi\sqrt{\dfrac{R2}{R1}}\,R1\,C} \qquad Q = \frac{\sqrt{\dfrac{R2}{R1}}}{3}$$

**Figure 10.3: A 2nd-order highpass Butterworth multiple-feedback (MFB) filter with the analysis equations. Cutoff frequency = 1 kHz, Q = 0.707 and passband gain is unity.**

As for Sallen & Key filters, these buffer amplifiers can be dispensed with if the loading effects are taken into account in the design; this does however make the process much more difficult and is likely to make component sensitivities worse. I have never used these filters myself, and I think caution should be the watchword. Clearly an amplifier is saved; however, you can also make both lowpass and highpass 3rd-order Sallen & Key filters in one stage and therefore using only one amplifier. Examples of 3rd-order Butterworth lowpass and highpass filters are given next, which can be scaled for different cutoff frequencies.

## Multiple-Feedback 3rd-Order Lowpass Filters

A 3rd-order lowpass MFB filter can be made by placing an additional 1st-order lowpass circuit R4-C4 just before the 2nd-order MFB filter. In Figure 10.4 R4 = R1 = R2 = R3/2.

## Multiple-Feedback 3rd-Order Highpass Filters

A 3rd-order highpass MFB filter can be made by placing an additional first-order highpass circuit C4-R4 just before the 2nd-order MFB filter. In Figure 10.5, C4 = C1 = C2 = 2C3. There is a clear temptation to make C3 the preferred value of 47 nF. If you do, the cutoff frequency and the later roll-off are unaffected, but the gain in the passband is increased from 0 dB to +0.54 dB; you can probably live with this. Although there are four capacitors, this is still a 3rd-order filter, in the same way that Figure 10.3 has three capacitors but is only a 2nd-order filter. Given the relatively high cost of capacitors, this is not an economic circuit.

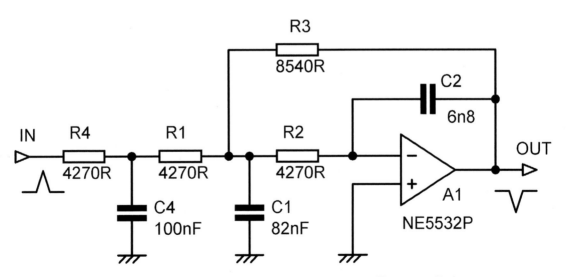

**Figure 10.4: Third-order Butterworth lowpass MFB filter. Cutoff 1kHz.**

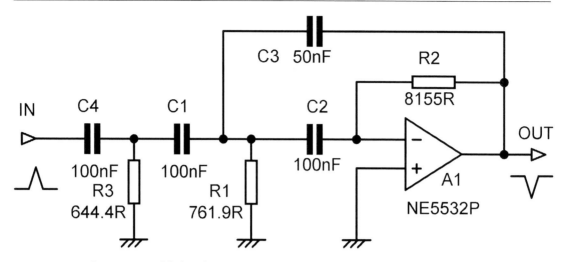

**Figure 10.5: Third-order Butterworth highpass MFB filter. Cutoff 1kHz.**

## Biquad Filters

Most biquad filters are composed of three amplifier stages in a loop. It is also possible to make biquads using just one amplifier with passive components around it, and they are unsurprisingly known as single-amplifier biquad (SAB) filters. Interest in SAB filters was high when the amplifiers were by a long way the most expensive parts in the circuit; that is no longer the case—it is now normally the capacitors which are the most expensive parts, unless for some reason you are using exotic opamps. SABs have their drawbacks and are not considered here.

## Akerberg-Mossberg Lowpass Filter

The Akerberg-Mossberg filter [2] uses three amplifiers. It is best suited to lowpass applications. It is a combination of an Antoniou active compensated non-inverting integrator (A1 and A2) and a leaky integrator (A3). The point of the Antoniou integrator [3] is that it does not invert, and so A3 can be a simple inverting stage, and that routing the negative feedback through A2 actively compensates for the finite bandwidth of A1, increasing their combined bandwidth. The inverted feedback via A2 means that the summing point (virtual earth) of the integrator is at the non-inverting input of A1, as in Figure 10.6, which shows an Akerberg-Mossberg 2nd-order Butterworth lowpass filter with a cutoff frequency of 1 kHz. It may look wrong in the diagram, but I assure you that's how it's supposed to be connected. A leaky integrator is an integrator with a resistance R5 placed across the capacitor C2.

The Akerberg-Mossberg filter is claimed to be superior to others in that it is very insensitive to amplifier shortcomings in gain and bandwidth. With modern opamps, this is not the big issue it used to be, but the configuration is also useful for generating higher $Q$'s than are practicable with the Sallen & Key or MFB configurations. $Q$'s of 400 or more are possible, if not obviously useful for crossover design. But . . . before you get too confident, read about the implementation problems I had.

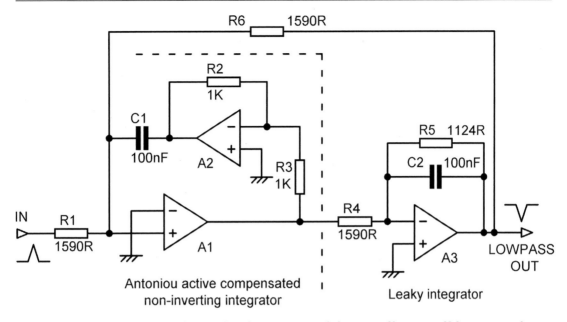

**Figure 10.6: Akerberg-Mossberg 2nd-order Butterworth lowpass filter; cutoff frequency 1 kHz.**

Multiple opamp filters sometimes yield a useful signal at every opamp output, the classic case being the state-variable filter. This is not the case here; the output of A1 is a −6 dB/octave roll-off, which seems unlikely to be useful in crossover design. This point also shows +3 dB of gain in the passband, which is very likely to cause headroom problems. The output of A2 is the same as the output at A1 but phase inverted; also of little or no use to us. The lowpass signal at A3 output is phase inverted with respect to the input in the passband, which is likely to be inconvenient unless you are cascading two of them.

As usual it is best to choose the capacitor values and let the resistors come out as they may. The usual design process for a Butterworth characteristic with $Q = 0.7071$ is:

| | | |
|---|---|---|
| Choose C | $C1 = C2 = C$ | 10.1 |
| Calculate R | $R = \dfrac{1}{2\pi f_0 C}$ | 10.2 |
| Then | $R4 = R6 = R$ | 10.3 |
| Calculate R5 | $R5 = \dfrac{R}{\sqrt{2}}$ | 10.4 |
| Choose R2 | $R3 = R2$ | 10.5 |

R2 and R3 have no link with the value of R; they just have to be equal.

$$\text{The passband gain A is: } A = \frac{R}{R1}$$
10.6

These equations give the 1 kHz lowpass filter seen in Figure 10.6, and it works very nicely except for a couple of non-obvious snags. First, when designing a filter with more than one amplifier it really is essential to check every amplifier output to see if it has a higher signal level than the actual filter output. It is very necessary here; the A-M filter in Figure 10.6 has a gain of +3.2 dB in the passband at the A1 and A2 outputs, seriously degrading headroom (or signal/to noise, if you drop the nominal level by 3.2 dB to accommodate this gain). This is illustrated in Figure 10.7, showing measured results from the circuit of Figure 10.6; A1 has a higher output level than A3 and only falls at −6 dB/octave. So far as I am aware this is a standard Akerberg-Mossberg lowpass filter.

I decided to try and do something about this level problem. I reasoned that it would be best to leave the Antoniou integrator alone, as any modifications would mess up the active compensation process. That leaves leaky integrator A3. It occurred to me that if R4 was reduced by a factor of √2 (3 dB) and R6 increased by √2, the level at A1 and A3 outputs would be same, but the Antoniou integrator would not know that anything had changed. Since R5 has not been altered, C2 remains the same value as C1, and an awkward capacitor value is avoided; see Figure 10.8. This process works very nicely, and I believe it is a new finding; I therefore intend to call this the Akerberg-Mossberg-Self filter. OK?

The second snag proved less tractable. All was going very nicely until I attempted to measure the noise and distortion of the filter. A version using 5532 sections showed odd jumps in THD with

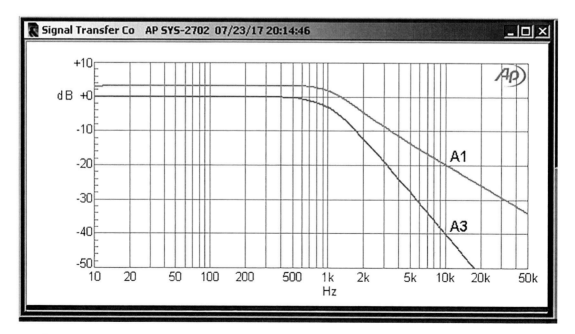

**Figure 10.7: Output levels referred to input of the Akerberg-Mossberg filter in Figure 10.6.**

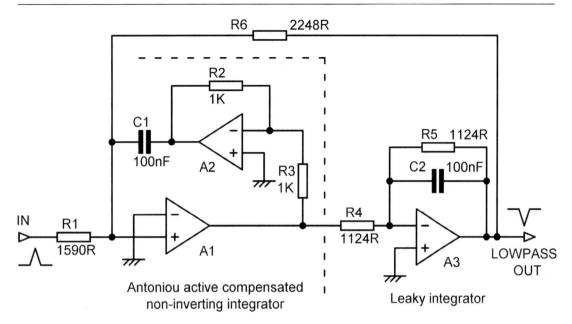

**Figure 10.8: Akerberg-Mossberg-Self 2nd-order Butterworth lowpass filter with no internal clipping problems; cutoff frequency 1 kHz.**

frequency, and generally too much THD altogether. The cause seems to be some sort of complicated VHF instability in the ingenious Antoniou integrator. Before anyone asks, A1 and A2 were in the same dual package to get amplifier matching. A TL072 was tried for A1, A2, in the hope that its lower open-loop gain would fix the situation, but the results were much the same. An LM4562 was also tried in this position, but again the same problems appeared. I have some experience in taming recalcitrant circuitry, but there didn't seem to be any way to fix this example. This was my first attempt at active compensation, and I am not impressed.

There the Akerberg-Mossberg situation rests for the present. Right now, if I needed to get a filter into production quickly, I wouldn't touch it with my worst enemy's bargepole.

The name is sometimes seen spelt "Ackerberg-Mossberg", but "Akerberg-Mossberg" is the correct usage.

## Akerberg-Mossberg Highpass Filters

The Akerberg-Mossberg is not much suited to making highpass filters. It can be done by replacing R1 with a third capacitor C3 connected to the inverting input of A3, as in Figure 10.9. There is now a bandpass response available at A1 and A2 outputs. A serious drawback is that the input is through C3 to a virtual ground, and the input impedance will be very low at high frequencies. Thus a 1 V input at 1 kHz gives input current of 628 uA, but at 20 kHz it is 12.5 mA. There are also likely to be stability issues with opamps driving large capacitive loads.

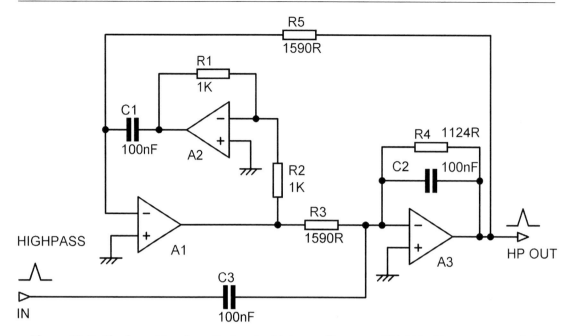

**Figure 10.9: Akerberg-Mossberg 2nd-order highpass filter; cutoff 1 kHz. Not recommended.**

There are also three capacitors instead of two, adding to the cost. However, errors in the value of C3 do not affect the cutoff frequency or $Q$ of the filter, just the gain, moving the response up or down without changing its shape. Depending on the application, this may mean that the accuracy of C3 can be relaxed compared with C1 and C2.

Overall, the highpass Akerberg-Mossberg filter is interesting but not recommended. I have therefore not given the detailed design process.

## Tow-Thomas Biquad Lowpass and Bandpass Filter

This lowpass and bandpass biquad filter configuration was introduced by Tow [4] and Thomas [5], [6] and is sometimes called the Ring-of-Three circuit. It should not be confused with the Ring-of-Two circuit, which is a clever way of making a voltage reference [7] and nothing to do with active filters. There are other multi-amplifier biquad configurations, but if the term "biquad" is used without qualification it normally means the Tow-Thomas biquad. All biquads are inherently 2nd-order, because the response is controlled by quadratic equations, but they can be cascaded into higher-order filters in the same way as other 2nd-order filters.

The Tow-Thomas biquad has advantages over other types of active filter. The component sensitivities are low. The distortion performance is good because all three amplifiers are working in shunt-feedback mode, so there is no common-mode signal on the opamp inputs and therefore no common-mode distortion. This is very much in contrast to the Sallen & Key filter, which usually has the full signal voltage on the inputs. The TT biquad was built and measured (see Chapter 11) and proved very tractable and easy to apply. Unlike the Akerberg-Mossberg filter . . .

The Tow-Thomas biquad gives lowpass and bandpass responses simultaneously; this might allow for some elegant fill-driver crossovers (see Chapter 4) so long as the bandpass output has the correct $Q$. The bandpass output has −6 dB/octave skirts. The TT does not directly give a highpass output. Configurations that give all three responses directly are often called universal filters rather than biquads; the KHN state-variable filter is the best-known example and is dealt with later in this chapter. The TT biquad is sometimes confused with the KHN state-variable filter because it has three amplifiers and apparently two integrators, but in the TT A1 is actually a leaky integrator, due to the presence of R3, and not a pure integrator like A2; the filter operation is quite different.

Figure 10.10 shows a Tow-Thomas 2nd-order Butterworth lowpass filter designed to give 0 dB gain in the passband of the lowpass output at A2. The lowpass output from A2 is in phase with the input in the passband. The output from A3 is the same signal as at A2 but phase inverted, and it has to be said that having two lowpass outputs in anti-phase is rarely if ever useful in crossover design. The inverter is essential to the circuit operation, but as I have said elsewhere it irks me to have an opamp doing just a phase inversion; it hardly seems to be earning its keep.

The design process is:

Decide the passband gain $A$ required.

$$\text{Choose a convenient value for C} \qquad \text{C1} = \text{C2} = \text{C} \qquad\qquad 10.7$$

(It is not essential to have C1 = C2, but it is obviously convenient and simplifies the design.)

Choose a convenient value for R2, such as 1 kΩ.

This however has deeper implications than mere convenience, which will be explained in a moment.

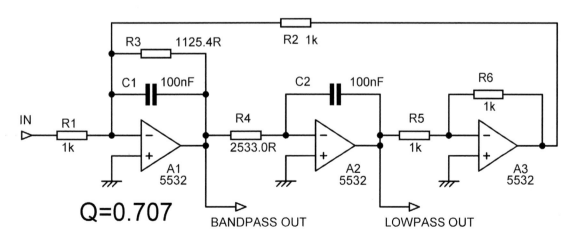

Figure 10.10: First version Tow-Thomas 2nd-order Butterworth lowpass biquad filter, centre frequency 1 kHz, +1 dB peak gain at A1 output.

Choose a convenient value for  R5 = R6, such as 1 kΩ.                                              10.8

$$\text{Then: } R3 = \frac{Q}{2\pi f_0 C}$$                                                        10.9

$$\text{and: } R4 = \frac{1}{(2\pi f_0 C)^2 R2}$$                                                   10.10

$$\text{Finally, R1 is set by the required passband gain A: } R1 = \frac{R2}{A}$$                  10.11

In any multi-amplifier filter, unity gain at one amplifier output can co-exist with more gain at the others, which will cause premature clipping internal to the filter. This needs to be checked for by simulation for each amplifier across the whole working frequency range. Here the output of A3 is clearly at the same level as A2, so no problem there. But . . . simulation shows a peak gain of +1.0 dB at the A1 bandpass output, which is not exactly disastrous but not ideal for maximum headroom; we can do better. The gain in the lowpass passband is always unity so long as R1 = R2.

The initial choice of R2 = 1000 Ω gave us Figure 10.10. The gain at A1 output can be reduced by increasing the initial value of R2. So we try setting R2 = 1100 Ω and find that recalculating the values of R3 and R4 gives a peak gain at A1 output of only +0.2 dB, which we can ignore. R1 is now also set to R2 = 1100 Ω to maintain unity gain in the lowpass passband. R3 remains at the same value, since R2 does not appear in Equation 10.9, but R4 does change; see Figure 10.11. The frequency response of this circuit is shown in Figure 10.12. We will call it TT Version 1; see Table 10.3.

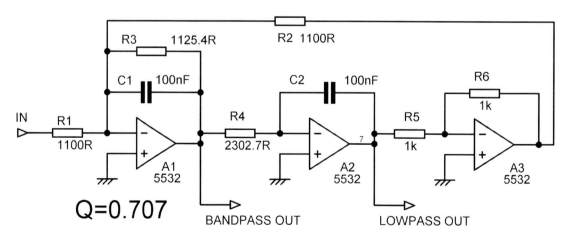

**Figure 10.11: Improved Tow-Thomas 2nd-order Butterworth lowpass biquad filter, centre frequency 1 kHz, +0.2 dB peak gain at A1 output. This is TT Version 1.**

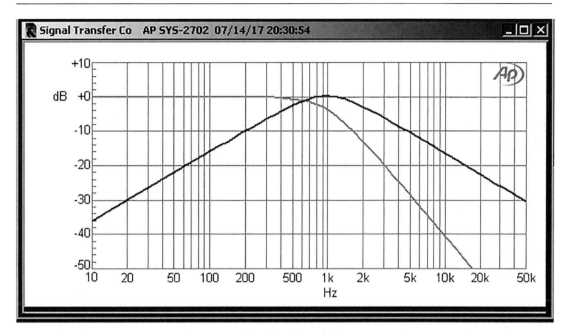

**Figure 10.12: Tow-Thomas 2nd-order biquad filter; centre frequency 1 kHz.**

The full analysis equations for the Tow-Thomas biquad, including non-equal capacitors, are:

$$\text{Cutoff frequency: } f_0 = \frac{1}{2\pi \cdot \sqrt{R2 \cdot R4 \cdot C1 \cdot C2}} \qquad 10.12$$

$$\text{For Q: } Q = \sqrt{\frac{R3^2 \cdot C1}{R2 \cdot R4 \cdot C2}} \qquad 10.13$$

If C1 = C2 = C and R2 = R4 = R, which is also likely to be true in many cases, the analysis equations are much simplified:

$$\text{Cutoff frequency: } f_0 = \frac{1}{2\pi RC} \qquad 10.14$$

$$\text{For Q: } Q = \frac{R3}{R} \qquad 10.15$$

$$\text{For passband gain A: } A = \frac{R2}{R1} \qquad 10.16$$

One of the advantages of the Tow-Thomas biquad over Sallen & Key configurations is that it allows what is called *orthogonal tuning*. This means that the three parameters of frequency $f_0$, Q, and gain can be adjusted independently if it is done in the right order; thus:

1.    Adjust R2 to get the required cutoff frequency $f_0$.
2.    Adjust R3 to get the required $Q$. This will not affect the $f_0$ previously set.
3.    Adjust R1 to get the desired passband gain A. This will not affect $f_0$ or $Q$.

As R1 is feeding into a virtual earth point, this is pretty obvious, unlike Step 2.

This will not be useful for production purposes, as presumably all parameters have been decided and fixed by the use of fixed resistors. It can however be very handy during crossover development work.

If the filter $Q$ is greater than 0.7071 (1/√2), then gain peaking will occur at the lowpass output around the cutoff frequency. If we assume a $Q$ of 1 is required, as is used to make up a two-stage 3rd-order Butterworth filter, then the peaking is +1.25 dB, which has a small effect on the headroom. We will call this TT Version 2, with the component values given in Table 10.3; they give 0 dB peak gain for the bandpass output, so there will be no headroom problems there. If you want to avoid any compromise on the headroom, the passband gain can be reduced as required by increasing the input resistor R1.

TT Version 3 is a TT example with a $Q$ of 2.56, the highest $Q$ required in a four-stage Butterworth 8th-order filter. If we try R2 = 1000 Ω we get R3 = 4074.4 Ω and R4 = 2533.0 Ω. SPICE simulation then shows us that the gain peaks at +8.2 dB at the lowpass output, which is what we expect with a $Q$ that is relatively high by crossover standards; the passband gain is still unity. However, as with all multi-stage filters, we must check the levels at every amplifier output. The A3 output is an inverted version of A2 output, so there are no problems there. But . . . at A1 the bandpass output peaks at +12.1 dB at the cutoff frequency, and that is going to restrict headroom.

As we saw in the first example, the gain at A1 output can be reduced by increasing the initial value of R2; this leads to a lower value for R4, so less gain is required in the A1 stage to give the same output at A2. R1 is then increased in the same ratio as R2 to keep the lowpass passband gain at unity. This process is illustrated in Table 10.2, which shows how increasing the initial value of R2 affects the other components and the gain distribution in the filter. Note that R3 does not change, as the $Q$ is not changing.

**Table 10.2: Component values for 1 kHz cutoff and $Q$ = 2.56. R3 = 4074.4 Ω.**
**R5 = R6 and C1 = C2 = 100 nF.**

| R2 | R4 | R1 | Peak gain at A1 out | Peak gain at A2 out |
|---|---|---|---|---|
| 1000 Ω | 2533.0 Ω | 1000 Ω | 12.1 dB | 8.2 dB |
| 1590 Ω | 1593.1 Ω | 1590 Ω | 8.2 dB | 8.4 dB |
| 2000 Ω | 1266.5 Ω | 2000 Ω | 6.2 dB | 8.3 dB |
| 2500 Ω | 1013.2 Ω | 2500 Ω | 4.2 dB | 8.3 dB |
| 3000 Ω | 844.34 Ω | 3000 Ω | 2.6 dB | 8.3 dB |
| 4000 Ω | 633.26 Ω | 4000 Ω | 0.16 dB | 8.3 dB |

**Table 10.3: Component values for 1 kHz cutoff at differing $Q$'s. R5 = R6 and C1 = C2 = 100 nF.**

| TT filter Version | Q | R1 | R2 | R3 | R4 |
|---|---|---|---|---|---|
| 1 | 0.7071 | 1100 Ω | 1100 Ω | 1125.4 Ω | 2302.7 Ω |
| 2 | 1 | 1590 Ω | 1590 Ω | 1590 Ω | 1590 Ω |
| 3 | 2.56 | 1590 Ω | 1590 Ω | 4074.4 Ω | 1590 Ω |
| 4 | 10 | 15.9 kΩ | 1590 Ω | 15.9 kΩ | 1590 Ω |

For crossover purposes R2 = 1590 Ω, giving equal peak gain at A1 and A2, gives probably the best noise/headroom compromise; we will call this TT Version 3. If for some reason you want near-unity gain at A1 output, you can see that R4 is getting too low; since it is driving a virtual earth, it is directly loading A1 output, and the distortion performance will suffer. In this case it would be better set C1 = C2 = 47 nF, which doubles the impedances. If then you set R1 = R2 = 8200 Ω, R3 emerges as 8668.9 Ω and R4 becomes 1398.4 Ω. R5 and R6 remain unaltered.

Suppose that for some reason you need a $Q$ that is high by crossover standards; I'll assume we want a lowpass with a $Q$ of 10. This obviously gives some serious gain peaking—ten times or +20 dBu at the cutoff frequency. If it's important to maintain full headroom at cutoff, then the only option is to reduce the passband gain by 20 dB, accepting there will be a noise penalty. This has been done in TT Version 4 in Table 10.3, so the gain at both lowpass and bandpass outputs is 0 dB at cutoff.

Table 10.3 summarises the component values for the versions of the Tow-Thomas biquad we have designed. In each case C1 and C2 are 100 nF and R5 and R6 are equal.

## Tow-Thomas Biquad Notch and Allpass Responses

The basic Tow-Thomas biquad gives only bandpass and lowpass outputs. However, by suitably summing these outputs and also the input signal, notch, allpass, and highpass responses can be obtained, as in Figure 10.13.

You will have spotted that a little bit of tweaking has been done to some of the resistors. R7 had to be adjusted to get a convincingly deep notch, simulation showing −58 dB. R10 had to be adjusted to get a suitably flat allpass output, simulation showing it flat to within ±0.01 dB. R13 had to be adjusted to keep the highpass output falling with frequency at −12 dB/octave down to −80 dB at 10 Hz; without the correct value the slope lessens to −6 dB/octave within the audio band. You will note that all the resistors were tweaked to exactly the same value, indicating that the design of the basic Tow-Thomas biquad is off by a fraction; the gain to the output of A1 is clearly 0.2 dB high. This is the gain value we decided to live with in the previous section, but now that we are making derived outputs, it is a decided nuisance. Putting the tweaked resistors back to 1 kΩ and recalculating the biquad with a fractionally higher initial value for R2 would give a more elegant result, but since there is a better way to make a highpass Tow-Thomas, as described in the next section, and that is more likely to be useful in active crossover design than notches or allpass filters, I have not pursued that further at present.

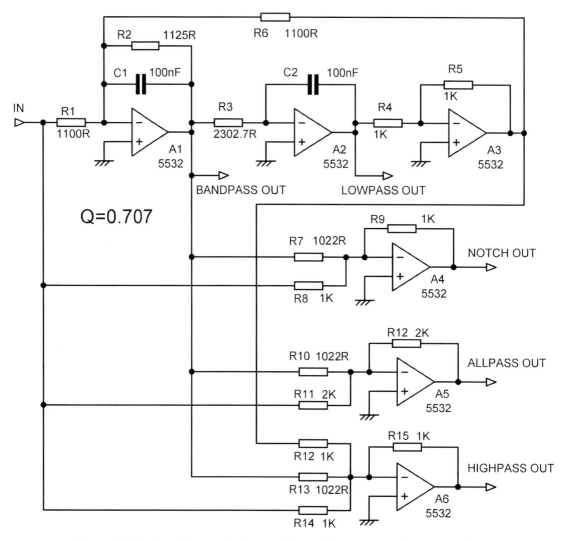

**Figure 10.13: Tow-Thomas 2nd-order biquad filter; centre frequency 1 kHz.**

In all derived outputs of this sort the summing resistors have to be accurate, as notch depth, allpass flatness, and the highpass −12 dB/octave slope are all critically dependent on precision summing.

## Tow-Thomas Biquad Highpass Filter

In the previous section we saw that deriving a highpass output from the standard Tow-Thomas biquad required four amplifiers and some critically accurate extra resistors. It is not an elegant solution, which is unfortunate, as active crossover design requires both lowpass and highpass filters. Mark Fortunato of Texas Instruments has reported a rearrangement of the Tow-Thomas configuration that gives a

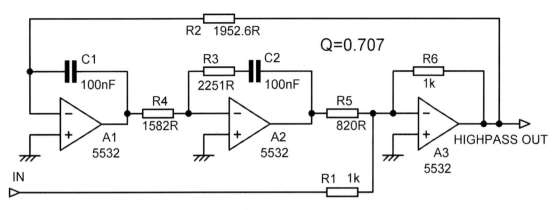

**Figure 10.14: Tow-Thomas/Fortunato 2nd-order highpass biquad filter; centre frequency 1 kHz.**

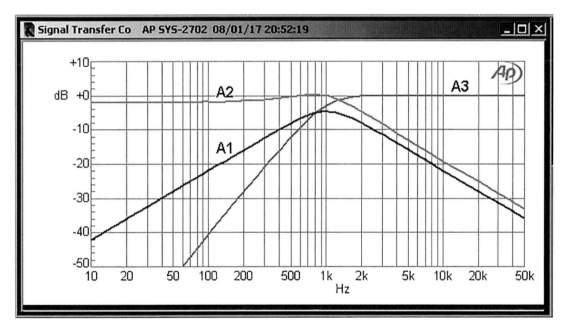

**Figure 10.15: Tow-Thomas/Fortunato 2nd-order highpass biquad filter: frequency response at each amplifier output.**

highpass output directly; [8] see Figure 10.14. The highpass output comes directly from A3. At A1 there is a bandpass output with−6 dB/octave skirts. At A2 there is a 1st-order roll-off with gain peaking just below the cutoff frequency; this is unlikely to be useful for anything.

As initially designed, R5 was set to 1 kΩ. For an input of 0 dB this gave a +2 dB peak at A2 output. It was decided that headroom was probably more desirable than the ultimate noise performance, so the circuit was recalculated with R5 set to 820R. This changes R2, and also R3 very slightly. R4 is unchanged, so this filter seems to have some of the orthogonal tuning features of the standard Tow-Thomas biquad. The alteration gives a gain peak at A2 of only +0.2 dB, which I think we can ignore. The level at A1 output now peaks at −5 dB, and there are no headroom issues there.

I have not derived the design equations for this filter. This is not as big a snag as you might think. Spreadsheets have facilities to set one quantity to a desired value by twiddling another; in Excel the feature is called Goalseek, and it is usually reliable, though I have known some equations to defeat it. This makes it fairly simple to work backwards from the analysis equations to get a design.

$$\text{Cutoff frequency} f_0 \quad f_0 = \frac{1}{2\pi \cdot \sqrt{R3 \cdot R6 \cdot \dfrac{R4}{R5} \cdot C1 \cdot C2}} \qquad\qquad 10.17$$

$$\text{For Q:} \quad Q = \frac{1}{R2}\sqrt{\frac{R3 \cdot R4 \cdot R6 \cdot C1}{R5 \cdot C2}} \qquad\qquad 10.18$$

$$\text{Passband gain A is:} \quad A = \frac{R6}{R1} \qquad\qquad 10.19$$

This is much sounder proposition for making an accurate TT highpass filter than the summation method; it uses one less amplifier, but perhaps more importantly, eliminates three precision resistors.

## State-Variable Filters

State-variable filters (SVFs) give highpass, bandpass, and lowpass outputs simultaneously. In crossover applications the bandpass output is not likely to be used, except perhaps for filler-driver applications. The Tow-Thomas biquad configuration looks similar, but is only a lowpass and bandpass filter with no highpass output. SVF filters show good component value sensitivity and are very simple to design. They are called "state-variable" filters because in the most common 2nd-order version, they are made up of two integrators and an amplifier, and signals from all three stages are used for feedback (the output of each integrator and the output of the amplifier). These three variables completely define the state of the circuit. Higher-order state-variable filters have more integrators but also a feedback signal from each of them. The state-variable filter is often called a KHN filter in textbooks, after Kerwin, Huelsman, and Newcombe, who introduced the configuration. Like the Tow-Thomas biquad, the SVF is a tractable and easy-to-use filter.

A basic 2nd-order SVF is shown in Figure 10.16. The input is applied to the inverting input of amplifier A1. As shown it has unity gain in the passbands of the highpass and lowpass outputs, and a gain of −3 dB at the passband centre of the bandpass output. The highpass output is phase inverted in the passband, the bandpass output is in-phase in the centre of the passband, and the lowpass output is phase inverted in the passband. The presence of two integrators indicates that it is a 2nd-order filter. Higher-order state variable filters are perfectly possible; for example a 4th-order state-variable filter has four integrators; higher-order SVFs are dealt with later in this chapter.

The amplitude responses of the three outputs can be seen in Figure 10.17, where the −3 dB points of the high and lowpass filters coincide with the band centre of the bandpass output at 1 kHz. The Q is set

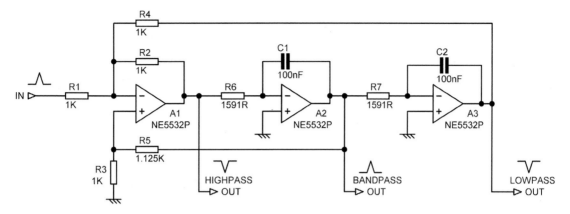

**Figure 10.16: Second-order state-variable Butterworth filter. Set R1 = R2 = R3 = R4, R6 = R7, and C1 = C2. Centre frequency is 1 kHz.**

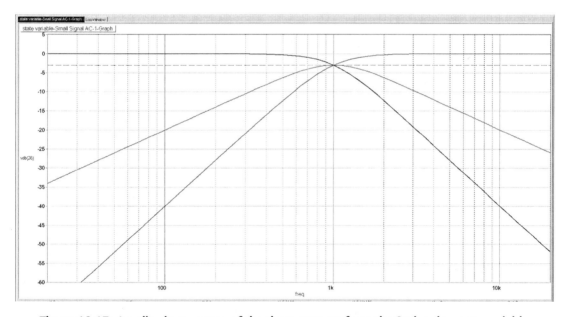

**Figure 10.17: Amplitude response of the three outputs from the 2nd-order state-variable Butterworth filter, centre frequency 1 kHz.**

at 0.7071 to give a Butterworth filter characteristic at the lowpass and highpass outputs, such as might be used in a 2nd-order crossover application. This low $Q$ gives a very broad bandpass characteristic, with no peaking at the centre. The lowpass and highpass outputs both have an ultimate slope of 12 dB/octave, as we would expect from a 2nd-order filter, while the bandpass output has slopes of 6 dB/octave on each side of its peak. Since the cutoff frequencies of the highpass and lowpass filters are inherently the same, this is obviously a very convenient way to make a 2nd-order crossover in one handy stage. This does however mean that it is not possible to employ frequency offset between the highpass and lowpass filters.

The state-variable filter may look more complex than the Sallen & Key or the MFB filter, but it is in fact delightfully simple to design. Choose a resistor value R = R1 = R2 = R3 = R4, and choose a capacitor value C = C1 = C2.

The required centre frequency *f* is then used to calculate R6 and R7:

$$R6 = R7 = \frac{1}{2\pi fC} \qquad\qquad 10.20$$

and Q is set by selecting an appropriate value for R5. For Butterworth it is 1.125 times the value of R.

To obtain other filter characteristics, it is necessary to set a different value of Q and use a frequency scaling factor to get the desired cutoff frequency, as we did with other forms of filter. So, to get a Bessel characteristic with a cutoff frequency of 1 kHz, R5 is set to 0.575 times R, and Equation 10.20 uses 1.273 times *f* as its input. The necessary values for the Q and scaling factor are given in Table 10.4

You will note that the lower the value of R5, the lower the Q; this is because the feedback path from the first integrator controls the damping of the circuit: the more feedback, the more damping.

Q can also be altered by changing the value of R3. Figure 10.18 shows the effect on the bandpass output Q of stepping R3 from 1000 Ω down to 100 Ω; as R3 is reduced, the filter Q increases, because the damping feedback signal is being reduced. It increases more rapidly at lower values of R3. If a variable Q control is required, a reverse-log type gives an acceptable law. The higher the Q, the higher the passband gain at the peak, which is usually very inconvenient in crossover and equalisation work; in Figure 10.18 the maximum Q gives a gain of +12 dB. Changes in Q have a minimal effect on the skirts of the bandpass plot away from the peak.

An alternative approach is to apply the input to the non-inverting input of amplifier A1, as in Figure 10.19. Now, as R3 is reduced, the Q is increased as before, but simultaneously the input signal is attenuated to a greater extent, giving a constant gain at the bandpass peak, as illustrated in Figure 10.20. This method is sometimes called a constant-amplitude state-variable filter, though that condition only applies to the bandpass peak. It is particularly useful for implementing parametric equalisers; see Chapter 14. Changes in Q now do affect the skirts of the bandpass plot, and as before, the rate of change of Q accelerates as Q increases. The gain at the passband peak is set by the ratio of

**Table 10.4: R5 scaling factor and frequency scaling factor for filter characteristics.**

| Filter type | R5 scaling factor | Frequency scaling factor |
|---|---|---|
| Bessel | 0.575 | 1.273 |
| Butterworth | 1.125 | 1.000 |
| 3 dB-Chebyshev | 3.47 | 0.841 |

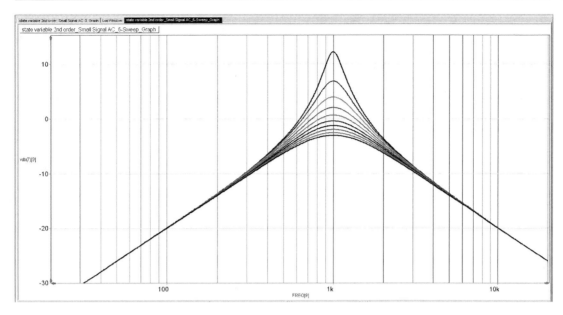

Figure 10.18: Effect of changing $Q$ on the state-variable Butterworth filter of Figure 10.16. Peak gain increases with $Q$. Centre frequency 1 kHz.

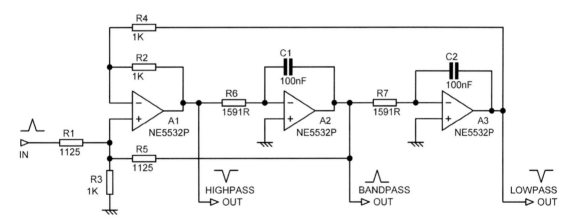

Figure 10.19: Second-order state-variable Butterworth filter with constant-amplitude input connection. Centre frequency is 1 kHz.

R1 and R5, and here is unity. Note that the phase relationships between input and outputs are in each case inverted with respect to that of Figure 10.16.

A notch output from a state-variable filter, with its centre at the same frequency as the bandpass output, can be obtained by the"1-bandpass" principle. The bandpass output is subtracted from the input signal by a fourth opamp. This is a relatively complicated way of making a notch filter, requiring precision

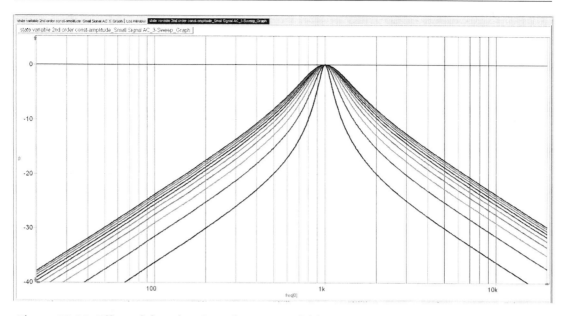

**Figure 10.20: Effect of changing *Q* on the state-variable Butterworth filter of Figure 10.19; peak gain is constant as *Q* changes. Centre frequency 1 kHz.**

resistors to get a good notch depth, and is not usually justified unless the ease of changing the centre frequency and *Q* which come with a state-variable filter is important. For fixed notches the Bainter filter is recommended, because its notch depth does not depend on the precise matching of resistors; see Chapter 12.

## Variable-Frequency Filters: State-Variable 2nd Order

The number of gangs in a crossover frequency control can be halved by abandoning the use of separate filters and instead using state-variable filters that produce outputs for two paths simultaneously; this technology is widely if not universally used in variable-frequency crossovers. We saw earlier that a two-gang variable resistor controls the cutoff frequencies of both the lowpass and highpass outputs simultaneously.

Figure 10.21 shows a variable-frequency 2nd-order state-variable filter based on the design of Figure 10.19. The frequency-control method is slightly more sophisticated than just using the frequency pots as variable resistors, because it aims to reduce the effect of control mismatches. The variable resistors are now being used as potentiometers, so to a first approximation the effect of differing total track resistances will cancel out. To make this effective the loading on the potentiometer needs to be light relative to the track resistance, and you will observe that R6, R7 have been increased in value by ten times, while C1 and C2 have been reduced in value by ten times, to raise the impedances without altering the centre frequency range. This will of course have some consequences in terms of increased noise. Log controls can be used to get a suitable frequency calibration law.

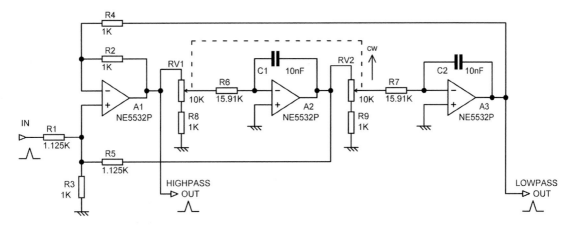

**Figure 10.21: Variable-frequency 2nd-order SVF crossover, with the crossover frequency variable from 85 Hz to 1 kHz.**

Altering the potentiometer setting alters the effective value of R6 and R7, and so the filter centre frequency. R8 and R9 are end-stop resistors to limit the frequency range at the lower end. Note that in this case the bandpass output of A2 is not used.

## Variable-Frequency Filters: State-Variable 4th-Order

While the 2nd-order variable-frequency SVF works very nicely, 2nd-order crossovers are not exactly the most popular choice, because of their relatively shallow slopes and considerable band overlap. What is much more desirable is a variable-frequency 4th-order Linkwitz-Riley state-variable crossover; precisely this was provided by Dennis Bohn in 1983, [9] and this very clever implementation has seen extensive use in the crossover business. Figure 10.22 shows a 4th-order variable filter with a crossover frequency range from 210 Hz to 2.10 kHz. Note that there are now four integrators plus the summing/differencing stage, to which feedback from all four integrators is returned. As usual the capacitors have been made preferred values, letting the resistor values fall where they may.

Figure 10.23 shows the lowpass and highpass outputs from the 4th-order filter. The two outputs are both 6 dB down at the crossover frequency, as expected for a Linkwitz-Riley crossover, and the ultimate slopes are at 24 dB/octave. The design of the filter for various frequencies is not hard. It follows the same process as the 2nd-order SVF described earlier. Ignoring the variable resistors, and treating R8, R9, R10, and R11 as the total resistance feeding each integrator from the previous stage, the centre frequency $f$ is then set by choosing R6 (= R7):

$$R8 = R9 = R10 = R11 = \frac{1}{2\pi fC} \qquad\qquad 10.21$$

Note that the resistors around A1 have to be in the ratio shown for correct operation and to get the correct $Q$ for a Linkwitz-Riley alignment. The required ratios are shown in brackets in Figure 10.22.

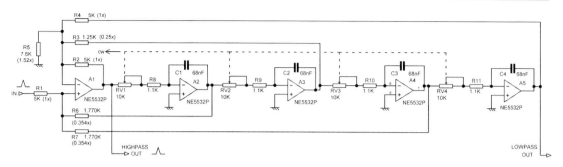

**Figure 10.22: Fourth-order variable-frequency SVF crossover, with the crossover frequency variable from 210 Hz to 2.10 kHz.**

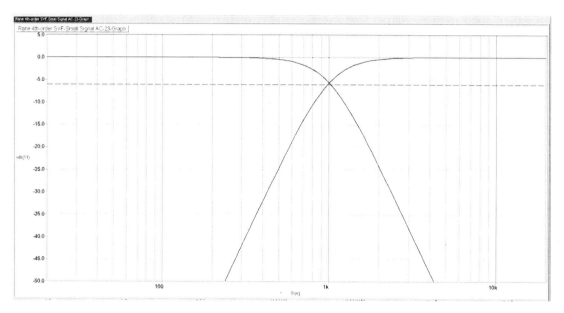

**Figure 10.23: Lowpass and highpass outputs from the 4th-order variable-frequency state-variable filter, crossing over at −6 dB for a Linkwitz-Riley alignment. Crossover frequency is set to 1 kHz.**

Figure 10.24 attempts to give some insight into how the filter works. The output from A1, the summing/differencing stage, is the highpass output, with an ultimate roll-off slope of 24 dB/octave below crossover as frequency decreases. The output of the first integrator A2 is the same but with a 6 dB/octave slope decreasing with frequency applied across the whole range, so the part of the response that was flat now slopes downward at 6 dB/octave slope with frequency, while the 24 dB/octave section has its slope reduced by 6 dB/octave to give 18 dB/octave. The second integrator A3

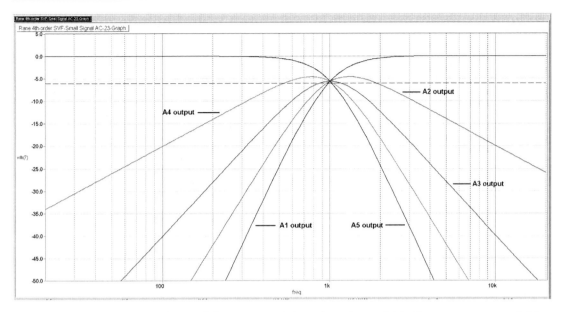

**Figure 10.24: The outputs from all five stages of the 4th-order variable-frequency state-variable filter, with crossover frequency set to 1 kHz.**

performs the same process again, so its output is a 4th-order bandpass response with skirt slopes of 12 dB/octave. The third integrator A4 does the same thing, its output having a 6 dB/octave slope in its low-frequency section, and a 18 dB/octave slope in its high-frequency section. The fourth integrator completes the process, and its output is the lowpass signal, with a flat low-frequency section and a 24 dB/octave roll-off above the crossover frequency. Note that all the plots in Figure 10.24 pass through the −6 dB point at the crossover frequency.

This 4th-order configuration could of course be adapted for fixed-frequency operation by removing the variable elements.

## Variable-Frequency Filters: Other Orders of State-Variable

You may at this point be wondering what happened to 3rd-order state-variable filters; there appears to be no reason why these should not be constructed in exactly the same way, using three integrators plus a summing/differencing stage. In view of the lesser desirability of 3rd-order crossovers compared with 4th-order, I have chosen not to investigate these at this time. It is certainly possible to make 8th-order variable-frequency state-variable filter crossovers by using the same approach, as demonstrated by Dennis Bohn in 1988, [10] and so there is every reason to assume that 5th-, 6th-, and 7th-order versions could be implemented without great difficulty. Eighth-order crossovers are sometimes used in sound-reinforcement applications but seem unlikely to become popular in domestic hi-fi.

# Other Filters

This chapter and the three preceding it by no means exhaust the filter types available. It is a fertile field for what Lord Rutherford would have called stamp-collecting. While I am not aware that the four filter configurations listed here have any special advantages for active crossover design, you may want to look them up. They are all multi-amplifier biquads.

> The Berka-Herpy biquad filter. [11]
> The Padukone-Mulawka-Ghausi (PMG) biquad filter. [12]
> The Mikhael-Bhattacharyya (MB) biquad filter. [13]
> The Natarajan biquad filter. [14] It is a modification of the standard Tow-Thomas biquad.

The first three use three amplifiers in the lowpass case, and the Natarajan lowpass uses four. The PMG and MB may or may not present technical challenges, but they certainly present spelling challenges and so are almost always referred to by their acronyms.

# References

[1] Deliyannis, T., Yichuang Sun and J. K. Fidler "Continuous-Time Active Filter Design" CRC Press, 1999, pp. 127–134

[2] Akerberg, D., and Mossberg, K. "A Versatile Active RC Building Block with Inherent Compensation for the Finite Bandwidth of the Amplifier" *IEEE Transactions on Circuits and System* 21, pp. 75–78, 1974.

[3] Ananda Mohan, P. V. "VLSI Analog Filters, Active RC, OTA-C, and SC" Springer 2023, Section 2.2 ISBN: 978-0-8176-8357-3 (Print) 978-0-8176-8358-0 (Online)

[4] Tow, J. "Active RC Filters: A State-Space Realization" Proc IEEE, 56, 1968, pp. 1137–1139

[5] Thomas, L. C. "The Biquad: Part-I—Some Practical Design Considerations" IEEE Transactions on Circuit Theory CT18, pp. 350–361, 1971

[6] Thomas, L. C. "The Biquad: Part II—a Multipurpose Active Filtering System" IEEE Transactions on Circuit Theory CT-18(3), 1971, pp. 358–361,

[7] Hartley Jones, Martin "A Practical Introduction to Electronic Circuits" Third Edn, Cambridge University Press, 1995, pp. 226–227

[8] Fortunato, Mark "A New Filter Topology for Analog High-Pass Filters" Texas Instruments Analog Applications Journal SLYT299, 2008

[9] Bohn, D. A. "A Fourth-Order State Variable Filter for Linkwitz-Riley Active Filters" 74th AES Convention, October 1983, preprint 2011 (B-2)

[10] Bohn, D. A. "An 8th-Order State Variable Filter for Linkwitz-Riley Active Crossover Design" 85th AES Convention, November 1988, preprint 2697 (B-11)

[11] Chen, Wai-Kai (ed.) "The Circuits and Filters Handbook" Second Edition, CRC Press, 2003, pp. 2491–2492

[12] ibid pp. 2486–2487

[13] ibid p. 2485

[14] Natarajan, S. "Modified Tow-Thomas Biquads with Improved Active Sensitivity" Proceedings of the 30th Midwest Symposium on Circuits and Systems, August 17–18, 1987, held in Syracuse, New York

# CHAPTER 11

# *Lowpass and Highpass Filter Performance*

## Aspects of Filter Performance: Noise and Distortion

There are several aspects to the performance of a filter, the most obvious being how accurately it follows the desired frequency response. That has been examined thoroughly in the last few chapters. Here we examine two other important aspects of performance; the noise and distortion behaviour of various types of filter. The distortion performance in particular is rather complicated. This situation has not been improved by the advent of a band called Filter Distortion, [1] which much impedes Google searches on the subject.

## Distortion in Active Filters

Distortion in active filters may come from the amplifier(s) or the capacitors. It is assumed that modern metal film resistors are used, which are perfectly linear as far as we are concerned at opamp levels. It may seem obvious that amplifiers have non-zero distortion, but the point is that when they are used in active filters, they may show much more distortion than if they were working as a simple unity-gain buffer. The basic amplifier distortion is will be increased in a Sallen & Key filter by common-mode distortion, due to significant series resistance in the lowpass case, and in both lowpass and highpass cases by the fact that the amplifier inputs see the full signal amplitude.

Another issue which has received some attention is distortion aggravation factor, (DAF) a term coined by Peter Billam. [2] The DAF is said to vary widely with frequency, filter type, and filter order; it can be as low as 1 or as high as 1000. There is more on this worrying prospect in the next section.

When they have a signal voltage across them, some capacitor types generate distortion. This unwelcome phenomenon is described in Chapter 15. It afflicts not only all electrolytic capacitors but also some types of non-electrolytic such as polyester. If electrolytics are being used as coupling capacitors, then the cure is simply to make them so large that they have a negligible signal voltage across them at the lowest frequency of interest; less than 80 mVrms is a reasonable criterion. This means they may have to be ten times the value required for a satisfactory frequency response; this not normally a problem, because in opamp circuitry they can usually have a low voltage rating such as 6V3.

However, when non-electrolytics are used in filters, they obviously must have a specific value and a substantial signal voltage across them if they are to actually do anything, so this simple fix is not usable. As described at length in Chapter 15 on passive components, some types of non-electrolytic capacitor such as polyester show easily measurable distortion when they have a significant signal voltage across them. Other types such as polypropylene show no detectable distortion but are more expensive than polyester. Polystyrene capacitors are also distortion free but are only economic for relatively small capacitances. This distortion behaviour works straightforwardly when single capacitors are used in simple 1st-order RC or CR filters, as described in Chapter 15, but things are more complicated when they are used in 2nd- or higher-order active filters. Unusually, this means complicated in a good way. In a 2nd-order lowpass Sallen & Key filter only one of the two capacitors—the *inner* capacitor C2—is the critical part and needs to be an expensive low-distortion type; this gives results just as good as with two expensive low-distortion capacitors. In a 2nd-order highpass Sallen & Key filter only the *outer* capacitor C1 is critical and needs to be a low-distortion type, which is not exactly intuitive.

In a 3rd-order lowpass single-stage filter only the inner capacitor C3 needs to be low-distortion, as for 2nd-order lowpass filters. Likewise, in a 3rd-order highpass single-stage filter only the outer capacitor C1 needs to be low distortion, In 4th-order single-stage filters things get a little more involved, but there are still savings to be made. This mysterious and intriguing behaviour also occurs in multiple-feedback (MFB) filters; it is described in detail in this chapter and summarised in Figure 11.22.

## Distortion in Sallen & Key Filters: The Distortion Aggravation Factor

The most common amplifier in Sallen & Key filters is a unity-gain stage; this is very often an opamp converted into a voltage-follower by applying 100% voltage feedback to the inverting input, but it can also be a discrete amplifier such as an emitter-follower, which inherently has 100% voltage feedback. This means that the filter can only show a gain of more than 1 (as do all S&K filters around the cutoff frequency when they have a Q greater than $1/\sqrt{2}$) if positive feedback is used selectively to increase the gain. This is equivalent to reducing the negative feedback, even if the opamp inverting input is directly connected to the output, and this will increase the distortion from the amplifier. This is why Sallen & Key filters are often referred to as "positive feedback filters", though the negative feedback always dominates the positive feedback.

In contrast, multiple-feedback (MFB) filters achieve their response by directly controlling the total amount of negative feedback around a high-gain amplifier. This is why MFB filters are often referred to as "negative feedback filters".

As I understand it the Billam position is that, in both cases, amplifier distortion is increased by the distortion aggravation factor (DAF), which can be determined from calculation or by performing a Billam Test, [2] either in simulation or in the real world. Other papers on DAF are by Bonello, [3] and there is another paper by Billam [4] which attempts to suggest ways of improving and optimising the DAF of a given filter, though the theoretical basis of this optimisation is not revealed.

Figure 11.1 shows how the Billam Test is performed. There are two identical S&K filters. The top filter is the standard 2nd-order Butterworth lowpass. The lower filter is the same configuration, with the same components, but here the positive feedback connection through C1 is broken and C1 connected to ground instead. The ratio of the gain to the output of each filter V1/V2 gives the DAF directly. It can be determined either by measurement or by simulation; in either case you can directly get a plot of

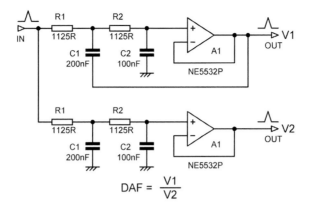

$$DAF = \frac{V1}{V2}$$

**Figure 11.1: Billam test on a 2nd-order Butterworth S&K lowpass filter. Cutoff 1 kHz.**

DAF against frequency by plotting V1/V2. Note this is simply a voltage ratio, and **not** the ratio of two dB figures.

The SPICE simulation results are shown in Figure 11.2. V1 (in dB) shows the usual Butterworth roll-off and passes through the −3 dB point at 1 kHz. V2 (also in dB) is the attenuation given by the grounded version of the feedback network. The DAF is shown here as a ratio, (not in dB) using the same y-axis as V1 and V2; like any ratio, it could be expressed in decibels, but I find that less useful when dealing with distortion in %. The DAF is 1 or close to it at most frequencies but peaks gently to a value of exactly 2 at the cutoff frequency.

A 2nd-order Butterworth S&K highpass stage has exactly the same behaviour as the lowpass, with unity DAF rising to a peak of exactly 2 at the cutoff frequency.

In practice a DAF of 2 times makes little difference to the overall distortion if a low-distortion opamp such as the 5532 is used. But . . . the DAF gets worse quickly with increasing $Q$. For a 2nd-order Butterworth Sallen & Key filter, the peak DAF can be exactly calculated as per Equation 11.1 from [4]. I have confirmed this by simulation.

$$DAF_{pk} = 1 + 2Q^2 \qquad\qquad 11.1$$

Thus a 2nd-order stage with a $Q$ of 1.932, as is used to construct a multi-stage 6th-order Butterworth filter, has a peak DAF of 8.46, which is rather more worrying. The DAF behaviour of 2nd-order filters is described in detail in the next section, because 2nd-order stages are particularly important as one of the building blocks for multi-stage higher-order filters. Table 11.1 gives the calculated and simulated DAFs for the most popular filter types. The Linkwitz-Riley filter has many advantages, particularly in its 4th-order form, and favourable DAF behaviour is another one of them.

Table 11.2 summarises the 2nd-order $Q$'s required for various order multi-stage Butterworth filters, and by referring to Equation 11.1 or Table 11.3, a forecast of the DAF that will be encountered can be made; and this may affect your choice of amplifier. If you are making an 8th-order Butterworth, then the DAF for its high-$Q$ stage peaks at more than 14 times, so it would appear that a low-distortion

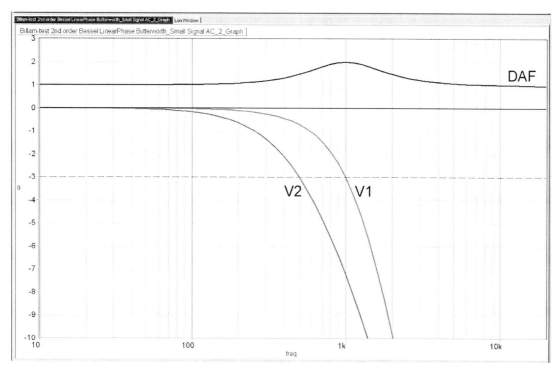

**Figure 11.2: DAF from Billam test for 2nd-order Butterworth S&K lowpass filter. Cutoff 1 kHz. Scale in dB for V1 and V2, scale linear for DAF.**

**Table 11.1: Second-order Sallen & Key unity-gain Qs and DAFs for various filter types.**

| Type | Q | DAF sim'd | DAF sim'd peak freq kHz | DAF calc'd |
|---|---|---|---|---|
| Linkwitz-Riley | 0.500 | 1.500 | 1.58 | 1.500 |
| Bessel | 0.577 | 1.66 | 1.30 | 1.666 |
| Linear-Phase 0.05deg | 0.600 | 1.72 | 1.21 | 1.720 |
| Linear-Phase 0.5deg | 0.640 | 1.82 | 1.12 | 1.819 |
| Butterworth | 0.707 | 2.00 | 1.00 | 2.000 |
| 0.5 dB-Chebyshev | 0.864 | 2.50 | 1.00 | 2.493 |
| 1.0 dB-Chebyshev | 0.956 | 2.83 | 1.03 | 2.828 |
| 2.0 dB-Chebyshev | 1.129 | 3.54 | 0.91 | 3.459 |
| 3.0 dB-Chebyshev | 1.306 | 4.40 | 0.85 | 4.411 |
| Billam example | 5 | 50.91 | 1.00 | 51.0 |

amplifier is very necessary. It is assumed in the table that the higher-$Q$ stages are placed last in the signal chain to optimise headroom, but this has no effect on DAF behaviour. Since DAF depends on $Q^2$, and in Table 11.2 there is in each case one stage with much higher $Q$ than the others, this suggests that in practice only this stage is likely to matter.

**Table 11.2: *Q*'s found in multi-stage Butterworth filters.**

|  | First-stage Q | Second-stage Q | Third-stage Q | Fourth-stage Q |
|---|---|---|---|---|
| 3rd-order Butterworth | 1.00 | - | - | - |
| 4th-order Butterworth | 0.514 | 1.306 | - | - |
| 5th-order Butterworth | 0.62 | 1.618 | - | - |
| 6th-order Butterworth | 0.52 | 0.71 | 1.932 | - |
| 7th-order Butterworth | 0.55 | 0.80 | 2.247 | - |
| 8th-order Butterworth | 0.51 | 0.60 | 0.900 | 2.563 |

**Table 11.3: Second-order Sallen & Key unity-gain Qs and maximum DAFs for multi-stage Butterworth filters.**

| Used in | Max Q | DAF sim'd | DAF sim'd peak freq kHz | DAF calc'd |
|---|---|---|---|---|
| 3rd-order Butterworth | 1.000 | 3.00 | 1.00 | 3.000 |
| 4th-order Butterworth | 1.306 | 4.41 | 1.00 | 4.411 |
| 5th-order Butterworth | 1.618 | 6.23 | 1.00 | 6.236 |
| 6th-order Butterworth | 1.932 | 8.45 | 1.00 | 8.465 |
| 7th-order Butterworth | 2.247 | 11.00 | 1.00 | 11.098 |
| 8th-order Butterworth | 2.563 | 14.12 | 1.00 | 14.138 |

**Table 11.4: Third-order Sallen & Key single-stage unity-gain lowpass; simulated DAFs for various filter types.**

| Type | DAF | DAF peak freq kHz |
|---|---|---|
| Linkwitz-Riley | 3.57 | 0.902 |
| Bessel | 3.39 | 1.282 |
| Linear-Phase 5% delay ripple | 5.35 | 1.132 |
| Butterworth | 7.07 | 0.922 |
| 0.5 dB-Chebyshev | 26.0 | 1.108 |
| 1.0 dB-Chebyshev | 41.1 | 0.940 |
| 2.0 dB-Chebyshev | 81.2 | 0.915 |
| 3.0 dB-Chebyshev | 144.1 | 0.895 |

Third-order Sallen & Key single-stage filters have similar DAF behaviour to the 2nd-order stages described earlier, with DAF being unity everywhere except for a rounded peak near the cutoff frequency. However, now the Butterworth DAF peaks not exactly at the cutoff frequency, but still near to it; this offset seems to have no significance for design. Table 11.4 summarises the DAFs for various types of filter; the Butterworth DAF of 7.07 is already cause for concern, but things get much worse when you get into Chebyshev territory. I am not aware of a closed-form equation for the DAF of a 3rd-order single-stage filter.

A 3rd-order Butterworth Sallen & Key single-stage unity-gain highpass has exactly the same behaviour as the matching lowpass, with a peak DAF of 7.07 at 1.068 kHz, the other side of the cutoff frequency.

As noted in Chapters 8 and 9, the design of 4th-order single-stage filters is not really an exact science, in that you have to solve a set of complex equations, and the answer you get after much work is not necessarily exactly what you wanted; it will certainly have an accurate Butterworth characteristic but may require slightly more than unity gain in the amplifier. If the gain is changed even slightly, the component values for the same characteristic will be rather different. For this reason the DAF results for these filters given in Tables 11.5 and 11.6 are rather less tidy than for the filters examined earlier, but they are all derived from simulation, and I believe them to be correct. I am not aware of an equation for the DAF of a 4th-order single-stage filter.

We have seen previously that 2nd- and 3rd-order single-stage highpass filters have the same sort of DAF peaking as 2nd- and 3rd-order single-stage lowpass filters. It would be very strange if this was not the case for 4th-order single-stage highpass filters, so I have not gone through the lengthy business of checking this for all 4th-order filter types. Instead I just tested two different solutions for a 4th-order single-stage Butterworth; they gave the same behaviour of DAF peaking around cutoff. Table 11.6 shows how the DAF is significantly affected by the gain.

**Table 11.5: Fourth-order Sallen & Key single-stage lowpass; simulated DAFs for various filter types.**

| Type | Gain | DAF | DAF peak freq kHz |
|---|---|---|---|
| Linkwitz-Riley | 1.13 | 8.77 | 0.909 |
| Bessel | 1.09 | 7.42 | 1.392 |
| Linear-Phase 5% delay ripple | 1.09 | 15.21 | 1.427 |
| Butterworth | 1.10 | 14.64 | 0.952 |
| 0.5 dB-Chebyshev | 1.09 | 55.55 | 0.983 |
| 1.0 dB-Chebyshev | 1.09 | 66.91 | 0.945 |
| 2.0 dB-Chebyshev | 1.09 | 94.58 | 0.922 |
| 3.0 dB-Chebyshev | 1.09 | 123.1 | 0.917 |

**Table 11.6: Fourth-order Sallen & Key single-stage highpass; DAFs for two different gains.**

| Type | Gain | DAF | DAF peak freq kHz |
|---|---|---|---|
| Butterworth | 1.05 | 21.67 | 1.078 |
| Butterworth | 1.10 | 16.70 | 1.045 |

## Distortion in Sallen & Key Filters: Looking for DAF

It is evident that if DAF theory is correct, we are going to see a large peak in distortion at some frequency. This will not occur at the cutoff frequency but at sub-multiples of it, which have harmonics falling at the cutoff frequency. If cutoff is at 1 kHz, and we have second-harmonic distortion in the amplifier, we will see a distortion peak at an input frequency of 500 Hz, as the second harmonic of 500 Hz is 1 kHz. Likewise, with third-harmonic distortion we will see a peak when the input frequency is 333 Hz.

Perhaps rather belatedly (after working out all the DAFs) I decided to see this effect for real. In Peter Billam's second paper on active filter distortion [4] he uses as an example an S&K lowpass filter with a $Q$ of 5, as shown in Figure 11.3a. The response has a +14 dB peak, shown in Figure 11.4. A $Q$ of 5 is higher than any we are likely use in crossover design; even an 8th-order Butterworth filter uses a maximum $Q$ of only 2.56. Note the high capacitor ratio of 100 needed to get $Q = 5$ with this type of filter. For the standard circuit, the DAF peaks at a frightening 50.9, labelled DAF 1 in Figure 11.4. I tried the Billam optimisation procedure, as described in [4], which gives the modified filter shown at Figure 11.3b, which has the same frequency response but a much reduced DAF peak of 11.0; it is labelled DAF 2 in Figure 11.4. The optimised filter has a much lower ratio of capacitors; 9.1 times instead of 100 times, which if nothing else will save a good deal of money on C1. But . . . to get the correct response the optimised filter has to be configured with a good deal more than unity gain; 1.91 times or +5.6 dB in fact. This is likely to be very inconvenient, causing significant problems with noise and headroom.

Trying this optimisation process with the rather more popular Butterworth 2nd-order lowpass, the peak DAF is reduced from 2 to 1.709, which is a less than impressive improvement. A gain of 1.336 times is required to get the correct response, and this is likely to be unhelpful.

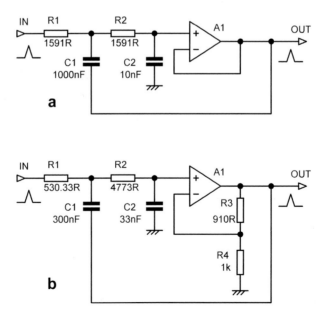

**Figure 11.3: Second-order lowpass Sallen & Key filters with $Q$ = 5. Standard filter at a, after Billam optimisation at b. Cutoff frequency is 1 kHz for both.**

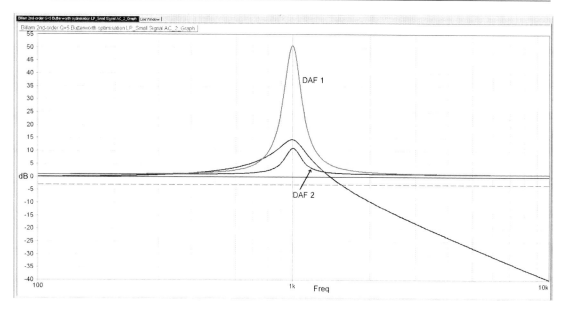

**Figure 11.4: Second-order lowpass Sallen & Key filters with _Q_ = 5; DAF 1 is for standard circuit, DAF 2 after Billam optimisation. Cutoff frequency is 1 kHz for both. DAF and dB share the same scale on the _y_-axis.**

Going back to the $Q = 5$ filter, the frequency response of the optimised version is the same as for the original filter, but the gain is higher. The question is, have we actually managed to reduce the filter distortion?

The filter in Figure 11.3a was built using a 5532 section and measured with input levels of 1 Vrms and 2 Vrms. Considering that S&K filters are regularly criticised as being unsuitable for anything other than low $Q$'s, I was pleasantly surprised at how well this $Q = 5$ version behaved. The cutoff frequency and response peaking were as calculated.

And now for the distortion behaviour. Figure 11.5 shows no trace of a big distortion peak. The 1 Vrms input shows no measurable distortion, but for a 2 Vrms input there is a small distortion peak at 500 Hz, indicating that some second harmonic is being produced. This only reaches 0.0007% at its peak. The distortion residual at this frequency could be seen to be fairly pure second harmonic as expected, while at all other frequencies only noise could be seen. Since the $Q$ of 5 gives a +14 dB (5x) gain peak at cutoff, and the clipping level is about 10 Vrms, a 2 Vrms input is about the most demanding test possible.

Having read the Billam and Bonello papers, it was rather a surprise to find in my measurements no evidence of DAF in action at all. The conclusion at this point must be that DAF theory need not be taken into account when designing crossover filters for low distortion, and the "optimisation" process is dubious if it requires extra filter gain. Choosing a low-distortion amplifier such as the 5532 seems to be all that is necessary.

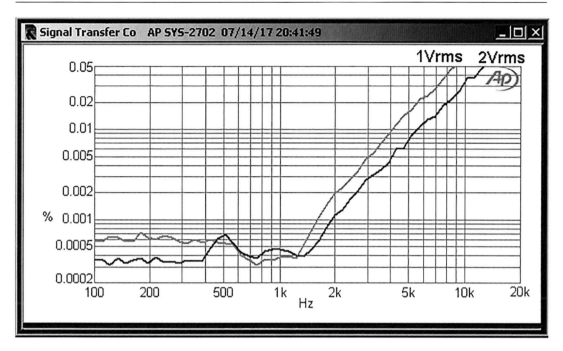

**Figure 11.5: Second-order lowpass Sallen & Key filters with $Q = 0.7071$ for Butterworth response. Cutoff frequency is 1 kHz.**

## Distortion in Sallen & Key Filters: 2nd-Order Lowpass

We now turn from the effect of the amplifiers to the effect of capacitor non-linearity on active filter distortion performance.

The capacitor-induced distortion behaviours of the 2nd-order lowpass and highpass Sallen & Key filters are rather different. The examples shown in Figure 11.6 both have a cutoff at 510 Hz, this frequency resulting from the convenient component values used. The frequency response of the Butterworth 510 Hz lowpass filter in Figure 11.6a is shown in Figure 11.7, while the distortion performance is shown in Figure 11.8. With 220 nF 100 V polyester capacitors there is a pronounced peak in distortion around 200 Hz, more than an octave below the cutoff frequency. All the distortion in this region is fairly pure third harmonic and clearly comes from the capacitors and not the opamp. This is demonstrated by replacing them with 220 nF 160 V polypropylene capacitors; the distortion is eliminated, as shown by the lower trace in Figure 11.8. Note that the test level is almost as high as opamps are capable of at 10 Vrms, and practical internal levels such as 3 Vrms will give much lower levels of distortion.

As expected, the Noise + THD reading in Figure 11.8 climbs rapidly as the cutoff frequency is exceeded and the output amplitude starts to fall at 12 dB/octave; in fact the Noise + THD reading climbs more rapidly than 12 dB/octave, which is a clear warning sign that the residual is not entirely composed of noise. At 1 kHz the reading is only 0.00037%, but crossover-ish distortion from the

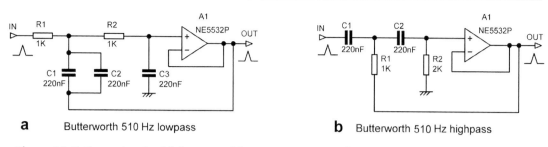

**a**    Butterworth 510 Hz lowpass             **b**    Butterworth 510 Hz highpass

**Figure 11.6: Second-order highpass and lowpass Butterworth Sallen & Key filters for distortion and noise tests. Cutoff frequency is 510 Hz for both.**

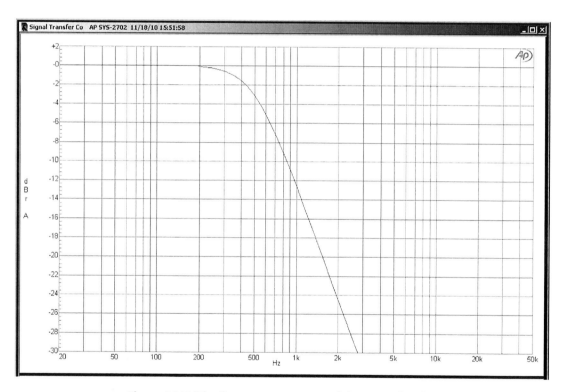

**Figure 11.7: The frequency response of the second order 510 Hz lowpass S&K filter in Figure 11.6a.**

opamp is clearly visible on the THD residual, well above the noise level. This condition persists as frequency rises, with opamp distortion still clear above the noise even at 10 kHz when the output has fallen below 24 mVrms. This is quite different behaviour from that of the highpass filter, where below the cutoff frequency the Noise + THD reading goes up at the 12 dB/octave rate, which would be expected if it was composed only of noise.

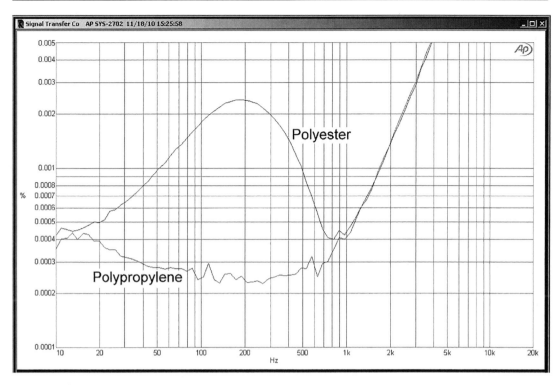

**Figure 11.8: THD plot from the 2nd-order 510 Hz S&K lowpass filter; input level 10 Vrms. The upper trace shows distortion from polyester capacitors; the lower trace, with polypropylene capacitors, shows noise and opamp distortion above 1 kHz.**

Removing the capacitors to get a flat frequency response much reduces the distortion seen at these frequencies, even though the output level is now much higher, which might perhaps be an example of the distortion aggravation factor at work. These are deep waters, Watson.

The signal voltage across the capacitors in these filters varies strongly with frequency, and I thought it might be instructive to see how this relates to their distortion behaviour. In a lowpass Sallen & Key filter, the voltage across C1 peaks at one half of the output voltage at the turnover frequency rolling off either side at 6 dB/octave. The voltage across C2 is the input voltage up to the turnover frequency; it then rolls off at 12 dB/octave, as illustrated in Figure 11.9. The capacitor voltages for the equivalent highpass filter are exactly the same, but with C1 and C2 swapped over.

Figure 11.9 tells us that the differing signal voltages across the capacitors are not the critical factor, because if it was C1 would be making a definite, if somewhat smaller, contribution to the distortion. Instead it contributes nothing. It is both one of the joys and one of the anguishes of electronics that a circuit made up of only five components can be so enigmatic in its behaviour.

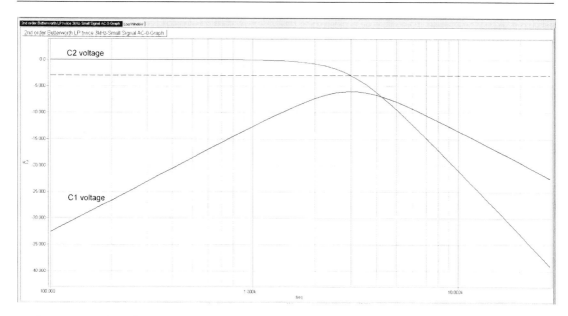

**Figure 11.9: The voltages across the two capacitors in a Sallen & Key Butterworth lowpass filter with a 3 kHz cutoff frequency. The voltage across C1 is never more than half the filter input voltage.**

## Distortion in Sallen & Key Filters: 2nd-Order Highpass

Figure 11.10 shows the frequency response of a conventional 2nd-order Butterworth Sallen & Key highpass filter with a −3 dB frequency of 510 Hz, as shown in Figure 11.6b. C1, C2 were 220 nF 100 V polyester capacitors, with R1 = 1 kΩ and R2 = 2 kΩ. The opamp was a Texas 5532. The distortion performance is shown by the upper trace in Figure 11.11; above 1 kHz the distortion comes from the opamp alone and is very low. However, you can see it rising rapidly below 1 kHz as the filter begins to act, and it has reached 0.015 % by 100 Hz, completely overshadowing the opamp distortion; it is basically 3rd-order. The output from the filter has dropped to −28 dB by 100 Hz, and so the amplitude of the harmonics generated is correspondingly lower, but it is still not a very happy outcome.

The test level used is high, at 10 Vrms. It is chosen to be about as much voltage swing as you are likely to encounter in an opamp system, so it will give worst-case distortion results but will be clear of any clipping effects. Most of the distortion tests in this book are run at 9 or 10 Vrms for this reason. The actual operating levels will naturally be significantly lower, to give some headroom, and the distortion levels produced will therefore be much lower. This is especially true in the case of capacitor distortion, which is essentially all third harmonic, and so the THD increases as the square of the signal voltage.

Thus if the test circuit here was run at a nominal voltage of 3 Vrms, which is probably as high as is advisable (see the discussion in Chapter 17 on elevated operating levels), the distortion levels would be lower by a factor of $(10/3)^2 = 11.1$ times. This is a big reduction, and explains why polyester capacitors are in practice acceptable in many applications where the highest possible quality is not being sought.

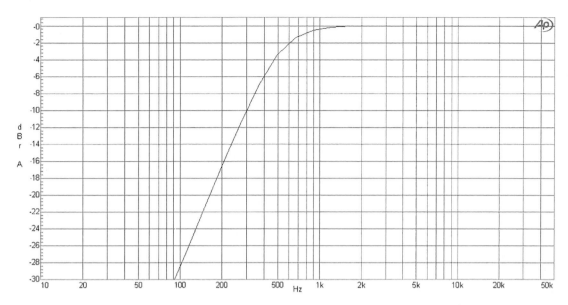

Figure 11.10: The frequency response of the 2nd-order
510 Hz highpass S&K filter in Figure 11.6b.

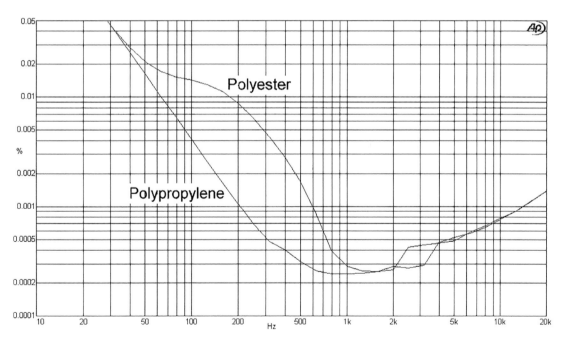

Figure 11.11: THD plot from the 2nd-order 510 Hz highpass S&K filter;
input level 10 Vrms. The upper trace shows distortion from polyester capacitors;
the lower trace, with polypropylene capacitors, shows noise only.

As explained in Chapter 15, polypropylene capacitors exhibit negligible distortion compared with polyester, and the lower trace in Figure 11.11 shows the improvement on substituting 220 nF 250 V polypropylene capacitors. The THD residual below 500 Hz is now pure noise, and the trace is only rising at 12 dB/octave, because circuit noise is constant but the filter output is falling. The important factor is the dielectric, not the voltage rating; 63 V polypropylene capacitors are also free from distortion. The only downside is that polypropylene capacitors are larger for a given CV product and more expensive.

The non-linearity of polyester capacitors, even those of the same type and voltage rating, appears to be rather variable, and if you repeat this experiment the results for the upper trace may be different but will still be much inferior to the polypropylene case. A further complication is that the non-linearity is time-dependent; if you set up a polyester capacitor in a simple RC circuit, with the frequency arranged so that there is a substantial voltage across the capacitor, the THD reading will slowly drift down over time, continuing to drop over 24 hours. [5] It will, however, never become as distortion free as a polypropylene capacitor.

## Mixed Capacitors in Low-Distortion 2nd-Order Sallen & Key Filters

But you have seen nothing yet. If you take the lowpass test filter and change the capacitor dielectric type one at a time, you find that *only C2 has to be polypropylene for low distortion.* If the filter has polyester in both the C1 and C2 positions, then you get the same peaked distortion trace seen in Figure 11.8; but if you replace just C2 with polypropylene, then the distortion performance is identical to the bottom trace obtained by making both capacitors polypropylene. This is, to the best of my knowledge, a novel observation, and it has the potential to much reduce capacitor costs in an active crossover; you will note it is the smaller of the two capacitors (C2) which has to be the more expensive polypropylene type, which is just the way we would like it for maximum economy. This arrangement is shown in Figure 11.12.

This is clearly a most interesting effect, and I attempted to find out how why the lowpass Sallen & Key circuit is so much more sensitive to the non-linearity of C2. Firing up the SPICE simulator and putting an AC voltage source in series with C1 shows a maximum gain of unity from the inserted source to the output at its peak (at the turnover frequency), falling off at 6 dB/octave on either side. The same

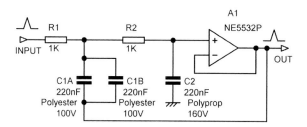

Butterworth 510 Hz lowpass

**Figure 11.12: Mixed-capacitor Sallen & Key lowpass filter has same distortion performance as if all capacitors were polypropylene.**

test on C2 shows there is actually a gain of 2 dB to the output just above the turnover frequency. This result appears to give some indication as to why C2 is the more critical component. It has a greater voltage across it, and the resulting distortion components appear amplified at the output. This does not however really answer the question, because repeated practical experiments show that using polyester capacitors for C1 does not generate a reduced amount of distortion—it produces none at all. More convincing SPICE results are obtained by altering the capacitor models so their value is voltage-dependent; this is dealt with later in this chapter.

But what you may ask, about the highpass filter? The components in the equivalent circuit positions to the capacitors are now resistors, which if metal film almost always have perfect linearity for our purposes, so there is little point in playing around with them. However, if you start swapping out polyester capacitors for polypropylene ones, it soon emerges that in this case C1 is the critical component for linearity. C2 has a very minor influence, but unless it is really bad, polypropylene for C1 only gives results very close to those obtained when both capacitors are polypropylene; the arrangement is shown in Figure 11.13. Just think of the money it saves!

## Distortion in Sallen & Key Filters: 3rd-Order Lowpass Single Stage

When higher-order filters are implemented in the conventional way by cascading 2nd- and 1st-order stages, the distortion behaviour and the effect of different capacitor types for the 2nd-order stages are as described earlier. The 1st-order stages behave as described in Chapter 15 on passive components. However, the intriguing and economical 3rd- and 4th-order single-stage filters have their own significantly different capacitor distortion behaviour. Since convenient component values do not occur, the higher-order filters examined have cutoff frequencies of 1 kHz, like most of the example filters in this book.

As we have seen, in the 2nd-order lowpass stage, only the inner capacitor C2 needs to be linear, while in the 2nd-order highpass stage only the outer capacitor C1 needs to be linear. The 3rd-order Butterworth lowpass single-stage filter tested here is shown in Figure 11.25c. Experiment shows that for low distortion, only the inner capacitor C3 needs to be linear. The outer capacitor C1 causes only a very small degradation in distortion behaviour when changed from polypropylene (PP) to polyester (PE), as shown in Figure 11.14; the lower trace is for all three capacitors as polypropylene; the upper one shows a very modest increase in distortion that peaks around 300 Hz.

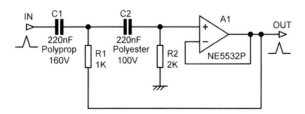

Butterworth 510 Hz highpass

**Figure 11.13: Mixed-capacitor Sallen & Key highpass filter has very nearly the same distortion performance as if both capacitors were polypropylene.**

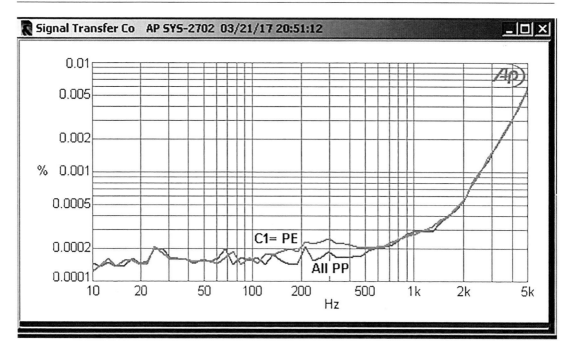

**Figure 11.14: Third-order Butterworth lowpass single-stage 1kHz cutoff: All PP, and C2, C3 = PP, C1 = PE; 9 Vrms in.**

The middle capacitor C2 has no measurable effect on distortion at all when it alone is changed to polyester, as seen in Figure 11.15. However, when C3 alone is made polyester the distortion increases dramatically, as in Figure 11.16, and it peaks around 400 Hz. The input voltage in all cases was 9 Vrms, at least twice the likely maximum internal level for an active crossover, and close to the opamp clipping level of about 10 Vrms.

This is extremely useful information. We now know that C2 can in every case be an economical polyester part, and it is always the largest capacitor in the filter. C1 is always the second-largest capacitor in the filter, and this too can be polyester except for the most demanding applications. The critical capacitor is C3, and in the worst case (Linkwitz-Riley) it is about a quarter of the size of C1; for a Butterworth filter it is about seven times smaller. The distortion behaviour is probably related to the fact that C3 has the highest component sensitivity for frequency, as explained in Chapter 8.

The 3rd-order lowpass single-stage filter distortion performance can therefore be much improved by a relatively small investment in polypropylene for C3. This is most convenient.

## Distortion in Sallen & Key Filters: 3rd-Order Highpass Single Stage

The 3rd-order Butterworth highpass single-stage filter tested here is shown in Figure 11.25d; the cutoff frequency is 1 kHz. In this case we find that only C1 has a significant effect on the distortion, as shown in Figure 11.17. C2 and C3 have no detectable effect, so their plots are not given.

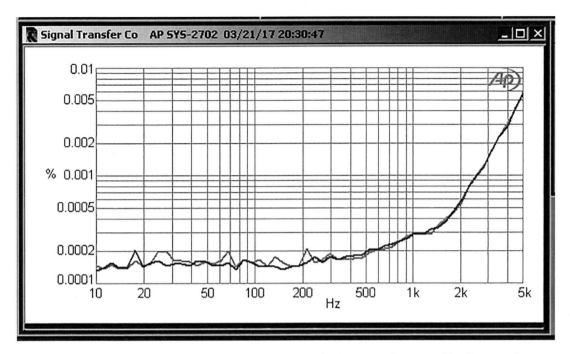

Figure 11.15: Third-order Butterworth lowpass single-stage 1kHz cutoff: All PP, and
C1, C3 = PP, C2 = PE; 9 Vrms in.

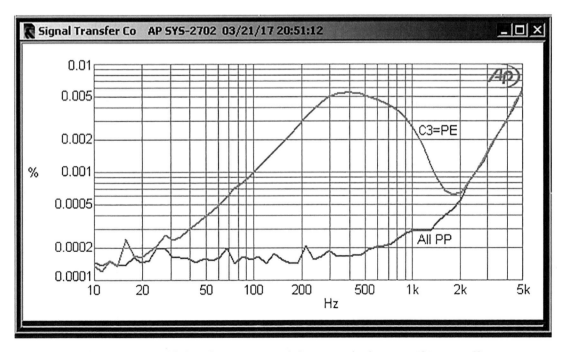

Figure 11.16: Third-order Butterworth lowpass single-stage 1kHz cutoff:
C1, C2 = PP, C3 = PE; 9 Vrms in.

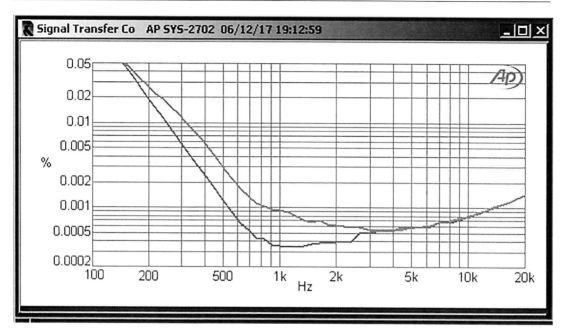

**Figure 11.17: Third-order Butterworth highpass single-stage 1kHz cutoff:**
**C2, C3 = PP, C1 = PE; 10 Vrms in.**

Because in a single-stage filter all of the filtering takes place *before* the amplifier, this allows large input voltages at frequencies where the filtering action reduces the output amplitude. This filter will not clip at 1 kHz, where the output is 3 dB down, until the input exceeds 15 Vrms.

## Distortion in Sallen & Key Filters: 4th-Order Lowpass Single Stage

The likeliest application for a 4th-order filter is the Linkwitz-Riley crossover, which, as described elsewhere in this book, has many advantages. The single-stage lowpass filter examined here has therefore a Linkwitz-Riley characteristic rather than a Butterworth characteristic; the circuit is shown in Figure 11.25e, which also shows a suitable 2xE24 parallel pair to give 1122.32 Ω. Figure 11.18 shows that making C1 alone polyester (PE) causes only a minor increase in distortion around 500 Hz; the lower trace is with all capacitors polypropylene (PP). The DAF for this filter peaks at 8.77 (see Table 11.5) Since this filter has a gain of 1.13 times (+1.06 dB), the maximum input before clipping is now 9 Vrms.

Figure 11.19 shows that making C2 alone polyester (PE) causes a serious increase in distortion around 400 Hz. The lower trace is with all capacitors polypropylene (PP).

Figure 11.20 shows that making C3 alone polyester (PE) causes only a barely measurable increase in distortion compared with the trace for all capacitors polypropylene (PP).

Figure 11.21 shows that making C4 alone polyester (PE) also causes no measurable increase in distortion compared with the trace for all capacitors polypropylene (PP).

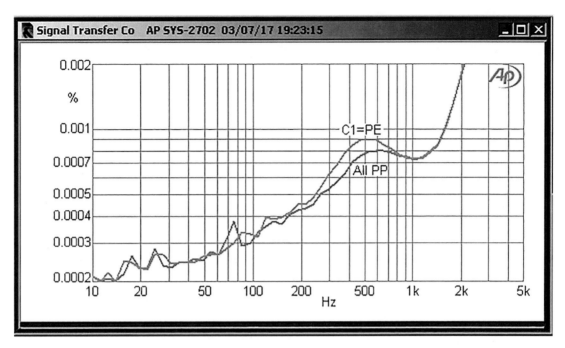

Figure 11.18: Fourth-order Linkwitz-Riley lowpass single-stage 1kHz cutoff: All PP, and
C1 only = PE; 9 Vrms in.

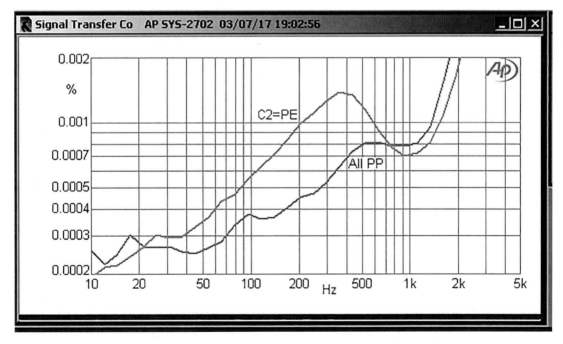

Figure 11.19: Fourth-order Linkwitz-Riley lowpass single-stage 1kHz cutoff: All PP, and
C2 only = PE; 9 Vrms in.

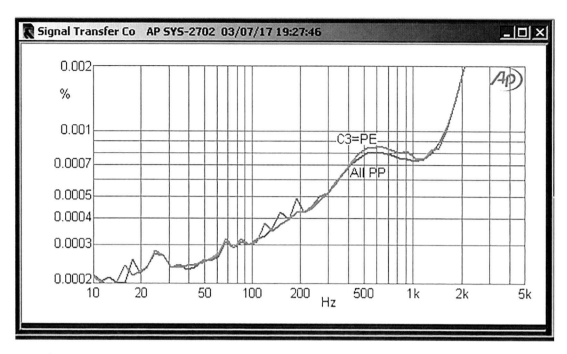

Figure 11.20: Fourth-order Linkwitz-Riley lowpass single-stage 1kHz cutoff: All PP, and C3 only = PE; 5532 9 Vrms in.

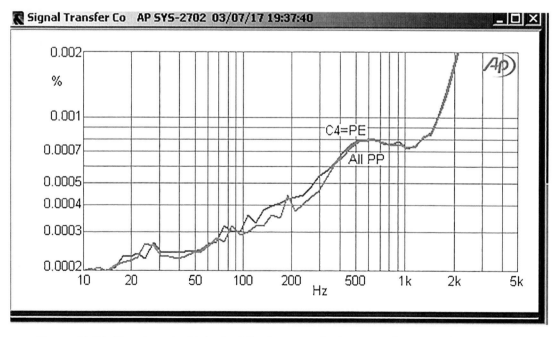

Figure 11.21: Fourth-order Linkwitz-Riley lowpass single-stage 1kHz cutoff: All PP, and C4 only = PE; 5532 9 Vrms in.

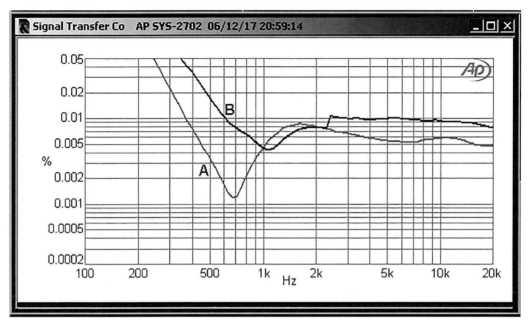

**Figure 11.22: Fourth-order Linkwitz-Riley highpass single-stage 1kHz cutoff: 5532 9 Vrms in. Trace A: all caps polyprop. Trace B: C1 only is polyester.**

The priority for low distortion at low cost is therefore to make C2 polypropylene first, possibly followed by C1 if the best distortion performance is required. Unlike the 3rd-order single-stage lowpass filter, in this case the capacitor with the worst sensitivity for frequency (C4) is not the critical capacitor for distortion (C2).

## Distortion in Sallen & Key Filters: 4th-Order Highpass Single Stage

As for the 4th-order lowpass, the 4th-order highpass filter tested here has a Linkwitz-Riley rather than a Butterworth characteristic; see Figure 11.25f for the schematic. In Figure 11.22, initially all four capacitors were polypropylene (PP), giving trace A. The distortion even in this condition is nevertheless much higher than for the previous filters; it is at a roughly constant level for frequencies above cutoff, and is third harmonic at all these frequencies. This is characteristic of common-mode distortion in the 5532, probably caused here by the large number of capacitors in series with the input and perhaps by the peak DAF of 8.77 near the cutoff frequency (see Table 11.5)

Changing C1 only to polyester (PE) gives trace B, which shows considerably increased distortion below 1 kHz.

Figure 11.23 shows much reduced distortion when the 5532 section is replaced by an LM4562 section, proving that most of the distortion (Trace A) in Figure 11.22 comes from the amplifier. Trace A in Figure 11.23 is the result with all PP capacitors and shows no measurable distortion, just an increase

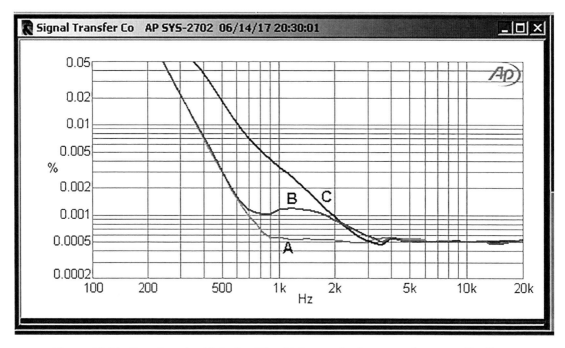

**Figure 11.23: Fourth-order Linkwitz-Riley highpass single-stage 1kHz cutoff: All PP, (A)C1 only = PE, (C) and C2 only = PE. (B)LM4562 9 Vrms in.**

in relative noise below the 1 kHz cutoff frequency. Changing C1 only to PE produces much increased distortion, giving trace C. Changing C2 only to PE produces a little extra distortion, giving trace B.

Figure 11.24 shows that changing either C3 or C4 to PE has very little effect on the distortion performance.

The interaction between capacitor type and distortion is more complex in the 4th-order single-stage filters than in the 2nd- and 3rd-order filters, because in both the lowpass and highpass cases there is not only a critical capacitor that creates most of the distortion but also a capacitor that causes a small but definite increase of distortion. Nevertheless it is still possible to build a filter with mixed capacitors that is just as good as one using all polypropylene capacitors but is considerably cheaper.

## Distortion in Sallen & Key Filters: Simulations

Given the very useful behaviour of active filters in requiring only some capacitors to be low distortion, it would be nice to know how it works. As noted earlier in the sections on 2nd-order filters, it is not explained by differing signal voltages across the capacitors.

As a first step in trying to understanding what is going on, I have done a large number of SPICE S&K simulations starting with default (i.e. perfect) capacitors and using a VCVS as a perfect amplifier. This establishes the numerical noise floor, which in PSPICE was gratifyingly low at around 0.000001%.

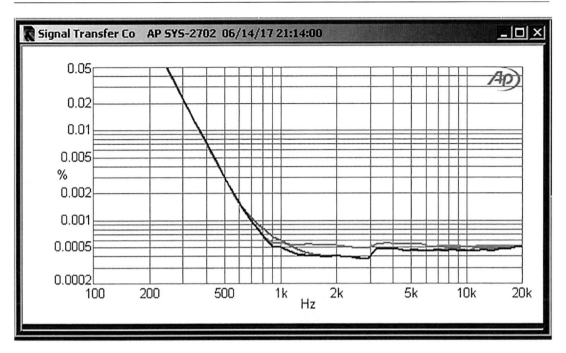

**Figure 11.24: Fourth-order Linkwitz-Riley highpass single-stage 1kHz cutoff: LM4562 9 Vrms in.**

So it should be, for all the circuit elements are mathematically linear. Then one of the capacitors was modified to make it non-linear using an analogue behavioural model of a voltage-controlled capacitor; a subcircuit with this function is included in most simulator packages, or can be found quickly on the Web. The non-linearity implies that the capacitance is in some way a function of the voltage across it, and the most plausible laws seemed to be either linear or square-law dependence on the voltage. Equation 11.2 shows a typical linear voltage-dependent model; C0 is the capacitance at zero voltage, and Vc the voltage across it. Another version of this did an absolute-value operation on Vc to make the law symmetrical about zero voltage.

$$C = C_0 \left(1 + 0.02 V_c\right)$$ 11.2

Equation 11.3 shows a typical square-law voltage-dependent model; note that the coefficient setting the non-linearity of the law is now much smaller to allow for the squaring process. This law is inherently symmetrical about zero voltage.

$$C = C_0 \left(1 + 0.0003(V_c)^2\right)$$ 11.3

Both linear models and the square-law model were tried during a *lot* of simulation, but the results were never quite right. In the 2nd-order lowpass S&K filter, C1 has no effect on the linearity; it is all down to C2. But the SPICE results showed that C1 did have an effect, though roughly ten times less than C2. This is far enough out of line to show that something is wrong with the assumptions in the simulation, and it cannot safely be used to try and explain what happens in the real world. Pretty much the only possibility seems to me to be inadequacies in the model used for the non-linear capacitor. There are no official or well-accepted models for capacitor non-linearity; it's not exactly a headline-grabbing issue in electronics. Possibly there is a model that would make the distortion generated by C1 vanish, but if that was true you would think it would need some careful fine tuning, somehow cancelling the distortion from C1. Even if it was possible, it seems highly unlikely to work across a lot of different capacitors and all the filter types shown in Figure 11.25.

I actually went through this simulation process twice, the attempts separated by several years. The second time I deliberately started again from scratch in case there was a subtle mistake buried somewhere in the first attempt. The results were however the same. Possibly a purely mathematical approach would be more revealing, but I make no claim to be a mathematician, and there the matter rests for the time being. Perhaps someone else would like to have a go, for it is very annoying to be unable to explain such a useful effect.

## Distortion in Sallen & Key Filters: Capacitor Conclusions

Figure 11.25 shows the schematics of all the filters examined earlier and the critical capacitors in each one; the 2nd-order filters have had their components recalculated to give a 1 kHz cutoff

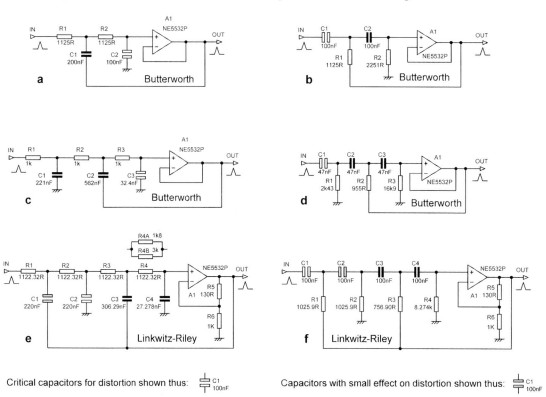

Critical capacitors for distortion shown thus:

Capacitors with small effect on distortion shown thus:

**Figure 11.25: Summary of the critical capacitors for distortion in Butterworth S&K filters: (a) 2nd-order lowpass, (b) 2nd-order highpass, (c) 3rd-order lowpass, (d) 3rd-order highpass, (e) 4th-order lowpass, (f) 4th-order highpass. All have 1 kHz cutoff.**

frequency to allow better comparison. The figure shows a truly remarkable state of affairs, with just one capacitor dominating the distortion behaviour of an active filter. For the 2nd- and 3rd-order lowpass filters, the innermost capacitor (C2 or C3) is critical. But this changes for the 4th-order lowpass, where C2 is the critical component. For the 2nd-, 3rd- and 4th-order highpass filters, things are quite different; it is the outer capacitor C1 that is critical. Capacitors having only a small effect on linearity are shown in grey. Just one shade of grey.

Since they are likely to be rather a minority interest, I have not so far studied the distortion behaviour of 5th- and 6th-order single-stage filters. I suspect it will be the same in that one particular capacitor will have a dominant effect on the distortion.

This filter behaviour is a quite unexpected gift from providence, as it means that very useful savings can be made on the capacitors, even when top-notch performance is sought. But . . . how does it work?

## Distortion in Multiple-Feedback Filters: The Distortion Aggravation Factor

As for S&K filters, the overall distortion performance depends partly on the amplifier distortion, (which, it is claimed, is exaggerated by the distortion aggravation factor) and the capacitor distortion. In multiple-feedback (MFB) filters the amplifier is working as a virtual earth, i.e. with no signal voltage on its inputs. This eliminates common-mode distortion from the amplifier.

According to Bonello [4] the DAF for either lowpass or highpass MFB filters is calculated from Equation 11.4:

$$\sqrt{(1+3Q^2)^2 + Q^2} \qquad 11.4$$

Some results from this equation are shown in Table 11.7, where they are compared with the DAFs for S&K filters with the same $Q$. The MFB DAF is always somewhat worse, the difference increasing as

**Table 11.7: Second-order MFB Qs and DAFs for various filter types, compared with S&K (from Table 11.1).**

| Type | Q | MFB DAF calc'd | S& K DAF calc'd |
|---|---|---|---|
| Linkwitz-Riley | 0.500 | 1.820 | 1.500 |
| Bessel | 0.577 | 2.080 | 1.666 |
| Linear-Phase 0.05deg | 0.600 | 2.165 | 1.720 |
| Linear-Phase 0.5deg | 0.640 | 2.319 | 1.819 |
| Butterworth | 0.707 | 2.598 | 2.000 |
| 0.5 dB-Chebyshev | 0.864 | 3.353 | 2.493 |
| 1.0 dB-Chebyshev | 0.956 | 3.862 | 2.828 |
| 2.0 dB-Chebyshev | 1.129 | 4.954 | 3.459 |
| 3.0 dB-Chebyshev | 1.306 | 6.255 | 4.411 |

$Q$ increases; this may be because a shunt-feedback stage has a higher noise gain than the equivalent series-feedback stage.

With the usual values of $Q$ used in crossovers, it seems likely that the absence of common-mode distortion will have more effect on the overall distortion performance than a slightly higher DAF.

## Distortion in Multiple-Feedback Filters: 2nd-Order Lowpass

Figure 11.26 shows lowpass and highpass Butterworth MFB filters, each with a cutoff frequency of 510 Hz and designed to work at a low impedance level. These filters were designed to allow direct comparison between them and the Sallen & Key 510 Hz filters of Figure 11.6 that were examined for distortion and noise earlier in this chapter.

Figure 11.27 shows the distortion performance of the 2nd-order 510 Hz MFB lowpass filter with an input level of 10 Vrms. As for the Sallen & Key lowpass filter, the use of polyester capacitors causes a broad peak of extra distortion, though here it is centred on 240 Hz rather than 200 Hz. As before the use of polypropylene capacitors eliminates the extra distortion.

Above 1 kHz, the Noise + THD reading climbs rapidly as the cutoff frequency is exceeded, and the output amplitude starts to fall at 12 dB/octave, but this time it rises at 12 dB/octave because it is composed of noise only; some opamp distortion is visible around 100–500 Hz when polypropylene capacitors are used. Comparing Figure 11.27 with the graph for the lowpass Sallen & Key version (Figure 11.8), here we have a Noise + THD reading of only 0.0008% at 2 kHz, compared with 0.0015% for S&K at the same frequency. The MFB lowpass filter has the better linearity, probably because of the absence of common-mode distortion.

You will recall that we discovered that the full benefits of capacitor distortion elimination could be obtained in the Sallen & Key 2nd-order lowpass filter by replacing only the second of the two capacitors with a polypropylene type. In the MFB lowpass filter only C2 needs to be polypropylene to get the full elimination of capacitor distortion. Extraordinary!

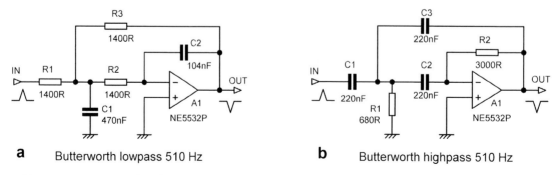

**a**   Butterworth lowpass 510 Hz          **b**   Butterworth highpass 510 Hz

**Figure 11.26: Second-order Butterworth highpass and lowpass MFB filters for distortion and noise tests. Cutoff frequency is 510 Hz for both.**

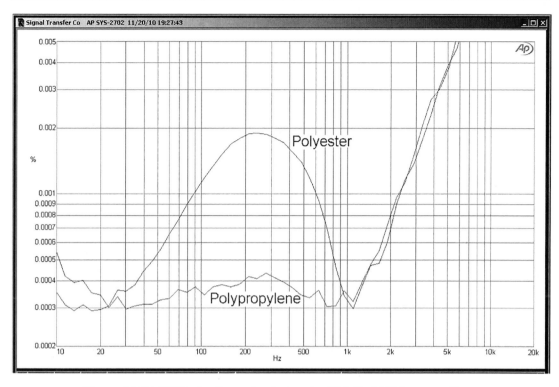

**Figure 11.27: THD plot from the 2nd-order 510 Hz MFB lowpass filter;
input level 10 Vrms. The upper trace shows distortion from polyester capacitors;
the lower trace, with polypropylene capacitors, shows only noise above 1 kHz.
Unlike the S&K filter, there is no visible opamp distortion.**

## Distortion in Multiple-Feedback Filters: 2nd-Order Highpass

It was only necessary to make one capacitor polypropylene in the Sallen & Key lowpass and highpass filters, and capacitor distortion was still eliminated. Figure 11.28 demonstrates that replacing polyester capacitors with polypropylene in the highpass case once again gives a dramatic reduction in distortion. I was initially not optimistic, because of the different ways in which the two kinds of filters perform their function, but it does work. In the highpass case both C1 and C3 need to be polypropylene, as both capacitors contribute distortion. C2 however, can be a cheaper and smaller polyester type.

There were some unexpected issues with the distortion performance above 3 kHz which were not resolved at the time of going to press, and at the moment the verdict has to be that the highpass MFB filter is not to be designed into a system without careful scrutiny. In contrast, the lowpass MFB filter works very well and gives less opamp distortion than the Sallen & Key equivalent.

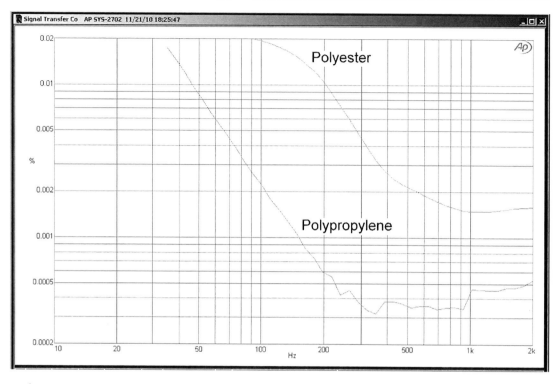

**Figure 11.28: THD plot from the 2nd-order 510 Hz MFB highpass filter; input level 10 Vrms. The upper trace shows distortion from polyester capacitors; the lower trace, with polypropylene capacitors, shows only noise.**

## Distortion in Tow-Thomas Filters: 2nd-Order Lowpass

The Tow-Thomas filter described in Chapter 10 uses three amplifiers rather than one; this extravagance suggests it should be expected give a good performance in other areas than accuracy of frequency response. The lowpass Tow-Thomas filter of Figure 10.10 was built and measured, using 5532 sections and initially with both capacitors as polypropylene. This gave the lower traces in Figures 11.29 and 11.30; the input level was 10 Vrms, which is about as much as you are going to see in an opamp-based system. The rise at HF is just noise increasing relative to the output level due to the lowpass filtering. Very small amounts of low-order distortion are just about visible in the THD residual below 400 Hz, but there are none of the grievous problems experienced with the Akerberg-Mossberg filter. Distortion is low partly because all three amplifiers are operating as virtual earth stages with shunt feedback, and so there is no common-mode signal on any of the amplifier inputs. In addition, the filter loop contains two integrators, and these notably have low distortion because the negative feedback increases with frequency, matching the increase of amplifier distortion with frequency.

Earlier in this chapter it was demonstrated that in 2nd-order Sallen & Key and MFB filters, only one of the capacitors needs to be of the expensive but linear polypropylene (PP) type, and the rest can be cheap polyester (PE) capacitors without degrading the distortion performance; exactly how this works

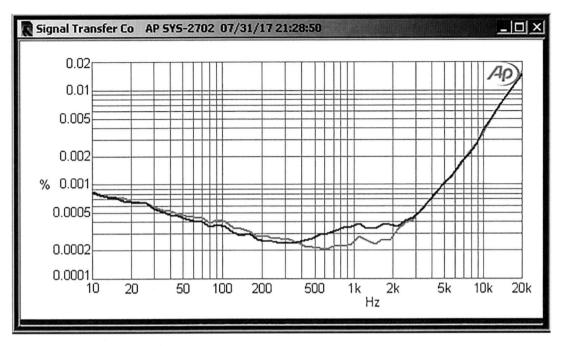

Figure 11.29: THD plot from Tow-Thomas 2nd-order highpass filter; input level 10 Vrms. Upper trace is THD with C1 PE; lower trace, with both capacitors PP.

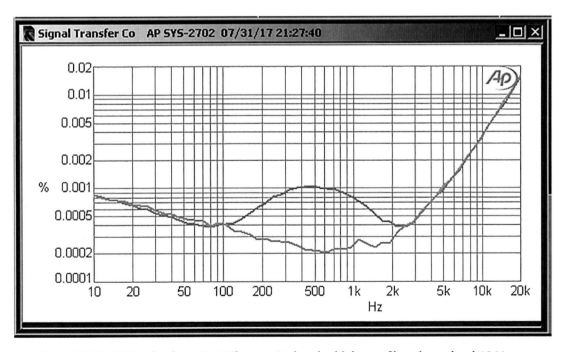

Figure 11.30: THD plot from Tow-Thomas 2nd-order highpass filter; input level 10 Vrms. Upper trace is THD with C2 PE; lower trace, with both capacitors PP.

remains a mystery. Something of the sort happens here, but unlike the previous cases the behaviour is simply explicable. Figure 11.29 shows that converting C1 from polypropylene to polyester only causes a very modest rise in THD residual between 400 Hz and 3 kHz. However, making C2 polyester (with C1 switched back to polyprop) causes a much greater rise between 100 Hz and 3 kHz, as in Figure 11.30. The same capacitor was used in each case, with precautions taken to minimise self-improvement (see Chapter 15). Reference to Figure 10.12 shows that the signal voltage across C2 is much greater than that across C1 (three times greater at 200 Hz), and this seems to be all that is needed to explain the fact that C2 is more critical for distortion than C1. At any rate, there is no doubt that if you can afford only one polyprop capacitor, it should be C2.

## Distortion in Tow-Thomas Filters: 2nd-Order Highpass

The highpass Tow-Thomas/Fortunato filter of Figure 10.14 was built and measured using 5532 sections, and initially with both capacitors as polypropylene. This gave the lower traces in Figures 11.31 and 11.32; the input level was again 10 Vrms, about as much as you can get from opamps. As it was for the lowpass Tow-Thomas filter, the general level of distortion is very low, because all three amplifiers are operating with shunt feedback and therefore have no common-mode signal on the amplifier inputs.

The distortion results are much the same as for the lowpass Tow-Thomas. Making C1 only polyester generates only a small amount of distortion, as in Figure 11.31, while making C2 only polyester

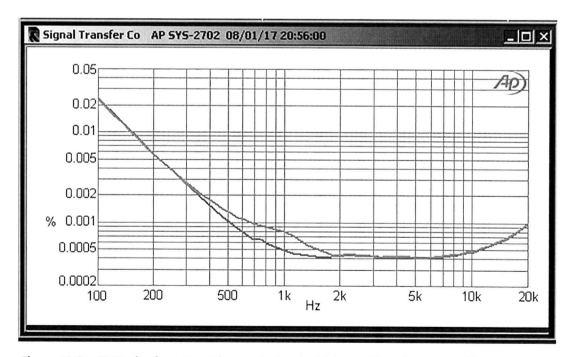

**Figure 11.31: THD plot from Tow-Thomas 2nd-order highpass filter; input level 10 Vrms. Upper trace is THD with polyester capacitor at C1; lower trace, with both capacitors polypropylene.**

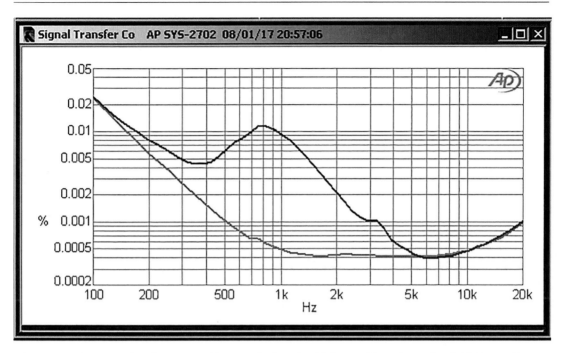

**Figure 11.32: THD plot from Tow-Thomas 2nd-order highpass filter; input level 10 Vrms. Upper trace is THD with polyester capacitor at C2; lower trace, with both capacitors polypropylene.**

gives much more, ten times as much as in the lowpass version; see Figure 11.32. The same capacitor was used in each case, with precautions taken to minimise self-improvement (see Chapter 15). Once more this seems to be completely explained by the differing signal voltages across the capacitors. The situation is slightly more complicated in Figure 10.14 (highpass) than in Figure 10.11 (lowpass) because of the presence of R3 in series with C2, but again there is no doubt that if you can afford only one polyprop capacitor, you should make it C2.

This chapter has not been able to give a completely worked-out account of distortion in active filters; some questions remain. Nonetheless I think it is the most comprehensive study ever published.

## Noise in Active Filters

In active filters distortion may come from the amplifiers and the capacitors. Noise, on the other hand, will come from the amplifiers and the resistors. Each amplifier generates its own voltage noise and current noise, while each resistor generates Johnson noise, which can be treated as a voltage or a current as convenient. Low-impedance design can be applied just as in other circuitry; it does not affect voltage noise but reduces the effect of current noise, because that is flowing through lower impedances and so produces less additional voltage noise. Reducing resistor values reduces their Johnson noise, but only slowly, as the noise voltage is proportional to the square root of the resistance.

The capacitors will be proportionally larger to obtain the same filter response, but a capacitor is a very close approximation to a pure reactance and so does not generate detectable Johnson noise.

Low-impedance design and the selection of appropriate amplifiers, such as the 5532 or LM4562, make it straightforward to design low-noise active filters. The limits of the low-impedance design approach are the need for amplifiers to drive the impedances chosen without excess distortion, and the need to keep the size of expensive precision capacitors within bounds.

All the noise measurements in the rest of this chapter were taken with the filter completely screened by metal plates to exclude hum from electric fields. All the noise results given here are for a bandwidth of 22 Hz–22 kHz, rms sensing. In each case the noise level for 22 Hz–22 kHz bandwidth was compared with that for 400 Hz–22 kHz bandwidth to ensure the hum component was negligible. Noise is of its very nature variable, and to deal with this each noise level given is the average of six readings. Where appropriate the noise of the testgear has been subtracted RMS fashion to give more accurate readings.

## Noise and Bandwidth

In assessing the noise performance of a system using filters, a difficulty is that the frequency response is of course very much not flat, and that must be allowed for. This is done by using the equivalent noise bandwidth (EQNBW)of the filter, this defined as the bandwidth of an infinitely sharp filter with infinite passband attenuation that will give the same results as our real filter when it filters white noise, including its own internally generated noise. In crossover work the greatest modification to the noise readings is usually made by the lowpass filters, as white noise has equal power in equal absolute bandwidth, *not* equal bandwidth ratios, and there is more absolute bandwidth in the upper half of the audio spectrum. The equivalent noise bandwidths for some common lowpass filters are given in Tables 11.8 to 11.10. For the Butterworth and Bessel filters, the cutoff frequency is as usual the −3 dB point, and for the Chebyshev as usual even-order versions have peaks above the 0 dB line and a response going through 0 dB at cutoff, while odd-order versions have dips below it and instead pass through the ripple value (e.g. −1 dB) at cutoff.

**Table 11.8: The equivalent noise bandwidth of Butterworth lowpass filters.**

| Filter order | EQNBW |
|:---:|:---:|
| 1 | 1.5708 |
| 2 | 1.1107 |
| 3 | 1.0472 |
| 4 | 1.0262 |
| 5 | 1.0166 |
| 6 | 1.0115 |
| 7 | 1.0084 |
| 8 | 1.0065 |

**Table 11.9: The equivalent noise bandwidth of Bessel lowpass filters.**

| Filter order | EQNBW |
|:---:|:---:|
| 1 | 1.57 |
| 2 | 1.56 |
| 3 | 1.08 |
| 4 | 1.04 |
| 5 | 1.04 |
| 6 | 1.04 |

**Table 11.10: The equivalent noise bandwidth of Chebyshev lowpass filters for various ripple amplitudes.**

| Filter order | 0.25 dB | 0.5 dB | 1.0 dB |
|:---:|:---:|:---:|:---:|
| 2 | 1.74486 | 1.48892 | 1.25316 |
| 3 | 1.28245 | 1.16659 | 1.04106 |
| 4 | 1.14049 | 1.06563 | 0.97348 |
| 5 | 1.07798 | 1.02083 | 0.94328 |
| 6 | 1.04484 | 0.99698 | 0.92717 |
| 7 | 1.02514 | 0.98276 | 0.91754 |
| 8 | 1.01246 | 0.97361 | 0.91133 |

For the higher-order Butterworth filters the EQNBW gets very close to unity. This is not the case for the Bessel filters in Table 11.9, because of their slower roll-off.

When assessing the noise performance of non-standard filters, it is not usually a good idea to plod through heavy mathematics to determine the EQNBW. Usually it is much quicker to simply build the circuit and measure it.

## Noise in Sallen & Key Filters: 2nd-Order Lowpass

It is difficult to give a comprehensive summary of the noise performance of even one type of filter in a reasonable space. There are a large number of variables, even if we restrict ourselves to the Sallen & Key configuration. There is the cutoff frequency, the Q, the impedance level at which it operates, whether it is highpass or lowpass, and whether it is 2nd-order or higher.

If we start with the 2nd-order 510 Hz lowpass filter in Figure 11.6a, it can be built with 1 kΩ resistors and 220 nF capacitors (using two for C1), as shown, in which case the noise output (corrected by subtracting the internal noise of the measuring system, as are all noise data in this section) is −117.8 dBu, which is pretty low. The measurement bandwidth was 22 Hz–22 kHz as usual. If we rebuild our lowpass filter with 10 kΩ resistors and 22 nF capacitors, so the cutoff frequency remains 500 Hz but the impedances are all ten times as great, we rather disconcertingly get the same figure, to within the limits of measurement, which are here about ±0.1 dB. This is because the cutoff frequency is towards the lower end of the audio bandwidth, and so the Johnson noise from the resistors never makes it to the filter output. What we are seeing is the basic voltage noise of the opamp, which is not altered by changing circuit impedances. The only way to reduce it is to choose a quieter opamp or to use multiple opamps for noise reduction, as explained in Chapter 20. This at first seems to make nonsense of the recommendations scattered throughout this book that resistance levels should be kept as low as practicable to minimise noise.

If however we are dealing with the same lowpass filter but with a ten times higher cutoff frequency of 5.1 kHz, things are very different. Our low-impedance version has 1 kΩ resistors and 22 nF capacitors; it produces −115.8 dBu of noise at its output, noticeably higher because of the higher cutoff frequency. If we once more increase the impedance level by ten times, by using 10 kΩ resistors and 2n2 capacitors, the output noise increases to −110.9 dBu, a sizable increase of 4.9 dBu. Clearly the higher the cutoff frequency of a lowpass filter, the greater the noise advantage to be gained by using low impedances. Raising the impedance ten times also makes the circuit noticeably more susceptible to electrostatically induced hum.

## Noise in Sallen & Key Filters: 2nd-Order Highpass

Turning our attention to the 2nd-order highpass Butterworth filter, the noise output for a low-impedance version with R1 = 1 kΩ, R2 = 2 kΩ, and 220 nF capacitors is a very low −119.3 dBu. Raising the impedance level by ten times, using R1 = 10 kΩ, R2 = 20 kΩ, and 22 nF capacitors, raises the noise output to −116.6 dBu, an increase of 2.7 dB. The low-impedance approach is therefore well worthwhile, though the capacitors will cost a bit more.

Filters of order three and higher are commonly built using 2nd-order and 1st-order stages as building blocks, and the noise output is simply the rms-sum of the cascaded stages. Filters that implement 3rd- and 4th-order responses in a single stage have rather different noise behaviours, in the same way that they have rather different distortion characteristics. Their noise behaviour is briefly described later.

## Noise in Sallen & Key Filters: 3rd-Order Lowpass Single Stage

The measured noise output of the 3rd-order lowpass Butterworth 1 kHz filter in Figure 11.25c. was −111.7 dBu, using a 5532 section as the amplifier. This is 6 dB noisier than the 2nd-order lowpass. There are now three series resistances; this increases the effect of the amplifier current noise and increases the Johnson noise generated.

## Noise in Sallen & Key Filters: 3rd-Order Highpass Single Stage

The measured noise output of the 3rd-order highpass Butterworth filter in Figure 11.25d was −109.2 dBu using a 5532 section as the amplifier. The three series capacitors are relatively small to prevent R2 becoming too low and loading the previous stage excessively at frequencies below cutoff, and so the effect of amplifier current noise is increased. An LM4562 section as amplifier did not give meaningful noise results because of the instability at unity-gain problem.

## Noise in Sallen & Key Filters: 4th-Order Lowpass Single Stage

The measured noise output of the 4th-order lowpass Linkwitz-Riley filter in Figure 11.25e. was −108.9 dBu using a 5532 section as the amplifier. This is 2.8 dB worse than the 3rd-order single-stage lowpass because of the extra series resistance generating more Johnson noise, the increased effect of opamp input current noise flowing through the increased series resistance, and the filter gain of +1.06 dB rather than unity.

## Noise in Sallen & Key Filters: 4th-Order Highpass Single Stage

The measured noise output of the 4th-order lowpass Linkwitz-Riley filter in Figure 11.25f. was −109.3 dBu using a 5532 section as the amplifier. Changing the amplifier to an LM4562 section made the noise 0.1 dB worse, and it therefore appears that in this case the higher current noise of the LM4562 flowing through the four series capacitors outweighs its lower voltage noise. The series capacitors do not generate Johnson noise, but the resistors in the circuit will do so. The overall gain is +1.1 dB rather than unity, and this increases the noise output by the corresponding amount.

The Sallen & Key filter noise results are summarised in Table 11.11.

## Noise in Multiple-Feedback Filters: 2nd-Order Lowpass

The low-impedance MFB lowpass filter in Figure 11.26a has a noise output of −117.8 dBu, which is quite satisfyingly low. As we did for the Sallen & Key filters, we will raise the circuit impedance level by ten times and see what effect that has on the noise performance. Plugging in R1 = R2 = R3 = 14 kΩ and C1 = 47 nF, C2 = 10.4 nF, the noise output rises to −114.0 dBu, an increase of 3.8 dBu (as before, the measurement bandwidth was 22 Hz–22 kHz, and the readings were corrected by subtracting the internal noise of the measuring system). This is quite different behaviour from the Sallen & Key lowpass filter, which showed no measurable change in noise output with a 510 Hz cutoff frequency. Once again, raising the impedance ten times makes the circuit noticeably more susceptible to electrostatically induced hum.

**Table 11.11: Summary of noise outputs from Sallen & Key filters;**
**22 Hz–22 kHz bandwidth, rms.**

|  | Lowpass noise out | Highpass noise out |
|---|---|---|
| 2nd-order | −117.8 dBu | −119.3 dBu |
| 3rd-order | −111.7 dBu | −109.2 dBu |
| 4th-order | −108.9 dBu | −109.3 dBu |

The low-impedance version of the MFB lowpass filter has a noise output of −117.8 dBu, which interestingly is exactly the same noise output as the Sallen & Key lowpass filter examined earlier. From the noise point of view there is nothing to choose between the two filter configurations.

## Noise in Multiple-Feedback Filters: 2nd-Order Highpass

With the values shown in Figure 11.26b, the noise output was −112.8 dBu, which is a hefty 6.5 dB worse than that of the Sallen & Key highpass filter, which was −119.3 dBu. This is believed to be due to the amplifier seeing a higher noise gain than for the lowpass case. This is not promising, and it reinforces the judgement that the highpass MFB filter is not to be recommended for use without further study.

## Noise in Tow-Thomas Filters

The noise output from the TT lowpass filter in Figure 10.11 was −114.4 dBu (22 Hz–22 kHz) and −115.2 dBu (400 Hz–22 kHz). Compare the 2nd-order Sallen & Key highpass filter described earlier with a noise output of −115.8 dBu (22 Hz–22 kHz). The more complex TT filter is only 1.4 dB noisier, which is gratifying.

The noise output from the TT/Fortunato highpass filter in Figure 10.14 was −107.4 dBu (22 Hz–22 kHz) and −107.5 dBu (400 Hz–22 kHz). Compare the 2nd-order Sallen & Key highpass filter described earlier with a noise output of only −119.3 dBu, some 12 dB better. This is very disappointing compared with the lowpass case and presumably results from the fact that there are three opamps to generate noise rather than one, and that the highpass filtering action does not reduce the noise as a lowpass TT would. There does not seem to be much scope for reducing noise by further lowering the impedance of the filter.

## References

[1]   www.filterdistortion.co.uk; Accessed August 2017
[2]   Billam, Peter "Harmonic Distortion in a Class of Linear Active Filter Networks" JAES26(6), June 1978, pp. 426–429 Free download from www.pjb.com.au/; Accessed March 2017
[3]   Billam, Peter" Practical Optimisation of Noise and Distortion in Sallen & Key Filter Sections" IEEE Journal of Solid-State Circuits: August 1979, SC-14(4), pp. 768–771 Free download from www.pjb.com.au/ (Accessed March 2017)
[4]   Bonello, Oscar "Distortion in Positive- and Negative-Feedback Filters" JAES 32(4), April 1984, pp. 239–245
[5]   Self, Douglas "Self-Improvement for Capacitors" Linear Audio, Volume 1, April 2011, p. 156, ISBN: 9-789490-929022

# Bandpass and Notch Filters

This chapter deals with bandpass and notch filters. Bandpass filters as such are rarely used in performing the basic band-splitting functions of a crossover, but they can be useful for equalisation purposes and are essential for putting together high-order allpass filters for time correction. Bandpass filters are essential for filler-driver crossover schemes. Notch filters can also be useful for equalisation, but their most important use is in the construction of notch crossovers, whether based on elliptical filters or other sorts of filtering.

## Multiple-Feedback Bandpass Filters

When a bandpass filter of modest $Q$ is required, the multiple-feedback or Rauch type shown in Figure 12.1 has many advantages. The capacitors are equal and so can be made any preferred value. The opamp is working with shunt feedback and so has no common-mode voltage on the inputs, which avoids one source of distortion. It does however phase-invert, which can be inconvenient. Phase inversions are no problem in passive crossovers—you simply swap over the wires to the speaker unit—but in an active crossover an extra inverting stage may be needed to undo the first inversion.

Figure 12.1 shows the normal configuration of an MFB or Rauch filter. The minimum $Q$ that can be normally achieved is 0.707 ($1/\sqrt{2}$), which requires R2 to be set to infinity, i.e. not fitted at all. If you want a lower $Q$ than that, you still omit R2, but the vital point is that a different set of design equations are used that give different values for the other components and allow lower $Q$'s to be realised. These lower $Q$'s are required for time-delay compensation allpass filters and for filler-driver crossovers. The design process is fully described in Chapter 13 on time-domain filtering, where low $Q$'s are especially necessary. This no-R2 variation on the standard MFB filter is sometimes called a Deliyannis filter.

Note that the Deliyannis filter is not suitable for high $Q$'s, because as the $Q$ is increased, the passband gain increases with it; it cannot be controlled independently, as it can in the standard MFB filter. This will lead to either headroom problems or, alternatively, noise problems if the signal is attenuated before it reaches the filter.

The filter response is defined by three parameters—the centre frequency $f_0$, the $Q$, and the passband gain (i.e. the gain at the response peak) A. The filter in Figure 12.1 was designed for $f_0 = 250$ Hz, $Q = 2$, and A = 1 using the design equations shown, and the usual awkward resistor values emerged. The resistors shown in Figure 12.1are the nearest E96 values, and the simulated results come out as $f_0 = 251$ Hz, $Q = 1.99$, and A = 1.0024, which, as they say, is good enough for rock 'n' roll. Using 2xE24 parallel pairs would give even more accurate values.

The $Q$ of the filter can be quickly checked from the response curve, as $Q$ is equal to the centre frequency divided by the −3 dB bandwidth, i.e. the frequency difference between the two −3 dB points

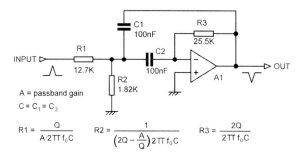

**Figure 12.1: An MFB or Rauch bandpass multiple-feedback filter with**
**$f_0$ = 250 Hz, $Q$ = 2 and a gain of 1.**

on either side of the peak. This configuration is not suitable for $Q$s greater than about 10, as the filter characteristics become unduly sensitive to component tolerances. If independent control of $f_0$ and $Q$ are required, the state-variable filter should be used instead.

Similar MFB configurations can be used for lowpass and highpass filters; see Chapter 10. The lowpass MFB filter does not depend on a low opamp output impedance to maintain stopband attenuation at high frequencies and so avoids the oh-no-it's-coming-back-up-again behaviour of Sallen & Key lowpass filters. The highpass MFB filter has some unresolved issues with noise and distortion.

## High-$Q$ Bandpass Filters

As we have just seen, the simple MFB/Rauch filter is not suitable for high $Q$'s. When these are required (which is not likely to be very often in crossover design), there are many, many kinds of active filter that can be used. We will just take a quick look at one of the most useful, the double-amplifier bandpass or DABP filter; this is a good example of the way that active filter performance can sometimes be transformed by adding one more inexpensive opamp section. Figure 12.2 shows the circuit with values for a centre frequency of 1 kHz and an impressively high $Q$ of 70, A2 being used to provide one of the feedback paths. The high value of R1 with respect to the rest of the circuit derives from the high $Q$.

The design equations are pleasantly straightforward; just choose a preferred value for C1 = C2 = C, calculate the value of R and then derive the values of R1, R2, R3 from it. R4 and R5 are non-critical so long as they are of equal value; as usual they should be made as low as possible without overloading A1 output, in order to keep down both their Johnson noise and the effect of the current noise flowing through them from the A2 non-inverting input. In this case R4 and R5 could be reduced to 1 k$\Omega$ so long as the external loading on A1 is not too great.

The gain at resonance is always two times (+6 dB), which is very often not wanted. The most convenient way to introduce a compensating 6 dB of attenuation is to split R1 into two resistors which have in parallel the same value as R1, and connect one of them to ground. This technique is also described in Chapter 8 on lowpass filters

Figure 12.3 shows the satisfyingly sharp resonance at a $Q$ of 70 obtained with the circuit of Figure 12.2. Away from the peak, the filter slopes slowly merge into straight lines at 6 dB/octave, as is

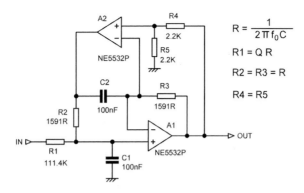

$$R = \frac{1}{2\pi f_0 C}$$

$$R1 = Q\,R$$

$$R2 = R3 = R$$

$$R4 = R5$$

Figure 12.2: Dual-amplifier bandpass (DABP) filter with a centre frequency of 1 kHz and a *Q* of 70. The gain at resonance is always +6 dB.

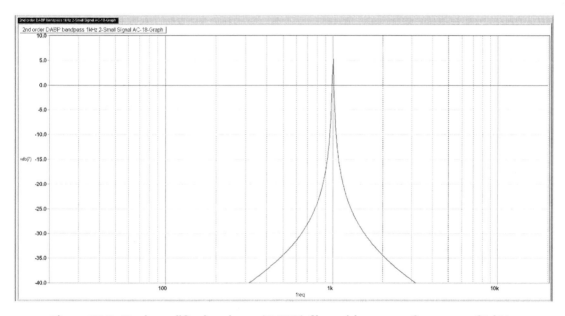

Figure 12.3: Dual-amplifier bandpass (DABP) filter with a centre frequency of 1 kHz and a *Q* of 70. Gain at resonance peak is +6 Db.

normal for 2nd-order bandpass responses. Fourth-order bandpass filters with 12 dB/octave skirt slopes can be made by cascading two 2nd-order bandpass filters.

## Notch Filters

There are a very large number of notch filters, each with their own advantages and disadvantages. The width of a notch is described by its *Q*. As for a resonance peak, the *Q* is equal to the centre frequency divided by the −3 dB bandwidth, i.e. the frequency difference between the −3 dB points either side of the notch. Figure 12.4 demonstrates this procedure for a notch with a centre frequency of 1.00 kHz

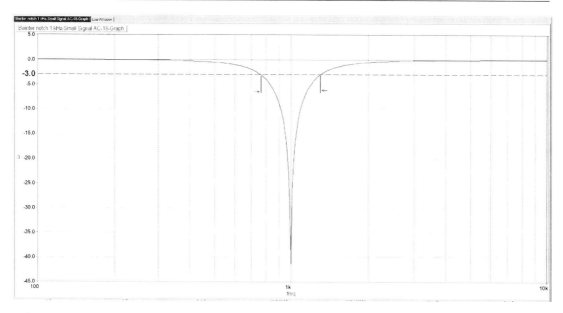

**Figure 12.4: Determining the $Q$ of a notch filter, which is equal to the centre frequency divided by the bandwidth between the two −3 dB points.**

and a $Q$ of 1.5; the bandwidth between the two −3 dB points is 0.667 kHz, so $Q = 1.00/0.667 = 1.5$. Be aware that the $Q$ of a notch has no special relation to the depth, but if the notch is really shallow— more of a dip than a notch—then the response at the −3 dB points may be affected, and this will affect the calculated $Q$.

The symmetrical notch shown in Figure 12.4 is the best-known sort, but there are also highpass notches and lowpass notches. A lowpass notch typically has unity gain on the low-frequency side of the notch and a lower gain on the high-frequency side. A highpass notch typically has unity gain on the high-frequency side of the notch and a lower gain on the low-frequency side. In both cases the responses are flat away from the notch. There is more information on the use of lowpass and highpass notches later in this chapter and in Chapter 5 on notch crossovers.

## The Twin-T Notch Filter

The best-known notch filter is the Twin-T notch network shown in Figure 12.5a, invented in 1934 by Herbert Augustadt. [1] It gives a symmetrical notch. The notch depth is infinite with exactly matched components, but with ordinary ones it is unlikely to be deeper than 40 dB. It requires ratios of two in component values such as 100 Ω–200 Ω, and there are only six such pairs in the E24 resistor series; see Chapter 15. Capacitors have sparser value series, and two will need to be paralleled to get the 1:2 ratio required.

The Twin-T notch when used alone has a $Q$ of only 1/4, which is not very useful. It is therefore normally used with positive feedback via an opamp buffer A2, as shown. The proportion of feedback

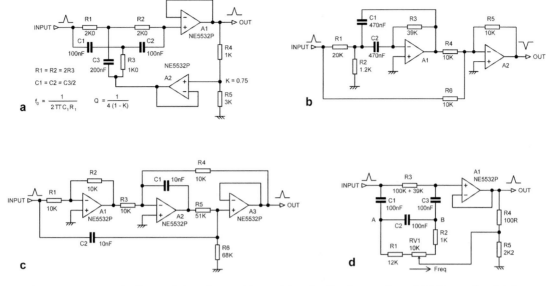

Figure 12.5: Notch filters: (a) Twin-T with positive feedback, notch at 795 Hz and a *Q* of 1; (b)"1-bandpass" filter with notch at 50 Hz, *Q* of 2.85; (c) Bainter filter with notch at 700 Hz, *Q* of 1.29; (d) bridged-differentiator notch filter tuneable 80 to 180 Hz.

K and hence the *Q*-enhancement is set by R4 and R5, which here give a *Q* of 1. The great drawbacks of the Twin-T are that it depends on matched components for good notch depth and that the notch frequency can only be altered by changing three components, keeping them carefully matched while doing so.

## The 1-Bandpass Notch Filter

Another way of making notch filters is the "one-minus-bandpass" principle, usually written "1-bandpass" or "1-BP". The input goes through a bandpass filter, typically the multiple-feedback type described earlier, and is then subtracted from the original signal. The accuracy of the cancellation and hence the notch depth is critically dependent on the midband gain of the bandpass filter. Figure 12.5b shows an example that gives a notch at 50 Hz with a *Q* of 2.85. The subtraction is performed by A2, as the output of the filter is phase inverted. The multiple-feedback filter is designed for unity passband gain, but the use of E24 values as shown means that the actual gain is 0.97, limiting the notch depth to −32 dB. The value of R6 can be tweaked to deepen the notch; the nearest E96 value is 10.2 kΩ, which gives a depth of −45 dB. The final output is inconveniently phase inverted in the passband.

## The Bainter Notch Filter

The Bainter notch filter is a very useful configuration, invented as recently as 1975, [2] making it the newest filter in this book. An example is shown in Figure 12.5c, where the values shown give a symmetrical notch at 700 Hz with a *Q* of 1.29. You will note that three amplifiers are required, but this is a small price to pay for such a capable filter. The first great advantage is that the depth of the notch does not depend in any way on the matching of components but is determined by the open-loop gain of the amplifiers, being roughly proportional to it; 5532 opamps give a notch depth of about −70 dB with no adjustment at all, while the old TL072 gives less impressive depths of about −40 to −50 dB. Having said that, deep notches are not normally essential for crossover design; even notches that look worryingly shallow usually have a negligible effect on the summed response.

Second, the Bainter notch filter can provide lowpass notches, symmetrical notches, and highpass notches depending on the design decisions (though only one of them at a time). Third, two out of the three amplifiers are working at virtual earth and will give no trouble with common-mode distortion. Fourthly, if it suffers internal clipping this is relatively benign, with only one-sixth of the distortion generated by the clipping amplifier reaching the filter output. Fifthly, the Bainter filter is non-inverting in the passband, which is very handy.

The only real disadvantage of the Bainter filter is that be it set for lowpass, symmetrical, or highpass notch, the gain asymptoted to above the notch frequency is always unity (0 dB), and so making a lowpass notch means having gain on the low-frequency side of the notch, which will often be highly inconvenient in a crossover signal path. About the only other thing I can think of to say against the Bainter is that it is one of those enigmatic configurations which on inspection give very little clue as to how they work. One thing is clear; at frequencies well above the notch, the filter must be unity gain and non-inverting, because then C2 is essentially a short-circuit from the input to the unity-gain buffer, which is connected directly to the output. The signal at the A2 output is then irrelevant, because the resistance of R5 will be much greater than the impedance of C1.

A Bainter filter consists of three opamps, the shunt-feedback stage A1, the integrator A2, and the output buffer A3. See Figure 12.5c. The gain of A1 is here set to unity. You can have any gain here so long as R3 is appropriate to inject the current amount of signal current into the summing point of A2, as noted later, but unity gain is often convenient.

To get a standard symmetrical notch with equal gain either side of the crevasse, R3 must equal R4. R4 greater than R3 gives a lowpass notch, while R3 greater than R4 gives a highpass notch; these responses are useful for making elliptical filters and for other applications in crossover design.

The Bainter filter is usually shown with equal values for C1 and C2. This leads to values for R5 and R6 that are a good deal higher than other circuit resistances, and I suggested in the first edition of this volume that this would impair the noise performance. A bit more thought shows that the junction of R5 and R6 is connected to the input by capacitor C1, and so in the audio band there is no high impedance to generate noise; low-impedance design here will give only a very small noise improvement, if any. The input impedance below the notch is the value of R1. Above the notch it falls to R1 in parallel with R5 and R6, which in most Bainter filters is a very small reduction.

The Bainter filter is set to give a lowpass, symmetrical, or highpass notch by the ratio of the notch frequency $f_z$ to the filter centre frequency $f_0$. The square of this ratio gives the parameter Z, which is used to calculate R3, R4, and R5 (please note this has nothing whatever to do with using Z to represent impedance). For Z > 1 you get a lowpass notch with R3 < R4. For Z = 1 you get a symmetrical notch with R3 = R4. For Z < 1 you get a highpass notch with R3 > R4. The gain *g* of the shunt-feedback stage A1 can be set arbitrarily, as it can be allowed for by the value of R3. Unity gain seems to give good results for noise and distortion, but I am not claiming each is always optimal. It occurs to me that suitable adjustment of R3 would allow both R1 and R2 to be preferred values.

The design procedure given here assumes the passband gain is unity. You can get gain by making A3 a series-feedback amplifying stage, but unity gain is usually what's wanted.

## Bainter Notch Filter Design

The design procedure is thus:

1. Choose     C= C1 = C2
2. Choose gain of shunt-feedback stage $g$
   (Different values can be used for C1 and C2, but it makes the calculation more complicated)
3. Choose R1
4. Calculate    $R2 = gR1$
5. Choose the centre frequency of the filter $f_0$
6. Choose the notch frequency $f_z$
   (this is only the notch frequency for a symmetrical notch)
7. Calculate:  $Z = \left(\dfrac{f_z}{f_0}\right)^2$
8. Calculate    $a = 2\pi f_0 C$
   For Z > 1 you get a lowpass notch
   For Z = 1 you get a symmetrical notch
   For Z < 1 you get a highpass notch
9. Choose **Q**
   The value of $Q$ that gives maximal flatness in the passband depends on the value of Z
   For Z = 1/2, $Q$ = 1 gives maximal flatness.
10. Calculate    $R3 = \dfrac{g}{2ZQa}$     $R4 = \dfrac{1}{2Qa}$     $R5 = R6 = \dfrac{2Q}{a}$

The effect of changing the parameter Z is illustrated in Figure 12.6. For Z > 1 you get a lowpass notch, for Z = 1 you get a symmetrical notch, and for Z < 1 you get a highpass notch, as already noted. The gain is always unity above the notch frequency because of the presence of C2. The symmetrical notch is centred on the low frequency of 10 Hz because these plots were originally part of a subsonic filter development program. The Bainter filter is an extremely convenient building block for highpass filters, but not so much for lowpass filters, because a lowpass notch means gain on the low-frequency side of the notch, and this will often be highly inconvenient.

Figure 12.7 shows the effect of different values of $Q$ with Z = 1/2 to give a highpass notch.

Looking at Figure 12.5c, we can attempt some deductions about the likely distortion performance. I never make predictions, and I never will, but I think we can say:

1. Opamp A1 is working in shunt-feedback mode, so there is no common-mode (CM) voltage on its inputs, and so no common-mode distortion.

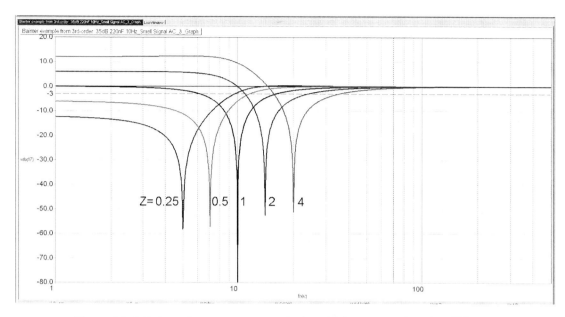

Figure 12.6: Bainter lowpass, symmetrical, or highpass notches for different values of the parameter Z. $Q = 1$.

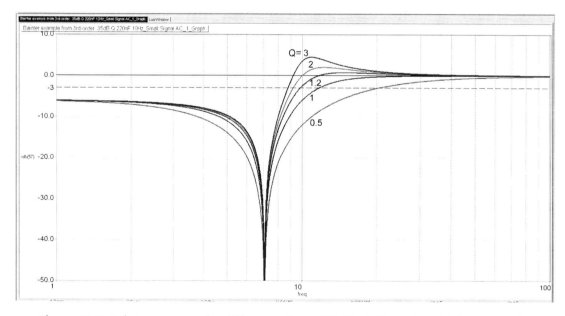

Figure 12.7: Bainter response for different values of $Q$. $Z = 1/2$ to give a highpass notch.

2. A1 is also working at a low noise gain so there will be plenty of negative feedback (NFB) to reduce distortion.
3. A2 is working in shunt-feedback mode, so there is no CM voltage on its inputs.
4. A2 is working as an integrator, so there will be a lot of NFB at high frequencies where open-loop gain is falling.
5. A3 is a voltage-follower, so it has maximal CM voltage on its inputs, which may cause increased distortion.
6. A3 is a voltage-follower, so it has maximal NFB, which should reduce distortion.
7. Perhaps most importantly, in the audio passband the signal only goes through C2 and A3, so distortion from A1 and A2 does not reach the output.

All these statements except (5) look promising. Now some thoughts about the likely noise performance:

1. A1 has low-value resistors around it, set by R2, which loads its output; 1 kΩ was chosen to give good THD results with 5532 opamps.
2. A1 is in shunt-feedback mode, so it is operating at a higher noise gain than its signal gain, increasing noise.
3. A2 is an integrator, and they are quiet because of low gain at high frequencies.
4. A3 is working at unity signal gain and noise gain, so noise output should be low.
5. Again, perhaps most importantly, over the audio band, A3 is directly connected to the input via C2, so only the input voltage noise of A3 will be seen at the output.

Both sets of predictions were very largely confirmed by the measurements made, confirming that the Bainter filter is an excellent configuration.

### Bainter Notch Filter Example

As an example of filter scaling and impedance reduction, a low-noise Bainter filter with a symmetrical notch at 1 kHz and a $Q$ of 2.3 is shown in Figure 12.8. Starting from the circuit of Figure 12.5c with the notch at 700 Hz, both capacitors were scaled by the same ratio to change the notch centre frequency to 1 kHz. The value of R6 was then changed to give the desired $Q$ of 2.3. The first stage of impedance reduction was to reduce R1 and R2 from 10 kΩ to 1 kΩ, which is about as far as you can take it without loading a previous stage excessively. The next stage was to then alter R3, R4 and C2 so that the resistance values were decreased by the same ratio as C2 was increased; this leaves the frequency-dependent behaviour of this network unchanged. The third stage of impedance reduction reduces R6 while increasing C1 by the same ratio, once again keeping the frequency-dependent behaviour the same, though as noted earlier this may not give much lower noise.

The limits of the first impedance reduction are set by the value of R3, which directly loads the output of A1, the value of R4 that loads A3, and the combined value of R5, R6 in series, which loads A2. The last condition is less critical; as you can see from Figure 12.6, the values of R5 and R6 are such as to present only a light load on A2 output, despite the fact that C1 is considerably bigger than C2. A further reduction in R5, R6 would be possible, but C1 then starts to get expensive, and the loading on whatever stage is driving this filter must also be considered.

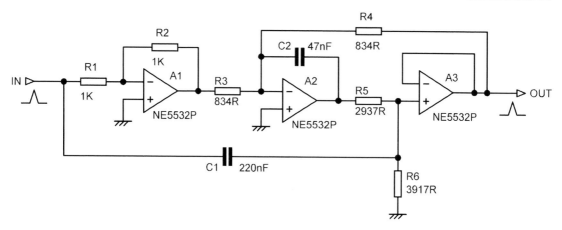

**Figure 12.8: Bainter filter with symmetrical notch at 1 kHz and a *Q* of 2.3.**

### *An Elliptical Filter Using a Bainter Highpass Notch*

The ability of the Bainter to give either a lowpass or highpass notch makes it a very useful building block for elliptical filters that have a zero or zeros in the stopband. I have made use of this in designing a comprehensive suite of elliptical highpass subsonic filters to assist with vinyl reproduction. [3] What I consider the most effective of these is shown in Figure 12.9, with 2xE24 pairs given for the awkward resistor values. It is a 4th-order elliptical filter with a single stopband notch centred on 10 Hz, where spurious vinyl signals are often at their maximum level due to arm/cartridge resonance. The complete filter consists of a 2nd-order highpass stage A4, placed at the front to reduce internal clipping problems (I call this the New Order) and followed by a Bainter highpass notch. The response is shown in Figure 12.10. The noise output was −113 dBu (22Hz–22kHz, rms sensing) with all opamps 5532. The filter easily handled 10 Vrms in/out at low distortion from 50 Hz to 50 kHz. My recent book *Electronics for Vinyl* [3] contains much more on subsonic filtering.

I like the Bainter filter. It is a very useful and well-behaved filter configuration. It may not be the simplest of filters, but its inherently deep notch and its good noise and distortion behaviour are a joy to work with. Mr Bainter, I salute you.

There are relatively few references to the Bainter filter in the literature; Valkenburg [4] is one.

## The Bridged-Differentiator Notch Filter

A notch filter that can be tuned with one control can be useful in development work. Figure 12.5d shows a bridged-differentiator notch filter tuneable from 80 to 180 Hz by RV1. R3 must theoretically be six times the total resistance between A and B, which here is 138 kΩ, but 139 kΩ gives a deeper notch, about −27 dB across the tuning range. The downside is that *Q* varies with frequency from 3.9 at 80 Hz to 1.4 at 180 Hz; clearly it is no substitute for a state-variable filter where frequency and *Q* can be changed independently. It may be an economical circuit in terms of amplifiers, but it uses three capacitors instead of the two in the 1-BP and Bainter filters.

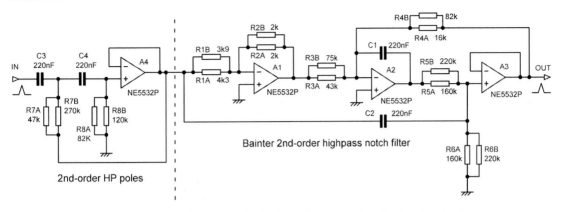

**Figure 12.9: Schematic of practical 4th-order A$_0$ = −35 dB classical highpass elliptical filter, using the New Order.**

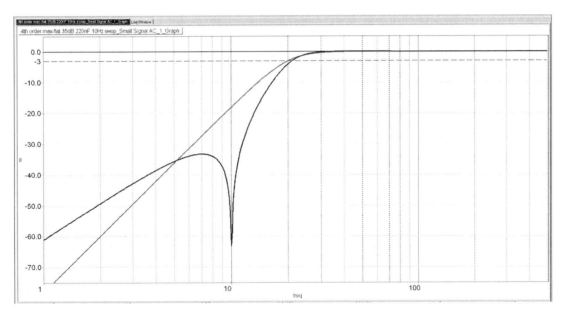

**Figure 12.10: Frequency response of 4th-order A$_0$ = −35 dB classic (non-MCP) highpass elliptical filter compared with 3rd-order Butterworth. Response is −3 dB at 21 Hz, almost the same as the Butterworth, which is −3 dB at 20 Hz.**

## Boctor Notch Filters

Another interesting notch filter is the Boctor circuit, which uses only one opamp. [5], [6] Versions exist that can give either a lowpass or highpass notch. The design equations are complicated but can be found in the two references given. Figure 12.11 shows the lowpass and highpass versions.

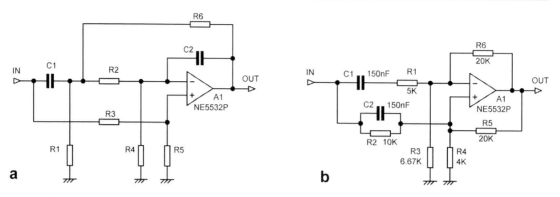

Figure 12.11: Lowpassnotch (a) and highpass notch (b) Boctor filters. The design at (b) has a notch at 150 Hz with a $Q$ of 1.

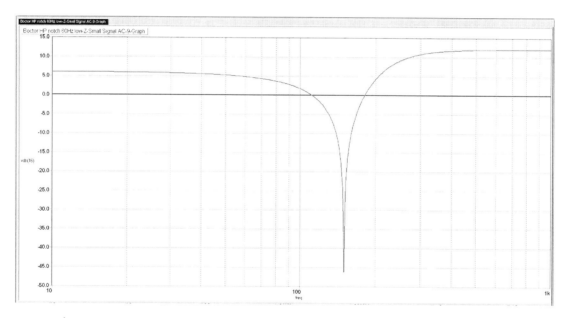

Figure 12.12: Response of Boctor filter in Figure 12.7b, giving a highpass notch at 150 Hz with a $Q$ of 1.

Figure 12.12 shows the highpass notch produced by the Boctor circuit of Figure 12.11b; the high-frequency passband gain is +12 dB, with a gain of +6 dB on the low-frequency side of the notch. The capacitor values may be scaled to change the notch frequency, but they must be the same. A lowpass notch is the mirror image of this sort of response, with the low-gain section on the high-frequency side.

Like Bainter filters, Boctor highpass or lowpass notch filters are frequently used as a component part of elliptical filters.

## Other Notch Filters

The field of active filters is rich in possibilities. Other notch filters that there is no space to examine here, but which can be found in the filter textbooks, can be derived from biquad filters such as the Fliege filter, [7] the Tow-Thomas filter, the KHN state-variable filter, the Berka-Herpy filter, the Akerberg-Mossberg filter, and the Natarajan filter; these give symmetrical notches. See Chapter 10 for more on them. Filters apart from the Bainter and the Boctor that can generate highpass or lowpass notches are Friend's SAB circuit [8] and the Scultety filter. The only reference I can find for the latter is [9], which at least gives the schematic; the same reference also gives schematics of the other filter mentioned in this section.

The Wien-Robinson filter is not biquad-based but uses a Wien network with feedback around it. [10] Notch depth depends on component accuracy, unlike the Bainter, and it seems to have no compensating advantages.

## Simulating Notch Filters

When simulating notch filters, assessing the notch depth can be tricky. You need a lot of frequency steps to ensure you really have hit bottom with one of them. For example, in one run, 50 steps/decade showed a −20 dB notch, but upping it to 500 steps/decade revealed it was really −31 dB deep. In most cases having a stupendously deep notch is pointless. If you are trying to remove an unwanted signal, then it only has to alter in frequency by a tiny amount and you are on the side of the notch rather than the bottom, and the attenuation is much reduced. The exception to this is the THD analyser, where a very deep notch (120 dB or more) is needed to reject the fundamental so very low levels of harmonics can be measured. This is achieved by continuously servo-tuning the notch, so it is kept exactly on the incoming frequency.

## References

[1] Augustadt, H. *Electric Filter* US Patent 2,106,785, February 1938, assigned to Bell Telephone Labs
[2] Bainter, J. R. "Active Filter Has Stable Notch, and Response Can Be Regulated" Electronics, pp. 115–117, 2 October 1975
[3] Self, Douglas "Electronics for Vinyl" Taylor & Francis, 2017
[4] Valkenburg, Van "Analog Filter Design", pp. 346–349
[5] Boctor, S. A. "Single Amplifier Functionally Tuneable Low-Pass Notch Filter" IEEE Trans Circuits & Systems, CAS-22, 1975, pp. 875–881
[6] Valkenburg, Van "Analog Filter Design", p. 349
[7] Carter and Mancini (ed.) "Opamps for Everyone" Third Edn, Newnes, 2009, Chapter 21, p. 447, ISBN: 978-1-85617-505-0
[8] Valkenburg, Van "Analog Filter Design", p. 356
[9] www.schematica.com/active_filter_resources/a_list_of_active_filter_circuit_topologies.html; Accessed July 2017
[10] Tietze and Shenk "Electronic Circuits" Second Edn, Springer, 2008, pp. 825–826

# CHAPTER 13

# *Time-Delay Filters*

## The Requirement for Delay Compensation

A lot of discussion on loudspeaker behaviour is based on mathematically simple point sources, but real loudspeaker drive units have a physical size and must be mounted with some space between them. If they are mounted on a flat baffle, as in Figure 13.1, their acoustic centres (the position from which the sound effectively radiates) have different horizontal distances from the baffle position. If we take our listening position as in-line with the tweeter axis, as is usually done, then there is a greater distance from the ear to the mid unit and an even greater one to the bass unit. To minimise the differences, the drive units are normally mounted as close together as physically possible. We therefore have two effects giving rise to different distances from drive unit to ear:

1. The vertical spacing of the drive units
2. Different distances between the drive unit acoustic centres and the front baffle

I am assuming here that the drive units are mounted in a vertical line. This is almost always the case, as it gives the best horizontal directivity.

These delays lead to frequency-dependent variations in the response, due to cancellation and reinforcement at a given point in space. The dimensions shown in Figure 13.1 are intended to be reasonably typical, but I will tell you now that they have been carefully chosen to require a compensation delay of 80 usec for the tweeter and 400 usec for the mid unit; these figures are used extensively as examples throughout this chapter. It must not be thought that a spacing of a few millimetres is insignificant. The 22 mm difference between the mid drive unit and the tweeter should be compared with the wavelength of sound at 4 kHz, which is 86 mm. At this frequency, 22 mm therefore gives a 90° phase-shift, turning a +6 dB fully in-phase reinforcement into a +3 dB partial reinforcement. This sort of thing is obviously going to cause frequency response irregularities.

This is why delay compensation is highly desirable for multi-way loudspeakers. The technique is sometimes called time equalisation or time alignment. Siegfried Linkwitz says that crossovers that are not time-compensated are only marginally usable, [1] and, for what it's worth, I agree with him. Figure 13.1 shows a simplified diagram of the situation; please be aware that the drawing is very much not to scale. First we have to decide how far away to put our reference listening point P, marked by that disembodied ear floating in space at the same level as the tweeter. This distance is usually set at 1 or 2 metres; I have here chosen 2 metres as the more realistic listening distance, but of course this will vary in practice, upsetting our careful calculations somewhat. We can but press on and do our best.

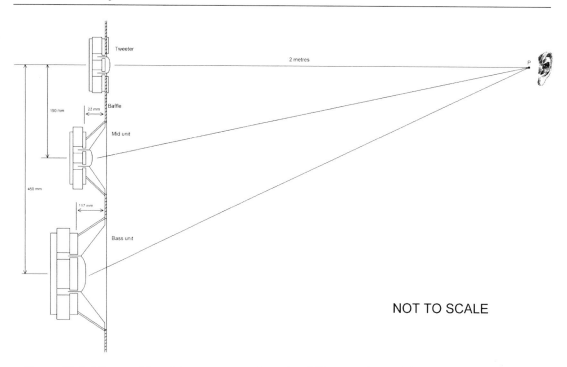

NOT TO SCALE

**Figure 13.1: The need for delay compensation; the differences in path lengths to the ear when three drive units are mounted in a common vertical plane.**

The vertical distance between the drive unit centre-lines is easy to determine, but the position of the acoustic centres is less easy, not least because it usually varies somewhat with frequency. Some people use the voice coil cap, but others claim that the centre of the voice coil itself should be regarded as the acoustic point of origin, as the speed of sound in the average coil assembly is roughly the same as in air; this seems more than a bit dubious, as the speed of sound is in fact quite different in solids, and there are actually two different speeds of sound, transmitted in one case by volumetric deformations and in the other by shear deformations.

Add to this the composite construction of the voice coil, with copper windings bonded to the coil former, and the whole situation gets very complicated and will probably have to be sorted out by experiment. The position of the acoustic centre is not normally part of a drive unit specification.

An important consideration is that we only need to consider the delay that must be applied to the tweeter with respect to the MID unit, and the delay to be applied to the MID unit with respect to the LF unit. It is not necessary to delay the tweeter to match the LF unit, because the LF unit output at a given frequency should be negligible when the tweeter is operating, and vice versa. This greatly reduces the amount of delay that needs to be used on the HF unit.

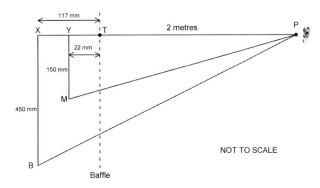

**Figure 13.2: The delay situation of Figure 13.1 reduced to its elements.
We need to calculate MP and BP.**

## Calculating the Required Delays

The initial step is to calculate what the path length differences are. We need to find the path lengths MP and BP in Figure 13.2. MP is the hypotenuse of the right-angled triangle MYP, and the two shorter sides are YM and (YT + TP). Using Pythagoras;

$$MP = \sqrt{YM^2 + (YT + TP)^2} \qquad\qquad 13.1$$

Likewise, BP is the hypotenuse of right-angled triangle BXP, and the two shorter sides are XB and (XT + TP). Flexing our Pythagoras again;

$$BP = \sqrt{XB^2 + (XT + TP)^2} \qquad\qquad 13.2$$

For the dimensions in Figure 13.2 and a listening distance TP of 2 metres, we find the path lengths are MP = 2.027 metres and BP = 2.165 metres. It is the difference between these path lengths and the tweeter listening distance that causes the delay we need to compensate for; the extra mid path length is thus 2 − 2.027 = 0.027 metres = 27 mm, and the extra LF path length is the difference between the mid path and the LF path, in other words 2.027 − 2.165 = 0.137 metres = 137 mm.

Having calculated the path length differences, the delays themselves are simply calculated by using the speed of sound. This is 343.2 metre/sec (equal to 767.7 mph or 1.13 feet per ms) at 20°C. It varies slightly with air temperature, being proportional to the square root of absolute temperature; this is measured from absolute zero at −273°C, which is why the speed variation is relatively small at the kind of temperatures we are used to. Table 13.1 shows this, together with the percentage variation. The speed of sound varies by ±2% between 10 and 30°C, so for this reason alone there really is no point in being overly precise with delay calculations. Sound velocity varies very slightly with barometric pressure, and humidity also has a small but measurable effect (it can cause an increase of 0.1%–0.6%), because some of the nitrogen and oxygen molecules in the air are replaced by lighter water molecules.

**Table 13.1: How the speed of sound varies with temperature.**

| Temp deg C | Speed m/sec | % ref 20°C |
|---|---|---|
| 35 | 351.9 | 2.5 |
| 30 | 349.0 | 1.7 |
| 25 | 346.1 | 0.8 |
| **20** | **343.2** | **0.0** |
| 15 | 340.3 | −0.9 |
| 10 | 337.3 | −1.7 |
| 5 | 334.3 | −2.6 |
| 0 | 331.3 | −3.5 |

Dividing the path length differences by the speed of sound, we find that the tweeter must be delayed by exactly 80 usec with respect to the mid unit, and the mid unit must be delayed by exactly 400 usec with respect to the LF unit. As I mentioned earlier, these nice round figures are no accident.

At the start of these calculations we had to choose a listening distance, and we picked 2 metres. In reality the listening distance will vary, and we'd better find out how much difference this makes to the delays required for exact delay compensation. A bit of automated Pythagoras on a spreadsheet gives us the rather worrying Table 13.2, which gives the delays required for listening distances between 1 and 10 metres, and for infinity (but not beyond).

Table 13.2 also gives the percentage variation of each delay with respect to the 2-metre listening distance. You can see at once that picking 2 metres as the reference rather than 1 metre has made big differences to the delays needed; 19% less for the tweeter delay and a thumping great 25% less for the mid delay. One metre is not a very practical distance for listening to a big hi-fi system of the sort that is likely to use active crossovers, and I would have thought that 1.5 to 5 metres is a more probable range for normal use. This assumption gives us a tweeter delay that varies from +6.6% to −12.1%, while the mid delay varies from +9.0% to −17.7%. This makes it clear that if fixed amounts of delay are used there is, once more, no call for extreme precision. I am not aware that any manufacturer has ever produced an active crossover with a "Listening distance" dial that would alter the two delays appropriately to match the listening position; it's an interesting idea. It would only work with one model of loudspeaker, as the variations required would depend on the horizontal and vertical offsets.

The required delays alter more slowly as the listening distance increases, and the vertical offsets become relatively less significant. The need to compensate for the horizontal offsets is however unchanged, and the delays required asymptote to these values at an infinite listening distance. The situation is illustrated in Figure 13.3, which plots how tweeter and mid delays change with the listening distance.

To put this into perspective, let's look at the effect of uncertainty in the position of the acoustic centre of a drive unit. We currently have to apply a delay of 400 usec to the mid unit to match the LF unit output. But if we hold the LF unit in our hands and solemnly contemplate it, we might conclude that

**Table 13.2: The variation of the delays required with listening-point distance.**

| Listening distance | Tweeter delay | Tweeter % | Mid delay | Mid % |
|---|---|---|---|---|
| Metres | usec | ref 2 metres | usec | ref 2 metres |
| 1.0 | 95.7 | 19.6% | 500.6 | 25.2% |
| 1.1 | 92.9 | 16.1% | 483.8 | 21.0% |
| 1.2 | 90.5 | 13.2% | 469.2 | 17.3% |
| 1.5 | 85.3 | 6.6% | 435.9 | 9.0% |
| **2.0** | **80.0** | **0.0%** | **400.0** | **0.0%** |
| 2.5 | 76.8 | −4.0% | 377.3 | −5.7% |
| 3.0 | 74.6 | −6.8% | 361.7 | −9.6% |
| 3.5 | 73.1 | −8.6% | 350.3 | −12.4% |
| 4.0 | 72.0 | −10.0% | 341.7 | −14.6% |
| 4.5 | 71.1 | −11.1% | 334.9 | −16.3% |
| 5.0 | 70.3 | −12.1% | 329.4 | −17.7% |
| 6.0 | 69.2 | −13.5% | 321.1 | −19.7% |
| 7.0 | 68.5 | −14.4% | 315.1 | −21.2% |
| 8.0 | 67.9 | −15.1% | 310.6 | −22.4% |
| 9.0 | 67.4 | −15.8% | 307.1 | −23.2% |
| 10.0 | 67.1 | −16.1% | 304.3 | −23.9% |
| Infinity | 63.8 | −20.3% | 278.4 | −30.4% |

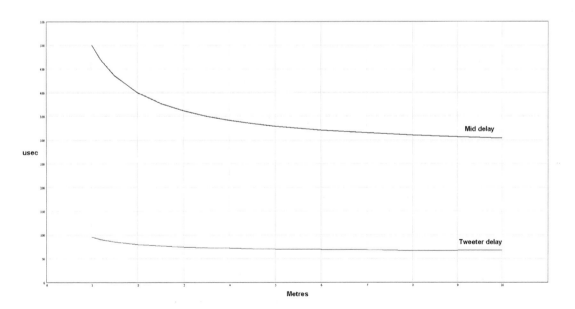

**Figure 13.3: The variation of the required tweeter delay (bottom) and the required mid delay (top) with listening distance, for the dimensions given in Figure 13.1.**

the true acoustic centre could be 30 mm either side of the position we have assumed. If it is 30 mm forward, the mid delay required falls to 313 usec; if it is 30 mm to the rear, we need 484 usec. This shows that errors in finding the true acoustic centre can have just as much effect as altering the listening distance, and is a bit worrying.

We have so far glibly assumed that the position of the acoustic centre is constant with frequency. I currently have no information on this, but if it does prove to vary with frequency then it should be possible to design a delay filter system that compensates for this.

Sound-reinforcement applications are rather different in the distances involved, but the delay calculations work the same way. The listening distances are much greater, but then the physical separation between banks of HF units and banks of mid units, for example, will also be greater.

Delay compensation is only really practicable with active crossovers. The electrical losses involved in LC delay lines in passive crossovers, and the cost of the components to implement, makes them an unattractive proposition. So far we have only dealt with 3-way loudspeakers. The same principles are of course applicable to four ways or more.

## Signal Summation

I shall shortly be showing you how the difficulty of designing a delay filter—and the cost of its components—depends very much on how much delay is required and over what range of frequency it must remain constant. Around the crossover frequency the delay must be constant at the required value, but at some frequency above this the amount of delay will begin to fall, and we need to know how much the acoustical summation of the two signals, which combine in the air in front of the speaker, is going to be affected by a fall in the time delay above the crossover frequency. Figure 13.4 gives the rules for adding sine waves of arbitrary magnitude and phase.

There is, as far as I am aware, no published work on the delay flatness required for loudspeaker delay compensation. I have here assumed that a reduction of not more than 10% in the group delay is acceptable so long as it is well away from the crossover frequency, at which point the contribution of the drive unit being compensated is negligible.

Let us assume that the delay falls by $-10\%$ at a frequency 2 octaves above the relevant crossover frequency. If we are using a Linkwitz-Riley 4th-order crossover, then the ultimate filter slopes will be

With amplitudes $a$, $b$ and phase shift $\alpha$

We get $\quad c \sin(x + \beta) = a \sin x + b \sin(x + \alpha)$

where magnitude $\quad c = \sqrt{a^2 + b^2 + 2ab \cos \alpha}$

and new phase $\beta = \arctan\left(\dfrac{b \sin \alpha}{a + b \cos \alpha}\right) + \begin{cases} 0 & \text{if } a + b \cos \alpha \geq 0 \\ \pi & \text{if } a + b \cos \alpha < 0 \end{cases}$

**Figure 13.4: The summation of two sinusoidal signals of the same frequency but arbitrary amplitude and phase to give a single sinusoidal output (Equation 13.3).**

**Table 13.3: The result of combining two in-phase sinewaves, with one at a lower amplitude.**

| Amplitude | Combined | Amplitude | Combined |
|---|---|---|---|
| dB | dB | dB | dB |
| 0 | 6.021 | −25 | 0.475 |
| −1 | 5.535 | −30 | 0.270 |
| −5 | 3.876 | −35 | 0.153 |
| −10 | 2.387 | −40 | 0.086 |
| −15 | 1.422 | −45 | 0.049 |
| −20 | 0.828 | −50 | 0.027 |

24 dB/octave. The lower frequency unit of the drive unit pair, the one with the compensation delay added, will therefore be something like 48 dB down, though this figure is likely to be affected by drive unit response irregularities. From the equations in Figure 13.4, we can quickly determine that, even using the worst-case phase-shift for the smaller signal (0°, 180°, 360°, etc), the output will only be altered in amplitude by ±0.035 dB, which is completely negligible compared with other possible errors. I therefore suggest that the 10% fall-off in time delay 2 octaves away from the crossover frequency will have no audible consequences.

Table 13.3 shows how the amplitude of the combined signal is increased for varying attenuations of the delayed signal, assuming a worst case of signals exactly in phase. It is clear that delayed signals below −40 dB will have an absolutely negligible effect, no matter what phase-shift is imposed on them by a fall-off of delay time.

## Physical Methods of Delay Compensation

Before we leap into the fascinating world of time-domain filters, we will take a quick look at some of the physical methods used for delay compensation. These are shown in 2-way loudspeaker form in Figure 13.5. Clearly it is best to mount the drive units as close together vertically as physically possible to minimise the path difference, but loudspeaker manufacturers do not appear to always do this, possibly because of the aesthetic drawbacks of two drive units that look "crammed together". Possibilities are:

1.   Tilting the front baffle. This compensates the delay for a listener on the horizontal axis. The problem is that the drive units will have been designed for optimal performance on the drive unit axis, and now you are listening to off-axis output. This method is relatively cheap to manufacture, but people are used to having the front of a loudspeaker facing them directly, and as a result there is a general feeling that it looks wrong. It may compensate for the horizontal offsets (though for it to work with a 3-way system you will have to have the three acoustic centres in a straight line, which may not be easy to arrange) but only reduces the effective vertical offsets slightly.

2.   Putting a step or steps in the front baffle. This enables you to listen on the intended axis but costs more to implement. There may be diffraction problems with the step or steps, which will need

to be carefully shaped or possibly compensated for by equalisation in the crossover. It makes the enclosure more complex and significantly more expensive to build. It can correct the horizontal offsets exactly but does nothing at all to reduce the vertical offsets—in fact they will have to be bigger to make room for the steps between the drive units.

3.  A separate box for the tweeter. More expensive again, and there may be more diffraction problems with the second box. A 3-way system will require three boxes, and once more the horizontal offsets may be corrected exactly, but the vertical offsets are likely to be increased.

While these methods can compensate for the differing distances of the drive unit acoustic centres from the front baffle—the horizontal offsets—none of them can help much with the other half of the problem—the vertical offsets. This appears to be a serious inherent problem in the vast majority of loudspeaker designs. The only way to eliminate the vertical offsets is to mount two loudspeakers in effectively the same physical position by using a dual-concentric design, where typically a tweeter is mounted in front of an LF unit voice coil or behind it. Current versions of this are the British KEF Uni -Q driver and the French Cabasse coaxial drive unit, which has an annular diaphragm surrounding a small horn-loaded dome tweeter. The famous Tannoy Monitor Red Loudspeaker had a rear-mounted tweeter that used the LF cone itself as a horn.

The dual-concentric approach has the great benefit that there are no vertical offsets that require delay compensation that varies with listening distance. The horizontal offsets may or may not remain, depending on the details of construction, but can be easily compensated for by constant delays. It would at first appear that a major limitation of dual-concentric technology that it is restricted to 2-way loudspeakers, as it is hard to mechanically fit LF, mid, and tweeter units *all* on the same axis. However, Cabasse have recently announced not only the TC23 3-way coaxial driver, which combines a tweeter and separate annular high-mid and low-mid diaphragms, but also the QC55 four-way unit that combines the TC23 with a 22-inch LF unit. This is built into a spherical enclosure, which is the ideal shape for minimising diffraction effects. This technology clearly wants watching.

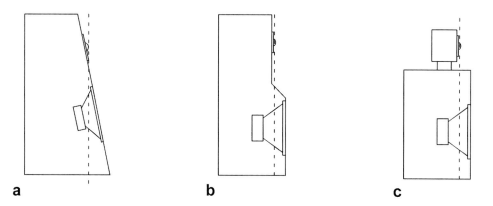

a                                        b                                        c

**Figure 13.5: Physical methods of compensating for drive unit delay: (a) tilting the front baffle; (b) stepping the front baffle; (c) separate box for the tweeter.**

## Delay Filter Technology

In the analogue domain, delay equalisation is performed by using allpass filters, which have a flat amplitude-frequency response but a phase-shift that varies with frequency. Allpass filters, like their more familiar lowpass and highpass relatives, come in 1st-order, 2nd-order, 3rd-order, and higher-order variants. The higher orders have increasing complexity but give a faster phase change at the turnover frequency, allowing the group delay to remain flatter for further up the frequency range. This is analogous to the frequency response of a filter for amplitude; the higher order the filter, the flatter the response will be until the turnover frequency is reached.

Allpass filters are sometimes used to flatten the delay curve of a high-order lowpass or highpass filter. This often means matching the combined delay of two or three allpass stages with a delay curve from the high-order filter that looks like a dog's back leg and typically requires computer optimisation. Here, mercifully, the situation is much simpler. We just want a constant delay with frequency.

You can also obtain delays with lowpass filters such as the Bessel sort, which is optimised for flat group delay, but the bandwidth is much lower, as you might expect. There is more on the topic of delay with lowpass filters at the end of this chapter.

## Sample Crossover and Delay Filter Specification

This is the basic specification for a loudspeaker crossover that we will use as an example for this chapter.

| | |
|---|---|
| Number of bands: | Three |
| Type: | Linkwitz-Riley 4th-order |
| Mid/HF crossover frequency: | 2.5kHz |
| LF/mid crossover frequency: | 400Hz |
| HF delay: | 80 usec, tolerance ±5 % |
| Mid delay: | 400 usec, tolerance ±5% |

Most of the delay filter examples in this chapter will be designed for an 80 usec delay.

## Allpass Filters in General

When I first encountered the phrase "allpass filter", my immediate reaction—and here I suspect I am not alone—was "What's the point of that, then?" The point is of course that while the frequency response is flat, the phase-shift varies with frequency. A less enigmatic name would be "variable-phase-shift-only filter", but I'm quite sure that won't catch on.

The general form of an allpass filter is shown in Figure 13.6. There are two signal paths, one of which goes through the box labelled T(s), which just means that it has a frequency response that varies with

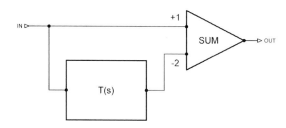

**Figure 13.6: The general form of an allpass filter. The path bypassing
T(s) must be inverted and have half the gain.**

frequency. It will be either a simple RC 1st-order filter or a 2nd-order bandpass filter; its output is amplified by two and subtracted from the original signal.

It is important to be aware that allpass filters do not inherently have an absolutely flat frequency response. Due to the finite gain-bandwidth-product of the opamps and components that do not have mathematically exact values, the magnitude response is likely to show small deviations from perfect flatness. These deviations are however usually very small and of no consequence compared with speaker unit tolerances.

## First-Order Allpass Filters

Figure 13.7 shows the two versions of the basic 1st-order allpass filter. This deceptively simple circuit is analogous to a 1st-order RC filter, and so gives a slow phase change as frequency changes.

The first version in Figure 13.7a is non-inverting at low frequencies; in other words the phase-shift is 0°. As the frequency rises, the phase-shift increases until it approaches 180° (inverting) at high frequencies, as shown in Figure 13.10. This changing phase-shift is equivalent to a delay. The sharp-eyed reader will note that in Figure 13.10 the phase has actually reached 180° and is clearly headed for more. This plot was produced by simulating the circuit with a model of a real opamp, rather than just evaluating a mathematical equation, and the extra phase-shift has accumulated because of the finite open-loop bandwidth of that opamp. This is what will happen when you build real crossovers, but with any modern opamp the effects are negligible. Looking at Figure 13.10 and Figure 13.11 we can see that the phase is still changing quite quickly in the 10 kHz–20 kHz range (from −135° to −158°, i.e. 21°) when the delay has fallen to a quarter of its maximum value. This is not going to give us a minimum-phase crossover.

When the positions of the resistor and capacitor are exchanged, the same phase change is obtained but phase inverted, so the second version in Figure 13.7b inverts at low frequencies but has an in-phase output at high frequencies. An allpass filter gives twice the maximum phase-shift of an ordinary filter of the same order. The non-inverting version of Figure 13.7a could also be called the RC version, as the resistor comes before the capacitor, forming a lowpass filter. This means the common-mode voltage will fall with frequency. The inverting version of Figure 13.7b is correspondingly the CR version, and the capacitor-resistor arm now acts as a highpass filter.

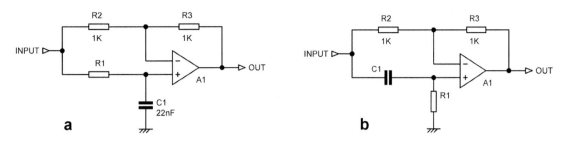

**Figure 13.7: Two versions of the 1st-order allpass filter: (a) non-inverting or RC type (b) inverting or CR type.**

A 1st-order allpass filter has only one parameter to set: the RC time-constant; the delightfully simple calculation for the low-frequency delay is shown in Equation 13.4. There is no such thing as the $Q$ of a 1st-order allpass. The output of an allpass filter does not have a "turnover" as such. According to some authors its operation is defined by the frequency at which the group delay has fallen to $1/\sqrt{2}$ of its low frequency value; while this corresponds in a way to "−3 dB" for an amplitude-frequency filter, it has no real significance in itself. A more useful description is the frequency $f_{90}$ at which the phase-shift reaches 90°, halfway between the extremes of 0° and 180°, because it is derived very easily from the circuit values, as in Equation 13.5. Note that these are *not* the same frequencies.

$$Delay = 2RC \qquad\qquad 13.4$$

$$f_{90} = \frac{1}{2\pi RC} \qquad\qquad 13.5$$

You will recall that in our example loudspeaker, the tweeter signal had to be delayed by 80 usec. The resulting RC allpass circuit is shown in Figure 13.8.

If you trustingly feed a positive-going step waveform into Figure 13.8, what emerges is an immediate negative spike followed by a long slow approach to the steady input voltage, as shown in Figure 13.9."Doesn't look much like a delay to me!" I hear you cry, which was pretty much what I cried when I first tried this experiment.

The reason for this unhelpful-looking output is that allpass filters do not give a constant time delay with frequency, and this one is no exception. Figure 13.11 shows how as the frequency goes up the group delay is indeed 80 usec initially, but begins to decrease early as the frequency increases. Figure 13.10 gives the phase response for comparison. The delay is down by 10% at 2.5 kHz and down to 50% at 9.3 kHz, slowly approaching zero above 100 kHz. This is clearly not much use for equalising the delay in the HF path. Since there is only one variable—the time-constant R3, C1—the only way to keep the delay constant to higher frequencies would be to reduce its value, which would make the filter useless for implementing the required delay compensation.

The fall-off in delay with frequency explains Figure 13.9; the high frequencies that make up the edge of the input step function are hardly delayed at all, and since they are subject to a 180° phase-shift, give the immediate inverted spike. The lower frequencies are delayed but get through eventually,

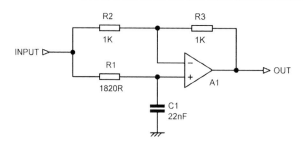

**Figure 13.8: A non-inverting RC 1st-order allpass filter designed for a group delay of 80 usec.**

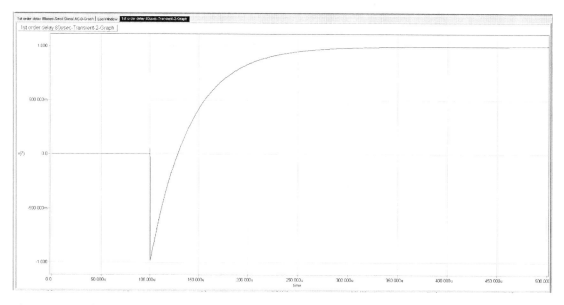

**Figure 13.9: The disconcerting response of the 80 usec 1st-order allpass filter to a 1 V step input.**

causing the slow rise. The mistake we have made is applying a stimulus waveform with a full frequency range.

Figure 13.12 shows the rather more convincing result obtained if the 1 V step input is band-limited before it is applied to our allpass filter. The step input (Trace 1) has been put through a 4th-order Bessel lowpass filter with a −3 dB frequency of1.2 kHz, the Bessel characteristic being chosen to prevent overshoot in the filtered waveform; a Butterworth filter would have given a bit of overshoot and confused matters. The Bessel filter output is the leisurely-rising waveform of Trace 2. When this goes through the allpass filter it emerges as Trace 3 with almost exactly the same shape, but delayed by 80 usec. That looks a bit more like a delay, eh?

If you look closely at Figure 13.12 you will see that the output of the allpass filter does show a very small amount of undershoot before it rises. This could have been further suppressed by reducing the

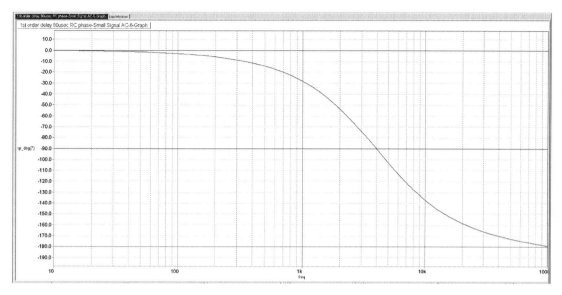

Figure 13.10: The phase response of the 80 usec 1st-order RC-type allpass filter. The phase-shift is still changing at the top of the audio spectrum.

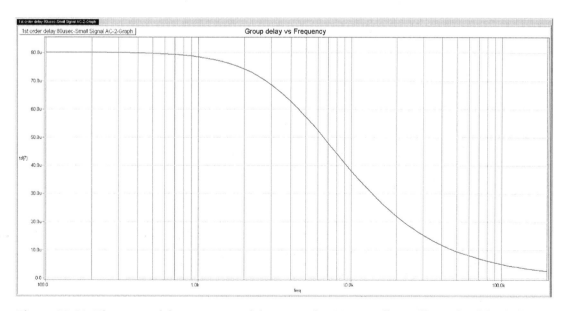

Figure 13.11: The group delay response of the 1st-order 80 usec allpass filter. The delay is down by 10% at 2.5 kHz and 50% at 9.3 kHz.

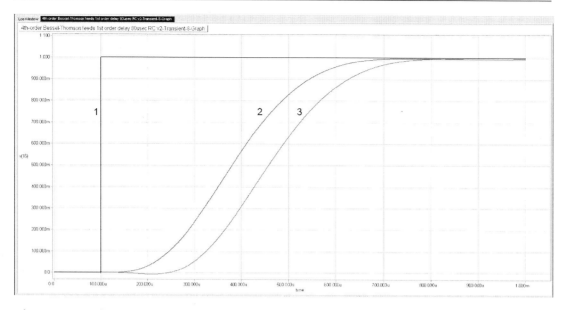

**Figure 13.12: The 1 V step input is Trace 1; putting it through a 4th-order Bessel-Thomson filter gives Trace 2, and the 80 usec 1st-order allpass filter delays it to give Trace 3.**

cutoff frequency of the lowpass filter. You will also note that the output of the 4th-order Bessel lowpass filter, Trace 2, has already been delayed a good deal compared with the step input Trace 1, in fact far more than by the simple 1st-order allpass filter, because the Bessel is 4th-order. The use of Bessel lowpass filters to delay signals is examined further at the end of this chapter.

The circuit in Figure 13.7a must be non-inverting at low frequencies, because when C1 is effectively an open circuit, we get a non-inverting stage because of the direct connection to the non-inverting input. Likewise, in Figure 13.7b, when C1 is effectively open circuit, the configuration is clearly a unity-gain inverter.

The input impedance of the RC version in Figure 13.8 is 2.7 kΩ at 10 Hz, falling to 646 Ω at 1 kHz, whereafter it remains flat (it happens to be 666 Ω at 100 Hz, but I don't think you should try to read too much into that). The equivalent CR version (with all component values the same) has an input impedance that is flat at 1 kΩ from 10 Hz to 1 kHz. Above that the impedance rises slowly until it levels off at 1.8 kΩ around 30 kHz. It is pretty clear that the CR version will be an easier load for the preceding stage, especially at high audio frequencies, and this will have its effect on the distortion performance of that stage. There is more on that in the later section on the performance of 3rd-order allpass filters.

Figure 13.11 shows an elegantly sinuous curve, quite unlike the tidy straight-line approximation roll-offs we are used to when we look at filter amplitude responses. These latter of course are plotted logarithmically with dB on the vertical axis, whereas here we have group delay time as a linear vertical axis. This is how it is normally done; dB are useful for measuring amplitude, not least because they

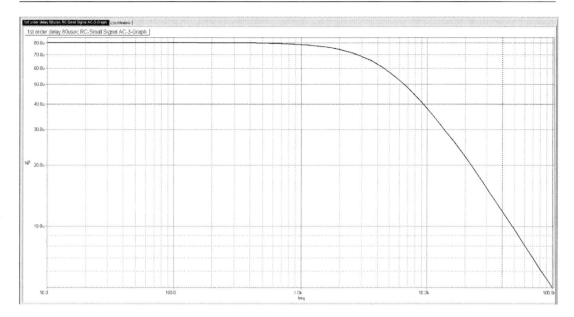

**Figure 13.13: The group delay response of the 1st-order 80 usec allpass filter plotted on a log time axis. The delay is down by 10% at 2.5 kHz.**

follow the logarithmic nature of how we perceive loudness, but there is no perceptual analogue for how we experience small time delays. Still, I thought it might be interesting to plot group delay with the log-of-time as the vertical axis, and the result is seen in Figure 13.13. I don't recall seeing this done before.

You can see that it looks very much like the amplitude plot of a 6 dB per octave roll-off, with a linear roll-off to the right. Is this a more useful way of plotting group delay? Probably not, because it tends to de-emphasise variations in the region of maximum group delay, which is where we are most concerned. Interesting picture, though.

## Distortion and Noise in 1st-Order Allpass Filters

While the two versions of the 1st-order allpass may appear to be functionally identical, apart from the phase inversion, they do in fact have different distortion behaviour at high frequencies. The non-inverting version or RC version (Figure 13.7a) has the resistor before the capacitor, forming a lowpass filter, so the common-mode voltage will fall with frequency. The inverting or CR version (Figure 13.7b) has a capacitor-resistor arm that acts as a highpass filter, so the CM voltage will rise with frequency. This makes a significant difference to the HF distortion when using the 5532 opamp, as shown in Figure 13.14; at 20 kHz the CR version gives 0.0026% as opposed to 0.0007% for the RC version. The allpass filter measured was the 80 usec design, the non-inverting version of which is shown in Figure 13.8. Polypropylene capacitors were used to eliminate any capacitor distortion. Note

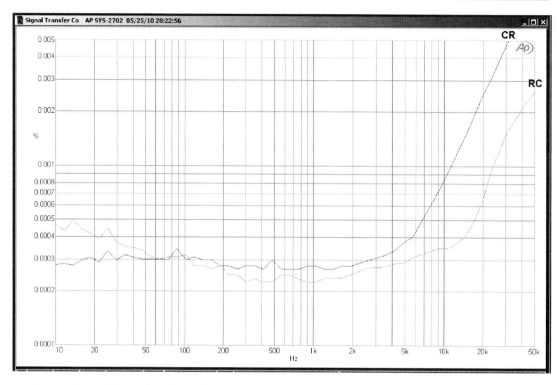

**Figure 13.14: Distortion plots for RC and CR 1st-order 80 usec allpass filters using 5532s. The CR version has more HF distortion because the common-mode voltage is high at high frequencies, and vice versa for the RC version. Polypropylene capacitors, input 9 Vrms.**

that the 9 Vrms test level is close to the maximum possible, and practical internal levels such as 3 Vrms will give significantly lower levels of distortion.

When the input level is 9 Vrms, the RC version has a CM voltage of 1.67 Vrms at 20 kHz. The CR version has a CM voltage of 8.84 Vrms at 20 kHz, more than five times as much, leading to much-increased HF distortion. First-order allpass filters with longer delays have larger capacitors, and so the difference is even more marked. The measurements in Figure 13.14 were taken using 63 V polypropylene capacitors; however, replacing these with 100 V polyester microbox capacitors in both RC and CR versions made no measurable difference to the THD plots. This was somewhat unexpected.

It is therefore well worthwhile using the non-inverting (RC) version wherever possible. If there is a need for a phase inversion, then it may be worth looking for some other part of the signal path in which to place it, where it will not cause degradation of the distortion performance.

The noise output for the non-inverting 80 usec version is −110.3 dBu (22 Hz–22 kHz) The noise output for the inverting 80 usec version is −110.7 dBu (22 Hz–22 kHz). The difference is small but real.

## Cascaded 1st-Order Allpass Filters

The delay required in a typical crossover is too long to be effectively provided by a single 1st-order allpass filter, because the longer the delay used, the lower the frequency at which that delay starts to roll off. See Chapter 23 for a real example. The Siegfried Linkwitz crossover design in [2] addresses this problem by using three 1st-order filters in series, as shown in Figure 13.15. This spreads the delay over three sections, allowing each one to be set to a lower delay which can be sustained up to higher frequencies. The RC version of the filter is used here because it is non-inverting and has a markedly superior distortion performance.

As Figure 13.16 shows, this approach greatly improves the delay flatness, which is now 10% down at 7.2 kHz and down 50% at 28 kHz, outside the audio band. However, 7.2 kHz is only 1.5 octaves

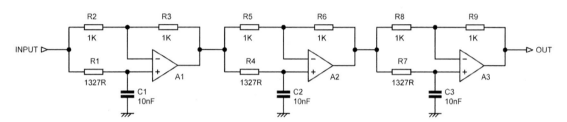

**Figure 13.15: Three 1st-order allpass filters in series, designed for an overall group delay of 80 usec.**

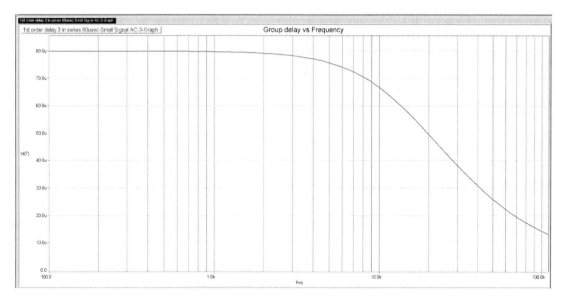

**Figure 13.16: The group delay response of the three 1st-order allpass filters in series. The 80 usec delay is now down 10% at 7.2 kHz, a higher frequency by a factor of three.**

clear of the 2.5 kHz mid-HF crossover frequency. The signal to the mid drive unit will, allowing for the initially shallower filter slope, be 35 dB down at that point. Its acoustic contribution will probably be less than that, as it is hardly likely to have a flat response 1.5 octaves above the crossover frequency, and may be significantly down; on the other hand it might have an ugly peak. In view of this uncertainty is hard to judge if we have sufficient delay flatness to avoid audible effects, but as from Table 13.3 we can see that a −35 dB signal is only going to increase the total amplitude by 0.15 dB, I am inclined to think we are fairly safe. It would of course be easy to divide the required group delay over more stages, to further extend the flat-delay response, but since every stage adds a certain amount of noise, distortion, and cost, and requires power, this is not that attractive a route.

While the amplitude perturbations resulting from this solution may be tiny, a further potential worry is the varying phase-shift of the filter above the frequencies at which we are trying to keep the group delay constant.

Figures 13.10 and 13.11, for a single 1st-order section, showed that the phase is still changing quite quickly in the 10 kHz–20 kHz range. Our triple-allpass scheme has flattened the delay/frequency curve, but has it helped with the phase? No. This now changes considerably more, from 10 kHz to 20 kHz, going from −240° to −357°, a change of 157°. This very large increase is not just because we now have three filters all phase-shifting, but also because the part of the phase curve with the most rapid changes is now moved up over the 10 kHz–20 kHz range. As demonstrated in Chapter 3, this sort of thing is actually not audible, but in any case it's not clear there's very much we can do about it. The only solution would be to make the group delay effectively constant up to a very high frequency, well out of the audio band. Let's assume we grit our teeth and accept a phase-shift that goes up to 10° at 20 kHz. That implies raising the group delay −10% frequency by no less than 57 times, to 410 kHz. This is hardly practical—here we need three stages just to achieve 7.2 kHz—and so it looks like we are going to have to live with frequency-dependent phase-shifts if we use this sort of delay technology.

The iterated nature of the multiple-1st-order circuit means we are using three times as many components. A more efficient approach is to use a higher-order allpass filter to create the desired delay.

## Second-Order Allpass Filters

Second-order allpass filters have the advantage that you can get a flatter delay response using fewer components than the cascaded-1st-order approach. The disadvantage is that they are conceptually more complex.

The usual method of making a 2nd-order allpass filter is the 1–2BP configuration, where the signal is fed to a conventional 2nd-order bandpass filter with a passband gain (i.e. gain at the response peak) of two times and then subtracted from the original signal, as in Figure 13.17. It is not exactly intuitively obvious, but this process gives a flat amplitude response and a 2nd-order allpass phase response. (This method should not be confused with the 1-BP configuration, where the bandpass path has unity gain; that gives a notch amplitude response.) To obtain a maximally flat delay response, the $Q$ of the bandpass filter must be 0.5 (note that this differs from the amplitude response of highpass and lowpass filters, where a $Q$ of $1/\sqrt{2}$, i.e. 0.707, is required to get a maximally flat Butterworth response).

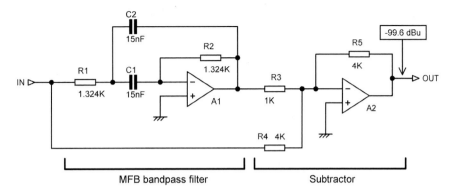

**Figure 13.17: A 2nd-order allpass filter using the1–2BP principle, designed for a group delay of 80 usec. The MFB filter has a gain of −6 dB, so R3 = R4/4.**

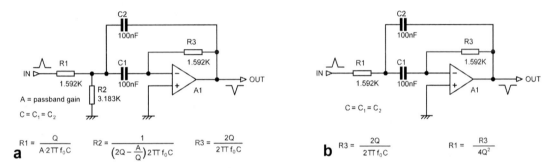

**Figure 13.18: Two multiple-feedback (MFB) bandpass filters with centre frequencies of 1 kHz: (a) the more usual version, with design equations; passband gain = 1 and Q = 1; (b) the low-Q Deliyannis version with design equations; passband gain = 0.5 and Q = 0.5. Note the similarity in component values.**

A 2nd-order allpass has two parameters to set instead of one—we now have both frequency and $Q$ for the bandpass filter. A convenient configuration is the multiple-feedback (MFB) type, as shown in two forms in Figure 13.18, which has the useful property of inverting the signal, so the required subtraction from the original signal can be implemented simply by summing original and filter output together. A vital point to be aware of is that if you use the more common form of MFB filter in Figure 13.18a, with its associated design equations, it is not possible to get a $Q$ of less than 0.707, and we need here a $Q$ of 0.5. If you put a $Q$ value of 0.707 into these equations, then R2 comes out as infinite. This is not a problem—you just leave it out altogether—but it also means that putting in values of $Q$ less than 0.707 yield negative values for R2, which is a little harder to implement (though I hasten to add I have tried it, and it does work).

A simpler approach is to use the Deliyannis version of the MFB filter shown in Figure 13.18b; the circuit is the same for infinite R2, but the vital point is that different design equations are used. It is now not possible to set passband gain independently of $Q$. The values shown give a $Q$ of 0.5, which also sets the passband gain at 0.5 (−6 dB). This is only true in this case, and in general the passband gain is not equal to $Q$.

To design the filter, first choose a value for C (C1 and C2 are equal) and put that and the required centre frequency into Equation 13.6 to get R3, then use Equation 7 to calculate R1.

$$R3 = \frac{2Q}{2\pi fC}$$

13.6

$$R1 = \frac{R3}{4Q^2}$$

13.7

A practical implementation of this method is shown in Figure 13.17, which gives an 80 usec delay at low frequencies. To achieve this, the MFB filter, consisting of R1, R2, C1, C2, and A1, is designed for a centre frequency of 7.91 kHz and a $Q$ of 0.5. Its inverted output allows the required subtraction from the original signal to be done simply by summing original and filter output together by the shunt-feedback stage A2. Note that R3 is half the value of R4; this implements the 2 in the 1–2BP. Because of the low $Q$ required in the MFB for a maximally flat delay, the passband gain is −6 dB. This means that to achieve 1–2BP, the resistor R3 must be a quarter of the value of R4 and R5. In this case R4, R5 can conveniently be made up of two 2.0 kΩ resistors. The group delay is shown in Figure 13.19; it is down 10% at 4.79 kHz, which is predictably better than the 1st-order filter (10% down at 2.5 kHz) but worse than the triple 1st-order filter (10% down at 7.2 kHz)

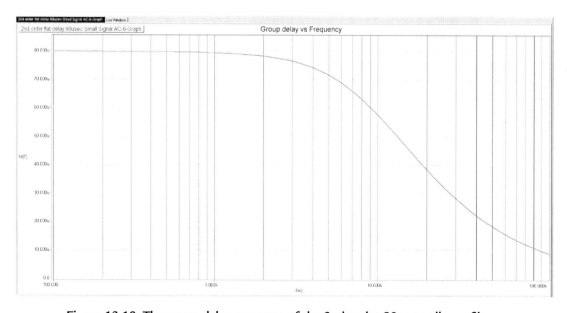

**Figure 13.19: The group delay response of the 2nd-order 80 usec allpass filter. The delay is now down 10%, at 4.79 kHz.**

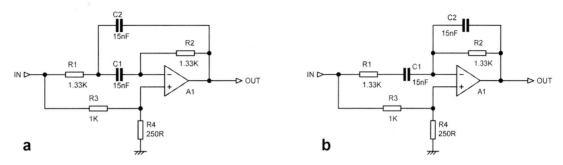

**Figure 13.20: Two 2nd-order allpass filters using only one opamp:
(a) the Delyiannis circuit (b) the Lloyd circuit.**

Second-order allpass filters of this type give a phase inversion at low frequencies and also at high frequencies; as mentioned before, an allpass filter gives twice the phase-shift of a conventional filter of the same order. The only time when the phase is non-inverting (0° phase-shift) is as the phase plot goes through 0° as it moves smoothly from 180° to −180°.

It is possible to make a 2nd-order allpass filter that only requires one opamp, but there are compromises on noise, gain, and component sensitivity that make this unattractive for our purposes. Two kinds of one-opamp 2nd-order allpass filter are shown in Figure 13.20.

The first version in Figure 13.20a is also generally known as the Delyiannis circuit [3] and is essentially an MFB bandpass filter (R1, R2, C1, C2) with an additional subtraction path via the attenuator R3, R4; the basic aim is to implement the circuit of Figure 13.17 in one stage. The values shown give an 80 usec delay that is 10% down at 4.8 kHz with a maximally flat delay response. While this circuit is undoubtedly minimalist, it has the great drawback that the lower the Q, the lower the gain. Setting the $Q$ to the required value of 0.5 results in a stage gain of 1/5, or −14 dB, and the considerable amount of amplification required to get the output back to the input level will introduce a lot of additional noise; the extra amplifier stage required to do it also undoes the cost advantage of a one-amplifier allpass filter.

Figure 13.20b shows another 2nd-order allpass filter, known as the Lloyd circuit. [4] R1, C1, R2, C2 form a 2nd-order bandpass response; this configuration is only capable of a low Q, but that is all we need. Once more there is a subtraction path via R3, R4. The values shown give an 80 usec delay that is 10% down at 4.8 kHz. It is notable that the circuit values are the same as for the Delyiannis circuit, despite the different configuration. This circuit has a delay response equivalent to two cascaded 1st-order stages but uses only opamp. However, once again the stage shows a considerable loss of −14 dB, which must be made up somewhere, and my conclusion is that pursuing economy in this way is not a very hopeful quest.

Two other 2nd-order allpass configuration were thoroughly investigated by Robert Orban in 1991:[5] these were the Bruton circuit [6] using two opamps and the Steffen circuit using one. [7] Neither seems to have any great advantages over the 1–2BP method, but the Bruton beats the Steffen if a very flat amplitude response is required.

## Distortion and Noise in 2nd-Order Allpass Filters

The 2nd-order filter in Figure 13.17 was built using 5532 opamps and polypropylene capacitors to eliminate any capacitor distortion; the measured THD results are shown in Figure 13.21. The THD falls rapidly from 0.0022% at 20 kHz as frequency decreases, but then peaks in the range 2 to 10 kHz, as this is the only region where the MFB filter output, with its own distortion content, makes a significant contribution. Distortion from the MFB filter is low because of its shunt configuration, which eliminates CM voltages, and because the output voltage is low due to the −6 dB passband gain; THD does not exceed 0.0004%. The high-frequency distortion comes solely from the second opamp, as the MFB filter output is 10 dB down at 20 kHz.

The circuit was then remeasured using 63 V polyester microbox capacitors, to find out how much extra distortion they introduced. The answer is: not much. The THD in Figure 13.20 now peaked at 0.0009% at 5 kHz instead of 0.0007% for the polypropylene capacitors. The modest increase is due to the low signal level in the MFB filter.

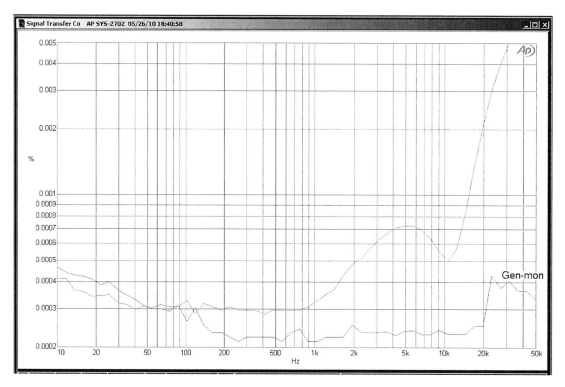

**Figure 13.21: Distortion plot for the 2nd-order 80 usec allpass filter in Figure 13.17, using 5532s. The THD drops rapidly from 0.0022% at 20 kHz as frequency falls, but peaks around 2 to 10 kHz, where the MFB makes a significant contribution. High-frequency distortion is solely from the second opamp. Polypropylene capacitors, input 9 Vrms.**

The noise output is −99.6 dBu (22 Hz–22 kHz), which is more than 10 dB noisier than the 1st-order stage measured earlier. The greatest proportion of this extra noise comes from the second stage A2, which because of the low value of R3 is working at a noise gain of 15.6 dB. When R3 is disconnected, the noise output drops precipitately to −109.5 dBu; when, however, R3 is connected to ground instead of the MFB filter output, noise only increases to −102.1 dBu, showing that some of the extra noise is produced by the filter.

The high noise gain in the second stage is inherent in the 1–2BP mode of operation, because of the subtraction involved, and it is not easy to see a way round that.

## Third-Order Allpass Filters

A 3rd-order allpass filter is made up of a 2nd-order allpass cascaded with a 1st-order allpass, in the same way that 3rd-order frequency/amplitude domain filters are constructed. The 2nd-order allpass is arranged to peak in its delay, but the slow delay roll-off of the 1st-order allpass cancels out this peaking, giving a maximally flat delay overall. The arrangement is shown in Figure 13.22; note that the 1st-order stage is now of the inverting or CR type, which undoes the inversion performed by the 2nd-order stage (as we noted earlier in the chapter, this version has an inferior distortion performance to its non-inverting(RC) equivalent; however, the overall outcome for distortion is not obvious, as related in the next section, on the performance of this filter).

The output of the 2nd-order allpass is the middle trace in Figure 13.23, peaking just above 10 kHz. When this is combined with the slow roll-off of the following 1st-order stage (the lower trace), the final result is a flat response that then rolls off quickly. This is the top trace in Figure 13.23, where the group delay remains flat and then rolls off rapidly, being 10% down at 12.7 kHz and 50% down at 31.9 kHz. This gives almost twice the frequency range of the three cascaded 1st-order networks in Figure 13.15 but actually uses one less resistor. This 3rd-order filter is clearly the better solution.

The delay −10% point is now 2.3 octaves above the 2.5 kHz crossover frequency, so assuming a 24 dB/octave slope, the mid speaker output will be about 55 dB down, and so I suggest that the fall-off in time delay will have no audible consequences.

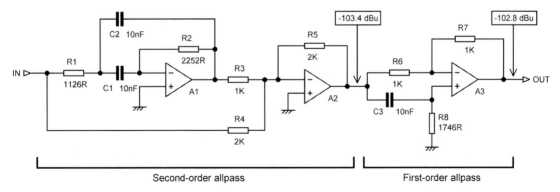

**Figure 13.22: A 3rd-order 80 usec allpass filter, made up of a 2nd-order allpass followed by a 1st-order allpass. Noise at each stage output is shown for 5532s.**

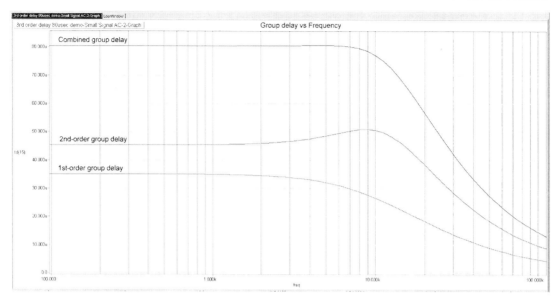

**Figure 13.23: The group delay response of the 3rd-order 80 usec allpass filter.
The delay is now down 10% at 12.7 kHz.**

Looking at the low-frequency area to the left, you will see that the 2nd-order stage gives a flat delay of 45 usec; to this is added the 35 usec of the 1st-order stage, giving a total delay of 80 usec.

The 3rd-order allpass filter has input and output in phase at the low-frequency end. At the high-frequency end, the phase-shift reaches a total of 540° lagging (180° × 3).

## Distortion and Noise in 3rd-Order Allpass Filters

The 3rd-order filter in Figure 13.22, with its CR final stage, was built using 5532 opamps and polypropylene capacitors. Figure 13.24 shows the distortion at the 2nd-order stage output and the final output. An interesting observation is that between 200 Hz and 6 kHz the final output shows a lower THD reading than the intermediate point in the circuit. This is because there is some cancellation of the second harmonic from the 2nd-order filter in the final stage. Both points show 0.002% at 20 kHz.

Figure 13.25 shows the result of swapping the CR 1st-order stage for its RC equivalent, which as we saw earlier has in itself lower distortion. We might hope that that would give us a better overall distortion performance, but actually it is markedly worse; we now suffer 0.003% THD at 20 kHz, and perhaps more worryingly, we have THD around 0.0014% in the audible region from 6 kHz to 10 kHz, whereas before we were well below 0.001% in this region.

The alert reader (and I trust you all are) will have noticed that the THD plots for the intermediate 2nd-order output in Figures 13.24 and 13.25 are not the same, and you may very well wonder why this is

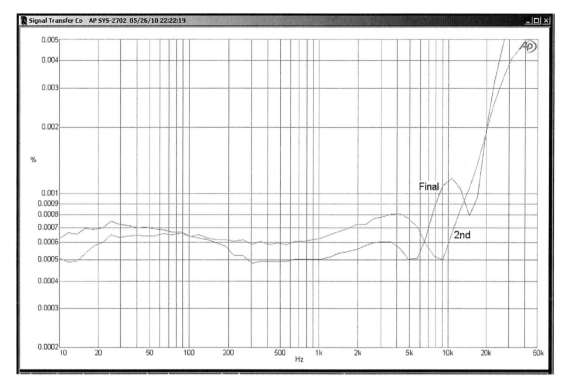

**Figure 13.24: Distortion plots for 3rd-order 80 usec allpass filter in Figure 13.22, showing the 2nd-order stage output and the final output. Polypropylene capacitors, 5532s, input 9 Vrms.**

the case when the 2nd-order stage has not been altered. The answer is that the CR and RC 1st-order stages differ considerably in how much loading they are putting on the previous stage. The RC version has a three times lower input impedance at high frequencies, and in this circuit that outweighs the fact that it has lower distortion of its own. The input impedances of 1st-order allpass filters were described earlier in this chapter.

I hope you don't think that I am belabouring a trivial point here. There is in fact a fundamental message—that you must never neglect the loading that one stage imposes on its predecessor if you are aiming for the lowest practicable distortion. To ram the point home I will show one more THD plot. Figure 13.26 shows the THD at the 2nd-order stage output for the RC and CR cases, and also for the unloaded condition (NL), with the 1st-order stage disconnected to remove its loading altogether.

You can see that the traces for the RC and CR cases are the same as those in Figures 13.24 and 13.25, but the unloaded (NL for no load) trace is a good deal lower than either of them, except around 5 to 10 kHz, where the CR plot gains from some cancellation effect. Only the CR case shows elevated LF distortion from 10 Hz to 5 kHz; this is characteristic of loading on a 5532 output, and it is caused by the relatively low input impedance of the CR version at low frequencies. At higher frequencies the input impedance rises, which is why the CR trace takes a nose-dive at 5 kHz before coming back up again.

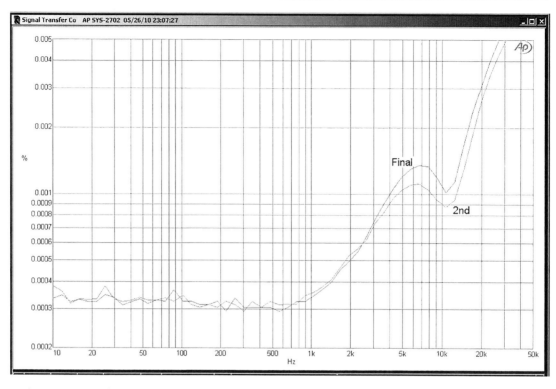

**Figure 13.25: Distortion plot for the 3rd-order 80 usec allpass filter in Figure 13.22, but with an RC-type1st-order filter as the final stage. The final distortion output is unexpectedly worse. Polypropylene capacitors, 5532s, input 9 Vrms.**

The 20 kHz THD at the 2nd-order stage output (A2) in Figure 13.26 is really not very good in either the CR or RC case. There are two reasons for this:

1.  Opamp A2 in Figure 13.22 is heavily loaded by the input impedance of the 1st-order stage, as described earlier.
2   A2 is working at a relatively high noise gain and so has less negative feedback than usual available for linearisation.

In Chapter 16, it is described how the distortion performance of a heavily loaded 5532 can be improved by biasing the output. This looks like a good place to apply that technique. Adding 3k3 to V+ on the 2nd-order output reduces the LF distortion from around 0.0006% to around 0.0005%; not exactly a giant improvement, but then it is almost free. See Figure 13.27, where there is a general reduction in LF distortion, but we have lost the dip around 8 kHz.

The noise at the output of the 2nd-order stage is −103.4 dBu (22 Hz–22 kHz), and the noise at the final output = −102.8 dBu.

Why is this circuit 3.2 dB quieter than the −99.6 dBu we get from the 2nd-order allpass filter, when we have more opamps in the signal path? It's a subtle chain of reasoning. First, since the 2nd-order stage

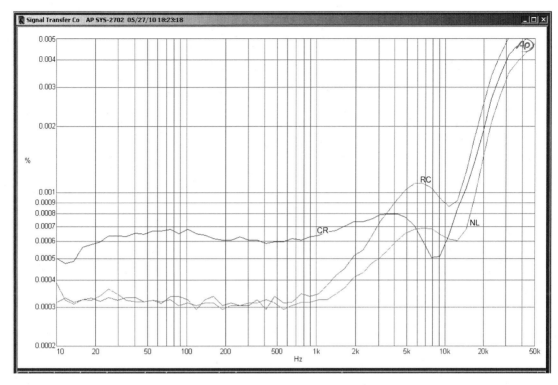

**Figure 13.26: Distortion plot for the 2nd-order stage output (A2) of the 3rd-order 80 usec allpass filter in Figure 13.22, when it is loaded by the RC or CR final stage, and also with no load (NL). Polypropylene capacitors, 5532s, input 9 Vrms.**

is now peaking in delay, the MFB filter has a higher $Q$ and therefore a higher passband gain, here 0 dB. Thus R3 is half the value of R4, R5 instead of a quarter. Therefore the noise gain in the subtractor stage A2 is 3.6 dB lower at 12.0 dB.

This is another reason why a 3rd-order allpass is better than a 2nd-order allpass delay filter; it is not only flatter, it is quieter. Is this an original observation? I suspect so.

## Higher-Order Allpass Filters

If 3rd-order filters do not provide a sufficiently extended flat-delay characteristic, then higher-order filters might be considered. We have just seen that a 3rd-order allpass filter provides the most extended flat-delay response for a given component count and also has usefully lower noise. If greater delays or flatter delay characteristics than this configuration can provide are required, higher-order filters may be the answer.

The basic principle, as we saw for the 3rd-order filter, is to take one or more 2nd-order stages with a high Q, giving a peak in the delay response, and then cancel out the peak by cascading it with one or

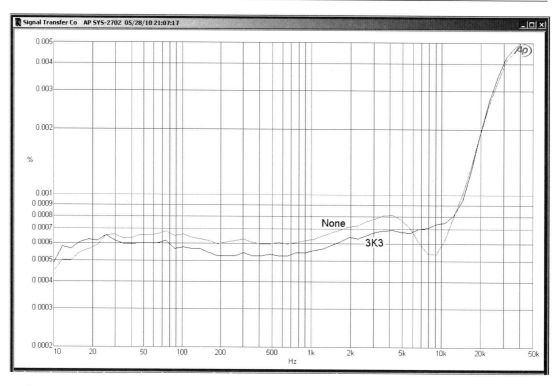

**Figure 13.27: Distortion plot for the 2nd-order-stage output of the 3rd-order 80 usec allpass filter in Figure 13.22 loaded only by the CR final stage. Adding the 3k3 output biasing resistor reduces distortion below 6 kHz. Polypropylene capacitors, 5532s, input 9 Vrms.**

more stages, either 1st- or 2nd-order, with a low $Q$. Figure 13.28 shows how higher-order filters are built up. For each stage the $Q$ is that of the overall allpass function and *not* the $Q$ of the MFB filter, which is a different quantity. For example, in a 2nd-order allpass filter the $Q$ for maximal flatness is 0.58, but the $Q$ of the MFB filter is 0.50. Here the 1st-order sections have been put at the end of the chain; this makes no difference, as all the stages have unity gain.

Designing high-order filters is not something to try by hand, and the usual design procedure is to consult a table of allpass filter coefficients, as in Table 13.4. This assumes you are using the 1–2BP approach to making allpass filters.

To design your filter, pick an order and the cutoff frequency $f$ required. For each stage of the complete filter, first select the value of $C = C1 = C2$ in the MFB filter (here I am using the component designations in the first stage of Figure 13.20); if this results in resistor values that are undesirably low or high, then you will have to go back and rethink the value of C. R1 and R2 are then calculated using the coefficients $a_i$ and $b_i$ thus:

$$R1 = \frac{a_i}{4\pi fC}$$

13.8

$$R2 = \frac{b_i}{a_i \pi fC} \qquad 13.9$$

This sets the frequency and $Q$ of the stage. We now need to work out the parameter $\alpha$:

$$\alpha = \frac{(a_i)^2}{b_i} \qquad 13.10$$

which lets us set R3 at $R/\alpha$, where R is the value of R4 and R5, to get the correct subtraction to create an allpass filter; parameter $\alpha$ includes the 2 in the 1–2BP. Job done, but your very next move should be to run the complete filter on a simulator to make sure you get what you want—it's all too easy to make an arithmetical error.

Figure 13.29 shows the delay response of 1st- to 6th-order allpass filters all set to the same cutoff frequency. This is the sort of diagram you often see in filter textbooks, but here it is simulated rather than derived mathematically.

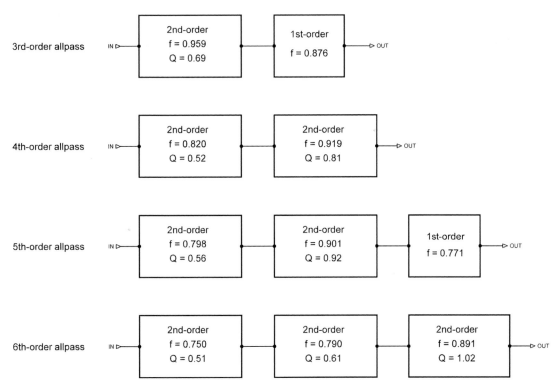

Figure 13.28: Constructing allpass filters from 3rd- to 6th-order by cascading 1st- and 2nd-order stages. The relative cutoff frequency and the *Q* are shown for each stage.

### Table 13.4: Filter coefficients for allpass filters (after Kugelstadt).

| Filter Order | Filter Section i | Coeff $a_i$ | Coeff $b_i$ | Section Q | R3 factor | Stage gain |
|---|---|---|---|---|---|---|
| | | | | Q | | |
| 1st | 1 | 0.6436 | n/a | | | 1.000 |
| 2nd | 1 | 1.6278 | 0.8832 | 0.58 | 3.000 | 0.667 |
| 3rd | 1 | 1.1415 | n/a | | | 1.000 |
| | 2 | 1.5092 | 1.0877 | 0.69 | 2.094 | 0.955 |
| 4th | 1 | 2.3370 | 1.4878 | 0.52 | 3.671 | 0.545 |
| | 2 | 1.3506 | 1.1837 | 0.81 | 1.541 | 1.298 |
| 5th | 1 | 1.2974 | n/a | | | 1.000 |
| | 2 | 2.2224 | 1.5685 | 0.56 | 3.149 | 0.635 |
| | 3 | 1.2116 | 1.2330 | 0.92 | 1.191 | 1.680 |
| 6th | 1 | 2.6117 | 1.7763 | 0.51 | 3.840 | 0.521 |
| | 2 | 2.0706 | 1.6015 | 0.61 | 2.677 | 0.747 |
| | 3 | 1.0967 | 1.2596 | 1.02 | 0.955 | 2.095 |

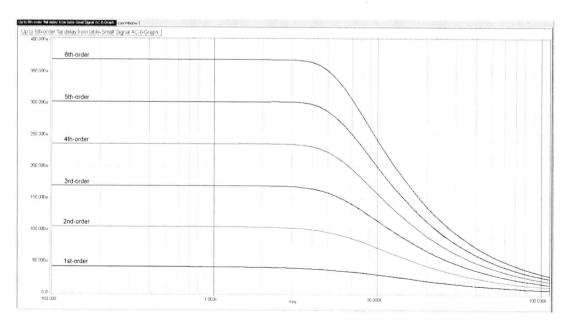

Figure 13.29: Delay response of 1st- to 6th-order allpass filters, all set to same cutoff frequency. The delay increases with filter order.

More relevant to our needs is to set the group delay to the same value (80 usec again) for each order of filter, and you can see how much more extended the flat portion of the delay curve is with the higher-order filters, as in Figure 13.30. It is also painfully obvious that the extra flat-delay bandwidth you gain with each increase of filter order decreases; this is true on a log-frequency plot because with each higher order you actually gain a constant extra bandwidth of about 5 kHz. There is clearly a limit to how high an order it is sensible to use. A powerful consideration here is that component sensitivity (i.e. how much the filter behaviour changes with component value tolerances) gets worse for higher-order filters. There are also issues with dynamic range, as higher-order allpass filters use at least one section with high Q—this means its MFB filter has gain well above unity, and this is a potential clipping point. The 6th-order allpass filter described later illustrates this problem.

I will now give some practical designs for 4th-, 5th-, and 6th-order allpass filters, all set for a 80 usec LF delay; these were the circuits used to generate the curves labelled 4, 5, and 6 in Figure 13.30. The capacitor values have been chosen to keep the resistor values in the MFB filters above 1 kΩ to minimise loading problems. The resistor values given are exact and not part of any series of preferred values.

Figure 13.31 shows a 4th-order allpass filter made by cascading two 2nd-order allpass stages. The delay is now 10% down at 17.2 kHz, rather than the 12.7 kHz we got with the 3rd-order filter examined earlier in this chapter, a useful increase in flat-delay bandwidth.

The lower the Q of the stage, the lower the gain of the MFB filter, and so the lower the value of the resistor feeding into the summing stage must be to get the correct gain. It must be remembered that this resistor constitutes a load to ground on the output of the MFB opamp, and its value must not be

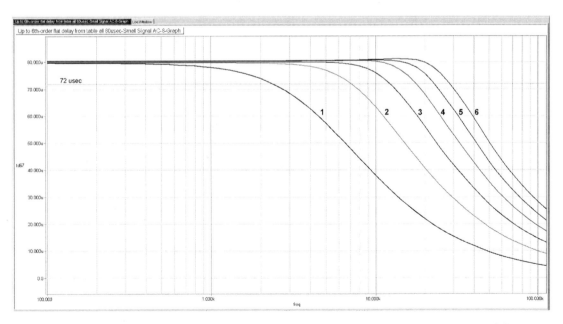

**Figure 13.30: Delay response of first to 6th-order allpass filters, all set to 80 usec delay. The 72 usec line shows where each response is 10% down.**

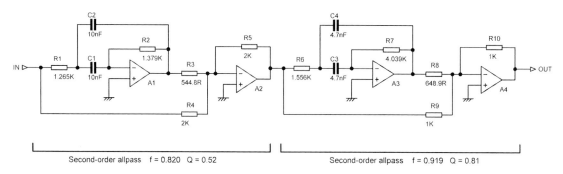

**Figure 13.31: Fourth-order allpass filter built by cascading two 2nd-order stages.
Delay = 80 usec.**

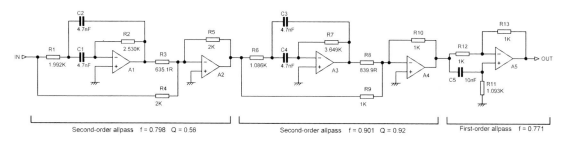

**Figure 13.32: Fifth-order allpass filter built by cascading two 2nd-order stages and
one 1st-order stage. Delay = 80 usec.**

allowed to fall below 500 Ω if 5532 or similar opamp types are being used. In the first stage here this means that R4 and R5 have to be 2 kΩ to get R4 up to 544.8 Ω. In the second stage the Q is higher and the MFB has more gain, so R9, R10 can be reduced to 1 kΩ for better noise performance, while R8 is a respectably high 648.9 Ω.

The values of C1, C2 and C3, C4 can be selected independently for each 2nd-order stage, and C3, C4 have therefore been set lower in value, at 4n7, to keep R6, R7 above 1 kΩ in value.

Figure 13.32 shows a 5th-order allpass filter made by cascading two 2nd-order allpass stages and one 1st-order stage. Because of the higher cutoff frequency, both C1, C2 and C3, C4 have been set to 4n7 to keep the associated resistors above 1 kΩ. C5 can be set independently and has been made 10 nF to give a suitable value for R11: as low as possible for noise purposes, but not so low that it puts an excessive load on A4.

Figure 13.33 shows a 6th-order allpass filter made by cascading three 2nd-order allpass stages. In the third high-Q stage it has been necessary to reduce C5, C6 to 3n3 to keep the associated resistors above 1 kΩ.

The final 2nd-order section has a Q of 1.02, which means a passband gain of 2.095 times in the MFB filter. This could cause headroom problems. It is another reason not to embrace higher-order allpass filters without careful consideration.

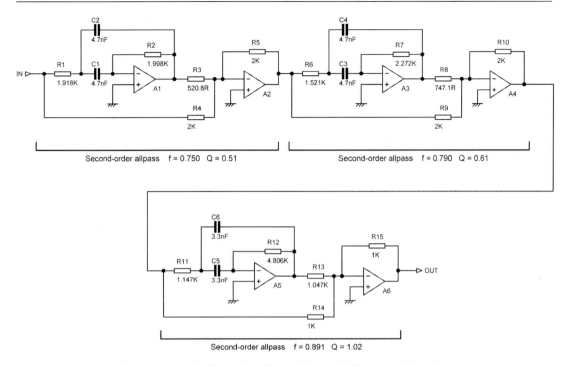

**Figure 13.33: Sixth-order allpass filter built by cascading three
2nd-order stages. Delay = 80 usec.**

Since higher-order all pass filters are made up of combinations of 1st- and 2nd-order stages, the
distortion and noise characteristics that apply have been dealt with earlier in this chapter. Noise will
clearly increase as more stages are cascaded, and life being what it is, it is highly likely that distortion
will also increase additively, though you may benefit from some degree of cancellation.

## Delay Lines for Subtractive Crossovers

There is another use for delay filtering in crossover design; it can be a fundamental part of the
crossover operation rather just compensating for mechanical displacements. Lipshitz and Vanderkooy
[8] published in 1983 a crossover topology in which a time delay in one signal path allowed the
construction of linear-phase crossovers with high slopes; a constructional project based on this was
put forward by Harry Baggen in *Elektor* in 1987. [9] Both are described in Chapter 6 on subtractive
crossovers. Lipshitz and Vanderkooy put forward a specimen crossover design, in which a delay of
289.2 usec was required, and they suggested the following possibilities:

1. Three cascaded ninth-order Bessel allpass filters, each giving a delay of 96.4 usec, accurate to 1%
   up to 20 kHz. Total delay 289.2 usec.
2. Four cascaded 6th-order Bessel allpass filters, each giving a delay of 72.3 usec, accurate to 1%
   up to 15 kHz and to 10% up to 20 kHz. This will have less component sensitivity because of the
   lower-order filters. Total delay 289.2 usec.

3.  Three cascaded 6th-order Bessel allpass filters, each giving a delay of 96.4 usec, accurate to 1% up to 10 kHz and to 20% up to 20 kHz. Total delay 289.2 usec.
4.  Four cascaded 4th-order Bessel allpass filters, each giving a delay of 72.3 usec, accurate to 1% up to 8 kHz and to 50% up to 20 kHz. Total delay 289.2 usec.

The trade-off here is between how far up the audio band you want linear-phase behaviour to extend and how complex and costly a delay line you are prepared to pay for (when you have many cascaded stages it is more usual to talk of a "delay line" rather than a "delay filter"). It is clearly impractical to implement any of these options in a passive crossover—the cost, the resistive losses, and the non-linearity of the many inductors required would be crippling. Such a delay-based crossover can only be realised by active circuitry, and even then it presents some serious challenges in the analogue domain. In a digital crossover, of course, manipulating time delays is very simple indeed.

I decided to have a go at constructing such a delay line, adopting the simplest option, Option 4. I used maximally flat allpass filters rather than Bessel types to explore how much more cost-effective a solution it was in terms of greater bandwidth for the same number and order of filters. The resulting schematic is shown in Figure 13.34; it consists of four cascaded identical 4th-order filters similar to that in Figure 13.31, but with the resistor and capacitor values in the MFBs altered to theoretically give a delay of 72.3 usec each.

The actual delay came out at 291 usec, showing that I should have used a bit more precision in my calculations; there is also a very small amount of delay peaking around 10 kHz for the same reason. The delay response is shown in Figure 13.35, and it is down by 1% at 12.5 kHz rather than 8 kHz, and down 11% at 20 kHz rather than 50%; compare Option 4. The 50% point is not reached until 47.8 kHz. These results would appear to be notably superior to the figures put forward by Lipshitz and Vanderkooy for the Bessel approach.

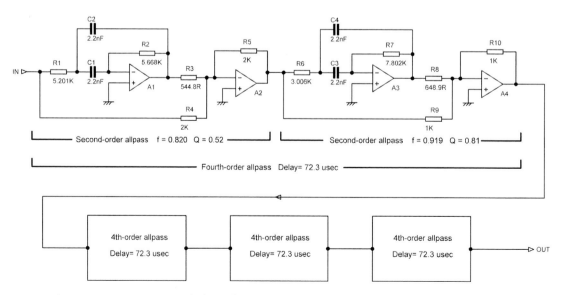

**Figure 13.34: Four cascaded maximally flat 4th-order allpass filters for Lipshitz and Vanderkooy time-delay crossover.**

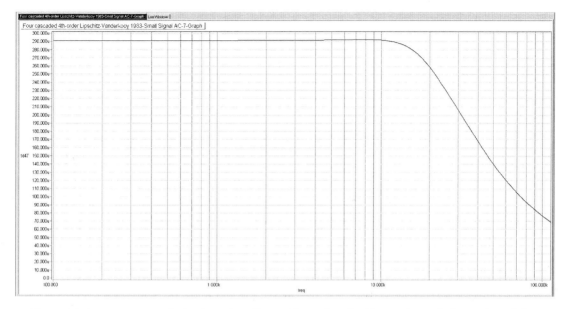

**Figure 13.35: Four cascaded 4th-order allpass filters for Lipshitz and Vanderkooy time-delay crossover. Delay = 291 usec, 1% down at 12.5 kHz, 11% down at 20 kHz.**

This delay-line design may hopefully be of use to those wishing to explore delay-based crossovers. The component cost is not excessive; you need 16 capacitors (all of the same size in this case), 40 resistors, and eight opamps. If a greater delay is needed then more stages could be added on.

It is a good question as to whether the delay line could be implemented more efficiently by using 3rd-order allpass filters. As we saw earlier in the chapter, a 3rd-order allpass is some 3 dB quieter if everything else is equal. A possible design solution would be to use five cascaded 3rd-order filters (using 15 amplifiers)instead of four cascaded 4th-order filters (using 16 amplifiers). We should therefore get a noise improvement as well as a slight saving on parts.

Please note that there are some questions about the practicality of subtractive crossovers, because they do seem to require very precise delays to maintain the desired slope steepness. This issue is looked into in Chapter 6.

## Variable Allpass Time Delays

Variable time delays are often required for subwoofer applications (see Chapter 18) because the distance between the main loudspeakers and the subwoofer is not fixed, and so the delay needs to be variable to get the best integration of responses.

It can also be useful to have a time-delay configuration which allows tweaking during the design stage—it would be particularly useful in cases where the effective acoustic centre of a drive unit is not easy to determine.

Unfortunately, making variable the most efficient delay filters, such as the 3rd-order 1–2BP configuration, requires multiple components to be changed in various ways and is not easy to implement. A more promising approach is to use cascaded 1st-order allpass filters, as described earlier, in which a two- or four-gang pot with equal-value sections (anything else will be hard to come by) can be used to alter the frequency-setting resistor in each stage. This method has been used in commercial crossovers such as those by Rane.

Figure 13.36 shows two cascaded 1st-order allpass filters with a two-gang control pot. The delay is variable from 40 usec (down 10% at 10 kHz) to 440 usec (down 10% at 900 Hz).

## Lowpass Filters for Time Delays

Lowpass Bessel filters are often used to preserve the waveform of a signal, typically being used to remove out-of-band noise before A-to-D conversion. Because in the analogue world any real-life filter must be causal (in other words you can get an output after the input, but never *before* the input), the output is inevitably a delayed version of the input. It seems reasonable to take a look at how effective Bessel lowpass filters are at implementing delays. There is at least the possibility that this sort of filter could implement delay compensation at the same time as giving a crossover roll-off for a mid drive unit.

Figure 13.37 shows a typical 4th-order Bessel lowpass filter with a −3 dB point at 4.2 kHz, which corresponds with an 80 usec delay, as used before. A similar lowpass filter (with the capacitors scaled up to give −3 dB at 1.2 kHz) was used to demonstrate 1st-order allpass delay action earlier in this chapter. Like almost all 4th-order filters, it is composed of two 2nd-order stages, with the later stage having the higher *Q*. The design of Bessel filters is described in Chapters 7, 8, and 9.

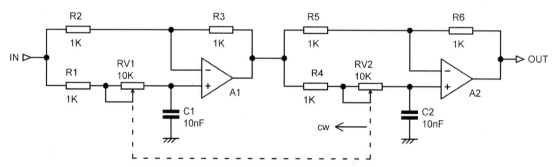

**Figure 13.36: Variable time-delay using cascaded 1st-order allpass filters. Moving the control clockwise increases the delay.**

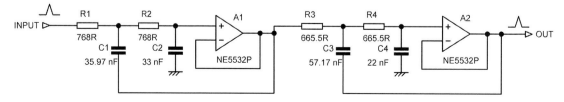

**Figure 13.37: Fourth-order Bessel filter giving a delay of 80 usec and an amplitude roll-off of −3 dB at 4.2 kHz.**

Figure 13.38 shows that the delay is 10% down at 8.5 kHz, compared with 10% down at 17.2 kHz for a 4th-order allpass filter. This is less than half the bandwidth and indicates that Bessel filters are not an efficient way of obtaining a flat delay.

The frequency response is shown in Figure 13.39. As with all Bessel filters, the roll-off is relatively slow. The amplitude response is perceptibly drooping at 2 kHz, but the −3 dB point does not occur until 4.2 kHz, more than an octave away.

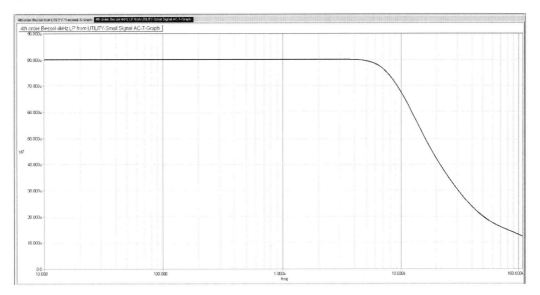

**Figure 13.38: Group delay of 4th-order Bessel filter of Figure 13.28. The −10% point is at 8.5 kHz.**

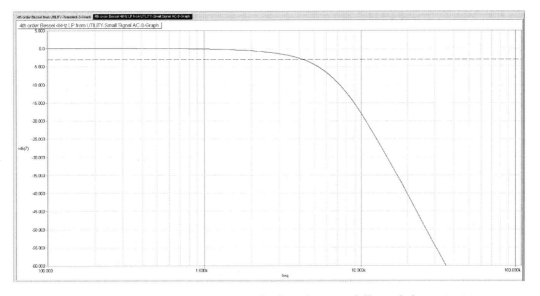

**Figure 13.39: Frequency response of 4th-order Bessel filter of Figure 13.28.**

# References

[1]  Linkwitz, Siegfried  www.linkwitzlab.com/crossovers.htm; Accessed August 2017
[2]  Linkwitz, Siegfried "Loudspeaker System Design" Electronics World, May, June 1978
[3]  Delyiannis, T., Yichuang Sun and J. Kel Fidler "Continuous-Time Active Filter Design" CRC Press, 1999, Chapter 4, p. 134
[4]  Lloyd, A. G. "Here's a Better Way to Design a 90° Phase-Differencing Network" Electronic Design, No. 15, p. 78, 1971
[5]  Orban, R. "Active All-Pass Filters for Audio" JAES, April 1991
[6]  Bruton, L. "RC-Active Circuits: Theory & Design" Prentice-Hall, NJ, 1980, p. 351
[7]  Haykin, S. S. "Synthesis of Active RC Circuits" McGraw-Hill, London, 1969, Chapter 3
[8]  Lipshitz and Vanderkooy "A Family of Linear-Phase Crossover Networks of High Slope Derived by Time-Delay" JAES January/February 1984
[9]  Baggen, Harry "Active Phase-Linear Crossover Network" Elektor, September 1987, p. 61

# CHAPTER 14

# *Equalisation*

## The Need for Equalisation

Equalisation to attain a desired flat frequency response may be applied to correct problems in the loudspeaker itself or, moving along the audio chain, to modify the interaction of the loudspeaker with the room it is operating in. Moving along the audio chain still further, equalisation can also be used to modify the response of the room itself, by cancelling resonances with dips or notches in the overall amplitude response. However, it is not normally considered a good idea to try to combine an active crossover with a room equaliser, not least because they are doing quite different jobs. Moving the loudspeakers from one listening space to another will not require adjustment of the crossover, except in so far as the loudspeaker placement with regard to walls and corners has changed, but would almost certainly require a room equaliser to be re-adjusted, unless the room dimensions, which determine its resonances, happen to be the same.

At low audio frequencies, normal rooms (i.e. not anechoic chambers with enormous sound absorption) have resonances at a series of frequencies where one dimension of the space corresponds to a multiple number of half-wavelengths of the sound being radiated. The half-wavelength is the basic unit because there must be a node, that is a point of zero amplitude, at each end. Sound travels at about 345 metres/second, so a room with a maximum dimension of 5 metres will have resonances from 34.5 Hz upwards. This is simply calculated from velocity/frequency = wavelength, bearing in mind that it is the half-wavelength that we are interested in. We might therefore expect a resonance at 34.5 Hz and another at about 69 Hz; this is twice the frequency, because we now need to fit in two half-wavelengths between the two reflecting surfaces. This continues for three and four times the lowest frequency, and so on. These "resonant modes" cause large peaks and dips in response, the height of which depends on the amount of absorbing material. A room with big soft sofas, thick carpeting, and heavy curtains will be acoustically fairly "dead", and the peaks and dips of the frequency response will typically vary by something like 5–10 dB. A bare room with hard walls and an uncarpeted floor will be much more acoustically "live", and the peaks and dips and dips are more likely to be in the range of 10 to 20 dB, though larger excursions are possible. Resonant modes at low frequencies cause the greatest problems, because they cannot be effectively damped by convenient absorption material such as curtains or wall hangings. Room equalisation that attempts to deal with this situation is a very different subject from active crossover design and is not dealt with further here.

This chapter deals only with the equalisation of the frequency response, but there is another very important form of equalisation; this is commonly called time equalisation, time-delay compensation, or phase equalisation, and while its results can be seen in the form of an improved frequency response, the circuitry involved essentially works in the time domain, and in itself has a flat frequency response. This branch of active crossover design is dealt with in detail in Chapter 13 and is not referred to further here.

# What Equalisation Can and Can't Do

When the word "equalisation" is used without qualification, it almost always refers to correcting the amplitude/frequency response, without attempting to simultaneously correct the phase/frequency response. Correcting both is much more difficult, but it can be done. It is important to realise that if you use the right sort of equaliser, you can put a peak into the frequency response and then cancel it out completely by using another equaliser with the reciprocal characteristic. The high-$Q$ peak/dip equaliser described later in this chapter can perform this, as it is reciprocal—in other words its boost and cut curves are exact mirror-images of each other about the 0 dB line. Figure 14.18 shows the symmetrical 6 dB peak and dip at 1 kHz that this circuit creates. If one equaliser is set to maximum boost, and it is then followed by an otherwise identical equaliser set to maximum cut, the final frequency response is exactly flat, as one might hope. What is much less obvious, but equally true, is that the phase-shifts introduced by the first equaliser are also cancelled out by the second equaliser, so a square wave input will emerge as a square wave output. This process is demonstrated in Figure 14.1, where the square wave with added ringing is the output from the first (peaking) equaliser.

The visually perfect square wave is not the input waveform; it is the output from the second (dip) equaliser. The rise and fall times of the input square wave were deliberately slowed down to 10 usec to avoid distracting effects due to the finite bandwidth of the opamps used. You will note that the height of the ringing takes a little while to settle down after the start at 0 ms. When applying square waves and the like to filters and other circuits with energy-storage elements, you need to allow enough cycles for the circuit to reach equilibrium; otherwise you may draw some false conclusions.

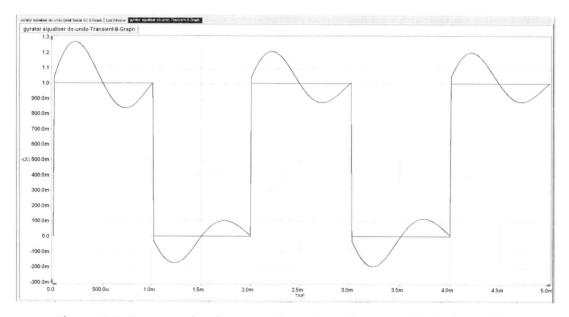

**Figure 14.1: Demonstrating that two reciprocal equalisers cancel their phase-shifts as well as amplitude/frequency response changes.**

Having reassured ourselves on this point, things are generally very much otherwise in crossover equalisation. If a drive unit has, say, a 2 dB peak or hump in its amplitude/frequency response, this will probably be due to some underdamped mechanical resonance or other electro-acoustic phenomenon, and while it is, in principle at least, straightforward to cancel this out, it is very unlikely that the associated phase-shifts will also cancel, because the physics behind the drive unit peak and the equaliser dip are completely different.

## Loudspeaker Equalisation

Active crossovers may consist simply of frequency-dividing filters but very often also include equaliser circuits with a carefully tailored frequency response which are used to counteract irregularities in the overall response of the system. Since the active crossover and its power supply, etc, is already present, relatively little extra circuitry is needed, and the cost is low. A large number of different equaliser circuits are described later in this chapter.

There are several reasons why equalisation might be necessary:

1.  Correcting irregularities in the frequency response of the drive units themselves.
2.  Correcting frequency response features inherent in the driver/enclosure system, for example 6 dB/ octave equalisation of dipole LF loudspeaker units
3.  Using equalisation to evade the physical limits of driver/enclosure system; most commonly extending the bass response of any kind of loudspeaker.
4.  Compensation for other unwanted interactions between the loudspeaker and its enclosure. The most common application is compensation for enclosure front panel diffraction effects.
5.  Compensation to deal with interactions between the loudspeaker and the listening space, such as when a loudspeaker is mounted against a wall.

## 1 Drive Unit Equalisation

Drive unit equalisation usually involves the correction of minor humps and dips in the frequency response. For example, a consistent 2 dB dip in drive unit response might be cancelled out by a 2 dB peak introduced by an equalisation circuit. Exact cancellation will not be possible, either because the shape of the dip is too complex to be mimicked by practical amounts of circuitry or because of variation in the shape of the dip due to driver tolerances or aging. This kind of correction is commonly performed by peak/dip equalisers.

A classic example of drive unit equalisation is constant directivity horn equalisation. Following the work of Don Keele at Electro-Voice, [1] constant directivity (CD) high-frequency horn loudspeakers appeared in the late 1970s, Basically an initial exponential section was combined with a final conical flare, causing the shorter high-frequency wavelengths to be dispersed more effectively off axis; purely conical horns, as seen on old gramophones and phonographs, are not satisfactory because of their poor low-frequency response. Because CD horns direct more high-frequency energy off axis, the amount of high-frequency energy available directly on axis is reduced. Therefore the CD horn no longer measures flat directly on axis unless it is given equalisation that is the inverse of the horn high-frequency roll off response. This is called constant directivity horn equalisation or CDEQ.

Constant directivity horns roll off the higher frequencies at about −6 dB per octave from around 2 to 4 kHz, continuing to 20 kHz and beyond, so the equalisation therefore takes the form of a +6 dB per octave boost starting at the appropriate frequency. The equalisation curve is arranged to shelve to prevent excessive boost at frequencies above 20 kHz, which could lead to inappropriate amplification of ultrasonic signals and imperil the sound system stability. The frequency at which the equalisation begins may be derived from measurements or from the −3 dB point of the CD horn frequency response, as provided by the manufacturer. The equalisation is typically performed by a HF-boost equaliser, as described later in this chapter.

## 2  6 dB/octave Dipole Equalisation

The operation of dipole loudspeakers was described in Chapter 2. To summarise it quickly, a dipole loudspeaker has a drive unit mounted on a flat panel, usually called a baffle, which prevents sound from simply sliding straight round from the front to the back of the drive unit and cancelling out. The larger the baffle, the lower the frequency down to which this is effective. The panel may be folded in various ways to save space, so it remains large acoustically but is physically smaller. The name "dipole" is derived from the way that the polar response consists of two lobes, which have equal radiation forwards and backwards (in opposite polarities) and none at right angles to the front-back axis of the drive unit. Because the baffle is of finite size, there is a frequency at which the response begins to fall off as sound goes around the edge of the baffle (loudspeakers in sealed boxes are sometimes referred to as being mounted in an "infinite baffle", but that does not of course mean that their response goes down to infinitely low frequencies).

Dipole speakers are therefore commonly regarded as needing equalisation in the form of a 6 dB/octave boost as frequency falls, starting from an appropriate frequency. This has to be done with great caution, as even apparently modest amounts of equalisation greatly increase both cone excursion and the amplifier power required; a dipole loudspeaker is particularly vulnerable to this because, unlike a drive unit with a sealed box, there is no loading on the cone at low frequencies. It is essential that the bass boost ceases at a safe frequency, and it is wise to arrange for an effective subsonic roll-off if drive unit damage is to be avoided. There is often also a need to equalise away a peak in the loudspeaker response where the LF roll-off begins. The biquad equaliser, described later, can do the task effectively and economically but may need to be supplemented with further subsonic filtering to make sure that no accidents occur.

## 3  Bass Response Extension

As described in Chapter 2, the bass response of a loudspeaker is essentially that of a highpass filter whose characteristics are determined by the type of enclosure and the parameters of the LF drive unit. A sealed-box loudspeaker consists of a mass-compliance-damping system that gives a classical 2nd-order highpass response and can for some purposes be simulated by a highpass active filter of the Sallen & Key or other configuration.

Much thought has been given to extending the low-frequency response of such systems by adding bass boost in a controlled fashion. The presence of an active crossover makes adding this facility straightforward, though its design is not so simple. The basic principle is to counteract the downward slope of the loudspeaker response with the upward slope of equaliser boost, but there are two major complicating factors. First, when the LF response of a loudspeaker is rolling off, the cone excursion is at its greatest. Adding equalisation increases the excursion further, and it is all too easy to exceed the safe limits of the drive unit. Second, a much increased amplifier power capability is required.

It has been claimed that not only is the LF response extended, but the time response may also be improved, because the ringing and overshoot caused by an underdamped LF response can be cancelled by a matching dip in the response of the equalisation circuit, since we are dealing with a simple minimum-phase system. The final LF roll-off is determined by that of the equaliser. It is questionable if this can really be counted as "equalisation" as such, because the intent is not so much to correct an error as to extend the performance beyond what would otherwise be physically possible. The biquad equaliser is a good choice for this form of equalisation, giving great freedom of parameter variation.

Because of the low frequencies at which this kind of equaliser operates, large capacitor values are required if impedance levels are to be kept suitably low, and this puts up the cost.

# 4 Diffraction Compensation Equalisation

We saw back in Chapter 2 that the diffraction of sound from the corners of a loudspeaker enclosure can have profound effects on the frequency response. A somewhat impractical spherical loudspeaker is free from most response irregularities but still has a gentle 6 dB rise with frequency, because at low frequencies, the long sound wavelengths diffract around the sides and rear of the enclosure, so radiation occurs into "full-space", while at high frequencies, the drive unit cone radiates mostly forwards, into what is called "half-space". This rise in response is sometimes called the "6 dB baffle step" though it is actually a very smooth transition between the two modes. Since the frequencies that define it are constant, being derived from the enclosure dimensions, it can easily be cancelled out by 1st-order shelving equalisers, such as those described later in this chapter.

Spherical loudspeakers are, however, very rare, and for practical reasons the vast majority of loudspeaker enclosures are rectangular boxes, as seen in Figure 14.2, which is derived from the all-time classic paper by Olson in 1969. [2] The lengths of the edges of the rectangular box were 2 feet and 3 feet, the drive unit being mounted midway between the two side edges and 1 foot from the top short edge. Olson comments that the dips at 1 kHz and 2 kHz are induced by diffraction from the top and side edges, the frequencies being relatively high because the distances from the drive unit to these edges are small; the broader response minimum just below 600 Hz is due to the longer distance to the bottom edge. It is obvious that equalising away these three minima. as well as compensating for the 6 dB baffle-step rise, is a fairly ambitious undertaking, requiring at least four equaliser stages. There are also minor dips of about 1 dB at 1.5 kHz and 2.5 kHz, which are at multiples of the frequency of the minimum due to the distance to the bottom edge and are due to the path containing whole wavelengths.

Olson suggested a more sophisticated enclosure, as in Figure 14.3, that would give blunter edges and much reduce the response irregularities, while still fitting into a living room better than a sphere would. The graph shows response deviations reduced to about 2 dB. This is a great improvement, and this sort of box has had some popularity, though it is of course more difficult to make and not everybody likes the shape. The response could be made almost flat by correcting the inherent 6 dB rise and then applying a little cut at 1 kHz and 2 kHz.

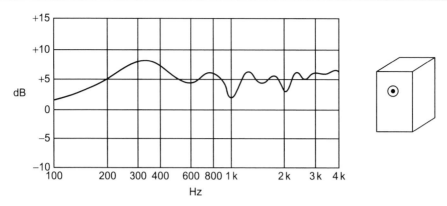

**Figure 14.2: Response disturbances, due to the sharp corners of a rectangular box, are superimposed on the inherent 6 dB response rise (after Olson, 1969).**

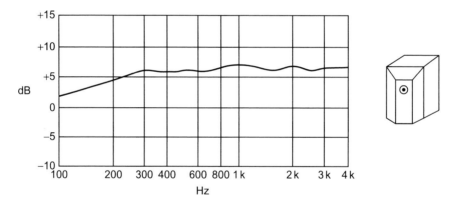

**Figure 14.3: Much reduced response disturbances due to the blunter corners at the front of the box, added to the inherent 6 dB rise (after Olson, 1969).**

It should be pointed out that these famous graphs show a rather smoothed version of the actual frequency response. There must have been many other minor irregularities that were not reproduced, as they were irrelevant to the central argument about the importance of diffraction effects. You cannot assume that the equalisation options suggested would result in a ruler-flat frequency response. Loudspeakers just don't work that way. In addition, diffraction effects vary according to the listening or measuring position, because the "virtual" sources are displaced from the position of the actual driver. Correcting the on-axis response may make the off-axis colouration worse.

## 5  Room Interaction Correction

In Chapter 2, we saw how the placement of a loudspeaker with respect to walls and corners could have a significant effect on the frequency response. A loudspeaker in free air (perhaps on a high pole in the middle of a large field, which can be useful for measurements but is less so for actual listening) is said to be working into "whole-space". As we saw in the previous section, the low-frequency output will tend to diffract around the loudspeaker enclosure and travel backwards away from the listener, while the high-frequency output will mostly be radiated forwards.

If we now mount the loudspeaker in the middle of a large vertical wall, with the front baffle flush with the wall, it is working in "half-space", and the low-frequency output can no longer travel backwards—it has to go forwards, and so more low-frequency energy reaches the listener. The high-frequency output is unchanged because it was all going forwards anyway, so the relative rise in LF level is about 6 dB. If we put the loudspeaker at the junction between a vertical wall and the floor, the effect is enhanced, and this is called "quarter-space" operation. Finally, we can put our loudspeaker in a corner, where the floor and two walls at right angles meet, and this is known as "eighth-space" operation. If this progression was taken further by adding more enclosing surfaces, we would end up with something like a horn loudspeaker.

From the point of view of crossover design, the important thing is that the low-frequency acoustic output is boosted relative to the high-frequency output each time we move from whole-space to half-space to quarter-space and then to eighth-space. A loudspeaker/crossover system designed to give a flat "free-space" response, typically for an outdoor sound-reinforcement application, will sound very bass-heavy indoors. Some studio monitor speakers have "half-space" and "quarter-space" settings which switch in a low-frequency roll-off, the frequency at which it starts depending on the enclosure size. A suitable equaliser might be the LF-cut circuit described later in this chapter.

While studio monitor speakers are mounted very carefully, often flush with a surface to give true "half-space" operation, more compromise is usually required in the domestic environment, and it is quite possible to encounter a situation where one loudspeaker is against a wall (quarter-space) while the other has to be in a corner (eighth-space). This is obviously undesirable, but sometimes in life one must make the best of a non-optimal situation, and providing separate open/wall/corner equalisation switches for the left and right channels of a crossover might be worth considering.

Things get more complicated when we contemplate a loudspeaker standing on the floor and not flush (or nearly so) with a wall but some distance from it. If the loudspeaker is one quarter of a wavelength away from a reflective wall at a given frequency, the low-frequency energy that diffracts backwards is reflected, so that its total path length is one half-wavelength, and it will reach the loudspeaker again in anti-phase so that cancellation occurs. If the front of a loudspeaker is 1 metre from a wall, the first cancellation notch will be at about 86 Hz, as this frequency has a quarter-wavelength of 1 metre. How complete the cancellation is depends on how accurately the speaker-wall distance is one quarter-wavelength and on the reflection coefficient of the wall. A lower frequency means a greater distance for a quarter-wavelength, and the amplitude of the reflected signal reaching the speaker will be lower because there is more opportunity for diffusion, and so the cancellation will be less effective.

If the loudspeaker is half a wavelength away from a the reflective wall, the total path length is a whole wavelength, and it will in arrive in-phase and reinforce the direct sound, theoretically giving a +6 dB increase in level. The effect on the sound reaching the listener will vary from complete cancellation to +6 dB reinforcement depending both on the relationship between the sound wavelength and the total path length via reflection and on the wall reflection coefficient. This effect is not confined to one frequency, because there can be any number of whole wavelengths in the go-and-return path; Figure 14.4 shows how this gives a "comb-filter" response, with the reinforcement peaks and the

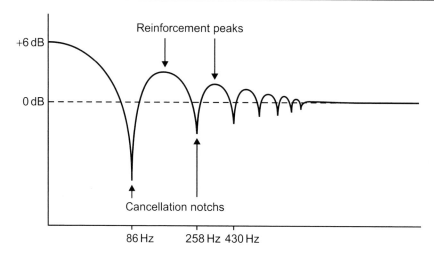

**Figure 14.4: Comb-filter effect produced by reflection from wall behind loudspeaker. The amplitude of the peaks and the notches reduces with increasing frequency because more sound is radiating forward and less is diffracting to the rear.**

cancellation notches reducing in amplitude with increasing frequency because more sound energy is being radiated forward and less is diffracting to the rear. The peaks and notches get closer together because the graph is drawn in the usual way with a logarithmic frequency axis, but the reinforcement/cancellation process is dependent on a linear function of frequency.

Calculating the actual path length is complicated by the fact that the low-frequency radiation has to go round a 180-degree corner, so to speak, as it diffracts around the front of the enclosure, and it is a question as to how sharp a turn it makes in a given situation.

This comb-filtering loudspeaker-room interaction has been examined here because it is a very good example of a mechanism that it is *not* possible to correct completely by equalisation, for both theoretical and practical reasons. Practically, boosting the gain at all the notch frequencies would be very hard to do, because of the need to line multiple boost equalisers up with multiple narrow notches, an alignment that would become grotesquely incorrect as soon as the loudspeaker was moved by a few inches. You will however note that the gain variations become less as the frequency increases, so what can be done is to make some compensation for the really big response variations at the LF end.

The other important point that stands out here is that the loudspeaker-spaced-from-a-wall situation is extremely common, giving rise to the sort of frequency response shown in Figure 14.4, and yet we still listen quite happily to the result. It is not necessary to have a ruler-flat frequency response to enjoy music.

Another important property of a listening space is the amount of high-frequency absorption it contains. A room with hard walls and floors will reflect high-frequency energy, and a proportion of this will reach the listener as reverberation. On the other hand a room with wall hangings, thick carpets, and comfy sofas will absorb some of high-frequency energy, and the effect will be less treble. The more absorbent room will give the more accurate sound, as a greater proportion of the energy at the listener will be directly from the loudspeaker, with an accurate frequency response, and will be subjected to less room colouration. Correcting for high-frequency absorption is a job for the preamplifier rather than the active crossover, and this is just one reason why preamplifiers without tone-controls are a daft idea.

## Equalisation Circuits

There are a many ways of obtaining a desired equalisation response, and any attempt to examine them all would probably fill up the whole book, so I have had to be very selective in picking those looked at here. I have aimed to provide circuits that are easy to design, predictable in their response, easy to configure for good noise and distortion performance, and well-adapted for dealing with common loudspeaker problems. I have included some where the fixed resistors that set the response can be temporarily replaced with variable controls to speed the optimisation of a crossover design.

## HF-Boost and LF-Cut Equaliser

This is also known as a shelving highpass equaliser. It gives a frequency response that at low frequencies is flat but begins to rise when the frequency passes the boost frequency $f_b$. It continues to rise at a basic rate of 6 dB/octave until the shelf frequency $f_s$ is reached, at which point it shelves or levels out to a fixed gain. Unless the boost frequency and the shelf frequency are spaced by several octaves, the transition slope between the gain regimes will not have time to develop an actual 6 dB/octave slope. The Figure 14.5 shows the inverting form of the circuit. A non-inverting version also exists but is less flexible, because at no frequency can it have a gain of less than unity.

The circuit shown here is essentially a shunt-feedback amplifier with unity gain at low frequencies, because R1 = R3. To make it an HF-boost equaliser, the extra network R2, C1 is added. As the

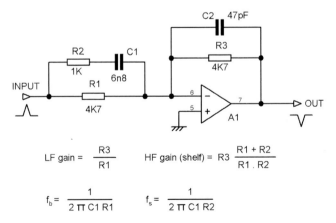

$$\text{LF gain} = \frac{R3}{R1} \qquad \text{HF gain (shelf)} = R3\,\frac{R1 + R2}{R1 \cdot R2}$$

$$f_b = \frac{1}{2\pi\,C1\,R1} \qquad f_s = \frac{1}{2\pi\,C1\,R2}$$

**Figure 14.5: Typical application of HF-boost equaliser for constant directivity horn equalisation; with the values shown, the boost starts around 2 kHz and begins to shelve to + 15 dB above 20 kHz.**

frequency rises the impedance of C1 falls and allows a greater input current to flow into the virtual earth point at the inverting input of A1. As the frequency increases further, this current is limited by R2, causing the gain at high frequencies to reach a maximum of R3 divided by the value of R1 and R2 in parallel. The stabilisation capacitor C2 across R3 has no effect on the response at audio frequencies and is included only to emphasise that it is always good practice to include such a measure. The shunt-feedback configuration has no common-mode signal voltage; this is handy if you are using an opamp prone to common-mode distortion. The downside is that it introduces a phase inversion which will need to be reversed somewhere else in the crossover system.

The design equations for the circuit are given in Figure 14.5, and the frequency response with the values given is shown in Figure 14.6. The equations for the LF gain and the HF or shelf gain are straightforward, but the expressions for the two frequencies require a little explanation. The boost frequency $f_b$ is the frequency at which the gain would have increased by 3 dB, just as the cutoff frequency of a lowpass filter is usually specified as the frequency at which the amplitudes response has fallen by 3 dB. Likewise, the shelf frequency $f_s$ is the frequency at which the gain is 3 dB below its final shelving value. As with the response slope, in practice the interaction between the boost and shelving actions is such that the equaliser response will only show these 3 dB figures in its response if the boost frequency and the shelf frequency are a long way apart—much further apart than is likely in any practical crossover design. This needs to be kept in mind when examining simulator outputs and measured frequency responses. It is perfectly possible to use component values that give the same boost and shelving frequencies; this does not mean the equaliser is doing nothing; it means that the LF and shelf gains are 6 dB apart.

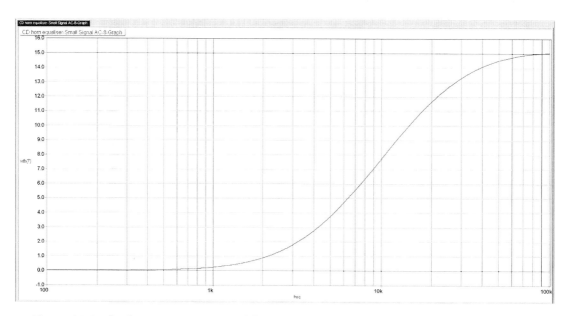

**Figure 14.6: The frequency response of the HF-boost equaliser for constant directivity horn equalisation shown in Figure 14.5.**

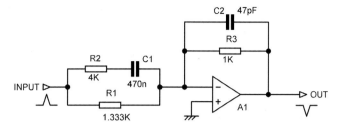

**Figure 14.7: The same circuit treated as an LF-cut rather than an HF-boost equaliser.**

If a basic gain of unity is not what is required, it can be set to any value above or below by altering the value of R3. As with active filters and other frequency-dependent circuits, it is best to decide on a preferred capacitor value first and derive the resistor values from that, given the much greater variety of resistor values available and the ease with which non-standard values can be obtained by combining two (2xE24 format) or three (3xE24 format) of them.

In choosing component values, the resistors should be kept as low as possible to minimise Johnson noise and the effects of current noise flowing through them, but they must not be so low that opamp distortion is increased, either in A1 or in the preceding stage. Particular care is needed with the latter because the input impedance of the circuit falls to R1 in parallel with R2 at high frequencies, where opamp distortion is most troublesome. The circuit values shown give an HF input impedance of 824 $\Omega$, and if 5532 opamps are being used, you will not want to go much lower than this.

This type of HF-boost equaliser can be used to deal with response irregularities of many kinds. A common application in the sound-reinforcement field is for constant directivity horn equalisation; the requirement for this is explained earlier in this chapter. Siegfried Linkwitz also recommends this configuration to smooth the transition between a floor-mounted woofer and a free-standing midrange/tweeter assembly. [3]

It is important to understand that while this circuit has so far been described as an HF-boost equaliser, it can also be regarded as an LF-cut equaliser; it is simply a matter of how you look at it. Figure 14.7 shows a circuit that has unity gain across most of the audio spectrum but gives a gentle cut from about 200 Hz down, as in Figure 14.8. The important point is that to set the normal or unequalised gain, where the impedance of C1 is so low it has no effect, to unity you must choose R1 and R2 so their combined value in parallel is equal to that of R3 (other unequalised gains greater or lesser than unity can be chosen). As the frequency falls, the impedance of C1 increases until the current flow through R2 is negligible and the shelving gain of the circuit is set by R1 alone, to −2.5 dB. In this case the boost frequency is *higher* than the shelving frequency, not lower, as it was with the CD horn example.

## HF-Cut and LF-Boost Equaliser

This is also known as a shelving lowpass equaliser. It gives a frequency response that at low frequencies is flat but begins to fall as the frequency rises and approaches the cut frequency $f_c$. It continues to fall at a basic rate of 6 dB/octave until the shelf frequency $f_s$ is reached, at which point

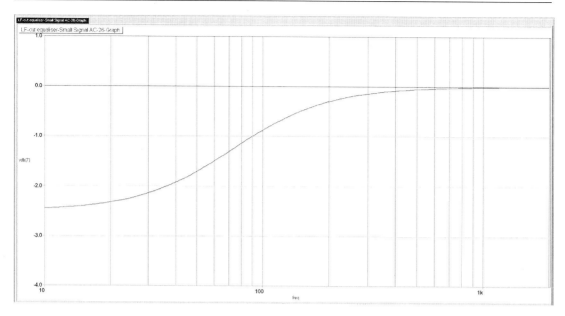

**Figure 14.8: The frequency response of the LF-cut equaliser in Figure 14.7.**

it shelves or levels out to a fixed gain. As with the HF-boost equaliser, unless the boost and the shelf frequency are spaced by several octaves, the slope between the gain regimes will be much less than 6 dB/octave. Figure 14.9 shows the inverting form of the circuit, with its design equations. A non-inverting version of this equaliser also exists but is less flexible because it cannot have a gain of less than 1.

The circuit is a shunt-feedback amplifier with unity gain at low frequencies because R1 = R3. To make it a HF-cut equaliser, the extra network R2, C1 is added. As the frequency rises, the impedance of C1 falls and allows a greater feedback current to flow into the virtual earth point at the inverting input of A1, reducing the gain. As the frequency increases further, this feedback current is limited by R2, causing the gain at high frequencies to reach a minimum of R2 and R3 in parallel, divided by the value of R1. A small stabilisation capacitor C2 is placed across R3, as before. Once again, the shunt-feedback configuration has the advantage of no common-mode signal voltage, but it introduces a phase inversion which will need to be reversed elsewhere.

A typical use of this kind of equaliser is compensating for the high-frequency boost resulting from diffraction around the edges of the front panel of a loudspeaker. With the values shown in Figure 14.9, the basic gain at low frequencies is unity, and the fall in response starts at around 200 Hz, with the gain shelving to −6 dB around 5 kHz, as seen in Figure 14.10. The middle −3 dB point is at 1 kHz.

As for the previous HF-boost/LF-cut equaliser, this HF-cut equaliser can also be regarded as a LF-boost circuit. To obtain a normal gain of unity at high frequencies, R1 is set equal to the value of the parallel combination of R2 and R3. Then, as frequency falls, the point is reached where the impedance of C1 becomes significant and reduces the feedback current through R2; the gain therefore rises, and shelves when the impedance of C1 becomes large compared with R3.

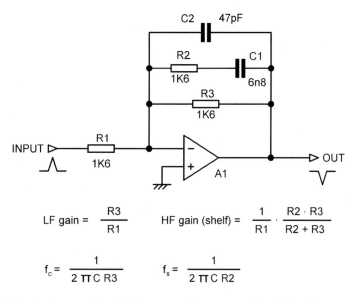

$$\text{LF gain} = \frac{R3}{R1} \qquad \text{HF gain (shelf)} = \frac{1}{R1} \cdot \frac{R2 \cdot R3}{R2 + R3}$$

$$f_c = \frac{1}{2\pi C \, R3} \qquad f_s = \frac{1}{2\pi C \, R2}$$

**Figure 14.9: Typical example of HF-cut equaliser set up for diffraction compensation; with the values shown the response falls from about 200 Hz and shelves to −6 dB around 5 kHz. The middle −3 dB point is at 1 kHz.**

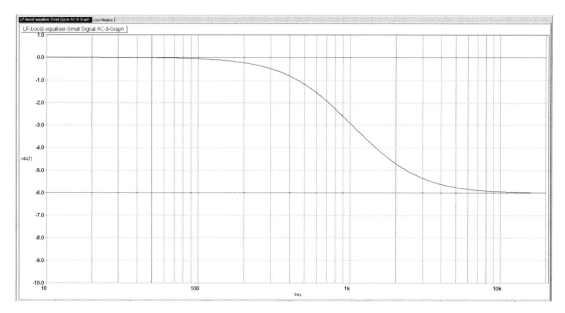

**Figure 14.10: The frequency response of the HF-cut equaliser in Figure 14.9, showing a typical curve for diffraction correction.**

This form of equaliser is useful for compensating for the low-frequency roll-off from a loudspeaker fitted to an open baffle; the smaller the baffle, the greater is the loss of low-frequency response and the greater the amount of boost required. Since the drive unit cone is not loaded at low frequencies, care must be taken to avoid excessive cone excursions. The biquad equaliser provides a more complex but much more versatile alternative.

## Combined HF-Boost and HF-Cut Equaliser

The first of these equaliser types has its frequency-dependent network in the input arm of the shunt-feedback amplifier; the second has its frequency-dependent network in the feedback arm. It is therefore possible to put a frequency-dependent network in both input and feedback arms, thus getting two equalisers for the cost of one opamp. Since the inverting opamp input is at virtual earth, there is no interaction between the two networks.

## Adjustable Peak/Dip Equalisers: Fixed Frequency and Low $Q$

It is often desirable to include an equaliser that can put a peak (like that of a resonant circuit) or a dip (essentially a broad notch)in the frequency response. This is particularly useful for correcting amplitude response irregularities in drive units. The circuit shown in Figure 14.11 is based on the Baxandall tone control concept and is commonly used in low-end mixing consoles. It can implement a low-$Q$ peak or dip of variable height, giving a flat response if the control is set centrally. The centre frequency can only be altered by changing the capacitor values. This equaliser is described as "adjustable" because it can be set to either peak or dip by any desired amount within its limits. It is very unlikely you would want to incorporate a peak/dip control potentiometer in a production crossover, though it could be extremely useful during the development phase. In manufacture RV1 would be replaced by a pair of fixed resistors that give the desired response.

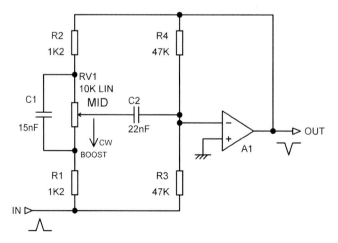

**Figure 14.11: A adjustable peak/dip equaliser based on the Baxandall tone control concept.**

The operation of this versatile circuit is very simple. As frequency increases from the low end of the audio band, the impedance of C2 falls, and the position of the pot wiper begins to take effect. When RV1 is in the boost position, more of the input signal is passed to the inverting input of A1 and adds to the output, giving a peaking response. When RV1 is in the cut position, more of the output signal is passed to the inverting input, giving more negative feedback, and the gain is reduced, causing a dip in the response.

At a still higher frequency, the impedance of C1 becomes low enough to effectively tie the two ends of the pot together, and the position of the wiper no longer has any effect, the circuit reverting to a fixed gain of unity at high frequencies. Thus the circuit only acts over a limited band of frequencies, giving the pleasingly symmetrical response curves in Figure 14.12 for varying control settings. It must be said that the benefits of symmetry are here visual rather than audible. More information on this type of equaliser can be found in my book *Small Signal Audio Design*; get the latest edition for maximum information. [4]

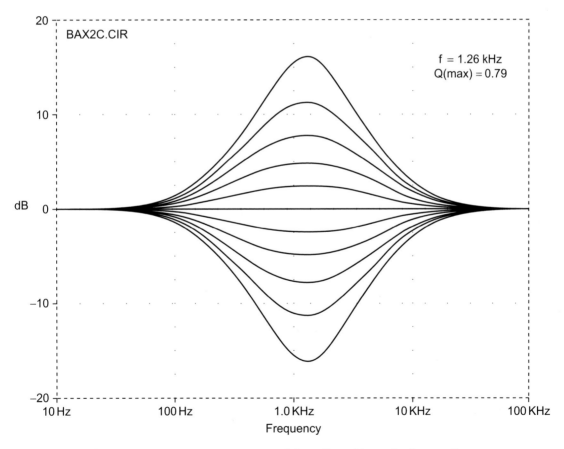

**Figure 14.12: Frequency response of the adjustable peak/dip equaliser, for different boost/cut settings.**

The component values shown in Figure 14.11 give a centre frequency of 1.26 kHz, which can be simply altered by scaling the values of C1 and C2 while keeping them in the same ratio. This apparently random frequency is a consequence of the fact that both potentiometers and capacitors come in relatively few values. The maximum $Q$ is 0.79, though this is only obtained at maximum boost or cut. At intermediate settings the curves are flatter and the $Q$ considerably lower; with a boost of +8 dB the $Q$ is 0.29. The boost/cut limits are ±15 dB, though this is hopefully a much greater range than will be required for equalisation in practice; the range can be reduced by increasing R1, R2.Note that R4 and R5 are needed to maintain negative feedback for unity gain at DC and to keep the stage biased properly. They must be high in value compared with the impedances in the rest of the circuit.

It is not possible to obtain high values of $Q$ with this configuration, and that is probably its major drawback. The capacitor ratio in Figure 14.11 gives the maximum possible $Q$. This equaliser is therefore mainly useful for dealing with large-scale trends in the frequency response.

As with the previous equalisers, the shunt-feedback configuration used here means there is no common-mode voltage on the opamp inputs, but with that comes an inconvenient phase inversion which must be taken into account in the system design.

## Adjustable Peak/Dip Equalisers: Variable Centre Frequency and Low $Q$

The fixed-frequency peak/dip equaliser we have just looked at gives excellent control over the amount of boost or cut applied, but the centre frequency can only be altered by changing the capacitors. In the development phase of crossover design it is extremely useful to be able to temporarily include an equaliser that also has continuously variable control of its centre frequency as well as the amount of peak or dip. When optimisation of the crossover is complete, it is replaced in the final design by a fixed equaliser like that in the previous section.

The circuit shown in Figure 14.13 also uses mixing console technology, where it is usually called a "sweep-middle EQ". It is based on a modified Wien-bridge network of the sort sometimes used in oscillators; this acts as a low-$Q$ bandpass filter, so that only a selected band of frequencies reach the non-inverting input of A1. When RV1 is in the boost position, the input signal passes through the Wien network and adds to the output, giving a peaking response. When RV1 is in the cut position, the signal through the Wien network constitutes extra negative feedback, and the gain is reduced, causing a dip in the response.

The variable load that the Wien network puts on the cut/boost pot RV1 and the variable source impedance from its wiper cause a small amount of interaction between boost/cut and centre frequency settings. This is not likely to cause any significant problem, but if necessary it could be eliminated by putting a unity-gain buffer stage between the RV1 wiper and the Wien bandpass network. There will also be minor inaccuracies due to imperfect matching of the two sections of the frequency control.

The combination of 100k pot sections and a 6k8 end-stop resistor gives a theoretical centre frequency range of 15.7 to 1, which is about as much as can be usefully employed when using reverse-log Law C pots. Greater ranges will give excessive cramping of the frequency calibrations at the high-frequency end of the scale. These calibrations should only be used as a guide; it is more accurate to measure

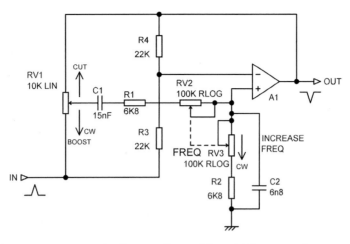

**Figure 14.13: A peak/dip equaliser with variable centre frequency, intended for crossover development work.**

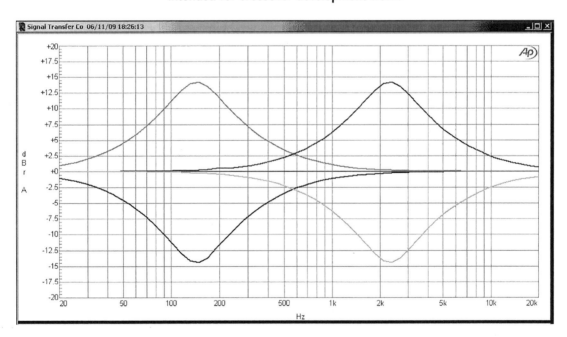

**Figure 14.14: Frequency response of the variable-frequency peak/dip equaliser, showing extremes of frequency setting.**

the response of the circuit after you have completed the optimisation of the crossover system. The measured frequency responses at the control limits are shown in Figure 14.14. The frequency range is from 150 Hz to 2.3 kHz; the ratio is slightly adrift from theory due to component tolerances. To obtain different frequencies, scale C1 and C2, keeping the ratio between them the same. More information on this kind of equaliser can be found in *Small Signal Audio Design*. [5]

This variable-frequency circuit is relatively complex compared with fixed equalisers and is noisier because of the extra Johnson noise from the high-value frequency-determining resistors. In production it would probably be replaced by a fixed equaliser like that in the previous section, with component values set to give the desired response.

## Adjustable Peak/Dip Equalisers With High $Q$

The two peak/dip equalisers we have just examined have low Qs and are not suitable for dealing with relatively narrow response irregularities; obtaining higher Qs requires more complex circuitry. There are several different approaches that might be taken; for example a state-variable filter would give the most flexibility, with control over centre frequency and $Q$ as well as the amount of peak or dip. While it could be very useful for optimisation, it would however be excessively complex and costly for permanent inclusion in a crossover design. The approach I have chosen here is based on using a gyrator to simulate a series LC resonant circuit.

The essence of the scheme is shown in Figure 14.15, which the alert reader will spot as the basic concept behind graphic equalisers. [6] L1, C1 and R3 make up an LCR series-resonant circuit; this has a high impedance except around its resonant frequency; at this frequency the reactances of L1, C1 cancel each other out, and the impedance to ground is that of R3 alone (a parallel LC circuit works in the opposite way, having a low impedance at all frequencies except at resonance). At the resonant frequency, when the wiper of RV1 is at the R1 end of its track, the LCR circuit forms the lower leg of an attenuator of which R1 is the upper arm; this attenuates the input signal, and a dip in the frequency response is therefore produced. When the RV1 wiper is at the R2 end, an attenuator is formed with R2 that reduces the amount of negative feedback at resonance and so creates a peak in the response. It is not exactly intuitively obvious, but this process does give absolutely symmetrical cut/boost curves. At frequencies away from resonance the impedance of the RLC circuit is high and the gain of the circuit is unity.

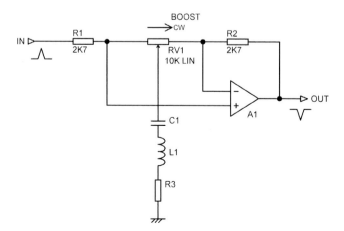

**Figure 14.15: The basic idea behind the peak/dip equaliser; gain is unity with the wiper central.**

Inductors are always to be avoided if possible, They are relatively expensive, often because they need to be custom-made. Unless they are air-cored (which limits their inductance to low values), the ferromagnetic core material will cause non-linearity. They can crosstalk to each other if placed close together and can be subject to the induction of interference from external magnetic fields. In general they deviate from being an ideal circuit element much more than resistors or capacitors do.

Gyrator circuits are therefore extremely useful, as they take a capacitance and "gyrate" it, so it acts in some respects like an inductor. This is simple to do if one end of the wanted inductor is grounded—which fortunately is the case here. Gyrators that can emulate floating inductors do exist but are far more complex.

Figure 14.16 shows how it works; C1 is the normal capacitor, as in the series LCR circuit, while C2 is made to act like the inductor L1. As the applied frequency rises, the attenuation of the highpass network C2, R1 falls, so that a greater signal is applied to unity-gain buffer A1 and it more effectively bootstraps the point X, making the impedance from X to ground increase. Therefore we have a part of the circuit where the impedance rises proportionally to frequency—which is just how an inductor behaves. There are limits to the $Q$ values that can be obtained with this circuit because of the inevitable presence of R1 and R2. The remarkably simple equation for the inductor value is shown; note that this includes R2 as well as R1.

The gyrator example in Figure 14.16 has values chosen to synthesise a grounded inductor of 100 mH in series with a resistance of 2 kΩ; that would be quite a hefty component if it was a real coil, but it would have a much lower series resistance than the synthesised version.

Figure 14.17 shows a gyrator-based high-$Q$ peak/dip equaliser, with a centre frequency of 1 kHz. The $Q$ at the maximum boost or cut of 6.3 dB is 2.2, considerably higher than that of the previous peak/dip equaliser we looked at and much more suitable for correcting localised response errors. The maximal cut and boost curves and some intermediate boost values are seen in Figure 14.18. The +4 dB peak results from the values R2 = 9 kΩ and R3 = 1 kΩ. The +1.5 dB peak results from the values R2 = 7 kΩ and R3 = 3 kΩ. For development work R2 and R3 can be replaced by a 10 kΩ pot RV1. To obtain different centre frequencies scale C1 and C2, keeping the ratio between them the same.

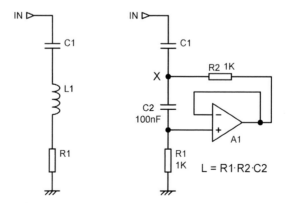

**Figure 14.16: Synthesising a grounded inductor in series with a resistance using a gyrator.**

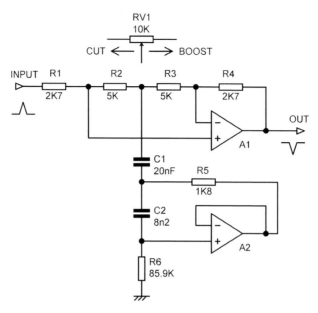

Figure 14.17: Gyrator-based high-Q peak/dip equaliser, with centre frequency fixed at 1 kHz. For development work R2 and R3 can be replaced by a 10 kΩ linear pot RV1.

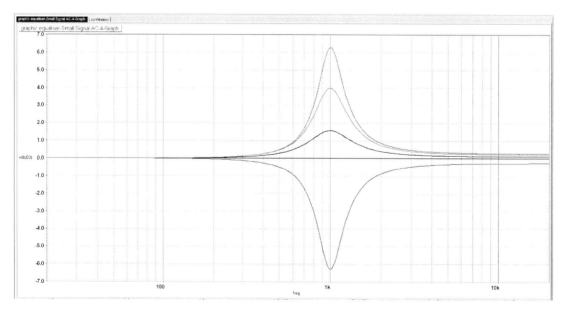

Figure 14.18: Frequency response of gyrator high-Q equaliser with various boost/cut settings.

The beauty of this arrangement is that two, three, or more LCR circuits, with associated cut/boost resistors or pots, can be connected between the two opamp inputs, giving us an equaliser with pretty much as many bands as we want. It is, after all, based on a classic graphic equaliser configuration.

This configuration can produce a response dip with well-controlled gain at the deepest point, but it is not capable of generating very deep and narrow notches of the sort required for notch crossovers (see Chapter 5). However, notches for equalisation purposes are not normally required to be particularly deep or narrow. The implementation of filters that do have deep and narrow notches is thoroughly dealt with in Chapter 12.

## Parametric Equalisers

The most complicated equalisers that are likely to be of use in crossover design are parametric equalisers. Since a standard 2nd-order resonance curve is completely defined by the three parameters of amplitude, centre frequency, and Q, they are called parametric equalisers. They are most commonly encountered in mixing console input channels, where four equalisers covering LF, LOW-MID, HI-MID and HF are fitted to high-end models. The LF and HF equaliser are usually switchable to give shelving responses like the Baxandall tone control.

Given their flexibility, parametric equalisers are relatively complex, and it is not likely that anyone would want to build them permanently into a crossover. However, even more so than other variable equalisers, they can be extremely useful during the development period for determining exactly what correction gives the best results, and this can then be implemented in simpler and cheaper fixed circuitry.

Parametric equalisers are almost always based around state-variable filters because of their ability to alter centre frequency and $Q$ independently. State-variable filters are described in Chapter 10; for equalisers the constant-amplitude version is used so that the gain at the peak or dip does not vary with the $Q$ setting. In Figure 14.19, the state-variable filter A1, A2, A3 only provides a bandpass boost resonance curve, and for equalisation is combined with the cut/boost stage A1.

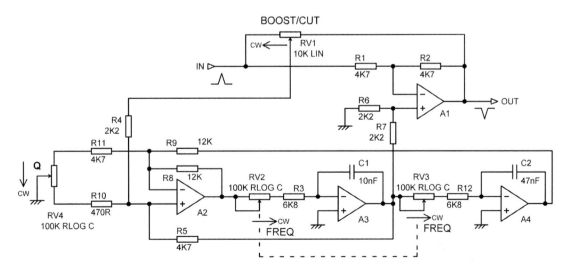

Figure 14.19: Low-noise parametric equaliser giving ±15 dB of boost or cut at 70 Hz–1.2 kHz, with $Q$ 0.7–5.

The arrangement of Figure 14.19 is that commonly used in mixing consoles. The boost/cut stage A1 gives a maximum of ±15 dB of boost or cut at the centre frequency. The state-variable filter shown has component values for typical LO-MID use, with centre frequency variable over a wide range from 70 Hz to 1.2 kHz and Q variable from 0.7 to 5. The frequency range is best altered by scaling C1 and C2 by the same ratio.

The circuit includes a cunning feature which is an important improvement on the standard circuit topology. Most of the noise in a parametric equaliser comes from the state-variable filter. In this design the filter path signal level is set to be 6 dB higher than usual by reducing the value of R4. After the filter, the signal returned to the non-inverting input of the cut/boost stage is reduced by 6 dB by the attenuator R6, R7. This attenuates the filter noise by 6 dB as well, and the final result is a parametric section approximately 6 dB quieter than the industry standard. The circuit is configured so that despite the doubled level in the filter, clipping cannot occur in it with any combination of control settings. If excess boost is applied, clipping can only happen at the output of A1, as would be expected.

Other features are the Q control RV4, which is configured to give a wide control range without affecting the peak gain. Note the relatively low values for the DC feedback resistors R1 and R2, chosen to minimise Johnson noise and the effect of opamp current noise without causing excessive opamp loading. If bipolar opamps are used, then arrangements for DC blocking are likely to be needed to stop the pots rustling when moved.

## The Bridged-T Equaliser

There are many types of bridged-T equaliser, [7] but the configuration shown in Figure 14.20 is probably the best known; it is described by Linkwitz in [8].This equaliser is potentially useful for modifying the LF roll-off of loudspeakers, but it has serious limitations compared with the biquad equaliser described in the next section.

One of these limitations is that no design equations are in common use. Figure 14.20 shows the values used for the attempt at LF equalisation in Figure 14.21, using the constraint R1a = R1b = R1, and I quite happily admit that they were obtained by twiddling values on a simulator and not by any more sophisticated process. The LF response of the loudspeaker shows a fairly high Q of 1.20 and a −3 dB cutoff point of 74 Hz. This apparently random frequency derives from the fact that if the Q was reduced to 0.7071 the −3 dB point would be at the nice round number of 100 Hz. As you can see, it is possible to convert the peaking response to something approximating maximally flat, but in the process

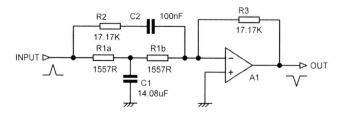

**Figure 14.20: Bridged-T equaliser circuit designed for loudspeaker LF extension.**

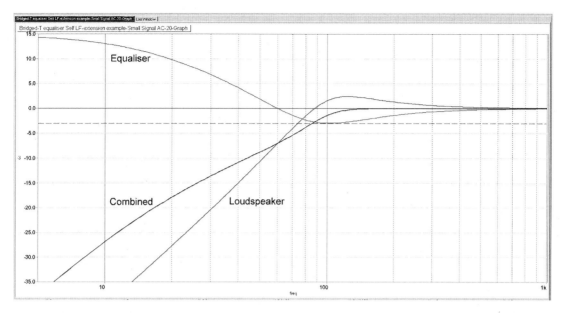

**Figure 14.21: Low-frequency extension using a biquad equaliser for a loudspeaker with an unequalised *Q* of 1.20. The dotted line is at −3 dB.**

the −3 dB frequency has actually increased from 74 Hz to 85 Hz, and there is a rather unhappy-looking bend in the combined response around 60 Hz. Not until 50 Hz does the equalised response exceed the unmodified loudspeaker response. Despite this, the equaliser gain at 40 Hz is +4.0 dB, and so the amplifier power will have to be more than doubled to maintain the maximum SPL down to this frequency. All in all, the result is not very satisfactory.

This bridged-T configuration has the disadvantage of high noise gain—in other words the gain of the circuit for the opamp noise is greater than the gain for the signal. This is because at high frequencies C1 has negligible impedance, so we have R1b connected from the virtual earth point at the opamp inverting input to ground. R2 also affects the noise gain, but by a small amount. In this case the noise gain is (R3 + R1b)/R1b, which works out at a rather horrifying 12 times or +21.8 dB. This stage will be much noisier than the active filters and other parts of the crossover system, and this is a really serious drawback.

Neville Thiele described several different bridged-T networks for loudspeaker LF extension in 2004, [7] but these work rather differently, as they are placed in the feedback path of an opamp. He noted that he was surprised that he could find no earlier analysis of bridged-T networks. I can testify that there appeared to be no useful material on these configurations on the Internet in 2010.

## The Biquad Equaliser

An especially effective equaliser is a combination of two bridged-T equalisers, as shown in Figure 14.22. It is much better adapted to equalising drive unit errors, being able to simultaneously

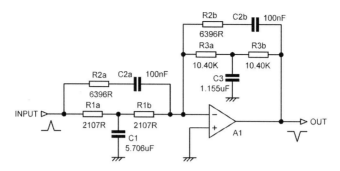

**Figure 14.22: Biquad equaliser circuit designed for f0 = 100 Hz, Q0 = 1.5, and fp = 45 Hz, Qp = 1.3.**

generate a peak and an independently controlled dip in the response. It is referred to as a biquad equaliser because its mathematical description is a fraction with one quadratic equation divided by another.

While it is possible to alter all of the ten passive components independently, in practice it is found that the easiest way to get manageable design equations is to set:

$$R1a = R1b = R1$$

$$R2a = R2b = R2$$

$$C2a = C2b = C2$$                           14.1, 14.2, 14.3, 14.4

$$R3a = R3b = R3$$

This approach is illustrated in Figure 14.22; it reduces the number of degrees of freedom to four, and the amplitude response is conveniently defined by f0, Q0 (the frequency and $Q$ of the dip in the response) and fp, Qp (the frequency and $Q$ of the peak in the response). Note that $Q$ values greater than 0.707 (1/√2) are required to give peaking or dipping. The circuit in Figure 14.22, derived from a crossover design I did for a client, has f0 = 100 Hz, Q0 = 1.5, and fp = 45 Hz, Qp = 1.3, and so helpfully shows both a peak and a dip with different $Q$'s; its response is shown in Figure 14.23. You will see that the gain flattens out +13.9 dB at the LF end, considerably greater than the HF gain of 0 dB. The further apart the f0 and fp frequencies are, the greater in difference the gain will be, and care is needed to make sure it does not become excessive. The actual peak frequency in Figure 14.23 is 36 Hz and the dip frequency 115 Hz, differing from the design values because f0 and fp are sufficiently close together for their responses to interact significantly.

The value of R2a and R2b (i.e. R2) has been kept reasonably low to reduce noise, but as a consequence of the low frequencies at which the equaliser acts, the capacitors C1 and C3 are already sizable, and it is not very feasible to reduce R2 further. C1 in particular will be expensive and bulky if it is a non-electrolytic component.

The gain of this circuit always tends to unity at the HF end of the response, so long as R2a = R2b, because at high frequencies the impedance of C2a, C2b becomes negligible and R2a, R2b set the gain. At the LF end all capacitors may be regarded as open circuit, so gain is set by R3/R1 and may be either

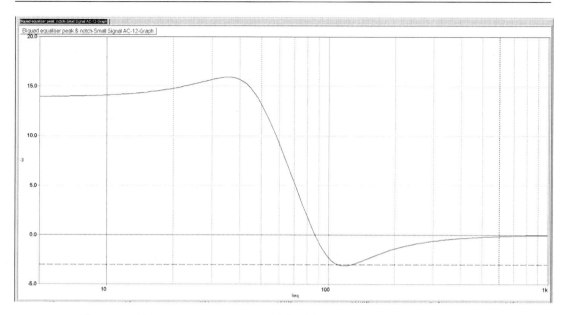

**Figure 14.23: Frequency response of the biquad equaliser in Figure 14.22.**

much higher or much lower than unity, depending on the values set for f0 and fp. Note that the stage is phase-inverting, and this must be allowed for in the system design of a crossover.

We saw in the previous section that the bridged-T configuration has the disadvantage of high noise gain because at high frequencies C1 has negligible impedance and R1b is effectively connected from the virtual earth point to ground. In this case the value of R2a cannot be neglected, as it is comparable with R1b, so the parallel combination of the two is used to give a more accurate equation for the noise gain:

$$Noisegain = \frac{(R2a + R1b)\, R2b + (R2a \cdot R1b)}{R2a \cdot R1b} \qquad 14.5$$

Thus the noise gain for Figure 14.22 comes to 5.04 times or +14.0 dB, which is certainly uncomfortably high but nothing like as bad as the +21.8 dB given by the bridged-T equaliser in the previous section.

The design procedure given here is based on the design equations introduced by Siegfried Linkwitz. [8] The procedure is thus:

1.  First calculate the design parameter $k$.

$$k = \frac{\dfrac{f_0}{f_p} - \dfrac{Q_0}{Q_p}}{\dfrac{Q_0}{Q_p} - \dfrac{f_p}{f_0}} \qquad 14.6$$

The parameter $k$ must come out as positive, or the resistor values obtained will be negative.

2.  Choose a value for C2 (a preferred value is wise, because it's probably the only one you're going to get, and it occurs twice in the circuit), then calculate R1:

$$R1 = \frac{1}{2\pi \cdot f_0 \cdot C2 \cdot \left(2Q_0\left(1+k\right)\right)}$$

14.7

3.  Calculate R2, C1, C3 and R3

$$R2 = 2 \cdot k \cdot R1$$

14.8

$$C1 = C2\left(2Q_0\left(1+k\right)\right)^2$$

14.9

$$C3 = C1\left(\frac{f_p}{f_0}\right)^2$$

14.10

$$R3 = R1\left(\frac{f_0}{f_p}\right)^2$$

14.11

The HF gain is always unity, because R2a = R2b, but the LF gain in dB is variable and is given by:

$$Gain_{LF} = 40\log\left(\frac{f_0}{f_p}\right)$$

14.12

These equations can be very simply automated on a spreadsheet, which is just as well, as several iterations may be required to get satisfactory answers.

The topology in Figure 14.22 appears to have been first put forward by Siegfried Linkwitzin 1978; [3] it is not clear if he invented it, but it seems he was certainly the first to develop useful design equations. Its use for equalising the LF end of loudspeaker systems was studied by Greiner and Schoessow in 1983. [9] Rather surprisingly, Greiner and Schoessow laid very little emphasis on the flexibility and convenience of this topology. The shunt-feedback configuration means that the two halves of the circuit are completely independent of each other and there is no interaction of the peak and dip parameters. However, it is important to understand that if f0 and fp are close together, the summation of their responses at the output may make the centre frequencies appear to be wrong at a first glance.

We will now see how this equaliser can be used for LF response extension. Figure 14.24 shows the low-end response of the loudspeaker we used in the previous section, with the relatively high Q of 1.20 and a −3 dB cutoff point of 74 Hz, which corresponds to a −3 dB point that would be at 100 Hz if the Q was reduced to 0.7071 for maximal flatness. The loudspeaker response peaks by 2.4 dB at 124 Hz. We now design a biquad equaliser with f0 = 100 Hz—note that is the frequency for the Q = 0.7071 version of the loudspeaker—and a Q of 1.20. This will cancel out the peaking in the loudspeaker response. The choice of fp and Qp depends on how ambitious we are in our plan to extend the LF response, but in this case fp has been set to 40 Hz and Qp to 0.5.

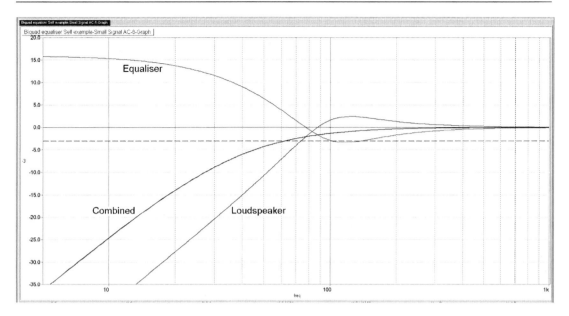

**Figure 14.24: Low-frequency extension using a biquad equaliser for a loudspeaker with an unequalised *Q* of 1.20. The dotted line is at −3 dB.**

Figure 14.24 shows that the response peak from the original loudspeaker has been neatly cancelled, and the overall LF response extended. The −3 dB point has only moved from 74 Hz to 62 Hz, which does not sound like a stunning improvement, but the gentler roll-off of the new response has to be taken into account. The −6 dB point has moved from 63 Hz to 40 Hz, a more convincing change. The LF roll-off takes on the *Q* of Qp, which at 0.5 would be considered overdamped by some critics, but I adopted it deliberately, as it shows how equalisation can turn an underdamped response into an overdamped one.

It is worth noting that it is not possible to set fp to 50 Hz instead of 40 Hz while leaving the other setting unchanged. This gives a negative value for the parameter k and a negative value of −3537 Ω for R2. While it is possible to construct circuitry that emulates floating negative resistors, it is not a simple business. If you run into trouble with this you should use another kind of equaliser.

The improvements may not be earth-shattering, but bear in mind that the equaliser has a gain of 9.0 dB at 40 Hz, so if the maximum SPL is going to be maintained down to that frequency, the output of the associated power output amplifier will have to increase by almost eight times. Driver cone excursion increases rapidly—quadrupling with each octave of decreasing frequency for a constant sound pressure—so great care is required to prevent it becoming excessive, leading to much-increased non-linear distortion and possibly physical damage. Thermal damage to the voice coil must also be considered; a sobering thought.

The equaliser circuit for Figure 14.24 is shown in Figure 14.25. The noise gain is now a much more acceptable 2.1 times or +6.4 dB, which emphasises that the biquad equaliser is much superior to the bridged-T. C3 could be a 100 nF capacitor in parallel with 1.5 nF without introducing significant error.

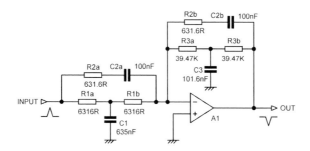

Figure 14.25: Schematic of a biquad equaliser designed for LF extension with exact component values.

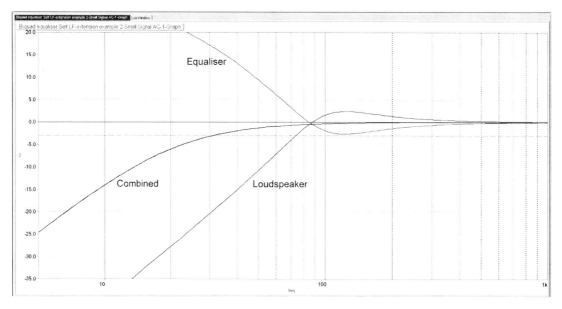

Figure 14.26: An ambitious attempt at low-frequency extension with the equaliser fp = 20 Hz and Qp = 0.5. The combined response is now −3 dB at 31 Hz, but the equaliser gain to achieve this is excessive. The dotted line is at −3 dB.

C1 is an awkward value, but if you are restricted to the E6 capacitor series it could be made up of 470 nF +150 nF + 15 nF in parallel. This is a relatively expensive solution, especially if you are using polypropylene capacitors to avoid distortion, and will also use up a lot of PCB area. A solution to this problem is provided in the next section.

It is instructive to try some more options for LF extension. Once the design equations have been set up on a spreadsheet and the circuit built on a simulator (including a block to simulate the response of the loudspeaker alone, which can in many circumstances be a simple 2nd-order filter of appropriate cutoff frequency and Q), variations on a theme can be tried out very quickly.

Figure 14.26 shows an optimistic attempt to further extend the LF response further by setting the equaliser fp to 50 Hz and keeping the Qp at 0.5. The −3 dB point is now moved from 74 Hz to 31 Hz, which would be a considerable improvement were it practical, but note that the equaliser gain is now heading off the top of the graph and does not begin to level out until around 5 Hz, at which point the gain has reached 27 dB. At the 31 Hz −3 dB point, equaliser gain is +17.4 dB, which would require no less than 55 times as much power from the amplifier and a truly extraordinary drive unit to handle it. It is important to realise that there is only so much you can do with LF extension by equaliser.

Figure 14.27 shows a more reasonable approach. The aim is to make the combined response maximally flat ($Q = 0.707$) rather than overdamped and keep the demands for extra amplifier power and increased cone excursion within practicable limits. We therefore set the equaliser fp to 50 Hz and the Qp at 0.707. The −3 dB point is now moved from 74 Hz to 50 Hz, which is well worth having, but the equaliser gain never exceeds 12 dB, and so the amplifier power increase is limited to a somewhat more feasible but still very substantial 16 times. If it is accepted that maximum SPL will only be maintained down to the new −3 dB frequency, only six times as much power is required. In practice the drive unit cone excursion constraints and the thermal performance of the voice coil assembly will probably set the limits of what is achievable.

An example of the use of the biquad equaliser for LF response extension is shown in the Christhof Heinzerling subtractive crossover. [10] This design also includes a boost/cut control for the low LF based on the 2-C version of the Baxandall tone control.

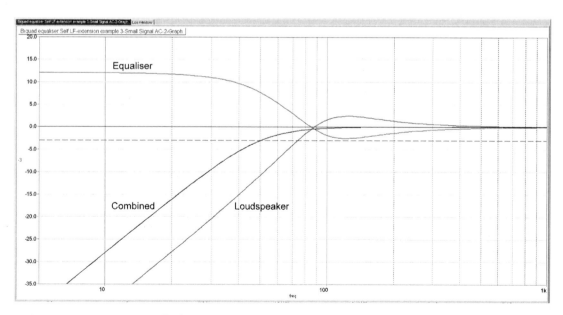

**Figure 14.27: A more realistic plan for low-frequency extension, with the equaliser fp = 50 Hz and $Q$ = 0.707. The modified response is now −3 dB at 50 Hz, and the equaliser gain is now acceptable. The dotted line is at −3 dB.**

## Capacitance Multiplication for the Biquad Equaliser

Since this equaliser is commonly used at the LF end of the spectrum and must work at low frequencies, if the resistors are to be kept low to minimise noise, then the capacitors can become inconveniently large in value, physical size, and cost. This is particularly true if polypropylene capacitors are used to prevent capacitor distortion. The use of a capacitance-multiplier architecture in an LF biquad equaliser means that the capacitors can be kept to a reasonable size, and any desired capacitance value can be obtained without paralleling components, simply by varying the multiplication factor.

The capacitance multiplication is very straightforward because the capacitors in question are grounded at one end in the basic circuit. Applying it to a floating capacitor would be much more difficult. The basic plan is to connect the normally grounded end of the capacitor to an opamp output, which is then driven so that as the voltage at the top of the capacitor (point A in Figure 14.28) rises, then the voltage at the bottom of the capacitor falls, increasing the current through the capacitor for a given voltage at point A. If the fall is equal in voltage to the rise, then as far as the rest of the circuit is concerned, the value of the capacitor has doubled. Varying the amount by which the bottom of the capacitor is driven allows the multiplication factor to be set to any desired value by varying a single resistor.

This technique is demonstrated in Figure 14.28, which implements the circuit of Figure 14.25 without the need to find and pay for a 635 nF capacitor. A3 is a shunt-feedback inverting stage that drives the bottom of capacitor C1 in anti-phase to the voltage at it top end. R4 and R5 are kept low in value to minimise current noise and Johnson noise, and the resulting low input impedance of the A3 stage is therefore buffered from the rest of the circuit by the voltage-follower A2. The multiplication factor, determined by R5/R4, is 1.924 times. so the 330 nF capacitor will appear to the rest of the equaliser circuit as 635 nF.

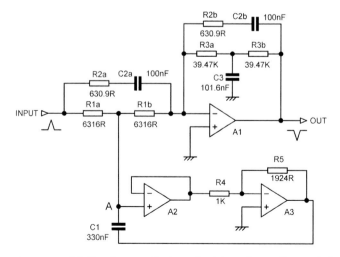

**Figure 14.28: Capacitance multiplication applied to the biquad equaliser of Figure 14.25. Here a 330 nF capacitor acts like a 635 nF capacitor, as its lower terminal is driven in anti-phase.**

Any ingenious new circuitry that adds more active devices needs to be carefully scrutinised to make sure that neither the noise or distortion performance is unduly compromised. The capacitance-multiplier method shown here does not significantly degrade either noise or distortion. The lack of extra noise is explained by the fact that if a signal is injected into the inverting input of A3 (which is a virtual earth point) through a resistor of equal value to R5, the gain to the output of the stage is −11 dB, and so the extra noise contribution from A2 and A3 is of little significance.

The technique does require careful consideration of signal levels. Given the condition R1a = R1b referred to earlier, the signal voltage at the top of C1 is half that at the input of the equaliser. Therefore, if multiplication factors greater than two are used, clipping may occur in the capacitance multiplier at the output of A3 before it does in any other part of the circuit.

## Equalisers With Non-Standard Slopes

Since lowpass and highpass filter slopes come only in multiples of 6 dB/octave, creating any other slope requires some approach other than a straightforward filter. Lesser slopes can be made over small frequency ranges by putting single highpass and lowpass time-constants close together, but this will not work over larger ranges. There are more sophisticated ways of combining time-constants to get a non-standard response slope, which I shall illustrate with a classic non-loudspeaker example. Pink noise is much more useful than white noise for audio measurement because it gives a flat line on a spectrum analyser, but the various methods of noise generation all give white noise. White noise is therefore converted into pink noise by what is called a "pinkening filter", which has a −3 dB/octave slope over the whole audio band from 20 Hz to 20 kHz.

### Equalisers With −3 dB/Octave Slopes

A −3 dB/octave slope is proverbially difficult, or at least non-obvious, to obtain, as the ultimate slopes of filters come in multiples of −6 dB/octave only. The standard solution when you need a pinkening filter is a series of overlapping lowpass and highpass time-constants that approximate to the required slope. This gives a response that wiggles up and down around a −3 dB/octave line, and the more pairs of poles and zeros used, the less the wiggle and the more accurately the response approximates to the line.

Two possible versions are shown in Figure 14.29; in both cases capacitor values have been restricted to the E6 series. The simpler version uses three RC networks with a final unmatched pole introduced by C4. Its response is shown in Figure 14.30, with an exact −3 dB/octave line (dotted) added for comparison; the error trace at the top has been multiplied by ten times to make the deviations more visible. The worst errors in the 100 Hz–10 kHz range are +0.22 dB at 259 Hz, −0.09 dB at 2.5 kHz, and +0.36 dB at 8.2 kHz, which I hope you will agree is not bad for a simple circuit.

This 3xRC circuit was originally published in *Small Signal Audio Design*, [11] but with R2 = 12k, R3 = 3k9, and R4 = 1k2. This had errors of up to ± 0.65 dB because of a historical requirement to use

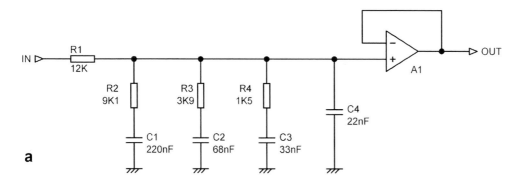

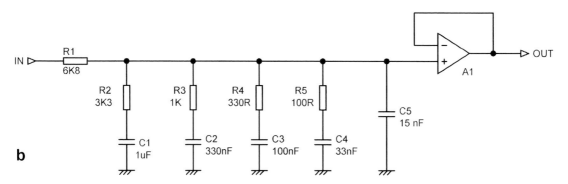

**Figure 14.29: Two −3 dB/octave equalisers built from repeated RC networks. The version at (b) is more accurate over a wider frequency range.**

only E12 resistor values. The 3x RC circuit is not the best option unless you are really closely counting the cost of every component, for reasons that will now appear.

Figure 14.31 shows the response of the version with four RC networks shown in Figure 14. 29b. It has four pole-zero pairs, with a final unmatched pole. It gives rather better performance for two reasons: one obvious, the other less so. Clearly, the more RC networks are used, the closer is the pole-zero spacing, and so the less the wobble on the frequency response. Less obvious is the fact that four RC networks allow the pole-zero frequencies required to match E6 capacitor values much better—you will note the tidy pattern of component values in Figure 14.29b, where $100/33 = 3.03$ and $33/10 = 3.3$, giving near-equal spacings on a log-frequency scale.

The error plot now shows a pleasing sine wave undulation around zero, with one cycle per decade of frequency, indicating that the errors ($\pm 0.28$ dB) are as low as they can be with this number of RC networks. The frequency range over which the errors are well controlled is also much greater; a very useful 20 Hz–20 kHz span. This is a good return from adding just one resistor and one capacitor. Finally, you will note that our "premium" four RC network has an overall lower impedance to reduce Johnson noise and the effect of device current noise in the resistors. This means that C1 is relatively

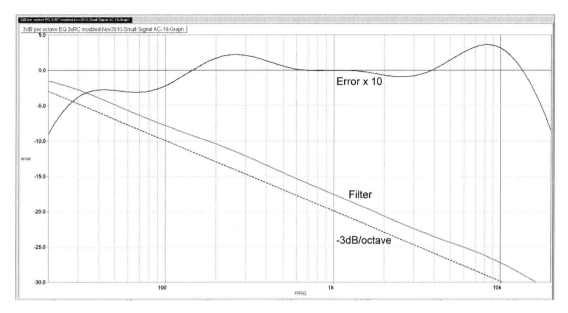

Figure 14.30: The response of the 3x RC −3 dB/octave equaliser in Figure 14.29a. The error trace at top (multiplied by ten) shows a maximum error of +0.36 dB in the 100 Hz–20 kHz range. The dotted line shows an exact −3 dB/octave slope.

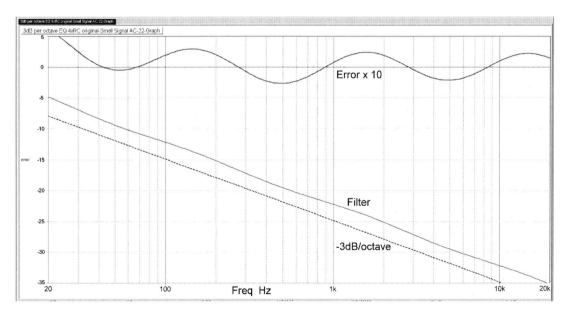

Figure 14.31: The response of the 4x RC −3 dB/octave equaliser in Figure 14.29b. The error trace at top (multiplied by ten) shows a maximum error of +0.28 dB over the 20 Hz–20 kHz range.

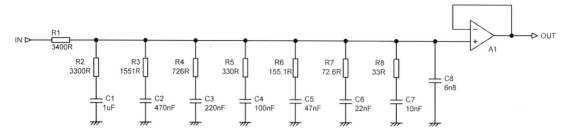

**Figure 14.32: A −3 dB/octave equaliser with better accuracy using seven RC networks that fit in with the E6 capacitor series. Accuracy is within +0.1 dB over a 10 Hz–20 kHz frequency range.**

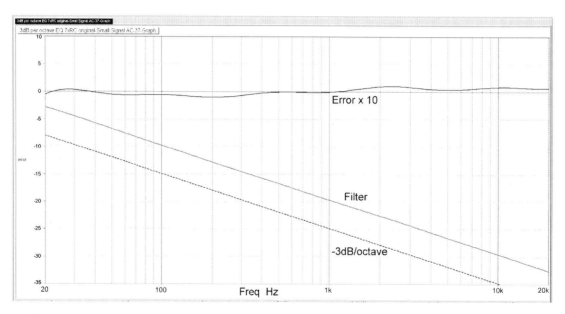

**Figure 14.33: The response of the 7x RC −3 dB/octave equaliser in Figure 14.32. The error is now within ±0.1 dB over a 10 Hz–20 kHz frequency range.**

large at 1 uF, but I see no reason why the capacitance multiplication concept described for biquad equalisers should not work nicely in this application; I should however say that I have not yet tried it.

As we have just seen, it makes sense to go with the flow of the E6 capacitor series. The 4xRC network we have just looked at uses just two values from the series; 10 and 33. This principle can be extended by using three values instead of two; 10, 22, and 47. This gives ratios of $100/47 = 2.10$, $47/22 = 2.14$, and $22/10 = 2.20$. The resulting 7xRC circuit is seen in Figure 14.32, where exact resistor values derived from the capacitor ratios are shown rather than preferred values.

This gives considerably improved accuracy over the 4x RC network. The error is now within ±0.1 dB over a frequency range extended to 10 Hz–20 kHz, though the error curve does not have the same symmetry; see Figure 14.33. In each stage both the resistor and capacitor values halve. It will of course be necessary to combine resistors, preferably in a 2xE24 format, to obtain those awkward values. With the E6 capacitor series the ultimate accuracy could be obtained by using all the values in a decade, with the ratios 100/68 = 1.47, 68/47 = 1.45, 47/33 = 1.42. 33/22 = 1.50, 22/15 = 1.47 and 15/10 = 1.50. For crossover design this would give much greater accuracy than that permitted by normal transducer variations.

Since the equaliser network is just a passive attenuator, quite large amounts of attenuation are introduced in creating the desired slope; in Figure 14.33 the attenuation at 1 kHz is 20 dB and at 10 kHz 30 dB. This implies that amplification is likely to be needed to restore the desired nominal level, and this will introduce noise. One approach to minimising noise is to split the equaliser into two or more halves, with amplifiers introduced to prevent the signal level ever falling too low.

### Equalisers With −3 dB/Octave Slopes Over Limited Range

While this example gives a clear insight into the process of implementing non-standard slopes, it is unlikely that speaker equalisation will require such a slope over the whole audio bandwidth. A more limited frequency range allows us to economise on the number of time-constants used while maintaining the accuracy of the slope.

In a real-life example, I needed an accurate slope of −3 dB/octave over the range 10 Hz to 500 Hz. Below 10 Hz the response was required to level out to flat. Above 500 Hz the response was of little importance, as other filtering was bringing in attenuation, but I certainly did not want it to go *up*; levelling out to flat would be acceptable, but transitioning to a −6 dB/octave slope was preferred. Figure 14.34 shows the resulting circuit, which does just that. The number of time-constants is reduced to three, plus a final shunt capacitor. As for the full-range equaliser of Figure 14.32, in each stage both the resistor and capacitor values halve.

The error is comfortably less than ±0.1 dB from 10 to 500 Hz, reaching a maximum of +0.074 dB at 10 Hz. Figure 14.34 shows the exact component values, but clearly some of the resistors are temptingly close to preferred values. If we set R1 = 27 kΩ, R2 = 27 kΩ, R3 = 12 kΩ and R4 = 5k6, the accuracy is still ±0.25 dB from 50 Hz to 500 Hz but is worsened to +1 dB at lower frequencies. In many applications that may be fine, but for mine it was not, so more exact resistor values using two E24 parts in parallel were used.

At low frequencies the response flattens out to level, and so the error becomes increasingly negative, as seen in Figure 14.35. Above 1 kHz the presence of C4 causes the response to steepen to −6 dB/octave,

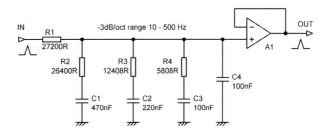

**Figure 14.34: Limited frequency range 3 dB/octave equaliser working over 10–500 Hz.**

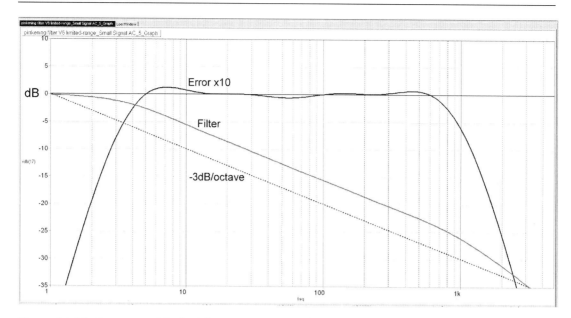

**Figure 14.35: Response of limited frequency range 3 dB/octave equaliser working over 10–500 Hz.**

as desired; the filter output starts to turn downwards, causing the error with respect to a −3 dB/octave slope to head negative again.

## Equalisers With −4.5 dB/Octave Slopes

In the first edition of this book, I glibly asserted that: "This approach can be used to create a filter with any desired slope by altering the spacing of the poles and zeros," though I had never actually attempted it. This sort of confident assertion has a nasty habit of coming back to bite you, so it is with some relief that after checking, I can tell you that the statement really is true.

A slope of 4.5 dB /octave is halfway between 3 dB/octave and 6 dB/octave, so seems like a good test. The network shown in Figure 14.36 has its time-constants arranged to give this result. The frequency plot of Figure 14.37 shows the result, plus the error compared with a mathematically derived perfect 4.5 dB/octave straight line. The error is within ±0.5 dB from 20 Hz to 20 kHz, and its generally symmetrical nature suggests that this is the best result that can be obtained with this number of time-constants.

In this case, in each stage both the capacitor values halve, as before, but the resistors are reduced by a factor of ten.

## Equalisers With Other Slopes

If a slope greater than 6 dB/octave but less than 12 dB/octave is required, a standard 6 dB/octave filter or integrator can be cascaded, with one having a lesser slope. For example, cascading a −6 dB/octave filter with a −3 dB/octave filter gives a −9 dB/octave slope.

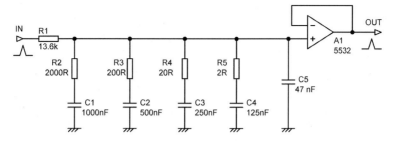

**Figure 14.36: The 4.5 dB/octave filter network.**

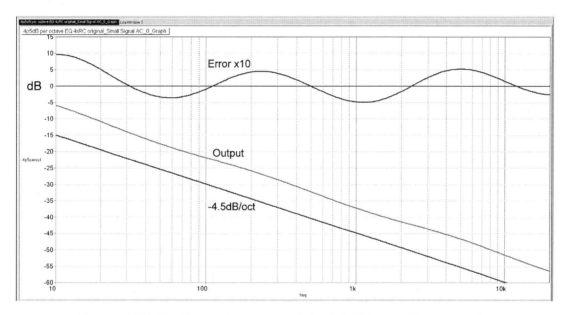

**Figure 14.37: The frequency response of the 4.5 dB/octave filter network.
The upper error trace is multiplied by 10 for clarity.**

All the filters examined give a response that falls with frequency; to get a response that rises with frequency, the networks can be inserted into the feedback network of an opamp. Stability will need to be checked and may be problematic. You don't want a situation where there is no feedback at high frequencies.

## Equalisation by Filter Frequency Offset

In Chapter 4 we saw that peaks and dips in the crossover region between a pair of filters can be manipulated and minimised by offsetting the frequencies of the two filters. If there is a peak at the crossover frequency then it can very often be reduced to a much smaller peak with two flanking dips by separating the filter frequencies. This technique can also be used for the purposes of response equalisation, though since there are only two variables (the highpass and lowpass filter cutoff frequencies) it is rather inflexible, and it is going to take a good deal of luck to get a response irregularity of just the right size and shape so it can be corrected simply by frequency offset. Even if this does happen, if you are manufacturing you might need to change one of the drive units for a different type, with different response irregularities, and you will then need to re-design the crossover to add extra equalisation stages that can cope with them.

## Equalisation by Adjusting All Filter Parameters

Much greater flexibility is possible if other filter parameters apart from the cutoff frequencies are manipulated. If we assume we have a 4th-order Linkwitz-Riley crossover, then the lowpass path has two cascaded Butterworth 2nd-order filters. Both the cutoff frequency and the $Q$ of each Butterworth filter could be altered to meet specific requirements, giving us four variables. The highpass path of the crossover similarly has four variables, so there are many more degrees of freedom.

The problem is that all these variables have an interactive effect on the final response, and tweaking them to get a desired response is going to be a deeply tiresome business; some sort of automatic means of optimisation is highly desirable. The Linear-X Systems Filtershop CAD software [12] contains optimisation routines that can take measured drive unit amplitude response as input and optimise the parameters of a chosen filter configuration to get the desired final response. It is a most impressive software package.

This approach has the advantage that it has much more ability to correct response irregularities than simple frequency-offsetting. It also requires no extra circuitry to perform the equalisation; on the other hand, adding dedicated equalisation stages will only have a small impact on the price of a typical active crossover. It does have the serious drawback that the resulting circuitry is likely to be wholly opaque, with no obvious indication of what parameters have been tweaked to compensate for what response irregularities. Good documentation is essential, and if specialised software is used to perform the optimisation, it will need to be kept available so that design changes can be made, if necessary.

Passive crossovers must stringently minimise component count and power losses, so equalisation often has to be performed by manipulating filter cutoff frequencies and $Q$'s.

## References

[1] Keele, Don "What's So Sacred About Exponential Horns?" AES Preprint #1038, May 1975
[2] Olson, H. F. "Direct Radiator Loudspeaker Enclosures" JAES 17(1), January 1969, pp. 22–29
[3] Linkwitz, Siegfried "Loudspeaker System Design" Wireless World, May, June 1978
[4] Self, Douglas "Small Signal Audio Design Handbook" Second Edn, Focal Press, 2015, pp. 431–434, ISBN 978-0-415-70974-3 hbk
[5] Self, Douglas Ibid, pp. 437–442
[6] Self, Douglas Ibid, pp. 447–450
[7] Thiele, Neville "An Active Biquadratic Filter for Equalising Overdamped Loudspeakers" AES Convention Paper 6153, 116th convention, May 2004

[8] Linkwitz, Siegfried "Loudspeaker System Design: Part 2" Wireless World, June 1978, p. 71

[9] Greiner and Schoessow "Electronic Equalisation of Closed-Box Loudspeakers" JAES 31(3), March 1983, pp. 125–134

[10] Heinzerling, Christhof "Adaptable Active Speaker System" (subtractive) Electronics World, February 2000, p. 105

[11] Self, Douglas Ibid, pp. 705–706

[12] Strahm, Chris "Linear-X Systems Filtershop Application Manual", Application Note 5, p. 223

# CHAPTER 15

# Passive Components for Active Crossovers

In this chapter we will look at some passive component properties that are especially important in the design of small-signal audio equipment in general and active crossovers in particular. All passive components differ from the ideal mathematical models—resistors have series inductance, capacitors have series resistance, and so on. There are also well-known issues with the accuracy of the component value and the way that the value changes with temperature. What is less publicised is that some can show significant non-linearity.

It is most unwise to assume that all the distortion in an electronic circuit will arise from the active devices. This is pretty clearly not true if transformers or other inductors are in the audio path, but it is also a very unsafe assumption if the only passive components you are using are resistors and capacitors. I recall that I was horrified when I first began designing active filters; the distortion from the polyester capacitors completely obliterated the quite low THD from the discrete transistor unity-gain buffers I was using. More on this problem later.

## Component Values

Active filters are the basic building blocks of electronic crossovers, and their proper operation depends on having accurate ratios between resistors and capacitors. Setting these up is complicated by the fact that capacitors are available in a much more limited range of values than resistors, often restricted to the E6 series, which runs 10, 15, 22, 33, 47, 68. Fortunately resistors are widely available in the E24 series (24 values per decade) and the E96 series (96 values per decade). There is also the E192 series (you guessed it, 192 values per decade), but this is less freely available. Using the E96 or E192 series means you have to keep a lot of different resistor values in stock, and when non-standard values are required it is usually more convenient to use a series or parallel combination of two E24 resistors. More on this later, too.

It is common in USA publications to see something like: "Resistor values have been rounded to the nearest 1% value", which usually means the E96 series has been used. The value series from E3 to E96 is given in Appendix 5.

## Resistors

In the past there have been many types of resistor, including some interesting ones consisting of jars of liquid, but only a few kinds are likely to be met with now. These are usually classified by the kind of material used in the resistive element, as this has the most important influence on the fine details of performance. The major materials and types are shown in Table 15.1.

Table 15.1: Characteristics of resistor types.

| Type | Resistance tolerance | Temperature coefficient | Voltage coefficient |
|---|---|---|---|
| Carbon composition | ±10% | +400 to −900 ppm/° | 350 ppm |
| Carbon film | ±5% | −100 to −700 ppm/°C | 100 ppm |
| Metal film | ±1% | +100 ppm/°C | 1 ppm |
| Metal oxide | ±5% | +300 ppm/°C | variable but too high |
| Wirewound | ±5% | ±70% to ±250% | 1 ppm |

These values are illustrative only, and it would be easy to find exceptions. As always, the official data sheet for the component you have chosen is the essential reference. The voltage coefficient is a measure of linearity (lower is better), and its sinister significance is explained later.

It should be said that you are most unlikely to come across carbon composition resistors in modern signal circuitry, but they frequently appear in vintage valve equipment, so they are included here. They also live on in specialised applications such as switch-mode snubbing circuits, where their ability to absorb a high peak power in a mass of material rather than a thin film is very useful.

Carbon film resistors are currently still sometimes used in low-end consumer equipment but elsewhere have been supplanted by the other three types. Note from Table 15.1 that they have a significant voltage coefficient.

Metal film resistors are now the usual choice when any degree of precision or stability is required. These have no non-linearity problems at normal signal levels. The voltage coefficient is usually negligible.

Metal oxide resistors are more problematic. Cermet resistors and resistor packages are metal oxide and are made of the same material as thick-film SM resistors. Thick-film resistors can show significant non-linearity at opamp-type signal levels and should be kept out of high-quality signal paths.

Wirewound resistors are indispensable when serious power needs to be handled. The average wirewound resistor can withstand very large amounts of pulse power for short periods, but in this litigious age component manufacturers are often very reluctant to publish specifications on this capability, and endurance tests have to be done at the design stage; if this part of the system is built first, then it will be tested as development proceeds. The voltage coefficient is usually negligible.

Resistors for general PCB use come in both through-hole and surface-mount types. Through-hole (TH) resistors can be any of the types in Table 15.1; surface-mount (SM) resistors are always either metal film or metal oxide. There are also many specialised types; for example, high-power wirewound resistors are often constructed inside a metal case that can be bolted down to a heatsink.

## Through-Hole Resistors

These are too familiar to require much description; they are available in all the materials mentioned earlier: carbon film, metal film, metal oxide, and wirewound. There are a few other sorts, such as metal foil, but they are restricted to specialised applications. Conventional through-hole resistors are now almost always 250 mW 1% metal film. Carbon film used to be the standard resistor material, with the expensive metal film resistors reserved for critical places in circuitry where low tempco and an absence of excess noise were really important, but as metal film got cheaper, so it took over many applications.

TH resistors have the advantage that their power and voltage rating greatly exceed those of surface-mount versions. They also have a very low voltage coefficient, which for our purposes is of the first importance. On the downside, the spiral construction of the resistance element means they have much greater parasitic inductance; this is not a problem in audio work.

## Surface-Mount Resistors

Surface-mount resistors come in two main formats, the common chip type and the rarer (and much more expensive) MELF format.

Chip surface-mount (SM) resistors come in a flat tombstone format, which varies over a wide size range; see Table 15.2.

MELF surface-mount resistors have a cylindrical body with metal endcaps, the resistive element is metal film, and the linearity is therefore as good as conventional resistors, with a voltage coefficient of less than 1 ppm. MELF is apparently an acronym for "Metal ELectrode Face-bonded" though most people I know call them "Metal Ended Little Fellows" or something quite close to that.

Surface-mount resistors may have thin-film or thick-film resistive elements. The latter are cheaper and so more often encountered, but the price differential has been falling in recent years. Both thin-film and thick-film SM resistors use laser trimming to make fine adjustments of resistance value during the manufacturing process. There are important differences in their behaviour.

Thin-film (metal film) SM resistors use a nickel-chromium (Ni-Cr) film as the resistance material. A very thin Ni-Cr film of less than 1 um thickness is deposited on the aluminium oxide substrate by sputtering under vacuum. Ni-Cr is then applied onto the substrate as conducting electrodes. The use of a metal film as the resistance material allows thin-film resistors to provide a very low temperature coefficient, much lower current noise and vanishingly small non-linearity. Thin-film resistors need only low laser power for trimming (one-third of that required for thick-film resistors) and contain no glass-based material. This prevents possible micro-cracking during laser trimming and maintains the stability of the thin-film resistor types.

Thick-film resistors normally use ruthenium oxide ($RuO2$) as the resistance material, mixed with glass-based material to form a paste for printing on the substrate. The thickness of the printing material is usually 12 um. The heat generated during laser trimming can cause micro-cracks on a thick-film resistor containing glass-based materials, which can adversely affect stability. Palladium/silver (PdAg) is used for the electrodes.

The most important thing about thick-film surface-mount resistors from our point of view is that they do not obey Ohm's law very well. This often comes as a shock to people who are used to TH resistors, which have been the highly linear metal film type for many years. They have much higher voltage coefficients than TH resistors, at between 30 and 100 ppm. The non-linearity is symmetrical about zero voltage and so gives rise to third-harmonic distortion. Some SM resistor manufacturers do not specify voltage coefficient, which usually means it can vary disturbingly between different batches and

**Table 15.2: The standard surface-mount resistor sizes with typical ratings.**

| Size L × W | Max power dissipation | Max voltage |
|---|---|---|
| 2512 | 1 W | 200 V |
| 1812 | 750 mW | 200 V |
| 1206 | 250 mW | 200 V |
| 0805 | 125 mW | 150 V |
| 0603 | 100 mW | 75 V |
| 0402 | 100 mW | 50 V |
| 0201 | 50 mW | 25 V |
| 01005 | 30 mW | 15 V |

different values of the same component, and this can have dire results on the repeatability of design performance.

Chip-type surface-mount resistors come in standard formats with names based on size, such as 1206, 0805, 0603, and 0402. For example, 0805, which used to be something like the "standard" size, is 0.08 inches by 0.05 inches; see Table 15.2. The smaller 0603 is now more common. Both 0805 and 0603 can be placed manually if you have a steady hand and a good magnifying glass.

The 0402 size is so small that the resistors look rather like grains of pepper; manual placing is not really feasible. They are only used in equipment where small size is critical, such as mobile phones. They have very restricted voltage and power ratings, typically 50 V and 100 mW. The voltage rating of TH resistors can usually be ignored, as power dissipation is almost always the limiting factor, but with SM resistors it must be kept firmly in mind.

Recently, even smaller surface-mount resistors have been introduced; for example several vendors offer 0201, and Panasonic and Yageo offer 01005 resistors. The latter are truly tiny, being about 0.4 mm long; a thousand of them weigh less than a twentieth of a gram. They are intended for mobile phones, palmtops, and hearing aids; a full range of values is available from 10 Ω to 1 MΩ (jumper inclusive). Hand placing is really not an option.

Surface-mount resistors have a limited power-dissipation capability compared with their through-hole cousins, because of their small physical size. SM voltage ratings are also restricted, for the same reason. It is therefore sometimes necessary to use two SM resistors in series or parallel to meet these demands, as this is usually more economic than hand-fitting a through-hole component of adequate rating. If the voltage rating is the issue, then the SM resistors will obviously have to be connected in series to gain any benefit.

## Resistors: Values and Tolerances

Resistors come in a rich variety of values compared with other components such as capacitors, and this gives a much needed flexibility to crossover design and active filter design generally.

Active filters very often require both precise resistor values and precise resistor ratios. When designing circuit blocks such as Sallen & Key highpass filters, where resistor ratios of exactly two are required, it is useful to keep in mind the options in the E24 series, as in Table 15.3, and the E96 series, as in Tables 15.4 to 15.7. The E24 series offers six options for a ratio of two, while the E96 series offers 16. In both cases the 1:2 pairs are closely spaced at the bottom of the decade, and this will have to be taken into account when choosing capacitor values.

**Table 15.3: There are six E24 resistor value pairs in a 1:2 ratio.**

| R | 2R |
|---|---|
| 75 | 150 |
| 100 | 200 |
| 110 | 220 |
| 120 | 240 |
| 150 | 300 |
| 180 | 360 |

**Table 15.4: There are 16 E96 resistor value pairs in a 1:2 ratio.**

| R | 2R |
|---|---|
| 100 | 200 |
| 105 | 210 |
| 113 | 226 |
| 137 | 274 |
| 140 | 280 |
| 147 | 294 |
| 158 | 316 |
| 162 | 324 |
| 174 | 348 |
| 187 | 374 |
| 196 | 392 |
| 221 | 442 |
| 232 | 464 |
| 590 | 1180 |
| 732 | 1464 |
| 750 | 1500 |

**Table 15.5: There are four E24 resistor value pairs in a 1:3 ratio.**

| R | 3R |
|---|---|
| 100 | 300 |
| 110 | 330 |
| 120 | 360 |
| 130 | 390 |

**Table 15.6: There are only two E24 resistor value pairs in a 1:4 ratio.**

| R | 4R |
|---|---|
| 300 | 1200 |
| 750 | 3000 |

**Table 15.7: There are six E24 resistor value pairs in a 1:5 ratio.**

| R | 5R |
|---|---|
| 150 | 750 |
| 200 | 1000 |
| 220 | 1100 |
| 240 | 1200 |

Other resistor ratios are less commonly used, but it is useful to be aware of them. In the E24 series there are four pairs of values in the ratio 1:3, two pairs in the ratio 1:4, and six pairs in the ratio 1:5; see Tables 15.5 to 15.7. In the E96 series there are no pairs in the ratio 1:3 or 1:4. This may sound paradoxical, but it is simply because the E96 series does not include many values from the E24 series. E12 is a subset of E24, but E24 is *not* a subset of E96. This less-than-obvious fact needs to be remembered when you are selecting resistors.

This information is also summarised in Appendix 5.

The standard tolerance for TH resistors is ±1%. In the direction of increasing accuracy, the next popular tolerance is ±0.1%; in the E24 series they cost between 10 and 15 times as much as the 1% parts but nowadays are freely available from the usual component distributors and are a good option when real precision is required in a few places in a crossover.

In many cases you will aim to use a capacitor value or values in the relatively sparse E6 series, and this will almost invariably result in non-standard resistor values. These are easily obtained by putting two or more resistors in series or parallel, and multiple resistors will be much cheaper than multiple capacitors. There are other advantages to this, as I shall now explain.

# Improving Accuracy With Multiple Components: Gaussian Distribution

Using two or more resistors to make up a desired value has a valuable hidden benefit. If it is done correctly it will actually increase the average accuracy of the total resistance value so it is *better* than the tolerance of the individual resistors; this may sound paradoxical, but it is simply an expression of the fact that random errors tend to cancel out if you have a number of them. This works for any parameter that is subject to random variations, but for the time being we will focus on the concrete example of multiple resistors.

Resistor values are usually subject to a Gaussian distribution, also called a normal distribution. It has a familiar peaked shape, not unlike a resonance curve, showing that the majority of the values lie near the central mean and that they get rarer the further away from the mean you look. This is a very common distribution in statistics, cropping up wherever there are many independent things going on that all affect the value of a given component. The distribution is defined by its mean and its standard deviation, which is the square root of the sum of the squares of the distances from the mean—the RMS-sum, in other words. Sigma ($\sigma$) is the standard symbol for standard deviation. A Gaussian distribution will have 68.3% of its values within $\pm 1\ \sigma$, 95.4% within $\pm 2\ \sigma$, 99.7% within $\pm 3\ \sigma$, and 99.9% within $\pm 4\ \sigma$. This is illustrated in Figure 15.1, where the *x*-axis is calibrated in numbers of standard deviations on either side of the central mean value.

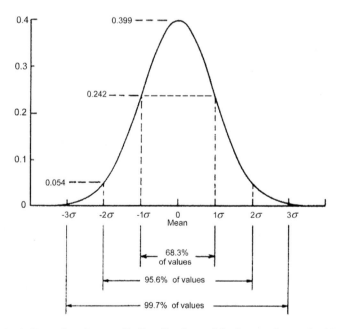

**Figure 15.1: A Gaussian (normal) distribution with the *x*-axis marked in standard deviations on either side of the mean. The apparently strange value for the height of the peak results because the area under the curve is 1.**

If we put two equal-value resistors in series, or in parallel, the total value has a narrower distribution than that of the original components. The standard deviation of summed components is the sum of the squares of the individual standard deviations, as shown in Equation 15.1; σsum is the overall standard deviation, and σ1 and σ2 are the standard deviations of the two resistors in series.

$$\sigma_{sum} = \sqrt{(\sigma_1)^2 + (\sigma_2)^2}$$ 15.1

Thus if we have four 100 Ω 1% resistors in series, the standard deviation of the total resistance increases only by the square root of 4, that is 2 times, while the total resistance has increased by 4 times; thus we have inexpensively made an otherwise costly 0.5% close-tolerance 400 Ω resistor. There is a happy analogue here with the use of multiple amplifiers to reduce electrical noise; we are using essentially the same technique to reduce "statistical noise". Note that this equation (Equation 15.1) only applies to resistors in series and cannot be used when resistors are connected in parallel to obtain a desired value. The improvement in accuracy nonetheless works in the same way.

You may object that putting four 1% resistors in series means that the worst-case errors can be four times as great. This obviously true—if all the components are 1% low, or 1% high, the total error will be 4%. But the probability of this occurring is actually very, very small indeed. The more resistors you combine, the more the values cluster together in the centre of the range.

Perhaps you are not wholly satisfied that this apparently magical improvement in average accuracy is genuine. I could reproduce here the standard statistical mathematics, but it is not very exciting and can be easily found in the standard textbooks. I have found that showing the process actually at work on a spreadsheet makes a much more convincing demonstration.

In Excel, the standard random numbers with a uniform distribution are generated by the function RAND(), but random numbers with a Gaussian distribution and specified mean and standard deviation can be generated by the function NORMINV(). Let us assume we want to make an accurate 20 kΩ resistance. We can simulate the use of a single 1% tolerance resistor by generating a column of Gaussian random numbers with a mean of 20 and a standard deviation of 0.2; we need to use a lot of numbers to smooth out the statistical fluctuations, so we generate 400 of them. As a check we calculate the mean and standard deviation of our 400 random numbers using the AVERAGE() and STDEV() functions. The results will be very close to 20 and 0.2, but not identical, and will change every time we hit the F9 recalculate key, as this generates a new set of random numbers. The results of five recalculations are shown in Table 15.8, demonstrating that 400 numbers are enough to get us quite close to our targets.

**Table 15.8: Mean and standard deviation of five batches of 400 Gaussian random resistor values.**

| Mean | Standard deviation |
| --- | --- |
| 20.0017 | 0.2125 |
| 19.9950 | 0.2083 |
| 19.9910 | 0.1971 |
| 19.9955 | 0.2084 |
| 20.0204 | 0.2040 |

To simulate the series combination of two 10 kΩ resistors of 1% tolerance we generate two columns of 400 Gaussian random numbers with a mean of 10 and a standard deviation of 0.1. We then set up a third column which is the sum of the two random numbers on the same row, and if we calculate the mean and standard deviation of that using AVERAGE() and STDEV() again, we find that the mean is still very close to 20, but the standard deviation is reduced on average by the expected factor of $\sqrt{2}$. The result of five trials is shown in Table 15.9. Repeating this experiment with two 40 kΩ resistors in parallel gives the same results.

If we repeat this experiment by making our 20 kΩ resistance from a series combination of four 5 kΩ resistors of 1% tolerance, we have to generate four columns of 400 Gaussian random numbers with a mean of 5 and a standard deviation of 0.05. We sum the four numbers on the same row to get a fifth column and calculate the mean and standard deviation of that. The result of five trials is shown in Table 15.10. The mean is very close to 20, but the standard deviation is now reduced on average by a factor of $\sqrt{4}$, which is 2.

I think this demonstrates quite convincingly that the spread of values is reduced by a factor equal to the square root of the number of the components used. The principle works equally well for capacitors or indeed any quantity with a Gaussian distribution of values. The downside is the fact that the improvement depends on the square root of the number of equal-value components used, which means that big improvements require a lot of parts and the method quickly gets unwieldy. Table 15.11 demonstrates how this works; the rate of improvement slows down noticeably as the

**Table 15.9: Mean and standard deviation of five batches of 400 Gaussian resistors with two in series.**

| Mean | Standard deviation |
|---|---|
| 19.9999 | 0.1434 |
| 20.0007 | 0.1297 |
| 19.9963 | 0.1350 |
| 20.0114 | 0.1439 |
| 20.0052 | 0.1332 |

**Table 15.10: Mean and standard deviation of five batches of 400 Gaussian resistors with four in series.**

| Mean | Standard deviation |
|---|---|
| 20.0008 | 0.1005 |
| 19.9956 | 0.0995 |
| 19.9917 | 0.1015 |
| 20.0032 | 0.1037 |
| 20.0020 | 0.0930 |

**Table 15.11: The improvement in value tolerance with number of equal-value parts.**

| Number of equal-value parts | Tolerance reduction factor |
|---|---|
| 1 | 1.000 |
| 2 | 0.707 |
| 3 | 0.577 |
| 4 | 0.500 |
| 5 | 0.447 |
| 6 | 0.408 |
| 7 | 0.378 |
| 8 | 0.354 |
| 9 | 0.333 |
| 10 | 0.316 |

number of parts increases. The largest number of components I have ever used in this way for a production design is five. Constructing a 0.1% resistance from 1% resistors would require a hundred of them and would not normally be considered a practical proposition, due to the PCB area and assembly time required.

The spreadsheet experiment is easy to do, though it does involve a bit of copy-and-paste to set up 400 rows of calculation. It takes only about a second to recalculate on my PC, which is by no stretch of the imagination state of the art.

You may be wondering what happens if the series resistors used are *not* equal. If you are in search of a particular value, the method that gives the best resolution is to use one large resistor value and one small one to make up the total, as this gives a very large number of possible combinations. However, the accuracy of the final value is essentially no better than that of the large resistor. Two equal resistors, as we have just demonstrated, give a $\sqrt{2}$ improvement in accuracy, and near-equal resistors give almost as much, but the number of combinations is very limited, and you may not be able to get very near the value you want. The question is, how much improvement in accuracy can we get with resistors that are some way from equal, such as one resistor being twice the size of the other?

The mathematical answer is very simple; even when the resistor values are not equal, the overall standard deviation is still the RMS-sum of the standard deviations of the two resistors, as shown in Equation 15.1; $\sigma 1$ and $\sigma 2$ are the standard deviations of the two resistors in series. Note that this equation is only correct if there is no correlation between the two values whose standard deviations we are adding; this is true for two separate resistors but would not hold for two film resistors fabricated on the same substrate.

Since both resistors have the same percentage tolerance, the larger of the two has the greater standard deviation and dominates the total result. The minimum total deviation is thus achieved with equal resistor values. Table 15.12 shows how this works, and it is clear that using two resistors in the ratio 2:1 or 3:1 still gives a worthwhile improvement in average accuracy.

**Table 15.12: The improvement in value tolerance with number of equal-value parts.**

| Series resistor values Ω | Resistor ratio | Standard deviation kΩ |
|---|---|---|
| 20k single | | 0.2000 |
| 19.9k + 100 | 199 : 1 | 0.1990 |
| 19.5k + 500 | 39 : 1 | 0.1951 |
| 19k + 1k | 19 : 1 | 0.1903 |
| 18k + 2k | 9 : 1 | 0.1811 |
| 16.7k + 3.3k | 5 : 1 | 0.1700 |
| 16k + 4k | 4 : 1 | 0.1649 |
| 15k + 5k | 3 : 1 | 0.1581 |
| 13.33k + 6.67k | 2 : 1 | 0.1491 |
| 12k + 8k | 1.5 : 1 | 0.1442 |
| 11k + 9k | 1.22 : 1 | 0.1421 |
| 10k + 10k | 1 : 1 | 0.1414 |

The entries for 19.5k +500 and 19.9k +100 demonstrate that when one large resistor value and one small are used to get a particular value, its accuracy is very little better than that of the large resistor alone.

## Resistor Value Distributions

At this point you may be complaining that this will only work if the resistor values have a Gaussian (also known as normal) distribution with the familiar peak around the mean (average) value. Actually, it is a happy fact this effect does *not* assume that the component values have a normal (Gaussian) distribution, as we shall see in a moment. An excellent account of how to handle statistical variations to enhance accuracy is in [1]. This deals with the addition of mechanical tolerances in optical instruments, but the principles are just the same.

Also we have assumed that the mean is absolutely accurate. This is actually completely reasonable because controlling the mean value emerging from a manufacturing process is usually easy compared with controlling all the variables that lead to a spread in values. In the examples given next, it is plain that the mean is very well controlled, and the spread is pretty much under control as well.

You sometimes hear that this sort of thing is inherently flawed, because, for example, 1% resistors are selected from production runs of 5% resistors. If you were using the 5% resistors, then you would find there was a hole in the middle of the distribution; if you were trying to select 1% resistors from them, you would be in for a very frustrating time, as they have already been selected out, and you wouldn't find a single one. If instead you were using the 1% components obtained by selection from the 5% population, you would find that the distribution would be much flatter than Gaussian and the accuracy improvement obtained by combining them would be reduced, although there would still be a definite improvement.

However, don't worry. In general this is not the way that components are manufactured nowadays, though it may have been so in the past. A rare contemporary exception is the manufacture of carbon composition resistors, [2] where making accurate values is difficult, and selection from production runs, typically with a 10% tolerance, is the only practical way to get more accurate values. Carbon composition resistors have no place in audio circuitry because of their large temperature and voltage coefficients and high excess noise, but they live on in specialised applications such as switch-mode snubbing circuits, where their ability to absorb high peak power in bulk material rather than a thin film is useful, and in RF circuitry where the inductance of spiral-format film resistors is unacceptable.

So, having laid that fear to rest, what is the actual distribution of resistor values like? It is not easy to find out, as manufacturers are not very forthcoming with this sort of information, and measuring thousands of resistors with an accurate DVM is not a pastime that appeals to all of us. Any nugget of information in this area is therefore very welcome; Hugo Kroeze [3] reported the result of measuring 211 metal film resistors from the same batch with a nominal value of 10 kΩ and 1% tolerance. He concluded that:

1.  The mean value was 9.995 k Ω (0.05% low).
2.  The standard deviation was about 10 Ω, i.e. only 0.1%. This spread in value is surprisingly small (the resistors were all from the same batch, and the spread across batches widely separated in manufacture date might have been less impressive).
3.  All the resistors were within the 1% tolerance range.
4.  The distribution appeared to be Gaussian, with no evidence that it was a subset from a larger distribution.

I have added my own modest contribution of data to what we know. I measured 100 ordinary metal film 1kΩ resistors of 1% tolerance from Yageo, a reputable Chinese manufacturer, and very tedious it was too. If you think you can achieve some sort of Zen ecstasy by repeatedly measuring resistors, I can assure you it did not happen to me. I used a recently-calibrated 4.5 digit multimeter.

1.  The mean value was 997.66 Ohms (0.23% low).
2.  The standard deviation was 2.10 Ohms, i.e. 0.21%.
3.  All resistors were within the 1% tolerance range. All but one was within 0.5%, with the outlier at 0.7%.
4.  The distribution appeared to be Gaussian, with no evidence that it was a subset from a larger distribution.

These results are not as good as those obtained by Hugo Kroeze but reassuringly demonstrate that these ordinary commercial resistors were well within their specification, and using multiple parts to increase nominal accuracy should work well.

Now these are only two reports, and it would be nice to have more confirmation, but there seems to be no reason to doubt that the distribution of resistance values is Gaussian, though the range of standard deviations you are likely to meet remains enigmatic. Whenever I have attempted this kind of statistical improvement in accuracy, I have always found that the expected benefit really does appear in practice.

# Improving Accuracy With Multiple Components: Uniform Distribution

As I mentioned earlier, improving average accuracy by combining resistors does not depend on the resistance value having a Gaussian distribution. Even a batch of resistors with a uniform distribution gives better accuracy when two of them are combined. A uniform distribution of component values may not be likely, but the result of combining two or more of them is highly instructive, so stick with me for a bit.

Figure 15.2a shows a uniform distribution that cuts off abruptly at the limits L and−L and represents 10 kΩ resistors of 1% tolerance. We will assume again that we want to make a more accurate 20 kΩ resistance. If we put two of the uniform-distribution 10 kΩ resistors in series, we get not another uniform distribution, but the triangular distribution shown in Figure 15.2b. This shows that the total resistance values are already starting to cluster in the centre; it is possible to have the extreme values of 19.8 kΩ and 20.2 kΩ, but it is very unlikely.

Figure 15.2c shows what happens if we use more resistors to make the final value; when four are used, the distribution is already beginning to look like a Gaussian distribution, and as we increase the number of components to eight and then 16, the resemblance becomes very close.

Uniform distributions have a standard deviation just as Gaussian ones do. It is calculated from the limits L and−Las in Equation 15.2. Likewise, the standard deviation of a triangular distribution can be calculated from its limits L and −L, as in Equation 15.3.

$$\sigma = \frac{1}{\sqrt{3}} L \qquad\qquad 15.2$$

$$\sigma = \frac{1}{\sqrt{6}} L \qquad\qquad 15.3$$

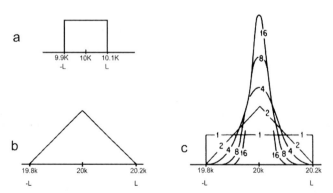

**Figure 15.2: How a uniform distribution of values becomes a Gaussian (normal) distribution when more component values are summed (after Smith).**

Applying Equation 15.2 to the uniformly-distributed 10 kΩ 1% resistors in Figure 15.2a, we get a standard deviation of 0.0577. Applying Equation 15.3 to the triangular distribution of 20 kΩ resistance values in Figure 15.2b, we get 0.0816. The mean value has doubled, but the standard deviation has less than doubled, so we get an improvement in average accuracy; the ratio is √2, just as it was for two resistors with a Gaussian distribution. This is also easy to demonstrate with the spreadsheet method described earlier.

## Obtaining Arbitrary Resistance Values

If you are making up a particular resistance value, the best resolution is obtained by using one large resistor value and one small one to make up the total, so for example 3100 Ω would be made up of 3000 Ω and 100 Ω; this is what you might call the asymmetrical solution. On the other hand, if it is possible to make up the required value with two resistors of roughly the same value, there is an advantage in terms of increased precision, for the statistical reasons described previously. In the optimal case, where the two resistors are equal, the average accuracy is improved by √2, but resistors in the ratio 2:1 or 3:1 still give a useful improvement.

The best procedure is therefore to determine at the beginning how accurate the resistor value needs to be, start with two near-equal resistor values, and if no combination can be found within our error window, we try values that are more and more unequal until we get a satisfactory result. As an example, we will assume that we are using E24 values, and our calculations show that 2090 Ω is the exact value required. The closest we can get with near-equal resistors in series is 1000 Ω + 1100 Ω = 2100 Ω, an error of 0.48%, which you may well be able to live with. If not, we try again using near-equal resistors in parallel, and we come up with 4300 Ω in parallel with 3900 Ω = 2045 Ω; this is in error by −2.2%, which is rather less appealing. Clearly the series option comes up with the closer answer in this case, but if 2100 Ω is not close enough, one option is to abandon completely the "near-equal" constraint and go for one large value and one small value. The obvious answer is 2000 Ω + 91 Ω = 2091 Ω, which is an error of only +0.048%, small enough to be utterly lost in other component tolerances. However, there is no improvement in precision. As we saw in Table 15.12, intermediate conditions between "near equal" and "one large and one small" will give intermediate improvements in precision.

The equivalent asymmetrical parallel-combination best approach is 2200 Ω in parallel with 43 k = 2092.9 Ω, which has an error of +0.14%, and so the series solution is in this case the better one.

A step-by step example of choosing parallel resistor combinations for the best possible increased precision while achieving a result within a specified percentage error window is given in Chapter 23, where a complete active crossover is designed.

Trying out the various resistor combinations on a hand calculator rapidly becomes life-threateningly tedious, even on my much-prized Casio FX-19 scientific calculator, now more than 30 years old but still working beautifully. [4] A spreadsheet is somewhat quicker once you've set it up, but the best answer is clearly some software that automatically comes up with the best answers. One example is resistor combination calculator utility called "ResCalc" by Mark Lovell and Morgan Jones, [5] which is superior to most as it allows mixed E24 and E96 series, but like all the other resistor-calculator applications I have seen, it does not appear to prioritise near-equal values to minimise tolerance errors.

A closer approach to the desired value while keeping the values near equal is possible by using combinations of three resistors rather than two. The cost is still low because resistors are cheap, but seeking out the best combination is an unwieldy process. By the time this book appears I hope to have a Javascript solution on my website. [6]

All of the statistical features described here apply to capacitors as well but are harder to apply because capacitors come in sparse value series, and it is also much more expensive to use multiple parts to obtain an arbitrary value. It is worth going to a good deal of trouble to come up with a circuit design that uses only standard capacitor values; the resistors will then almost certainly be all non-standard values, but this is easier and cheaper to deal with. A good example of the use of multiple parallel capacitors, both to improve accuracy and to make up values larger than those available, is the Signal Transfer RIAA preamplifier; [7] [8] see also the end of this book. This design is noted for giving very accurate RIAA equalisation at a reasonable cost. There are two capacitances required in the RIAA network, one being made up of four polystyrene capacitors in parallel and the other of five polystyrene capacitors in parallel. This also allows non-standard capacitance values to be used.

## Other Resistor Combinations

So far we have looked at serial and parallel combinations of components to make up one value, as in Figure 15.3a and 15.3b. Other important combinations are the resistive divider at Figure 15.5c, which is frequently used as the negative feedback network for non-inverting amplifiers. A two-output divider is shown at Figure 15.3d and a three-output divider at Figure 15.3e. Another important configuration is the inverting amplifier at Figure 15.5f, where the gain is set by the ratio R2/R1. All resistors are assumed to have the same tolerance about an exact mean value.

I suggest it is not obvious whether the divider ratio of Figure 15.5c, which is R2/(R1 + R2), will be more or less accurate than the resistor tolerance, even in the simple case with R1 = R2 giving a divider

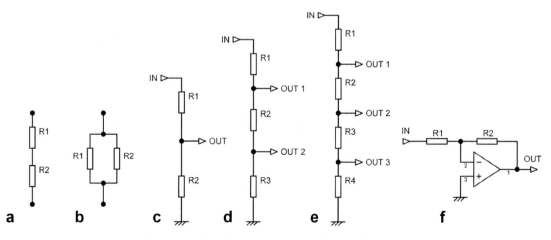

**Figure 15.3: Resistor combinations: (a) series; (b) parallel; (c) one-tap divider; (d) two-tap divider; (e) three-tap divider; (f) inverting amplifier.**

ratio of 0.5. However, the Monte Carlo method shows that in this case partial cancellation of errors still occurs, and the division ratio is more accurate by a factor of √2.

This factor depends on the divider ratio, as a simple physical argument shows:

- If the top resistor R1 is zero, then the divider ratio is obviously one with complete accuracy, the bottom resistor value is irrelevant, and the output voltage tolerance is zero.
- If the bottom resistor R2 is zero, there is no output and accuracy is meaningless, but if instead R2 is very small compared with R1, then R1 completely determines the current through R2, and R2 turns this into the output voltage. Therefore the tolerances of R1 and R2 act independently, and so the combined output voltage tolerance is worse by a factor of √2.

Some more Monte Carlo work, with 8000 trials per data point, revealed that there is a linear relationship between accuracy and the "tap position" of the output between R1 and R2, as shown in Figure 15.4. Plotting against division ratio would not give a straight line. With R1 = R2 the tap is at 50%, and accuracy improved by a factor of √2, as noted earlier. With a tap at about 30% (R1 = 7 kΩ,

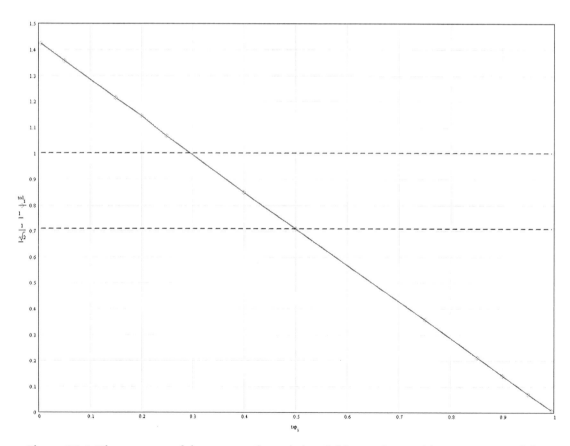

**Figure 15.4: The accuracy of the output of a resistive divider made up with components of the same tolerance varies with the divider ratio.**

R2 = 3 kΩ) the accuracy is the same as the resistors used. This assessment is *not* applicable to potentiometers, as the two sections of the pot are not uncorrelated in value—they are highly correlated.

The two-tap divider (Figure 15.3d) and three-tap divider(Figure 15.3e) were also given a Monte Carlo work-out, though only for equal resistors. The two-tap divider has an accuracy factor of 0.404 at OUT 1 and 0.809 at OUT 2. These numbers are very close to $\sqrt{2}/(2\sqrt{3})$ and $\sqrt{2}/(\sqrt{3})$ respectively. The three-tap divider has an accuracy factor of 0.289 at OUT 1, of 0.500 at OUT 2, and of 0.864 at OUT 3. The middle figure is clearly 1/2 (twice as many resistors as a one-tap divider, so $\sqrt{2}$ times more accurate), while the first and last numbers are very close to $\sqrt{3}/6$ and $\sqrt{3}/2$ respectively. It would be helpful if someone could prove analytically that the factors proposed are actually correct.

For the inverting amplifier of Figure 15.3f, the values of R1 and R2 are completely uncorrelated, and so the accuracy of the gain is always $\sqrt{2}$ worse than the tolerance of each resistor, assuming the tolerances are equal. The nominal resistor values have no effect on this. We therefore have the interesting situation that a non-inverting amplifier will always be equally or more accurate in its gain than an inverting amplifier. So far as I know this is a new result. Maybe it will be useful.

## Resistor Noise: Johnson and Excess Noise

All resistors, no matter what their method of construction, generate Johnson noise. This is white noise, which has equal power in equal absolute bandwidth, i.e. with the bandwidth measured in Hz, not octaves. There is the same noise power between 100 and 200 Hz as there is between 1100 and 1200 Hz. The level of Johnson noise that a resistor generates is determined solely by its resistance value, the absolute temperature (in degrees Kelvin), and the bandwidth over which the noise is being measured. For our purposes the temperature is 25 °C and the bandwidth is 22 kHz, so the resistance is really the only variable. The level of Johnson noise is based on fundamental physics and is not subject to modification, negotiation, or any sort of rule-bending . . . however, you might want to look up "load synthesis", in which a high-value resistor is made to act like a low-value resistor but still having the low Johnson current noise of the high-value resistor. [9] Sometimes Johnson noise from resistors places the limit on how quiet a circuit can be, though more often the noise from the active devices is dominant. It is a constant refrain in this book that resistor values should be kept as low as possible, without introducing distortion by overloading the circuitry, in order to minimise the Johnson noise contribution and the effects of opamp current noise.

The rms amplitude of Johnson noise is calculated from the classic equation:

$$v_n = \sqrt{4kTRB} \qquad\qquad 15.4$$

Where:

$v_n$ is the rms noise voltage      $T$ is absolute temperature in °K      $B$ is the bandwidth in Hz

$k$ is Boltzmann's constant      $R$ is the resistance in Ohms

The thing to be careful with here is to use Boltzmann's constant ($1.380662 \times 10^{-23}$), and NOT the Stefan-Boltzmann constant ($5.67\ 10^{-08}$), which relates to black-body radiation, has nothing to do with resistors, and will give some impressively wrong answers. The voltage noise is often left in its squared form for ease of RMS-summing with other noise sources.

The noise voltage is inseparable from the resistance, so the equivalent circuit is of a voltage source in series with the resistance present. Johnson noise is usually represented as a voltage, but it can also be treated as a Johnson noise current, by means of the Thevenin-Norton transformation, [10] which gives the alternative equivalent circuit of a current source in shunt with the resistance. The equation for the noise current is simply the Johnson voltage divided by the value of the resistor it comes from:

$$i_n = v_n/R. \hspace{4cm} 15.5$$

Excess resistor noise refers to the fact that some resistors, with a constant voltage drop across them, generate extra noise in addition to their inherent Johnson noise. This is a very variable quantity, but is essentially proportional to the DC voltage across the component; the specification is therefore in the form of a "noise index" such as "1 uV/V". The uV/V parameter increases with increasing resistor value and decreases with increasing resistor size or power dissipation capacity. Excess noise has a 1/f frequency distribution. It is usually only of interest if you are using carbon or thick-film resistors—metal film and wirewound types should have little or no excess noise. A rough guide to the likely range for excess noise specs is given in Table 15.13

One of the great benefits of opamp circuitry is that it allows simple dual-rail operation, and so it is noticeably free of resistors with large DC voltages across them; the offset voltages and bias currents involved are much too low to cause measurable excess noise. If you are designing an active crossover on this basis, then you can probably forget about the issue. If, however, you are using discrete transistor circuitry, it might possibly arise; specifying metal film resistors throughout, as you no doubt would anyway, should ensure you have no problems.

To get a feel for the magnitude of excess resistor noise, consider a 100 kΩ 1/4 W carbon film resistor with a steady 10 V across it. The manufacturer's data gives a noise parameter of about 0.7 uV/V, and so the excess noise will be of the order of 7 uV, which is −101 dBu. That could definitely be a problem in a low-noise preamplifier stage.

**Table 15.13: Resistor excess noise.**

| Type | Noise Index uV/V |
|---|---|
| Metal film TH | 0 |
| Carbon film TH | 0.2–3 |
| Metal oxide TH | 0.1–1 |
| Thin film SM | 0.05–0.4 |
| Bulk metal foil TH | 0.01 |
| Wirewound TH | 0 |

(Wirewound resistors are normally considered to be completely free of excess noise.)

# Resistor Non-Linearity

Ohm's law is, strictly speaking, a statement about metallic conductors only. It is dangerous to assume that it invariably applies exactly to resistors simply because they have a fixed value of resistance marked on them; in fact resistors—whose *raison d'etre* is packing a lot of controlled resistance in a small space—sometimes show significant deviation from Ohm's law in that current is not exactly proportional to voltage. This is obviously unhelpful when you are trying to make low-distortion circuitry. Resistor non-linearity is normally quoted by manufacturers as a voltage coefficient, usually the number of parts per million (ppm) that the resistance changes when 1volt is applied. The measurement standard for resistor non-linearity is IEC 6040.

The common through-hole metal film resistors show effectively perfect linearity at opamp signal levels. It can be a different matter when they have power amplifier output levels across them and temperature changes cause cyclic resistance variations, but this is not going to occur in active crossovers.

Wirewound resistors, both having voltage coefficients of less than 1 ppm, are also distortion free but are unlikely to be used in active crossovers, except perhaps for Subjectivist marketing purposes with cost no object.

Carbon film resistors, now almost totally obsolete, tend to be quoted at around 100 ppm; in many circumstances this is enough to generate more distortion than that produced by the active devices. Carbon composition resistors are of historical interest only so far as audio is concerned and have rather variable voltage coefficients in the area of 350 ppm, something that might be pondered by connoisseurs of antique amplifying equipment.

Today the most serious concern over resistor non-linearity is about thick-film surface-mount resistors, which have high and rather variable voltage coefficients; more on this later.

Table 15.14 (produced by SPICE simulation) gives the THD in the current flowing through the resistor for various voltage coefficients when a pure sine voltage is applied. If the voltage coefficient is significant, this can be a serious source of non-linearity.

**Table 15.14: Resistor voltage coefficients and the resulting distortion at +15 and +20 dBu.**

| Voltage | THD at | THD at |
|---|---|---|
| Coefficient | +15 dBu | +20 dBu |
| 1 ppm | 0.00011% | 0.00019% |
| 3 ppm | 0.00032% | 0.00056% |
| 10 ppm | 0.0016% | 0.0019% |
| 30 ppm | 0.0032% | 0.0056% |
| 100 ppm | 0.011% | 0.019% |
| 320 ppm | 0.034% | 0.060% |
| 1000 ppm | 0.11% | 0.19% |
| 3000 ppm | 0.32% | 0.58% |

The resistor is assumed to be symmetrical, and as a result it only generates odd harmonics, with level decreasing quickly as harmonic order increases. All of these harmonics rise proportionally as level increases. There is much more on this, and other related forms of distortion, in my power amplifier book. [11]

My own measurement setup is shown in Figure 15.5. The resistors are usually of equal value, to give 6 dB attenuation. A very low-distortion oscillator that can give a large output voltage is necessary; the results in Figure 15.6 were taken at a 10 Vrms (+22 dBu) input level. Here thick-film SM and through-hole resistors are compared. The gen-mon trace at the bottom is the record of the analyser reading the oscillator output and is the measurement floor of the AP System I used. The THD plot is higher than this floor, but this is not due to distortion. It simply reflects the extra Johnson noise generated by two 10 kΩ resistors. Their parallel combination is 5 kΩ, and so this noise is at −115.2 dBu. The SM plot, however, is higher again, and the difference is the distortion generated by the thick-film component.

For both thin-film and thick-film SM resistors non-linearity increases with resistor value and also increases as the physical size (and hence power rating) of the resistor shrinks. The thin-film versions are much more linear; compare Figures 15.7 and 15.8.

Sometimes it is appropriate to reduce the non-linearity by using multiple resistors in series. If one resistor is replaced by two with the same voltage coefficient in series, the THD in the current flowing is halved. Similarly, using three resistors reduces THD to a third of the original value. There are obvious

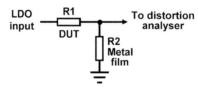

**Figure 15.5: Test circuit for measuring resistor non-linearity. The not-under-test resistor R2 in the potential divider must be a metal film type with negligible voltage coefficient.**

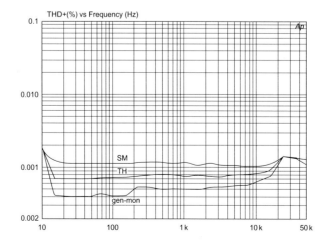

**Figure 15.6: SM resistor distortion at 10 Vrms input, using 10 kΩ 0805 thick-film resistors.**

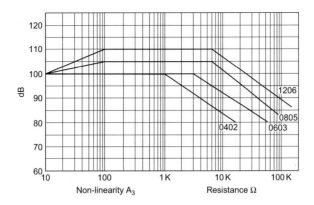

**Figure 15.7: Non-linearity of thin-film surface-mount resistors of different sizes. THD is here in dB rather than percent.**

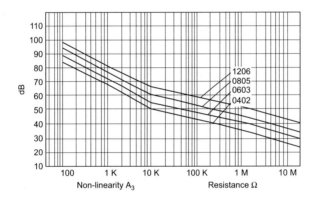

**Figure 15.8: Non-linearity of thick-film surface-mount resistors of different sizes.**

economic limits to this sort of thing, and it takes up PCB area, but it can be useful in specific cases, especially where the voltage rating of the resistor is a limitation.

## Capacitors: Values and Tolerances

The need for specific capacitor ratios creates problems, as capacitors are available in a much more limited range of values than resistors, usually the E6 series, running 10, 15, 22, 33, 47, 68. If other values are needed, then they often have to be made up of two capacitors in parallel; this puts up the cost and uses significantly more PCB area, so it should be avoided if possible, but for applications like Sallen & Key lowpass filters where a capacitance ratio of 2 is required, it is often the only way. Sallen & Key filters with equal component values can be used (see Chapters 8 and 9), but they must be configured to give voltage gain, which is often unwanted and will force compromises over either noise performance or headroom. Selecting convenient capacitor values will almost invariably lead to a need for non-standard resistance values, but this is much less of a problem, as combining two resistors to get the right value is much cheaper and uses less PCB area.

As for the case of resistors, when contriving a particular capacitor value, the best resolution is obtained by using one large value and one small one to make up the total, but two capacitors of approximately the same value give better average accuracy. When the two capacitors used are equal in nominal value, the accuracy is improved by $\sqrt{2}$. This effect is more important for capacitors because the cost premium for 1% parts is considerable, whereas for resistors it is very small, if it exists at all. When using multiple components like this to make up a value or improve precision, it is important to keep an eye on both the extra cost and the extra PCB area occupied.

The tolerance of non-electrolytic capacitors is usually in the range ±1% to ±10%; anything more accurate than this tends to be very expensive. Electrolytic capacitors used to have *much* wider tolerances, but things have recently improved, and ±20% is now common. This is still wider than any other component you are likely to use in a crossover, but this is not a problem, for as described later, it is most unwise to try to define frequencies or time-constants with electrolytic capacitors.

## Obtaining Arbitrary Capacitance Values

The problem of making up an arbitrary capacitance value is very different from making up a resistance values. Two things are different:

1.  Capacitors are often only available in the E3 or E6 series. Sometimes you can get E12, but rarely E24.
2.  Capacitors are significantly more expensive, so using them in multiples is less favourable economically.

There are often different options for capacitor values when designing active crossover filters. For example, you might start your calculations with 1 kΩ resistors, and you find that the biggest capacitor comes out at, say, 450 nF. That can be changed to give the same response using the E3 preferred value of 470 nF by scaling the resistors; here 1 kΩ scaled by 450/470 is 957 Ω. The 2xE24 parallel combination of 1600 Ω and 2400 Ω is only 0.31% high in nominal value. Since the resistor value is slightly reduced, you need to check that the loading on the previous stage has not become excessive.

Alternatively, you might decide to make up the capacitance with two 220 nF capacitors in parallel, to give an improvement in effective tolerance of $\sqrt{2}$. Now 440 nF is less than 450 nF, so to get the same response the resistors must be scaled up by 450/440, which works out as 1.022 kΩ. The 2xE24 parallel combination of 1300 Ω and 4700 Ω is only 0.36% low in nominal value.

Decisions like this can only be made intelligently if you know the cost of the various options. Resistors cost the same whatever their value, so that simplifies matters. The cost of a capacitor, however, is a strong function of its value. I have tried to give a handle on this in Table 15.15, which gives the relative cost of capacitors at 63 V DC rating and 2% tolerance. Prices were obtained for production quantities as close as possible to 1000-off. These figures are distinctly approximate, as it proved difficult to cross-reference various manufacturers, price breaks, and so on. Some things are, however, clear; for polyester, cost increases more slowly than proportionally with capacitance, very roughly in a square root law. Polypropylene cost increases faster with capacitance but still significantly more

**Table 15.15: Relative cost of polyester and polypropylene capacitors, relative to 100 nF polyester.**

|  | Manufacturer A | Manufacturer B | Manufacturer C | Manufacturer A |
|---|---|---|---|---|
| Dielectric | Polyester | Polyester | Polyester | Polypropylene |
| 100 nF | 1.00 | 1.00 | 1.00 | 9.42 |
| 220 nF | 1.18 | 1.76 | 1.55 | 25.66 |
| 470 nF | 1.55 | 2.25 | 2.38 | 31.58 |

slowly than proportionally. Therefore, making up a capacitance with two equal values, as in our example of 440 nF = 2 × 220 nF, is going to cost more than using a single 470 nF capacitor. The table also demonstrates that polypropylene capacitors cost a lot more than polyester, so the cost difference is much greater. In making your design decisions you will of course be working with prices from your intended supplier.

Polypropylene capacitors are much more expensive than polyester, but their great advantage is that they do not generate distortion (see later). In many filter types only one of the capacitors needs to be polypropylene for capacitor distortion to be eliminated, and this can save some serious money. This intriguing state of affairs is described in detail in Chapter 11.

## Capacitor Shortcomings

Capacitors fall short of being an ideal circuit element in several ways, notably leakage, equivalent series resistance (ESR), equivalent series inductance (ESL), dielectric absorption, and non-linearity: Capacitor leakage is equivalent to a high value resistance across the capacitor terminals, which allows a trickle of current to flow when a DC voltage is applied. Leakage is usually negligible for non-electrolytics, but is much greater for electrolytics. It is not normally a problem in audio design.

Equivalent series resistance (ESR) is a measure of how much the component deviates from a mathematically pure capacitance. The series resistance is partly due to the physical resistance of leads and foils and partly due to losses in the dielectric. It can also be expressed as tan-$\delta$ (tan-delta), which is the tangent of the phase angle between the voltage across and the current flowing through the capacitor. Once again it is rarely a problem in the audio field, the values being small fractions of an ohm and very low compared with normal circuit resistances.

Equivalent series inductance (ESL) is always present. Even a straight piece of wire has inductance, and any capacitor has lead-out wires and internal connections. The values are normally measured in nanohenries and have no effect in normal audio circuitry.

Dielectric absorption is a well-known effect; take a large electrolytic, charge it up, and then fully discharge it. Over a few minutes the charge will partially reappear. This "memory effect" also occurs in non-electrolytics to a lesser degree; it is a property of the dielectric and is minimised by using polystyrene, polypropylene, NP0 ceramic, or PTFE dielectrics. Dielectric absorption is invariably modelled by adding extra resistors and capacitances to an ideal main capacitor, such as Figure 15.9 for a 1

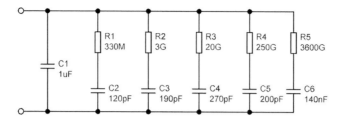

**Figure 15.9: A model of dielectric absorption in a 1 uF polystyrene capacitor.
All components are linear.**

uF polystyrene capacitor, which does not hint at any source of non-linearity. [12] However, the dielectric absorption mechanism does seem to have some connection with capacitor distortion, since the dielectrics that show the least dielectric absorption also show the lowest non-linearity. Dielectric absorption is a major consideration in sample-and-hold circuits but of no account in itself for normal linear audio circuitry; you can see from Figure 15.9 that the additional components are relatively small capacitors with very large resistances in series, and the effect these could have on the response of a filter is microscopic, and far smaller than the effects of component tolerances. Note that the model is just an approximation and is not meant to imply that each extra component directly represents some part of a physical process.

Capacitor non-linearity is the least known but by far the most troublesome of capacitor shortcomings. An RC lowpass filter to demonstrate the effect can be made with a series resistor and a shunt polyester capacitor, as in Figure 15.10, and if you examine the output with a distortion analyser, you will find to your consternation that the circuit is not linear. If the capacitor is a non-electrolytic type with a dielectric such as polyester, then the distortion is relatively pure third harmonic, showing that the effect is symmetrical. For a 10 Vrms input, the THD level may be 0.001% or more. This may not sound like much, but it is substantially greater than the midband distortion of a good opamp. The definitive work on capacitor distortion is a magnificent series of articles by Cyril Bateman in *Electronics World*. [13] The authority of this is underpinned by Cyril's background in capacitor manufacture.

Capacitors are used in audio circuitry for four main functions, where their possible non-linearity has varying consequences:

1.  Coupling or DC-blocking capacitors. These are usually electrolytics, and if properly sized have a negligible signal voltage across them at the lowest frequencies of interest. The non-linear properties of the capacitor are then unimportant unless current levels are high; power amplifier output capacitors can generate considerable midband distortion. [14] This makes you wonder what sort of non-linearity is happening in those big non-polarised electrolytics in passive crossovers. A great deal of futile nonsense has been talked about the mysterious properties of coupling capacitors, but it is all total twaddle. How could a component with negligible voltage across it put its imprint on a signal passing through it? For small-signal use, as long as the signal voltage across the coupling capacitor is kept low, non-linearity is not detectable by the best THD methods. The capacitance value is non-critical, as it has to be, given the wide tolerances of electrolytics.

2.  Supply filtering or decoupling capacitors. Electrolytics are used for filtering out supply rail ripple, etc, and non-electrolytics, usually around 100 nF, are used to keep the supply impedance low at high frequencies and thus keep opamps stable. The capacitance value is again non-critical.

3. For active crossover purposes, by far the most important aspect of capacitors is their role in active filtering. This is a much more demanding application than coupling or decoupling, for first, the capacitor value is now crucially important, as it defines the accuracy of the frequency response. Second, there is by definition a significant signal voltage across the capacitor, and so its non-linearity can be a serious problem. Non-electrolytics are always used in active filters, though sometimes a time-constant involving an electrolytic is ill-advisedly used to define the lower end of the system bandwidth; this is a very bad practice, because it is certain to introduce significant distortion at the bottom of the frequency range.

4. Small value ceramic capacitors are used for opamp compensation purposes, and sometimes in active filters in the HF path of a crossover. So long as they are NP0 (C0G) ceramic types, their non-linearity should be negligible. Other kinds of ceramic capacitor, using the XR7 dielectric, will introduce copious distortion and must never be used in audio paths. They are intended for high-frequency decoupling, where their linearity or otherwise is irrelevant.

## Non-Electrolytic Capacitor Non-Linearity

It has often been assumed that non-electrolytic capacitors, which generally approach an ideal component more closely than electrolytics and have dielectrics constructed in a totally different way, are free from distortion. It is not so. Some non-electrolytics show distortion at levels that is easily measured and can exceed the distortion from the opamps in the circuit. Non-electrolytic capacitor distortion is essentially third harmonic, because the non-polarised dielectric technology is basically symmetrical. The problem is serious, because non-electrolytic capacitors are commonly used to define time-constants and frequency responses (in RIAA equalisation networks, for example) rather than simply for DC blocking.

Very small capacitances present no great problem. Simply make sure you are using the C0G (NP0) type, and so long as you choose a reputable supplier, there will be no distortion. I say "reputable supplier" because I did once encounter some allegedly C0G capacitors from China that showed significant non-linearity. [15]

Middle-range capacitors, from 1 nF to 1 uF, present more of a problem. Capacitors with a variety of dielectrics are available, including polyester, polystyrene, polypropylene, polycarbonate, and polyphenylene sulphide, of which the first three are the most common (note that what is commonly called "polyester" is actually polyethylene terephthalate, PET).

Figure 15.10 shows a simple lowpass filter circuit which, with a good THD analyser, can be used to measure capacitor distortion. The values shown give a measured pole frequency, or −3 dB roll-off

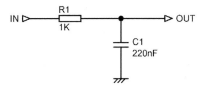

**Figure 15.10: Simple lowpass test circuit for non-electrolytic capacitor distortion; −3 dB at 723 Hz.**

point, at 723 Hz. We will start off with polyester, the smallest, most economical, and therefore the most common type for capacitors of this size.

The THD results for a microbox 220 nF 100 V capacitor with a polyester dielectric are shown in Figure 15.11, for input voltages of 10, 15 and 20 Vrms. They are unsettling.

The distortion is all third-harmonic. It peaks at around 300 to 400 Hz, well below the −3 dB frequency, and even with the input limited to 10 Vrms will exceed the non-linearity introduced by opamps such as the 5532 and the LM4562. Interestingly, the peak frequency changes with applied level. Below the peak, the voltage across the capacitor is roughly constant, but distortion falls as frequency is reduced, because the increasing impedance of the capacitor means it has less effect on a circuit node at a 1 kΩ impedance. Above the peak, distortion falls with increasing frequency because the lowpass circuit action causes the voltage across the capacitor to fall.

The level of distortion varies with different samples of the same type of capacitor; six of the type in Figure 15.11 were measured, and the THD at 10 Vrms and 400 Hz varied from 0.00128% to 0.00206%. This puts paid to any plans for reducing the distortion by some sort of cancellation method.

The distortion can be seen in Figure 15.11 to be a strong function of level, roughly tripling as the input level doubles. Third-harmonic distortion normally quadruples for doubled level, so there may well be an unanswered question here. It is however clear that reducing the voltage across the capacitor reduces the distortion. This suggests that if cost is not the primary consideration, it might be useful to put two capacitors in series to halve the voltage and the capacitance, and then double up this series combination to restore the original capacitance, giving the series-parallel arrangement in Figure 15.12. The results are shown in Table 15.16, and once more it can be seen that halving the level has reduced distortion by a factor of three rather than four.

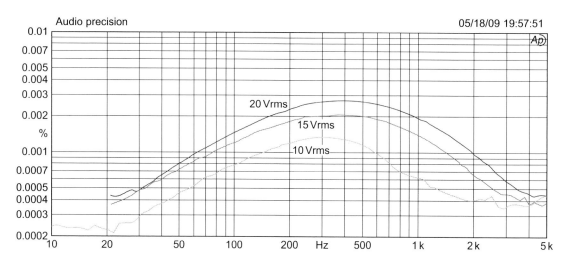

**Figure 15.11: Third-harmonic distortion from a 220 nF 100 V polyester capacitor, at 10, 15, and 20 Vrms input level, showing peaking around 400 Hz.**

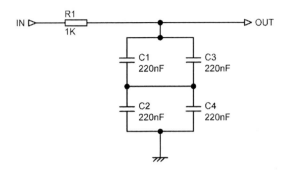

**Figure 15.12: Reducing capacitor distortion by series-parallel connection.**

**Table 15.16: The reduction of polyester capacitor distortion by series-parallel connection.**

| Input level Vrms | Single capacitor | Series-parallel capacitors |
|---|---|---|
| 10 | 0.0016% | 0.00048% |
| 15 | 0.0023% | 0.00098% |
| 20 | 0.0034% | 0.0013% |

The series-parallel arrangement has obvious limitations in terms of cost and PCB area occupied but might be useful in some cases. It has the advantage that, as described earlier in the chapter, using multiple components improves the average accuracy of the total value.

An unexpected complication in these tests was that every time a polyester sample was remeasured, the distortion was lower than before. I found a steady reduction in distortion over time; if a test signal was applied continuously, 9 Vrms at 1 kHz would roughly halve the THD over 11 hours. This change consists of both temporary and permanent components. When the signal is removed, over days the distortion increases again but never rises to its original value. When the signal is reapplied, the distortion slowly falls again. This self-improvement effect is therefore of little practical use, but if nothing else it demonstrates that polyester capacitors are more complicated than you might think. I do not believe this has anything to do with dim-witted audio commentators claiming that circuitry has to be left switched on for months or years before it attains its optimum sound; for one thing, nothing happens unless you also continuously apply a signal of much higher amplitude than normal, which would be hard to arrange and has never been touted as a good way to run-in equipment. For fuller details of this strange self-improvement business, see my *Linear Audio* article. [16]

Clearly polyester capacitors can generate significant distortion, despite their extensive use in audio circuitry of all kinds. The next dielectric we will try is polystyrene. Capacitors with a polystyrene dielectric are extremely useful for some filtering and RIAA equalisation applications because they can be obtained at a 1% tolerance at up to 10 nF at a reasonable price. They can be obtained in larger sizes but at much higher prices.

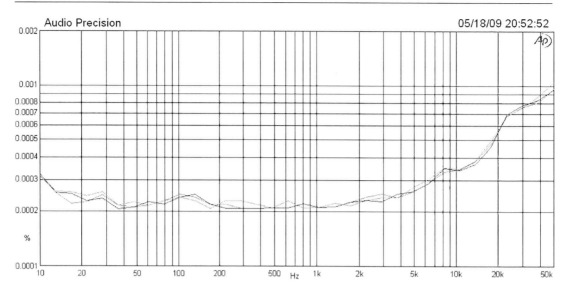

**Figure 15.13: The THD plot with three samples of 4n7 2.5% polystyrene capacitors, at 10 Vrms input level. The reading is entirely noise.**

The distortion test results are shown in Figure 15.13 for three samples of a 4n7 2.5% capacitor; the series resistor R1 has been increased to 4.7 kΩ to keep the −3 dB point inside the audio band, and it is now at 7200 Hz. Note that the THD scale has been extended down to a subterranean 0.0001%, because if it was plotted on the same scale as Figure 15.11 it would be bumping along the bottom of the graph. Figure 15.13 in fact shows no distortion at all, just the measurement noise floor, and the apparent rise at the HF end is simply due to the fact that the output level is decreasing because of the lowpass action, and so the noise floor is relatively increasing. This is at an input level of 10 Vrms, which is about as high as might be expected to occur in normal opamp circuitry. The test was repeated at 20 Vrms, which might be encountered in discrete circuitry, and the results were the same, yielding no measurable distortion.

The tests were done with four samples of 10 nF 1% polystyrene from LCR at 10 Vrms and 20 Vrms, with the same results for each sample. This shows that polystyrene capacitors can be used with confidence; this finding is in complete agreement with Cyril Bateman's results. [17]

Having resolved the problem of capacitor distortion below 10 nF, we need now to tackle it for larger capacitor values. Polyester having proven unsatisfactory, the next most common capacitor type is polypropylene, and I am glad to report that these are effectively distortion free in values up to 220 nF. Figure 15.14 shows the results for four samples of a 220 nF 250 V 5% polypropylene capacitor from RIFA. The plot shows no distortion at all, just the noise floor, with the apparent rise at the HF end being increasing relative noise due to the lowpass roll-off, as in Figure 15.13. This is also in agreement with Cyril Bateman's findings. [18] Rerunning the tests at 20 Vrms gave the same result— no distortion. This is very pleasing, but there is a downside. Polypropylene capacitors of this value and voltage rating are much larger physically than the commonly used 63 or 100 V polyester capacitor, and more expensive.

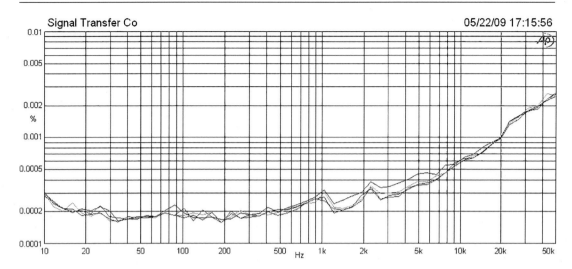

**Figure 15.14: The THD plot with four samples of 220 nF 250 V 5% polypropylene capacitors, at 10 Vrms input level. The reading is again entirely noise.**

It was therefore important to find out if the good distortion performance was a result of the 250 V rating, and so I tested a series of polypropylene capacitors with lower voltage ratings from different manufacturers. Axial 47 nF 160 V 5% polypropylene capacitors from Vishay proved to be THD-free at both 10 Vrms and 20 Vrms. Likewise, microbox polypropylene capacitors from 10 nF to 47 nF with ratings of 63 V and 160 V from Vishay and Wima proved to generate no measurable distortion, so the voltage rating appears not to be an issue. This finding is particularly important, because the Vishay range has a 1% tolerance, making them very suitable for precision filters and equalisation networks. The 1% tolerance is naturally reflected in the price.

The higher values of polypropylene capacitors (above 100 nF) appear to be currently only available with 250 V or 400 V ratings, and that means a physically big component. For example, the EPCOS 330 nF 400 V 5% part has a footprint of 26 mm by 6.5 mm, with a height of 15 mm, and capacitors like that take up a lot of PCB area. One way of dealing with this is to use a smaller capacitor in a capacitance multiplication configuration, so a 100 nF 1% component could be made to emulate 330 nF. It has to be said that this is only straightforward if one end of the capacitor is connected to ground; see Chapter 14 for an example of this concept applied to a biquad equaliser.

When I first started looking at capacitor distortion, I thought that the distortion would probably be lowest for the capacitors with the highest voltage rating. I therefore tested some RF-suppression X2 capacitors, rated at 275 Vrms, equivalent to a peak or DC rating of 389 V. The dielectric material is unknown. An immediate snag is that the tolerance is 10 or 20%, not exactly ideal for precision filtering or equalisation. A more serious problem, however, is that they are far from distortion free. Four samples of a 470 nF X2 capacitor showed THD between 0.002% and 0.003% at 10 Vrms. A high-voltage rating alone does not mean low distortion.

For more information on how capacitor non-linearity affects filters see Chapter 11.

## Electrolytic Capacitor Non-Linearity

Cyril Bateman's series in *Electronics World* [17][18] included two articles on electrolytic capacitor distortion. It proved to be a complex subject, and many long-held assumptions, such as "DC biasing always reduces distortion", were shown to be quite wrong (my own results confirm this—DC biasing is at best pointless and can increase distortion). The distortion levels Cyril measured were in general a good deal higher than for non-electrolytic capacitors, and I can confirm that too.

My view is that electrolytics should never, ever, under any circumstances, be used to set important time-constants in audio. There should be a dominant time-constant early in the signal path, based on a non-electrolytic capacitor, that determines the lower limit of the system bandwidth; preferably proper bandwidth definition should be implemented with a subsonic filter. All the electrolytic-based time-constants should be much longer so that the electrolytic capacitors can never have significant signal voltages across them and so never generate detectable distortion. Electrolytics have large tolerances and cannot be used to set accurate time-constants anyway.

However, even if you think you are following this plan, you can still get into trouble. Figure 15.15 shows a simple highpass test circuit representing an electrolytic capacitor in use for coupling or DC blocking. The load of 1 kΩ is the sort of value that can easily be encountered if you are using low-impedance design principles. The calculated −3 dB roll-off point is 3.38 Hz, so the attenuation at 10 Hz, at the very bottom of the audio band, will be only 0.47 dB; at 20 Hz it will be only 0.12 dB, which is surely a negligible loss. As far as frequency response goes, we are doing fine. But . . . examine Figure 15.16, which shows the measured distortion of this arrangement. Even if we limit ourselves to a 10 Vrms level, the distortion at 50 Hz is 0.001%, already above that of a good opamp. At 20 Hz it has risen to 0.01%, and at 10 Hz is a most unwelcome 0.05%. The THD is increasing by a ratio of 4.8 times for each octave fall in frequency, in other words increasing faster than a square law. The distortion residual is visually a mixture of second and third harmonic, and the levels proved surprisingly consistent for a large number of 47 uF 25 V capacitors of different ages and from different manufacturers.

Figure 15.16 also shows that the distortion rises rapidly with level; at 50 Hz, going from an input of 10 Vrms to 15 Vrms almost doubles the THD reading. To underline the point, consider Figure 15.17, which shows the measured frequency response of the circuit with 47 uF and 1 kΩ; note the effect of the capacitor tolerance on the real versus calculated response. The roll-off that does the damage, by allowing an AC voltage to exist across the capacitor, is very modest indeed, less than 0.2 dB at 20 Hz.

Having demonstrated how insidious this problem is, how do we fix it? As we have seen, changing capacitor manufacturer is no help. Using 47 uF capacitors of higher voltage also does not work—tests showed there is very little difference in the amount of distortion generated. An exception was the sub-miniature style of electrolytic, which was markedly more non-linear than standard types.

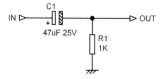

**Figure 15.15: Highpass test circuit for examining electrolytic capacitor distortion.**

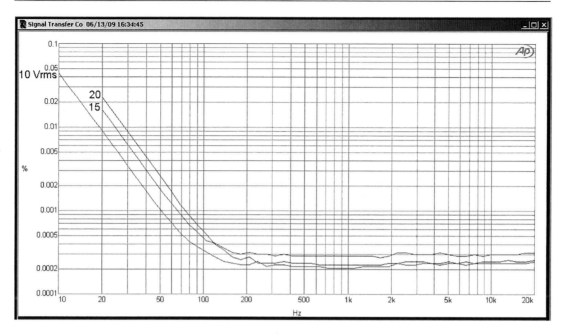

**Figure 15.16: Electrolytic capacitor distortion from the circuit in Figure 15.15.
Input level 10, 15, and 20 Vrms.**

The answer is simple—just make the capacitor bigger in value. This reduces the voltage across it in the audio band, and since we have shown that the distortion is a strong function of the voltage across the capacitor, the amount produced drops more than proportionally. The result is seen in Figure 15.18, for increasing capacitor values with a 10 Vrms input.

Replacing C1 with a 100 uF 25 V capacitor drops the distortion at 20 Hz from 0.0080% to 0.0017%, an improvement of 4.7 times; the voltage across the capacitor at 20 Hz has been reduced from 1.66 Vrms to 790 mVrms. A 220 uF 25 V capacitor reduces the voltage across itself to 360 mV and gives another very welcome reduction to 0.0005% at 20 Hz, but it is necessary to go to 1000 uF 25 V to obtain the bottom trace, which is the only one indistinguishable from the noise floor of the AP-2702 test system. The voltage across the capacitor at 20 Hz is now only 80 mV. From this data, it appears that the AC voltage across an electrolytic capacitor should be limited to below 80 mVrms if you want to avoid distortion. I would emphasise that these are ordinary 85°C rated electrolytic capacitors, and in no sense special or premium types.

This technique can be seen to be highly effective, but it naturally calls for larger and somewhat more expensive capacitors and larger footprints on a PCB. This can be to some extent countered by using capacitors of lower voltage, which helps to bring back down the CV product and hence the can size. I tested 1000 uF 16 V and 1000 uF 6V3 capacitors, and both types gave exactly the same results as the 1000 uF 25 V part in Figure 15.18, which seems to indicate that the maximum allowable signal voltage across the capacitor is an absolute value and not relative to the voltage rating. Naturally, 1000 uF 16 V and 1000 uF 6 V3 capacitors gave very useful reductions in CV product, can size, and

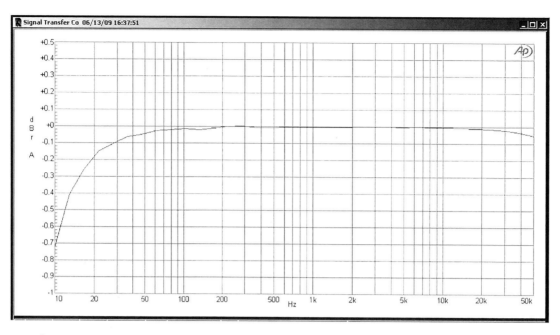

Figure 15.17: The measured roll-off of the highpass test circuit for examining electrolytic capacitor distortion.

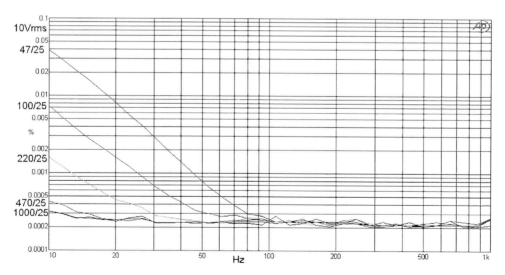

Figure 15.18: Reducing electrolytic capacitor distortion by increasing the capacitor value. Input 10 Vrms.

PCB area occupied. This does of course assume that the capacitor is, as is usual, being used to block small voltages from opamp offsets to prevent switch clicks and pot noises rather than for stopping a substantial DC voltage.

The use of large coupling capacitors in this way does require a little care, because we are introducing a long time-constant into the circuit. Most opamp circuitry is pretty much free of big DC voltages, but if there are any, the settling time after switch-on may become undesirably long.

# References

[1] Smith, Warren J. "Modern Optical Engineering" McGraw-Hill, 1990, ISBN: 0-07-059174-1, p. 484

[2] Johnson, Howard  www.edn.com/article/509250-7_solution.php, June 2010

[3] Kroeze, Hugo  www.rfglobalnet.com/forums/Default.aspx?gposts&m61096 (inactive) March 2002

[4] www.voidware.com/calcs/fx19.htm; Accessed January 2017

[5] Lovell, Mark and Jones, Morgan  www.pmillett.com/rescalc.htm

[6] Self, Douglas  www.douglas-self.com/

[7] Self, Douglas "Small Signal Audio Design" Second Edn, Focal Press, 2015, Chapter 8, pp. 265–268, ISBN 978-0-415-70974-3 hardback

[8] The Signal Transfer Company  www.signaltransfer.freeuk.com/

[9] Self, Douglas "Small Signal Audio Design Handbook" Second Edn, Focal Press, 2015, pp. 321–324, ISBN: 978-0-415-70974-3 hbk

[10] https://en.wikipedia.org/wiki/Th%C3%A9venin's_theorem; Accessed January 2017

[11] Self, Douglas "Audio Power Amplifier Design" Sixth Edn, Newnes, 2013, pp. 39–45, ISBN: 978-0-240-52613-3

[12] Kundert, Ken "Modelling Dielectric Absorption in Capacitors" www.designers-guide.org/Modeling/da.pdf, 2008

[13] Bateman, Cyril "Capacitor Sound?" Parts 1–6, Electronics World, July 2002–March 2003

[14] Self, Douglas "Audio Power Amplifier Design" Sixth Edn, Newnes, 2013, pp. 107–108, (amp output cap), ISBN: 978-0-240-52613-3

[15] Self, Douglas "Audio Power Amplifier Design" Sixth Edn, Newnes, 2013, pp. 299–300 (C0G cap), ISBN: 978-0-240-52613-3

[16] Self, Douglas "Self-Improvement for Capacitors" Linear Audio, Volume 1, April 2011, p156 ISBN 9–789490–929022

[17] Bateman, Cyril "Capacitor Sound?" Part 3. Electronics World, October 2002, pp. 16, 18

[18] Bateman, Cyril "Capacitor Sound?" Part 4, Electronics World, November 2002, p. 47

# CHAPTER 16

# *Opamps for Active Crossovers*

## Active Devices for Active Crossovers

It is a truth universally acknowledged, that if you are designing an active crossover, you will need active devices. In this day and age that almost always means opamps. Active crossovers can be built with discrete transistor circuitry if this is desirable for performance benefits or marketing reasons. This is relatively straightforward for Sallen & Key filters that require only a unity-gain buffer or voltage-follower, which can be implemented as some form of emitter-follower. It is a bit more complex for MFB filters and some equaliser circuits, which require a high-open-loop-gain inverting amplifier, and more complex again for configurations such as state-variable filters, which require multiple differential amplifiers (the first active crossover I designed for production was in fact a discrete transistor system, solely on the grounds of performance, for the affordable opamps of the time were really not very good). However, even if the crossover architecture is confined to Sallen & Key filters, matching the distortion performance of the newer opamps is going to be no easy matter. The noise performance of discrete circuitry should be slightly better, but in a typical application the differences will be marginal. Discrete transistor circuitry undoubtedly scores on the issue of headroom, because you can use supply rails that are pretty much as high as you like; the downside is that the rail voltages have to be increased considerably to get a meaningful increase in headroom when it is expressed in dB, and equipment capable of producing very high output voltages can be dangerous to other parts of the system if the output levels are mismanaged. While the study of suitable discrete circuitry for active crossovers would be fascinating, its doubtful utility means that space cannot be given to it here, and this chapter concentrates solely on opamps.

It is of course also possible to make active crossovers using valves. Given the large number of active elements required in a high-quality crossover and the necessity of working at high impedance levels with accompanying higher noise, this is for most of us a truly unattractive proposition. If you are one of those who insist on using directly heated triodes left over from WWI, then the prospect becomes quite surreal.

An active crossover is an optional part of an audio system, in the sense that you could use passive crossovers instead. If you are inserting an extra piece of equipment into the audio path, especially one that requires a good deal of expenditure on additional power amplifiers and so on, it really makes sense that it should be of high quality. The cost of a mediocre active crossover is not going to be radically different from that of a really good one, the main differences being the cost of the opamps, the cost of the capacitors (there is a great deal on this issue in Chapters 8, 9, and 15), and perhaps the cost of balanced outputs with their associated connectors. In active crossovers there are likely to be four or more filter stages in succession, so the quality of each one must be high. For these reasons this chapter focuses on achieving the best possible performance rather than cutting the last penny off the costing sheet.

## Opamp Types

You might be questioning how a discourse on opamps for active crossovers will differ from one on opamps for general use in audio circuitry. One issue is that active crossover design tends to make great use of Sallen & Key filters, which rely on voltage-followers as the active part of the circuitry. Opamps in voltage-followers work under the most demanding conditions possible as regards common-mode (CM) distortion, and this topic, which gets little attention in most textbooks, is therefore examined in detail here.

General audio design has for a great many years relied on a very small number of opamp types; the dual-opamp TL072, the single-opamp 5534, and the dual-opamp 5532 dominated the audio small-signal scene for a long time. The TL072, with its JFET inputs, was used wherever its negligible input bias currents, lower power consumption, and lower cost were important. For a long time the 5534 and 5532 were much more expensive than the TL072, so the latter was used wherever feasible in an audio system, despite its markedly inferior noise, distortion, and load-driving capabilities. The 5534 or 5532 was reserved for critical parts of the circuitry. Although it took many years, the price of the 5534 or 5532 is now down to the point where it is usually the cheapest opamp you can buy, and its cost/performance ratio is outstanding. This is believed to be due to its massive use in CD player output stages. You need a very good reason to choose any other type of opamp for audio work.

The TL072 and the 5532 are dual opamps; the single equivalents are TL071 and 5534. Dual opamps are used almost universally, as the package containing two is usually cheaper than the package containing one, simply because it is more popular. The 5534 also requires an external compensation capacitor for closed-loop gains of less than 3, which adds to the cost.

It took a long time for better opamps for audio to come along. Some were marketed specifically for audio applications, such as the idiosyncratic OP275, which had both BJT- and JFET-input devices. Unfortunately it had higher noise and higher distortion than a 5532 and cost six times as much, so it made little headway. It is only in the last few years that opamps have appeared that have better performance than the 5532. Notable examples are the LM4562 and the LME49990, and these are examined in this chapter.

There are many opamps on the market which could be applied to audio purposes, and to deal with them all in detail would fill this book, so only a selected range is covered here. Samuel Groner [1] has measured a wide range of opamp types, and his published measurements should be your first recourse if your favourite opamp is not included here.

## Opamp Properties: Noise

Table 16.1 ranks the opamps most commonly used for audio in order of voltage noise. The great divide is between JFET-input opamps and BJT (bipolar junction transistor) input opamps. The JFET opamps have more voltage noise but less current noise than bipolar input opamps, the TL072 being particularly voltage noisy. The BJT opamps have the lowest voltage noise, but the current noise is much higher. The difference in voltage noise between a modern JFET-input opamp such as the OPA2134 and the old faithful 5532 is only 4 dB, but the JFET part is a good deal more costly.

It is important to realise that the voltage noise, like the poor, is always with you, but the effect of the current noise is negotiable in that it only becomes measurable and audible voltage noise when it flows through an impedance. Current noise can therefore be reduced by using suitably low circuit impedances, and here the ability of the opamp to drive heavy loads without increased distortion becomes very important. This noise advantage is one reason why BJT opamps are a better choice for high-quality circuitry where the impedance levels can be chosen.

Looking at the range of BJT opamps, the current noise increases as the voltage noise decreases, as both depend on the standing current in the input devices, so parts like the AD797 and LME49990 are definitely best suited to low-impedance circuitry. The LM4562 has almost 6 dB less voltage noise than the 5532, while both the LME49990 and the AD797 are almost 15 dB quieter, given sufficiently low impedance levels that their high current noise does not intrude.

A further complication in this noise business is that the OP27, LT1115, and LT1028 devices have bias-cancellation systems which can cause unexpectedly high noise, for reasons described later.

Opamps with bias-cancellation circuitry are often unsuitable for audio use due to the extra noise the circuitry creates. The amount depends on circuit impedances and is not taken into account in Table 16.1.

**Table 16.1: Opamps ranked by typical voltage noise density.**

| Opamp | $e_n$ nV/rtHz | $i_n$ pA/rtHz | Input device type | Bias cancel? |
|---|---|---|---|---|
| TL072 | 18 | 0.01 | JFET | No |
| OPA604 | 11 | 0.004 | JFET | No |
| NJM4556 | 8 | Not spec'd | BJT | No |
| OPA2134 | 8 | 0.003 | JFET | No |
| LME49880 | 7 | 0.006 | JFET | No |
| OP275 | 6 | 1.5 | BJT+FET | No |
| OPA627 | 5.2 | 0.0025 | DIFET | No |
| 5532A | 5 | 0.7 | BJT | No |
| LM833 | 4.5 | 0.7 | BJT | No |
| MC33078 | 4.5 | 0.5 | BJT | No |
| 5534A | 3.5 | 0.4 | BJT | No |
| OP270 | 3.2 | 0.6 | BJT | No |
| OP27 | 3 | 0.4 | BJT | **YES** |
| LM4562 | 2.7 | 1.6 | BJT | No |
| LME49710 | 2.5 | 1.6 | BJT | No * |
| LME49990 | 0.9 | 2.8 | BJT | No * |
| AD797 | 0.9 | 2 | BJT | No |
| LT1115 | 0.85 | 1 | BJT | No * |
| LT1028 | 0.85 | 1 | BJT | **YES** |

*Not directly stated on data sheet but inferred from specifications.
NB: A DiFET is a dielectrically isolated FET, i.e. a MOSFET rather than a JFET.

## Opamp Properties: Slew Rate

Slew rates vary more than most parameters; a range of 23:1 is shown here in Table 16.2. A maximum slew rate greatly in excess of what is required appears to confer no benefits whatever.

The 5532 slew rate is typically ±9 V/us. This version is internally compensated for unity-gain stability, not least because there are no spare pins for compensation when you put two opamps in an 8-pin dual package. The single-amp version, the 5534, can afford a couple of compensation pins and so is made to be stable only for gains of 3x or more. The basic slew rate is therefore higher, at ±13 V/us.

Compared with power amplifier specs, which often quote 100 V/us or more, these speeds may appear rather sluggish. In fact they are not; even ±9 V/us is more than fast enough. Assume you are running your opamp from ±18 V rails and that it can give a ±17 V swing on its output. For most opamps this is distinctly optimistic, but never mind. To produce a full-amplitude 20 kHz sine wave you only need 2.1 V/us, so even in the worst case there is a safety margin of at least four times. Such signals do not of course occur in actual use, as opposed to testing. More information on slew-limiting is given in the section on opamp slew-limiting distortion.

Table 16.2: Opamps ranked by typical slew rate.

| Opamp | V/us |
| --- | --- |
| OP270 | 2.4 |
| OP27 | 2.8 |
| NJM4556 | 3 |
| MC33078 | 7 |
| LM833 | 7 |
| 5532A | 9 |
| LT1028 | 11 |
| TL072 | 13 |
| 5534A | 13 |
| LT1115 | 15 |
| LME49880 | 17 |
| LME49710 | 20 |
| OPA2134 | 20 |
| LM4562 | 20 |
| AD797 | 20 |
| OP275 | 22 |
| LME49990 | 22 |
| OPA604 | 25 |
| OPA627 | 55 |

## Opamp Properties: Common-Mode Range

This is simply the range over which the inputs can be expected to work as proper differential inputs. It usually covers most of the range between the rail voltages, with one notable exception. The data sheet for the TL072 shows a common-mode (CM) range that looks a bit curtailed at −12 V. This bland figure hides the deadly trap this IC contains for the unwary. Most opamps, when they hit their CM limits, simply show some sort of clipping. The TL072, however, when it hits its negative limit, promptly inverts its phase, so your circuit either latches up or shows nightmare clipping behaviour, with the output bouncing between the two supply rails. The positive CM limit is in contrast trouble-free. This behaviour can be especially troublesome when TL072s are used in highpass Sallen & Key filters; they are definitely not recommended for this role.

## Opamp Properties: Input Offset Voltage

A perfect opamp would have its output at 0 V when the two inputs were exactly at the same voltage. Real opamps are not perfect, and a small voltage difference—usually a few millivolts—is required to zero the output. These voltages are large enough to cause switches to click and pots to rustle, and DC-blocking capacitors are very often required to keep them in their place.

The typical offset voltage for the 5532A is ±0.5 mV typical, ±4 mV maximum at 25 °C; the 5534A has the same typical spec but a lower maximum, at ±2 mV. The input offset voltage of the new LM4562 is only ±0.1 mV typical, ±4 mV maximum at 25 °C.

## Opamp Properties: Bias Current

Bipolar-input opamps not only have larger noise currents than their JFET equivalents, they also have much larger bias currents. These are the base currents taken by the input transistors. This current is much larger than the input offset current, which is the difference between the bias current for the two inputs. For example, the 5532A has a typical bias current of 200 nA, compared with a much smaller input offset current of 10 nA. The LM4562 has a lower bias current of 10 nA typical, 72 nA maximum. In the case of the 5532/4 the bias current flows into the input pins, as the input transistors are NPN.

Bias currents are a considerable nuisance when they flow through variable resistors; they make them noisy when moved. They will also cause significant DC offsets when they flow through high-value resistors.

It is often recommended that the effect of bias currents can be cancelled out by making the resistance seen by each opamp input equal. Figure 16.1a shows a shunt-feedback stage with a 22 kΩ feedback resistor. When 200 nA flows through this, it will generate a DC offset of 4.4 mV, which is a good deal more than we would expect from the input offset voltage error.

If an extra resistance Rcompen of the same value as the feedback resistor is inserted into the non-inverting input circuit, then the offset will be cancelled. This strategy works well and appears to be done almost automatically by some designers. However, there is a snag. The resistance Rcompen

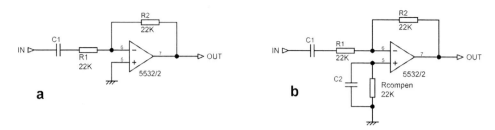

**Figure 16.1: Compensating for bias current errors in a shunt-feedback stage. The compensating resistor must be bypassed by a capacitor C2 to prevent it adding Johnson noise to the stage.**

generates extra Johnson noise, and to prevent this it is necessary to shunt the resistance with a capacitor, as in Figure 16.1b. This extra component costs money and takes up PCB space, so it is questionable if this technique is actually very useful for audio work. It is usually more economical to allow offsets to accumulate in a chain of opamps and then remove the DC voltage with a single output blocking capacitor. This assumes that there are no stages with a large DC gain and that the offsets are not large enough to significantly reduce the available voltage swing. Care must also be taken if controls are involved, because even a small DC voltage across a potentiometer will cause it become crackly, especially as it wears.

FET-input opamps have very low bias current at room temperature; however, it doubles for every 10°-Centigrade rise. This is pretty unlikely to cause trouble in most audio applications, but a combination of high internal temperatures (such as inside a big mixing console)and high-value pots can lead to some unexpected rustling and crackling noises.

## Opamp Properties: Cost

While it may not appear on the data sheet, the price of an opamp is obviously a major factor in deciding whether or not to use it. Table 16.3 lists the opamps commonly used in audio, and was derived from the averaged prices for 25+ quantities across a number of UK distributors. At the time of writing (2010) the 5532 was not only by far the most popular audio opamp but also the cheapest, so its price was taken as unity and used as the basis for the price ratios given.

The table is ranked by cost per package, but of course some are dual and some are single packages. The column on the right shows cost per opamp, with the singles bolded. This shows that some parts are relatively very expensive indeed, costing almost 100 times as much per opamp as a 5532.

Table 16.3 was compiled using prices for DIL packaging and the cheapest variant of each type. It is obviously only a rough guide. Purchasing in large quantities or in different countries may change the rankings somewhat, but the basic look of things will not alter too much. One thing is obvious—the 5532 is one of the great opamp bargains of all time.

Active crossovers are not going to be used in low-cost systems where every penny counts; they are found in high-end hi-fi systems and professional PA rigs, and so they are expected to have

**Table 16.3: Opamps ranked by price per package (2010) relative to the 5532.**

| Opamp | Format | Price ratio per package 25+ | Price ratio per Opamp 25+ |
|---|---|:---:|:---:|
| 5532 | Dual | **1.00** | **1.00** |
| TL072 | Dual | 1.00 | 1.00 |
| LM833 | Dual | 1.12 | 1.12 |
| MC33078 | Dual | 1.27 | 1.27 |
| 5534A | Single | 1.52 | **3.04** |
| TL052 | Dual | 2.55 | 2.55 |
| OP275GP | Dual | 3.42 | 3.42 |
| OPA2134PA | Dual | 4.45 | 4.45 |
| OPA604 | Dual | 5.03 | 5.03 |
| OP27 | Single | 6.76 | **13.52** |
| LM4562 | Dual | 9.06 | 9.06 |
| LME49990 | Single | 9.58 | 19.15 |
| LT1115 | Single | 12.73 | **25.45** |
| AD797 | Single | 13.09 | **26.18** |
| LT1028 | Dual | 17.88 | 17.88 |
| OPA270 | Single | 24.42 | **48.85** |
| LME49710 | Single | 30.58 | **61.15** |
| OPA627 | Single | 48.42 | **96.85** |

(No representative UK distributor was found for the LME49880 at time of writing.)

appropriately good performance. There is usually no point in compromising this to make minor cost savings. However, as we shall see, there is in fact very little in the way of agonising dilemmas about cost/performance tradeoffs. The 5532/5534 may be the cheapest opamp, but it is also an exceptionally good one. When first introduced it was very much an expensive premium part, but its popularity has driven prices down. When used with a little care (and I shall be going into that in detail in this chapter) it is capable of a performance that can only be beaten by using parts that cost ten times as much, such as the LM4562.

## Opamp Properties: Internal Distortion

This is what might be called the basic distortion produced by the opamp you have selected. Sam Groner calls it "transfer distortion". [1] Even if you scrupulously avoid clipping, slew-limiting, common-mode issues, and excessive output loading, opamps are not completely distortion free, though some types such as the 5532, the LM4562, and the LME49990 do have very low levels indeed. If distortion appears when the opamp is run with shunt feedback, to prevent common-mode voltages on the inputs, and with very light output loading, then it is probably wholly internal, and all you can do is

(a) run the opamp at the maximum safe supply rails (this will improve linearity but not usually by very much) or (b) pick a better opamp.

If the distortion is higher than expected, the cause may be internal instability provoked by putting a capacitive load directly on the output, or neglecting the supply decoupling. The classic example of the latter effect is the 5532, which shows high distortion if there is not a capacitor across the supply rails close to the package; 100 nF is always adequate in my experience. No actual HF oscillation is visible on the output with a general-purpose oscilloscope, so the problem is likely be instability in one of the intermediate gain stages.

## Opamp Properties: Slew Rate Limiting Distortion

This is essentially an overload condition, and it is the designer's responsibility to make sure it never happens. If users crank up the gain until the signal is within a hair of clipping, they should still be able to assume that slew-limiting will never occur, even with aggressive material full of high frequencies.

Arranging this is not too much of a problem. If the rails are set at the usual maximum voltage, i.e. ±18 V, then the maximum possible signal amplitude is 12.7 Vrms, ignoring the saturation voltages of the output stage. To reproduce this level cleanly at 20 kHz requires a minimum slew rate of only 2.3 V/usec. Most opamps can do much better than this, though with the OP27 (2.8 V/usec) you are sailing rather close to the wind. This calculation obviously assumes that the incoming signal is free of ultrasonic signals at any significant level; these may not be audible, but if they are large enough to provoke slew-limiting, there will be severe intermodulation distortion in the audio band. Bandwidth definition filters at the start of the audio chain will stop such unwanted signals.

Horrific as it may now appear, audio paths full of LM741s were quite common in the early 1970s. Entire mixers were built with no other active devices, and what complaints there were tended to be about noise rather than distortion. The reason for this is that full-level signals at 20 kHz simply do not occur in reality; the energy at the HF end of the audio spectrum is well known to be lower than that at the bass end.

This assumes that slew-limiting has an abrupt onset as level increases, rather like clipping. This is in general the case with opamps. As the input frequency rises and an opamp gets closer to slew-limiting, the input stage is working harder to supply the demands of the compensation capacitance. There is an absolute limit to the amount of current this stage can supply, and when you hit it the distortion shoots up, much as it does when you hit the supply rails and induce voltage clipping. Before you reach this point, the linearity may be degraded, but usually only slightly until you get close to the limit. It is not normally necessary to keep big margins of safety when dealing with slew-limiting. If you are employing the Usual Suspects in the audio opamp world—the 5532 and TL072, with maximal slew rates of 9 and 13 V/usec respectively—you are most unlikely to suffer any slew rate non-linearity.

## Opamp Properties: Distortion Due to Loading

Output stage distortion is always worse with heavier output loading because the increased currents flowing exacerbate the gain changes in the Class-B output stage. These output stages are not in general individually trimmed for optimal quiescent conditions (as are audio power amplifiers), and so the crossover distortion produced by opamps both tends to be higher and can be more variable between different specimens of the same chip. On the other hand, the intimate contact between biasing circuits and the output transistors means that it is possible to make the quiescent conditions very stable against temperature, especially when compared with discrete-component power amplifiers, where the thermal losses and lags are much greater. Distortion increases with loading in different ways for different opamps. It may rise only at the high-frequency end, or there may be a general rise at all frequencies. Often both effects occur, as in the TL072 and the 5532.

The lowest load that a given opamp can be allowed to drive is an important design decision. It will typically be a compromise between the distortion performance required and opposing factors such as the number of opamps in the circuit, cost of load-capable opamps, and so on. It even affects noise performance, for the lower the load resistance an amplifier can drive, the lower the resistance values in the negative feedback can be, and hence the lower the Johnson noise they generate. There are limits to what can be done in noise reduction by this method, because Johnson noise is proportional to the square root of circuit resistance and so improves only slowly as opamp loading is increased. Voltage noise from the opamps is not reduced at all, but the effect of current noise falls proportionally to the value of the circuit resistances. Overall the sum of these contributions decreases at a fairly gentle rate. Opamp distortion, however, at least in the case of the ubiquitous 5532, tends to rise rapidly when the loading exceeds a certain amount, and a careful eye needs to be kept on this issue. More modern devices such as the AD797 and the LM4562 handle heavy loading better, with less increase in distortion, and the very recent LME49990 barely reacts at all to loads down to 500 Ω (see Figure 16.19).

## Opamp Properties: Common-Mode Distortion

This is the general term for extra distortion that appears when there is a large signal voltage on both the opamp inputs. The voltage difference between these two inputs will be very small, assuming the opamp is in its linear region, but the common-mode (CM) voltage can be a large proportion of the available swing between the rails, and in the case of the voltage-follower will equal it.

Common-mode distortion appears to be the least understood distortion mechanism, and it gets little or no attention in opamp books, but it is actually one of the most important influences on opamp linearity. It is simple to separate this effect from the basic forward-path distortion by comparing THD performance in series and shunt-feedback modes; this should be done at the same noise gain. Bear in mind that a voltage-follower has a noise gain of 1, while a unity-gain inverting stage has a noise gain of 2. The distortion is often a good deal lower for the shunt-feedback case, where there is no common-mode voltage. BJT- and JFET-input opamps show rather different behaviour as regards common-mode distortion, and this is an important difference between the two types. This fact seems to be very little appreciated.

A BJT opamp requires both a CM voltage and a significant source resistance driving an input for it to generate CM distortion. While this is somewhat speculative, my hypothesis is that this is due to Early effect occurring in the input stage when there is a large CM voltage, modulating the high input bias currents, and this is the cause of the distortion (Early effect occurs in a bipolar transistor when changes in its collector-emitter voltage cause changes in the collector current, even though the base-emitter voltage is constant). The signal input currents, which are in general non-linear, are much smaller due to the high open-loop gain of the opamp and appear to have a negligible effect. There is more on this in the section on the CM distortion behaviour of the 5532.

With JFET inputs the problem is not the input bias currents of the input devices themselves, which are quite negligible, but the currents drawn by the non-linear junction capacitances inherent in field-effect devices. These capacitances are effectively connected to one of the supply rails. For P-channel JFETs, as used in the input stages of most JFET opamps, the important capacitances are between the input JFETs and the substrate, which is normally connected to the V-rail. See for example Jung. [2] According to the Burr-Brown data sheet for the OPA2134,"The P-channel JFETs in the input stage exhibit a varying input capacitance with applied CM voltage." It goes on to recommend that the input impedances should be matched if they are above 2 kΩ, to cancel out the non-linearity.

The amount of CM distortion generated by a given type of opamp is very important for our purposes here because active crossover circuitry very often uses voltage-followers, primarily in Sallen & Key filters but also for general buffering purposes. This configuration is the worst case for CM distortion because the full output voltage appears as a CM signal on the inputs. With BJT opamps, CM distortion can be rendered negligible by keeping the source impedances low, which is also a very good idea as it reduces noise and susceptibility to electrostatic interference.

## Opamps Surveyed

As we have seen, opamps with JFET inputs tend to have higher voltage noise but lower current noise than BJT-input types and therefore give a better noise performance with high source resistances. Their very low bias currents often allow circuitry to be simplified by omitting DC blocking. These advantages are however of little use in active crossover circuitry. Their CM distortion performance also tends to be much worse, and for this reason, and to economise on space, I shall only look at one JFET opamp before examining some BJT types in detail.

Figure 16.2 shows some of the test circuits used in this chapter.

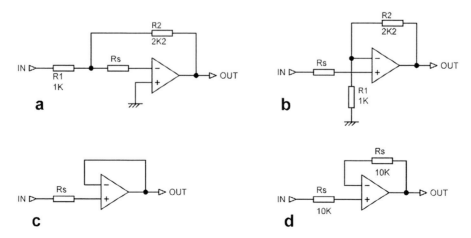

**Figure 16.2: Opamp test circuits with added source resistance Rs: (a) shunt; (b) series; (c) voltage follower; (d) voltage-follower with cancellation resistor in feedback path.**

# The TL072 Opamp

The TL072 is, or perhaps was, one of the most popular opamps, having very high-impedance inputs with effectively zero bias and offset currents. The JFET-input devices give their best noise performance at medium impedances, in the range 1 kΩ–10 kΩ. It has a modest power consumption, at typically 1.4 mA per opamp section, which is significantly less than the 5532. The slew rate is higher than for the 5532, at 13 V/us against 9 V/us. The TL072 is a dual opamp; there is a single version called the TL071 which has offset null pins. Nowadays the TL072 is regarded as somewhat obsolescent, as a result of its high voltage noise and mediocre load-driving capabilities, but it still gives a good illustration of JFET opamp issues.

Figure 16.3 shows the distortion performance of the TL072 in voltage-follower mode as in Figure 16.2c, with varying extra source resistance in the input path. With a low driving impedance the THD at 10 kHz is 0.0025%, but with a 10 kΩ source resistance inserted, this rises to an alarming 0.035%. Note that the flat parts of the traces to the left are not the noise floor, as would be the case with more modern opamps; it is real distortion at about 0.0006%. The noise floor is equivalent to 0.00035%. Be aware that the test level is almost as high as possible at 10 Vrms, to get the low-frequency distortion clear of the noise floor; practical internal levels such as 3 Vrms will give much lower levels of distortion. This applies to all the distortion tests in this chapter, and indeed to the whole book.

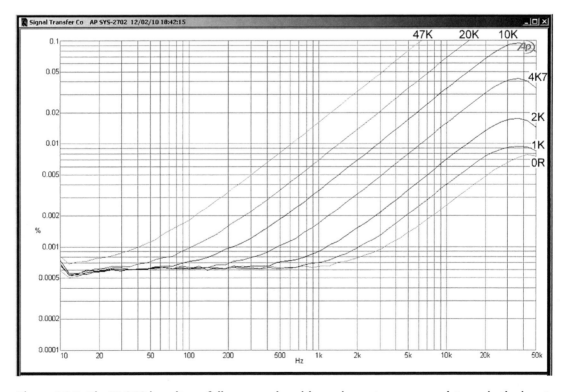

**Figure 16.3: The TL072 in voltage-follower mode, with varying extra source resistance in the input path. CM distortion is much higher than for the BJT opamps. 10 Vrms out, ±18 V supply rails.**

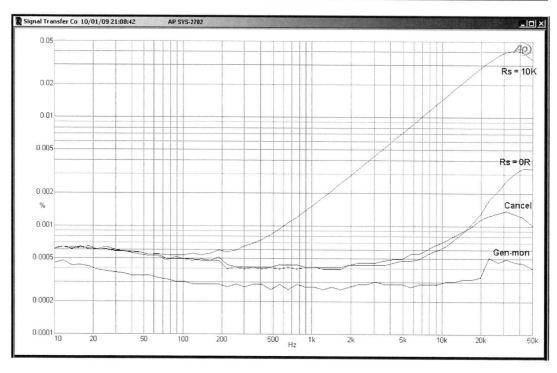

**Figure 16.4: A TL072 voltage-follower driven from a low source resistance produces reasonably low distortion (Rs = 0R), but adding a 10 kΩ source resistance makes things much worse (Rs = 10k). Putting a 10 kΩ resistance in the feedback path as well gives complete cancellation of this extra distortion (Cancel). Output 5 Vrms, supply ±18 V.**

Figure 16.3 should be compared with the same test carried out on BJT opamps; see Figure 16.10 (5532) where for Rs = 10 kΩ at 10 kHz the THD is 0.0015% against 0.035% for the TL072. Figure 16.19 (LM4562) shows 0.0037%, Figure 16.24 (LME49990) shows 0.024%, and Figure 16.27 (AD797) shows 0.0025%. Clearly the TL072 has far worse CM distortion than all except the LME49990, and that opamp is redeemed by its very low voltage noise.

Since the inputs of an opamp are nominally identical, it is possible to greatly reduce the effects of CM distortion by making both inputs see the same source impedance. In the voltage-follower case, this can be done by inserting an equal resistance in the feedback path, as in Figure 16.2d. Figure 16.4 shows how the high level of CM distortion can be radically reduced—observe the "Cancel" trace. However, adding resistances for distortion cancellation in this way has the obvious disadvantage that they introduce extra Johnson noise into the circuit. Taking that with the higher voltage noise of JFET opamps, this approach is definitely going to be noisy.

This cancellation technique is not pursued further in the sections on BJT opamps, because there the effects of extra resistance noise are relatively more serious due to the lower opamp voltage noise.

It would clearly be desirable to examine more JFET opamps, but for reasons of space we need to cut to the chase and look in detail at the far more promising BJT opamps.

# The NE5532 and 5534 Opamps

Since the 5534/5532 is by far the most popular audio opamp I make no apology for describing its behaviour in quite a bit of detail. This will also illuminate issues that apply to all the other opamps examined here.

The 5532 is a low-noise, low-distortion bipolar dual opamp with internal compensation for unity-gain stability. The 5534 is a single version internally compensated for gains down to 3, and an external compensation capacitor can be added for unity-gain stability; 22 pF is the usual value. The dual 5532 is used much more than the single 5534 as it is cheaper per opamp and does not require an external compensation capacitor when used at unity gain; the 5534 is however significantly quieter. The common-mode range of the inputs is a healthy ±13 V, with no phase-inversion problems if this is exceeded. It has a distinctly higher power consumption than the TL072, drawing approximately 4 mA per opamp section when quiescent. The DIL version runs perceptibly warm when quiescent on ±17 V rails.

The internal circuitry of the 5532 has never been publicly explained but appears to consist of nested Miller loops that permit high levels of internal negative feedback.

The 5534/5532 has BJT input devices. It therefore gives low noise with low source resistances, but the downside is that it draws relatively high bias currents through the input pins. The input transistors are NPN, so the bias currents flow into the chip from the positive rail. If an input is fed through a significant resistance, then the input pin will be more negative than ground due to the voltage drop caused by the bias current. The inputs are connected together with back-to-back diodes for reverse-voltage protection and therefore should not be forcibly pulled to different voltages.

The 5532 and 5534 type opamps require adequate supply decoupling if they are to remain stable; otherwise they appear to be subject to some sort of internal instability that seriously degrades linearity without being visible as oscillation on a normal oscilloscope. The essential requirement is that the +ve and −ve supply rails should be decoupled with a 100 nF capacitor between them, not more than a few millimetres from the opamp; normally one such capacitor is fitted per package as close to it as possible. It is *not* necessary, and often not desirable, to have two capacitors going to ground; every capacitor between a supply rail and ground carries the risk of injecting rail noise into the ground.

The 5532 is a robust opamp, but it is possible to damage one so that it keeps working but shows high distortion. This seems to be associated with faults where one supply rail fails; see Chapter 22 for ways of guarding against this. Obviously such specimens should be disposed of at once to prevent confusion in the future.

## The 5532 With Shunt Feedback

Figure 16.5 shows the distortion from a 5532 working in shunt mode with low-value feedback resistors of 1 kΩ and 2k2 setting a gain of 2.2 times, at an output level of 5 Vrms. This is the circuit of Figure 16.2a with Rs set to zero. This is the simplest situation, as the use of shunt mode means there is no CM voltage and hence no extra CM distortion. The THD is well below 0.0005% up to 20 kHz; this underlines what a superlative bargain the 5532 is.

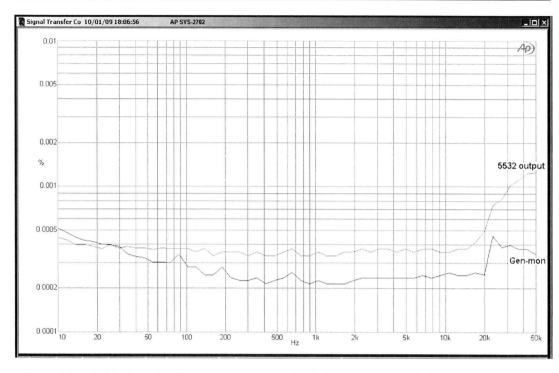

**Figure 16.5: 5532 distortion in a shunt-feedback circuit (as in Figure 16.2 a) at 5 Vrms out. This shows the AP SYS-2702 output (lower trace) and the opamp output (upper trace). Supply ±18 V.**

Figure 16.6 shows the same situation but with the output increased to 10 Vrms (the clipping level on ±18 V rails is about 12 Vrms), and there is now some significant distortion above 10 kHz, though it only exceeds 0.001% when the frequency reaches 18 kHz. This remains the case when Rs in Figure 16.2a is increased to 10 kΩ and 47 kΩ—the noise floor is higher but there is no real change in the distortion behaviour. The significance of this will be seen later when we look at CM distortion.

## 5532 Output Loading in Shunt-Feedback Mode

When a significant load is placed on a 5532 output, the distortion performance deteriorates in a predictable way. Figure 16.7 shows the effect on a shunt-feedback amplifier (to eliminate the possibility of input common-mode distortion). The output of the opamp is of course always loaded by the feedback resistor, which is effectively connected to ground at its other end. The circuit is as Figure 16.2a with a load resistor to ground added.

There are two differing regimes visible. At low frequencies (below 10 kHz) the distortion is flat with frequency. Above this, the THD rises rapidly, at approximately 12 dB/octave. This is very different from Blameless power amplifier behaviour, where the distortion normally rises at only 6 dB/octave when it emerges from the noise floor, typically around 2 kHz. The increase in THD in the flat region (at 1 kHz) is summarised in Table 16.4.

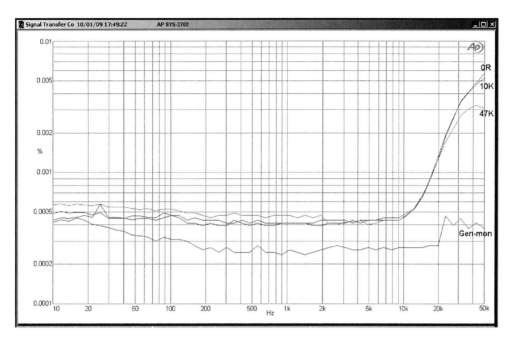

Figure 16.6: 5532 distortion in the shunt-feedback circuit of Figure 16.2a. Adding extra resistances of 10 kΩ and 47 kΩ in series with the inverting input does not degrade the distortion at all but does bring up the noise floor a bit. Test level is now 10 Vrms out, supply ±18 V.

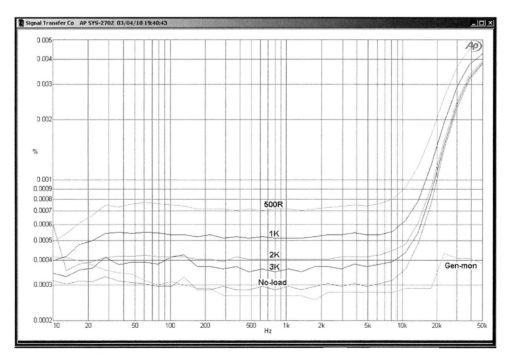

Figure 16.7: The effect of output loading on a shunt-feedback 5532 stage; 500 Ω, 1 kΩ, 2 kΩ, 3 kΩ, and no load, plus gen-mon (bottom trace). Feedback resistors 1 kΩ and 2k2, noise gain 3.2 times. Output 9 Vrms, supply ±18 V.

**Table 16.4: THD at 1 kHz for varying external
loads on a shunt-feedback 5532.**

| External load resistance Ω | THD at 1 kHz |
|---|---|
| No external load | 0.00030 % |
| 3 kΩ | 0.00036 % |
| 2 kΩ | 0.00040 % |
| 1 kΩ | 0.00052 % |
| 500 Ω | 0.00072 % |

With external no load, the THD trace is barely distinguishable from the Audio Precision output until we reach 10 kHz, where the THD rises steeply. In this condition the opamp is of course still loaded by the 2k2 feedback resistor.

### The 5532 With Series Feedback

Figure 16.8 shows the distortion from a 5532 working in series-feedback mode, as in Figure 16.2b. Note that the stage gain is greater at 3.2 times, but the opamp is working at the same noise gain and so has the same amount of negative feedback available to reduce distortion. The working conditions are however less favourable, as we shall see in the next section on common-mode problems, and the distortion now begins to rise from 5 kHz and reaches 0.0025% at 20 kHz, as opposed to 0.0014% at 20 kHz for the shunt version, as in Figure 16.7.

Figure 16.8 also shows the effect of loading the output. As for the shunt case, the increased distortion is mostly apparent in the flat LF section of the traces.

### Common-Mode Distortion in the 5532

In a series-feedback amplifier with a gain of 3.2 times the CM voltage is 3.1 Vrms for a 10 Vrms output. See Figure 16.9. The trace labelled "0R" (i.e. no source resistance) is the same as the "No load" trace of Figure 16.8. The source resistance seen by the inverting input is not zero, because of the impedance of the feedback network, but this is only 2k2 in parallel with 1 kΩ; in other words 687 Ω. Figure 16.9 implies that this will have only a very small effect, but more on that later. When we add some source resistance Rs, the picture is radically worse, with serious midband distortion rising at 6 dB/octave and roughly proportional to the amount of resistance added. We will note it is 0.0015% at 10 kHz with Rs = 10 kΩ. The horizontal low-frequency parts of the traces are raised by the Johnson noise from the source resistances and also by the opamp current noise flowing in those resistances.

The worst case for CM distortion is the voltage-follower configuration, as in Figure 16.2c, where the CM voltage is equal to the output voltage. Voltage-followers typically give low distortion because they have the maximum possible amount of negative feedback, and Figure 16.8 shows that even with a CM voltage of 10 Vrms, the distortion when driven from a low impedance is no greater than for the shunt

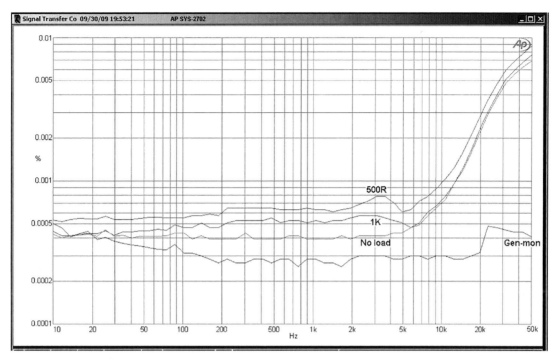

**Figure 16.8: 5532 distortion in a series-feedback stage with 2k2 and 1 kΩ feedback resistors (gain 3.2 times) and zero source resistance. The output level is 10 Vrms with 500 Ω, 1 kΩ loads, and no load. The gen-mon trace is the output of the distortion analyser measured directly. Supply ±18 V.**

mode. However, when source resistance is inserted in series with the input, the distortion mixture of second, third, and other low-order harmonics increases markedly; in the 3.2 times stage 10 kΩ of source resistance caused 0.0015% THD at 10 kHz, but with a voltage-follower we get 0.0035%. The distortion increases with output level, approximately quadrupling as the level doubles. Figure 16.10 shows what happens with source resistances of 10 kΩ and below; when the source resistance is below 2k2, the distortion is close to the zero source resistance trace.

Close examination reveals the intriguing fact that a 1 kΩ source actually gives *less* distortion than no source resistance at all, reducing THD from 0.00065% to 0.00055% at 10 kHz. Minor variations around 1 kΩ make no measurable difference. This effect is due to the 1 kΩ source resistance cancelling the effect of the source resistance of the feedback network, which is 1 kΩ in parallel with 2k2, i.e. 688 Ω. The improvement is small, but if you are striving for the very best linearity, the result of deliberately adding a small amount of source resistance appears to be repeatable enough to be exploited in practice. A large amount will compromise the noise performance, as seen in Figure 16.9, and this is another argument for keeping feedback network impedances as low as practicable without impairing distortion.

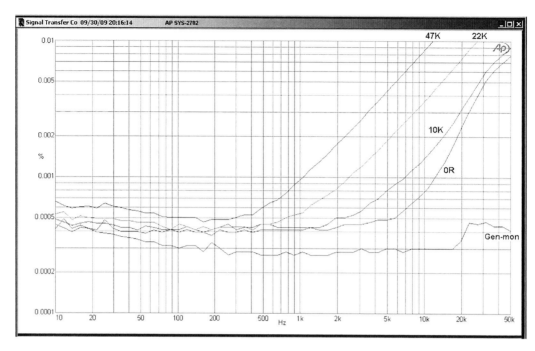

Figure 16.9: 5532 distortion in a series-feedback stage with 2k2 and 1 kΩ feedback resistors (gain 3.2 times) and varying source resistances; 10 Vrms output.

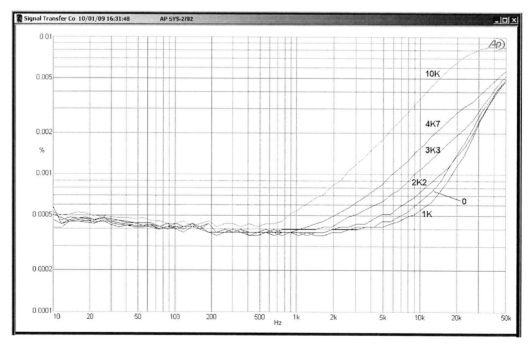

Figure 16.10: 5532 distortion in a voltage-follower circuit with a selection of source resistances; a 1 kΩ source resistance actually gives less distortion than no resistance, due to cancellation. The gen-mon THD was as before. Test level 10 Vrms, supply ±18 V.

This CM distortion behaviour is unfortunate because voltage-followers are very frequently required in active crossover design, primarily in Sallen & Key filters but also for general buffering purposes. The only cure is to keep the source impedances low, which is obviously also a good idea from the noise point of view. A total of 2 kΩ is about as high as you want to go, and this is why the lowpass Sallen & Key filters in Chapter 8 have been given series resistors that do not exceed this figure.

So, what exactly is going on here with common-mode distortion? Is it simply due to non-linear currents being drawn by the opamp inputs? Audio power amplifiers have discrete input stages which are very simple compared with those of most opamps and draw relatively large input currents. These currents show appreciable non-linearity even when the output voltage of the amplifier is virtually distortion free, and if they flow through significant source resistances will introduce added distortion. [3]

If this was the case with the 5532, then the extra distortion would manifest itself whenever the opamp was fed from a significant source resistance, no matter what the circuit configuration. But as we saw earlier, it only occurs in series-feedback situations; increasing the source resistance in a shunt-feedback amplifier does not perceptibly increase distortion.

The only difference is that the series circuit has a CM voltage of about 3 Vrms, while the shunt circuit does not, and the conclusion is that with a bipolar input opamp, you must have *both* a CM voltage and a significant source resistance to see extra distortion. The input stage of a 5532 is a straightforward long-tailed pair, with a simple tail current source and no fancy cascoding, and I suspect that Early effect is significant when there is a large CM voltage, modulating the quite high input bias currents, and this is what causes the distortion. The signal input currents are much smaller, due to the high open-loop gain of the opamp, and as we have seen appear to have a negligible effect.

### Reducing 5532 Distortion by Output Stage Biasing

There is a useful though relatively little-known (and where it is known, almost universally misunderstood) technique for reducing the distortion of the 5532 opamp. While the general method may be applicable to some other opamp types, here I concentrate on the 5532, and it must not be assumed that the relationships or results will be emulated by any other opamp type.

The principle is that if a biasing current of the right polarity is injected into the opamp output, then the output stage distortion can be significantly reduced. This technique is sometimes called "output stage biasing", though it must be understood that the DC voltage conditions are not significantly altered; because of the high level of voltage feedback the actual DC potential at the output is shifted by only a tenth of a millivolt or so.

You may have recognised that this scheme is very similar to the Crossover Displacement (Class XD) system I introduced for power amplifiers, which also injects an extra current, either steady or signal-modulated, into the amplifier output. [4] It is not however quite the same in operation. In power amplifiers the main aim of Crossover Displacement is to prevent the output stage from traversing the crossover region at low powers. In the 5532 at least the crossover region is not easy to spot on the distortion residual, the general effect being of second- and third-harmonic distortion rather than spikes or edges; it appears that the 5532 output stage is simply more linear when it is pulling down rather than pulling up, and the biasing current is compensating for this.

For the 5532, the current *must* be injected from the positive rail; currents from the negative rail make the distortion emphatically worse. This confirms that the output stage of the 5532 is in some way asymmetrical in operation, for if it was simply a question of suppressing crossover distortion by Crossover Displacement, a bias current of either polarity would be equally effective. The continued presence of the crossover region, albeit displaced, would mean that the voltage range of reduced distortion would be quite small and centred on 0 V. It is rather the case that there is a general reduction in distortion across the whole of the 5532 output range, which seems to indicate that the 5532 output stage is better at sinking current than sourcing it, and therefore injecting a positive current is effective at helping out.

Figure 16.11a shows a 5532 running in shunt-feedback mode with a moderate output load of 1 kΩ; the use of shunt feedback makes it easier to see what's going on by eliminating the possibility of common-mode distortion. With normal operation we get the upper trace in Figure 16.12, labelled "No bias". If we then connect a current-injection resistor between the output and to the V+ rail, we find that the LF distortion (the flat bit) drops almost magically, giving the trace labelled "3k3", which is only just above the gen-mon trace. Since noise makes a significant contribution to the THD residual at these levels, the actual reduction in distortion is greater than it appears.

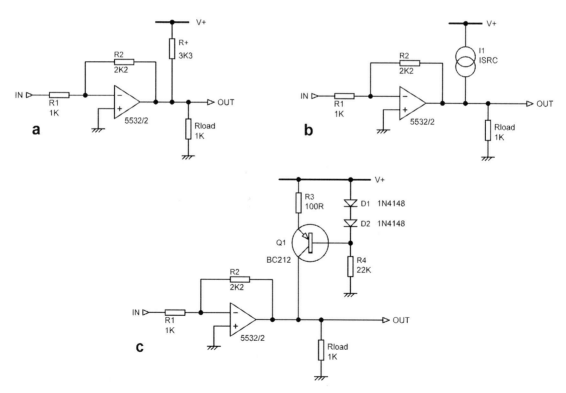

**Figure 16.11: Reducing 5532 distortion in the shunt-feedback mode by biasing the output stage with a current injected through a resistor R+ or a current source.**

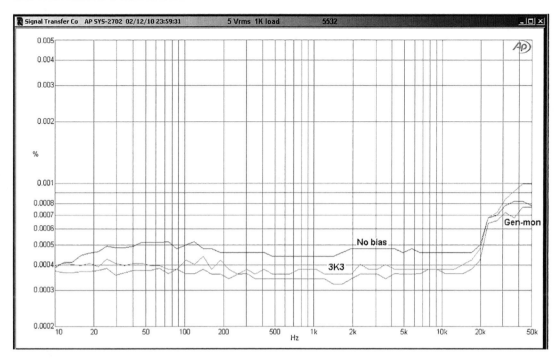

**Figure 16.12: The effect of output biasing, with a 3k3 resistor to V+, on a unity-gain shunt-feedback 5532 stage. Output load 1 kΩ, input and feedback resistors are 2k2, noise gain 2.0 times. Output 5 Vrms, supply ±18 V.**

The optimum resistor value for the conditions shown (5 Vrms and 1 kΩ load) is about 3k3, which injects a 5.4 mA current into the output pin. A 2k2 resistor gives greater distortion than 3k3, no doubt due to the extra loading it imposes on the output; in AC terms the injection resistor is effectively in parallel with the output load. In fact, 3k3 seems to be close to the optimal value for a wide range of output levels and output loadings.

The extra loading that is put on the opamp output by the injection resistor is a disadvantage, limiting the improvement in distortion performance that can be obtained. By analogy with the canonical series of Class-A power amplifier outputs, [5] a more efficient and elegant way to inject the required biasing current is by using a current source connected to the V+ rail, as in Figure 16.11b. Since this has a very high output impedance, the loading on the opamp output is not increased. Figure 16.11c shows a simple but effective way to do this; the current source is set to the same current as the 3k3 resistor injects when the output is at 0 V (5.4 mA), but the improvement in distortion is greater. There is nothing magical about this figure; however, increasing the injection current to, say, 8 mA, gives only a small further improvement in the THD figure, and in some cases may make it worse; also the circuit dissipation is considerably increased, and in general I would not recommend using a current source value of greater than 6 mA. Here in Table 16.5 are typical figures for a unity-gain shunt amplifier as before, with the loading increased to 680 Ω to underline that the loading is not critical; output biasing is effective with a wide range of loads.

**Table 16.5: Output biasing improvements with unity-gain shunt feedback, 5 Vrms out, load 680 Ω, supply ±18 V.**

| Injection method | THD at 1 kHz (22 kHz bandwidth) |
|---|---|
| None | 0.00034% |
| 3k3 resistor | 0.00026% |
| 5.4 mA current source | 0.00023% |
| 8.1 mA current source | 0.00021% |

**Table 16.6: Output biasing improvements with 3.2 times gain, series feedback, 9.6 Vrms out, load 1 kΩ, supply ±18 V.**

| Injection method | THD at 1 kHz (22 kHz bandwidth) |
|---|---|
| None | 0.00037% |
| 3k3 resistor | 0.00033% |
| 5.4 mA current source | 0.00027% |
| 8.1 mA current source | 0.00022% |

As mentioned before, at such low THD levels the reading is largely noise, and the reduction of the distortion part of the residual is actually greater than it looks from the raw figures. Viewing the residual shows a dramatic difference.

You might be concerned about the Cbc of the transistor, which is directly connected to the opamp output. The 5532/5534 is actually pretty resistant to HF instability caused by load capacitance, and in the many versions of this configuration I tested I have had no problems whatever. The presence of the transistor does not reduce the opamp output swing.

Output biasing is also effective with series-feedback amplifier stages in some circumstances. Table 16.6 shows it working with a higher output level of 9.6 Vrms, and a 1 kΩ load. The feedback resistors were 2k2 and 1 kΩ to keep the source resistance to the inverting input low.

The output biasing technique is in my experience only marginally useful with voltage-followers, as the increased feedback factor with respect to a series amplifier with gain reduces the output distortion below the measurement threshold. Table 16.7 demonstrates this.

As a final example, Figure 16.13 shows that the output biasing technique is still effective with higher gains, here 14 times. The distortion with the 5.4 mA source is barely distinguishable from the testgear output up to 2 kHz. The series-feedback stage had its gain set by 1k3 and 100 Ω feedback resistors, their values being kept low to minimise common-mode distortion. It also underlines the point that in some circumstances an 8.1 mA current source gives worse results than the 5.4 mA version.

When extra common-mode distortion is introduced by the presence of a significant source resistance, this extra distortion is likely to swamp the improvement due to output biasing. In a 5532 amplifier stage

**Table 16.7: Output biasing improvements for voltage-follower,
9.6 Vrms out, load 680 Ω, supply ±18 V.**

| Injection method | THD at 1 kHz (22 kHz bandwidth) |
| --- | --- |
| None | 0.00018% (almost all noise) |
| 3k3 resistor | 0.00015% (all noise) |
| 5.4 mA current source | 0.00015% (all noise) |
| 8.1 mA current source | 0.00015% (all noise) |

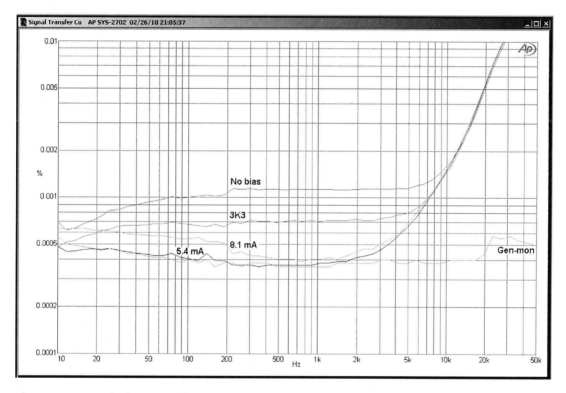

**Figure 16.13: Reducing 5532 distortion with series feedback by biasing the output stage with
a 3k3 resistor, or 5.2 or 8.1 mA current sources. Gain 14 times, no external load.
Test level 5 Vrms out, supply ±18 V.**

with a gain of 3.2 times and a substantial source resistance, the basic output distortion with a 1 kΩ load at 9.6 Vrms, 1 kHz out was 0.0064%. A 3k3 output biasing resistor to V+ reduced this to 0.0062%, a marginal improvement at best, and an 8.1 mA current source could only reduce it to 0.0059%.

Earlier I said that the practice of output stage biasing appears to be pretty much universally misunderstood, judging by how it is discussed on the Internet. The evidence is that every application of it that my research has exposed shows a resistor (or current source) connected between the opamp output and the *negative* supply rail. This is very likely based on the assumption that displacing the crossover region in either direction is a good idea, coupled with a vague feeling that a resistor to the negative rail is somehow more "natural". However, the assumption that the output stage is symmetrical is usually incorrect; as we have seen, it is certainly not true for the 5532/5534. For the 5532—which surely must be the most popular audio opamp by a long way—a pulldown resistor would be completely inappropriate, as it *increases* rather than decreases the output stage distortion.

You may be thinking that this is an ingenious method of reducing distortion but rather clumsy compared with simply using a more linear opamp like the LM4562. This is true, but on the other hand, if the improvement from output biasing is adequate, it will be much cheaper than switching to a more advanced opamp that costs ten times as much.

### Which 5532?

It is an unsettling fact that not all 5532s are created equal. The part is made by a number of manufacturers, and there are definite performance differences. While the noise characteristics appear to show little variation in my experience, the distortion performance does vary noticeably from one manufacturer to another. Although, to the best of my knowledge, all versions of the 5532 have the same internal circuitry, they are not necessarily made from the same masks, and even if they were there would inevitably be process variations between manufacturers.

The main 5532 sources at present are Texas Instruments, Fairchild Semiconductor, ON Semiconductor (was Motorola), NJR (New Japan Radio), and JRC (Japan Radio Company). I took as wide a range of samples as I could, ranging from brand-new devices to parts over 20 years old, and it was reassuring to find that without exception, every part tested gave the good linearity we expect from a 5532. But there were differences. I did THD tests on six samples from Fairchild, JRC, and Texas, plus one old Signetics 5532 for historical interest.

All tests were done using a shunt-feedback stage with a gain of unity, both input and feedback resistors being 2 kΩ; the supply rails were ±18 V. The output level was high, at 10.6 Vrms, only slightly below clipping. No external load was applied, so the load on the output was solely the 2 kΩ feedback resistor; applying extra loading will make the THD figures worse. The test instrumentation was an Audio Precision SYS-2702.

For most of the manufacturers, distortion only starts to rise very slowly above the measurement floor at about 5 kHz, and remains well below 0.0007% at 20 kHz. The exceptions were the six Texas 5532s, which surprisingly consistently showed somewhat higher distortion at high frequencies. Distortion at 20 kHz ranged from 0.0008% to 0.0012%, showing more variation than the other samples as well as being generally higher in level. The low-frequency section of the plot, below 10 kHz, was at the measurement floor, as for all the other devices, and distortion is only just visible in the noise. Compared with other makers' parts, the THD above 20 kHz is much higher—at least three times greater at 30 kHz. Fortunately, this should have no effect unless you have very high levels of ultrasonic signals that could cause intermodulation. If you do, then you have bigger problems than picking the best opamp manufacturer.

All of the measurements given in this book were performed using the Texas version of the 5532, to ensure worst-case results. If you use 5532s from one of the other manufacturers, then your high-frequency distortion results should be somewhat better.

## The 5534 Opamp

The single-opamp 5534 is somewhat neglected compared with the 5532, often being regarded as the same thing but inconveniently packaged with only one opamp per 8-pin DIL and in many cases requiring the expense of an external compensation capacitor. However, it can in fact be extremely useful when a somewhat better performance than the 5532 can give needs to be achieved economically, as it is between two and three times as expensive per opamp as the 5532 but about five times cheaper than the still somewhat exotic LM4562. The price ratio is likely to be nearer two when purchasing large quantities. (There was once a 5533, which was basically a dual 5534 with external compensation and offset null facilities in a 14-pin DIL package, but it seems to have achieved very little market penetration—the only place I have ever seen one is on the equaliser board of the famous EMT turntable. However, Philips were still putting out spec sheets for it in 1994.)

The 5534 has a very significant distortion advantage when gains of greater than unity are required, because the lighter internal compensation means that an NFB factor three times greater can be applied, with distortion reduced proportionally.

The 5534 also has lower noise than the 5532, its input voltage noise density being 3.5 uV/√Hz rather than 5 uV/√Hz. This means that in situations where the voltage noise dominates over the current noise—the sort of situation where you would want to use a BJT input opamp—the noise performance is potentially improved by 3.0 dB. Johnson noise from the circuit resistances is likely to reduce this difference in some cases.

Samuel Groner points out [1] that the higher voltage noise of the 5532 is almost certainly because the input pair has emitter degeneration resistors added to make unity-gain compensation easier. These are not shown on any internal schematic I have ever seen, but their presence is suggested by the fact that the 5532 has more voltage noise but much the same current noise and has a faster slew rate than a unity-gain compensated 5534 because its compensation capacitor can be smaller.

Figure 16.14 demonstrates the excellent linearity of an uncompensated 5534 with shunt feedback for 3.2 times gain and no external loading; distortion is 0.0005% at 20 kHz and 10 Vrms out. Compare this with Figure 16.6, which shows 0.0015% at 20 kHz for a 5532 section in the same situation. The 5534 distortion is reduced proportionally to the increased NFB factor, being three times less.

Figure 16.15 shows an uncompensated 5534 with series feedback for 3.2 times gain and no external loading; distortion is 0.00077% at 20 kHz and 10 Vrms out. Compare this with the "No-load" trace in Figure 16.8, which shows 0.0022% at 20 kHz for a 5532 section in the same situation. The reduction is again proportional to the increased NFB factor, being three times lower once more. Note that there is no source resistance added, so the 5534 is fed from a low impedance and will not exhibit CM distortion.

If the 5534 is used with external compensation to allow unity-gain stability, then the distortion advantage is lost, but the better noise performance remains. The generally used value is 22 pF for unity-gain stability, though this does not as far as I know have any official seal of approval from a manufacturer. In fact I have often found that uncompensated 5534s were stable and apparently quite happy at unity gain, but this is not something to rely on.

For a unity-gain shunt-feedback stage the opamp noise gain is 2 times. There appears to be no generally accepted value for the external compensation capacitor required in this situation, or other circuits requiring a 5534 to work at a gain of 2, but I have always found that 10 pF does the job.

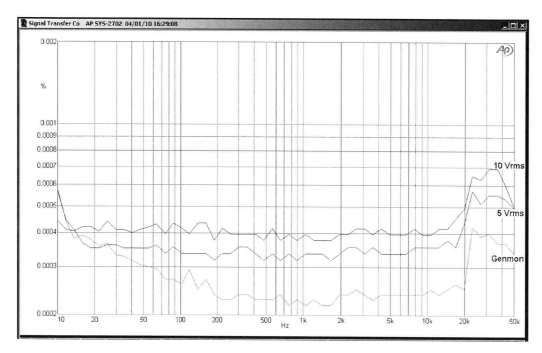

Figure 16.14: Distortion from an uncompensated 5534 with shunt feedback with 3.2 times gain and no external loading, at 5 Vrms and 10 Vrms out. The gen-mon (testgear output) is also shown. Supply ±18 V.

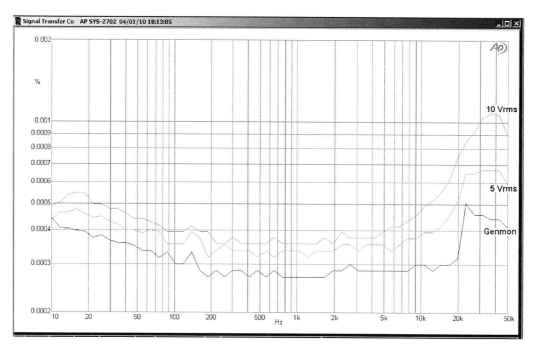

Figure 16.15: Distortion from an uncompensated 5534 with series feedback and 3.2 times gain, no external loading, at 5 Vrms and 10 Vrms out. Supply ±18 V.

# The LM4562 Opamp

The LM4562 is a relatively new opamp which first become freely available at the beginning of 2007. It is a National Semiconductor product. It is a dual opamp—there is no single or quad version. It costs about ten times as much as a 5532.

The input noise voltage is typically 2.7 nV/√Hz, which is substantially lower than the 5 nV/√Hz of the 5532. For suitable applications with low source impedances this translates into a useful noise advantage of 5.3 dB. The bias current is 10 nA typical, which is very low and would normally imply that bias cancellation, with its attendant noise problems, was being used. However, in my testing I have seen no sign of excess noise, and the data sheet is silent on the subject. No details of the internal circuitry have been released so far, and quite probably never will be. The LM4562 is not fussy about decoupling, and as with the 5532, 100 nF across the supply rails close to the package seems to ensure HF stability. The slew rate is typically ±20 V/us, more than twice as fast as the 5532.

The first THD plot in Figure 16.16 shows the LM4562 working at a closed-loop gain of 2.2 times in shunt-feedback mode, at a high level of 10 Vrms. The top of the THD scale is now a very low 0.001%, and the plots look a bit jagged because of noise. The no-load trace is barely distinguishable from the AP SYS-2702 output, and even with a heavy 500 Ω load driven at 10 Vrms there is only a very small

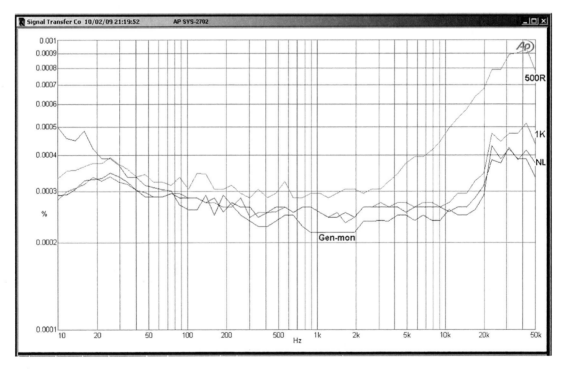

**Figure 16.16: The LM4562 in shunt-feedback mode, with 1 kΩ, 2k2 feedback resistors giving a gain of 2.2x. Shown for no load (NL) and 1 kΩ, 500 Ω loads. Note the vertical scale ends at 0.001% this time. Output level is 10 Vrms. ±18 V supply rails.**

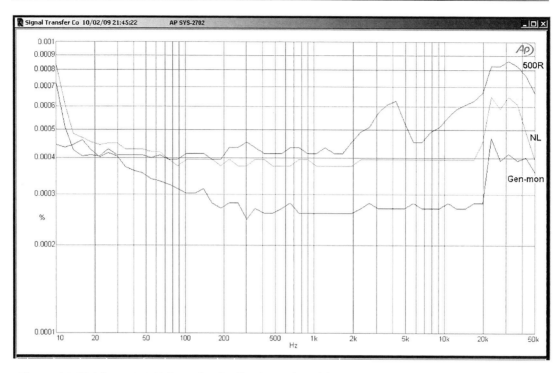

**Figure 16.17: The LM4562 in series-feedback mode, with 1 kΩ, 2k2 feedback resistors giving a gain of 3.2x. No load (NL) and 500 Ω load. 10 Vrms output. ±18 V supply rails.**

amount of extra THD, reaching 0.0007% at 20 kHz. Compare this with the 5532 performance in Figure 16.7, where loading brings up the distortion in the flat region below 10 kHz to 0.0008% and gives a THD of 0.0020% at 20 kHz with a 500 Ω load, as opposed to only 0.0007% for the LM4562.

Figure 16.17 shows the LM4562 working at a gain of 3.2x in series-feedback mode, both modes having a noise gain of 3.2 times. The extra distortion from the 500 Ω loading is very low.

### Common-Mode Distortion in the LM4562

For Figures 16.16 and 16.17 the feedback resistances were 2k2 and 1 kΩ, so the minimum source resistance presented to the inverting input is 688 Ω. In Figure 16.18 extra source resistances were then put in series with the input path (as was done with the 5532 in the section on common-mode distortion), and this revealed a remarkable property of the LM4562—with moderate levels of CM voltage it is much more resistant to common-mode distortion than the 5532. At 10 Vrms and 10 kHz, with a 10 kΩ source resistance, the 5532 generates 0.0014% THD (see Figure 16.9), but the LM4562 gives only 0.00046% under the same conditions. I strongly suspect that the LM4562 has a more sophisticated input stage than the 5532, probably incorporating cascoding to minimise the effects of common-mode voltages. Note that only the rising curves to the right represent actual distortion. The

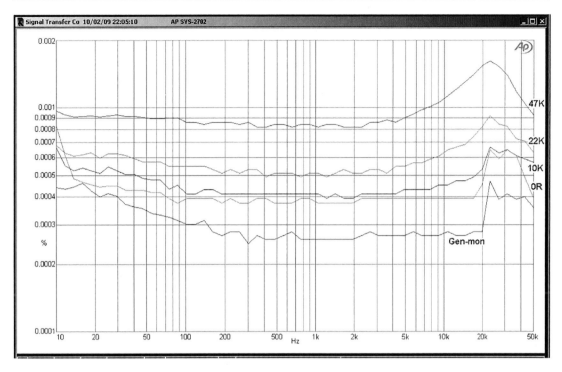

**Figure 16.18: The LM4562 in series-feedback mode, gain 3.2x, with varying extra source resistance in the input path. The extra distortion is much lower than for the 5532. 10 Vrms out, ±18 V supply rails.**

raised levels of the horizontal traces at the LF end are due to Johnson noise from the added series resistances, plus opamp current noise flowing in them.

As we saw with the 5532, the voltage-follower configuration is the most demanding test for CM distortion, because the CM voltage is at a maximum. Now the LM4562 does not work quite so well. At 10 kHz the distortion with a 10 kΩ source resistance has leapt up from 0.00046% in Figure 16.18 to 0.0037% in Figure 16.19, which allowing for the noise component in the former must be an increase of some ten times. This is a surprising and unwelcome result, because it means that despite its much greater cost the performance of the LM4562 in this configuration is no better than that of the 5532; compare Figure 16.10.

The reason for this rapid increase in distortion with CM voltage appear to be a non-linearity mechanism that is activated quite suddenly when the CM voltage exceeds 4 Vrms. This is illustrated in Figure 16.20, which shows distortion against level for the voltage-follower at 10 kHz, with different source resistances. The left side of the plot shows only noise decreasing relatively as the test level increases; the steps in the lowest trace are measurement artefacts. Since the CM voltage for the series-feedback configuration is only about 3 Vrms, the non-linearity mechanism is not activated, and CM distortion in that case is very low.

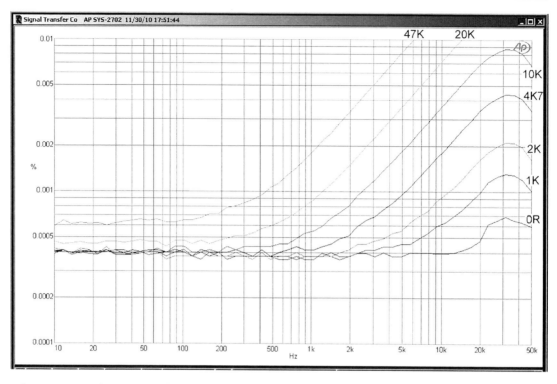

**Figure 16.19: The LM4562 in voltage-follower mode, with varying extra source resistance in the input path. CM distortion is much higher than for the series-feedback amplifier in Figure 16.16. 10 Vrms out, ±18 V supply rails.**

The conclusion has to be once more that if you are using voltage-followers and want low distortion, it is well worthwhile to expend some time and trouble on getting the source impedances as low as possible.

It has taken an unbelievably long time—nearly 30 years—for a better audio opamp than the 5532 to come along, but at last it has happened. The LM4562 is superior in just about every parameter, apart from having more than twice as much current noise and no better CM distortion in voltage-follower use. These issues should not be serious problems if the low-impedance design philosophy is followed. At present the LM4562 has a much higher price; hopefully that will change, but it may take a very long time, based on the price history of other opamps. It has already begun to appear in hi-fi equipment, one example being the Benchmark DAC1 HDR, a combined DAC and preamplifier.

There is, however, or at any rate appears to be, a slight problem with using the LM4562 as a voltage-follower. On several occasions I have built such a voltage-follower, supplied it with a low source impedance such as 40 Ω, and attempted to measure the noise performance. This has been rendered impossible by the stage picking up and demodulating radio stations. In the same circuit position a 5532 works fine. I assume some internal oscillation is causing heterodyning; no oscillation is visible on the output. A cure I found helpful is connecting the inverting input to the opamp output with a 100 Ω

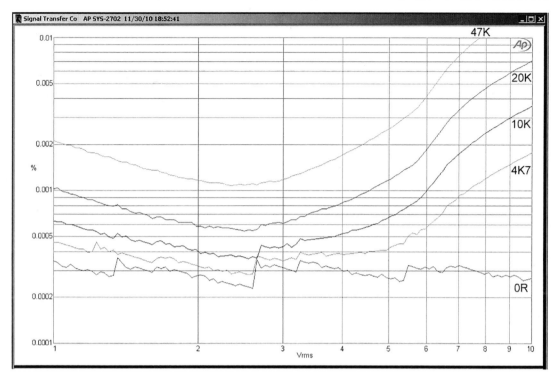

**Figure 16.20: The LM4562 in voltage-follower mode, showing CM distortion versus signal voltage with varying extra source resistance. The frequency is 10 kHz. A non-linearity is kicking in at about 4 Vrms. 1 to 10 Vrms out, ±18 V supply rails.**

resistor rather than a direct connection. This is of course not ideal if you are seeking the best possible noise performance.

## The LME49990 Opamp

The LME49990 from National Semiconductor is a single opamp available only as an 8-lead narrow body SOIC surface-mount package. It was released in early 2010. It is part of their "Overture" series, which the data sheet describes as an "ultra-low distortion, low noise, high slew rate operational amplifier series optimized and fully specified for high performance, high fidelity applications", and from my measurements on the LME49990 I'll go along with that. The Overture series also includes the LME49880, which is a dual JFET-input opamp. The LM49710 is another BJT opamp with very low noise and distortion specs, but for unknowable reasons it does not appear to be part of the Overture series.

Figure 16.21 shows the distortion performance in the shunt-feedback configuration, which prevents any common-mode distortion issues. The input and feedback resistors are 1 kΩ and 2k2, giving a gain

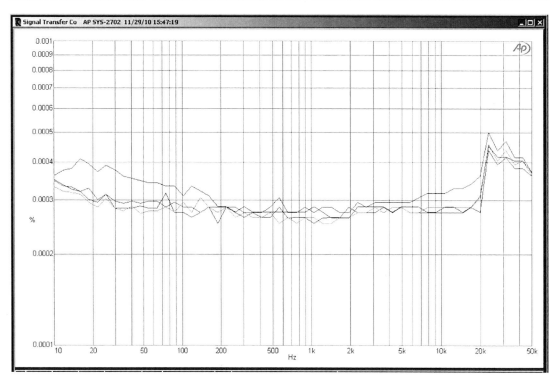

**Figure 16.21: The LME49990 in shunt-feedback mode, with a 1 kΩ input resistor and a 2k2 feedback resistor giving a gain of 2.2x. Shown for no load (NL) and 1 kΩ, 500 Ω loads. The generator output is also plotted. Note the top of the vertical scale is at only 0.001%. The output level is 9 Vrms, with ±17 V supply rails.**

of 2.2 times (and a noise gain of 3.2 times, as for the series version of this test). The traces are for no load (apart from the feedback resistor) and 1 kΩ, and 500 Ω loads at an output of 9 Vrms, and also the AP 2722 output for reference. As you can see, these traces are pretty much piled up on top of each other, with no distortion visible on the residual except for a small amount between 10 kHz and 20 kHz with the 500 Ω load; clearly the LME49990 is very good at driving 500 Ω loads. The step at 20 kHz is an artefact of the Audio Precision SYS-2702 measuring system.

If we compare this plot with Figure 16.5, we can see that the LME49990 is somewhat superior to the 5532, but since we are down in the noise floor most of the time, the differences are not great.

In the series configuration, with 1 kΩ and 2k2 feedback resistors giving a gain of 3.2 times and a significant common-mode mode voltage of 3 Vrms, things are not quite so linear; there is now clearly detectable distortion at high frequencies, as shown in Figure 16.22. Even so, the distortion is less than half that of a 5532 in the same situation—compare Figure 16.6.

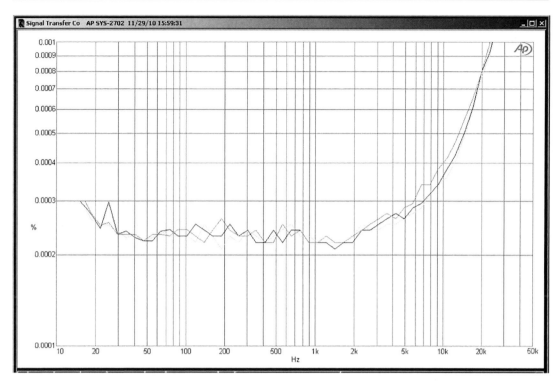

**Figure 16.22: The LM4562 in series-feedback mode, with 1 kΩ, 2k2 feedback resistors giving a gain of 3.2x. No load, 1 kΩ, and 500 Ω loads. 9 Vrms output; ±17 V supply rails.**

### Common-Mode Distortion in the LME49990

It looks as though common-mode distortion may be more of an issue with the LME49990 than it was with the LM4562. As we saw earlier, BJT input opamps do not show common-mode distortion unless the configuration has both a significant common-mode voltage *and* a significant source impedance. If we repeat the series-feedback test with no external load, but increasing source resistance, we get Figure 16.23, where as usual more source resistance means more high-frequency distortion. The exception is for a 1 kΩ source resistance, where the distortion actually decreases; this is because the 1 kΩ is partially cancelling the 688 Ω source resistance of the feedback network. We saw exactly the same effect with the 5532; see Figure 16.10. The horizontal low-frequency parts of the traces are raised by the Johnson noise from the added source resistances and also by the opamp current noise flowing in those resistances. There is no distortion visible in this region.

The voltage-follower configuration has the worst working conditions for CM distortion because there is no amplification, and so the CM voltage on the inputs is as large as the output voltage. Figure 16.24 shows that in this case the CM distortion is much worse. With a 2 kΩ source resistance, the THD at 10 kHz has increased from 0.0015 % to 0.0042 %, and all the other figures show a similar increase. The conclusion has to be that if you are working with a large CM voltage and a significant source

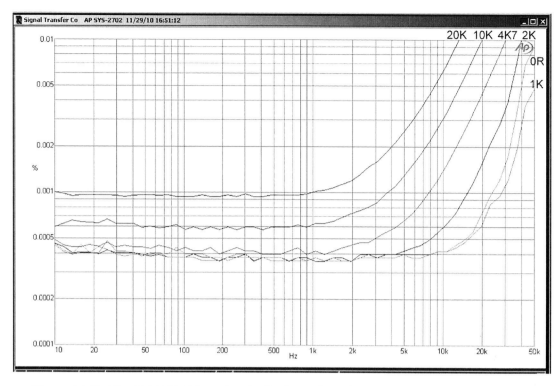

**Figure 16.23: The LME49990 in series-feedback mode, gain 3.2x, with varying extra source resistance in the input path; note that a 1 kΩ source resistance actually gives less distortion than none. The CM distortion is lower than for the 5532 but higher than for the LM4562. 10 Vrms out, ±18 V supply rails.**

resistance, the LME49990 is not the best choice, and either the 5532, the LM4562, or the AD797 will give considerably lower distortion.

The SOIC-only format of the LME4990 is not helpful for experimentation or home construction. I soldered my samples to an 8-pin DIL adaptor and plugged that into a prototype board without any problems. The LME49990 samples were kindly supplied by Don Morrison.

## The AD797 Opamp

The AD797 (Analog Devices) is a single opamp with very low voltage noise and distortion. It appears to have been developed primarily for the cost-no-object application of submarine sonar, but it works very effectively with normal audio—if you can afford to use it. The cost is something like 20 times that of a 5532. No dual version is available, so the cost ratio per opamp section is 40 times. The AD797 incorporates an ingenious error-correction feature for internal distortion cancellation, the operation of which is described on the manufacturer's data sheet. The measurements presented here seem to show that it works effectively.

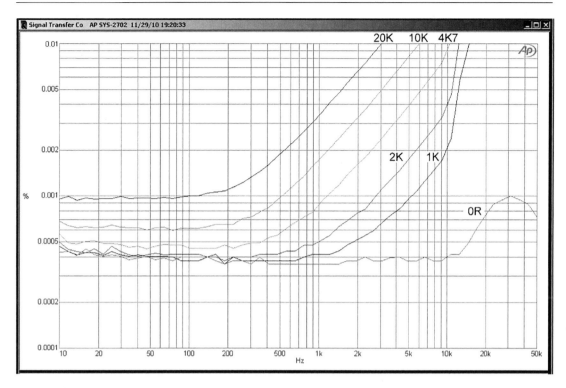

**Figure 16.24: LME49990 distortion in a voltage-follower circuit with a selection of source resistances; test level 10 Vrms, supply ±18 V.**

The AD797 is a remarkably quiet device in terms of voltage noise, but current noise is correspondingly high due to the large standing currents in the input devices. Early versions appeared to be quite difficult to stabilise at HF; the current product seems to be easier to handle but still a bit harder to stabilise than the 5532 or the LM4562. Possibly there has been a design tweak, or on the other hand my impression may be wholly mistaken. It has taken quite a long time for the AD797 to make its way into audio circuitry, possibly because of discouraging early reports on instability but more likely because of the high cost. At the time of writing (2010) AD797s are showing up in commercial audio equipment, a contemporary example being the Morrison E.L.A.D. line-level preamplifier, and they are also appearing in DIY designs. Audio Precision use this opamp in their 2700 series state-of-the-art distortion analysers.

Figure 16.25 shows the AD797 in shunt-feedback mode, to assess how it copes with output loading; as you can see, the effect is very small, with a very modest rise above 10 kHz with the 500 Ω load. Below 10 kHz the loading has no detectable effect at all, which makes it a better opamp than the 5532—see Figure 16.5. The step at 20 kHz is a measurement artefact. When the loads were applied, a 33 pF capacitor had to be put across the 2k2 feedback resistor to obtain stability—this was not required with either the 5532, the LM49990, or the LM4562, and shows that the AD797 does require a little more care to get dependable HF stability.

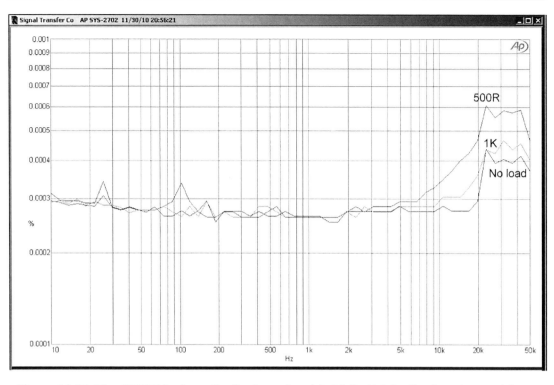

**Figure 16.25: The AD797 in shunt-feedback mode, with 1 kΩ, 2k2 feedback resistors giving a gain of 2.2x. No load and 1 kΩ, 500 Ω loads. Note top of vertical scale is at 0.001%. Output level is 10 Vrms. ±18 V supply rails.**

Figure 16.26 shows the AD797 in series-feedback mode, with no output loading apart from its own 2k2 feedback resistor, to test its CM distortion with moderate CM voltages, here 3.1 Vrms. The series-feedback test also needed 33 pF across the feedback resistor to ensure HF stability.

### Common-Mode Distortion in the AD797

The AD797 is next tested in voltage-follower mode with a 10 Vrms signal level and various source impedances. The traces in Figure 16.27 do not fall out very neatly because there appears to be some sort of cancellation effect going on, but from the difference between the Rs = 4k7 and Rs = 10 kΩ traces it is abundantly clear that, as usual, source impedances need to be kept low when there is a large CM signal. The results are significantly better than the 5532 in voltage-follower mode—compare Figure 16.10.

## The OP27 Opamp

The OP27 from Analog Devices is a bipolar input single opamp primarily designed for low noise and DC precision. It was not intended for audio use, but in spite of this it is frequently recommended for such applications as RIAA and tape head preamps. This is unfortunate, because while at first sight it

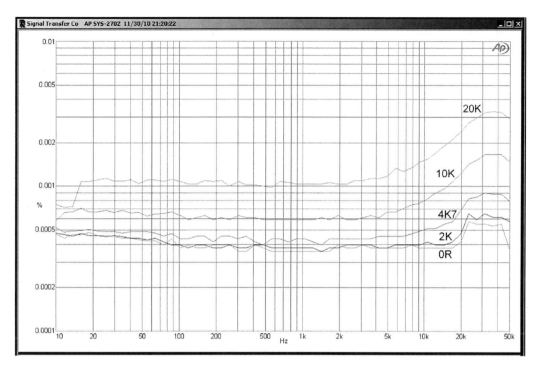

Figure 16.26: AD797 with series-feedback THD with no external load, at 10 Vrms.
Gain = 3.2x, ±18 V supply rails.

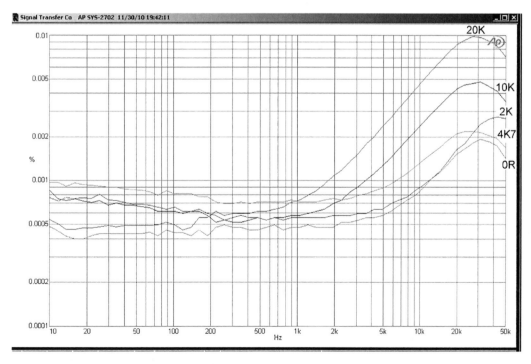

Figure 16.27: AD797 voltage-follower THD at 10 Vrms. Output is virtually
indistinguishable from input.

appears that the OP27 is quieter than the 5534/5532, as the $e_n$ is 3.2 nV/rtHz compared with 4 nV/rtHz for the 5532, in practice it is usually slightly noisier. This is because the OP27 is in fact optimised for DC characteristics and so has input bias-current cancellation circuitry that generates common-mode noise. When the impedances on the two inputs are very different the CM noise does not cancel; this can degrade the overall noise performance significantly, and certainly to the point where the OP27 is noisier than a 5532.

For a bipolar input opamp, there appears to be a high level of common-mode input distortion, enough to bury the output distortion caused by loading; see Figures 16.28 and 16.29. It is likely that this too is related to the bias-cancellation circuitry, as it does not occur in the 5532.

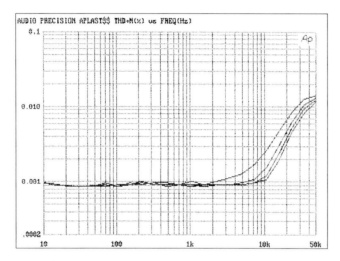

**Figure 16.28: OP27 THD in shunt-feedback mode with varying loads. This opamp accepts even heavy (1 kΩ) loading relatively gracefully.**

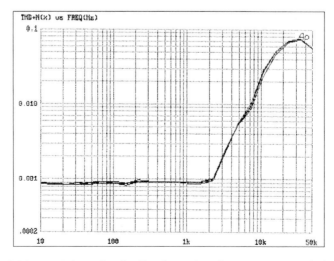

**Figure 16.29: OP27 THD in series-feedback mode. The common-mode input distortion completely obscures the output distortion.**

The maximum slew rate is low compared with other opamps, being typically 2.8 V/us. However, this is not the problem it may appear. This slew rate would allow a maximum amplitude at 20 kHz of 16 Vrms, if the supply rails permitted it. I have never encountered any particular difficulties with decoupling or stability of the OP27.

## Opamp Selection

In audio work, the 5532 is pre-eminent. It is found in almost every mixing console and in a large number of preamplifiers. Distortion is very low, even when driving 600 Ω loads. Noise is very low, and the balance of voltage and current noise in the input stage is well matched to low-impedance audio circuitry. Large-quantity production has brought the price down to a point where a powerful reason is required to pick any other device.

The 5532 is not, however, perfect. It suffers common-mode distortion. It has high bias and offset currents at the inputs as an inevitable result of using a bipolar input stage (for low noise) without any sort of bias-cancellation circuitry.

With tiresome inevitability, the very popularity and excellent technical performance of the 5532 has led to it being savagely criticised by Subjectivists who have contrived to convince themselves that they can tell opamps apart by listening to music played through them. This always draws a hollow laugh from me, as there is probably no music on the planet that has not passed through a hundred or more 5532s on its way to the consumer.

In some cases, such as variable-frequency crossover filters, bipolar-style bias currents are a nuisance, because keeping them out of pots to prevent scratching noises requires the addition of blocking capacitors. JFET-input opamps have negligible bias currents and so do not need these extra components, but there is no obviously superior device that is the equivalent of the 5532 or 5534. The TL072 has been used in the $EQ$ sections of low-cost preamplifiers and mixers application for this purpose for many years, but its HF linearity is poor by modern standards, and distortion across the band deteriorates badly as output loading increases. It is also more subject to common-mode distortion than bipolar types. There are of course more modern JFET opamps such as the OPA2134, but their general linearity is not much better, and they also suffer from common-mode distortion.

If you are looking for something better than the 5532, the newer opamps (LM4562, LME49990, AD797) have definite advantages in noise performance. The LM4562 has almost 6 dB less voltage noise, while both the LME49990 and the AD797 are nearly 15 dB quieter, given sufficiently low impedance levels. Both have better load-driving characteristics than the 5532. The LME49990 has really excellent load-driving capabilities, but its CM distortion is disappointingly high. The 3 dB noise advantage of the 5534A should not be forgotten.

To summarise some related design points:

* The trade-off between noise and distortion-reducing circuit impedances will reduce noise relatively slowly. The effect of opamp current noise is proportional to impedance, but Johnson noise is proportional to square root of impedance.
* Keep circuit impedances as low as possible to minimise their Johnson noise and the effect of current noise flowing through them. This will also reduce common-mode distortion and capacitive crosstalk.

- If distortion is more important than noise, and you don't mind the phase inversion, use the shunt configuration.
- If noise is more important than distortion, or if a phase inversion is undesirable, use the series configuration. In this case keeping circuit impedances low has an extra importance, as it minimises common-mode distortion arising in the feedback network.
- To minimise distortion, keep the output loading as light as possible. This of course runs counter to the need to keep feedback resistances as low as possible.
- If the loading on a 5532/5534 output is heavy, consider using output biasing to reduce distortion. If the improvement is sufficient, this will be cheaper than switching to a more advanced opamp.

# References

[1]   Groner, Samuel "Operational Amplifier Distortion" October 2009,
      www.sg-acoustics.ch/analogue_audio/index.html
[2]   Jung, Walt (ed.) "Op-Amp Applications Handbook" Newnes, 2006, Chapter 5, p. 399
[3]   Self, Douglas "Audio Power Amplifier Design" 6th Edn, Newnes, 2013, pp. 142–150 (i/p current distn)
      ISBN: 978-0-240-52613-3
[4]   Self, Douglas Ibid, pp. 449–462 (XD)
[5]   Self, Douglas Ibid, pp. 423–426 (Class-A)

# CHAPTER 17

# Active Crossover System Design

So far we have looked at the design of the functional blocks that go together to make an active crossover. It is now time to look at putting them together into a complete system. This involves not only selecting the optimal types of circuit block to perform the required crossover functions but also considerations about what order in which to place them in the signal path and what nominal signal levels to operate them at. This chapter also covers minor crossover functions such as output level controls and channel muting switches.

## Crossover Features

The actual crossover functions of filtering and equalisation make up the main part of a crossover's circuitry, but other functions are required to make a device that is practically useful. Not all crossovers have all of them. These are:

### Input Level Controls

These allow varying incoming levels to be normalised to the nominal internal level of the crossover. They frequently have a deliberately restricted gain range, typically not going down to zero; they are not intended to be used as volume controls. As described later in this chapter, better signal-to-noise performance can be obtained by using active gain controls that actually vary the gain of an amplifier stage, as opposed to combinations of fixed-gain amplifiers and variable passive attenuation. This often means designing a variable-gain balanced input stage, which requires some subtlety if a good CMRR is to be obtained at all gain settings. This issue is thoroughly explored in Chapter 20 on line inputs.

### Subsonic Filters

Highpass filters are used to protect loudspeakers from subsonic signals. Some loudspeaker types have no cone loading at very low frequencies and are especially vulnerable to this sort of damage. Subsonic filters are typically of the 2nd-order or 3rd-order Butterworth (maximally flat) configuration, rolling off at rates of 12 dB/octave and 18 dB/octave respectively. Fourth-order 24 dB/octave filters are less common, presumably due to worries about the possible audibility of rapid phase changes at the very bottom of the audio spectrum; when they are used, the cutoff frequency is lower. There seems to be a general consensus that a 3rd-order Butterworth filter with a cutoff frequency around 20 to 25 Hz is the correct approach, though at least one manufacturer has used a 4th-order filter starting at 15 Hz. A 3rd-order Butterworth filter with a −3 dB cutoff at 20 Hz will be −18.6 dB down at 10 Hz and −36.0 dB down at 5 Hz. The 30 Hz response is down by−0.37 dB, and if you feel that is too much, the filter cutoff frequency can be moved down a little.

Subsonic filters are placed as early in the signal path as possible, typically immediately after a balanced input amplifier. This is not so much to avoid the generation of intermodulation distortion (which should be negligible at low frequencies in a decent opamp) but to prevent headroom being eroded by large subsonic signals. Standard Sallen & Key filters are normally used.

### Ultrasonic Filters

These are also intended for speaker protection, in the event of HF instability somewhere upstream in the audio system. A typical ultrasonic filter would be a 2nd-order Butterworth with a cutoff frequency around 40 kHz; this will be only−0.24 dB at 20 kHz. If you feel that is too much of an intrusion on the audio spectrum, moving the cutoff frequency up to 50 kHz gives a 20 kHz loss of only −0.09 dB. Ultrasonic filters are rarely steeper than 2nd-order, presumably because of a fear that phase changes are more audible at the very top of the audio spectrum than the very bottom. The ultrasonic filter may be placed just after the balanced input amplifier to minimise HF intermodulation distortion in the following circuitry or, alternatively, placed only in the HF crossover path, the assumption being that the lowpass filters in the LF and MID paths will very effectively remove any ultrasonic signals and that it is better to put as little of the audio spectrum as possible through as few stages as possible, to minimise signal degradation. Standard Sallen & Key filters are commonly used.

The combination of a subsonic filter and an ultrasonic filter is often called a bandwidth definition filter. There is much information on how to make them economically in Chapter 9.

### Output Level Trims

Output level controls tend also to have a limited range; once again they are not intended to be used as volume controls. Allowing for power amplifier gain variations in a nominally identical set of power amplifiers may require as little as ±1 dB, but it is usual to provide at least ±3 dB to allow for drive unit variations, and ±6 dB is more usual. If the crossover is intended to work with a wide range of power amplifiers of differing sensitivities, the range may need to be considerably greater than this. There is more on output level controls later in this chapter.

### Output Mute Switches, Output Phase-Reverse Switches

These are dealt with later in this chapter.

### Control Protection

Active crossovers are not some sort of glorified tone-control—they do a quite different job and are meant to be carefully set up to match the power amplifiers and loudspeakers and then left alone. They are not meant to be tweaked, twiddled or frobbed by every casual passer-by. It is possible to cause serious damage by maladjustment—if the highpass cutoff frequency for the HF loudspeakers is adjusted radically downwards, then expensive and gig-cancelling damage is virtually a certainty.

It is therefore common to cover up crossover controls with a panel to prevent tampering; this is sometimes called a "security cover". This might be a substantial piece of clear plastic (the look-but-don't-touch approach), or a solid piece of metal which is more robust against impact. It is usually fixed with so-called security screws, but since you can buy sets of drivers for those very easily, we're not exactly talking Fort Knox here.

## Features Usually Absent

There are also some features that, while appearing in many kinds of audio equipment, are rarely if ever found in active crossovers. These include:

### Metering

Active crossovers are not usually fitted with comprehensive level metering, as the assumption is that they will be installed in some secluded spot where a visual display will not be seen. Peak-detect or clip-detect indicators can however be useful if there is a possibility of internal clipping. These are dealt with later in this chapter. Signal-present indicators that illuminate some way below the nominal signal level (often at −20 dB) can be useful for fault finding.

### Relay Output Muting

Active crossovers are not normally expected to have relay output muting, the function of which is to avoid sending out unpleasant transients at power-up and power-down. Since there are likely to be six outputs, probably balanced, the extra cost of six good-quality two-changeover relays is significant. In sound-reinforcement work, putting mute relays on every piece of equipment is very much not favoured, as they present one more place for things to go wrong and stop the signal. Thump suppression is normally considered to be the job of the power amplifier output muting relays, which have to be fitted, as one of their most important functions is protection of the loudspeakers from DC fault conditions.

Manual mute switches for each output are however often fitted to sound-reinforcement crossovers to simplify checking and fault finding.

## Switchable Crossover Modes

Crossovers intended for sound reinforcement are often constructed so they can be used in several different modes. Figure 17.1 shows the block diagram of a stereo 2-way crossover that can be switched to act as a mono 3-way crossover or as a 2-way crossover with a single mono LF output for a subwoofer. Alternatively a crossover might be switchable between stereo 2-way and mono 3-way crossover, with a wholly separate mono subwoofer output always available.

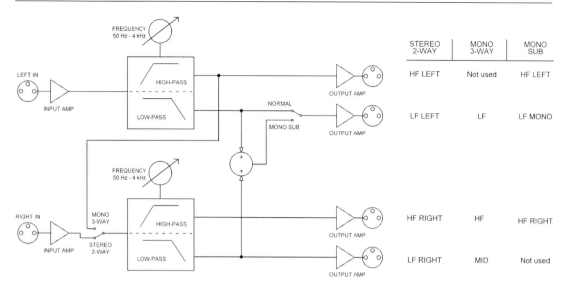

**Figure 17.1: Mode switching in an electronic crossover: it can be used as a stereo 2-way crossover, a mono 3-way crossover, or a 2-way crossover with a mono subwoofer output.**

Figure 17.1 has the mode switches in the normal stereo 2-way crossover position, and 2-way outputs are obtained as shown in the left column of the text in Figure 17.1. If the 2-way/3-way switch is operated, then the right input is not used and the input to the right filter block now comes from the highpass output of the left filter block. If the left filter block is set to the LF/MID crossover frequency and the right filter block set to the MID/HF crossover frequency, 3-way outputs are obtained as shown in the middle column of the text in Figure 17.1. This kind of mode switching requires a very wide range of filter frequency variation, as the filter block must be able to cover both LF/MID and MID/HF crossover points.

For mono-sub operation the 2-way/3-way switch is left in its normal position, and the mono-sub switch operated instead. This causes the left LF output to be fed with the mono sum of the two LF outputs from the filter blocks. The right LF output is not used. In this case the left and right filter blocks are set to the same frequency—the crossover point between the two HF outputs and the single LF output. The summing function has to be implemented carefully if crosstalk between the two channels feeding it is to be avoided.

Each filter block is shown with a single frequency control to emphasise that the cutoff frequencies of the highpass and lowpass sections move together; this is usually implemented with a state-variable filter (SVF) that simultaneously gives highpass and lowpass outputs.

More complex mode switching schemes are possible. A stereo 3-way crossover could be switchable to work as a mono four-way or five-way crossover. This is all very ingenious, but it does require a lot of complicated switching and a very clear head when you are setting up all those crossover frequencies.

Manufacturers often warn that mode switches should not be operated while the whole system is active, stating that this can lead to damaging transients; presumably they are worried that you might get a level you don't expect rather than concerned about minor DC clicks.

# Noise, Headroom, and Internal Levels

The choice of the internal signal level in a piece of audio equipment is a serious matter, as it controls both the signal-to-noise ratio and the headroom available before clipping occurs. A vital step in any design is the determination of the optimal signal level at each point in the circuit; there is no reason why you have to stick to the same level in every section. Obviously a real signal, as opposed to a test sine wave, continuously varies in amplitude, and the signal level chosen is purely a nominal level. One must steer a course between two evils:

1. If the signal level is too low, it will be contaminated unduly by noise. The absolute level of noise in a circuit is not of great significance in itself—what counts is how much greater the signal is than the noise—in other words, the signal to noise ratio.
2. If the signal level is too high, there is a risk it will clip and introduce severe distortion.

You will note that the first evil is a certainty, in that there will always be some addition of noise, even if it is insignificant, while the second is a statistical risk.

The wider the gap between the noise level and the clipping level, the greater is the dynamic range. If the best possible signal-to-noise is required for hi-fi use, then the internal level should be high, and if there is an unexpected overload it's not the end of the world. In sound-reinforcement applications it will often be preferable use a lower internal level, sacrificing some noise performance to reduce the risk of clipping. Heavy clipping, which in an active crossover system can be surprisingly hard to detect by ear, is likely to imperil HF speaker units, though not for the frequently quoted but quite untrue reason that excess harmonics are generated; the real problem is the general rise in level. [1] Later in this chapter we will look at some ways of detecting and indicating clipping.

The internal level chosen depends on the purpose of the equipment. For example, suppose you are designing a mixing console. If it is intended for studio recording, you only have to get the performance right once, but you do have get it exactly right, i.e. with the best possible signal-to-noise ratio, so the internal level is relatively high, very often −2 dBu (615 mVrms), which gives a headroom of about 24 dB. If it is broadcast work intended to air you only have one chance to get it right, and a mildly impaired signal-to-noise ratio is much more acceptable than a crunching overload, so the internal levels need to be significantly lower. The Neve 51 Series broadcast consoles used −16 dBu (123 mVrms), which gives a much increased headroom of 38 dB. Apart from this specialised application, general audio equipment might be expected to have a nominal internal level in the range −6 dBu (388 mVrms) to 0 dBu (775 mVrms), with −2 dBu probably being the most popular choice.

If you have a given dynamic range and you're not happy with it, you can either increase the maximum signal level or lower the noise floor. The maximum signal levels in opamp-based equipment are set by the voltage capabilities of the opamps used, and this usually means a maximum signal level of about 10 Vrms or +22 dBu. Discrete transistor technology removes this absolute limit on supply voltage and allows the voltage swing to be at least doubled before the supply rail voltages get inconveniently high. For example, ±40 V rails are quite practical for small-signal transistors and permit a theoretical voltage swing of 28 Vrms or +31 dBu. However, in view of the complications of designing your own discrete circuitry and the greater space and power it requires, the extra 9 dB of headroom is bought at a high price. You will need a lot more PCB area, and of course the specialised knowledge of how to design discrete transistor stages. My book *Small Signal Audio Design* will be of considerable help with the latter. [2] If you plan to use signal levels as high as 28 Vrms, you might want to consider what that will do to opamp circuitry downstream if applied directly; it is likely to end badly. Attenuators at the very outputs of the crossover can remove this danger by reducing the high internal maximum level of 28 Vrms to a safe 10 Vrms; since the noise will be attenuated by the same amount as the signal, the advantage of the high internal level is preserved. This technique is also useful with all-opamp crossover designs and is described in detail later.

A current example of a crossover with all-discrete circuitry in the signal path is the Bryston 10B. [3]

## Circuit Noise and Low-Impedance Design

Increasing the dynamic range by reducing the noise levels in the circuitry is more practical (and in general a good deal cheaper), but there are some quite restrictive limits on how much you can do this. Adopting low-impedance design—in other words using the lowest resistor values you can without creating extra distortion by overloading the opamps—will reduce the Johnson noise the resistors generate and the effects of opamp current noise flowing in them. It also makes the circuit more immune to capacitive crosstalk and interference pickup. However, Johnson noise is proportional to the square root of the resistance, and so moving from 10 k$\Omega$ to 1 k$\Omega$ will only reduce the noise by 10 dB ($\sqrt{10}$ times) rather than 20 dB (10 times). A reduction of 10 dB is nevertheless very well worth having. Things get more difficult if you want to reduce the impedance levels further, as opamp distortion will start to increase due to the heavier loading.

This can be countered by using opamps in parallel to increase the drive capability, assuming you're not designing to the absolute minimum cost. Two opamps working together allow the circuit impedances to be halved, giving us another 3 dB improvement, while four opamps allow them to be halved again, giving 6 dB less noise. This is going to be about as far as it is economical to go unless you're designing really gold-plated gear, so we have a total Johnson noise improvement from low-impedance design of 16 dB.

Johnson noise is however only one component of the circuit noise, the other two important contributions coming from the voltage noise and the current noise of the active devices. Reducing the circuit impedances reduces the effect of current noise—proportionally this time, as the current noise only manifests itself when it causes a voltage drop across an impedance. Voltage noise is a tougher proposition to reduce, the options being (a) shell out for quieter and more expensive active devices; (b) make use of opamps in parallel again. If two opamp stages of the same gain are connected together by low-value resistors (say 10 $\Omega$), then at their junction you get the average of the two outputs, so the signal level is unchanged, but the noise drops by 3 dB ($1/\sqrt{2}$), as the two noise components are uncorrelated and so partially cancel. Four opamp stages give a 6 dB improvement. This technique obviously goes extremely well with using opamps in parallel to allow circuit impedance reduction and can make for some very neat and effective circuitry.

## Using Raised Internal Levels

When setting the internal levels of an active crossover, a great deal depends on the way that it is going to be built into the overall system. If the crossover is running directly into power amplifiers with no intermediate level control, it can be guaranteed that the output of the power amplifier will clip long before the crossover outputs, as even the most insensitive amplifiers are unlikely to need more than +8 dBu (2 Vrms) to drive them fully. It therefore occurred to me that it would be quite safe to raise the crossover internal level to 2 Vrms, which would theoretically give a 10 dB better signal-to-noise than the often-used −2 dBu level. That is a significant improvement.

As an example, look at Figure 17.2a, where the crossover is operating with an internal level of 0 dBu throughout, which is also the input required by the power amp. Only one of the paths through the crossover is shown. There is a unity-gain input amplifier A1 which has a noise level of −100 dBu (the input is assumed to be completely noise-free; keeping down the noise level from the preamplifier is someone else's problem). The filters, etc of the crossover have a noise level of −85 dBu, and when this is summed with the −100 dBu from A1 we get −84.9 dBu at the crossover output; the noise from A1 makes only a tiny contribution. The signal is then passed directly to the power amplifier, and our signal-to-noise ratio is 84.9 dB.

An elevated internal level +8 dBu (2 Vrms) is used in Figure 17.2b, input amplifier A1 having a gain of +8 dB. The noise out from A1 is now −92 dBu, and with the −85 dBu of noise from the crossover filters added, the total is −84.2 dBu. A passive 8 dB output attenuator R1, R2 then reduces the signal level back to the 0 dBu required by the power amplifier, and the noise is also reduced by 8 dB, giving us −92.5 dBu at the amplifier input. The signal-to-noise ratio has increased from 84.9 dB to 92.5 dBu, an improvement of 7.6 dB. We do not get the whole 8 dB because of the increased noise from input amplifier A1.

An elevated internal level not only makes the signal more proof against noise as it passes through the signal chain but also against hum and other interference, though a good design should have negligible levels of these last two anyway.

You will note that I specified a passive output attenuator, so that the very low noise output is not compromised by an opamp stage after it. The attenuation can be made variable to give an output level trim control, working over perhaps a ±3 dB range. Given opamps with good load-driving capability, it is possible to make the passive attenuator with low resistance values so the output impedance is still acceptably low for driving long cables. The attenuator values shown in Figure 17.2b give an output impedance of 166 Ω, which is not perhaps to the highest professional standards but quite good enough for a domestic installation with limited cable runs. The load on the last opamp in the crossover is 700 Ω, which is high enough to prevent significantly increased distortion from a 5532 stage. There is more on such output networks, and how they can be combined with balanced outputs, later in this chapter.

There is an assumption here that the crossover is mostly composed of unity-gain circuitry such as the standard Sallen & Key filters. This is not always true—if you are using equal-C Sallen & Key Butterworth filters, then each 2nd-order stage has a gain of +4.0 dB, and this will have to be dealt with somehow. Equalisation circuitry may have gains greater than this at some frequencies, with a corresponding reduction in headroom; this applies particularly to equalisers intended to extend the LF speaker unit response, which may have gains of +6 dB or more.

However, let's take it a little further. If our power amplifier clips with an input of 0 dBu, which corresponds to a crossover internal level of +8 dBu (2 Vrms), then as far as the crossover is concerned there is a range of signal levels from 2 Vrms to 10 Vrms (opamp maximum output) that is unusable. That is a 14 dB range. The effective signal-to-noise ratio could be further improved if the crossover ran

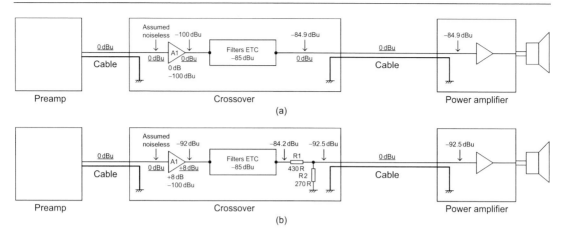

**Figure 17.2: Gain structure for the preamplifier-crossover-power amplifier chain: (a) internal crossover level of 0 dBu gives an S/N ratio of 84.9 dB; (b) raising the internal level to +8 dBu gives an S/N ratio of 92.5 dB, an improvement of 7.6 dB.**

at a still higher level, say 5 Vrms (we want to keep a little safety margin), with the signal more heavily attenuated at the output to reduce it to 0 dBu; the internal crossover signals would then be another 8 dB higher, and we should now get some 16 dB more signal-to-noise ratio than with the 0 dBu internal level.

If there are full-range level controls between the crossover outputs and the power amplifiers, to allow level trimming, then the operator may unwisely set them for a considerable degree of attenuation; they will probably then crank up the input into the crossover to compensate, so there is now a higher nominal level in the crossover to obtain the same final output level. This means that the headroom in the crossover is reduced, and the option of running it at a deliberately elevated internal level to improve the signal-noise ratio now looks less attractive. The question here is whether the noise performance should be compromised by increasing the headroom to allow for maladjustment.

## Placing the Output Attenuator

Let's stick for the moment with the situation that there are no full-range level controls on the power amplifiers, just an output level trim, as described earlier. We therefore have our passive attenuator at the crossover outputs. The signal has therefore been brought back down to something of the order of 1 Vrms before it passes along the interconnection between crossover and power amplifier. However—if the attenuator is placed at the destination end of the interconnect cable, as in Figure 17.3a, any hum and interference picked because of currents flowing through the cable ground will also be attenuated.

Ideally the output attenuator should be actually inside the power amplifier, as in Figure 17.3b, as this would also deal with any voltages caused by ground current flowing through the connector earth pins, but this requires a specialised amplifier design which would have a low input impedance and low overall gain. Its input parameters would have to be defined by a Domestic Electronic Crossover Standard, and unless there was some sort of input switching, it would not be usable as an ordinary power amplifier. A better idea is to put the attenuator inside the connector that plugs into the power amp.

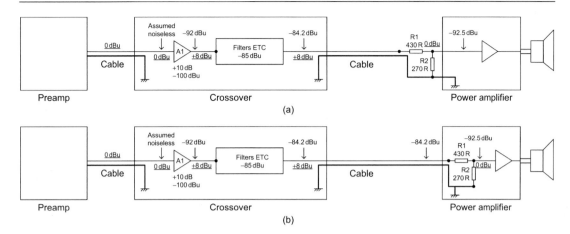

**Figure 17.3:** The preamplifier-crossover-power amplifier chain: (a) attenuator placed in the connector at the far end of the crossover-power amplifier; (b) attenuator placed in the power amplifier itself.

By now you are probably thinking: "Why not use a balanced interconnection? They are designed to discriminate against ground voltages"—which is of course absolutely correct. However, if you peruse the chapter on line inputs and outputs (Chapter 20), you will note that a practical real-life balanced input has a fairly modest common-mode rejection ratio (CMRR) of the order of 40 to 50 dB, and the most straightforward way of improving that is to have a preset CMRR trim. On a six-channel amplifier that is six adjustments to make, and while they are set-and-forget adjustments, this won't appeal to all manufacturers. Putting the passive attenuator at the power amplifier end of the cable gives another 10 to 16 dB of immunity against ground noise on top of the basic CMRR. This plan obviously makes for a more complicated cable, but it may be worth it, because ground-loop problems are a never-ending cause of distress in the audio business.

## The Amplitude/Frequency Distribution of Musical Signals and Internal Levels

In considering the headroom requirements, it is important to remember that the energy contained in the HF, MID, and LF bands of the typical crossover are very different. The energy in the HF band is much lower, and this prompts me to suggest the possibility of running the HF channel at a higher nominal level than the others so noise performance can be improved without significantly compromising headroom. This is particularly appropriate for the HF channel because it contains no lowpass filters and therefore may be expected to have a higher noise output.

So how much should the HF channel level be raised? Statistics on the distribution of amplitude with frequency are surprisingly hard to find, given how fundamental this information is to audio design; some information is given in [4], but it is not presented in a very convenient form for our purposes. A good source is Greiner and Eggars, [5] who derived a lot of statistical data from 30 CDs of widely varying musical genres. The data is presented as the level in each of ten octave bands which is exceeded 90% of the time, 50% of the time, 10% of the time, and so on. This is a great deal of very useful information, but it does not directly tell us what we want to know, which might typically be phrased as: "If I have a 3-way system with crossover frequencies at 500 Hz and 3 kHz, how much level can I expect in each of the three crossover bands?" Greiner and Eggars do however at the end of their paper summarise the spectral energy levels in each octave band, which is more helpful. See Table 17.1, which gives the levels in dB with respect to full CD level for eight different musical genres, and the average of them all. The frequencies are the centres of the octave bands.

Figure 17.4 shows how the average levels are distributed across the audio band. The maximal levels are the region 100 Hz–2 kHz, with roll-offs of about 10 dB per octave at each end of the audio spectrum.

All we have to do is to decide which of the octave bands fit into our crossover bands and sum the levels in those bands to arrive at a composite figure for each crossover band. A puzzling difficulty is to decide how to sum them—should they be treated as correlated (so two −6 dB levels sum to 0 dB) or uncorrelated (two −3 dB levels sum to 0 dB)? At first it seems unlikely that there would be much correlation between the octave bands, but on the other hand harmonics from a given musical instrument are likely to spread over several of them, giving some degree of correlation. I tried both, and for our purposes here the results are not very different.

Let's assume that our LF crossover band includes the bottom four octave bands (31.5 to 250 Hz), the MID crossover band includes the middle three octave bands (500 Hz to 2 kHz), and the HF crossover band includes the top three octave bands (4 kHz to 16 kHz). Uncorrelated summing gives us Table 17.2, where the actual levels in each crossover band are on the left and the relative levels of the MID and HF bands compared with the LF band are in the two columns on the right.

Looking at the bottom line, the average of all the genres, we see that in general the MID crossover band will have similar levels in it to the LF crossover band. This is pretty much as expected, though it's always good to have confirmation from hard facts. We also see that the average HF level is a heartening 11 dB below the other two crossover bands, so it looks as if we could run the HF channel at an increased level, say +10 dB, and get a corresponding improvement in signal-to-noise ratio.

**Table 17.1: Spectral energy levels in octave bands for different genres (after Greiner and Eggars).**

| Octave centre Hz | 31.5 | 63 | 125 | 250 | 500 | 1k | 2k | 4k | 8k | 16k |
|---|---|---|---|---|---|---|---|---|---|---|
| Piano B | −63 | −47 | −34 | −28 | −27 | −33 | −38 | −46 | −58 | −63 |
| Organ A | −32 | −30 | −29 | −31 | −30 | −29 | −32 | −37 | −50 | −69 |
| Orchestra B | −34 | −33 | −29 | −29 | −28 | −30 | −32 | −39 | −48 | −58 |
| Orchestra C | −26 | −26 | −30 | −32 | −32 | −33 | −35 | −38 | −45 | −54 |
| Chamber music | −78 | −62 | −45 | −39 | −41 | −46 | −49 | −51 | −65 | −78 |
| Jazz A | −48 | −36 | −33 | −31 | −29 | −27 | −29 | −32 | −39 | −49 |
| Rock A | −39 | −35 | −34 | −32 | −31 | −32 | −30 | −36 | −48 | −57 |
| Heavy metal | −45 | −31 | −27 | −27 | −33 | −37 | −31 | −29 | −36 | −46 |
| AVERAGE | −45.6 | −37.5 | −32.6 | −31.1 | −31.4 | −33.4 | −34.5 | −38.5 | −48.6 | −59.3 |

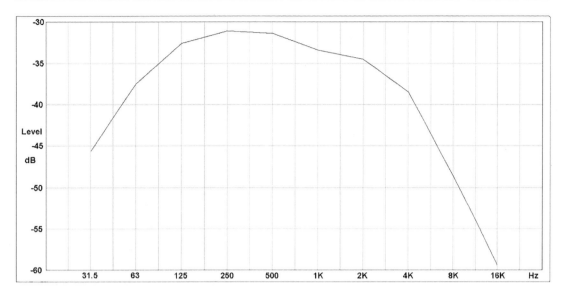

**Figure 17.4: Average spectral levels in musical signals from CDs (after Greiner and Eggars).**

**Table 17.2: Uncorrelated sums of levels in the three crossover bands; HF band = top three octaves. The two columns on the right are levels relative to the 31.5–250 Hz LF band.**

| Octave centre Hz | LF dB 31.5–250 | MID dB 500–2k | HF dB 4k–16k | MID dB 500–2k Relative to LF | HF dB 4k–16k Relative to LF |
|---|---|---|---|---|---|
| Piano B | −27.0 | −25.8 | −45.7 | 1.2 | −18.7 |
| Organ A | −24.3 | −25.4 | −36.8 | −1.1 | −12.4 |
| Orchestra B | −24.7 | −24.9 | −38.4 | −0.3 | −13.8 |
| Orchestra C | −21.8 | −28.4 | −37.1 | −6.6 | −15.4 |
| Chamber music | −38.0 | −39.3 | −50.8 | −1.3 | −12.8 |
| Jazz A | −28.1 | −23.5 | −31.1 | 4.6 | −3.1 |
| Rock A | −28.3 | −26.2 | −35.7 | 2.2 | −7.4 |
| Heavy Metal | −23.2 | −28.3 | −28.1 | −5.1 | −5.0 |
| AVERAGE | −26.9 | −27.7 | −38.0 | −0.8 | −11.1 |

However . . . look a little deeper than the average. Table 17.2 shows that the HF level is significantly lower in all cases except for "Heavy Metal", where the MID and HF levels are the same. This is so unlike all the other data that I am not convinced it is correct. It looks as though we could indeed run the HF channel a good 10 dB hotter if it were not for the existence of Heavy Metal—a sobering thought.

If we change our assumptions, moving the upper crossover frequency higher so that the MID crossover band now includes the middle four octave bands (500 Hz to 4 kHz) and the HF crossover

Table 17.3: Uncorrelated sums of levels in the three crossover bands; HF band = top two octaves. The two columns on right are levels relative to the 31.5–250 Hz LF band.

| Octave centre Hz | LF dB 31.5–250 | MID dB 500–4k | HF dB 8k–16k | MID dB 500–4k Relative to LF | HF dB 8k–16k Relative to LF |
|---|---|---|---|---|---|
| Piano B | −27.0 | −25.7 | −56.8 | 1.3 | −29.8 |
| Organ A | −24.3 | −25.1 | −49.9 | −0.8 | −25.6 |
| Orchestra B | −24.7 | −24.8 | −47.6 | −0.1 | −22.9 |
| Orchestra C | −21.8 | −27.9 | −44.5 | −6.2 | −22.7 |
| Chamber music | −38.0 | −39.0 | −64.8 | −1.0 | −26.8 |
| Jazz A | −28.1 | −22.9 | −38.6 | 5.2 | −10.5 |
| Rock A | −28.3 | −25.7 | −47.5 | 2.6 | −19.2 |
| Heavy metal | −23.2 | −25.6 | −35.6 | −2.4 | −12.4 |
| AVERAGE | −26.9 | −27.1 | −48.2 | −0.2 | −21.2 |

band includes only the top two octave bands (8 kHz to 16 kHz), then things look rather better. Uncorrelated summing now gives us Table 17.3, where the HF levels are now down by some 20 dB with respect to the LF and MID channels. Now even Heavy Metal can have the level in the HF path pumped up by 10 dB.

Nevertheless, what we can only call The Heavy Metal Problem is messing up the nice clear results from this study. The HM track used by Greiner and Eggars is only identified as "Metallica", but a little detective work revealed it to be the track "Fight Fire with Fire" from the CD "Ride The Lightning". I am listening to it as I write this, and I cannot honestly say that the HF content is obviously greater than that of the normal run of rock music. Are we justified in regarding this Heavy Metal data as an outlier that can be neglected? Hard to say.

At this point the only conclusion can be that you can run the HF channel hotter, but how much depends on where the upper crossover point is. If it is around 3 kHz, then elevating the HF channel level by 6 dB looks pretty safe, and this is the value adopted in my crossover design example in Chapter 23.

Another area where the distribution of amplitude with frequency is important is that of transformer balanced input and outputs. Transformers are notorious for generating levels of third-harmonic distortion that rise rapidly as frequency falls. It is fortunate that Table 17.1 shows that in almost all cases (church organ music being a notable exception), the signal levels in the bottom audio octave are roughly 10–12 dB lower than the maximum amplitudes, which occur in the middle octaves. This reduces the transformer distortion considerably.

## Gain Structures

There are some very basic rules for putting together an effective gain structure in a piece of audio equipment. Breaking them reduces the dynamic range of the circuitry, either by raising the noise floor or lowering the headroom. Three simple rules are:

1. DON'T AMPLIFY THEN ATTENUATE. It is all too easy to thoughtlessly add a bit of gain to make up for a loss later in the signal path, and immediately a few dB of precious headroom are gone for good. This assumes that each stage has the same power rails and hence the same clipping point, which is usually the case. Figure 17.5a shows a fragment of a system with a gain control designed to have +10 dB of gain at maximum. There is assumed to be no noise at the input. A and B are unity-gain buffers, and each contributes−100 dBu of its own noise; Amplifier 1 has a gain of +10 dB and an EIN of −100 dBu. The expectation is that the level control will spend most of its time set somewhere near the "0 dB" position, where it introduces 10 dB of attenuation. To keep the nominal signal level at 0 dBu we need 10 dB of gain, and Amplifier 1 has been put before the gain control. This is a bad decision, as this amplifier will clip 10 dB before any other stage before it in the system, and this introduces what one might call a headroom bottleneck. On the positive side, the noise output is only −96.8 dBu because the signal level never falls below 0 dBu and so is relatively robust against the noise introduced by the stages.

2. DON'T ATTENUATE THEN AMPLIFY. Since putting our 10 dB of amplification before the gain control has disastrous effects on headroom, it is more usual to put it afterwards, as shown in Figure 17.5b. Now noise performance rather than headroom suffers; the amount of degradation depends on the control setting, but as a rule it is much more acceptable than a permanent 10 dB reduction in headroom. The signal-to noise ratio is impaired for all gain control settings except maximum; if we dial in 10 dB of attenuation as shown, then the signal reaching Amplifier 1 is 10 dB lower, at −10 dBu. The noise generated by Amplifier 1 is unchanged, and almost all the noise at its output is its own EIN amplified by 10 dB. This is slightly degraded by the noise of block B to give a final noise output of −89.2 dBu, worse than Figure 17.5a by 7.6 dB. If there are options for the amplifier stages in terms of a noise/cost trade-off and you can only afford one low-noise stage, then it should clearly be Amplifier 1.

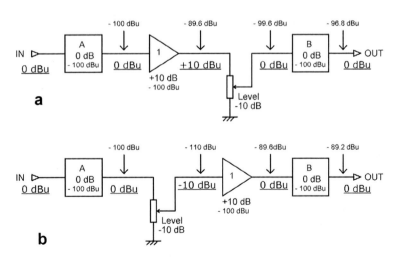

**Figure 17.5: Gain structures: (a) amplification then attenuation. Amplifier 1 always clips first, reducing headroom. (b) Attenuation then amplification. Noise from Amplifier 1 degrades the S/N ratio at low-gain settings. Noise levels along the signal path indicated by arrows; signal levels are underlined.**

3.    AMPLIFY AS SOON AS POSSIBLE. Get the signal up to the nominal internal level as soon as it can be done, preferably in the first stage, to minimise its contamination with noise from later stages. Consider the signal path in Figure 17.6a, which has a nominal input level of −10 dBu and a nominal internal level of 0 dBu. It has an input amplifier with 10 dB of gain followed by two unity-gain buffers A and B. As before, all circuit stages are assumed to have an equivalent input noise level of −100 dBu, and the incoming signal is assumed to be entirely noise-free. The noise output from the first amplifier is therefore −100 dBu + 10 dB = −90 dBu. The second stage A adds in another −100 dBu, but this is well below −90 dBu, and its contribution is very small, giving us −89.6 dBu at its output. Block B adds another −100 dBu, and the final noise output is −89.2 dBu.

Compare that with a second version of the signal path in Figure 17.6b that has an input amplifier with 5 dB of gain, followed by block A, a second amplifier with another 5 dB of gain, then block B. The signal has received exactly the same amount of gain, but the noise output is now 1.7 dB higher at −87.5 dB because the signal passed through block A at −5 dBu rather than 0 dBu. There is also an extra amplifier stage to pay for, and the second version is clearly an inferior design.

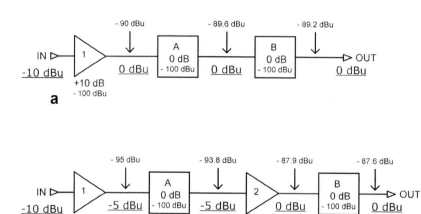

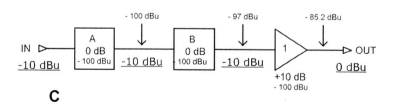

Figure 17.6: Why you should amplify as soon as possible: (a) all amplification in first stage gives best noise performance of −89.2 dBu. (b) Amplification split over two stages. Noise from Stage A degrades the S/N ratio. (c) Amplification late in chain. Now both Stages A and B degrade the S/N ratio. Noise levels indicated by arrows; signal levels are underlined.

An even worse architecture is shown in Figure 17.6c, where both the two unity-gain blocks A and B are now ahead of a +10 dB amplifier. The low-level signal at −10 dBu is now more vulnerable to the noise from block B, and the output noise is now −85.2 dBu, 4.0 dB worse than the optimal arrangement in Figure 17.6a, where all the amplification is placed at the start of the signal path.

In active crossover design the problem is not usually quite so marked as in the examples in Figure 17.6, because any level controls are usually trims rather than full-range volume controls, but the principles still stand. Let us examine a typical scenario; assume that you want to make a 4th-order Linkwitz-Riley filter, and you are planning to use two cascaded 2nd-order Sallen & Key Butterworth filters of the equal-C type to make component procurement simpler. For a 2nd-order Sallen & Key filter to be equal-C, the stage gain has to be 1.586 times (+4 dB), and when you cascade two to make the 4th-order Linkwitz-Riley filter, you have a total gain of 2.5 times (+8 dB)to deal with. There are three possibilities: First, you can attenuate the signal by 8 dB *before* the filters, but the low level in the first filter will almost certainly degrade the noise performance.

Second, you can attenuate the signal by 8 dB *after* both filters, but now there are higher signal levels, in particular 8 dB higher at the output of the second filter, and headroom problems may occur at this point.

Third, and usually best, is some compromise between these two extremes. For example, putting 4 dB of attenuation after the first filter and then another 4 dB of attenuation after the second filter will reduce the headroom problems by 4 dB. Alternatively, putting 4 dB of attenuation before the first filter, followed by 4 dB of attenuation after it, will reduce the noise contribution of the first stage.

The aforegoing assumes you do not need to add buffers between the attenuators and the filters to give the latter a low source impedance. There is some useful information on avoiding buffers and economically combining attenuators with filters in Chapter 8.

If you are seeking the best possible performance, then probably your best option is to not use Sallen & Key equal-C filters in the first place but stick to the more usual unity-gain sort.

## Noise Gain

Before we dig deeper into the details of getting the cleanest possible signal path, we need to look at the concept of noise gain. Look at Figure 17.7, which shows two inverting gain stages, both with unity gain. The only difference is the presence of an innocent-looking resistor R3 in the (b) version of the circuit. Since the inverting input of the opamp is at virtual ground (i.e. it has a negligible signal voltage on it but is not actually connected to ground) because of the very high negative feedback factor, the gain will be very nearly exactly equal to −1 for both circuits. So are they functionally identical? No indeed. The (b) version will be 3.5 dB noisier, simply because R3 is lurking in there.

This is nothing to do with Johnson noise from R3, which I have ignored; even if R3 was some sort of magical resistor with no Johnson noise, (which is not possible in this universe) the (b) version would be 3.5 dB noisier. This is because although it has the same signal gain, it has a higher *noise gain*.

Opamp voltage noise is accurately modelled by assuming a noiseless opamp and putting a voltage noise generator between the two inputs (I am ignoring current noise, as it has negligible effects with

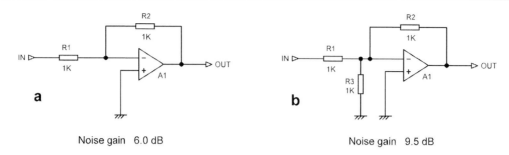

Figure 17.7: Demonstration of the concept of noise gain: (a) unity-gain inverting stage with noise gain of 6.0 dB; (b) unity-gain inverting stage with noise gain of 9.5 dB.

low resistances like those shown). One side of this generator is grounded via the non-inverting input, and as far as the opamp is concerned, the input is also effectively at ground, because it is fed from a low source impedance. Therefore, as far as noise is concerned, we have not an inverting amplifier but a non-inverting amplifier, with the noise generator as the input and R1, R2 in Figure 17.7a as the feedback network. R1 equals R2 so the gain from noise generator to output is 2 times; (+6 dB) Figure 17.7a has unity gain for signal but 6 dB of gain for its own noise.

For Figure 17.7b the story is different; so far as noise is concerned, the upper arm of the feedback network is unchanged, but the lower arm now consists of two equal resistors, so their combined resistance is half the value and the gain from noise generator to output is 3 times (+9.5 dB).

When concocting amplifier circuits you should always be aware of the possibility of accidentally building in R3 or some less obvious equivalent. Noise gain is a really major issue in mixing consoles, where the large number of mix resistors feeding the virtual earth summing bus [6] are effectively in parallel and represent a very low value for R3; a 32-input console mix system has a noise gain of +30.4 dB, and you can see why summing amplifiers with the lowest possible noise are required. In active crossovers, noise gain issues are most likely to arise in equaliser circuits such as the bridged-T and the biquad equaliser; see Chapter 14. As a general rule any amplifier or filter will amplify its own noise by more than an incoming signal, even if there is nothing as obvious as R3. The only obvious exception is the voltage-follower, which has a signal gain of unity and a noise gain of unity.

## Active Gain Controls

So far we have dealt only with combinations of fixed-gain amplifiers and passive attenuation controls, and we have seen that either noise performance or headroom must be compromised when a gain control function is included in the signal path. If however we move beyond the idea of a fixed-gain amplifier, this compromise can be to a large extent avoided. If we have an amplifier stage with variable gain, we can set it to give just the gain we want and no more. Using less gain when the maximum is not required will reduce the noise generated compared with a fixed-gain amplifier placed after a level control, because the fixed-gain amplifier has to have the maximum gain required and is hissing away all the time with no following level control to turn it down.

Simply making a stage with variable gain is straightforward; you just vary the amount of negative feedback. Achieving a given gain law requires a little more thought (see Chapter 20 on line inputs). In a crossover it will not normally be necessary for the gain to be variable right down to zero, as it would be for a preamplifier volume control. [7]

Another advantage of active gain controls is that they can give much better channel balance by eliminating the uncertainties of log pots. The Baxandall active gain configuration gives excellent channel balance, as it depends solely on the mechanical alignment of a dual linear pot—all mismatches of its electrical characteristics are cancelled out, and there are no quasi-log dual slopes to cause heartache. [8]

To demonstrate the advantages of active gain controls, take a look at Figure 17.8. The circuit at (a) is a conventional passive level control followed by a +10 dB amplifier stage. At (b) is an active gain control also giving a maximum of +10 dB gain. In this case we have a shunt-feedback circuit that can be adjusted all the way down to zero gain, but that is not an essential feature. If we assume both opamps have the same voltage noise (the actual level does not matter) and neglect current noise and Johnson noise, which make little contribution with such low impedances, then it is straightforward to calculate the signal-to-noise (S/N) ratio for each circuit. This is expressed as a the difference between the two circuits in Table 17.4, as this makes the actual signal level irrelevant. The level control setting is recorded as dB down from fully up, whatever the maximum gain.

The S/N advantage of the active gain control is greatest at intermediate and low-level settings, and this advantage increases as the maximum gain increases. However, when the control is set near or at maximum gain, the S/N is marginally worse. This is because the noise gain of Figure 17.8b is always one times more than the signal gain because of its shunt-feedback configuration, whereas for Figure 17.8a the signal gain and noise gain are always equal. The higher the maximum gain required, the less significant this effect is; for the 20 dB maximum gain case with the control fully up, the active version is 0.8 dB noisier, but backing it off by only 1 dB reverses the situation, and the passive version is now 0.1 dB noisier. At lower control settings the advantage of the active version increases rapidly, being a useful 7.6 dB at the −10 dB setting and asymptotically increasing to 20 dB.

If we compare our active stage not with the "passive control then amplifier" structure of Figure 17.8a but with the rather impractical "amplifier first then passive gain control" of Figure 17.5a, the active

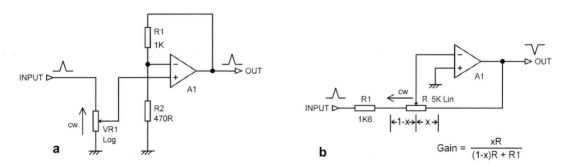

**Figure 17.8: Passive and active gain controls, both with a maximum gain of +10 dB.**

Table 17.4: The S/N improvement with an active gain control for various maximum gains.

| Max gain = | S/N ratio compared with passive circuit in Figure 14.8a | | |
| | 10 dB | 15 dB | 20 dB |
| Level control dB | dB | dB | dB |
| 0 | −2.4 | −1.4 | −0.8 |
| −1 | −1.6 | −0.6 | 0.1 |
| −2 | −0.9 | 0.2 | 1.0 |
| −5 | 1.1 | 2.6 | 3.6 |
| −7 | 2.4 | 4.1 | 5.2 |
| −10 | 4.0 | 6.1 | 7.6 |
| −20 | 7.6 | 11.1 | 14.0 |
| −30 | 9.2 | 13.6 | 17.6 |
| −40 | 9.7 | 14.5 | 19.2 |
| −50 | 9.9 | 14.8 | 19.7 |
| −60 | 10.0 | 15.0 | 19.9 |
| −70 | 10.0 | 15.0 | 20.0 |
| −80 | 10.0 | 15.0 | 20.0 |

noise performance is inferior at low-gain settings because the noise gain of the shunt-feedback stage cannot fall below unity, whereas a final passive control can attenuate the noise fed to it indefinitely. However, as we have seen, "amplifier first then passive gain control" is rarely if ever a good choice in system design because of the crippling loss of headroom. There is never any issue of reduction of headroom with the active gain control, as only the amount of gain required is actually in use.

Whenever you need a level control you should carefully consider the possibility of using active gain control circuitry.

## Filter Order in the Signal Path

The audio performance is affected by the order in which filters and other circuit blocks are placed. If you have any lowpass filters in a signal path, then put them last, so they will attenuate noise from the other circuit blocks ahead of them.

As an example of this, Figure 17.9 shows a typical crossover MID path composed of 4th-order Linkwitz-Riley lowpass (400 Hz) and highpass (3 kHz) filters. Each Linkwitz-Riley filter is made up of two identical 2nd-order Butterworth filters; note that "2 x 47 nF" means two 47 nF capacitors in parallel. The lowpass filters are here placed *before* the highpass filters in the signal path. The noise levels in the signal path were measured, and the results are shown in Figure 17.10.

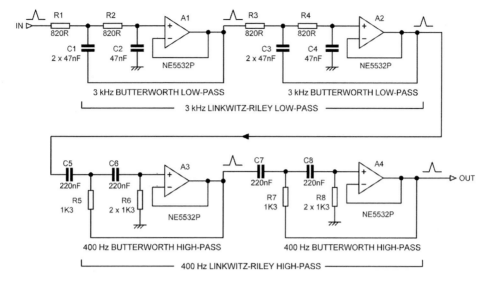

**Figure 17.9: A typical Linkwitz-Riley MID path with the lowpass filters placed before the highpass filters. Reversing the order reduces the noise output.**

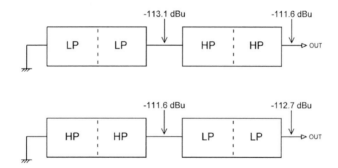

**Figure 17.10: How changing the filter order so the lowpass filters come after the highpass filters reduces the noise output.**

The order of the filters was then reversed so the lowpass filters came at the end of the chain, and Figure 17.10 shows that we reduce the noise by 1.1 dB by this simple change. I do not pretend this is a radical improvement, but it has the advantage that it costs *absolutely nothing*. If there is more circuitry in the signal path than just the crossover filters, such as frequency/amplitude equalisers or delay-compensation allpass filters, these also should be placed upstream of the lowpass filters, and the improvement in noise performance will be rather greater, as these circuits tend to be noisier than simple highpass filters. Bridged-T equalisers in particular have a reputation for being noisy due to their high internal noise gain.

You might wonder if the distortion performance is in some way impaired. It isn't. It is only possible to make meaningful measurements between about 200 Hz and 5 kHz, as the rapid 4th-order amplitude roll-offs mean that you are very quickly just measuring noise rather than distortion, but in this region the distortion is in fact slightly lower with the lowpass filters at the end of the path. This is because they tend to filter out distortion harmonics from the preceding highpass filters.

One possibility to consider there might be increased HF intermodulation in the highpass filters, as they are now handling a more extended range of the audio spectrum. Since the amount of negative feedback is reduced at HF, this would be of greater concern than at LF. I have seen no sign that this could be a significant problem.

When settling the arrangement of the filter blocks in the crossover, you should be aware of the amplitude response irregularities that can be caused by filter phase-shift when crossover frequencies are not widely spaced. This very important issue is dealt with at the end of Chapter 4.

## Output Level Controls

The first question is how much gain control should be available on the output of a crossover. We are not trying to make a volume control that can be adjusted from full up to zero, first because the place for the volume control is on the preamplifier, where you can get at it—the crossover will usually be tucked away out of side and possibly hidden in an equipment cabinet. Second, if we did want to put a volume control in this position, a 3-way crossover would require a six-gang pot, which would be expensive and probably have some unsettling channel balance errors, unless a switched attenuator or active gain control was used, and an unpleasant mechanical feel.

If we are simply intending to allow for gain variations in a nominally identical set of power amplifiers, then a vernier control of as little as ±1 dB may be sufficient. If in addition the gain controls must allow for production variations in transducers, then a wider range of ±3 dB or ±6 dB may be appropriate, but it is essential to realise that simply altering the gain of each channel will probably not give anything like enough control; making really accurate allowances for transducer tolerances will probably require tuning of crossover frequencies and several equalisation parameters.

This sort of calibration is quite feasible if it is computer-based, and each crossover example is permanently allocated to its matched loudspeaker. There is no guarantee that loudspeakers will have identical units in a stereo pair, and so it is essential not to swap left and right channels; it is unlikely that the crossover settings will be identical.

As we saw earlier in this chapter, the use of raised internal levels in the crossover circuitry followed by passive attenuation can give significant improvements in signal-to-noise ratio. The passive attenuator can be conveniently combined with an output trim network as shown in Figure 17.11, where the component values are chosen to give a ±6 dB adjustment range. This is a fragment from one of my active crossover designs, where the internal signal path was running at an elevated level of 3 Vrms and the nominal output was 880 mVrms.

The design considerations are:

1.  That the network resistances should be high enough to not excessively load the last opamp in the signal path, to ensure no excess distortion is generated. The load imposed by Figure 17.11 is 625 Ω, which is about as heavy as you would want to make it.

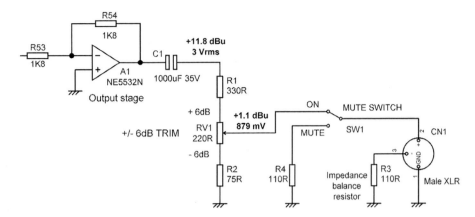

**Figure 17.11: Low-impedance passive output attenuator with gain trim, mute switch, and impedance-balancing resistor.**

2.  That the network resistances should be low enough to give a suitably low output impedance to drive reasonable lengths of cable without HF losses. The network in Figure 17.11 has a maximum output impedance of 156 Ω, which occurs when the control is fully up. The output impedances with the control central and fully down are 130 Ω and 66 Ω respectively. In the worst case (fully up), this arrangement could drive up to 53 metres of 150 pF/metre cable before the loss at 20 kHz reached 0.1 dB.

You can of course only have a ±6 dB trim range if the internal signal level is at least 6 dB above the nominal output level. It is very likely that the preset pot track resistance will have a wider tolerance than the fixed resistors, and this can cause channel balance errors. Note that no separate drain resistor is required to remove any charges left on C1 by external DC voltages.

The unbalanced output uses an XLR to allow impedance-balanced operation; R3 is connected between the cold output pin and ground to balance the impedances seen on the output pins and so maximise the CMRR when driving a balanced input. The value of 110 Ω is the average of the maximum and minimum output impedances, as the setting of the gain trim is varied. This variation inevitably compromises the balancing of the impedances, and picking an average figure for R3 is the best we can do. There is more on impedance balancing in Chapter 21 on line outputs and in Chapter 23. Note that C1 is rather big at 1000 uF; this is not to extend the frequency response but to make sure it does not generate capacitor distortion as it drives the relatively low impedance of the attenuator network. It is a non-polarised component, so it is immune to external fault voltages of either polarity.

## Mute Switches

Crossovers intended for sound reinforcement are often fitted with mute switches for each output to simplify system checking and debugging. The straightforward method shown in Figure 17.11 connects the output to ground and gives very good attenuation indeed, as almost all the crossover circuitry is disconnected. The "impedance-balanced" feature is retained because R4 = R3. Very slightly lower output noise might be obtained by replacing R4 with a short-circuit, but it is more important to retain the best possible CMRR, even in the test-only mute mode. Illuminated switches are often used to make operation easier in shadowy backstage areas.

## Phase-Invert Switches

This is another feature which is pretty much restricted to crossovers for sound reinforcement. The ability to invert the phase of just one output can be useful for checking and for the rapid correction of a phase error somewhere else in the system. Some crossovers (e.g. Behringer Super-XPRO CX2310) have separate phase-invert switches on all the main outputs and also the subwoofer output.

A phase-invert switch is easy to arrange if you have a balanced output; the outputs are just swapped over, as shown in Figure 17.12. With an unbalanced output it will be necessary to switch in a unity-gain inverting stage. If you are doing that, then you might as well make use of the added opamp section all the time and press it into service to implement a balanced output. Once again, illuminated switches are useful.

## Distributed Peak Detection

An electronic crossover is not normally fitted with level meters in the way that mixing consoles or power amplifiers are, it being assumed that it will be carrying out its function in some out-of-the-way corner where no one will see it. Nevertheless, some indication of grossly erroneous signal levels is useful.

When an audio signal path consists of a series of circuit blocks, each of which may give either gain or attenuation, it is something of a challenge to make sure that excessive levels do not occur anywhere along the chain. Simply monitoring the level at the end of the chain is no use, because a circuit block that gives gain, leading to clipping, may be followed by one that attenuates the clipped signal back to

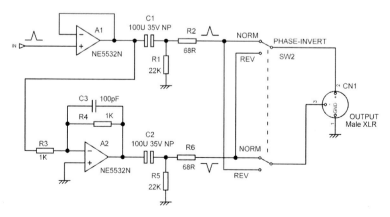

**Figure 17.12: A phase-invert switch added to a balanced output.**

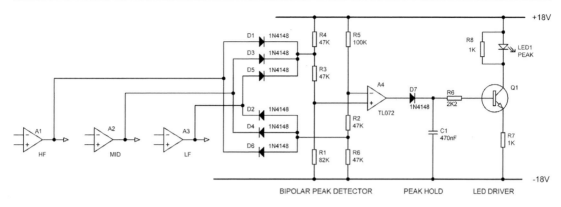

**Figure 17.13: A multi-point bipolar peak detector, monitoring three circuit blocks.**

a lower level that does not trip a final peak-detect circuit. It is not, however, necessary to add a bipolar peak-detection circuit at the output of every opamp to be sure that no clipping is happening anywhere along the path.

If a stage is followed by another stage with a flat gain, say of +4 dB, then there is of course no need to monitor the first stage. as the one following it will always clip first. If, however, the gain of the second stage is not flat, as in an electronic crossover it will very often not be, it may be necessary to monitor both stages. For example, if the second stage is a lowpass filter with a cutoff frequency of 400 Hz, then a signal at 10 kHz may cause the first stage to clip heavily without the second stage getting anywhere near its clipping point, even if it has gain in its passband..

A multi-point or distributed peak-detection circuit that I have made extensive use of is shown in Figure 17.13, where it is shown monitoring a circuit block in each path of a 3-way crossover. It can detect when either a positive or negative threshold is exceeded, at any number of points desired; to add another stage to its responsibilities you need only add another pair of diodes, so it is very economical. If a single peak detector is triggered by too many points in the signal path, it can be hard to determine at which of them the excessive level exists, but the basic message is still clear—turn it down a bit.

The operation is as follows. Because R5 is greater than R1, normally the non-inverting input of the opamp is held below the inverting input and the opamp output is low. If any of the inputs to the peak system exceed the positive threshold set at the junction of R4, R3, one of D1, D3, D5 conducts and pulls up the non-inverting input, causing the output to go high. Similarly, if any of the inputs to the peak system exceed the negative threshold set at the junction of R2, R6, one of D2, D4, D6 conducts and pulls down the inverting input, once more causing the opamp output to go high. When this occurs, C1 is rapidly charged via D7. The output-current limiting of the opamp discriminates against very narrow noise pulses. When C1 charges, Q1 turns on and illuminates D8 with a current set by the value of R7. R8 ensures that the LED stays off when A4 output is low, as it does not get close enough to the negative supply rail for Q1 to be completely turned off.

Each input to this circuit has a non-linear input impedance, and so for this system to work without introducing distortion into the signal path, it is essential that the diodes D1–D6 are driven directly from the output of an opamp or an equivalently low impedance. Do not try to drive them through a coupling capacitor, as asymmetrical conduction of the diodes can create unwanted DC-shifts on the capacitor.

The peak-detect opamp A4 must be a FET-input type to avoid errors due to its bias currents flowing in the relatively high-value resistors R1–R6, and a cheap TL072 works very nicely here; in fact, the resistor values could probably be raised significantly without any problems.

You will note all the circuitry operates between the two supply rails, with no connection to ground, preventing unwanted currents from finding their way into the ground system and causing distortion.

## Power Amplifier Considerations

When selecting the power amplifiers that are to work with active crossovers, the first consideration is how much power is required in each band. Earlier in this chapter we looked at the amplitude/frequency distribution of musical signals and found that while the LF and MID bands may have roughly the same amplitude, the HF band may be between 5 and 20 dB lower, depending on where the MID/ HF crossover frequency is set. Converting this into watts (with their square-law relationship with volts) tells us that if the LF and MID bands require 100 W amplifiers, that for the HF channel need be capable of no more than 31 watts in the worst case. There is some useful information on amplifier power requirements in Penkov. [4]

Another important factor is the efficiency of the speaker drive units for each band. Getting good low-frequency extension from an enclosure of limited size means low efficiency, and for this reason more power is generally required for LF channels than for MID channels.

In general there is little to be gained by designing different amplifiers for the frequency bands, apart from the power differences described earlier. You might be able to cut down on the size of the power supply reservoir capacitors in the MID amplifier compared with the LF amplifier because of the absence of sustained transients, but the cost saving is hardly worth the trouble.

The exception to that statement is the HF power amplifier. A DC fault in this amplifier can vaporise a tweeter in the twinkling of a voice coil, as unlike a tweeter connected to a passive crossover, it is not protected by a series capacitor. This would seem to demand that DC-offset protection circuitry for HF amplifiers would need to be especially fast-acting to be effective; since such protection would inevitably be triggered by large bass signals, we have created an amplifier that cannot be used for anything other than tweeter-driving duties. This runs completely counter to the desirability of the user being able to connect up any amplifiers he feels are the best for the job. The vulnerability of HF drive units is an excellent reason to not use HF amplifiers that are any more powerful than is necessary to do the job.

The DC-offset protection could be much improved by adding a capacitor in series with the tweeter, but this re-introduces a large and expensive component, of the sort we thought we had left behind when we adopted an active crossover; there should however be no need for it to maintain a precise value, as would be required if it was part of a passive crossover. The series capacitor is likely to introduce its own shortcomings in the form of non-linearity, especially if it is an electrolytic, when it may introduce distortion even in the midband, where it is not implementing any kind of roll-off. [9] There is also the point that in the world of hi-fi, where almost any technology that can make some sort of noise has its adherents, it seems that no one is prepared to advocate capacitor-coupled power amplifiers.

# References

[1] Ross, M. "Power Amplifier Clipping and Its Effects on Loudspeaker Reliability" RaneNote 128, Rane Corporation

[2] Self, Douglas "Small Signal Audio Design Handbook" Second Edn, Focal Press, 2015, Chapter 3, ISBN: 978-0-415-70974-3 hardback

[3] Bryston http://bryston.com/10b_m.html

[4] Penkov "Power & Real Signals in an Audio System" JAES, June 1983, p. 423

[5] Greiner, R. and J. Eggars "The Spectral Amplitude Distribution of Selected Compact Discs," Journal of the Audio Engineering Society, 37, pp. 246–275, April 1989

[6] Self, Douglas Ibid, pp. 626–627

[7] Self, Douglas Ibid, Chapter 13 (Volume controls)

[8] Self, Douglas Ibid, pp. 361–368 (Baxandall vol control)

[9] Self, Douglas "Audio Power Amplifier Design" Sixth Edn, Newnes, 2013, pp. 106–107, ISBN: 978-0-240-52613-3

# *Subwoofer Crossovers*

## Subwoofer Applications

A subwoofer loudspeaker is a separate enclosure from the main loudspeakers that is specifically designed to reproduce low bass. They are normally used singly rather than in stereo pairs, the assumption being that low frequencies can be regarded as non-directional. Another justification for mono subwoofers is that bass information is commonly pan-potted to the centre of the stereo stage. This is essential if the recorded material is likely to be used to cut vinyl discs, as big differences in low-frequency information between left and right channels creates large up-and-down, as opposed to lateral, contours in the groove that increase the possibility of mistracking and can in bad cases throw the stylus out of the groove altogether. Another good reason for centralising the bass is that it needs large amounts of amplifier power to reproduce it, and it is therefore desirable to make full use of both channels.

In home entertainment the subwoofer is normally of modest size and is often installed under the television screen, in front of the listeners, though other placements are possible, exploiting the non-directional characteristics of its output. In automotive use subwoofers are installed in the boot (the trunk, to some of you) or the rear cabin space.

The typical frequency range for subwoofer operation is between 20 and 200 Hz. Sound-reinforcement subwoofers typically operate below 100 Hz, and THX-approved systems in cinemas operate below 80 Hz.

## Subwoofer Technologies

A wide variety of types of loudspeaker enclosure are used in subwoofers. The relatively small frequency range that has to be covered allows techniques such as bandpass enclosures to be used when they would not be usable with LF or mid drive units. The main types are:

*   Sealed (infinite baffle)
*   Reflex (ported)
*   Transmission line
*   Bandpass
*   Isobaric
*   Dipole
*   Horn loaded

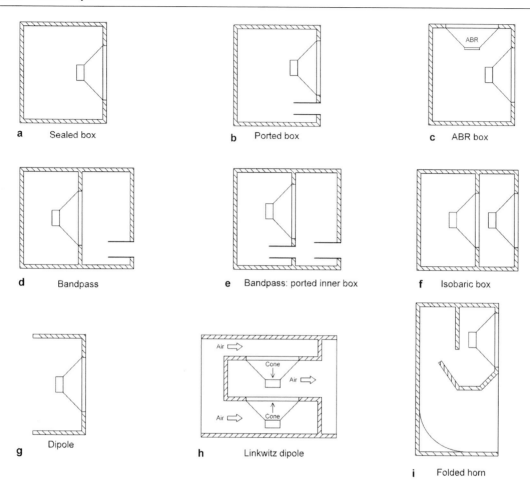

**Figure 18.1: The common types of subwoofer enclosure.**

The first two of these are by far the most popular. While I have not done a comprehensive survey, it seems clear that sealed enclosures are more popular than ported types, which is perhaps surprising given the greater bass extension possible with ported enclosures. Brief descriptions of the operation of each type are given next; I have kept them fairly short, as many aspects are covered in Chapter 2 on loudspeakers.

## Sealed-Box (Infinite Baffle) Subwoofers

The general characteristics of a sealed subwoofer enclosure, as shown in Figure 18.1a, are much the same as for normal loudspeakers, as examined in Chapter 2. For subwoofer use the driver may be mounted on one of the sides of the box or on the top or bottom; sometimes there are multiple drivers mounted on different sides of the box. It has a good transient response, good LF power handling because the drive unit is always loaded by the box air, and lower sensitivity to parameter misalignment than other approaches. Sealed-box enclosures have higher low-frequency cutoff points and lower sensitivity than other subwoofer types for the same box volume. Despite this they appear to be the most popular configuration for subwoofers.

The efficiency of a drive unit in a sealed box is proportional to the cube of the drive unit resonance frequency (the Thiele-Small parameter $F_s$), [1] so quite small improvements to low-frequency extension with the same drive unit and box volume lead to big reductions in the efficiency of converting electrical energy into sound energy. Subwoofers of reasonable size are therefore typically very inefficient by the standards of normal loudspeakers and require both powerful amplifiers and drivers with considerable power handling capability. For this reason enclosures other than a sealed box (e.g. bass reflex designs) are often used for subwoofers to increase the efficiency of the driver/ enclosure system and reduce the amplifier power requirements.

### Reflex (Ported) Subwoofers

Reflex or ported subwoofer enclosures work exactly as described for normal loudspeakers in Chapter 2. At low frequencies the port output is in phase with the forward radiation of the driver and allows the bass response to be extended without the lower efficiency that occurs with a sealed box. See Figure 18.1b.

With everything else being equal, a ported subwoofer can have lower drive unit distortion, higher power handling, and a lower cutoff frequency than a sealed-box system with the same drive unit. Distortion rapidly increases below the cutoff frequency, however, as the driver becomes unloaded and its excursion increases. The transient response of a ported subwoofer is usually worse than that of a sealed-box system using the same driver. The drivers used in ported subwoofers usually have a Qts value in the range 0.2 to 0.5.

Because of the low frequencies involved, large amounts of air are in motion, and the port should have the largest practicable diameter and be flared at both ends to minimise chuffing noises from turbulent air flow. It is often recommended that air velocity should be kept below 25 metre/sec, though maxima between 15 and 34 metre/sec have also been quoted.

### Auxiliary Bass Radiator (ABR) Subwoofers

In auxiliary bass radiator subwoofers (also called passive radiator subwoofers) the mass of air in a port is replaced by the mass of the ABR cone. The response is therefore very similar to that of a ported subwoofer using the same driver, but with a notch in the response corresponding to the resonance frequency of the ABR. The larger the ABR, the more mass its cone will have and the lower its resonance frequency will be for the same target Fb (the resonance frequency of speaker and box combined), pushing the notch further out of the subwoofer passband. The large size of suitable ABR subwoofer units means that they are commonly mounted on a different face of the box from the main drive unit, as shown in Figure 18.1c, where the ABR is mounted on the top. Two ABR units are sometimes combined with a single active driver, occupying three sides of the box.

## Transmission Line Subwoofers

The transmission line approach is not a very practical technology for subwoofers. To get worthwhile reinforcement the transmission line needs to be a quarter of a wavelength long at the frequency of interest, and at subwoofer frequencies that means an impractically long duct. The wavelength of sound at 25 Hz is 13.8 metres, and a quarter of that is 3.4 metres; that is obviously impractical as a straight duct, unless you live in a cathedral, and even if folded, as is usual in transmission line speakers, it will take up a lot of space. This means that transmission line subwoofers are relatively rare, but it seems that no audio technology is without its supporters. The Wisdom Audio transmission line subwoofer [2] has a duct folded once and is 90 inches high (2.3 metres). Drivers used with transmission line loudspeakers normally have a low Qts in the region of 0.25 to 0.4.

## Bandpass Subwoofers

This kind of enclosure is specific to subwoofers because of its very limited bandwidth. The bandpass enclosure in Figure 18.1d consists of a sealed chamber mounting the drive unit with a second ported chamber in front of it. This has a 2nd-order lowpass roll-off (12 dB/octave) at low frequencies and a 2nd-order highpass roll-off at its upper frequency limit and is sometimes called a 4th-order bandpass enclosure. I think this is unnecessarily confusing, as there is no actual 4th-order filtering or response, and it would be far better to call it a 2nd-order +2nd-order system, or something similar. Here I will use 2nd + 2nd order to describe it. The 2nd-order lowpass roll-off gives a more gently falling bass response than that shown by ported or ABR subwoofers but gives a lower cutoff frequency for the same driver and volume than the sealed-box approach; it is often claimed that the low-frequency response can be extended by about an octave compared with the sealed-box alignment. The box volume required for a given cutoff frequency can be further reduced by using two drivers in an isobaric configuration. The bandpass principle is not a new idea; it was first patented in 1934. Variations on the concept can be made by replacing the ports with auxiliary bass radiators (ABRs).

Because the driver cone never becomes unloaded, 2nd +2nd order bandpass subwoofers have good power handling characteristics at low frequencies. It is generally considered that the transient response is second only to that of sealed-box subwoofers. In a bandpass subwoofer of this sort, all of the output is via the port, which therefore needs to have the largest practicable diameter to minimise air-flow noise. The port ends are often flared to reduce turbulence; the larger the flare radius the better the results.

Figure 18.1e shows a 3rd +3rd order bandpass subwoofer. Now both the front and rear chambers are tuned by ports. As for a straightforward ported subwoofer, power handling is poor at frequencies below the passband due to lack of loading on the rear of the drive unit cone. The transient performance of a 3rd +3rd order bandpass subwoofer is usually inferior to either sealed, ported, or 2nd +2nd order bandpass systems, so this type of bandpass subwoofer is more appropriate to sound-reinforcement applications than to hi-fi. Once again the entire acoustical output is by way of the outer port, which must have the largest practicable diameter. The inner port also needs to be designed with the issue of port noise in mind, but it is less critical because its noise output will be attenuated by the lowpass action of the front chamber.

It is also possible to design a 4th + 4th order bandpass subwoofer, but there is a general consensus that the transient response is so poor that any improvements, such as in bass extension, are irrelevant.

## Isobaric Subwoofers

In an isobaric subwoofer there are two drive units acoustically in series, driven by the same electrical signal. One version is shown in Figure 18.1f. At low frequencies the two cones move together and are equivalent to a single drive unit with twice the cone mass. This reduces the resonant frequency to about 70% of what it would have been with one drive unit of the same type working in a sealed box with the same volume. The same cutoff frequency can be achieved with half the box volume, though the volume of the chamber between the two drive units must be added; the air volume in this chamber has no acoustic function beyond coupling the drive units, so it is desirable to make it as small as the physical constraints allow.

At higher frequencies cone break-up will occur, and the drive units no longer act as one unit. Other configurations are possible with the drive units mounted to front-to-front instead of front-to-back; this requires the phase of one drive unit to be reversed. There seems to be a general feeling that isobaric loudspeakers are a new idea, but this is not so. Once again, the technique is older than you might think. It was introduced by Harry F. Olson in the early 1950s.

## Dipole Subwoofers

The concept of the dipole subwoofer is quite different from the other subwoofer configurations we have looked at. They all aspire to be monopole radiators—ideally a single point source of sound that has the same polar response in all directions. A large area source or a multiple source causes reinforcements and cancellations in the sound pressure level that cause major amplitude response irregularities. To do this the rear radiation from the drive unit must be dealt with in some way, either by suppressing it (sealed boxes) or getting it into phase with the front radiation to reinforce the output (ported boxes and transmission lines).

A dipole subwoofer does neither of these things; instead the rear radiation is allowed out into the room without modification. Typically one (or quite often, more) large-diameter drive unit is mounted on a relatively small baffle, as in Figure 18.1g; this means that the bass will begin a 6 dB/octave roll-off relatively early, because of cancellation around the edges of the baffle. To counteract this, drivers with a high Qts are sometimes used to give a peaky underdamped response that lifts the overall response before roll-off; adding dipole equalisation to an active crossover is a far better approach, as it gives complete controllability and eliminates the need for special drivers that may be hard to get. This usually consists of a low-frequency shelving characteristic, where the amount of boost plateaus at low frequencies to avoid excessive cone excursions. More details are given in Chapter 14 on equalisation.

When the path around the baffle edge from front to rear of the drove unit equals a wavelength, front and back radiation are in phase and will reinforce, giving a 6 dB peak in the response; the frequency at which this happens can be simply calculated from the baffle dimensions. However, reinforcement only happens at a single frequency when the baffle is circular, so all paths are of equal length, and this is not likely to be a practical shape. The open-backed folded baffle shown in Figure 18.1g gives a more complex response but is far more compact and structurally rigid.

Figure 18.1h shows a dipole subwoofer design put forward by Siegfried Linkwitz [3] which gives greater baffle path lengths and is intended for stereo subwoofer operation. It uses two drive units working in anti-phase, which should offer at least the possibility of cancelling some even-order distortion products.

Since a dipole subwoofer has effectively two point sources, these interact to give a highly non-uniform polar response with a "figure of eight" shape. The output is greatest directly in front and behind the baffle, decreasing to a zero null at each side, where the front and rear waveforms cancel each other out. The presence of these polar response nulls is held to be the major advantage of a dipole bass system, it being claimed that the much smaller sideways radiation excites fewer room modes and so leads to a smoother overall amplitude response in the room.

### Horn-Loaded Subwoofers

As with horn-loaded loudspeakers for higher frequency ranges, this type of subwoofer uses a horn as an acoustic transformer to match the impedance of the drive unit cone to that of the air; it makes the driver cone appear to have a much greater surface area than it actually does. This increases the efficiency of electric-acoustic considerably; the efficiency of a sealed-box design may only be 1%, but a well-designed horn loudspeaker can give 10% or more. Obviously this makes a radical difference to the number of power amplifiers and speakers you have to haul to the gig.

The size of horn required depends on the wavelength of sound being handled, and so subwoofer horns are very big—impractically so for most home entertainment purposes. The low-frequency cutoff of a horn depends on both its mouth diameter and the rate at which it expands along its length, known as the "flare rate". A cutoff frequency of 40 Hz requires something like a mouth diameter of 2.5 metres and a length of some 4 metres. Even for sound-reinforcement applications, such a straight horn would be infeasibly huge, and so the horn is normally folded, as shown in Figure 18.2i (there are many ways to fold a horn. and this is just one example). This can lead to amplitude response irregularities at the upper end of the operating range due to resonances and reflections as the sound output makes its way round the corners in the horn.

## Subwoofer Drive Units

As we have noted, subwoofers are typically very inefficient and require driver units with considerable power handling ability. Driver cone excursion increases at 12 dB/octave with decreasing frequency for a constant sound pressure (SPL) because four times as much air has to be moved, so a large $X_{max}$ (maximum linear excursion of the cone) and $X_{mech}$ (maximum physical excursion before physical damage) are required. The voice coil must also be designed to withstand considerable thermal stress.

## Hi-Fi Subwoofers

When the main format for music delivery was the vinyl disc, the first problem to be overcome in reproducing loud and deep bass was to get the information off the wretched disc. The lower the frequency, the greater the amplitude of the groove deviations for a constant level, and the greater the chance that mistracking of the stylus would occur. From subsonic up to about 1 kHz, a limit on groove amplitude is the constraint on the maximum level that can be cut on the disc. The welcome appearance of the CD format meant that much greater levels of clean, low bass could be accessed, and this gave a great stimulus to the development of subwoofers and the pursuit of an extended bass response in general.

If the subwoofer approach is applied to upscale music-listening rather than an audio-visual experience, it is normal not to take chances with the possibility of losing low-frequency stereo information, and two subwoofers are used, for left and right in the usual way. The subwoofers are often placed under the main speakers, or very close to them, to preserve what stereo cues can be extracted from their output. They are not placed almost at random in the listening room in the way that mono subwoofers often appear to be. A classic application of stereo subwoofers is the extension of the bass response of electrostatic loudspeakers, notably those by Quad, such as the ESL-57, introduced in 1955, and the later ESL-63.

Since the technology of the hi-fi and the home entertainment subwoofer are similar, they are dealt with together in the next section.

## Home Entertainment Subwoofers

When the emphasis is watching television rather than listening to music, it is more common to use a single subwoofer. In multi-channel formats the extra directional information from rear and centre channels means that any lack of stereo in the deep bass is more likely to go unnoticed, and a single subwoofer takes up less space and is easier to fit into a room. In this application the drive units are typically between 4 and 15 inches in diameter,

Table 18.1 gives the vital statistics of a handful of home entertainment subwoofers picked pretty much at random from those on the market now (2010). This does not in any way claim to be a representative

**Table 18.1: Specs for some current subwoofer designs on the market.**

| Model | Driver diameter cm | Driver orientation | Box size H × W × D cm | Box type | Amplifier power W rms |
|---|---|---|---|---|---|
| Monitor Audio Vector VW-8 | 20 | Forward | 32 × 28 × 28 | Ported | 100 |
| Velodyne Impact-Mini | 16.5 | Forward | 25 × 25 × 30 | Sealed | 180 |
| B&W ASW610 | 25 | Forward | 31 × 31 × 31 | Sealed | 200 |
| Wilson Benesch Torus | 36 | Upward | 45 × 90 × 30 | Sealed | 200 |
| Energy ESW-M6 | 1 × 16.5 active<br>2 × 16.5 passive | Forward & sides | 20 × 20 × 20 | ABR | 200 |
| Audio Pro B1.36 | 25 | Forward | 45 × 35 × 38 | Ported | 200 |
| Wharfedale Diamond SW250 | 25 | Downward | 42 × 42 × 38 | Sealed | 250 |
| Mordaunt-Short Mezzo 9 | 2 × 20 | Forward | 32 × 34 × 35 | Sealed | 375 |
| Velodyne SPL-1500R | 38 | Forward | 47 × 46 × 44 | Sealed | 1000 |

selection, but it does give some feel for the basic subwoofer format. Note that ABR stands for Auxiliary Bass Radiator.

Domestic considerations require the subwoofer to use as small a box as possible, while at the same time being capable of reproducing deep bass. This means that efficiency is inevitably low, and powerful amplifiers are needed to generate the desired sound levels—considerably more powerful than those driving the main loudspeakers. It is common for the subwoofer amplifier to have ten times the power capability in Watts compared with the main amplifiers.

The facilities offered on a subwoofer are subject to some variation, but typical features you might expect to find are:

- Low-level inputs (unbalanced)
- Low-level inputs (balanced)
- High-level inputs
- High-level outputs
- Mono summing
- LFE input
- Level control
- Crossover in/out switch (LFE/normal)
- Crossover frequency control (lowpass filter)
- Highpass subsonic filter
- Phase switch (normal/inverted)
- Variable phase (delay) control
- Signal activation out of standby

### Low-Level Inputs (Unbalanced)

The low-level inputs are intended to be driven from a preamplifier or AV processor. In many cases they are phono (RCA) connectors, and so are inherently unbalanced. The input impedance should not be less than 10 kΩ.

### Low-Level Inputs (Balanced)

More upscale subwoofers are likely to have balanced inputs which will reject ground noise caused by ground loops, etc. This means paying for an XLR connector, but the cost of the electronics to implement the balanced function is small. Much more information on balanced inputs can be found in Chapter 20 on line inputs.

### High-Level Inputs

The high-level inputs are designed to be connected directly to the amplifier outputs that feed the main loudspeakers. They drive a resistive attenuator, usually with high resistor values to reduce power dissipation, which reduces the incoming level to that of the low-level inputs. Input protection clamping diodes are often fitted to prevent damage from excessive input levels. The circuitry is often arranged to sum the low-level and high-level inputs so that a switch between low-level and high-level inputs is not necessary; this can be done by the same circuitry that sums the left and right halves of the incoming signal. This approach is made practical by making the input impedance of the high-level inputs quite low, so they are not liable to pick up external noise. A typical value is 100 $\Omega$; the downside to this is that such a value resistor will wastefully dissipate a lot of power when connected to a powerful amplifier, and it needs to be a substantial wirewound component.

An important objection to this method is that the signals reaching the high-level inputs have passed through the main power amplifiers and will be degraded by whatever noise, hum and distortion those power amplifiers introduce. The signal has then to go through the subwoofer amplifier, so it is degraded twice instead of once. For these reasons, the use of high-level inputs should be avoided if possible, and manufacturers that provide them state in their instruction manuals that the use of the low-level inputs is preferred.

### High-Level Outputs

High-level outputs are sometimes also provided, so the subwoofer and main loudspeakers can be daisy-chained via high-level inputs instead of being connected in parallel. This means that the power amplifier signal has to go first to the subwoofer and then out again to main loudspeakers, passing through twice as many connectors and extra lengths of cable. If the subwoofer is installed away from the rest of the system, then the extra length of speaker cable may be considerable, and this will increase the impedance seen by the loudspeakers and reduce the so-called damping factor of the system. As is now well established, the actual effect of speaker cable resistance on loudspeaker damping is very small, because most of the resistance is in the voice coils, but a real and more worrying effect is irregularities in the frequency response caused by variations in loudspeaker impedance interacting with cable resistance.

### Mono Summing

Both the low-level and high-level inputs are in stereo and so must be summed to mono before they are presented to the crossover and power amplifier. As mentioned earlier, the same summing circuit is often used to sum the low-level and high-level inputs, to save on a selector switch.

### LFE Input

LFE stands for low-frequency effects. There will only be one LFE input connector, as the LFE channel is already in mono. The input is typically unbalanced; the input impedance should not be less than 10 k$\Omega$.

### Level Control

This adjusts the volume of the subwoofer relative to the rest of the system.

### Crossover In/Out Switch

This is sometimes labelled "Normal/LFE", as when the input is the LFE channel the internal subwoofer crossover is not required.

### Crossover Frequency Control (Lowpass Filter)

The filtering in a subwoofer is not strictly speaking a crossover, because instead of splitting the input between two or more outputs, it simply rejects all high frequencies; it is more accurate to simply call it a lowpass filter. The cutoff frequency is always adjustable, and a typical range is 50 Hz–150 Hz. The filter is normally a simple 2nd-order Butterworth lowpass using the Sallen & Key configuration. See Chapter 8 for more information on variable-frequency lowpass filters.

### Highpass Subsonic Filter

Many subwoofers use reflex (ported) enclosures to obtain more bass extension. At very low frequencies these enclosures put no restraint on cone movement, and when this factor is combined with the high output capability of subwoofer power amplifiers, you have a recipe for disaster: large subsonic signals (oops, dropped the needle) will almost certainly cause excess cone displacement and serious damage. For this reason most subwoofers—and especially those of the reflex type—include subsonic filters in the signal path. These are usually 2nd-, 3rd-, or 4th-order and are usually based on the Sallen & Key configuration.

### Phase Switch (Normal/Inverted)

This control puts the subwoofer output in phase or 180° out of phase with the incoming signal material. The intention is to try to cope with the fact that the difference between the distance from the main loudspeakers to the listener and the distance from the subwoofer to the listener are unpredictable, and the resulting time delay must be compensated to give good subwoofer integration. A phase-invert switch can only do this very crudely, as what is really required is a continuously variable control. A passage like this appears in some form in most subwoofer user manuals:

> There is no correct or incorrect setting of the phase switch. The proper setting depends on many variables such as subwoofer placement, room acoustics, and listener position. Set the phase switch to maximize bass output at the listening position.

This sort of thing is a bit disingenuous. There certainly will be a correct phase-shift which gives the flattest and best bass response, but it is unlikely to coincide with either of the two arbitrary settings provided by a phase-invert switch.

## *Variable Phase Control*

As just explained, a phase-invert switch is very much a gesture at achieving correct phase-compensation. What is really required is a variable amount of phase-shift. See Chapter 13 on variable delays in time-domain filtering to see how this can be accomplished.

## *Signal Activation Out of Standby*

Increasing attention is being paid to economy in the use of energy, and it is now common for power amplifiers to have a standby mode where the main transformer is disconnected from the supply when the unit is not in use, but a small transformer remains energised to power circuitry that will bring the unity out of standby when either an incoming signal is detected or a +12 V trigger voltage is applied to a control input. This also applies to the powerful amplifiers found in subwoofers, and they commonly include a signal activation facility. It is convenient to have the unit wakeup automatically when a signal is applied. It is now only necessary to pop in a CD or whatever and start it playing; there is no need to push a button on the subwoofer. This has particular force because subwoofers are likely to be hidden away out of sight where they cannot easily be got at.

A full discussion of the technical challenges presented by signal activation, for example the need to avoid switch-on in response to isolated noise clicks, is given in my book on power amplifiers. [4]

# Home Entertainment Crossovers

So far as the crossover is concerned there are two modes of subwoofer operation. In 5.1 systems there is a dedicated channel for the lowest frequencies called the LFE (low-frequency effects) channel. This is the "0.1"in 5.1 and 7.1 discrete surround systems; it is important to realise that it contains low-frequency bass material not included in the other channels and so cannot be derived from the other channels by filtering. AV receivers and processors therefore have a mono output labelled "Subwoofer

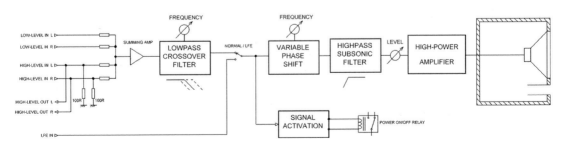

**Figure 18.2: Block diagram of a typical home entertainment subwoofer.**

Out" or "LFE Out". The LFE channel has a digital brickwall lowpass filter at 120 Hz, and to keep clear of this producers normally use their own gentler lowpass roll-off at around 80 Hz.

AV processors have so-called bass management facilities which can redirect bass from any channel to a subwoofer output (this output is *not* the same as an LFE output, because, as I said, the LFE channel contains unique information). This is useful if no LFE channel exists, as in a normal stereo signal in Format 2.0. This redirection is done by the LFE crossover.

AV processors classify the main loudspeakers into "Large" and "Small", with large being considered capable of handling the full audio bandwidth, while small speakers cannot cope with the lowest frequencies, so their signal is subjected to a highpass roll-off, as part of the LFE crossover function. The unused low frequencies are mixed and sent to the LFE output; the phase-shifts between the low-frequency signals in the satellite channels are generally small, so simple summation of the signals is all that is required. Processors differ in the amount of control they offer over the crossover frequency. The possibilities are fixed, variable, and multiple variable, which work as follows:

### Fixed Frequency

The only choice is between large or small main loudspeakers. When "Small" is selected, the crossover frequency is fixed, usually in the range 80 to 90 Hz. This restricted approach is considered obsolescent.

### Variable Frequency

There is the choice of large or small for the main loudspeakers, plus ability to alter a single crossover frequency for optimal subwoofer and main loudspeaker integration. This facility is available on most modern AV receivers and processors. A typical range of crossover frequency is 80 to 160 Hz.

### Multiple Variable

There is the choice of large or small for the main loudspeakers, plus the extra flexibility of being able to set different crossover frequencies for multiple speakers, such as the front, centre, and rear channels.

## Power Amplifiers for Home Entertainment Subwoofers

Because of the powerful amplifiers required, more efficient operating modes than Class-B are commonly used. When quality is the first consideration, Class-G is preferred. [5] When keeping down heatsink size and overall weight are the priorities, Class-D amplifiers are used. Despite considerable development effort, Class-D amplifiers remain much inferior in their distortion performance to other types of power amplifier, and it is often stated that they are only acceptable for subwoofer applications because their high-order distortions are not reproduced by the drive unit or units. This does overlook the possibility of intermodulation generating non-linearity products in the lower frequencies.

Class-G amplifiers, on the other hand, are capable of linearity competitive with all but the very best Class-B power amplifiers, so long as modern design approaches are used to minimise distortion. [5]

## Subwoofer Integration

The output of a subwoofer must combine properly with the bass output of the main loudspeakers. This issue is often ignored simply because it is complex and the problems are not easy to solve. [3] There are many factors to consider in integrating the LF and mid outputs in a conventional loudspeaker, where the drive units are in a fixed physical relationship to each other. Subwoofer integration adds the complications of summing stereo to mono and the variable placing of the subwoofer relative to the main loudspeakers. The very presence of a phase-invert switch, and often a variable phase control, indicates the uncertainty in the situation.

Subwoofers can integrate with the main loudspeakers in two ways—augmentation and crossover. In augmentation mode the main loudspeakers handle the full frequency range, down to the bottom of the bass end, but are limited in their ability to reproduce deep bass at high levels without distortion. The subwoofer supplements their output, typically below 80 Hz. There is a body of opinion that some rooms can benefit from smoother bass by getting deep bass from three room locations, i.e. the subwoofer plus the main left and right speakers, rather than from the subwoofer alone.

In crossover mode, the subwoofer replaces the deep bass output from the main loudspeakers. Here is a crossover from the main loudspeaker LF drive units to the subwoofer, just as there is from the mid to the LF drive units within the main loudspeakers. A highpass filter is therefore applied to the main loudspeaker signal, removing the low bass.

A very important consideration is the choice of the crossover frequency between the main loudspeakers and the subwoofer. In current variable crossovers the frequencies available range from 10 Hz to 4.2 kHz, though it is not easy to see how the latter figure counts as subwoofer territory. It is usually best to make the crossover frequency between the subwoofer and the main loudspeakers as low as possible; certainly below 90 Hz, and preferably below 60 Hz if this can be done without overtaxing the main loudspeakers, as this reduces the chance of the subwoofer becoming audible as a separate sound source. This is less important if the subwoofer is at the front of the room between the main loudspeakers, as that is usually where the bass is placed in the stereo stage. Let us examine the various factors affecting the choice of crossover frequency.

The advantages of higher crossover frequencies for subwoofer integration are:

1. Less output is required from the main loudspeakers, which should give lower distortion and less thermal compression where it is more audible in the audio spectrum; this does of course assume that the subwoofer is itself capable of reproducing low frequencies cleanly, which it certainly should be.
2. Since a subwoofer has a greater flexibility of placement than the main loudspeakers, it may be possible to position it so that room standing waves are excited less. The higher the crossover frequency, the more likely this is to be successful.
3. For a given crossover alignment, the group delay of the crossover filters will be inversely proportional to the crossover frequency. Thus the group delay for a 50 Hz crossover frequency will be twice that of a 100 Hz crossover frequency.

4. When the main channels have speaker size set to "Small", a typical AV processor sums all these channels with the LFE channel, and this combined signal goes through a lowpass crossover filter. If this crossover frequency is lower than 80 Hz (the practical upper limit of the LFE channel), then information at the upper end of the LFE bandwidth will be lost.
5. In the right conditions, the use of a subwoofer can give a more consistent low-frequency response. When the bass from the main loudspeaker channels is routed to a subwoofer, the frequency response over the subwoofer operating range will be identical between channels (though not necessarily good in itself). As the frequency range of the subwoofer becomes wider, this advantage grows in significance. In contrast, the bass signals radiated by the main loudspeakers will have differing amplitude responses because they are in different physical locations in the listening space.

The disadvantages of higher crossover frequencies for subwoofer integration are:

1. As higher frequencies are more audible than lower frequencies, a higher crossover frequency will place the crossover transitions in a more sensitive part of the audio spectrum. Any response irregularities induced by the crossover will therefore be more obvious.
2. It is normal to use a single subwoofer, and so the higher the crossover frequency, the greater the chance there is that stereo bass information will be lost. Stereo low-frequency information will most probably be in the form of phase differences between channels as opposed to level differences.
3. It is clearly important not to operate the subwoofer above its intended frequency range. This will almost certainly lead to increased distortion and greater frequency response anomalies.
4. The crossover needs to have a suitably steep lowpass slope so that the higher bass frequencies that are easier to localise (say above 100 Hz) are not rendered in mono. A 4th-order Linkwitz-Riley configuration is recommended.

The greatest problems with subwoofer integration arise when a mono subwoofer is placed some distance from the main speakers. As we have seen in Chapter 13, time alignment of the drivers is crucial for obtaining the desired amplitude/frequency response. It is difficult enough when all the drivers are mounted in the same enclosure, but separate stereo subwoofers may be metres away, requiring considerable time-delay compensation, and a mono subwoofer makes things worse because it may not be symmetrically placed between the two main speakers, and therefore different time-compensation delays would have to be applied to each main channel.

With a metre or more spacing the delays required are much longer, but implementing them is not too hard because the allpass filters only need to maintain a constant delay up to a relatively low frequency—that at which the subwoofer output has become negligible compared with the output of the LF drive units of the main loudspeakers.

## Sound-Reinforcement Subwoofers

Subwoofers are now pretty much standard in quality sound-reinforcement systems. The requirements are very different from those for home entertainment; the output levels are much higher, and the need to have the smallest possible box volume is less pressing (though not jettisoned altogether), and so they are usually much bigger. Sound-reinforcement subwoofers are usually of the sealed-box, ported-box, or horn-loaded types. The drive units are typically between 8 and 21 inches in diameter,

It is normal to use multiple subwoofers (up to a hundred at a time have been used) to get the sound output required, and this multiplicity can be exploited to beam the sound in the desired direction. This not only increases efficiency but reduces the likelihood of bitter complaints from outside the venue. Several techniques for steering the sound energy are used; most of them require signal delays to do this, and this is where the crossover setup comes in. For reasons of space only the two most important methods are described here; it is a big and a most fascinating subject.

### Line or Area Arrays

Installing multiple subwoofers in a vertical line array focuses the sound energy into a narrow beam so that a relatively small amount is sent up into the air or down towards the floor; most of it is focused at the audience. The reduction in the amount of low-frequency sound reflected from the ceiling (if working indoors) reduces frequency response and feedback problems. The longer the array, the greater the directional effect; note that this technique is implemented simply by the physical placement of the subwoofers, and signal delays are not required.

Giving the speaker array greater numbers horizontally as well gives an area array, which also focuses the sound radiation into a beam in the horizontal plane; this effect can become a problem rather than an advantage if a long line of subwoofers is require to get enough output, as the beam becomes too narrow to cover the whole audience. If however the signals to the outer subwoofers are delayed by a few milliseconds prior relative to those at the centre, considerable control can be exercised over the beam width. This technique is sometimes called a "delay-shaded array".

### Cardioid Subwoofer Arrays

The various types of cardioid subwoofer array (CSA) are, as the name implies, more concerned with altering the front/back ratio of output power rather than controlling beam width. The polar response is very similar to that of a cardioid microphone. One method of creating a cardioid pattern is applicable to a horizontal subwoofer array across the front of the stage. The polarity of every third subwoofer is reversed (this can be done simply by turning the cabinet around so it's firing backwards), and the signals to these are delayed. The result is that the radiation pattern is no longer quasi-omnidirectional; instead the sound energy being sent backward to the stage is much reduced, reducing feedback problems and making the lives of the performers more tolerable. The amount of delay can be manipulated to maximise the cancellation of the most troublesome frequencies in stage area. The technique is only effective over a limited frequency range, but the delays can usually be adjusted to give a significant improvement over slightly more than an octave.

## Aux-Fed Subwoofers

Subwoofers are sometimes used in sound reinforcement quite separately from the main loudspeaker system with its active crossovers. In what is called "aux-fed" operation, the subwoofer is fed via a lowpass filter (which may or may not be part of the main crossover system, but is at any rate handling only the subwoofer signal) from a dedicated auxiliary send on the mixing console. This send is used to make a mix specifically of those instruments with the greatest amount of bass output, such as bass guitar, kick drum, and keyboards. Aux-fed operation is claimed to give a cleaner sound, because microphones tend to pick up bass information that is not intended for them, with unpredictable phase delays. This low-frequency rubbish goes into a conventional stereo mix and is fed to the subwoofers, producing what is usually described as a "muddy" effect; the bass-cut filters normally present on mixer channels are generally not considered effective at controlling this because of their limited slope. If the subwoofer is aux-fed, then the unwanted low frequencies are filtered out by the main crossover system and never reach the subwoofer.

A disadvantage is that in most mixing consoles, fading down the entire audio system will require an aux master to be turned turn to fade the subwoofers at the same time as the main faders are pulled back. This is obviously undesirable, and consoles designed for aux-feeding have an extra fader placed next to the main output faders so easy simultaneous control is possible.

## Automotive Audio Subwoofers

The loudspeakers fitted as standard to cars have a very limited low-frequency capability because of their small size; they are installed in car doors or dashboards, so very little space is available. If some serious bass is required, one or more subwoofers are installed in the boot or back seat space.

Getting decent sound in a car by any means is a serious challenge. The cabin volume is much smaller than the average listening room, and consequently the effects of resonances and reflections are much more severe. Vance Dickason [6] describes the listening space as a "lossy pressure field". A true pressure field would have perfectly rigid walls, but the thin metal panels of a car are a long way from rigid, and Vance points out that this leads to unpredictable variation in the low-frequency response of the space over a ±3 to ±6 dB range. Simply opening a window (more optional in these days of widespread air-conditioning) has a radical effect on the response of the space.

The resonances and reflections in the listening space can give a considerable lift to low frequencies; this is sometimes called "cabin gain". Vance reports one test that yielded a boost of 7 to 8 dB between 40 and 50 Hz and a frightening 20 dB boost at 20 Hz. This sort of thing has obvious implications for crossover design; an unexpected hefty bass note from a CD could pop your windows out, so some effective subsonic filtering is an extremely good idea. If you are implementing variable equalisation, bear in mind that while it very often comes only in the form of bass boost, in this case the ability to cut the bass is very necessary. Circuitry that can give 20 dB of attenuation at 20 Hz, 8 dB at 40 Hz, and very little at, say, 60 Hz is going to need at least 12 dB/octave slopes and will need something more sophisticated than the conventional Baxandall configuration. See Chapter 14 on equalisation.

Automotive subwoofer enclosures are either sealed boxes or ported designs. Transmission line and horn-loaded subwoofers are not generally used, mostly because the physical size require for low-frequency operation is quite impossible to fit into any normal-sized car.

One of the main problems of car audio is that only a nominal 12 V is available to power amplifiers. This, even assuming a lossless amplifier, only allows 2.25 W into an 8 Ω loudspeaker. Speaker impedances of 4 Ω give only 4.5 W, and lower impedances than this do not give worthwhile improvements, because the losses in amplifiers and wiring resistances become large. Using a bridged pair of power amplifiers, driving each side of the speaker in anti-phase, doubles the voltage swing available and so theoretically quadruples the power, giving 9 W into 8 $\Omega$ and 18 W into 4 $\Omega$, though because of losses the real increase will be significantly less than 4 times. When higher powers than this are required, the normal practice is to use a switch-mode power supply to convert 12 V into whatever higher voltage is required.

Many after-market automotive amplifiers are of high power, to drive relatively small subwoofer enclosures that are inevitably inefficient. A typical model might be in two-channel format, giving 2 × 350 W into 4 $\Omega$ and 2 × 700 W into 2 $\Omega$, with a bridging facility to give 1 × 1400 W into 4 $\Omega$. Built-in variable-frequency crossovers are often provided, providing an HF output to the main loudspeakers as well as the lowpass feed to the subwoofer. These are usually based on 2nd-order Butterworth filters. Subsonic filters are usually included for drive unit protection; variable equalisation (invariably in the form of bass boost) is sometimes also provided.

Power amplification of this sort obviously makes heavy demands on a car's electrical system. The alternator fitted as standard will typically be capable of generating 80 amps maximum, and this is not adequate for high-power audio systems; they are commonly replaced with special high output alternators that can give up to 200 amps. These are not normally direct physical replacements, and some mechanical engineering is required to fit them. An alternative approach is the "split" system, where the original vehicle electrical system is left alone and a second alternator of high output is installed that charges a separate set of batteries dedicated to the audio system. Split systems are commonly used on emergency vehicles, though in that case the extra batteries run lighting, defibrillators, etc.

Capacitor banks, which are large numbers of high-value capacitors connected in parallel, are sometimes wired across the power supply when it is feared that the batteries, with their associated wiring, will not be able to respond quickly enough to a sudden demand for current. The enormous capacitances used are measured in farads rather than microfarads, and the values used range from 1 F to at least 50 F.

There is a distinctive genre of competition in car audio, where constructors contend simply to create the most awesome installations. Car audio competitions started in the early 1980s, with the first known event in 1981 at Bakersfield in California. While some competitions focus on sound quality and neat installation, the majority appear to be held simply to find the highest sound pressure levels; this is sometimes called "dB drag racing". In recent years the two aims, sound quality vs SPL, appear to have become almost mutually exclusive.

Cars built for dB-drag-racing are barely driveable, as the interior is almost completely filled with extra batteries, capacitor banks, amplifiers, and loudspeakers, but they must be driven 20 feet to prove that some movement is possible. Sound pressure levels, with all cabin openings sealed, of 155 or 160 dB above threshold are commonly reached. The current world record of 180.5 dB is held by Alan Dante. This was achieved with a single 18-inch subwoofer driven by four amplifiers totalling 26 kW, powered by fifteen 16 volt batteries. This equipment was installed in a Volvo that appears to have been weighted down with concrete. It should perhaps be mentioned that people do not sit in the cars during testing. It would not be a survivable experience.

# References

[1]   https://en.wikipedia.org/wiki/Thiele/Small; Accessed January 2017

[2]   Anon "Regenerative Transmission Line™ Subwoofers" www.wisdomaudio.com/pdfs/RTL.pdf

[3]   Linkwitz, Siegfried  www.linkwitzlab.com/frontiers_4.htm#S

[4]   Self, Douglas "Audio Power Amplifier Design" Sixth Edn, Newnes, 2013, pp. 679–682, ISBN: 978-0-240-52613-3 (signal activation)

[5]   Self, Douglas Ibid, Chapter 19 (Class-G)

[6]   Dickason, Vance "Loudspeaker Design Cookbook" Fifth Edn, Audio Amateur Press, p. 159, ISBN: 1-882580-10-9

# CHAPTER 19

# *Motional Feedback Loudspeakers*

## Motional Feedback Loudspeakers

The use of motional feedback to improve loudspeaker operation is closely related to active crossover technology. The amplifier is usually physically part of the loudspeaker because it is even more closely coupled than an active crossover system. As described in Chapter 1, you can buy an active crossover and matched loudspeaker from Company A and between them put a power amplifier from Company B; all you have to do is get the gain right. Motional feedback (MFB), on the other hand, depends critically on loudspeaker characteristics not only to work at all but in some cases to avoid catastrophic instability. Only one company (Yamaha) has released stand-alone amplifiers with a motional feedback capability.

Many loudspeaker problems are caused by the fact that a drive unit cone does not follow exactly the voltage driving it but has a life of its own due to cone inertia and suspension non-linearities; this is further complicated by cone break-up modes, where different parts of the cone are moving independently. Negative feedback can work miracles in controlling precisely the output voltage of a power amplifier, and it is a logical step to attempt to gain the same iron control over a drive unit cone by sensing its movement and using that to close a negative feedback loop; this promises both a better frequency response and lower distortion. It is however much more difficult to control a speaker cone with negative feedback than it is a simple output voltage. Motional feedback (MFB) for anything other than a full-range driver unit requires the use of an active crossover, because one amplifier can only control the movement of one drive unit. MFB is typically only applied to the LF unit, as this has the largest excursions, creates the greatest amount of distortion, and is generally further removed from being an infinitely rigid zero-mass piston than the MID and HF units. MFB is also relevant to this book because equalisation of the drive units is often necessary.

Despite this promising prospectus, motional feedback has had little success in the marketplace. The best-known example is the Philips 22RH544 loudspeaker, which derived its motional feedback from a small accelerometer mounted on the dust-cap of the LF unit voice coil. This was made and sold for ten years and was undoubtedly a good product, but it failed to trigger a wave of enthusiasm for MFB.

One of the main challenges of MFB is sensing the motion of the drive unit cone without putting any restraint on its movement. The basic idea is shown in Figure 19.1. There are three basic ways to do this:

1. Measure the cone position.
2. Measure the cone velocity.

The velocity can be integrated to get the cone position.

3.   Measure the cone acceleration.

This can be done with a small and light accelerometer. Acceleration can be integrated to get cone velocity and then integrated again to get cone position. The prime example of this is the method used by the Philips 22RH544.

## History

As usual, the history of the concept goes back a surprisingly long way. The first reference found was [1] from 1951. See also [2] from 1958 and [3] from 1963. None of these are easy to come by, but a very good early reference that is available is the 1964 *Philips Technical Review*, [4] which describes the origins of the accelerometer approach.

## Feedback of Position

This clearly needs to interfere with cone movement as little as possible; one obvious method that adds very little weight is capacitance sensing. It is not enough simply to fit a conductive dust-cap and place a fixed plate some distance from it; the capacitance is inversely proportional to the distance between the two capacitor plates, and so the position signal will be grossly distorted. An early attempt was by Brodie in 1958, [5] but as far as I can see it does not address the inverse proportionality. We need a linear change of capacitance with cone position. A neat solution was put forward by ServoSpeaker in 2004 [6][7] and is shown in Figure 19.2. The capacitance depends on the degree of overlap of the moving and fixed plates and is to a first approximation linear. A patent application was made in 2004, [8] and the technology was described in *Hifi-World* in 2008, [9] but as far as I can determine no product got to market, and ServoSpeaker no longer seems to exist as a company.

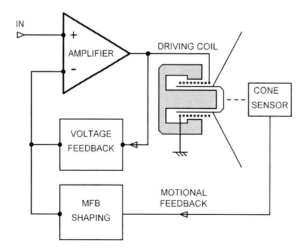

**Figure 19.1: The basic principle of motional feedback.**

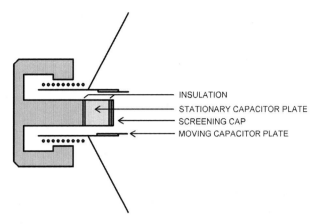

**Figure 19.2: Motional feedback of cone position using capacitive sensing: ServoSpeaker method.**

Ever since I first heard of MFB, it struck me that laser interferometry would be the ideal method; no moving parts added and no extra wires attached to the cone. Laser diodes can't be expensive, since there's one in every CD player—how hard can it be? However, I know little of such matters; one person consulted told me that the laser required to get the necessary coherence length would be pretty big—a metre or more long for a frequency stabilised helium-neon laser, which I was told was about the minimum hardware I could get away with. That doesn't sound too practical; I would be glad to hear if others agree on the possibilities of laser interferometry.

## Feedback of Velocity

At first it looks as if this might be easy; a secondary sensing coil is added to the cone primary drive coil and will produce a voltage proportional to its velocity in the magnetic field. The main problem with this approach is transformer coupling between the main voice coil and the sensing coil, which will completely swamp the desired velocity signal. A solution to this which was perhaps clumsy but has a certain no-nonsense appeal was that adopted by Panasonic MF-800 in the early 1960s (see Figure 19.3). [10] A second voice coil and magnet were mounted at the front of the drive unit, much reducing transformer action. On reading the user manual it is claimed that MFB reduces "distortion", but it is not clear if that refers to frequency response errors or non-linear distortion or both. The main aim is clearly to extend the LF response of the loudspeaker, and switches for controlling this were provided, labelled "Frequency" and "Damping".

There seems no doubt that this method worked as advertised, though according to the manual some control settings gave bass extension combined with a +6 dB resonance peak, which probably did not sound great. This could be suppressed by increasing the damping setting, but this much reduced the bass extension. There are very few references to this system anywhere, which suggests it was not a marketing success. No pictures of the actual drive units have so far been found.

A more subtle and apparently superior approach is to use the same voice coil for both driving the cone and measuring its velocity. The movement of the coil in the magnetic field generates a voltage signal

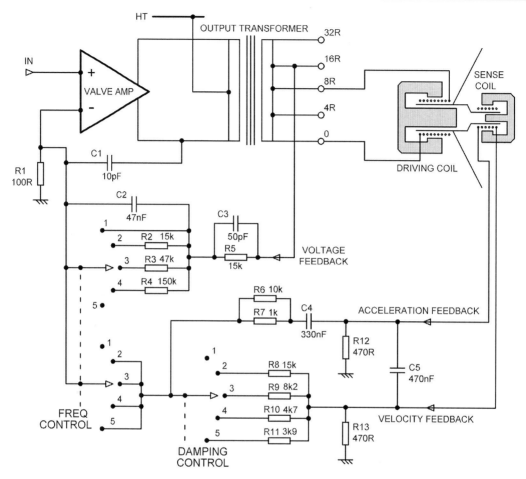

**Figure 19.3: Motional feedback of velocity using a separate sense coil and magnet: Panasonic MF-800.**

Vvel proportional to the velocity, but this has to be somehow separated out from the voltage driving the speaker. Figure 19.4 shows one way; a balancing impedance R1, L1 is put in series with the drive unit; typical values are shown. The transformer reads the voltage across this load and subtracts it from the main feedback signal so that the velocity signal is isolated and can be added to the feedback as required. Note that the balancing impedance is made equal to the *blocked* impedance of the drive unit, which is its impedance when the voice coil is fixed so that it cannot move.

This method has a disadvantage which I don't think I have ever seen explicitly pointed out; as shown it wastes half of the amplifier power that you trustingly feed into it, dissipating it in the balancing impedance. To make this approach practical the balancing impedance must be scaled so it dissipates a small fraction of the power in the drive unit, by raising the transformer ratio or otherwise modifying the feedback arrangements. See [4] for other versions of this circuit.

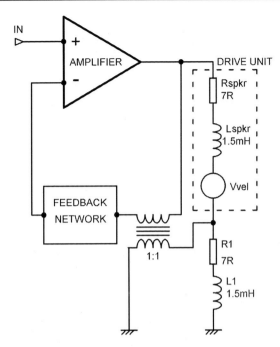

**Figure 19.4: Motional feedback of velocity from the voice coil with R1, L1 and the transformer cancelling the drive voltage.**

A transformer is not a welcome part in a negative feedback loop; it may have its own linearity problems at low frequencies, and HF phase-shifts are likely to cut down the stability margins. Figure 19.5 shows a more up-to-date version, where the required subtraction is performed by opamp A1. The balancing impedance has been scaled down by a factor of seven to demonstrate the process, but the power dissipated in R4 will still be substantial, and further scaling would be desirable. The scaling is embodied in the values of R1, R2, and R3.

This method can also be looked at as making the output impedance of the amplifier negative—it is equal and opposite to the blocked impedance of the drive unit. This should not be confused with current-drive of loudspeakers, where the output impedance is made very high.

This all seems very promising—you can use standard drive units and do the MFB with a few cheap electronic parts. But there are two big problems remaining with voice coil velocity feedback: First, the resistance of a drive unit voice coil changes significantly when it gets hot; a 40% increase is entirely possible with a 100°C temperature rise. This plays mayhem with the impedance balancing, and it would be all too easy to produce a design that became unstable as it heated up, banging the amplifier output against one of the rails. In practice this means that only a small amount of negative feedback can be applied safely.

Second, the uniformity of the magnetic field in which the voice coil moves is imperfect, and this reduces the accuracy of the velocity feedback signal.

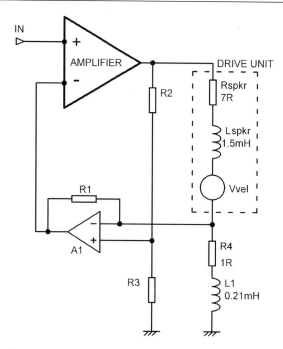

**Figure 19.5: Motional feedback of velocity using the voice coil, with R1, L1 scaled.**

Much work has been put into the first issue. Temperature compensation is difficult because the voice coil is a highly inaccessible position in the magnet gap. The only workable technique appears to be the addition of another coil on the former, made of the same grade copper. Depending on the details of construction, this coil will get close to the voice coil temperature. Note this is not a velocity-sensing coil; its proximity to the voice coil means the transformer action would be too strong. Instead the coil is wound non-inductively, so that transformer action is (mostly) prevented, and it responds only to temperature. A difficulty is that you have to put two windings in where but one grew before, and this is likely to require a bigger magnetic gap, leading to lower efficiency.

Variations on this technique were used by Servosound, a Belgian company operating in the early 1970s, and by Yamaha, with their Active Servo Technology (AST). One example is the AST A-90M, a 45 W/6 Ω integrated amplifier. This separate amplifier was matched to different Yamaha loudspeakers by means of a cartridge plugged into the front panel of the amplifier, which controlled both the impedance characteristics and the LF equalisation. Another was the AST-A10 power amplifier (1989), where the loudspeaker current was measured by 0.2 Ω resistors in the return connection. There was also the AST-P2601 professional audio amplifier. All three used the cartridge system, and Yamaha naturally only provided cartridges for their own loudspeakers. I have not been able to discover what happened if you plugged in the wrong cartridge for a given loudspeaker.

## Feedback of Acceleration

The best-known example of this technology is the Philips 22RH541 motional feedback loudspeaker, manufactured from 1975 to 1985. [11] It was a 2-way loudspeaker with the HF and LF drive units

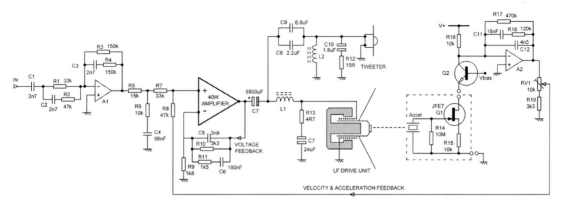

**Figure 19.6: Motional feedback of acceleration using an accelerometer: the Philips 22RH541.**

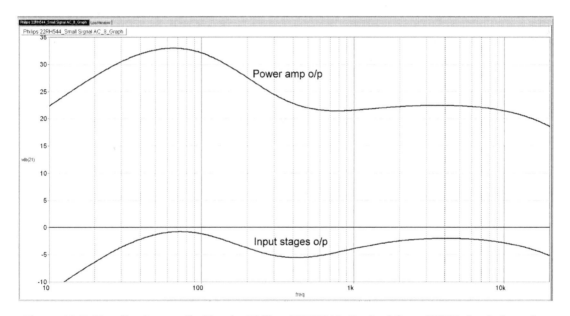

**Figure 19.7: Equalisation applied by the Philips 22RH541. Derived from SPICE simulation of simplified circuitry.**

fed from a passive crossover, with MFB applied only to the 8inch LF drive unit. Figure 19.6 shows a simplified version of the arrangement. A piezo-electric accelerometer with a preamplifier consisting of a JFET and two resistors was mounted inside the LF unit dust-cap (components inside dashed box). The JFET drain current then passed through cascode transistor Q2 to virtual earth amplifier A2.

A good deal of equalisation is applied around the input stage A1, as shown by the lower trace in Figure 19.7; three time-constants are employed. Further equalisation is applied by the components in the voltage feedback path of the power amplifier, giving the upper trace in Figure 19.7. The power amplifier is AC-coupled to the drive units via C7, so there is no possibility of them being damaged if the amplifier sticks to one rail. The passive 2nd-order crossover uses C8, C9 and L2 to direct the HF to the tweeter, while L1, R13 and C7 direct the LF to its driver. There is a Zobel network R12, C10 across the tweeter and another Zobel network R13, C7 across the LF unit.

There is no short-circuit protection, but then there is no access to the amplifier output, so it is not going to be short-circuited by the user.

The accelerometer preamp stage was cleverly arranged to work into the common-base stage Q2, so that only two wires to the accelerometer assembly were required. The signal then goes to A2, which implements the lowpass filter and integrator required to produce acceleration and velocity feedback signals. See [4]. The RC network involved contains a 4n3 capacitor; E24 capacitor values are unusual even today, and this suggests some fairly critical adjustment was being performed. There is a preset RV1 that controls the amount of MFB applied; I have been unable to determine exactly what this compensates for, but I suggest probably variations in the LF drive unit compliance. The 22RH541 also contained a significant amount of circuitry to switch the unit out of standby when a signal was applied.

There were eight models in the Philips MFB range. The first to be introduced was the RH532 in 1974, [12] reviewed in *Popular Electronics* in March 1975. [13] It was a 3-way bookshelf speaker, with very similar technology to the 22RH541 described earlier. A 20 W power amplifier drove the HF and MID units through a 2nd-order passive crossover (at 3.5 kHz) with Zobel networks across the drive units. A separate 40 W power amplifier drove the LF unit, which was the only one with MFB applied. The crossover frequency between the two amplifiers was 500 Hz.

The specifications for the various models are summarised in Table 19.1. All had MFB on the LF unit only. Cabinet volume is not given when I only have the external dimensions to work on, because the internal electronics take up a significant but unknown amount of space. More details on specific models are given later. The schematics are easily found on the Web.

The RH545 was the first MFB model to offer switchable correction filters for "Rear to wall", "Floor", and "Side to wall" loudspeaker placement. "Side to wall" used three gyrators to create a −3 dB dip

Table 19.1: Philips MFB loudspeakers.

| Model | Drive units | LF unit diam | Crossover freq | MID unit diam | Crossover freq | HF unit diam | Amplifiers | Cabinet volume | Introduced |
|---|---|---|---|---|---|---|---|---|---|
| RH532 | 3 | 200 mm | 500 active | 130 mm | 3500 passive | 25 mm | 40 W + 20 W | | 1974 |
| RH541 | 2 | 180 mm | n/a | n/a | 1400 passive | 25 mm | 30 W | | 1975 |
| RH544 | 3 | 200 mm | 500 active | 50 mm | 4000 passive | 25 mm | 40 W + 20 W | | 1975 |
| RH567 | 3 | 250 mm | 500 active | 50 mm | 3500 passive | 25 mm | 40 W + 20 W | | 1978 |
| RH545 | 3 | 300 mm | 500 active | 50 mm | 3000 active | 25 mm | 50 W + 35 W + 15 W | 70 litre | 1976 |
| AH585 | 2 | 180 mm | n/a | n/a | 1500 active | 20 mm | 30 W + 5 W | 9 litre | 1979 |
| AH586 | 2 | 200 mm | n/a | n/a | 1500 active | 20 mm | 40 W + 5 W | 14 litre | 1979 |
| AH587 | 3 | 200 mm | 650 active | 75 mm | 3500 active | 20 mm | 50 W + 20 W + 5 W | 25 litre | 1978 |
| F9638 | 3 | 200 mm | 700 | 75 mm | 4000 | ribbon | 50 W + 35 W | 26 litre | 1983 |

between 55 and 160 Hz, "Rear to wall" used a single gyrator to create a −5 dB dip at 60 Hz, and "Floor" used a single gyrator to create a −5 dB dip at 200 Hz. Each gyrator was made from three discrete transistors; they are not 'standard' gyrators, and their detailed operation is quite complex.

The AH586 was a 2-way speaker offering switchable correction filters for "Rear to wall", "Floor", and "Side to wall" placement, but the equalisation was simpler. "Rear to wall" used a single gyrator to create a −6 dB dip at 63 Hz, "Floor" used a single gyrator to create a −3.5 dB dip at 125 Hz, and "Side to wall" equalisation used two simple time-constants to create a very gentle 3.5 dB dip around 63 Hz. Thus only two gyrators were used instead of five.

The AH587 was a 3-way speaker. Notably, each of the three drive units had a different impedance. The HF drive unit had a 15 Ω impedance, the MID unit an 8 Ω impedance, and the LF unit a 4 Ω impedance, to suit the signal levels in each frequency range. "Rear to wall", "Floor", and "Side to wall" correction options were provided, exactly as for the AH586.

The F9638 was a 3-way speaker using flat-membrane LF and MID drive units, with a ribbon tweeter. It was presumably intended as the next generation of Philips MFB loudspeakers. MFB was only applied on the LF unit. It was introduced in 1983 but does not appear to have been successful, and went out of production in 1986. I think this was the only F-series MFB loudspeaker.

The Philips numbering system is somewhat opaque. For example AH468 and AH483 sound like they might be in the same MFB product series, but actually both were passive crossover loudspeakers.

Slightly off-topic, as this loudspeaker did not use MFB, but still of considerable interest is the Philips DSS930, which used DSP for its active crossover functions and a Class-G amplifier [14] for the LF drive unit.

Since the Philips loudspeakers were designed there have been major advances in accelerometer technology. When I first conceived this chapter, I thought a promising new approach would be the use of a MEMS accelerometer. [15] It did not take long to find out that someone was well ahead of me, using an Analog Devices ADXL001–250BEZ. [16] This is a single axis accelerometer IC that can handle up to ±250G and up to 10 kHz, and as far as I can see it is an excellent device for the job.

## Other MFB Speakers

The Sony FH-150R was a hefty boom-box weighting 14 kg that had MFB applied to its LF units. This was switchable with a button marked AMFB, meaning "Accurate Motional FeedBack"; the MFB path could be disabled by analogue switch ICs. No details are currently known, but there are two wires labelled MFB+ and MFB−coming back from the LF drive unit, which suggests a velocity-sensing coil. Other Sony loudspeakers, such as the SA-W2500, 3000, and 3800 subwoofers, sensed the drive unit current with a 0R22 series resistor.

Bang and Olufsen filed a US patent in 1977 [17] for an MFB system using acceleration feedback. The sensor was just described as an "acceleration transducer" with no further details. I am not aware that any B&O MFB speaker reached the marketplace.

Interest in MFB is not dead. In 2013 Harman International Industries filed a long and complex patent based on voice coil current sensing. [18] This patent is worth seeking out for the very long list of MFB references it gives.

## Published Projects

Given the mechanical difficulties of adding transducers to drive units, it is not surprising that few MFB build projects have been published. De Greef and Vandewege published an acceleration feedback design in *Wireless World* in 1981 [19] that was explicitly inspired by the RH532 and actually used Philips 200 mm drivers with integral accelerometers. Two were connected in series; it appears that MFB was only taken from one of them.

Ian Hegglun wrote on velocity feedback in 1996, [20] and an MFB design by Russel Breden appeared in *Electronics World* in 1997. [21] Breden's motional feedback was accomplished by using dual-coil drivers, one coil being driven by the amplifier and the other giving velocity feedback. No mention was made of transformer action between the two coils. In fact, two of these dual-coil drivers were used in parallel, but feedback was taken from only one of them, which does not sound ideal unless the drivers were exactly identical in every respect. There was very little in the way of measured results.

## Conclusions

Consensus of any kind in the audio business is extremely rare, but there seems to be agreement that the Philips MFB loudspeakers gave unusually deep and clean bass from boxes that were apparently impossibly small. The technology was adaptable to a wide range of two- and 3-way loudspeakers. Nonetheless, they lasted only ten years in the marketplace. I suggest that was partly because they were expensive, though of course you didn't have to buy a power amplifier. However, people want to choose their own power amplifiers, so this was a disadvantage rather than a feature. Perhaps Philips were simply not a cool brand—at that point Philips to most people meant inexpensive cassette recorders rather than the new technology of CDs.

Whatever the commercial outcome, there is no doubt that Philips' implementation of acceleration feedback was both elegant and technically sound, and there seems to be every reason to re-introduce it.

## References

[1] Tanner, R. L. "Improving Loudspeaker Response With Motional Feedback" Electronics 24(3), 1951, p. 142
[2] Werner, R. E. "Loudspeakers and Negative Impedances" IRE Trans AU-6, 1958, pp. 83–89 See also US patent 2,887.532, October 1956
[3] Holdaway, H. W. "Design of Velocity-Feedback Transducer Systems for Stable Low Frequency Behaviour" IEEE Trans AU-11, 1963, pp. 155–173
[4] Klassen and de Koning "Motional Feedback with Loudspeakers" Philips Technical Review 29, 1964, pp. 148–157, www.extra.research.philips.com/hera/people/aarts/_Philips%20Bound%20Archive/ PTechReview/PTechReview-29-1968-148.pdf; Accessed January 2017
[5] Brodie, G. H. *Speaker Diaphragm Controlled Capacitor for Negative Feedback Control* US Patent 2,857,461 October 1958
[6] www.servospeaker.com/index.html (inactive); Accessed Jan 2017
[7] www.servospeaker.com/pdf/ProtoDesc.pdf (inactive); Accessed Jan 2017
[8] Nuutinmaki, P. V. M. Patent application WO 2004/082330 Mar 2004

[9] Hi-Fi World March 2008

[10] www.proaudiodesignforum.com/forum/php/viewtopic.php?f=12&t=707 (Panasonic MF-800) Accessed Jan 2017

[11] Jarman & Howard Vintage Hifi: Philips 22RH544 Motional Feedback Loudspeaker *Hi-fi News* Nov 2016, pp 118–123

[12] Billboard, 13 July 1974, p. 34

[13] *Popular Electronics* March 1975, pp. 61–62

[14] Self, Douglas "Audio Power Amplifier Design" Sixth Edn, Newnes, 2013, pp. 150–157, ISBN: 978-0-240-52613-3 (Class G)

[15] https://en.wikipedia.org/wiki/Microelectromechanical_systems; Accessed February 2017

[16] Schneider et al. "Design and Evaluation of Accelerometer Based Motional Feedback Audio Engineering Society" 138th Convention, Warsaw, May 2015 (Paper 9293)

[17] Baakgaard, Knud *Loudspeaker Motional Feedback System.* US patent 4,180,706 May 1977

[18] Stanley, Gerald *Motional Feedback System* US patent 8,401,207 March 2013

[19] Greef, De and Vandewege "Acceleration Feedback Loudspeaker" Wireless World, September 1981, pp. 32–36

[20] Hegglun, Ian "Speaker Feedback" (i.e. motional feedback) Electronics World, May 1996, p. 378

[21] Breden, Russel "Roaring Subwoofer" (with motional feedback) Electronics World, February 1997, p. 104

# CHAPTER 20

# *Line Inputs*

## External Signal Levels

There are several standards for line signal levels. The −10 dBv standard is used for a lot of semi-professional recording equipment, as it gives more headroom with unbalanced connections—the professional levels of +4 dBu and +6 dBu assume balanced outputs which inherently give twice the output level for the same supply rails, as it is measured between two pins with signals of opposite phase on them. See Table 20.1.

Signal levels in dBu are expressed with reference to 0 dBu = 775 mVrms; the origin of this odd value is that it gives a power of 1 mW in a purely historical 600 Ω load. Signals in dBv (or dBV) are expressed with reference to 0 dB = 1.000 Vrms.

These standards are well established, but that does not mean that all equipment uses them as a nominal level. Many power amplifiers require more than 0 dBV for full output; the Yamaha P7000S power amplifier requires +8 dBu (1.95 Vrms) to give its full output of 750 W into 8 Ω. The "0 VU" on VU meters is nominally 1.23 volts.

## Internal Signal Levels

In any audio system it is necessary to select a suitable nominal level for the signal passing through it. This level is always a compromise—the signal level should be high so it suffers minimal degradation by the addition of circuit noise as it passes through the system but not so high that it is likely to suffer clipping before it reaches a gain control or generate undue distortion below the clipping level. (This last constraint is not normally a problem with modern circuitry, which gives very low distortion right up to the clipping point.)

The choice of internal levels for active crossovers is complicated by the fact that there are two or three parallel signal paths carrying signals with completely different spectral distributions. It is well known that signal levels at the upper end of the audio band are much lower than those at low and middle frequencies, which raises the question of whether an HF signal path will require more gain so that it can be run at a higher nominal internal level. This would give a better signal-noise-ratio, which is important, as noise is much more obvious in the HF path. The LF and MID paths both contain lowpass filters, which, if they are appropriately placed late in the processing chain, discriminate heavily against the most audible part of the noise spectrum.

This is a complicated issue and is dealt with in more detail in Chapter 17 on crossover system design.

Table 20.1: Nominal signal levels.

|  | Vrms | dBu | dBv |
|---|---|---|---|
| Semi-professional | 0.316 | −7.78 | −10 |
| Professional | 1.228 | +4.0 | +1.78 |
| German ARD | 1.55 | +6.0 | +3.78 |

If the incoming signal does have to be amplified, this should be done as early as you can in the signal path, to get the signal well above the noise floor as soon as possible. If the gain is in the input amplifier, balanced or otherwise, the signal will pass through later stages at a high level, and so their noise contribution will be less. However, if the input stage is configured with a fixed gain, this must be kept low, as it is not possible to turn it down to avoid clipping. Ideally any input stage should have variable gain. It is not straightforward to combine this feature with a good balanced input, but several ways of doing it are shown later in this chapter.

## Input Amplifier Functions

It is important that the very first thing an incoming signal meets is some form of RF filtering, to prevent RF breakthrough and other EMC problems. It must be done before the incoming signal encounters any semiconductors where RF demodulation could occur and can be regarded as a "roofing filter". Next, the low-end frequency response is given an early limit by DC-blocking capacitors, and in some cases overvoltage spikes are clamped by diodes. An input amplifier should have a reasonably high impedance: certainly not less than 10 kΩ, and preferably more. It must have a suitable gain— possibly switched or variable—to scale the incoming signal to the nominal internal level. Balanced input amplifiers also accurately perform the subtraction process that converts differential signals to single-ended ones, so noise produced by ground loops and the like is rejected. That's quite a lot of functionality for one stage.

## Unbalanced Inputs

The simplest unbalanced input would feed the incoming signal directly to the first stages of the crossover. This is not practical, because these stages will very likely be active filters that require a low source impedance to give the expected response. In addition, the input impedance would vary with frequency and could fall to rather low values. Some sort of input amplifier which can be fed from a significant impedance without ill effect is needed.

A typical unbalanced input amplifier is shown in Figure 20.1. The opamp U1:A is a unity-gain voltage-follower; it could be altered to give a fixed gain by adding two series-feedback resistors. A 5532 bipolar opamp is used here for its low distortion and low noise; with the low source impedances that are likely, an FET-input opamp would be noisier by 10 dB or more. R1 and C1 are a 1st-order lowpass filter to stop incoming RF before it reaches the opamp, where it would be likely to be demodulated into the audio band; once this has happened, any further attempts at RF filtering are pointless. R1 and

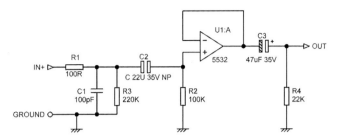

**Figure 20.1: An unbalanced input amplifier with RF filter, DC drain, and input and output DC blocking.**

C1 must be as physically close to the input socket as possible to prevent RF being radiated inside the equipment enclosure before it is shunted to ground, and this is why they should always be the first components in the signal path.

Selecting component values for input RF filters like this is always a compromise, because the output impedance of the source equipment is not known. If the source is an active preamplifier, then the output impedance ought to be around 50 Ω, but it could be 200 Ω or as high as 1 kΩ. If the source is one of those oxymoronic "passive preamplifiers"—in other words just an input selector switch, and a volume potentiometer or switched attenuator, and an improbably large price-tag—then the output impedance will be much higher in some circumstances. If you really must use a piece of equipment that blazons forth its internal contradictions in its very name, you will find that by far the most popular potentiometer value is 10 kΩ, with a maximum output impedance (when set for 6 dB of attenuation) of 2.5 kΩ, very much higher than the 50 Ω we might expect from a good active preamplifier. This is in series with R1 and affects the turnover frequency of the RF filter. Effective RF filtering is very desirable, but it is also important to avoid a frequency response that sags significantly at 20 kHz. Valve equipment is also likely to have a high output impedance.

Taking 2.5 kΩ as the worst-case source impedance and summing it with R1, we get 2.6 kΩ. With a 100 pF capacitor we would get −3 dB at 612 kHz; the loss at 20 kHz is a wholly negligible 0.005 dB, so we might decide that C1 could be usefully increased. If we make it 220 pF, then the 20 kHz loss is still a tiny 0.022 dB, but the−3 dB point is now 278 kHz, much improving the rejection of what used to be called The Medium Wave. If we take C1 as 220 pF and assume an active output with a 50 Ω impedance in the source equipment, then together with the 100 Ω of R1 we have 150 Ω, which in conjunction with 100 pF gives us −3 dB at 4.82 MHz. This is a bit higher than is really desirable, but it is not easy to see what to do about it. If there was a consensus that the output impedance of a respectable piece of audio equipment should not exceed 100 Ω, then things would be much easier in this area.

Another important consideration is that the series resistance R1 must be kept as low as practicable to minimise Johnson noise; but lowering this resistance means increasing the value of shunt capacitor C1, and if it becomes too big, then its impedance at high audio frequencies will become too low. Not only will there be too low a roll-off frequency if the source has a high output impedance, but there might be an increase in distortion at high audio frequencies because of excessive loading on the source output stage.

Replacing R1 with a small inductor to make an LC lowpass filter will give much better RF rejection at increased cost. This is justifiable in professional audio equipment, but it is much less common in hi-fi, one reason being that the unpredictable source impedance makes the filter design difficult, as we have just seen. In the professional world one *can* assume that the source impedance will be low. Adding more capacitors and inductors allows a three- or four-pole LC filter to be made. If you do use inductors, then it is important to check the frequency response to make sure it is as intended and there is no peaking at the turnover frequency, as you could have made a filter that does more harm than good.

C2 is a DC-blocking capacitor to prevent voltages from ill-conceived source equipment getting into the input amplifier. It is a non-polarised type, as voltages from outside are of unpredictable polarity, and it is rated at not less than 35 V, so that even if it gets connected to defective equipment with an opamp output jammed hard against one of its supply rails, the capacitor will not be damaged. It will give no protection against faulty valve equipment that may put out a couple of hundred volts. R3 is a DC drain resistor that prevents any charge put on C2 by external equipment from remaining there for a long time and causing a thud when connections are replugged; as with all input drain resistors, its value is a compromise between discharging the capacitor quickly and keeping the input impedance high. The input impedance here is R3 in parallel with R2, i.e. 220 kΩ in parallel with 100 kΩ, giving 68 kΩ. This is a suitably high value and should work well with just about any source equipment, including that valve-based stuff.

R2 provides the biasing for the opamp input; it must be a high value to keep the input impedance up, but bipolar input opamps draw significant input bias current. The Fairchild 5532 data sheet quotes 200 nA typical, and 800 nA maximum, and these currents would cause a voltage drop across R2 of 20 mV and 80 mV respectively. This offset voltage will be reproduced at the output of the opamp, with the input offset voltage added; this is only 4 mV maximum, much less than the offset due to the bias current. The 5532 has NPN input transistors, so the bias current flows into the input pins, and the voltage at Pin 3 and hence the opamp output may be negative with respect to ground by anything up to 84 mV.

Such offset voltages do not significantly affect the output voltage swing, but they will generate unpleasant clicks and pops if the input stage is followed by any sort of switching, and they are big enough to make potentiometers crackly; the DC voltage (you know perfectly well what I mean) is therefore blocked by C3. R4 is another DC drain to keep the output at zero volts; it can be made lower in value than the input drain R3, as the only requirement is that it should not significantly load the opamp output. 22 kΩ or 47 kΩ resistors are commonly used.

FET-input opamps have much lower input bias currents, so that the offsets they generate as they flow through biasing resistors are usually negligible, but they still have input offsets of a few millivolts, so DC blocking will still be needed if switches downstream are to work silently.

This input stage, with its input terminated by 50 Ω to ground, has a noise output of only −119.0 dBu over the usual 22 Hz–22 kHz bandwidth. This is very quiet indeed and is a reflection of the low voltage noise of the 5532, and the fact that R1, the only resistor in the signal path, has the low value of 100 Ω and so generates a very little Johnson noise, only −132.6 dBu, to be precise. This noise is wholly swamped by the voltage noise of the opamp, which is basically all we see; its current noise has negligible effect because of the low circuit impedances.

# Balanced Interconnections

Balanced inputs are used to prevent noise and crosstalk from affecting the input signal, especially in applications where long interconnections are used. They are standard on professional audio equipment and are quite quickly becoming more common in the world of hi-fi. Their importance is that they can render ground loops and other connection imperfections harmless. Since there is no point in making a wonderful piece of equipment and then feeding it with an impaired signal, making sure you have an effective balanced input really is of the first importance, and I will go into it in some detail.

The basic principle of balanced interconnection is to get the signal you want by subtraction, using a three-wire connection. In some cases a balanced input is driven by a balanced output, with two anti-phase output signals; one signal wire (the hot or in-phase) sensing the in-phase output of the sending unit, while the other senses the anti-phase output.

In other cases, when a balanced input is driven by an unbalanced output, as shown in Figure 20.2, one signal wire (the hot or in-phase) senses the single output of the sending unit, while the other (the cold or phase inverted) senses the unit's output-socket ground, and once again the difference between them gives the wanted signal. In either of these two cases, any noise voltages that appear identically on both lines (i.e. common-mode signals) are in theory completely cancelled by the subtraction. In real life the subtraction falls short of perfection, as the gains via the hot and cold inputs will not be precisely the same, and the degree of discrimination actually achieved is called the common-mode rejection ratio (CMRR), of which more later. I should also say of Figure 20.2 that the CMRR is often much improved by putting a resistance in series with the cold line, where it goes into the phono plug, of a value equal to the output impedance of the unbalanced output; more on that later, too.

It is deeply tedious to keep referring to non-inverting and inverting inputs, and so these are usually abbreviated to "hot" and "cold" respectively. This does *not* necessarily mean that the hot terminal carries more signal voltage than the cold one. For a true balanced connection, the voltages will be equal. The "hot" and "cold" input terminals are also often referred to as IN+ and IN−, and this latter convention has been followed in the diagrams here.

The subject of balanced interconnections is a large one, and a big book could be written on this topic alone; two of the classic papers on the subject are by Neil Muncy [1] and Bill Whitlock, [2] both of which are very well worth reading.

To make a start, let us look at the pros and cons of balanced connections:

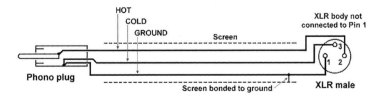

**Figure 20.2: Unbalanced output to balanced input interconnection, with cold joined to ground at the unbalanced end.**

## The Advantages of Balanced Interconnections

- Balanced interconnections discriminate against noise and crosstalk, whether they result from ground currents or electrostatic or magnetic coupling to signal conductors.
- Balanced connections make ground loops much less intrusive and usually inaudible, so people are less tempted to start "lifting the ground" to break the loop, with possibly fatal consequences. In the absence of a dedicated ground-lift switch that leaves the external metalwork firmly connected to mains safety earth, the foolhardy and the optimistic will break the mains earth (not quite so easy now that moulded mains plugs are standard), and this is highly dangerous, as a short-circuit from mains to the equipment chassis will then result in live metalwork but dead people.
- A balanced interconnection incorporating a true balanced output gives 6 dB more signal level on the line, which should give 6 dB more dynamic range. However, this is true only with respect to *external* noise—as is described later in this chapter, a standard balanced input using 10 kΩ resistors is about 14 dB noisier than the unbalanced input shown in Figure 20.1.
- Balanced connections are usually made with XLR connectors. These are a professional 3-pin format and are far superior to the phono (RCA) type normally used for unbalanced connections. More on this later.

## The Disadvantages of Balanced Interconnections

- Balanced inputs are inherently noisier than unbalanced inputs by a large margin, in terms of the noise generated by the input circuitry itself rather than external noise. This may appear paradoxical, but it is all too true, and the reasons will be fully explained in this chapter.
- More hardware means more cost. Small-signal electronics is relatively cheap; unless you are using a sophisticated low-noise input stage, of which more later, most of the extra cost is likely to be in the balanced input connectors.
- Balanced connections do not of themselves provide any greater RF immunity than an unbalanced input. For this to happen both legs of the balanced input would have to demodulate the RF in equal measure for common-mode cancellation to occur. The chances of this happening over any sort of frequency range are effectively zero. It remains vital to provide the passive RF filtering before the first active electronics.
- There is the possibility of introducing a phase error. It is all too easy to create an unwanted phase inversion by confusing hot and cold when wiring up a connector, and this can go undiscovered for some time. The same mistake on an unbalanced system interrupts the audio completely and leaves little room for doubt that something is amiss.
- Balanced connectors (usually XLRs) are inevitably more expensive. However, their security of connection and general quality make it money well spent, in my view.

## Balanced Cables and Interference

In a balanced interconnection two wires carry the signal, and the third connection is the ground wire, which has two functions.

First, it joins the grounds of the interconnected equipment together. This is not always desirable, and if galvanic isolation is required, a transformer balancing system will be necessary because the large common-mode voltages are likely to exceed the range of an electronic balanced input. A good transformer will also have a very high CMRR, which will be needed to get a clean signal in the face of large CM voltages.

Second, the presence of the ground allows electrostatic screening to shield the two signal wires, preventing both the emission and pickup of unwanted signals. In cheap cables this will mean a "lapped screen", with wires laid parallel to the central signal conductors. The screening coverage is not total and can be badly degraded, as the screen tends to open up on the outside of cable bends. A braided screen around the signal wires gives better coverage, but still not 100%. This is much more expensive, as it is harder to make.

The best solution is an overlapping foil screen, with the ground wire (often called the drain wire in this context) running down the inside of the foil and in electrical contact with it. This is usually the most effective, as the foil is a solid sheet and cannot open up on bends. It should give perfect electrostatic screening, and it is much easier to work with than either lap screen or braided cable.

There are three main ways in which an interconnection is susceptible to hum and noise:

1.  Electrostatic coupling

An interfering signal at significant voltage couples directly to the inner signal line, through stray capacitance. The stray capacitance between imperfectly screened conductors will be a fraction of a pF in most circumstances, as electrostatic coupling falls off with the square of distance. This form of coupling can be serious in studio installations with unrelated signals running down the same ducting.

The three main lines of defence against electrostatic coupling are effective screening, low impedance drive, and a good CMRR maintained up to the top of the audio spectrum. As regards screening, an overlapped foil screen provides complete protection.

Driving the line from a low impedance, of the order of 100 $\Omega$ or less, is also helpful, because the interfering signal, having passed through a very small stray capacitance, is a very small current and cannot develop much voltage across such a low impedance. This is convenient because there are other reasons for using a low output impedance, such as optimising the interconnection CMRR, minimising HF losses due to cable capacitance, and driving multiple inputs without introducing gain errors. For the best immunity to crosstalk, the output impedance must remain low up to as high a frequency as possible. This is definitely an issue, as opamps invariably have a feedback factor that begins to fall from a low and quite possibly sub-audio frequency, and this makes the output impedance rise with frequency as the negative feedback factor falls, as if an inductor were in series. Some line outputs have physical series inductors to improve stability or EMC immunity, and these should not be so large that they significantly increase the output impedance at 20 kHz. From the point of view of electrostatic screening alone, the screen does not need to be grounded at both ends or form part of a circuit. [3] It must of course be grounded at some point.

If the screening is imperfect, and the line impedance non-zero, some of the interfering signal will get into the hot and cold conductors, and now the CMRR must be relied upon to make the immunity acceptable. If it is possible, rearranging the cable-run away from the source of interference and getting some properly screened cable is more practical and more cost-effective than relying on very good common-mode rejection.

Stereo hi-fi balanced interconnections almost invariably use XLR connectors. Since an XLR can only handle one balanced channel, two separate cables are almost invariably used, and interchannel capacitive crosstalk is not an issue. Professional systems, on the other hand, use multi-way connectors that do not have screening between the pins, and there is an opportunity for capacitive crosstalk here, but the use of low source impedances should reduce it to below the noise floor.

### 2.   Magnetic coupling

If a cable runs through an AC magnetic field, an EMF is induced in both signal conductors and the screen, and according to some writers, the screen current must be allowed to flow freely, or its magnetic field will not cancel out the field acting on the signal conductors, and therefore the screen should be grounded at both ends, to form a circuit. [4] In practice the magnetic field cancellation will be very imperfect, and reliance is better placed on the CMRR of the balanced system to cancel out the hopefully equal voltages induced in the two signal wires. The need to ground both ends to possibly optimise the magnetic rejection is not usually a restriction, as it is rare that galvanic isolation is required between two pieces of audio equipment.

The equality of the induced voltages can be maximised by minimising the loop area between the hot and cold signal wires, for example by twisting them tightly together in manufacture. In practice most audio foil-screen cables have parallel rather than twisted signal conductors, but this seems adequate almost all of the time. Magnetic coupling falls off with the square of distance, so rearranging the cable-run away from the source of magnetic field is usually all that is required. It is unusual for it to present serious difficulties in a hi-fi application.

### 3.   Ground voltages

These are the result of current flowing through the ground connection and is often called "common-impedance coupling" in the literature. [1] This is the root of most ground-loop problems. The existence of a loop in itself does no harm, but it is invariably immersed in a 50 Hz magnetic field that induces mains-frequency currents plus harmonics into it. This current produces a voltage drop down non-negligible ground-wire resistances, and this effectively appears as a voltage source in each of the two signal lines. Since the CMRR is finite, a proportion of this voltage will appear to be a differential signal, and will be reproduced as such.

## Balanced Connectors

Balanced connections are most commonly made with XLR connectors, though it can be done with stereo (tip-ring-sleeve) jack plugs. XLRs are a professional 3-pin format, and are a much better connector in every way than the usual phono (RCA) connectors used for unbalanced interconnections. Phono connectors have the great disadvantage that if you are connecting them with the system active (inadvisable, but then people are always doing inadvisable things), the signal contacts meet before the grounds, and thunderous noises result. The XLR standard has Pin 2 as hot, Pin 3 as cold, and Pin 1 as ground.

As described in Chapter 1, in domestic crossover use there is a good case for using multi-way connectors that carry several three-wire balanced connections, in order to cut down the amount of visible cabling.

## Balanced Signal Levels

Many pieces of equipment, including preamplifiers and power amplifiers designed to work together, have both unbalanced and balanced inputs and outputs. The general consensus in the hi-fi world is that if the unbalanced output is, say, 1 Vrms, then the balanced output will be created by feeding the in-phase output to the hot output pin and also to a unity-gain inverting stage, which drives the cold output pin with 1 Vrms phase inverted. The total balanced output voltage between hot and cold pins is therefore 2 Vrms, and so the balanced input must have a gain of 1/2 or −6 dB relative to the unbalanced input to maintain consistent internal signal levels.

## Electronic vs Transformer Balanced Inputs

Balanced interconnections can be made using either transformer or electronic balancing. Electronic balancing has many advantages, such as low cost, low size and weight, superior frequency and transient response, and no low-frequency linearity problems. Transformer balancing has advantages of its own, particularly for work in very hostile RF/EMC environments, but serious drawbacks. The advantages are that transformers are electrically bullet-proof (and quite possibly physically bullet-proof), retain their high CMRR performance forever, and consume no power even at high signal levels. They are essential if galvanic isolation between ground is required. Unfortunately, transformers can generate LF distortion, particularly if they have been made with minimal core sizes to save weight and cost. They are liable to have HF response problems due to leakage reactance and distributed capacitance, and compensating for this requires a carefully designed Zobel network across the secondary. Inevitably they are heavy and expensive compared with an opamp and a few R's and C's. Transformer balancing is therefore relatively rare, even in professional audio applications, and the greater part of this chapter deals with electronically balanced inputs.

## Common-Mode Rejection Ratio (CMRR)

Figure 20.3 shows a balanced interconnection reduced to its bare essentials; hot and cold line outputs with source resistances Rout+, Rout−and a standard differential amplifier at the input end. The output resistances are assumed to be exactly equal, and the balanced input in the receiving equipment has two exactly equal input resistances to ground R1, R2. The ideal balanced input amplifier senses the voltage difference between the points marked IN+ (hot) and IN−(cold) and ignores any common-mode voltage which are present on both. The amount by which it discriminates is called the common-mode rejection ratio or CMRR and is usually measured in dB. Suppose a differential voltage input between IN+ and IN−gives an output voltage of 0 dB; now reconnect the input so that IN+ and IN−are joined together, and the same voltage is applied between them and ground. Ideally the result would be zero output, but in this imperfect world it won't be, and the output could be anywhere between −20 dB (for a bad balanced interconnection, which probably has something wrong with it) and −140 dB (for an extremely good one). The CMRR when plotted may have a flat section at low frequencies, but it very commonly degrades at high audio frequencies and may also deteriorate at very low frequencies. More on that later.

In one respect balanced audio connections have it easy. The common-mode signal is normally well below the level of the unwanted signal, and so the common-mode range of the input is not an issue. In other areas of technology, such as electrocardiogram amplifiers, the common-mode signal may be many times greater than the wanted signal.

The simplified conceptual circuit of Figure 20.3, under SPICE simulation, demonstrates the need to get the resistor values right for a good CMRR, before you even begin to consider the rest of the circuitry. The differential voltage sources Vout+, Vout−, which represent the actual balanced output, are set to zero, and Vcm, which represents the common-mode voltage drop down the cable ground, is set to 1 volt to give a convenient result in dBV. The output resulting from the presence of this voltage source is measured by a mathematical subtraction of the voltages at IN+ and IN−, so there is no actual input amplifier to confuse the results with its non-ideal performance.

Let us begin with Rout+ and Rout− set to 100 Ω and input resistors R1, R2 set to 10 kΩ. These are typical real-life values as well as being nice round figures. When all four resistances are exactly at their nominal value, the CMRR is in theory infinite, but if just one of the output resistors or one of the input resistors is then altered in value by 1%, then the CMRR drops sharply to −80 dB. If the deviation is 10%, things are predictably worse, and the CMRR degrades to −60 dB. The CMRR is here flat with frequency because our simple model has no frequency-dependent components.

Each line of the connection is a potential divider with Rout at the top and R1 at the bottom. With these values, the attenuation on each line is the very small figure of 0.0087 dB. If we can reduce this attenuation, then the gain on each line will be closer to unity, and minor changes in resistor values will have less effect. As an example, if we increase the input resistors R1, R2 to 100 kΩ with Rout+, Rout− kept at 100 Ω, then a 1% deviation only degrades the CMRR to −100 dB. This change is quite practical, so long as you buffer the inputs to the actual balanced amplifier—the technique is explained in more detail later. An even higher value for R1, R2 of 1 MΩ, which is still feasible but slightly more difficult, gives −120 dB for a single 1% resistance deviation and −100 dB for a single 10% deviation.

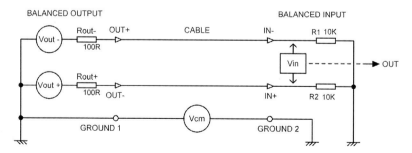

**Figure 20.3: A theoretical balanced interconnection showing how the output and input impedances influence CMRR.**

We could also improve CMRR by reducing the output impedances Rout+, Rout−. Dropping them to 10 Ω and using the conventional 10 kΩ input resistors gives −100 dB for a single 1% deviation. The problem is that these resistors are in the source equipment and usually outside our control. A 10 Ω output resistor is also too low to prevent HF instability caused by cable capacitance and would need to be supplemented by expensive output chokes. Alternatively, and much more economically, the output stage could be configured as a "zero-impedance output" as described in the section of the chapter on line outputs; an output impedance of a fraction of an ohm at 1 kHz combined with stability is very easy to achieve.

If, however, as usual we have to take the source equipment as it comes, we cannot assume the output resistors will be less than 100 Ω, and they may be a good deal higher. If the output is unbalanced, then we effectively have one output resistor at, say, 100 Ω, while the other is zero, as there is a direct connection between the cold line and the output ground. This messes things up dramatically, and the CMRR collapses to 43 dB.

Time for a reality check. You will have noticed that the CMRR figures we are dealing with, of 80 or 100 dB, are much better than we measure in reality. This is because we are only altering one of four resistances—in real life all four will be subject to a statistical distribution, and the CMRR results likewise come out as a statistical distribution.

There is no point in going into that level of detail here, because there are other and more important influences on CMRR, which we will look at shortly. The lesson to take away here is that we need the lowest possible output impedances (if we have any say in their value) and the highest possible input impedances to get the maximum common-mode rejection. This is highly convenient, because low output impedances are already needed to drive multiple inputs and cable capacitance, and high input impedances are needed to minimise loading and consequent signal losses. However . . . it will soon emerge that the influences on CMRR are such that what we should really conclude is that balanced line input impedances should be as high as possible *without compromising anything else*.

## The Basic Electronic Balanced Input

Figure 20.4 shows the basic balanced input amplifier. To achieve balance R1 must be equal to R3 and R2 equal to R4. The amplifier in Figure 20.4 has a gain of R2/R1 (= R4/R3). The standard one-opamp balanced input or differential amplifier is a very familiar circuit block, but its operation often appears somewhat mysterious. Its input impedances are *not* equal when it is driven from a balanced output; this has often been commented on, [5] and some confusion has resulted.

The source of the confusion is that a simple differential amplifier has interaction between the two inputs, so that the input impedance seen on the cold input depends on the signal applied to the hot input. Input impedance is measured by applying a signal and seeing how much current flows into the input, so it follows that the apparent input impedance on each leg varies according to how the cold input is driven. If the amplifier is made with four 10 kΩ resistors, then the input impedances on hot and cold are given in Table 20.2:

Some of these impedances are not exactly what you might expect, and require some explanation.

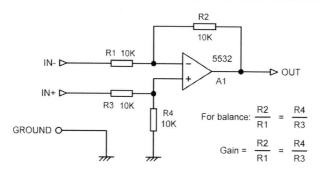

**Figure 20.4: The basic balanced input amplifier, with standard 10 kΩ resistors.**

**Table 20.2: The input impedances for different input drive conditions.**

| Case | Pins driven | Hot input res Ω | Cold input res Ω |
|------|-------------|-----------------|------------------|
| 1 | Hot only | 20k | Grounded |
| 2 | Cold only | Grounded | 10k |
| 3 | Both (balanced) | 20k | 6.66k |
| 4 | Both common-mode | 20k | 20k |
| 5 | Both floating | 10k | 10k |

## Case 1

The balanced input is being used as an unbalanced input by grounding the cold input and driving the hot input only. The input impedance is therefore simply R3 + R4. Resistors R3 and R4 reduce the signal by a factor of a half, but this loss is undone, as R1 and R2 set the amplifier gain to two times, and the overall gain is unity. If the cold input is not grounded, then the gain is 0.5 times. The attenuate-then-amplify architecture, plus the Johnson noise from the resistors, makes this configuration much noisier than the dedicated unbalanced input of Figure 20.1, which has only a single 100 Ω resistor in the signal path.

## Case 2

The balanced input is again being used as an unbalanced input, but this time by grounding the hot input and driving the cold input only. This gives a phase inversion, and it is unlikely you would want to do it except as an emergency measure to correct a phase error somewhere else. The important point here is that the input impedance is now only 10 kΩ, the value of R1, because shunt negative feedback through R2 creates a virtual earth at Pin 2 of the opamp. Clearly this simple circuit is not as symmetrical as it looks. The gain is unity, whether or not the hot input is grounded; grounding it is desirable because it not only prevents interference being pickup on the hot input pin but also puts R3 and R4 in parallel, reducing the resistance from opamp Pin 3 to ground and so reducing Johnson noise.

## Case 3

This is the standard balanced interconnection. The input is driven from a balanced output with the same signal levels on hot and cold, as if from a transformer with its centre-tap grounded, or an electronically balanced output using a simple inverter to drive the cold pin. The input impedance on the hot input is what you would expect; R3 + R4 add up to 20 kΩ. However, on the cold input there is a much lower input impedance of 6.66 kΩ. This at first sounds impossible, as the first thing the signal encounters is a 10 kΩ series resistor, but the crucial point is that the hot input is being driven simultaneously with a signal of the opposite phase, so the inverting opamp input is moving in the opposite direction to the cold input due to negative feedback, and what you might call anti-bootstrapping reduces the effective value of the 10 kΩ resistor to 6.66 kΩ. These are the differential input impedances we are examining, the impedances seen by the balanced output driving them. Common-mode signals see a common-mode impedance of 20 kΩ, as in Case 4.

You will sometimes see the statement that these unequal differential input impedances "unbalance the line". From the point of view of CMRR, this is not the case, as it is the CM input impedance that counts. The line is, however, unbalanced in the sense that the cold input draws three times the current from the output that the hot one does. This current imbalance might conceivably lead to inductive crosstalk in some multi-way cable situations, but I have never encountered it. The differential input impedances can be made equal by increasing the R1 and R2 resistor values by a factor of three, but this degrades the noise performance markedly and makes the common-mode impedances to ground unequal, which is a much worse situation, as it compromises the rejection of ground voltages, and these are almost always the main problem in real life.

## Case 4

Here both inputs are driven by the same signal, representing the existence of a common-mode voltage. Now both inputs shown an impedance of 20 kΩ. It is the symmetry of the common-mode input impedances that determines how effectively the balanced input rejects the common-mode signal. This configuration is of course only used for CMRR testing.

## Case 5

Now the input is driven as from a floating transformer with the centre-tap (if any) unconnected, and the impedances can be regarded as equal; they must be, because with a floating winding the same current must flow into each input. However, in this connection the line voltages are *not* equal and opposite: with a true floating transformer winding, the hot input has all the signal voltage on it while the cold has none at all, due to the negative feedback action of the balanced input amplifier. This seemed very strange when it emerged in SPICE simulation, but a sanity-check with real components proves it true. The line has been completely unbalanced as regards crosstalk to other lines, although its own common-mode rejection remains good.

Even if absolutely accurate resistors are assumed, the CMRR of the stage in Figure 20.4 is not infinite; with a TL072 it is about −90 dB, degrading from 100 Hz upwards, due to the limited open-loop gain of the opamp. We will now examine this effect.

## Common-Mode Rejection Ratio: Opamp Gain

In the earlier section on CMRR we saw that in a theoretical balanced line, choosing low output impedances and high input impedances would give very good CM rejection even if the resistors were not perfectly matched. Things are a bit more complex (i.e. worse) if we replace the mathematical subtraction with a real opamp. We quickly find that even if perfectly matched resistors everywhere are assumed, the CMRR of the stage is not infinite, because the two opamp inputs are not at exactly the same voltage. The negative feedback error-voltage between the inputs depends on the open-loop gain of the opamp, and that is neither infinite nor flat with frequency into the far ultra-violet. Far from it. There is also the fact that opamps themselves have a common-mode rejection ratio; it is high, but once more it is not infinite.

As usual, SPICE simulation is instructive, and Figure 20.5 shows a simple balanced interconnection, with the balanced output represented simply by two 100 Ω output resistances connected to the source equipment ground, here called Ground 1, and the usual differential opamp configuration at the input end, where we have Ground 2.

A common-mode voltage Vcm is now injected between Ground 1 and Ground 2, and the signal between the opamp output and Ground 2 measured. The balanced input amplifier has all four of its resistances set to precisely 10 kΩ, and the opamp is represented by a very simple model that has only two parameters; a low-frequency open-loop gain, and a single pole frequency that says where that gain begins to roll-off at 6 dB per octave. The opamp input impedances and the opamp's own CMRR are assumed infinite, as in the world of simulation they so easily can be. Its output impedance is set at zero.

For the first experiments, even the pole frequency is made infinite, so now the only contact with harsh reality is that the opamp open-loop gain is finite. That is however enough to give distinctly non-ideal CMRR figures, as Table 20.3 shows:

With a low-frequency open-loop gain of 100,000, which happens to be the typical figure for a 5532 opamp, even perfect components everywhere will never yield a better CMRR than −94 dB.

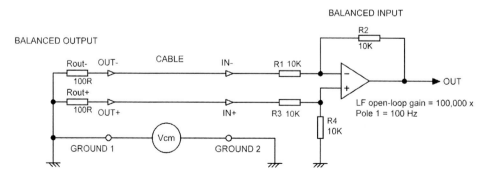

**Figure 20.5: A simple balanced interconnection for SPICE simulation to show the effect that opamp properties have on the CMRR.**

**Table 20.3: The effect of finite opamp gain on CMRR for the circuit of Figure 20.5.**

| Open-loop gain | CMRR dB | CMRR ratio |
|---|---|---|
| 10,000 | −74.0 | $19.9 \times 10^{-5}$ |
| 30,000 | −83.6 | $66.4 \times 10^{-6}$ |
| 100,000 | −94.0 | $19.9 \times 10^{-6}$ |
| 300,000 | −103.6 | $6.64 \times 10^{-6}$ |
| 1,000,000 | −114.1 | $1.97 \times 10^{-6}$ |

The CMRR is shown as a raw ratio in the third column so you can see that the CMRR is inversely proportional to the gain, and so we want as much gain as possible.

## Common-Mode Rejection Ratio: Opamp Frequency Response

To examine these we will set the low-frequency gain to 100,000 which gives a CMRR "floor" of −94 dB, and then introduce the pole frequency that determines where it rolls off. The CMRR now worsens at 6 dB/octave, starting at a frequency set by the interaction of the low-frequency gain and the pole frequency. The results are summarised in Table 20.4, which shows that as you might expect, the lower the open-loop bandwidth of the opamp, the lower the frequency at which the CMRR begins to fall off. Figure 20.6 shows the situation diagrammatically.

Table 20.5 gives the open-loop gain and pole parameters for a few opamps of interest. Both parameters, but especially the gain, are subject to considerable variation; the typical values from the manufacturers' data sheets are given here.

Some of these opamps have very high open-loop gains, but only at very low frequencies. This may be good for DC applications, but in audio line input applications, where the lowest frequency of CMRR interest is 50 Hz, they will be operating above the pole frequency, and so the gain available will be

**Table 20.4: The effect of opamp open-loop pole frequency on CMRR for the circuit of Figure 20.5.**

| Pole frequency | CMRR breakpoint frequency |
|---|---|
| 10 kHz | 10.2 kHz |
| 1 kHz | 1.02 kHz |
| 100 Hz | 102 Hz |
| 10 Hz | 10.2 Hz |

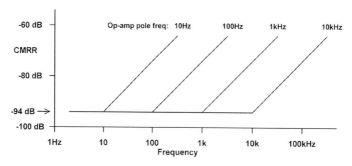

**Figure 20.6: How the CMRR degrades with frequency for different opamp pole frequencies. All resistors are assumed to be perfectly matched.**

**Table 20.5: Typical LF gain and open-loop pole frequency for some opamps commonly used in audio.**

| Name | Input device type | LF gain | Pole freq | Opamp LF CMRR dB |
|---|---|---|---|---|
| NE5532 | Bipolar | 100,000 | 100 Hz | 100 |
| LM4562 | Bipolar | 10,000,000 | below 10 Hz | 120 |
| LT1028 | Bipolar | 20,000,000 | 3 Hz | 120 |
| TL072 | FET | 200,000 | 20 Hz | 86 |
| OP27 | FET | 1,800,000 | 3 Hz | 120 |
| OPA2134 | FET | 1,000,000 | 3 Hz | 100 |
| OPA627 | FET | 1,000,000 | 20 Hz | 116 |

less—possibly considerably so, in the case of opamps like the OPA2134. This is not however a real limitation, for even if a humble TL072 is used, the perfect-resistor CMRR is about −90 dB, degrading from 100 Hz upwards. This sort of performance is not attainable in practice. We will shortly see why not.

## Common-Mode Rejection Ratio: Opamp CMRR

Opamps have their own common-mode rejection ratio, and we need to know how much this will affect the final CMRR of the balanced interconnection. The answer is that if all resistors are exactly correct, the overall CMRR is equal to the CMRR of the opamp. [6] Since opamp CMRR is typically very high (see the examples in Table 20.5), it is most unlikely to be the limiting factor.

The CMRR of an opamp begins to degrade above a certain frequency, typically at 6 dB per octave. This is (fortunately) at a higher frequency than the open-loop pole and is frequently around 1 kHz. For example, the OP27 has a pole frequency at about 3 Hz, but the CMRR remains flat at 120 dB until 2 kHz, and it is still greater than 100 dB at 20 kHz.

## Common-Mode Rejection Ratio: Amplifier Component Mismatches

We saw earlier in this chapter that when the output and input impedances on a balanced line have a high ratio between them and are accurately matched, we got a very good CMRR; this was compromised by the imperfections of opamps, but the overall results were still very good—and much higher than the real CMRRs that we measure in practice. There remains one place where we are still away in theory-land; we have so far assumed the resistances around the opamp were all exactly accurate. We must now face reality, admit that these resistors will not be perfect, and see how much damage to the CMRR they will do.

SPICE simulation gives us Table 20.6. The situation with LF opamp gains of both 100,000 and 1,000,000 is examined, but the effects of finite opamp bandwidth or opamp CMRR are not included. R1 in Figure 20.5 is varied, while R2, R3 and R4 are all kept at precisely 10 kΩ, and the balanced output source impedances are set to exactly 100 Ω.

Table 20.6 shows with glaring clarity that our previous investigations, which took only output and input impedances into account and determined that 100 Ω output resistors and 10 kΩ input impedances gave a CMRR of −80 dB for a 1% deviation in either, were definitely optimistic, and even adding in opamp imperfections left us with implausibly good results. Looking at component imbalances in the amplifier itself brings us down to earth; when a 1% tolerance resistor is used for R1 (and nowadays there is no financial incentive to use anything less accurate), the CMRR plummets to −46 dB; the same figure results from varying any other one of the four resistances by itself. If you are prepared to shell out for 0.1% tolerance resistors, the CMRR is a rather better −66 dB.

Table 20.6 also shows that there really is no point in getting anxious about the gain of the opamp you use in balanced inputs; unless you're planning to use 0.01% resistors (and I'm sure you're not), the effect of the opamp gain is negligible.

**Table 20.6: How resistor tolerances affect the CMRR for two realistic opamp open-loop gains.**

| R1 Ω | R1 deviation | Gain x | CMRR dB |
|---|---|---|---|
| 10k | 0% | 100,000 | −94.0 |
| 10.001k | 0.01% | 100,000 | −90.6 |
| 10.01k | 0.1% | 100,000 | −66.5 |
| 10.1k | 1% | 100,000 | −46.2 |
| 11k | 10% | 100,000 | −26.6 |
| 10k | 0% | 1,000,000 | −114.1 |
| 10.001k | 0.01% | 1,000,000 | −86.5 |
| 10.01k | 0.1% | 1,000,000 | −66.2 |
| 10.1k | 1% | 1,000,000 | −46.2 |
| 11k | 10% | 1,000,000 | −26.6 |

The results in Table 20.6 give an illustration of how resistor accuracy affects CMRR, but it is only an illustration, because in real life—a phrase that seems to keep cropping up, showing how many factors affect a practical balanced interconnection—all four resistors will of course be subject to a tolerance, and a more realistic calculation would produce a statistical distribution of CMRR rather than a single figure. One method is to use the Monte Carlo function in SPICE, which runs multiple simulations with random component variations and collates the results. However you do it, you must know (or assume) how the resistor values are distributed within their tolerance window. Usually you don't know, and finding out by measuring hundreds of resistors is not a task that appeals to all of us.

It is straightforward to assess the worst-case CMRR, which occurs when all resistors are at the limit of the tolerance in the most unfavourable direction. The CMRR in dB is then:

$$CMRR = 20\log\left(\frac{1 + R2/R1}{4T/100}\right)$$

20.1

Where $R1$ and $R2$ are as in Figure 20.5, and $T$ is the tolerance in %.

This rather pessimistic equation tells us that 1% resistors give a worst-case CMRR of only 34.0 dB, 0.5% parts give only 40.0 dB, and expensive 0.1% parts yield but 54.0 dB. Things are not however quite that bad in actuality, as the chance of everything being as wrong as possible is actually very small. I have measured the CMRR of more of these balanced inputs, built with 1% resistors, than I care to contemplate, but I do not recall that I ever saw one with an LF CMRR worse than 40 dB.

There are 8-pin SIL packages that offer four resistors that ought have good matching, if not accurate absolute values; be very, very wary of these, as they usually contain thick-film resistive elements that are not perfectly linear. In a test I did, a 10 kΩ SIL resistor with 10 Vrms across it generated 0.0010% distortion. Not a huge amount, perhaps, but in the quest for perfect audio, resistors that do not stick to Ohm's law are not a good start.

To conclude this section, it is clear that in practical use it is the errors in the balanced amplifier resistors that determine the CMRR, though both unbalanced capacitances (C1, C2 in Figure 20.7) and the finite opamp bandwidth are likely to cause further degradation at high audio frequencies. If you are designing both ends of a balanced interconnection and you are spending money on a few precision resistors, you should most definitely put them in the input amplifier, not the balanced output. The LF gain of the opamp and opamp CMRR have virtually no effect.

In fact, balanced input amplifiers like Figures 20.4 and 20.9, built with four ordinary 1% resistors, are used very extensively in the professional audio business and almost always prove to have adequate CMRR for the job; I have spent a lot of time designing mixing consoles, and I do not recall a single occasion when this was not the case. When more CMRR is wanted, for example in high-end mixing consoles, one of the resistances is made trimmable with a preset, as shown in Figure 20.7. This can mean a lot of tweaking in manufacture, as there might easily be three or four of these balanced inputs per channel, but looking on the bright side, it is a quick set-and-forget adjustment that will never need to be touched again unless one of the four fixed resistors needs replacing, and that is extremely unlikely. CMRRs at LF of more than 70 dB can easily be obtained by this method, but the CMRR at HF will degrade due to the opamp gain roll-off and stray capacitances.

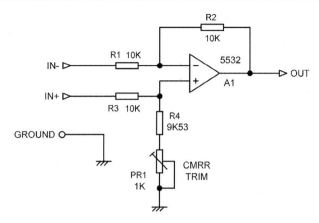

**Figure 20.7: A balanced input amplifier, with preset pot to trim for best LF CMRR. R4 is an E96 value.**

Figure 20.8 shows the CMRR measurements for a trimmable balanced input amplifier. The flat line at −50 dB was obtained from a standard balanced input using four 1% 10 kΩ resistors straight out of the box, while the much better (at LF, anyway) trace going down to −85 dB was obtained from Figure 20.7 by using a multi-turn preset for PR1. Note that R4 is an E96 value, so a 1k preset can swing the total resistance of that arm both above and below the nominal 10 kΩ.

The CMRR is dramatically improved by more than 30 dB in the region 50–500 Hz, where ground noise tends to intrude itself, and is significantly better across almost all the audio spectrum.

The sloping part of the trace in Figure 20.8 is partly due to the finite open-loop bandwidth of the opamp and partly due to unbalanced circuit capacitances. The CMRR is actually worse than 50 dB above 20 kHz, due to the stray capacitances in the multi-turn preset and the fact that I threw the circuit together on a piece of prototype board. In professional manufacture the value of PR1 would probably be much smaller and a small one-turn preset used with much less stray capacitance. Still, I think you get the point; for relatively small manufacturing quantities, trimming is both economic and effective.

## A Practical Balanced Input

The simple balanced input circuits shown in Figures 20.4 and 20.9 are not fit to face the outside world without additional components. Figure 20.9 shows a fully equipped version. First, and most important, C1 has been added across the feedback resistor R2; this prevents stray capacitances from Pin 2 to ground causing extra phase-shifts that lead to HF instability. The value required for stability is small, much less than that which would cause an HF roll-off anywhere near the top of the audio band. The values here of 10k and 27 pF give −3 dB at 589 kHz, and such a roll-off is only down by 0.005 dB at 20 kHz. C2, of equal value, must be added across R4 to maintain the balance of the amplifier, and hence its CMRR, at high frequencies.

C1 and C2 must not be relied upon for EMC immunity, as C1 is not connected to ground, and there is every chance that RF will demodulate at the opamp inputs. A passive RF filter is therefore added

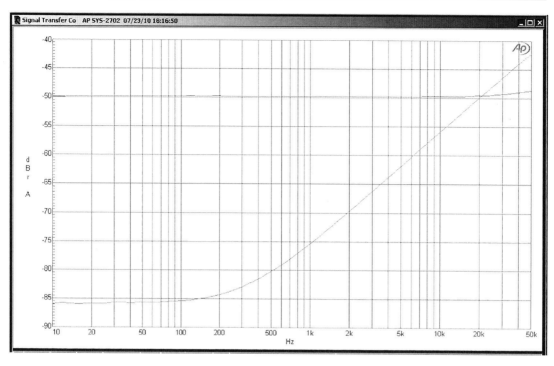

**Figure 20.8: CMRR results from a standard balanced amplifier as in Figure 20.4 and from trimmed Figure 20.7. The opamp was a 5532, and all resistors were 1%. The trimmed version is better than 80 dB up to 500 Hz.**

to each input, in the shape of R5, C3 and R6, C4, so the capacitors will shunt incoming RF to ground before it reaches the opamp. Put these as close to the input socket as possible to minimise radiation inside the enclosure. Ideally the connection from input socket terminal to PCB would be made with R5 and R6 to minimise the amount of potentially radiating wiring.

I explained earlier in this chapter when looking at unbalanced inputs that it is not easy to guess what the maximum source impedance will be, given the existence of "passive preamplifiers" and valve equipment. Neither is likely to have a balanced output unless implemented by transformer, but either might be used to feed a balanced input, and so the matter needs thought.

In the unbalanced input circuit resistances had to be kept as low as practicable to minimise the generation of Johnson noise that would compromise the inherently low noise of the stage. The situation with a standard balanced input is however different from the unbalanced case, as there have to be resistances around the opamp, and they must be kept up to a certain value to give acceptably high input impedances; this is why a balanced input like this one is much noisier. We could therefore make R5 and R6 much larger without a measurable noise penalty if we reduce R1 and R3 accordingly to keep unity gain. In Figure 20.9, R5 and R6 are kept at 100 $\Omega$, so if we assume 50 $\Omega$ output resistances in both legs of the source equipment, then we have a total of 150 $\Omega$, and 150 $\Omega$ and 100 pF give $-3$ dB at 10.6 MHz. Returning to a possible passive preamplifier with a 10 k$\Omega$ potentiometer, its maximum

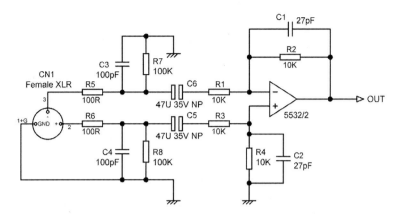

**Figure 20.9: Balanced input amplifier with the extra components required for DC blocking and EMC immunity.**

output impedance of 2.5k plus 100 Ω with 100 pF gives −3 dB at 612 kHz, which remains well clear of the top of the audio band.

As with the unbalanced input, replacing R5 and R6 with small inductors will give much better RF filtering but at increased cost. Ideally a common-mode choke (two bifilar windings on a small toroidal core) should be used, as this improves CMRR performance. Check the frequency response to make sure the LC circuits are well-damped and not peaking at the turnover frequency.

C5 and C6 are DC-blocking capacitors. They must be rated at no less than 35 V to protect the input circuitry and are the non-polarised type, as external voltages are of unpredictable polarity. The lowest input impedance that can occur with this circuit when using 10 kΩ resistors is, as described earlier, 6.66 kΩ when it is being driven in the balanced mode. The low-frequency roll-off is therefore −3 dB at 0.51 Hz. This may appear to be undesirably low, but the important point is not the LF roll-off but the possible loss of CMRR at low frequencies due to imbalance in the values of C5 and C6; they are electrolytics with a significant tolerance. Therefore they should be made large so their impedance is a small part of the total input impedance. 47 uF is shown here, but 100 uF or 220 uF can be used to advantage if there is the space to fit them in. The low-end frequency response must be defined somewhere in the input system, and the earlier the better, to prevent headroom or linearity being affected by subsonic disturbances, but this is not a good place to do it. A suitable time-constant immediately after the input amplifier is the way to go, but remember that capacitors used as time-constants may distort unless they are NP0 ceramic, polystyrene, or polypropylene. See the chapter on passive components (Chapter 15) for more on this.

R7, R8 are DC drain resistors to prevent charges lingering on C5 and C6. These can be made lower than for the unbalanced input, as the input impedances are lower, so a value of, say, 100 kΩ rather than 220 kΩ makes relatively little difference to the total input impedance.

A useful property of this kind of balanced amplifier is that it does not go mad when the inputs are left open-circuit—in fact it is actually *less* noisy than with its inputs shorted to ground. This is the opposite of the "normal" behaviour of a high-impedance unterminated input. This is because two things happen: open-circuiting the hot input doubles the resistance seen by the non-inverting input of the opamp,

Table 20.7: Noise output measured from simple balanced amps using a 5532 section.

| R value Ω | 50 Ω terminated inputs | Open circuit inputs | Terminated/open difference |
|---|---|---|---|
| 100k | −95.3 dBu | −97.8 dBu | 2.5 dBu |
| 10k | −104.8 dBu | −107.6 dBu | 2.8 dBu |
| 2k0 | −109.2 dBu | −112.0 dBu | 2.8 dBu |
| 820 | −111.7 dBu | −114.5 dBu | 2.8 dBu |

raising its noise contribution by 3 dB. However, opening the cold input makes the noise gain drop by 6 dB, giving a net drop in noise output of approximately 3 dB. This of course refers only to the internal noise of the amplifier stage, and pickup of external interference is always possible on an unterminated input. The input impedances here are modest, however, and the problem is less serious than you might think. Having said that, deliberately leaving inputs unterminated is usually bad practice.

If this circuit is built with four 10 kΩ resistors and a 5532 opamp section, the noise output is −104.8 dBu, with the inputs terminated to ground via 50 Ω resistors. As noted earlier, the input impedance of the cold input is actually lower than the resistor connected to it when working balanced, and if it is desirable to raise this input impedance to 10 kΩ, it could be done by raising the four resistors to 16 kΩ; this slightly degrades the noise output to −103.5 dBu. Table 20.7 gives some examples of how the noise output depends on the resistor value; the third column gives the noise with the input unterminated and shows that in each case the amplifier is about 3 dB quieter when open-circuited. It also shows that a useful improvement in noise performance is obtained by dropping the resistor values to the lowest that a 5532 can easily drive (the opamp has to drive the feedback resistor), though this usually gives unacceptably low input impedances. More on that at the end of the chapter.

## Variations on the Balanced Input Stage

I now give a collection of balanced input circuits that offer advantages or extra features over the standard balanced input configuration. These circuit diagrams often omit stabilising capacitors, input filters, and DC-blocking capacitors to improve the clarity of the basic principle. They can easily be added; in particular bear in mind that a stabilising capacitor like C1 in Figure 20.9 is often needed between the opamp output and the negative input to guarantee freedom from high-frequency oscillation.

## Combined Unbalanced and Balanced Inputs

If both unbalanced and balanced inputs are required, it is extremely convenient if it can be arranged so that no switching between them is required. Switches cost money, mean more holes in the metalwork, and add to assembly time. Figure 20.10 shows an effective way to implement this. In balanced

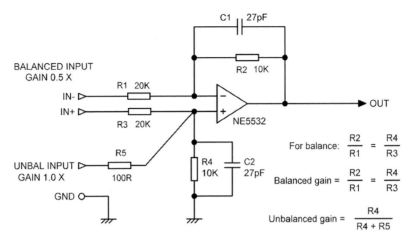

**Figure 20.10: Combined balanced and unbalanced input amplifier with no switching required but some performance compromises.**

mode, the source is connected to the balanced input and the unbalanced input left unterminated. In unbalanced mode, the source is connected to the unbalanced input and the balanced input left unterminated, and no switching is required. It might appear that these unterminated inputs would pick up extra noise, but in practice this is not the case. It works very well. and I have used it successfully in high-end equipment for two prestigious manufacturers.

As described earlier, in the world of hi-fi, balanced signals are at twice the level of the equivalent unbalanced signals, and so the balanced input must have a gain of 1/2 or −6 dB relative to the unbalanced input to get the same gain by either path. This is done here by increasing R1 and R3 to 20 kΩ. The balanced gain can be greater or less than unity, but the gain via the unbalanced input is always 1. The differential gain of the amplifier and the constraints on the component values for balanced operation are shown in Figure 20.10 and are not repeated in the text to save space. This applies to the rest of the balanced inputs in this chapter.

There are two minor compromises in this circuit which need to be noted. First, the noise performance in unbalanced mode is worse than for the dedicated unbalanced input described earlier in this chapter, because R2 is effectively in the signal path and adds Johnson noise. Second, the input impedance of the unbalanced input cannot be very high, because it is set by R4, and if this is increased in value all the resistances must be increased proportionally and the noise performance will be markedly worse. It is important that only one input cable should be connected at a time, because if an unterminated cable is left connected to an unused input, the cable capacitance to ground can cause frequency response anomalies and might in adverse circumstances cause HF oscillation. A prominent warning on the back panel and in the manual is a very good idea.

## The Superbal Input

This version of the balanced input amplifier, shown in Figure 20.11, has been referred to as the "Superbal" circuit because it gives equal impedances into the two inputs for differential signals. It

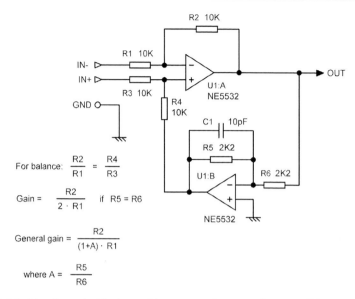

For balance: $\dfrac{R2}{R1} = \dfrac{R4}{R3}$

Gain $= \dfrac{R2}{2 \cdot R1}$   if R5 = R6

General gain $= \dfrac{R2}{(1+A) \cdot R1}$

where $A = \dfrac{R5}{R6}$

**Figure 20.11: The Superbal balanced input requires another amplifier but has equal input impedances.**

was originated by David Birt of the BBC; see [7]. With the circuit values shown, the differential input impedance is exactly 10 kΩ via both hot and cold inputs. The common-mode input impedance is 20 kΩ, as before.

In the standard balanced input R4 is connected to ground, but here its lower end is actively driven with an inverted version of the output signal, giving symmetry. The increased amount of negative feedback reduces the gain with four equal resistors to −6 dB instead of unity. The gain can be reduced below −6 dB by giving the inverter a gain of more than 1; if R1, R2, R3, and R4 are all equal, the gain is $1/(A+1)$, where $A$ is the gain of the inverter stage. This is of limited use, as the inverter U1:B will now clip before the forward amplifier U1:A, reducing headroom. If the gain of the inverter stage is gradually reduced from unity to zero, the stage slowly turns back into a standard balanced amplifier, with the gain increasing from −6 dB to unity and the input impedances becoming less and less equal. If a gain of less than unity is required, it should be obtained by increasing R1 and R3.

R5 and R6 should be kept as low in value as possible to minimise Johnson noise; there is no reason why they have to be equal in value to R1, etc. The only restriction is the ability of U1:A to drive R6 and U1:B to drive R5, both resistors being effectively grounded at one end. The capacitor C1 will almost certainly be needed to ensure HF stability; the value in the figure is only a suggestion. It should be kept as small as possible, because reducing the bandwidth of the inverter stage impairs CMRR at high frequencies.

## Switched-Gain Balanced Inputs

A balanced input stage that can be switched to two different gains while maintaining CMRR is very useful. Equipment often has to give optimal performance with both semi-pro (−7.8 dBu) and

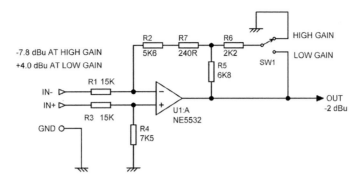

**Figure 20.12: A balanced input amplifier with gain switching that maintains good CMRR.**

professional (+4 dBu) input levels. If the nominal internal level of the system is in the normal range of −2 to −6 dBu, the input stage must be able to switch between amplifying and attenuating, while maintaining good CMRR in both modes.

The brute-force way to change gain in a balanced input stage is to switch the values of either R1 and R3, or R2 and R4, in Figure 20.4, keeping the pairs equal in value to maintain the CMRR; this needs a double-pole switch for each input channel. A much more elegant technique is shown in Figure 20.12. Perhaps surprisingly, the gain of a differential amplifier can be manipulating by changing the drive to the feedback arm (R2 etc) only and leaving the other arm R4 unchanged, without affecting the CMRR. The essential point is to keep the source resistance of the feedback arm the same but drive it from a scaled version of the opamp output. Figure 20.12 does this with the network R5, R6, which has a source resistance made up of 6k8 in parallel with 2k2, which is 1.662 kΩ. This is true whether R6 is switched to the opamp output (low-gain setting) or to ground (high-gain setting), for both have effectively zero impedance. For low gain the negative feedback is not attenuated but fed through to R2 and R7 via R5, R6 in parallel. For high-gain, R5 and R6 become a potential divider, so the amount of feedback is decreased and the gain increased. The value of R2 + R7 is reduced from 7k5 by 1.662 kΩ to allow for the source impedance of the R5, R6 network; this requires the distinctly non-standard value of 5.838 kΩ, which in Figure 20.12 is approximated by R2 and R7, which give 5.6 kΩ + 240 Ω = 5.840 kΩ. This is too high by 2 Ω (0.03%), but that is much less than a 1% tolerance on R2 and so will have only a vanishingly small effect on the CMRR. If we instead use a parallel pair of resistors in 2xE24 format, a good combination is 6.2 kΩ in parallel with 100 kΩ, which is only 0.004% high, and the effective tolerance for 1% parts is reduced to 0.94%, a small but helpful improvement.

Note that this stage can attenuate as well as amplify if R1, R3 are set to be greater than R2, R4, as shown here. The nominal output level of the stage is assumed to be −2 dBu; with the values shown, the two gains are −6.0 and +6.2 dB, so +4 dBu and −7.8 dBu respectively will give −2 dBu at the output. Other pairs of gains can of course be obtained by changing the resistor values; the important thing is to stick to the principle that the value of R2 + R7 is reduced from the value of R4 by the source impedance of the R5, R6 network. With the values shown, the differential input impedance is 11.25 kΩ via the cold and 22.5 kΩ via the hot input. The common-mode input impedance is 22.5 kΩ.

Switched-gain inputs like this one have the merit that there are no issues with balance between channels because the gain is defined by relatively precise fixed resistors rather than ganged pots, as used in the next section. This neat little circuit has the added advantage that nothing bad happens when the switch is moved with the circuit operating. When the wiper is between contacts, you simply get a gain intermediate between the high and low settings, which is pretty much the ideal situation. Make sure the switch is a break-before-make type (as most of them are) to avoid shorting the opamp output to ground.

## Variable-Gain Balanced Inputs

The beauty of a variable-gain balanced input is that it allows you to get the incoming signal up or down to the nominal internal level as soon as possible, minimising both the risk of clipping and contamination with circuit noise. The obvious method of making a variable-gain differential stage is to use dual-gang pots to vary either R1, R3 or R2, R4 together, to maintain CMRR. This is clumsy and gives a CMRR that is both bad and highly variable due to the inevitable mismatches between pot sections. For a stereo input the required four-gang pot is an unappealing proposition.

There is however a way to get a variable gain with good CMRR, using a single pot section. The principle is essentially the same as for the aforementioned switched-gain amplifier; keep constant the source impedance driving the feedback arm but vary the voltage applied. The principle is shown in Figure 20.13. To the best of my knowledge I invented this circuit in 1982; any comments on this point are welcome. The feedback arm R2 is driven by voltage-follower U1:B. This eliminates the variations in source impedance at the pot wiper, which would badly degrade the CMRR. R6 limits the gain range and R5 modifies the gain law to give it a more usable shape. When the pot is fully up (minimum gain), R5 is directly across the output of U1:A, so do not make it too low in value. If a centre-detent pot is used to give a default gain setting, this may not be very accurate, as it partly depends on the ratio of pot track resistance (no better than ±10% tolerance, and very often ±20%) to 1% fixed resistors.

This configuration is very useful as a general line input with an input sensitivity range of −20 to +10 dBu. For a nominal output of 0 dBu, the gain of Figure 20.13 is +20 to −10 dB, with R5 chosen for 0 dB gain at the central wiper position. An opamp in a feedback path may appear a dubious proposition for HF stability, because of the extra phase-shift it introduces, but here it is working as a voltage-follower, so its bandwidth is maximised and in practice the circuit is dependably stable.

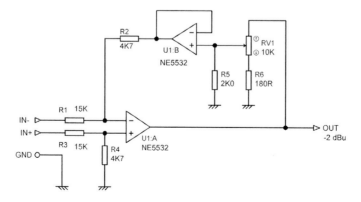

**Figure 20.13: Variable-gain balanced input amplifier.**

Circuitry like this is ideal for single-channel applications but can create difficulties when used in stereo or other multi-channel formats, because of matching problems in ganged potentiometers. This version of the circuit has a fairly wide gain range, and it is necessary to use an RD (anti-log 10%) law pot to get a reasonable linear-in-dB control law. This introduces gain errors between nominal identical channels because, first, the value of the pot track is not controlled anything like as closely as that of a 1% resistor, and so the effect of R5 on the pot law will vary, which leads to gain differences between the channels. Second, any log or anti-log pot law is made up of dual resistance sections, and this introduces more errors. If a more restricted gain range is being used, then it may be possible to use a linear pot and rely on the loading effect of R5 to give an acceptable control law. The whole problem of multi-channel gain control with potentiometers of limited accuracy is examined in detail in [8].

## The Self Variable-Gain Balanced Input

If the gain range is not great, an unloaded linear pot can be used, as it will give an acceptable control law. It is then possible to eliminate the effect of the pot track resistance (which will probably have a 20% tolerance) on gain accuracy.

This is done by changing the way the maximum gain is defined. In Figure 20.13 that is done by end-stop resistor R6; if it can be removed, then the pot really does works as a potentiometer; its output voltage depends only on its angular position and not on its track resistance. However, the maximum gain must then be defined in some other way. The answer is to add a resistor that gives a separate feedback path around the differential opamp only. R6 in Figure 20.14 provides negative feedback independently of the gain control and so limits the maximum gain. It is not intuitively obvious (to me, at any rate), but the CMRR is still preserved when the gain is altered, just as for Figure 20.13. You will note in Figure 20.14 that two resistors R4, R7 are paralleled to get a value exactly equal to R2 in parallel with R6. The resistor values shown give a minimum gain of −3.6 dB and a maximum gain of +5.8 dB.

Input buffers are used to make the input impedance pretty much as high as you like, while allowing low-value resistors to be used to reduce noise (more on this later). The gain with the pot wiper central

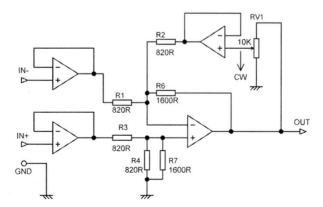

**Figure 20.14: The Self variable-gain balanced input amplifier, with no pot track dependence.**

is −0.1 dB, which I would suggest is close enough to unity for anyone and probably better than the mechanical accuracy of a pot centre-detent. This concept eliminates the effect of pot track resistance on gain, and I call it the Self Input. No-one has argued so far . . .

This circuit will be extremely useful for accurate crossover input gain trims with excellent stereo matching. It was used as the balanced input and active balance control in my Low-Noise Preamplifier, published in *Linear Audio* in 2013; [9] some 50-odd have been home-built without any hint of problems. The original article is reproduced in *Self On Audio*, [10] along with some extra circuit information.

## High Input Impedance Balanced Inputs

We saw earlier that high input impedances are required to maximise CMRR of a balanced interconnection, but the input impedances offered by the standard balanced circuit are limited by the need to keep the resistor values down to control Johnson noise. High-impedance balanced inputs are also useful for interfacing to valve equipment in the strange world of retro-hi-fi. Adding output cathode-followers to valve circuitry is expensive and consumes a lot of extra power, and so the output is often taken directly from the anode of a gain-stage, and even a so-called bridging load of 10 kΩ may seriously compromise the distortion performance and output capability of the source equipment.

Figure 20.15 shows a configuration where the input impedances are determined only by the bias resistances R1 and R2. They are shown here as 100 kΩ but may be considerably higher if opamp bias currents permit. A useful property of this circuit is that adding a single resistor Rg increases the gain but preserves the circuit balance and CMRR. This configuration cannot be set to attenuate, because the gain of an opamp with series feedback cannot be reduced below unity.

It is of course always possible to give a basic balanced input a high input impedance by putting unity-gain buffers in front of each input, but that uses three opamp sections rather than two. Sometimes, however, it is appropriate. Much more on that later.

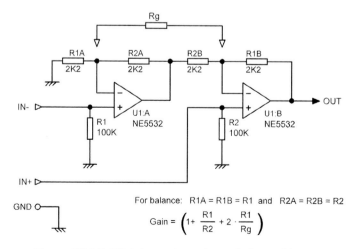

For balance:   R1A = R1B = R1  and  R2A = R2B = R2

$$\text{Gain} = \left(1 + \frac{R1}{R2} + 2 \cdot \frac{R1}{Rg}\right)$$

**Figure 20.15: High input impedance balanced input.**

We saw earlier that the simple balanced amplifier is surprisingly quiet and well-behaved when its inputs are unterminated. This is not the case with this configuration, which because of its high input impedances will be both noisy and susceptible to picking up external interference if either input is left open circuit.

## The Instrumentation Amplifier

Almost every book on balanced or differential inputs includes the three-opamp circuit of Figure 20.16 and praises it as the highest expression of the differential amplifier. It is usually called the instrumentation amplifier configuration because of its undoubted superiority for data-acquisition. (Specialised ICs exist that are sometimes also called instrumentation amplifiers or in-amps; these are designed for very high CMRR data-acquisition. They are expensive and in general not optimised for audio work.)

The beauty of the three-opamp configuration is that the dual input stage buffers the balanced line from the input impedances of the final differential amplifier; the four resistances around it can now be made much lower in value, reducing their Johnson noise by a significant amount while keeping the CMRR benefits of presenting high input impedances to the balanced line. The other feature, which is usually much more emphasised because of its unquestionable elegance, is that the dual input stage, with its shared feedback network R3, R4, R5, can be set to have a high differential gain by giving R4 a low value, but its common-mode gain is always unity; this property is *not* affected by mismatches between R3 and R5. The final amplifier then does its usual job of common-mode rejection, and the combined CMRR can be very good indeed if the first-stage gain is high.

This is all very well, but a high first-stage gain is not likely to be useful for audio balanced line inputs. A data-acquisition application like ECG monitoring may need a gain of thousands of times, which will allow a stunning CMRR without using precision resistors, but the cruel fact is that in audio use, gain in a line input amplifier is often simply not wanted. In a typical opamp signal path, the nominal internal

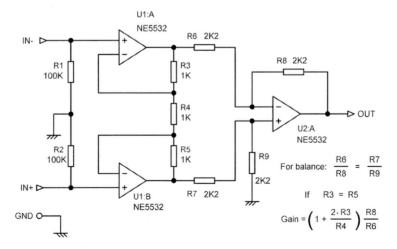

**Figure 20.16: The instrumentation amplifier configuration. Gain here is 3 times.**

level is usually between −6 and 0 dBu, and if the level of the incoming balanced signal is at the professional level of +4 dBu, what you need is 6 dB of attenuation rather than any gain. Gain now and attenuation later must introduce what can only be called a headroom bottleneck. If the incoming level was the semi-pro −7.8 dBu, a small amount of gain could be introduced, but then the CMRR advantage would be equally small and certainly not worth the cost of the extra circuitry.

However, active crossovers are a special case. Chapter 17 shows how running active crossover circuitry at a much higher level than is usual in audio equipment is entirely practical, and Chapter 23 demonstrates that the signal/noise ratio improvements are both real and impressive. The crossover design example in Chapter 23 has an input amplifier gain of 12 dB, and using an instrumentation amplifier there should give a very worthwhile CMRR advantage.

## Instrumentation Amplifier Applications

First, there are some applications where gain in a balanced input amplifier is either essential or can be made use of. Active crossovers and power amplifiers are probably the two most common.

Active crossovers are commonly used with no volume control or only a limited-range gain trim between them and the power amplifiers, so that it can always be assumed that the power amplifier will clip first. As described in Chapter 17, under these conditions the crossover can be run at an internal level of 3 Vrms or more, with the output level dropped back to, say, 1 Vrms at the very output of the crossover by a low-impedance passive attenuator. This can give a very impressive noise performance, because the signal goes through the chain of filters at a high level and the accumulated noise is attenuated with the signal at the output.

For a typical active crossover application like this, we therefore need an input stage with gain. If the nominal input is 0 dBu (775 mVrms), we want a gain of about four times to get the signal up to 3 Vrms (+11.8 dBu). We will not need a greater output than this because it is sufficient to clip the following power amplifiers, even after the passive attenuators. This demonstrates that you can break The Rule "Do not amplify then attenuate" so long as there is a more severe headroom restriction downstream and a limited gain control range.

If we use the strategy of input buffers followed by balanced amplifiers working at low impedances, as described later in this chapter (i.e. *not* an instrumentation amplifier, as there is no network between the two input opamps), the required gain can be obtained by altering the ratio of the resistors in the balanced amplifier of a 1–1 configuration. With suitable resistor values (R1 = R3 = 560R, R2 = R4 = 2k2) the output noise from this stage (inputs terminated by 50 Ω) is −100.9 dBu, so its equivalent input noise (EIN) is −100.9 − 11.8 = −112.7 dBu, which is reasonably quiet. All noise measurements are 22 Hz–22kHz, rms sensing, unweighted.

The other area for instrumentation amplifier techniques is that of power amplifiers with balanced inputs. The voltage gain of the power amplifier itself is frequently about 22 times (+26.8 dB) for its own technical reasons. If we take a power output of 100 W into 8 Ω as an example, the maximum output voltage is 28.3 Vrms, so the input level must be 28.3/22 = 1.29 Vrms. If the nominal input level is +4 dBu (1.23 Vrms), then a very modest gain of 1.05 times is required, and we will gain little from using an instrumentation amplifier with that gain in Stage 1. If however the nominal input level is

0 dBu (0.775 Vrms), a higher gain of 1.66 times is needed, but that is still only a theoretical CMRR improvement of 4.4 dB. However, we can again make use of the fact that a relatively small signal voltage is enough to clip the power amplifier, and we can amplify the input signal—perhaps by four times again—and then attenuate it, along with the noise from the balanced input amplifier.

## The Instrumentation Amplifier With 4x Gain

Since we need a significant amount of gain, the instrumentation amplifier configuration is a viable alternative to the input buffer–balanced amplifier approach, giving a CMRR theoretically improved by 11.8 dB, the gain of its first stage. It uses the same number of opamps- three. A suitable circuit is shown in Figure 20.17. This gives a gain of 3.94 times, which is as close as you can get to 4.00 with E24 values. My tests show that the theoretical CMRR improvement really is obtained. To take just one set of results, when I built the second stage, it gave a CMRR of −56 dB working alone, but when the first stage was added, the CMRR leapt up to −69 dB, an improvement of 13 dB. These one-off figures (13 dB is actually better than the 12 dB improvement you would get if all resistors were exact) obviously depend on the specific resistor examples I used, but you get the general idea. The CMRR was flat across the audio band.

I do realise that there is an unsettling flavour of something-for-nothing about this, but it really does work. The total gain required can be distributed between the two stages in any way, but it should all be concentrated in Stage 1, as shown in Figure 20.17, to obtain the maximum CMRR benefit.

What about the noise performance? Fascinatingly, it is considerably improved. The noise output of this configuration is usefully lower at −105.4 dBu, which is no less than 4.5 dB better than the −100.9 dBu we got from the more conventional 1–1 configuration described earlier. This is because the

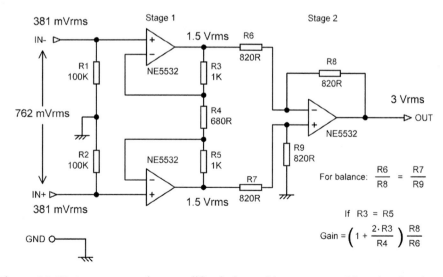

**Figure 20.17: Instrumentation amplifier balanced input stage with gain of 4 times.**

amplifiers in Stage 1 are working in better conditions for low noise than those in Stage 2. They have no significant resistance in series with the non-inverting inputs to generate Johnson noise or turn the opamp current noise into voltage noise. As they implement all the gain before the signal reaches Stage 2, the noise contribution of the latter is therefore less significant.

When noise at −105.4 dBu is put through a 10 dB passive attenuator to get back down to the output level we want, as might typically be done at the output of an active crossover, it will become −115.4 dBu, which I suggest is very quiet indeed.

The circuit in Figure 20.17 requires some ancillary components before it is ready to face the world. Figure 20.18 shows it with EMC RF filtering (R19, C1 and R20, C4), non-polarised DC-blocking capacitors (C2, C3), and capacitors C5, C6 to ensure HF stability. R21, R22 are DC drain resistors that ensure that any charge left on the DC-blocking capacitors by the source equipment when the connector is unplugged is quickly drained away. These extra components are omitted from subsequent diagrams to aid clarity.

So, here we are with all of the gain in the first stage, for the best CMRR, and this is followed by a unity-gain Stage 2. Now let's extend this thinking a little. Suppose Stage 2 had a gain of less than unity? We could then put more gain in Stage 1 and get a further free and guaranteed CMRR improvement while keeping the overall gain the same. But can we do this without causing headroom problems in Stage 1?

The instrumentation amplifier in Figure 20.19 has had the gain of Stage 1 doubled to 8 times (actually 7.67 times; 0.36 dB low, but as close as you can get with E24 values) by reducing R4, while the input resistors R6, R7 to Stage 2 have been doubled in value so that stage now has a gain of 0.5 times, to keep the total gain at approximately 4 times.

In this case the theoretical CMRR improvement is once more the gain of Stage 1, which is 7.67 times, and therefore 17.7 dB. Stage 2 gave a measured CMRR of −53 dB working alone, but when Stage 1 was added, the CMRR increased to −70 dB, an improvement of 17 dB, significantly greater than the 14 dB improvement obtained when Stage 1 had a gain of 4 times. The noise output was −106.2 dBu,

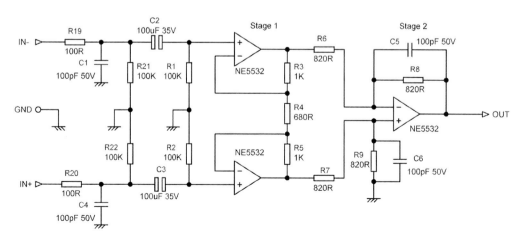

**Figure 20.18: Practical instrumentation amplifier balanced input stage with gain of 4 times.**

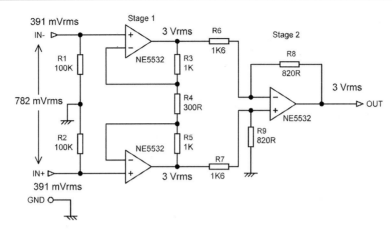

**Figure 20.19: Instrumentation amplifier balanced input stage with gain of 8 times in Stage 1 and 0.5 times in Stage 2.**

which is better than the four-times instrumentation amplifier by 0.8 dB and better than the somewhat more conventional 1–1 configuration by a convincing 5.3 dB. When the noise level of −106.2 dBu is put through the 10 dB passive attenuator to get back down to the level we want, this will become −116.2 dBu, which is even quieter than the previous version.

You may well be getting nervous about having a gain as high as 8 times followed by attenuation; have we compromised the headroom? If we assume that, as before, the maximum output we need is 3 Vrms, then with a total gain of 4 times we will have 0.775 Vrms between the two inputs, and the corresponding signal voltages at each output of Stage 1 are also 3 Vrms. These outputs are in anti-phase, and so when they are combined in Stage 2, with its gain of 0.5 times, we get 3 Vrms at the final output. This means that in balanced operation all three amplifiers will clip simultaneously, and the input amplifier can pass the maximum opamp output level before clipping. With the usual ±17 V supply rails this is about 10 Vrms.

So far so good; all the voltages are well below clipping. But note that we have assumed a balanced input, and unless you are designing an entire system, that is a dangerous assumption. If the input is driven with an unbalanced output of the same level, we will have 0.775 Vrms on one input pin and nothing on the other. You might fear we would get 6 Vrms on one output of Stage 1 and nothing on the other; that would still be some way below clipping, but the safety margin is shrinking fast. In fact things are rather better than that. The cross-coupling of the feedback network R3, R4, R5 greatly reduces the difference between the two outputs of stage 1 with unbalanced inputs. With 0.775 Vrms on one input pin and nothing on the other, the Stage 1 outputs are 3.39 Vrms and 2.61 Vrms, which naturally sum to 6 Vrms, reduced to 3 Vrms out by Stage 2. The higher voltage is at the output of the amplifier associated with the driven input. Maximum output with an unbalanced input is slightly reduced from 10 Vrms to 9.4 Vrms. This minor restriction does not apply to the earlier version with only 4 times gain in Stage 1.

This shows us that if we can assume a balanced input and stick with a requirement for a maximum output of 3 Vrms, we have a bit more elbow-room than you might think for using higher gains in Stage 1 of an instrumentation amplifier. Taking into account the possibility of an unbalanced input, we could

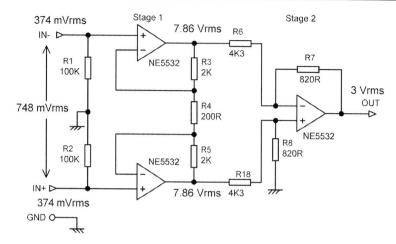

**Figure 20.20: Instrumentation amplifier balanced input stage with gain of 21 times in Stage 1 and 0.19 times in Stage 2. Maximum output 3 Vrms.**

further increase the gain by a factor of 10/3.39 = 2.95 times, on top of the doubling of Stage 1 gain we have already implemented. This gives a Stage 1 gain of 22.6 times. I decided to be conservative and go for a Stage 1 gain of 21 times; the gain of Stage 2 is correspondingly reduced to 0.19 times to maintain a total gain of 4 times. This version is shown in Figure 20.20; you will note that the feedback resistors R3, R4, R5 in Stage 1 have been scaled up by a factor of two. This is because Stage 1 is now working at a higher level; remember also that the resistors R3 and R5 are not the only load on Stage 1, because each of its opamp outputs is looking at the Stage 2 inputs driven in anti-phase, so its inverting input loading is heavier than it at first appears. Leaving R3, R5 at 1 kΩ would significantly degrade the distortion performance; I found it doubled the THD at 10 kHz from 0.0012% to 0.0019% for a 3 Vrms output.

When Stage 2 was tested alone, its CMRR was −61 dB; clearly we have been lucky with our resistors. Adding on Stage 1 increased it to a rather spectacular −86 dB, an improvement of no less than 25 dB. This degrades to −83 dB at 20 kHz; this is the first time it has not appeared flat across the audio band. The theoretical CMRR improvement is 26.4 dB, without trim presets or selected components.

The output noise level remains at −106.2 dBu; we have clearly gained all the noise advantage we can by increasing the Stage 1 gain.

Note that throughout this discussion I have assumed that there are no objections to running an opamp at a signal level not far from clipping, and in particular that the distortion performance is not significantly harmed by doing that. This is true for many modern opamps such as the LM4562 and also for the veteran 5532, when it is not loaded to the limit of its capability.

## The Instrumentation Amplifier at Unity Gain

If you want a unity-gain input stage but also some CMRR improvement, some compromise is required.

Putting R3, R5 = 1 kΩ and R4 = 2 kΩ gives a Stage 1 gain of 2 times, which is undone by setting R6, R7 to 1k6 in Stage 2 to give a gain of 0.5 times. This will pass the full 10 Vrms maximum opamp level, but only in balanced mode. In unbalanced mode Stage 1 will clip at the opamp connected to the driven input, because when unbalanced the gain to this point from the input is 1.5 times. The maximum signal that can be handled in unbalanced mode is therefore about 7 Vrms. The resulting input amplifier is shown in Figure 20.21. A heavy load is placed on the Stage 1 amplifiers in this situation, and for the best distortion behaviour it would be better to make R3, R5 = 2 kΩ and R4 = 4 kΩ.

When I did the measurements, Stage 2 alone had a CMRR of −57 dB; adding on Stage 1 brought this up to −63 dB (this exact correspondence of theoretical and measured improvement is just a coincidence). You can therefore have 6 dB more CMRR free, gratis, and for absolutely nothing, so long as you are prepared to tolerate a loss of 3.5 dB in headroom if the stage is run unbalanced but with the same input voltage. The noise output is −111.6 dBu.

A typical hi-fi output gives 1 Vrms in unbalanced use and 2 Vrms when run balanced, as the cold output is provided by a simple inverter, the presence of which doubles the total output voltage. Figure 20.21 shows the signal voltages when our unity-gain input stage is connected to a 2 Vrms balanced input. This sort of situation is more likely to occur in hi-fi than in professional audio; in the latter, simulated "floating transformer winding" outputs are sometimes found, where the total output voltage is the same whether the output is run balanced or unbalanced. These are described in Chapter 21.

In hi-fi use it may be possible to assume that the instrumentation amplifier will never be connected to a balanced output with a total signal voltage of more than 2 Vrms. If this is the case, we can once more put extra gain into Stage 1 and undo it with Stage 2, without compromising headroom. We are now dealing with a restriction on input level rather than output level. We will assume that the 2 Vrms input is required to be reduced to 1 Vrms to give adequate headroom in the circuitry that follows, so an overall gain of 0.5 times is required.

Finally we will look at how the range of gain control between the instrumentation amplifier and the power amplifier affects the headroom situation. Figure 20.22 shows one of the instrumentation amplifiers described earlier with a maximum output of 3 Vrms, the version in Figure 20.19 with a

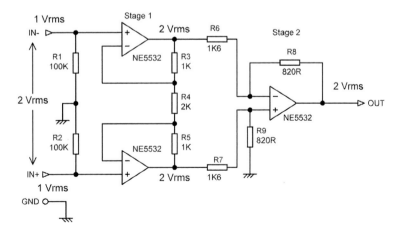

**Figure 20.21: Instrumentation amplifier balanced input stage with gain of 2 times in Stage 1 and 0.5 times in Stage 2 for overall unity gain.**

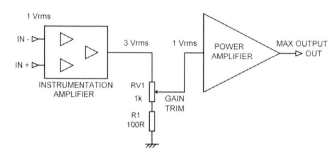

**Figure 20.22: Maximum gain trim range possible avoiding headroom restriction in the instrumentation amplifier.**

gain of 8 times in Stage 1 and 0.5 times in Stage 2. This is followed by a ±10 dB gain trim network that is normally expected to work at the 0 dB setting, i.e. giving 10 dB of attenuation and so reducing the noise from the input stage. I should point out that this simple network does not give "0 dB" at the centre of the pot travel; that point is more like three-quarters of the way down.

Since the maximum output of this stage is 10 Vrms with a balanced input (slightly less with an unbalanced input), clipping will always occur in the power amplifier first so long as the gain trim is set above −10 dB. At the −10 dB setting (which introduces 20 dB of actual attenuation) the input amplifier and the power amp will clip simultaneously; if more attenuation is permitted, then there is the danger that the input amplifier will clip first, and this cannot be corrected by the gain trim. Note the low resistor values in the gain trim network to minimise Johnson noise and also the effect of the power amplifier input noise current. A low source impedance is also required by most power amplifiers to prevent input current distortion (see my book *Audio Power Amplifier Design*).In the worst-case position (approximately wiper central) the output impedance of the network is 275 Ω, and the Johnson noise from that resistance is −128.2 dBu, well below the noise from the input amplifier, which is −106.2 dBu− 6 dB = −112.2 dBu, and also below the equivalent input noise of the power amplifier, which is unlikely to be better than −122 dBu.

## Transformer Balanced Inputs

When it is essential that there is no galvanic connection (i.e. no electrical conductor) between two pieces of equipment, transformer inputs are indispensable. They are also useful if EMC conditions are severe. Figure 20.23 shows a typical transformer input. The transformer usually has a 1:1 ratio and should have an inter-winding screen, which must be earthed to optimise the high-frequency CMRR and minimise noise pickup and EMC troubles. If the transformer is in a metal shielding can this needs to be grounded to reduce noise pickup; round cans can be held by a metal capacitor clip connected to ground.

The transformer secondary must see a high impedance, as this is reflected to the primary and represents the input impedance; here it is set by R2, and a buffer drives the circuitry downstream. In addition, if the secondary loading is too heavy, there will be increased transformer distortion at low frequencies. Line input transformers are built with small cores and are only intended to deliver very small amounts

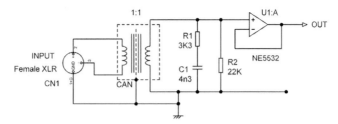

**Figure 20.23: A transformer balanced input. R1 and C1 are the Zobel network that damps the transformer secondary resonance.**

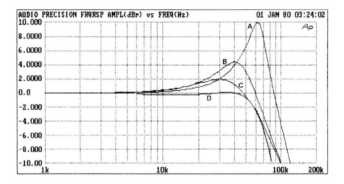

**Figure 20.24: Optimising the frequency response of a transformer balanced input by placing a Zobel network across the secondary winding.**

of power; they are *not* interchangeable with line output transformers. A most ingenious approach to dealing with this distortion problem by operating the input transformer core at near-zero flux was published by Paul Zwicky in 1986; [11] unfortunately, two transformers are required.

There is a bit more to correctly loading the transformer secondary. If it is simply loaded with a high-value resistor, there will be peaking of the frequency response due to resonance between the transformer leakage inductance and the winding capacitance. [12] This is shown in Figure 20.24, where a Sowter 3276 line input transformer (a high-quality component) was given a basic resistive loading of 100 kΩ. The result was Trace A, which has a 10 dB peak around 60 kHz. This is bad not only because it accentuates the effect of out-of-band noise but because it intrudes on the audio frequency response, giving a lift of 1 dB at 20 kHz. Reducing the resistive load R2 would damp the resonance, but it would also reduce the input impedance. The answer is to add a Zobel network, i.e. a resistor and capacitor in series, across the secondary; this has no effect except high frequencies. The first attempt used R1 = 2k7 and C1 = 1 nF, giving Trace B, where the peaking has been reduced to 4 dB around 40 kHz, but the 20 kHz lift is actually slightly greater. R1 = 2k7 and C1 = 2 nF gave Trace C, which is a bit better in that it only has a 2 dB peak. A bit more experimentation ended up with R1 = 3k3 and C1 = 4.3 nF (3n3 + 1 nF) and yielded Trace D, which is pretty flat, though there is a small droop around 10 kHz. The Zobel values are fairly critical for the flattest possible response and must certainly be adjusted if the transformer type is changed.

No discussion on transformer coupling is complete without pointing out that transformers have poor linearity at low frequencies—orders of magnitude worse than any electronic circuitry. A line input transformer is lightly loaded (by 22 kΩ in Figure 20.23), but this does not reduce the LF distortion, which comes as the inharmonious third harmonic. What does affect it is the source impedance feeding it—50 Ω can easily double the distortion compared with a negligible source impedance, and you're not likely to see lower than 50 Ω unless the source equipment has been deliberately designed with so-called zero-impedance output stages; these are described in Chapter 21 on line outputs. Fortunately, the signal levels in the bottom octave of the audio band are usually something like 10–12 dB lower than the maximum amplitudes, which occur in the middle frequencies, and this helps to ease the situation a bit; see Chapter 17 on system design for more on the amplitude/frequency distribution of musical signals.

Transformers also have the traditional disadvantages of size, weight, and cost. With these issues to grapple with, you can see why people don't design in transformers unless they really have to.

## Input Overvoltage Protection

Input overvoltage protection is not common in hi-fi applications but is regarded as essential in most professional equipment. The normal method is to use clamping diodes, as shown in Figure 20.25, that prevent certain points in the input circuitry from moving outside the supply rails.

This is straightforward, but there are two points to watch. First, the ability of this circuit to withstand excessive input levels is not without limit. Sustained overvoltages may burn out R5 and R6 or pump unwanted amounts of current into the supply rails; this sort of protection is mainly aimed at transients. Second, diodes have a non-linear junction capacitance when they are reverse-biased, so if the source impedance is significant, the diodes will cause distortion at high frequencies. To quantify this problem, here are the results of a few tests. If Figure 20.25 is fed from the low impedance of the usual kind of line output stage, the impedance at the diodes will be about 1 kΩ, and the distortion induced into an 11 Vrms 20 kHz input will be below the noise floor. However, if the source impedance is high so the impedance at the diodes is increased to 10 kΩ, with the same input level, the THD at 20 kHz was

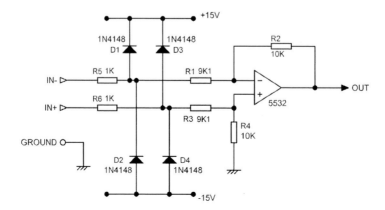

**Figure 20.25: Input overvoltage protection for a balanced input amplifier.**

degraded from 0.0030% to 0.0044% by adding the diodes. I have thought up a rather elegant way to eliminate this effect completely, but being a mercenary fellow I hope to sell it to someone.

## Noise and Balanced Inputs

So far we have not said much about the noise performance of balanced inputs, though on our way through the chapter we have noted that a standard balanced input amplifier constructed with four 10 kΩ resistors and a 5532 section has the relatively high noise output of −104.8 dBu with both its inputs terminated by 50 Ω to ground. That value of 50 Ω is in no way critical, because its value is so much lower than that of the 10 kΩ resistors.

When evaluating some sorts of input amplifier which have a well-defined input load, such as microphone preamplifiers or moving-magnet RIAA preamplifiers, it is useful to state the noise figure (NF). This is the difference in dB between what the Johnson noise from the input load would be if it was amplified by a theoretical noiseless amplifier and the real noise output with the real amplifier. It is a powerful tool, as it indicates at once how far short of perfection a design falls. Evaluating the NF of a microphone preamplifier is straightforward, as the input load is usually treated as a pure resistance of 200 Ω; NFs of 1 or 2 dB can be obtained easily using a combination of discrete BJTs as input devices, though usually only at high gains, because otherwise the Johnson noise from the gain pot resistance degrades things. [13] The situation with a moving-magnet RIAA preamplifier is more complex, as the input load has a high value of inductance as well as series resistance, and this has powerful effects on the noise generated both by the amplifier and the input loading resistance. Nonetheless, NFs of 3 dB or so are possible using 5532 or 5534 opamps for the input stage.

The situation is rather different if we try to apply this to a balanced input amplifier. Take the standard balanced input with its four 10 kΩ resistors, 5532 section, and both inputs terminated by 50 Ω to ground. The Johnson noise from each 50 Ω resistor is −135.2 dBu over a 22 kHz bandwidth at 25°C; the noise from both together is 3 dB more because the two noise sources are uncorrelated. This means that the noise voltage at the input is −132.2 dBu, a very low level in anyone's terms. Unfortunately, the noise output from the stage is −104.8 dBu, giving us a noise figure of 27.4 dB. In most fields of electronic endeavour this would be regarded as truly appalling and fit only for the dustbin. However, simple balanced input amplifiers of this type are widely used in the professional audio industry, so clearly noise figures are in this case not that useful a figure of merit.

What NFs do is to show us what room there is for improvement in our design: 27.4 dB is a lot of room, and in the next section we will attempt to cut the NF down to size.

## Low-Noise Balanced Inputs

I have remarked several times that the standard balanced input amplifier with four 10 kΩ resistors shown in Figure 20.26a is markedly noisier than an unbalanced input like that in Figure 20.1. The unbalanced input stage, with its input terminated by 50 Ω to ground, has a noise output of −119.0 dBu over the usual 22 Hz–22 kHz bandwidth. If the balanced circuit is built with 10 kΩ resistors and a 5532 section, the noise output is −104.8 dBu with the inputs similarly terminated. This is a big difference of 14.2 dB.

In the hi-fi world in particular, where an amplifier may have both unbalanced and balanced inputs, most people feel that this is the wrong way round. Surely the balanced input, with its professional XLR connector and its much-vaunted rejection of ground noise, should show a better performance in all departments? Well, it does—except as regards internal noise, and a 14 dB discrepancy is both clearly audible and hard to explain away. This section explains how to design it away instead.

We know that the source of the extra noise is the relatively high resistor values around the opamp (see Table 20.7 earlier in the chapter), but these cannot be reduced in the simple balanced amplifier without reducing the input impedances below what is acceptable. The answer is to lower the resistor values but buffer them from the input with a pair of voltage-followers; this arrangement is shown in Figure 20.26b. The 5532 is a good choice for this, combining low voltage noise with low distortion and good load-driving capability. Since the input buffers are voltage-followers with 100% feedback, their gain is very accurately defined at unity, and the CMRR is therefore not degraded; CMRR is still defined by the resistor tolerances and by the bandwidth of the differential opamp. In fact, the overall CMRR for the balanced link is likely to be improved, as we now have equal hot and cold input impedances in all circumstances, and we can make them much higher than the usual balanced input impedances. Figures 20.26b to 20.26d show 47 k$\Omega$ input resistors, but these could easily be raised to 100 k$\Omega$. The offset created by the input bias current flowing through this resistance needs to be watched, but it should cause no external troubles if the usual blocking capacitors are used.

There is a limit to how far the four resistors can be reduced, as the differential stage has to be driven by the input buffers, and it also has to drive its own feedback arm. If 5532s are used, a safe value that gives no measurable deterioration of the distortion performance is about 820 $\Omega$, and a 5532 differential stage alone (without the buffers) and 4 × 820 $\Omega$ resistors gives a noise output of −111.7 dBu, which is 6.6 dB lower than the standard 4 × 10 k$\Omega$ version. Adding the two input buffers degrades this only slightly to −110.2 dB, because we are adding only the voltage noise component of the two new opamps, and we are still 5.4 dB quieter than the original 4 × 10 k$\Omega$ version. It is an interesting point that we now have three opamps in the signal path instead of one, but we still get a significantly lower noise level.

This might appear to be all we can do; it is not possible to reduce the value of the four resistors around the differential amplifier any further without compromising linearity. However, there is almost always some way to go further in the great game that is electronics, and here are three possibilities. A step-up transformer could be used to exploit the low source impedance (remember we are still assuming the source impedances are 50 $\Omega$), and it might well work superbly in terms of noise alone, but transformers are always heavy, expensive, susceptible to magnetic fields, and of doubtful low-frequency linearity. We would also very quickly run into headroom problems; balanced line input amplifiers are normally required to attenuate rather than amplify.

We could design a discrete-opamp hybrid stage with discrete input transistors, which are quieter than those integrated into IC opamps, coupled to an opamp to provide raw loop gain; this can be quite effective, but you need to be very careful about high-frequency stability, and it is difficult to get an improvement of more than 6 dB. Thirdly, we could design our own opamp using all discrete parts; this approach tends to have less stability problems, as all circuit parameters are accessible, but it definitely requires rather specialised skills, and the result takes up a lot of PCB area.

Since none of those three approaches are appealing, now what? One of the most useful techniques in low-noise electronics is to use two identical amplifiers so that the gains add arithmetically, but the noise from the two separate amplifiers, being uncorrelated, partially cancels. Thus we get a 3 dB noise advantage each time the number of amplifiers used is doubled. This technique works very well with multiple opamps; let us apply it and see how far it may be usefully taken.

Since the noise of a single 5532-section unity-gain buffer is only −119.0 dBu and the noise from the 4 × 820 Ω differential stage (without buffers) is a much higher −111.7 dBu, the differential stage is clearly the place to start work. We will begin by using two identical 4 × 820 Ω differential amplifiers, as shown in the top section of Figure 20.26c, both driven from the existing pair of input buffers. This will give partial cancellation of both resistor and opamp noise from the two differential stages if their outputs are summed. The main question is how to sum the two amplifier outputs; any active solution would introduce another opamp and hence more noise, and we would almost certainly wind up worse off than when we started. The answer is, however, beautifully simple. We just connect the two amplifier outputs together with 10 Ω resistors; the gain does not change, but the noise output drops.

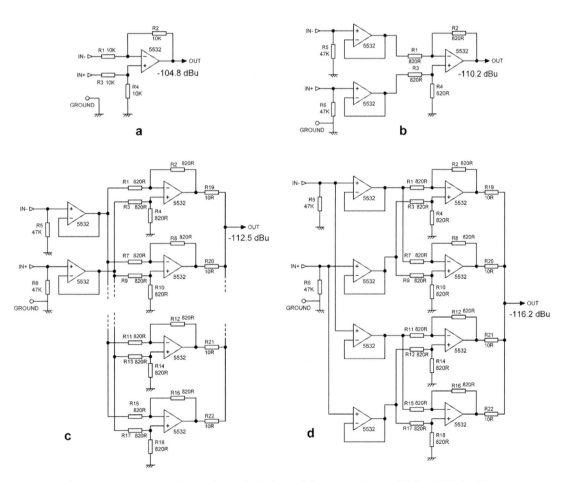

**Figure 20.26: Low-noise unity-gain balanced inputs using multiple 5532 buffers and differential amplifiers.**

The signal output of both amplifiers is nominally the same, so no current should flow from one opamp output to the other. In practice there will be slight gain differences due to resistor tolerances, but with 1% resistors I have never experienced any hint of a problem. The combining resistor values are so low at 10 Ω that their Johnson noise contribution is negligible.

The use of multiple differential amplifiers has another advantage—the CMRR errors are also reduced in the same way that the noise is reduced. This is also similar to the use of multiple resistors or capacitors to improve the accuracy of the total value, as explained in Chapter 15.

We therefore have the arrangement of Figure 20.26c, with single input buffers (i.e. one per input) and two differential amplifiers, and this reduces the noise output by 2.3 dB to −112.5 dBu, which is quieter than the original 4 × 10 kΩ version by an encouraging 7.4 dB. We do not get a full 3 dB noise improvement because both differential amplifiers are handling the noise from the input buffers, which is correlated and so is not reduced by partial cancellation. The contribution of the input buffer noise is further brought out if we take the next step of using four differential amplifiers. There is of course nothing special about using amplifiers in powers of two. It is perfectly possible to use three or five differential amplifiers in the array, which will give intermediate amounts of noise reduction. If you have a spare opamp section, then put it to work!

So, leaving the input buffers unchanged, we use them to drive an array of four differential amplifiers. These are added on at the dotted lines in the lower half of Figure 20.26c. We get a further improvement, but only by 1.5 dB this time. The output noise is down to −114.0 dBu, quieter than the original 4 × 10 kΩ version by 8.9 dB. You can see that at this point we are proceeding by decreasing steps, as the input buffer noise is starting to dominate, and there seems little point in doubling up the differential amplifiers again; the amount of hardware could be regarded as a bit excessive, and so would the PCB area occupied. The increased loading on the input buffers is also a bit of a worry.

A more fruitful approach is to tackle the noise from the input buffers by doubling them up as in Figure 20.26d, so that each buffer drives only two of the four differential amplifiers. This means that the buffer noise will also undergo partial cancellation and will be reduced by 3 dB. There is however still the contribution from the differential amplifier noise, and so the actual improvement on the previous version is 2.2 dB, bringing the output noise down to −116.2 dBu, which is quieter than the original 4 × 10 kΩ version by a thumping 11.1 dB. Remember that there are two inputs, and "double buffers" means two buffers per hot and cold input, giving a total of four in the complete circuit.

Since doubling up the input buffers gave us a useful improvement, it's worth trying again, so we have a structure with quad buffers and four differential amplifiers, as shown in Figure 20.27, where each differential amplifier now has its very own buffer. This improves on the previous version by a rather less satisfying 0.8 dB, giving an output noise level of −117.0 dBu, quieter than the original 4 × 10 kΩ version by 11.9 dB. The small improvement we have gained indicates that the focus of noise reduction needs to be returned to the differential amplifier array, but the next step there would seem to be using eight amplifiers, which is not very appealing. Thoughts about ears of corn on chessboards tend to intrude at this point.

This is a good moment to pause and see what we have achieved. We have built a balanced input stage that is quieter than the standard balanced input by 11.9 dB, using standard components of low cost. We have used increasing numbers of them, but the total cost is still small compared with enclosures, power supplies, etc. On the other hand, the power consumption is naturally several times greater. The

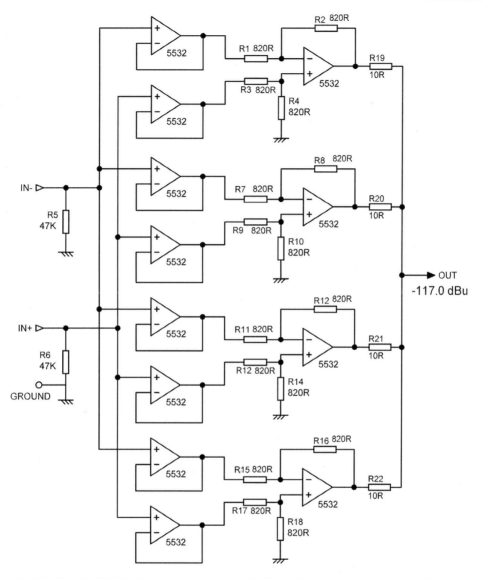

**Figure 20.27: The final 5532 low-noise unity-gain balanced input stage, with quad input buffers and four differential amplifiers. The noise output is only −117.0 dBu.**

technology is highly predictable and the noise reduction reliable; in fact it is bullet-proof. The linearity is as good as that of a single opamp of the same type, and in the same way there are no HF stability problems. The CMRR will also be improved.

What we have not done is build a balanced input that is quieter than the unbalanced one—we are still 2.0 dB short of that target, but at least we have reached a point where the balanced input is not obviously noisier. Earlier we evaluated the noise figure of the standard 4 × 10 kΩ balanced input and found it was a startling 27.4 dB; our new circuit here has a noise figure of 15.2 dB, which looks a good deal more respectable

The noise results, including noise figures, are all summarised in Table 20.9 at the end of this chapter.

## Low-Noise Balanced Inputs in Real Life

Please don't think that this examination of low-noise input options is merely a voyage off into pure theory. It has real uses and has been applied in practice: the Cambridge Audio 840 W power amplifier, a design of mine which, I might modestly mention in passing, won a CES Innovation Award in January 2008. This unit has both unbalanced and balanced inputs, and conventional technology would have meant that the balanced inputs would have been significantly the noisier of the two. Since the balanced input is the "premium" input, many people would think there was something amiss with this state of affairs. We therefore decided the balanced input had to be quieter than the unbalanced input. Using 5532s in an architecture similar to those outlined earlier, this requirement proved straightforwardly attainable, and the final balanced input design was both economical and quieter than its unbalanced neighbour by a dependable 0.9 dB. Two other versions were evaluated that made the balanced input quieter than the unbalanced one by 2.8 dB and by 4.7 dB, at somewhat greater cost and complexity. These were put away for possible future upgrades.

The Signal Transfer Company [14] manufactures a low-noise balanced input card based on these principles which has 47 kΩ input impedances, unity gain, and a noise output of only −115 dBu.

## Ultra-Low-Noise Balanced Inputs

In the section on low-noise balanced inputs earlier, we reclined briefly on our laurels, having achieved an economical balanced input stage with output noise at the extremely low level of −117.0 dBu. Regrettably this is still 2 dB noisier than a simple unbalanced input. It would be wrong to conclude from this that the resources of electronic design are exhausted. At the end of the noise-reduction sequence we were aware that the dominant noise source was currently the differential amplifier array, and we shrank from doubling up again to use eight amplifiers because of issues of cost and the PCB area occupied. We will take things another step by taking a much more relaxed view of cost (as one does at the "high end") and see how that changes the game. We will however retain some concern about PCB area.

An alternative way to make the differential amplifier array quieter is simply to use opamps that are quieter. These will inevitably be more expensive—much more expensive—than the ubiquitous 5532. Because of the low resistor values around the opamps, we need to focus on low voltage noise rather than low current noise, and there are several that are significantly better than the 5532, as shown by the typical noise density figures in Table 20.8.

Clearly, moving to the 5534A will give a significant noise reduction, but since there is only a single opamp per package and external compensation is needed, the board area used will be much greater. The new chip on the block is the LM4562, a bipolar opamp which has finally surpassed the 5532 in performance. The input voltage noise density is typically 2.7 nV/√Hz, substantially lower than

**Table 20.8: Voltage and noise densities for low-noise
balanced input opamp candidates.**

| Opamp | Voltage noise density nV√Hz | Current noise density pA√Hz |
|---|---|---|
| 5532 | 5 | 0.7 |
| 5534A | 3.5 | 0.4 |
| LM4562 | 2.7 | 1.6 |
| AD797 | 0.9 | 2 |
| LT1028 | 0.85 | 1 |

the 5 nV/√Hz of the 5532. For applications with low source impedances, this implies a handy noise advantage of 5 dB or more. The LM4562 is a dual opamp and will not take up any more space. At the time of writing it is something like 10 times more expensive than the 5532.

Step One—we replace all four opamps in the differential amplifiers with LM4562s. They are a drop-in replacement with no circuit adjustments required at all. We leave the quad 5532 input buffers in place. The noise output drops by an impressive 1.9 dB, giving an output noise level of −118.9 dBu, quieter than the original 4 × 10 kΩ version by 14.1 dB and only 0.1 dB noisier than the unbalanced stage.

Step Two—replace the quad 5532 buffers with quad LM4562 buffers. Noise falls by only 0.6 dB, the output being −119.5 dBu, but at last we have a balanced stage that is quieter than the unbalanced stage, by a small but solid 0.5 dB.

One of the pre-eminent low-noise-with-low-source-resistance opamps is the AD797 from Analog Devices, which has a remarkably low voltage noise at 0.9 nV/√Hz (typical at 1 kHz), but it is a very expensive part, costing between 20 and 25 times more than a 5532 at the time of writing. The AD797 is a single opamp, while the 5532 is a dual, so the cost per opamp is actually 40 to 50 times greater, and more PCB area is required, but the potential improvement is so great we will overlook that. . .

Step Three—we replace all four opamps in the differential amplifiers with AD797s, putting the 5532s back into the input buffers in the hope that we might be able to save money somewhere. The noise output drops by a rather disappointing 0.4 dB, giving an output noise level of −119.9 dBu, quieter than the original 4 × 10 kΩ version by 15.1 dB.

Perhaps putting those 5532s back in the buffers was a mistake? Our fourth and final move in this game of electronic chess is to replace all the quad 5532 input buffers with dual (not quad) AD797 buffers. This requires another four AD797s (two per input) and is once more not a cheap strategy. We retain the four AD797s in the differential amplifiers. The noise drops by another 0.7 dB, yielding an output noise level of −120.6 dBu, quieter than the original 4 × 10 kΩ version by 15.8 dB and quieter than the unbalanced stage by a satisfying 1.6 dB. You can do pretty much anything in electronics with a bit of thought and a bit of money.

If, however, we look at the noise figure for this final design, we feel a bit less happy. Despite deploying some ingenious circuitry and a lot of premium opamps, we still have an NF of 11.6 dB. We have not even managed to get the NF down to single figures, and so there is plenty of scope yet for some creative design. The relatively high NF is essentially because the reference input loads we are using are 50 Ω resistors, which naturally generate a very low level of Johnson noise (−136.2 dBu each). If we want to make more progress in this direction, we might start thinking about moving-coil preamp circuitry, which can achieve NFs of less than 7 dB with an input load as low as 3.3 Ω. [15] This can be achieved by using special low-Rb discrete transistors as input devices, with an opamp to provide open-loop gain. This sort of hybrid circuitry must be carefully designed to avoid HF stability problems, whereas simply plugging in more opamps always works.

You are probably wondering what happened to the LT1028 lurking at the bottom of Table 20.8. It is true that its voltage noise density is slightly better than that of the AD797, but there is a subtle snag. As described in Chapter 16, the LT1028 has bias-current cancellation circuitry which injects correlated noise currents into the two inputs. These will cancel if the impedances seen by the two inputs are the same, but in moving-magnet amplifier use the impedances differ radically, and the LT1028 is not useful in this application. The input conditions here are more benign, but the extra complication is unwelcome, and I have never used the LT1028 in audio work. In addition, it is a single opamp with no dual version.

This is not of course the end of the road. The small noise improvement in the last step we made tells us that the differential amplifier array is still the dominant noise source, and further development would have to focus on this. A first step would be to see if the relatively high current noise of the AD797s is significant with respect to the surrounding resistor values. If so, we need to see if the resistor values can be reduced without degrading linearity at full output. We should also check the Johnson noise contribution of all those 820 Ω resistors; they are generating −123.5 dBu each at room temperature, but of course the partial cancellation effect applies to them as well.

All these noise results are also summarised in Table 20.9.

**Table 20.9: A summary of the noise improvements made to the balanced input stage.**

| Buffer type | Amplifier | Noise output dBu | Improvement on previous version dB | Improvement over 4x10 kΩ diff amp dB | Noisier than unbal input by: dB | Noise figure Ref 2x 50 Ω dB |
|---|---|---|---|---|---|---|
| | 5532 voltage-follower | −119.0 | | | 0 dB ref | |
| None | Standard diff amp 10k 5532 | −104.8 | 0 | 0.0 dB ref | 14.2 | 27.4 |
| None | Single diff amp 820R 5532 | −111.7 | 6.9 | 6.9 | 7.3 | 20.5 |
| Single 5532 | Single diff amp 820R 5532 | −110.2 | 5.4 | 5.4 | 8.8 | 22.0 |
| Single 5532 | Dual diff amp 820R 5532 | −112.5 | 2.3 | 7.4 | 6.5 | 19.7 |

| Buffer type | Amplifier | Noise output dBu | Improvement on previous version dB | Improvement over 4x10 kΩ diff amp dB | Noisier than unbal input by: dB | Noise figure Ref 2x 50 Ω dB |
|---|---|---|---|---|---|---|
| Single 5532 | Quad diff amp 820R 5532 | −114.0 | 1.5 | 9.2 | 5.0 | 18.2 |
| Dual 5532 | Quad diff amp 820R 5532 | −116.2 | 2.2 | 11.4 | 2.8 | 16.0 |
| Quad 5532 | Quad diff amp 820R 5532 | −117.0 | 0.8 | 12.2 | 2.0 | 15.2 |
| Quad 5532 | Quad diff amp 820R LM4562 | −118.9 | 1.9 | 14.1 | 0.1 | 13.3 |
| Quad LM4562 | Quad diff amp 820R LM4562 | −119.5 | 0.6 | 14.7 | −0.5 | 12.7 |
| Quad 5532 | Quad diff amp 820R AD797 | −119.9 | 0.4 | 15.1 | −0.9 | 12.3 |
| Dual AD797 | Quad diff amp 820R AD797 | −120.6 | 0.7 | 15.8 | −1.6 | 11.6 |

# References

[1] Muncy, N. "Noise Susceptibility in Analog and Digital Signal Processing Systems" JAES 3(6), June 1995, p. 447

[2] Whitlock, Bill "Balanced Lines in Audio Systems: Fact, Fiction, & Transformers" JAES 3(6), June 1995

[3] Williams, T. "EMC for Product Designers" Newnes (Butterworth-Heinemann), pub 1992, p. 176, ISBN: 0-7506-1264-9

[4] Williams, T. As [3] above, p. 173.

[5] Winder, S. "What's the Difference?" Electronics World, November 1995, p. 989

[6] Graeme, Jerald "Amplifier Applications of Opamps" McGraw-Hill, 1999, p. 16, ISBN: 0-07-134643-0

[7] Birt, D. "Electronically Balanced Analogue-Line Interfaces" Proceedings of the Institute of Acoustics Conference, Windermere, UK, November 1990.

[8] Self, Douglas "Small Signal Audio Design Handbook" Second Edn, Focal Press, 2015, Chapter 13, ISBN 978-0-415-70974-3 hbk

[9] Self, Douglas Linear Audio Volume 5, April 2013, pp. 141–162

[10] Self, Douglas "Self on Audio (Audio Power Analysis)" Third Edn, Newnes, 2016, Chapter 19, ISBN: 978-1-138-85445-1 hbk

[11] Zwicky, P. *Low-Distortion Audio Amplifier Circuit Arrangement* US Patent No. 4,567,4431986

[12] Sowter, Dr G. A. V. "Soft Magnetic Materials for Audio Transformers: History, Production, and Applications" JAES 35(10), October 1987, p. 774

[13] Self, Douglas as [10] above, pp. 471–472

[14] www.signaltransfer.freeuk.com/ (low-noise balanced input PCB)

[15] Self, Douglas as [10] above, Chapter 12

# CHAPTER 21

# Line Outputs

## Unbalanced Outputs

There are only two electrical output terminals for an unbalanced output—signal and ground. However, the unbalanced output stage in Figure 21.1a is fitted with a three-pin XLR connector to emphasise that it is always possible to connect the cold wire in a balanced cable to the ground at the output end and still get all the benefits of common-mode rejection, if you have a balanced input. If a two-terminal connector is fitted, the link between the cold wire and ground has to be made inside the connector, as shown in Figure 20.2 in the chapter on line inputs.

The output amplifier in Figure 21.1a is configured as a unity-gain buffer, though in some cases it will be connected as a series-feedback amplifier to give gain. A non-polarised DC-blocking capacitor C1 is included; 100 uF gives a −3 dB point of 2.6 Hz with one of those notional 600 Ω loads. The opamp is isolated from the line shunt-capacitance by a resistor R2, in the range 47–100 Ω, to ensure HF stability, and this unbalances the hot and cold line impedances. A drain resistor R1 ensures that no charge can be left on the output side of C1; it is placed *before* R2, so it causes no attenuation. In this case the loss would only be 0.03 dB, but such errors can build up to an irritating level in a large system, and it costs nothing to avoid them.

If the cold line is simply grounded as in Figure 21.1a, then the presence of R2 degrades the CMRR of the interconnection to an uninspiring −43 dB, even if the balanced input at the other end of the cable has infinite CMRR in itself and perfectly matched 10 kΩ input impedances.

To fix this problem, Figure 21.1b shows what is called an impedance-balanced output. The cold terminal is neither an input nor an output but a resistive termination R3 with the same resistance as the hot terminal output impedance R2. If an unbalanced input is being driven, this cold terminal is ignored. The use of the word "balanced" is perhaps unfortunate, as when taken together with an XLR output connector it implies a true balanced output with anti-phase outputs, which is *not* what you are getting. The impedance-balanced approach is not particularly cost-effective, as it requires significant extra money to be spent on an XLR connector. Adding an opamp inverter to make it a proper balanced output costs little more, especially if there happens to be a spare opamp half available, and it sounds much better in the specification.

There is an instance of the use of impedance-balancing in the active crossover design example in Chapter 23; here the outputs come directly from low-impedance level trim controls with output impedances that vary somewhat with the level settings, so compromise values for the impedance-balancing resistances must be used.

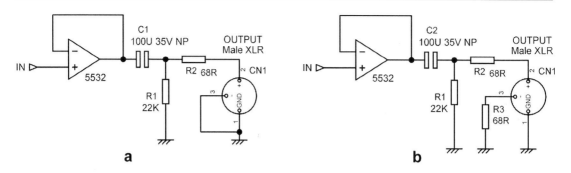

**Figure 21.1: Unbalanced outputs: (a) simple output and (b) impedance-balanced output for improved CMRR when driving balanced inputs.**

Active crossover output stages may also incorporate level trim controls, mute switches, and phase-invert switches. These features are covered in Chapter 17 on crossover system design.

## Zero-Impedance Outputs

Both the unbalanced outputs shown in Figure 21.1 have series output resistors to ensure stability when driving cable capacitance. This increases the output impedance and can lead to increased crosstalk in some situations, notably when different signals are being passed down the same signal cable. This is a particular problem when two or more layers of ribbon cable are laid together in a "lasagne" format for neatness or to save space. In some cases layers of grounded screening foil are interleaved with the cables, but this is rather expensive and awkward to do and does not greatly reduce crosstalk between conductors in the same piece of ribbon. The best way to deal with this is to reduce the output impedance.

A simple but very effective way to do this is the so-called "zero-impedance" output configuration. Figure 21.2a shows how the technique is applied to an unbalanced output stage with 10 dB of gain. Feedback at audio frequencies is taken from outside isolating resistor R3 via R2, while the HF feedback is taken from inside R3 via C2, so it is not affected by load capacitance and stability is unimpaired. Using a 5532 opamp, the output impedance is reduced from 68 ohms to 0.24 ohms at 1 kHz—a dramatic reduction that will reduce purely capacitive crosstalk by an impressive 49 dB. The output impedance increases to 2.4 ohms at 10 kHz and 4.8 ohms at 20 kHz, as opamp open-loop gain falls with frequency. The impedance-balancing resistor on the cold pin has been replaced by a link to match the near-zero output impedance at the hot pin.

Figure 21.2b shows a refinement of this scheme with three feedback paths. Electrolytic coupling capacitors can introduce distortion if they have more than a few tens of millivolts of signal across them, even if the time-constant is long enough to give a virtually flat LF response. (This is looked at in detail in Chapter 15 on passive components.) In Figure 21.2b, most of the feedback is now taken from outside C1 via R5, so it can correct capacitor distortion. The DC feedback goes via R2, now much higher in value, and the HF feedback goes through C2, as before, to maintain stability with capacitive loads. R2 and R5 in parallel come to 10 kΩ, so the gain is the same. Any circuit with separate DC and

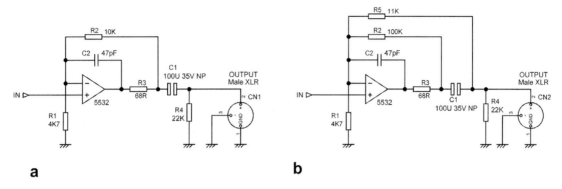

a                                                          b

**Figure 21.2: (a) Zero-impedance output; (b) zero-impedance output with NFB around output capacitor.**

AC feedback paths must be checked carefully for frequency response irregularities, which may happen well below 10 Hz.

## Ground-Cancelling Outputs

This technique, also called a ground-compensated output, appeared in the early 1980s in mixing consoles. It allows ground voltages to be cancelled out even if the receiving equipment has an unbalanced input; it prevents any possibility of creating a phase error by mis-wiring; and it costs virtually nothing in itself, though it does require a three-pin output connector.

Ground-cancelling (GC) separates the wanted signal from the unwanted ground voltage by addition at the output end of the link rather than by subtraction at the input end. If the receiving equipment ground differs in voltage from the sending ground, then this difference is added to the output signal so that the signal reaching the receiving equipment has the same ground voltage superimposed upon it. Input and ground therefore move together, and the ground voltage has no effect, subject to the usual effects of component tolerances. The connecting lead is differently wired from the more common unbalanced-out balanced-in situation, as now the cold line is joined to ground at the *input* or receiving end.

An inverting unity-gain ground-cancel output stage is shown in Figure 21.3a. The cold pin of the output socket is now an input and has a unity-gain path summing into the main signal going to the hot output pin to add the ground voltage. This path R3, R4 has a very low input impedance equal to the hot terminal output impedance, so if it *is* used with a balanced input, the line impedances will be balanced, and the combination will still work effectively. The 6 dB of attenuation in the R3-R4 divider is undone by the gain of two set by R5, R6. It is unfamiliar to most people to have the cold pin of an output socket as a low-impedance input, and its very low input impedance minimises the problems caused by mis-wiring. Shorting it locally to ground merely converts the output to a standard unbalanced type. On the other hand, if the cold input is left unconnected, then there will be a negligible increase in noise due to the very low input resistance of R3.

This is the most economical GC output, but obviously the phase inversion is not always convenient. Figure 21.3b shows a non-inverting GC output stage with a gain of 6.6 dB. R5 and R6 set up a gain

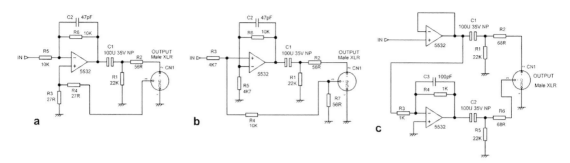

**Figure 21.3: (a) Inverting ground-canceling output; (b) non-inverting ground-canceling output; (c) a true balanced output.**

of 9.9 dB for the amplifier, but the overall gain is reduced by 3.3 dB by attenuator R3, R4. The cold line is now terminated by R7, and any signal coming in via the cold pin is attenuated by R3, R4 and summed at unity gain with the input signal. A non-inverting GC stage must be fed from a very low impedance such as an opamp output to work properly. There is a slight compromise on noise performance here because attenuation is followed by amplification.

Ground-cancelling outputs are an economical way of making ground loops innocuous when there is no balanced input, and it is rather surprising they are not more popular; perhaps people find the notion of an input pin on an output connector unsettling. In particular GC outputs offer the possibility of a quieter interconnection than the standard balanced interconnection, because a relatively noisy balanced input is not required (see Chapter 20 on line inputs). Ground-cancelling outputs can also be made zero impedance using the techniques described earlier.

## Balanced Outputs

Figure 21.3c shows a balanced output, where the cold terminal carries the same signal as the hot terminal but phase inverted. This can be arranged simply by using an opamp stage to invert the normal in-phase output. The resistors R3, R4 around the inverter should be as low in value as possible to minimise Johnson noise, because this stage is working at a noise gain of 2, but bear in mind that R3 is effectively grounded at one end, and its loading, as well as the external load, must be driven by the first opamp. A unity-gain follower is shown for the first amplifier, but this can be any other shunt or series-feedback stage as convenient. The inverting output, if not required, can be ignored; it must *not* be grounded, because the inverting opamp will then spend most of its time clipping in current-limiting, almost certainly injecting unpleasantly crunching distortion into the crossover grounding system. Both hot and cold outputs must have the same output impedances (R2, R6) to keep the line impedances balanced and the interconnection CMRR maximised.

It is vital to realise that this sort of balanced output, unlike transformer balanced outputs, by itself gives no common-mode rejection at all. It must be connected to a balanced input which can subtract one output from another if ground noise is to be cancelled.

The only advantages that this kind of balanced output has over an unbalanced output is that the total signal level on the interconnection is increased by 6 dB, which if correctly handled can improve the signal-to-noise ratio. It is also less likely to crosstalk to other lines, even if they are unbalanced, as the currents injected via the stray capacitance from each line will tend to cancel; how well this works depends on the physical layout of the conductors. All balanced outputs give the facility of correcting phase errors by swapping hot and cold outputs. This is however a two-edged sword, because it is probably how the phase got wrong in the first place.

There is no need to worry about the exact symmetry of level for the two output signals; ordinary 1% tolerance resistors are fine. Slight gain differences between the two outputs only affect the signal-handling capacity of the interconnection by a very small amount. This simple form of balanced output is the norm in hi-fi balanced interconnection but is less common in professional audio, where the quasi-floating output, which emulates a transformer winding, gives both common-mode rejection and more flexibility in situations where temporary connections are frequently being made.

## Transformer Balanced Outputs

If true galvanic isolation between equipment grounds is required, this can only be achieved with a line transformer, sometimes called a line isolating transformer; don't confuse them with mains isolating transformers. You don't, as a rule, use line transformers unless you really have to, because the much-discussed cost, weight, and performance problems are very real, as you will see shortly. However they are sometimes found in big sound-reinforcement systems and in any environment where high RF field strengths are encountered. They are unlikely to be used in active crossovers for domestic hi-fi. A basic transformer balanced output is shown in Figure 21.4a; in practice A1 would also have some other function such as providing gain or filtering. In good-quality line transformers there will be an inter-winding screen, which should be earthed to minimise noise pickup and general EMC problems. In most cases this does *not* ground the external can, and you have to arrange this yourself, possibly by mounting the can in a metal capacitor clip. Make sure the can is earthed, as this definitely does reduce noise pickup.

Be aware that the output impedance will be higher than usual because of the ohmic resistance of the transformer windings. With a 1:1 transformer, as normally used, both the primary and secondary winding resistances are effectively in series with the output. A small line transformer can easily have 60

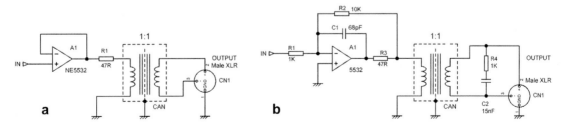

**Figure 21.4: Transformer balanced outputs: (a) standard circuit; (b) zero-impedance drive to reduce LF distortion, with Zobel network across secondary.**

Ω per winding, so the output impedance is 120 Ω plus the value of the series resistance R1 added to the primary circuit to prevent HF instability due to transformer winding capacitances and line capacitances. The total can easily be 160 Ω or more, compared with, say, 47 Ω for non-transformer output stages. This will mean a higher output impedance and greater voltage losses when driving heavy loads.

DC flowing through the primary winding of a transformer is bad for linearity, and if your opamp output has anything more than the usual small offset voltages on it, DC current flow should be stopped by a blocking capacitor.

## Output Transformer Frequency Response

If you have looked at the section in Chapter 20 on the frequency response of line input transformers, you will recall that they give a nastily peaking frequency response if the secondary is not loaded properly, due to resonance between the leakage inductance and the stray winding capacitances. Exactly the same problem afflicts output transformers, as shown in Figure 21.5; with no output loading there is a frightening 14 dB peak at 127 kHz. This is high enough in frequency to have very little effect on the response at 20 kHz, but such a high-$Q$ resonance isn't the sort of horror you want lurking in your circuitry. It could easily cause some nasty EMC problems.

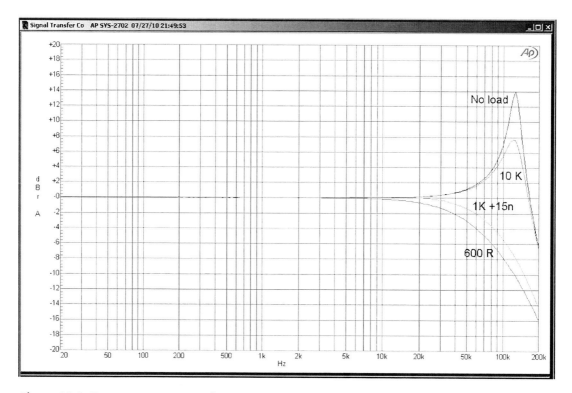

**Figure 21.5: Frequency response of a Sowter 3292 output transformer with various loads on the secondary. Zero-impedance drive as in Figure 21.4b.**

The transformer measured was a Sowter 3292 1:1 line isolating transformer. Sowter are a highly respected company, and this is a quality part with a mumetal core and housed in a mumetal can for magnetic shielding. When used as the manufacturer intended, with a 600 Ω load on the secondary, the results are predictably quite different, with a well-controlled roll-off that I measured as −0.5 dB at 20 kHz.

The difficulty is that there are very few if any genuine 600 Ω loads left in the world, and most output transformers are going to be driving much higher impedances. If we are driving a 10 kΩ load, the secondary resonance is not much damped, and we still get a thoroughly unwelcome 7 dB peak above 100 kHz, as shown in Figure 21.5. We could of course put a permanent 600 Ω load across the secondary, but that will heavily load the output opamp, impairing its linearity, and will give us unwelcome signal loss due in the winding resistances. It is also profoundly inelegant.

A better answer, as in the case of the line input transformer, is to put a Zobel network, i.e. a series combination of resistor and capacitor, across the secondary, as in Figure 21.4b. The capacitor required is quite small and will cause very little loading, except at high frequencies where signal amplitudes are low. A little experimentation yielded the values of 1 kΩ in series with 15 nF, which gives the much improved response shown in Figure 21.5. The response is almost exactly 0.0 dB at 20 kHz, at the cost of a very gentle 0.1 dB rise around 10 kHz; this could probably be improved by a little more tweaking of the Zobel values. Be aware that a different transformer type will require different values.

## Transformer Distortion

Transformers have well-known problems with linearity at low frequencies. This is because the voltage induced into the secondary winding depends on the rate of change of the magnetic field in the core, and so the lower the frequency, the greater the change in field magnitude must be for transformer action. [1] The current drawn by the primary winding to establish this field is non-linear, because of the well-known non-linearity of iron cores. If the primary had zero resistance and was fed from a zero source impedance, as much distorted current as was needed would be drawn, and no one would ever know there was a problem. But . . . there is always some primary resistance, and this alters the primary current drawn so that third-harmonic distortion is introduced into the magnetic field established and so into the secondary output voltage. Very often there is a series resistance R1 deliberately inserted into the primary circuit, with the intention of avoiding HF instability; this makes the LF distortion problem worse. An important point is that this distortion does not appear only with heavy loading—it is there all the time, even with no load at all on the secondary; it is not analogous to loading the output of a solid-state power amplifier, which invariably increases the distortion. In fact, in my experience transformer LF distortion is slightly better when the secondary is connected to its rated load resistance. With no secondary load, the transformer appears as a big inductance, so as frequency falls the current drawn increases, until with circuits like Figure 21.4a, there is a sudden steep increase in distortion around 10–20 Hz as the opamp hits its output-current limits. Before this happens the distortion from the transformer itself will be gross.

To demonstrate this I did some distortion tests on the same Sowter 3292 transformer. The winding resistance for both primary and secondary is about 59 Ω. It is quite a small component, 34 mm in diameter and 24 mm high and weighing 45 gm, and is obviously not intended for transferring large amounts of power at low frequencies. Figure 21.6 shows the LF distortion with no series resistance, driven directly from a 5532 output (there were no HF stability problems in this case, but it might be different with cables connected to the secondary), and with 47 and 100 Ω added in series with the primary. The flat part to the right is the noise floor.

Taking 200 Hz as an example, adding 47 Ω in series increases the THD from 0.0045% to 0.0080%, figures which are in exactly the same ratio as the total resistances in the primary circuit in the two cases. It's very satisfying when a piece of theory slots right home like that. Predictably, a 100 Ω series resistor gives even more distortion, namely 0.013% at 200 Hz, and once more proportional to the total primary circuit resistance.

If you're used to the near-zero LF distortion of opamps, you may not be too impressed with Figure 21.6, but this is the reality of output transformers. The results are well within the manufacturer's specifications for a high-quality part. Note that the distortion rises rapidly to the LF end, roughly tripling as frequency halves. It also increases fast with level, roughly quadrupling as level doubles. Having gone to some pains to make electronics with very low distortion, this non-linearity at the very end of the signal chain is distinctly irritating.

The situation is somewhat eased in actual use, as signal levels in the bottom octave of audio are normally about 10–12 dB lower than the maximum amplitudes at higher frequencies; see Chapter 17 for more on this.

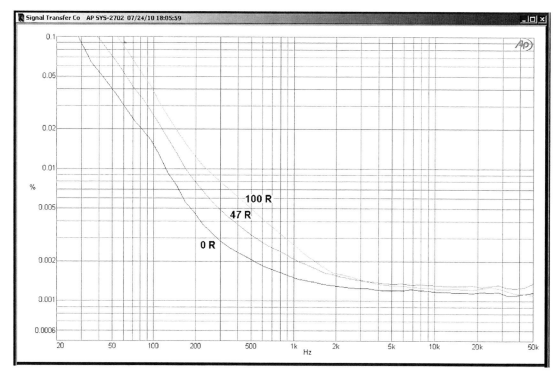

**Figure 21.6: The LF distortion rise for a 3292 Sowter transformer, without (0R) and with (47 Ω and 100 Ω) extra series resistance. Signal level 1 Vrms.**

## Reducing Transformer Distortion

In electronics, as in so many other areas of life, there is often a choice between using brains or brawn to tackle a problem. In this case "brawn" means a bigger transformer, such as the Sowter 3991, which is still 34 mm in diameter but 37 mm high, weighing in at 80 gm. The extra mumetal core material improves the LF performance, and thicker wire may be helping by lowering winding resistance, but you still get a distortion plot very much like Figure 21.6 (with the same increase of THD with series resistance), except now it occurs at 2 Vrms instead of 1 Vrms. Twice the metal, twice the level—I suppose it makes sense. You can take this approach a good deal further with the Sowter 4231, a much bigger open-frame design tipping the scales at a hefty 350 gm. The winding resistance for the primary is 12 Ω and for the secondary 13.3 Ω, both a good deal lower than the previous figures.

Figure 21.7 shows the LF distortion for the 4231 with no series resistance, and with 47 and 100 Ω added in series with the primary. The flat part to the right is the noise floor. Comparing it with Figure 21.5, the basic distortion at 30 Hz is now 0.015%, compared with about 0.10% for the 3292 transformer. While this is a useful improvement, it is gained at considerable expense. Now adding 47 Ω of series resistance has dreadful results—distortion increases by about five times. This is because the lower winding resistances of the 4231 mean that the added 47 Ω has increased the total resistance in the primary circuit to five times what it was. Predictably, adding a 100 Ω series resistance approximately doubles the distortion again. In general bigger transformers have thicker wire in the windings, and this in itself reduces the effect of the basic core non-linearity, quite apart from the improvement due to more core material. A lower winding resistance also means a lower output impedance.

The LF non-linearity in Figure 21.7 is still most unsatisfactory compared with that of the electronics. Since the "My policy is copper and iron!" [2] approach does not solve the problem, we'd better put brawn to one side and try what brains we can muster.

We have seen that adding series resistance to ensure HF stability makes things definitely worse, and a better means of isolation is a low-value inductor of, say, 4 uH paralleled with a low-value damping resistor of around 47 Ω. However, inductors cost money, and a more economic solution is to use a zero-impedance output as shown in Figure 21.4b. This gives the same results as no series resistance at all but with dependable HF stability. However, the basic transformer distortion remains because the primary winding resistance is still there, and its level is still too high. What can be done?

The LF distortion can be reduced by applying negative feedback via a tertiary transformer winding, but this usually means an expensive custom transformer, and there may be some interesting HF stability problems because of the extra phase-shift introduced into the feedback by the tertiary winding; this approach is discussed in [3]. What we really want is a technique that will work with off-the-shelf transformers.

A better way is to cancel out the transformer primary resistance by putting in series an electronically generated negative resistance; the principle is shown in Figure 21.8, where a zero-impedance output is used to eliminate the effect of the series stability resistor. The 56 Ω resistor R4 senses the current through the primary and provides positive feedback to A1, proportioned so that a negative output resistance of twice the value of R4 is produced, which will cancel out both R4 itself and the primary winding resistance. As we saw earlier, the primary winding resistance of the 3292 transformer is approximately 59 Ω, so if R4 was 59 Ω we should get complete cancellation. But . . .

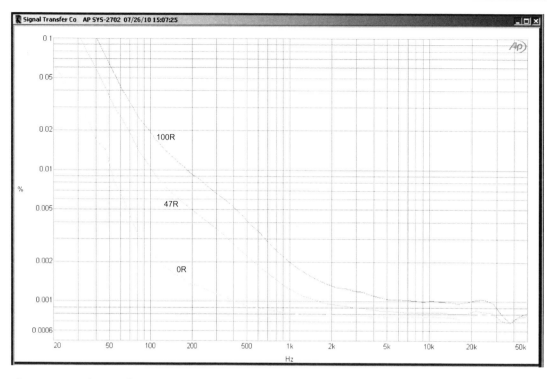

**Figure 21.7: The LF distortion rise for a much larger 4231 Sowter transformer, without and with extra series resistance. Signal level 2 Vrms.**

It is always necessary to use positive feedback with caution. Typically it works, as here, in conjunction with good old-fashioned negative feedback, but if the positive exceeds the negative (this is one time you do *not* want to accentuate the positive), then the circuit will typically latch up solid, with the output jammed up against one of the supply rails. R4 = 56 Ω in Figure 21.8 worked reliably in all my tests, but increasing R4 to 68 Ω caused immediate problems, which is precisely what you would expect. No input blocking capacitor is shown in Figure 21.8, but it can be added ahead of R1 without increasing the potential latch-up problems.

This circuit is only a basic demonstration of the principle of cancelling primary resistance, but as Figure 21.9 shows it is still highly effective. The distortion at 100 Hz is reduced by a factor of five, and at 200 Hz by a factor of four. Since this is achieved by adding one resistor, I think this counts as a triumph of brains over brawn and indeed is confirmation of the old adage that size is less important than technique.

The method is sometimes called "mixed feedback", as it can be looked at as a mixture of voltage and current feedback. The principle can also be applied when a balanced drive to the output transformer is used. Since the primary resistance is cancelled, there is a second advantage, as the output impedance of the stage is reduced. The secondary winding resistance is however still in circuit, and so the output impedance is usually only halved.

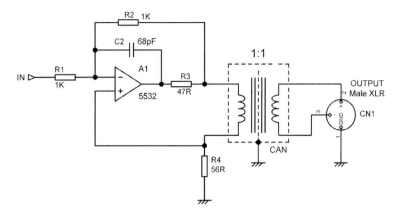

Figure 21.8: Reducing LF distortion by cancelling out the primary winding resistance with a negative resistance generated by current-sensing resistance R4. Values for Sowter 3292 transformer.

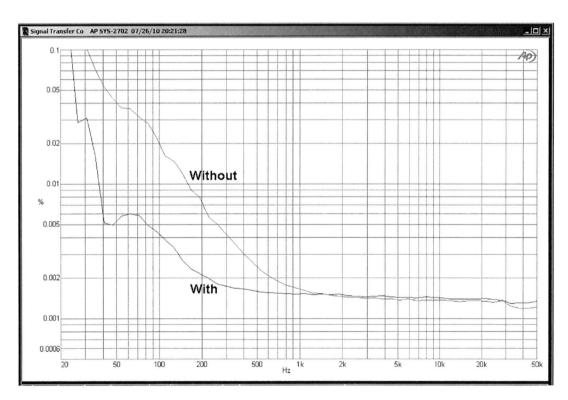

Figure 21.9: The LF distortion rise for a 3292 Sowter transformer, without and with winding resistance cancellation as in Figure 21.7. Signal level 1 Vrms.

If you want better performance than this—and it is possible to make transformer non-linearity effectively invisible down to 15 Vrms at 10 Hz—there are several deeper issues to consider. The definitive reference is Bruce Hofer's patent, which covers the transformer output of the Audio Precision measurement systems. [4] There is also more information in the *Analog Devices Opamp Applications Handbook*. [5]

# References

[1]  Sowter, Dr G. A. V. "Soft Magnetic Materials for Audio Transformers: History, Production, and Applications" JAES 35(10), October 1987, p. 769

[2]  Bismarck, Otto von Speech Before the Prussian Landtag Budget Committee, 1862 (actually, he said *blood* and iron).

[3]  Finnern, Thomas "Interfacing Electronics and Transformers" AES preprint #2194 77th AES Convention. Hamburg, March 1985

[4]  Hofer, Bruce *Low-Distortion Transformer-Coupled Circuit* US Patent 4,614,9141986

[5]  Jung, Walt (ed.) "Opamp Applications Handbook" Newnes, 2004, Ch 6, pp. 484–491, ISBN 13: 978-0-7506-7844-5

# Power Supply Design

*"We thought, because we had power, we had wisdom."*

Stephen Vincent Benet: Litany for Dictatorships, 1935

## Opamp Supply Rail Voltages

Running opamps at ±17 V rather than ±15 V gives an increase in headroom and dynamic range of 1.1 dB for virtually no cost and with no reliability penalty. This assumes that the opamps concerned have a maximum supply voltage rating of ±18 V, which is the case for the old Texas TL072, the new LM4562, and many other types.

The 5532 is (as usual) in a class of its own. Both the Texas and Fairchild versions of the NE5532 have an absolute maximum power supply voltage rating of ±22 V (though Texas also gives a "recommended supply voltage" of ±15 V), but I have never met any attempt to make use of this capability. The 5532 runs pretty warm on ±17 V when it is simply quiescent, and my view is that running it at any higher voltage is pushing the envelope. Moving from ±17 V rails to ±18 V rails only gives 0.5 dB more headroom, while stretching things to ±20 V would give a further 1.4 dB. Running at the full ±22 V would yield a more significant 2.2 dB improvement over the ±17 V case, but that is just asking for trouble. Anything above ±17 V is also going to cause difficulties if you want to run opamps with maximum supply ratings of ±18 V on the same supply rails.

We will therefore concentrate here on ±17 V supplies. They can be conveniently built with TO-220 regulators; this usually means an output current capability that does not exceed 1.5 amps, but that is plenty for even complicated active crossovers.

An important question is: how low does the noise and ripple on the supply output rails need to be? Opamps in general have very good power supply rejection ratios (PSRR), and some manufacturer's specs are given in Table 22.1.

The PSRR performance is actually rather more complex than the bare figures given in the table imply; PSRR is typically frequency dependent (deteriorating as frequency rises) and different for the +V and −V supply pins. It is however rarely necessary to get involved in this degree of detail. Fortunately even the cheapest IC regulators like the classic 78xx/79xx series have low enough noise and ripple outputs that opamp PSRR performance is rarely an issue.

There is however another point to ponder; if you have a number of electrolytic-sized decoupling capacitors between rail and ground, enough noise and ripple can be coupled into the non-zero ground resistance to degrade the noise floor. Intelligent placing of the decouplers can help—putting them near

**Table 22.1: PSRR specs for common opamps.**

| Opamp type | PSRR minimum dB | PSRR typical dB |
|:---:|:---:|:---:|
| 5532 | 80 | 100 |
| LM4562 | 110 | 120 |
| TL072 | 70 | 100 |

**Table 22.2: Typical additional supply rails for opamp-based systems.**

| Supply voltage | Function |
|:---:|:---:|
| +5 V | Housekeeping microcontroller |
| +9 V | Relays |
| +24 V | LED bar-graph metering systems, discrete audio circuitry, relays |

where the ground and supply rails come onto the PCB means that ripple will go straight back to the power supply without flowing through the ground tracks on the rest of the PCB.

Apart from the opamp supply rails, audio electronics may require additional supplies, as shown in Table 22.2.

It is often convenient to power relays from a +9 V unregulated supply that also feeds the +5 V microcontroller regulator.

## Designing a ±15 V Supply

Making a straightforward ±15 V 1 amp supply for an opamp-based system is very simple, and has been ever since the LM7815/7915 IC regulators were introduced (which was a long time ago). They are robust and inexpensive parts with both overcurrent and over-temperature protection and give low enough output noise for most purposes. We will look quickly at the basic circuit because it brings out a few design points which apply equally to more complex variations on the theme. Figure 22.1 shows the schematic, with typical component values; a centre-tapped transformer, a bridge rectifier, and two reservoir capacitors C1, C2 provide the unregulated rails that feed the IC regulators. The secondary fuses must be of the slow-blow type. The small capacitors C7–C9 across the input to the bridge reduce RF emissions from the rectifier diodes; they are shown as X-cap types not because they have to withstand 230 Vrms but to underline the need for them to be rated to withstand continuous AC stress. The capacitors C3, C4 are to ensure HF stability of the regulators, which like a low AC impedance at their input pins, but these are only required if the reservoir capacitors are not adjacent to the regulators, i.e. more than 10 cm away. C5, C6 are not required for regulator stability with the 78/79 series—they are there simply to reduce the supply output impedance at high audio frequencies.

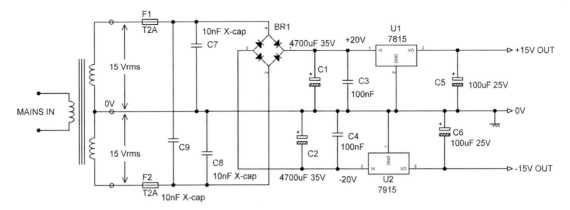

**Figure 22.1: A straightforward ±15 V power supply using IC regulators.**

There are really only two electrical design decisions to be made; the AC voltage of the transformer secondary and the size of the reservoir capacitors. As to the first, you must make sure that the unregulated supply is high enough to prevent the rails dropping out (i.e. letting hum through) when a low mains voltage is encountered, but not so high that either the maximum input voltage of the regulator is exceeded or it suffers excessive heat dissipation. How low a mains voltage it is prudent to cater for depends somewhat on where you think your equipment is going to be used, as some parts of the world are more subject to brown-outs than others. You must consider both the minimum voltage drop across the regulators (typically 2 V) and the ripple amplitude on the reservoirs, as it is in the ripple troughs that the regulator will first "drop out" and let through unpleasantness at 100 Hz.

In general, the RMS value of the transformer secondary will be roughly equal to the DC output voltage.

The size of reservoir capacitor required depends on the amount of current that will be drawn from the supply. The peak-to-peak ripple amplitude is normally in the region of 1 to 2 volts; more ripple than this reduces efficiency, as the unregulated voltage has to be increased to allow for unduly low ripple troughs, and less ripple is usually unnecessary and gives excessive reservoir capacitor size and cost. The amount of ripple can be estimated with adequate accuracy by using Equation 22.1:

$$Vpkpk \; = \; \frac{I \, \Delta t \; 1000}{C} \qquad\qquad 22.1$$

where:

    *Vpk-pk* is the peak-to-peak ripple voltage on the reservoir capacitor.
    *I* is the maximum current drawn from that supply rail in Amps.
    *Δt* is the length of the capacitor discharge time, taken as 7 milliseconds.
    *C* is the size of the reservoir capacitor in microfarads.
    The "1000" factor simply gets the decimal point in the right place.

Note that the discharge time is strictly a rough estimate and assumes that the reservoir is being charged via the bridge for 3 ms and then discharged by the load for 7 ms. Rough estimate it may be, but I have always found it works very well.

The regulators must be given adequate heatsinking. The maximum voltage drop across each regulator (assuming 10% high mains) is multiplied by the maximum output current to get the regulator dissipation in watts, and a heatsink selected with a suitable thermal resistance to ambient (in °C per watt) to ensure that the regulator package temperature does not exceed, say, 90 °C. Remember to include the temperature drop across the thermal washer between regulator and heatsink.

Under some circumstances it is wise to add protective diodes to the regulator circuitry, as shown in Figure 22.2. The diodes D1, D3 across the regulators are reverse-biased in normal operation, but if the power supply is driving a load with a large amount of rail decoupling capacitance, it is possible for the output to remain higher in voltage than the regulator input as the reservoir voltage decays, and this can damage the regulator. D1, D3 prevent this effect from putting a reverse voltage across the regulators.

The shunt protection diodes D2, D4 are once again reverse-biased in normal operation. D2 prevents the +15 V supply rail from being dragged below 0 V if the −15 V rail starts up slightly faster, and likewise D4 protects the −15 V regulator from having its output pulled above 0 V. This can be an important issue if rail-to-rail decoupling such as C9 is in use; such decoupling can be useful because it establishes a low AC impedance across the supply rails without coupling supply rail noise into the ground, as C7, C8 are prone to do. However, it also makes a low-impedance connection between the two regulators. D2, D4 will prevent damage in this case but leave the power supply vulnerable to start-up problems; if its output is being pulled down by the −15 V regulator, the +15 V regulator may refuse to start. This is actually a very dangerous situation, because it is quite easy to come up with a circuit where start-up will only fail one time in 20 or more, the incidence being apparently completely random but presumably controlled by the exact point in the AC mains cycle where the supply is switched on and other variables such as temperature, the residual charge left on the reservoir capacitors, and the

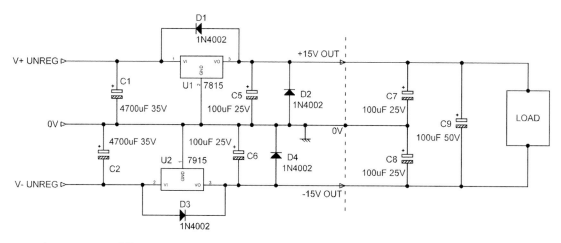

**Figure 22.2: Adding protection diodes to a±15 V power supply. The load has decoupling capacitors both to ground (C7, C8) and between the rails (C9); the latter can cause start-up problems.**

phase of the moon. If even one start-up failure event is overlooked or dismissed as unimportant, then there is likely to be serious grief further down the line. *Every power supply start-up failure must be taken seriously.*

## Designing a ±17 V Supply

There are 15 V IC regulators (7815, 7915), and there are 18 V IC regulators (7818, 7918), but there are no 17 V IC regulators. This problem can be effectively solved by using 15 V regulators and adding 2 volts to their output by manipulating the voltage at the REF pin. The simplest way to do this is with a pair of resistors that divide down the regulated output voltage and apply it to the REF pin, as shown in Figure 22.3a (the transformer and AC input components have been omitted in this and the following diagrams, except where they differ from those shown earlier). Since the regulator maintains 15 V between the OUT and REF pin, with suitable resistor values the actual output with respect to 0 V is 17 V.

The snag with this arrangement is that the quiescent current that flows out of the REF pin to ground is not well controlled; it can vary between 5 and 8 mA, depending on both the input voltage and the device temperature. This means that R1 and R2 have to be fairly low in value so that this variable current does not cause excessive variation of the output voltage, and therefore power is wasted.

If a transistor is added to the circuit as in Figure 22.3b, then the impedance seen by the REF pin is much lower. This means that the values of R1 and R2 can be increased by an order of magnitude, reducing the waste of regulator output current and reducing the heat liberated. This sort of manoeuvre is also very useful if you find that you have a hundred thousand 15 V regulators in store, but what you actually need for the next project is an 18 V regulator, of which you have none.

What about the output ripple with this approach? I have just measured a power supply using the exact circuit of Figure 22.3b, with 2200 uF reservoirs, and I found −79 dBu (87 uVrms) on the +17 V output rail and −74 dBu (155 uVrms) on the 17 V rail, which is satisfyingly low for inexpensive regulators and should be adequate for almost all purposes; note that these figures include regulator noise as well

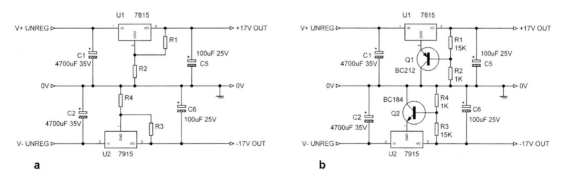

**Figure 22.3: Making a ±17 V power supply with 15 V IC regulators. (a) Using resistors is inefficient and inaccurate; (b) adding transistors to the voltage-determining resistor network makes the output voltage more predictable and reduces the power consumed in the resistors.**

as ripple. The load current was 110 mA. If you *are* plagued by ripple troubles, the usual reason is a rail decoupling capacitor that is belying its name by coupling rail ripple into a sensitive part of the ground system, and the cure is to correct the grounding rather than design an expensive ultra-low ripple PSU. Note that doubling the reservoir capacitance to 4400 uF only improved the figures to −80 dBu and −76 dBu respectively; just increasing reservoir size is not a cost-effective way to reduce the output ripple.

## Using Variable-Voltage Regulators

It is of course also possible to make a ±17 V supply by using variable output voltage IC regulators such as the LM317/337. These maintain a small voltage (usually 1.2 V) between the OUTPUT and ADJ (shown in figures as GND) pins and are used with a resistor divider to set the output voltage. The quiescent current flowing out of the ADJ pin is a couple of orders of magnitude lower than for the 78/79 series, at around 55 uA, and so a simple resistor divider gives adequate accuracy of the output voltage, and transistors are no longer needed to absorb the quiescent current. A disadvantage is that this more sophisticated kind of regulator is somewhat more expensive than the 78/79 series; at the time of writing they cost something like 50% more. The 78/79 series with transistor voltage-setting remains the most cost-effective way to make a non-standard-voltage power supply at the time of writing.

It is clear from Figure 22.4 that the 1.2 V reference voltage between ADJ and out is amplified by many times in the process of making a 17 V or 18 V supply; this not only increases output ripple but also output noise, as the noise from the internal reference is being amplified. The noise and ripple can be considerably reduced by putting a capacitor C7 between the ADJ pin and ground. This makes a dramatic difference; in a test PSU with a 650 mA load, the output noise and ripple were reduced from −63 dBu (worse than the 78xx series) to −86 dBu (better than the 78xx series), and so such a capacitor is usually fitted as standard. If it is fitted, it is then essential to add a protective diode D1 to discharge C7, C8 safely if the output is short-circuited, as shown in Figure 22.5.

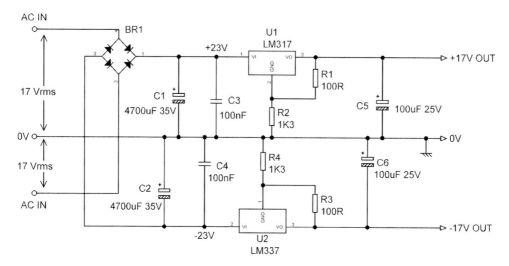

**Figure 22.4: Making a ±17 V power supply with variable-voltage IC regulators.**

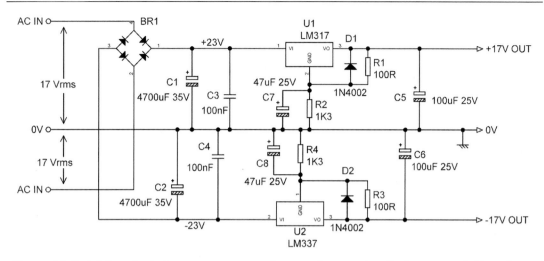

**Figure 22.5: Adding ripple-improvement capacitors and protective diodes to a variable-voltage IC regulator.**

**Table 22.3: Comparing the noise and ripple output of various regulator options.**

|  | 7815 + transistor | LM317 |
|---|---|---|
| No C on LM317 ADJ pin | −73 dBu (all ripple) | −63 dBu (ripple & noise) |
| 47 uF on LM317 ADJ pin | −73 dBu(all ripple) | −86 dBu (ripple & noise) |
| Input filter 2.2 Ω & 2200 uF | −78 dBu (ripple & noise) | −89 dBu (mostly noise) |
| Input filter 2.2 Ω & 4400 uF | −79 dBu (mostly noise) | −90 dBu (all noise) |

The ripple performance of the aforementioned test PSU, with a 6800 uF reservoir capacitor and a 650 mA load, is summarised for both types of regulator in Table 22.3. Note that the exact ripple figures are subject to some variation between regulator specimens.

## Improving Ripple Performance

Table 22.3 shows that the best noise and ripple performance that can be expected from a simple LM317 regulator circuit is about −86 dBu (39 uVrms), and this still contains a substantial ripple component. The reservoir capacitors are already quite large at 4700 uF, so what is to be done if lower-ripple levels are needed? The options are:

1.  Look for a higher-performance IC regulator. They will cost more, and there are likely to be issues with single sourcing.
2.  Design your own high-performance regulator using discrete transistors or opamps. This is not a straightforward business, especially if all the protection that IC regulators have is to be included. There can be distressing issues with HF stability.

3.   Add an RC input filter between the reservoir capacitor and the regulator. This is simple and pretty much bullet-proof and preserves all the protection features of the IC regulator, though the extra components are a bit bulky and not that cheap. There is some loss of efficiency due to the voltage drop across the series resistor; this has to be kept low in value, so the capacitance is correspondingly large.

The lower two rows of Table 22.3 show what happens. In the first case the filter values were 2.2 Ω and 2200 uF. This has a −3 dB frequency of 33 Hz and attenuates the 100 Hz ripple component by 10 dB. This has a fairly dramatic effect on the visible output ripple, but the dB figures do not change that much, as the input filter does not affect the noise generated inside the regulator. Increasing the capacitance to 4400 uF sinks the ripple below the noise level for both types of regulator.

## Dual Supplies From a Single Winding

It is very convenient to use third-party "wall-wart" power supplies for small pieces of equipment, as they come with all the safety and EMC approvals already done for you, though admittedly they do not look appropriate with high-end equipment.

The vast majority of these supplies give a single AC voltage on a two-pole connector, so a little thought is required to derive two supply rails. Figure 22.6 shows how it is done in a ±18 V power supply; note that these voltages are suitable only for a system that uses 5532s throughout. Two voltage-doublers of opposite polarity are used to generate the two unregulated voltages. When the incoming voltage goes negative, D3 conducts and the positive end of C1 takes up approximately 0 V. When the incoming voltage swings positive, D1 conducts instead, and the charge on C1 is transferred to C3. Thus the whole peak-to-peak voltage of the AC supply appears across reservoir capacitor C3. In the same way, the peak-to-peak voltage, but with the opposite polarity, appears across reservoir C4.

Since voltage-doublers use half-wave rectification, they are not suitable for high current supplies. When choosing the value of the reservoir capacitor values, bear in mind that the discharge time in

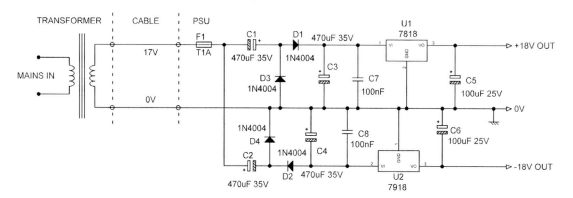

**Figure 22.6: A ±18 V power supply powered by a single transformer winding.**

Equation 22.1 above must be changed from 7 ms to 17 ms. The input capacitors C1, C2 should be the same size as the reservoirs.

## Mutual Shutdown Circuitry

The 5532 opamp is in general a tough item, but it has an awkward quirk. If one supply rail is lost and collapses to 0 V, while the other rail remains at the normal voltage, a 5532 can under some circumstances get into an anomalous mode of operation that draws large supply currents, and it ultimately destroys itself by over-heating. Other opamps may suffer the same problem, but I am not currently aware of any. To prevent damage from this cause opamp supplies should be fitted with a mutual shutdown system. This ensures that if one supply rail collapses because of overcurrent, over-temperature, or any other cause, the other rail will be promptly switched off. A simple way to implement this is shown in Figure 22.7, which also demonstrates how to make a ± supply using only positive regulators. Two separate transformer secondary windings are required; a single centre-tapped secondary cannot be used.

The use of a second positive regulator to produce the negative output rail looks a little strange at first sight, but I can promise you it works. The regulators in this case are the high-current Linear Tech LT1083, which can handle 7.5 amps but is not available in a negative version. That should be more than enough current for even the most complex active crossover . . .

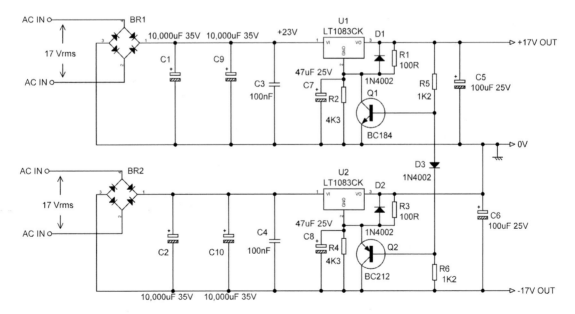

**Figure 22.7: A high-current ±17 V power supply with mutual rail shutdown, and using only positive regulators.**

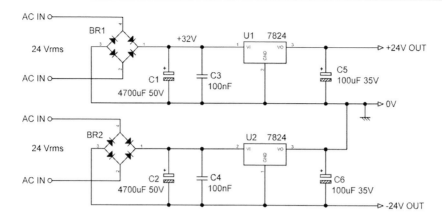

**Figure 22.8: A ±24 V power supply using only positive regulators.**

The extra circuitry to implement mutual shutdown is very simple; R5, D3, R6 and Q1 and Q2. Because R5 is equal to R6, D3 normally sits at around 0 V in normal operation. If the +17 V rail collapses, Q2 is turned on by R6, and the REF pin of U2 is pulled down to the bottom rail, reducing the output to the reference voltage (1.25 V). This is not completely off, but it is low enough to prevent any damage to opamps.

If the −17 V rail collapses, Q1 is turned on by R5, pulling down the REF pin of U1 in the same way. Q1 and Q2 do not operate exactly symmetrically, but it is close enough for our purposes.

Note that this circuit can only be used with variable output voltage regulators, because it relies on their low reference voltages to achieve what is effectively switch-off.

## Power Supplies for Discrete Circuitry

One of the main reasons for using discrete audio electronics is the possibility of handling larger signals than can be coped with by opamps running off ±17 V rails. The use of ±24 V rails allows a 3 dB increase in headroom, which is probably about the minimum that justifies the extra complications of discrete circuitry. A ±24 V supply can be easily implemented with 7824/7924 IC regulators.

A slightly different approach was used in my first published preamplifier design. [1] This preamp in fact used two LM7824 +24 V positive regulators, as shown in Figure 22.8, because at the time the LM7924 −24 V regulator had not yet reached the market. This configuration can be very useful in the sort of situation where you have a hundred thousand positive regulators in the stores but no negative regulators.

Note that this configuration does however require two separate transformer secondary windings; it cannot be used with a single centre-tapped secondary.

## Reference

[1]  Self, D. R. G. "An Advanced Preamplifier Design" Wireless World, November 1976

# CHAPTER 23

# *An Active Crossover Design*

Here I present an active crossover design which will illustrate a good number of the principles and techniques put forward in this book. A design that demonstrated all of them would be a cumbersome beast, so I have aimed instead to show the most important while also providing a practical and adaptable design which will be of use in the real world.

The design is a generic crossover that can be adapted to a wide range of applications by changing its parameters and its configuration. The crossover frequencies can be changed simply by altering component values, and circuit blocks for equalisation or time delay can be added or removed, with due care to preserve absolute phase, of course. Since widely varying drive units may be used, I have made no attempt to specify equalisation or integrate the driver response into the filter operation.

The crossover design is primarily aimed at hi-fi applications rather than sound reinforcement. It does not, for example, in its basic form have variable crossover frequencies or balanced outputs, though instructions are given for adding the latter.

## Design Principles

The aim of this chapter is not just to provide a complete design but to demonstrate the use of various design principles expounded in the body of this book.

- Low-impedance design for low noise          Chapter 17
- Elevated internal levels          Chapter 17
- Further elevated internal level for the HF path          Chapter 17
- Low-noise balanced inputs          Chapter 20
- Optimised filter order for best noise          Chapter 17
- Mixed-capacitor types in filters          Chapter 11
- Use of multiple resistors to improve accuracy          Chapter 15
- Delay compensation using 3rd-order allpass filters   Chapter 13

## Example Crossover Specification

This is the basic specification for an active crossover that we will use as an example.

| | |
|---|---|
| Number of bands: | Three |
| Type | Linkwitz-Riley 4th-order |
| Mid/HF crossover frequency: | 3 kHz |
| LF/Mid crossover frequency: | 400 Hz |

| HF path time-delay compensation | 80 usec, tolerance ±5 % (path length 27 mm) |
| Mid path time-delay compensation | 400 usec, tolerance ±5 % (path length 137 mm) |

(As noted in Chapter 13, the path lengths to be compensated for were carefully chosen to give nice round figures for the delay times required.)

The crossover has balanced inputs, but in its basic form the outputs are unbalanced, the assumption being that if a balanced interconnection is required, this will be provided at the power amplifier inputs. The possibility of adding balanced output stages is examined at the end of the chapter. If the overall audio system can be satisfactorily designed with balanced crossover inputs but unbalanced crossover outputs, as opposed to unbalanced crossover inputs but balanced crossover outputs, then the former is clearly more economical, as for a stereo crossover it requires two balanced input stages rather than six balanced output stages.

Low-impedance design is used throughout; the circuit impedances are designed to be as low as possible without causing extra distortion by overloading opamp outputs. This reduces the effect of opamp current noise and resistor Johnson noise but has no effect on opamp voltage noise.

The design assumes that there is no level control between the crossover and power amplifier, or if there is, it is set to maximum and left there; this allows us to use elevated internal levels in the crossover to improve the noise performance, as described in Chapter 17.

No equalisation stages are included in the signal paths.

The block diagram is shown in Figure 23.1.

## The Gain Structure

In Chapter 17 I explained how with suitable system design the internal levels of an active crossover could be significantly raised to reduce the effect of circuit noise. Here I have decided to go for an internal level of 3 Vrms, 12 dB higher than the assumed input voltage of 775 mVrms (0 dBu). This

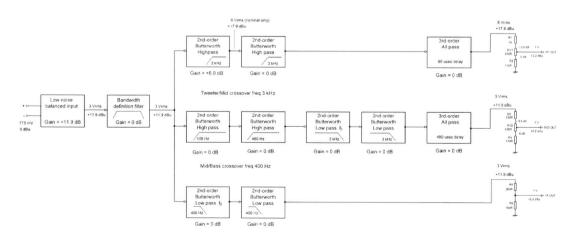

**Figure 23.1: Block diagram of 3-way Linkwitz-Riley 4th-order crossover.**

retains 10 dB of headroom to accommodate maladjusted input levels. We also saw in Chapter 17 that the distribution of amplitude with frequency in music is such that the levels in the top few octaves are much lower than in the rest of the audio spectrum. We decided conservatively that with an upper crossover point around 3 kHz, it was safe to raise the HF channel level by a further 6 dB.

It is therefore necessary to have an input amplifier with a gain of close to four times (+11.9 dB) to raise the internal level to a nominal 3 Vrms as soon as possible, and another +6 dB of amplification will be required as early as possible in the HF path.

The bandwidth definition filter is shown in Figure 23.1 as having unity gain for simplicity at this point. In fact, as we shall shortly see, it has a small loss of 0.29 dB, which has to be recovered elsewhere in the circuitry to keep the levels spot-on.

## Resistor Selection

The basic policy for component selection is that followed throughout this book—to use as few capacitors as possible, which means sticking to the sparse E3 or E6 series, and let the resistor values come out as they may. I decided that no more than two resistors from the E24 series would be used in series or parallel to get as close as possible to the desired value. Where it could be done the resistors would be of near-equal value to get the best possible improvement in accuracy by using multiple resistors, as explained in Chapter 15. An error window for the nominal value is set at ±0.5%, and 1% tolerance resistors are assumed.

Resistors in parallel rather than in series are generally to be preferred, as it makes the PCB layout easier; assuming they are lying side by side, you just have to join the adjacent pads with very short tracks. It could also be argued that the parallel connection makes for better reliability, as a dry solder-joint that develops into an open circuit is less likely to cause oscillation or stop the signal altogether. In some sound-reinforcement circumstances, any sound is better than none at all.

## Capacitor Selection

I decided to use polypropylene capacitors wherever they would have a measurable effect on the performance. They get bulky and expensive quickly as their value increases, and so I further concluded that 220 nF should be the largest value employed. It is assumed that only the E3 series (10, 22, 47) values are available. The number of different capacitor values (not counting small ceramics) has been kept down to four: 2n2, 10 nF, 47 nF and 220 nF. It makes sense to put some effort into this, as they are probably the most expensive electronic components in the crossover, and so you want to gain as much cost advantage from purchasing in quantity as possible. No particular assumptions are made about capacitor tolerance.

## The Balanced Line Input Stage

It was explained in Chapter 20 that the conventional unity-gain balanced input stage made with four 10 kΩ resistors is relatively noisy, especially when compared with an unbalanced input using a simple voltage-follower; the balanced stage noise output is about −104.8 dBu. This problem can be addressed by reducing the value of the 10 kΩ resistors around the balanced stage and driving the balanced stage with 5532 unity-gain buffers to keep the input impedance up. In the unity-gain case this reduces the noise output to −110.2 dBu, a very useful 5.4 dB quieter.

For our crossover design we do not want a unity-gain stage but one that gives a gain of 4 times (+11.9 dB) so that a nominal input of 775 mVrms is raised to 3 Vrms. We use the same strategy of input buffers and a balanced amplifier working at low impedances, but the required gain is obtained by altering the ratio of the resistors R7, R8 to R9, R10 in Figure 23.2. The output noise from this stage with the inputs terminated by 50 Ω resistors is −100.9 dBu, so its equivalent input noise (EIN) is −100.9 − 11.9 = −112.8 dBu. For what it's worth, this is the same as the Johnson noise from a 9.6 kΩ resistor.

This is a rare case where a significant amount of gain is required in a balanced input amplifier, so the instrumentation amplifier configuration might be an interesting alternative, giving a CMRR theoretically improved by 11.9 dB, the gain of the stage.

## The Bandwidth Definition Filter

The bandwidth definition stage is a combined filter consisting of a 3rd-order Butterworth highpass with a cutoff frequency of 20 Hz to remove subsonic signals and protect loudspeakers and a 2nd-order Butterworth lowpass with cutoff at 50 kHz. Unlike the example in Chapter 8, E24 resistor values have been used in the subsonic section to get the cutoff frequency exactly at 20 Hz. The physically large 220 nF capacitors in the filter are susceptible to electrostatic hum pickup, and this must be considered in the physical layout of the crossover.

The bandwidth definition filters are placed as early as possible in the signal path, immediately after the balanced input amplifier, to prevent headroom being eaten up by large subsonic signals. This should also minimise the generation of intermodulation distortion by large ultrasonic signals.

The combined filter has a small midband loss of −0.29 dB when designed with 220 nF capacitors. Redesigning it to use 470 nF capacitors (keeping the capacitors in the lowpass section the same) would reduce this to an even more negligible loss of−0.15 dB, but it's hard to argue that the result is worth the significant extra cost of the capacitors. The only real problem with the 0.29 dB loss is that when tracking a test signal through the HF path, you will encounter 2.90 Vrms at the combined filter output, instead of a nice round 3.0 Vrms. Rather than propagate this inelegance through the rest of the HF path, the signal is restored to 3.0 Vrms in the next stage. This attenuate-then-amplify process inevitably incurs a noise penalty, but in this case it is very small indeed. If an extra stage was used to recover the loss, this would be inelegant and uneconomic, but since the next stage in the HF path is already configured to give gain, there is no cost penalty at all.

## The HF Path: 3 kHz Linkwitz-Riley Highpass Filter

The HF signal path in Figure 23.3 includes two 2nd-order Butterworth filters that make a 4th-order Linkwitz-Riley filter. Both are of the Sallen & Key type. The first filter is configured for a gain of

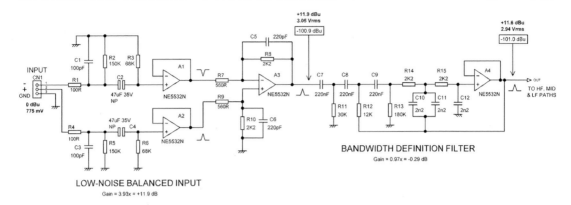

Figure 23.2: Schematic diagram of the input circuitry: the balanced input stage and bandwidth definition filter.

+6.29 dB, to raise the nominal level to make the signal less vulnerable to circuit noise and also to recover that 0.29 dB loss in the combined filter. The nominal level in the block diagram of Figure 23.1 is shown as "6 Vrms", which if it was a real level would be too high, but the relatively low amount of HF energy in musical signals gives a level of more like 3 Vrms in practical use. A level of 6 Vrms will however be obtained if you apply a signal at 775 mVrms to the input, given a suitably high frequency above the MID-HF crossover point.

The design of this stage is straightforward—see Chapter 8 for more information on designing Sallen & Key filters with whatever gain you want. The value of R20 comes very conveniently as the E24 preferred value of 1k3, but the value of the second resistance in the filter comes out as 970.7 $\Omega$, well away from either 910 $\Omega$ or 1 k$\Omega$. R2 is therefore made up of a parallel pair R21 and R22. The optimal method for selecting these was described in Chapter 15. If the two resistors are approximately equal in value, the accuracy of their combined resistance is improved by a factor of $\sqrt{2}$, as the errors tend to cancel. Thus a 1% tolerance becomes a 0.707% tolerance. As the values become more unequal, the improvement decreases until it is effectively 1%, as the tolerance is decided by only one resistor. The procedure is therefore to start with near-equal values and pick a resistor pair to give a result just above the target value, and then pick a resistor pair to give a result just below. If neither result is within the desired error window, one of the resistors is incremented, and we try again. As this algorithm proceeds, the resistor values become more and more unequal and the improvement in tolerance diminishes, so the sooner we find a satisfactory answer the better. In some respects you have to balance the improvement in tolerance against the accuracy of the combined resistance value obtained. The process is mechanical and tedious and would be best implemented in some procedural programming language, but once set up on a spreadsheet, it is fairly quick to do.

In this case our target value is 970.7 $\Omega$ and the error window is ±0.5%. Our progress is set out in Table 23.1; as Step 1 we start with 1800 $\Omega$ in parallel with 2000 $\Omega$, the combined resistance of which is 2.4% low. The next value above 2000 $\Omega$ is 2200 $\Omega$, which with 1800 $\Omega$ gives a combined resistance 1.99% high. Our target value is therefore irritatingly about halfway between these two resistor combinations. In each step we find the two resistor values that "bracket" the target: in other words one value gives a combined resistance that is too low and the other a combined resistance that is too high.

Step 2: We reduce the 1800 Ω resistor to 1600 Ω and try again. Putting 2400 Ω in parallel gives a combined resistance that is only 1.1% low, our best shot so far, but nowhere near good enough. Putting 2700 Ω in parallel gives a result 3.5% high. Are we downhearted? Yes, but faint heart never built fair crossover, so we will persist.

Step 3: We reduce the 1600 Ω resistor to 1500 Ω and find that 2700 Ω in parallel is only 0.66% low, which is temptingly close to 0.5%, but there is no point in setting error windows if you're not going to stick to them. We proceed, but with the nagging awareness that each step is reducing the accuracy improvement we will get from using multiple resistors.

Step 4: Reduce the 1500 Ω resistor to 1300 Ω and with 3600 Ω in parallel the result is 1.61% low. However, with 3900 Ω in parallel, bingo! The combined value is only 0.44%, inside our ±0.5% error window, but not, it must be said, a very long way inside it.

Our task is done. Or is it? There is always the nagging doubt that you might get a much better result if you go just a step or two further. In this case that doubt is very much justified.

Step 5: Reduce the 1300 Ω resistor to 1200 Ω, and with 4700 Ω in parallel the result is 1.52% low. But with 5100 Ω in parallel—cracked it! The result is only 0.08% high and more than good enough, considering the tolerance of the components. The accuracy improvement is much impaired, as one resistor is more than four times the other, but the 1300 Ω and 3900 Ω resistor combination in Step 4 is not much better in that respect.

We therefore select R21 as 1200 Ω and R22 as 5100 Ω. I realise that this process sounds a bit long-winded when every step is described, but a spreadsheet version is quick and convenient. I should warn you at this point that the Goalseek function in Excel apparently can't cope with the mathematical expression for the value of parallel resistors and tends to unhelpfully offer "solutions" up in the peta-ohm regions. However, this doesn't slow things down much; just don't use it.

**Table 23.1: Selecting the best parallel combination of R21 and R22 to get the target value of 970.7 Ω.**

| Step | R21 Ohms | R22 Ohms | Combined Ohms | Error % |
|------|----------|----------|---------------|---------|
| 1 | 1800 | 2000 | 947.368 | −2.40% |
|   | 1800 | 2200 | 990.000 | 1.99% |
| 2 | 1600 | 2400 | 960.000 | −1.10% |
|   | 1600 | 2700 | 1004.651 | 3.50% |
| 3 | 1500 | 2700 | 964.286 | −0.66% |
|   | 1500 | 3000 | 1000.000 | 3.02% |
| 4 | 1300 | 3600 | 955.102 | −1.61% |
|   | 1300 | 3900 | 975.000 | 0.44% |
| 5 | 1200 | 4700 | 955.932 | −1.52% |
|   | 1200 | 5100 | 971.429 | 0.08% |

The next step is to check by calculation or simulation (not measurement because the tolerances of the real components used will confuse things) that the filter cutoff frequency falls within the required error window.

In The Case of the Second Filter (Quick, Watson, the game's afoot!) we are much luckier. Using the process just described, we find that taking the value of the first resistor as 800 Ω gives an error of −0.23% in the filter cutoff frequency, which is less than the resistor tolerance (even after the full √2 accuracy improvement, which we get in this case) and probably *much* less than the capacitor tolerances. Thus R26 and R27 are 1k6, and R28 is also 1k6. It happens to work out very nicely, though in terms of component count we only save one resistor.

In each filter, only the first capacitor (C20, C22) is a polypropylene type, while the second (C21, C23) is polyester. Only one capacitor needs to be polypropylene to avoid capacitor distortion, but it must be the first one in the filter. This intriguing state of things is described more fully in Chapter 11.

## The HF Path: Time-Delay Compensation

When we looked at the question of time-delay compensation in Chapter 13, we noted that mercifully it is not necessary to maintain an absolutely accurate delay over the whole audio spectrum. Instead the delay only has to be constant over each crossover region. The HF delay needs only to be maintained around the 3 kHz crossover point for so long as both drive units are radiating significantly, and likewise the MID delay needs only to be constant around the 400 Hz crossover. One of the many advantages of the Linkwitz-Riley crossover configuration is that the slopes are steep at 24 dB/octave, and so these regions of overlap are relatively narrow, simplifying the problem.

When allpass filters are used as delay elements they give a constant group delay at low frequencies, but it begins to fall off at high frequencies. This makes the design of the HF delay more complicated than that of the MID delay. A 1st-order allpass filter designed for a delay of 80 usec unfortunately starts to roll-off quite early; as frequency rises the delay is down by 10% at 2.4 kHz, before we even reach the 3 kHz crossover point, and sinks to 50% at 9.3 kHz, slowly approaching zero above 100 kHz. A 1st-order allpass filter clearly won't do the job.

One solution is to cascade three 1st-order filters in series. The delay is now spread out over three sections, with each one set to a 80 usec/3 = 26.7 usec delay. The total 80 usec delay is now sustained up to three times the frequency, being 10% down at 7.5 kHz and not down 50% until 28 kHz, well outside the audio spectrum. So is this good enough? The MID and HF drivers will be contributing equally at 3 kHz, both of them being 6 dB down. While 7.5 kHz is only 1.3 octaves away from the 3 kHz MID-HF crossover frequency, the high slope of the Linkwitz-Riley crossover means that the signal to the MID drive unit will be 32 dB down, though its acoustical contribution is less certain, as it depends on the driver frequency response.

Another solution would be a 2nd-order allpass filter; the delay of an 80 usec version falls by 10% at 4.79 kHz, better than the 1st-order filter (10% down at 2.5 kHz) but worse than the triple 1st-order filter (10% down at 7.2 kHz). The 2nd-order filter as described in Chapter 13 also has the disadvantage of phase-inverting in the low-frequency range where its delay is constant. It is also 3.2 dB noisier than the 3rd-order solution we are about to look at. All in all, a 2nd-order allpass filter does not look promising.

The last delay filter examined in detail in Chapter 13 is the 3rd-order allpass filter, which is made up of a 2nd-order allpass cascaded with a 1st-order allpass. When designed for an 80 usec delay, it is 10% down at 12.7 kHz, giving almost twice the flat-delay frequency range of the three cascaded 1st-order filters, at a fractionally lower cost, as it actually uses one less resistor. It uses the same number of potentially expensive capacitors. The −10% point for the delay is now 2.1 octaves above our 3 kHz crossover frequency, and the signal sent to the MID driver will be down by 50 dB, so whatever the response of the driver itself we can be pretty sure that the fall-off in time delay will have no audible consequences. When designed for 80 usec, it is 3.2 dB quieter than the equivalent 2nd-order filter, and it does not inconveniently phase-invert in the low-frequency range. The 3rd-order filter is clearly the better solution, and so it is chosen for the HF path delay.

The 80 usec 3rd-order allpass filter is studied in detail in Chapter 13, with full consideration of its quite subtle noise and distortion characteristics, so I will not repeat that here. Suffice it to say that the circuit of Figure 13.22 is cut and pasted into our schematic in Figure 23.3. In the MFB filter, both capacitors are polypropylene, though the mixed-capacitor phenomenon applies to MFB filters as well as Sallen & Key filters, as described in Chapter 11.

The parallel resistor combinations for the awkward resistance values in the HF allpass filter are given in Table 23.2. We will get a good improvement in precision with R31 and R32, as the values are near equal, not much for R29 and R30, and virtually none for R37 and R38, where our luck was *right* out.

The schematic of the complete HF path is shown in Figure 23.3. Note that the component numbering starts at 20 for resistors and capacitors and 10 for opamps; this is to allow additions to other parts of the schematic without renumbering everything. The signal and noise levels are given for the output of each stage.

The allpass stage can of course be omitted if time-delay compensation is done by other means, such as the physical cabinet construction.

## The MID Path: Topology

The MID path contains a 400 Hz 4th-order Linkwitz-Riley highpass filter and a 3 kHz 4th-order Linkwitz-Riley lowpass filter. In this situation the order of the filters in the signal path needs to be considered. In Chapter 17 I demonstrated that a noise advantage can be gained by putting the lowpass filter second in the path, as it removes some of the noise from the previous highpass filter. With the filter frequencies we are using here, the noise advantage is 1.1 dB; not an enormous amount, but achieved at no cost whatsoever. Let's do it.

**Table 23.2: The best parallel combinations for non-preferred resistance values in the HF allpass filter.**

| Resistor | Target Ohms | Ra Ohms | Rb Ohms | Combined Ohms | Error % |
|---|---|---|---|---|---|
| R29 & R30 | 1126 | 1800 | 3000 | 1125.0 | −0.09% |
| R31 & R32 | 2251 | 4700 | 4300 | 2245.6 | −0.24% |
| R37 & R38 | 1746 | 1800 | 56000 | 1743.9 | −0.12% |

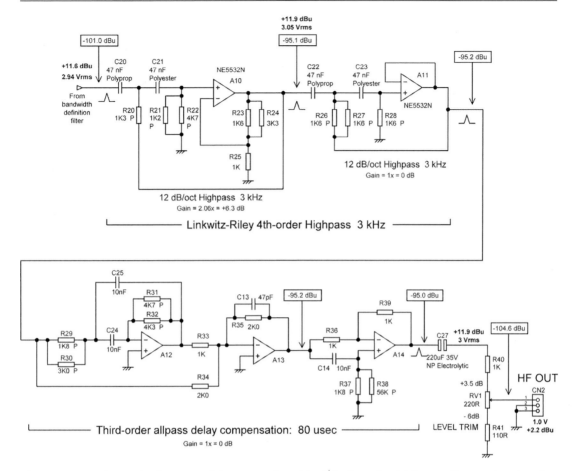

**Figure 23.3: Schematic diagram of the HF path, with Linkwitz-Riley highpass filter and 3rd-order time-delay compensator. Noise levels shown at each stage.**

You will recall that in the HF path the first filter was configured to give +6.3 dB of gain, recovering the 0.3 dB loss in the bandwidth definition filter. While it would be possible to configure the first MID filter to give +0.3 dB of gain (we do not want the +6.0 dB in this path), it does not seem worth the extra complication as +0.3 dB represents only a very small change in the position of the output level control.

## The MID Path: 400 Hz Linkwitz-Riley Highpass Filter

The MID path begins with two 400 Hz 2nd-order Butterworth highpass filters that make a 4th-order Linkwitz-Riley highpass filter. Both are of the unity-gain Sallen & Key type, since there is no need for gain, and in fact they are identical.

We might have been lucky with the resistor values of the second filter in the HF path, but things pan out differently here. The target value for the exact 400 Hz cutoff frequency is 1278.9 $\Omega$. That is temptingly close to 1300 Ω, but shoving that value in alone gives a cutoff frequency error of −1.6%; clearly we could simply add quite a high value in parallel to tweak the resistance just a little, but there will be effectively no improvement in accuracy. Perhaps there are two more nearly equal values that will give the combined value we want? Regrettably not. Playing the combinations, we find that the first pair of values that puts the cutoff frequency within the ±0.5%. error window is R50 = 1300 Ω in parallel with R51 = 82 k Ω, so that's what we have to use. The error in the cutoff frequency is only −0.07%.

The second resistance in the filter has a target value of 1278.9 Ω times two, which is 2557.7 Ω. Our first attempt at a parallel combination is R52 = 5100 Ω in parallel with R53 = 5100 Ω, which gives 2550 Ω, which is only 0.30% low. Looks like there's no need to need to look any further. And yet there is the nagging doubt that there might be a better solution . . . and there is. The very next step is 5600 Ω in parallel with 4700 Ω, giving 2555.3 Ω, which is only 0.09% low and still gives us almost all the accuracy improvement possible. In this case there is not much point in looking further yet. The same resistor values are used in both filters.

As for the filters in the HF path, only the first capacitors (C50, C52) need to be polypropylene for low distortion; the second capacitors (C51, C53) can be polyester.

## The MID Path: 3 kHz Linkwitz-Riley Lowpass Filter

After the 400 Hz highpass filter come the two 2nd-order Butterworth lowpass filters that make up a 4th-order Linkwitz-Riley lowpass filter. Both are of the unity-gain Sallen & Key type. Using 47 nF capacitors, the target value for the two resistors is 798.15 Ω. We can see at once that two 1600 Ω resistors in parallel is going to be very close, and the combined resistance of 800 Ω is in fact only 0.23% low. We also get the maximum possible improvement in accuracy, so we look no further.

## The MID Path: Time-Delay Compensation

The time delay required in the MID path is five times longer at 400 usec, but on the other hand it covers the 400 Hz crossover point, so the delay does not need to be constant up to such a high frequency. Let's see if that makes thing any easier.

Since a 1st-order allpass filter designed for a delay of 80 usec has the delay down by 10% at 2.4 kHz, we would expect the same filter designed for a 400 usec delay to be down by 10% at 480 Hz, and simulation confirms that this is so; the relationship is just simple proportion. Because 480 Hz is much too close to the 400 Hz crossover point, once more we are going to have to look at a more sophisticated solution. We will aim for 10% down at two octaves above crossover, in other words at 1.6 kHz, because this gave pretty convincing results in the HF path.

Looking at the options, we can say that: Three 1st-order filters in series will be 10% down at 7.5 kHz/5 = 1.5 kHz. This is 1.9 octaves away from crossover and is very close to our target, but this configuration uses a relatively large number of components. Four cascaded 1st-order filters would certainly do the job, but the component count is looking excessive compared with the other options, because four capacitors and four opamp sections are required, while the 3rd-order allpass filter only uses three of each.

A 2nd-order allpass filter designed for 400 usec will be down 10% at 4.79 kHz/5 = 958 Hz. This is not even close to our two-octave target, and, as we have seen, the 2nd-order allpass filter has performance issues compared with the 3rd-order, and an unwanted phase-inversion. We can rule this approach out.

A 3rd-order allpass filter designed for 400 usec will be 10% down at 12.7 kHz/5 = 2.54 kHz, and simulation confirms this figure is correct. It is 2.7 octaves away from crossover and gives us a healthy safety margin—in fact it looks almost a bit too healthy, as all the evidence is that a two-octave spacing is ample. However, it is the only alternative that meets our requirements, and there is no obvious way to save a few parts by cutting the spacing down a bit. It is therefore once more selected for our design.

The 400 usec 3rd-order allpass filter is very similar to that of the 80 usec version in the HF path, the main difference being that the capacitors have increased from 10 nF to 100 nF and the resistor values adjusted accordingly to get the desired delay time. The schematic of the complete MID path is shown in Figure 23.4. Note that the component numbering starts at 50 for resistors and capacitors and 20 for opamps; this is to allow additions to other parts of the schematic without renumbering everything. The signal and noise levels are given for the output of each stage.

The parallel resistor combinations for the three non-preferred resistance values in the MID allpass filter are given in Table 23.3. We will get a reasonable improvement in precision with R67 and R68, as the values are near equal, not much for R75 and R76, and virtually none for R69 and R70.

The allpass stage can be omitted if time-delay compensation is done by alternative means, such as the physical construction of the enclosure.

## The LF Path: 400 Hz Linkwitz-Riley Lowpass Filter

The LF path is the simplest of the three paths in the crossover. It consists of two 400 Hz 2nd-order Butterworth lowpass filters that make a 4th-order Linkwitz-Riley lowpass filter. Both are of the unity-gain Sallen & Key type. The target resistance value for the two resistors is 1278.9 $\Omega$, which not surprisingly is the same value as the first resistor in the MID highpass filter, as both use a capacitance value of 220 nF. As before, the first pair of values that puts the cutoff frequency within the ±0.5%. error window is 1300 $\Omega$ in parallel with 82 k$\Omega$. The error in the cutoff frequency is only −0.07%.

### Table 23.3: The best parallel combinations for non-preferred resistance values in the HF allpass filter.

| Resistor | Target Ohms | Ra Ohms | Rb Ohms | Combined Ohms | Error % |
|----------|-------------|---------|---------|---------------|---------|
| **R67, R68** | 1189 | 1600 | 4700 | 1193.7 | 0.39% |
| **R69, R70** | 2378 | 2700 | 20000 | 2378.9 | 0.04% |
| **R75, R76** | 1868 | 3300 | 4300 | 1867.1 | −0.05% |

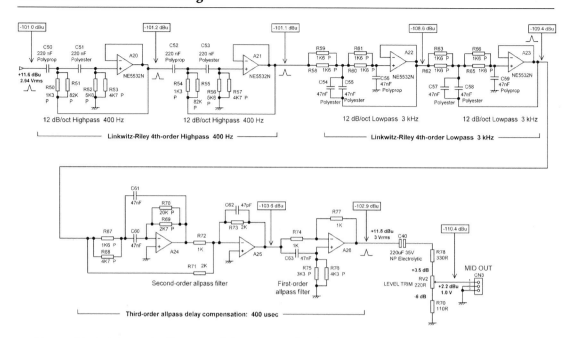

Figure 23.4: Schematic diagram of the MID path, with Linkwitz-Riley highpass and lowpass filters, followed by the 3rd-order time-delay compensator. Noise levels shown at each stage.

## The LF Path: No Time-Delay Compensation

No time-delay compensation is required in the LF path because the physical distance between the acoustic centres of the LF and MID drive units is compensated for by the delay in the MID path. In fact, if the delay compensation was in the LF path, it would have to have a negative delay; in other words the output would emerge before the input arrived. Such circuits, though they would be extremely useful for predicting lottery numbers if you wired enough in series, are notoriously hard to design.

The schematic of the complete LF path is shown in Figure 23.5. The component numbering starts at 80 for resistors and capacitors and 40 for opamps to allow additions to other parts of the schematic without renumbering. The signal and noise levels are given for the output of each stage.

## Output Attenuators and Level Trim Controls

The output level controls used here have a deliberately limited range because they are not intended to be used as volume controls. The range about the nominal 1 Vrms output is +3.5 dB to −6.0 dB, and this should be more than enough to allow for power amplifier gain tolerances (assuming a nominally identical set of power amplifiers) or drive unit sensitivity variations. If, however, the crossover is intended to work with a wide range of power amplifiers of varying sensitivities, the range may need to be extended by reducing the end-stop resistors at the top and bottom of the preset control.

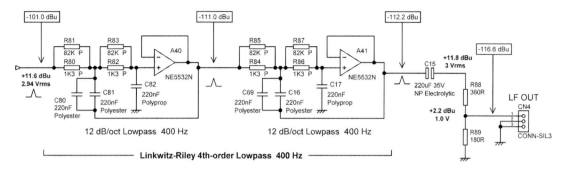

**Figure 23.5: Schematic diagram of the LF path, which consists simply of a 400 Hz Linkwitz-Riley lowpass filter and a fixed output attenuator.**

The values chosen give the nominal output with the preset wiper central in its track, so the system can be lined up pretty well just by eye; this won't of course work with multi-turn preset pots. The exactness of this unfortunately depends on the track resistance of the preset in relation to the fixed end-stop resistors above and below it, and the tolerance on these components is unlikely to be better than 10%, and it may be 20%. These tolerances are horribly wide compared with the 1% of fixed resistors. If really precise levels are required, the output will have to be measured during the trimming operation.

The output networks are configured with the lowest possible resistances that will not load the opamp upstream excessively, to keep the final output impedance low enough to drive a reasonable amount of cable without HF losses. The assumption in this design is that the power amplifiers will not be too far away, and their inputs will be driven directly. Putting a unity-gain buffer after the level trim network would allow a lower output impedance, especially if the "zero-impedance" type of buffer is used, but this will compromise the noise performance.

An alternative approach would be to reduce the resistor values in the output networks so the maximal output impedance is as low as desired and then drive this with a suitable number of 5532 unity-gain buffers connected in parallel via 10 Ω current-sharing resistors. The impairment of the noise performance due to another amplifier stage will in this case be negligible, especially since the noise contributions of the added buffers will partially cancel as they are uncorrelated. You might however need to keep an eye on the power dissipation in the output network resistors, especially the preset.

An unbalanced output will give better common-mode rejection if it is configured as impedance-balanced, by using a three-pin output connector with the cold (−) pin connected to ground through an impedance that approximates as closely as possible to the output impedance driving the hot (+) output pin. This improves the balance of the interconnection, as fully described in Chapter 20. In our circumstances here the output impedance of the level trim network varies as the wiper of the preset is moved from one end of its travel to the other, and so the impedance from the cold pin to ground is inevitably a compromise.

In the case of the HF path, the output impedance varies from 248 Ω with the trimmer at maximum and 101 Ω with the trimmer at minimum. The middle setting, which gives the nominal output level, gives an output impedance of 184 Ω. An impedance-balance resistor of 180 Ω will give negligible common-mode error.

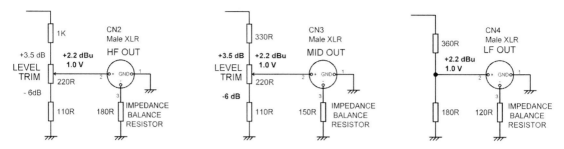

**Figure 23.6: Reconfiguring the output networks for impedance-balanced operation with power amplifiers fitted with balanced inputs.**

The MID path output impedance varies from 165 Ω at maximum to 92 Ω at minimum, with a middle value of 147 Ω, so an impedance-balance resistor of 150 Ω will give a very small error.

The LF path has no output level trim, and its output impedance is fixed at 120 Ω, so here the impedance-balancing can be made exact. The impedance-balanced outputs for the three paths are shown in Figure 23.6.

## Balanced Outputs

In the domestic applications for which this crossover is intended, it is doubtful if any real advantages will be gained by adding true balanced outputs. If placed after the output attenuator and trim networks, as in Figure 23.7a, they will compromise the noise performance, while if placed *before* the output networks, as in Figure 23.7b, a dual-gang trim-pot is required; good luck with sourcing that. A solution to this dilemma is to separate the gain trim and output attenuation functions, as shown in Figure 23.7c.

If the prospective power amplifiers have unbalanced inputs, consideration should be given to the possibility of using ground-cancelling outputs, as described in Chapter 21.

## Crossover Programming

Many of the resistors in the schematics in this chapter have the letter "P" next to their resistance value. These are the components that need to be altered to change the characteristics of the crossover. The idea is that these resistor positions on a prototype PCB can be fitted with single-way turned-pin sockets like those that ICs are plugged into. It is then possible to very easily plug resistors in and out during the development of the complete crossover-loudspeaker system. This is much quicker than pulling the PCB out of the case, desoldering one resistor and then re-soldering another one in, and PCBs will only take so much of this before the pads are damaged. When the design is finalised, the PCBs will have resistors soldered into the same positions in the usual way.

Single-pin sockets can be obtained by cutting up a strip of sockets. Make sure you use the kind of socket strip intended for this sort of thing—deconstructing IC sockets does not usually work well, as the plastic is more brittle and tends to shatter.

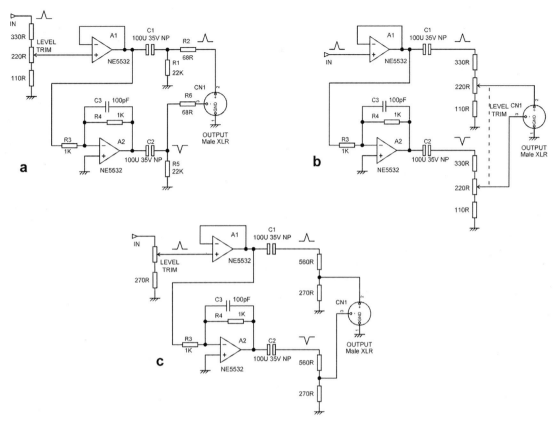

**Figure 23.7: Options for balanced outputs for crossovers with elevated internal levels and output attenuators.**

## Noise Analysis: Input Circuitry

The schematics of the input circuitry, the HF path, the MID path, and the LF path all have the measured noise levels at the output of each stage indicated by rectangular boxes with arrows. This information is essential for performing a noise analysis, in which the noise contribution of each stage is assessed to see if it is what we expect and how it relates the noise generated by the other stages in the path we are examining. Because of the way that uncorrelated noise sources add in a RMS fashion, the largest noise source tends to dominate the final result—the noise at the end of the path—and so we want to identify that source and see if it is worthwhile to make it quieter and so improve the overall noise performance.

As throughout this book, all the noise measurements given are unweighted and in a 22 kHz bandwidth.

The measured noise output from the balanced input stage with its +11.9 dB of gain is −100.9 dBu. When we terminate both input pins with 50 Ω to ground, the Johnson noise from each resistor is −135.2 dBu (22 kHz bandwidth at 25°C). The noise from both together is 3 dB more, not 6 dB, because the two noise sources are uncorrelated. The noise voltage at the input is thus a very very low −132.2 dBu. When amplified by +11.9 dBu this becomes −120.3 dBu, which is almost 20 dB below the stage output noise and is therefore making a negligible contribution to the total. Despite its special low-noise design, virtually all the output noise is generated by the balanced input amplifier.

The equivalent input noise (EIN) of the balanced input stage is −100.9 dBu − 11.9 dB = −112.8 dBu, giving us a noise figure of 22.4 dB with reference to 50 Ω. This figure would usually be considered pretty poor, but as is discussed in Chapter 20 on line inputs, noise figures are not a very useful figure of merit for balanced inputs. What is unquestionably true is that −100.9 dBu is a very low value of noise to find in audio circuitry, so we seem to have made a good start.

The noise level at the output of the bandwidth definition filter might be expected to be a bit higher, because of its noise contribution, but it is actually a tad lower, at −101.0 dBu. This is because of the lowpass action of the filter. If the 22 kHz measurement filter was a brickwall job this would not happen, but it has in fact only a 18 dB/octave roll-off, so the bandwidth definition filtering above 22 kHz does have enough effect to outweigh the noise contribution of the stage.

At this point we cannot say that either stage is in urgent need of improvement, though we note that effectively all the output noise from the input circuitry is generated internally by the balanced input amplifier.

## Noise Analysis: HF Path

The measured noise output from the input circuitry and the first 3 kHz highpass filter is −95.1 dBu. This filter has a gain of +6.3 dB, and the noise level at its input is −101.0 dBu, so if the stage was noiseless we would expect a noise output of −101.0 dBu + 6.3 dB = −94.6 dBu; in fact the measured noise output is *less* than this, at −95.1 dBu. This is of course because we are dealing with a 3 kHz highpass filter which is rejecting a substantial chunk of the audio spectrum. Calculating the noise contribution from this stage is therefore quite complicated. For the moment we will simply note that this stage does not appear to be excessively noisy.

The measured noise output from the second 3 kHz highpass filter is −95.2 dBu. This filter has a gain of 0 dB, but again the output noise level is fractionally less than the input noise, due to the filtering action.

The final stage in the HF path is the 80 usec 3rd-order allpass filter time-delay compensator, which we might expect to be relatively noisy compared with the filters because of its greater complexity. However, the noise at its output is only 0.2 dB higher than at its input, measuring −95.0 dBu. If we subtract the input noise from the output noise we get an estimate for the stage contribution of −108.5 dBu; unfortunately, we are subtracting two figures with only a small difference between them. That difference is fairly near the limit of measurement, and so the result will be very inaccurate. If you want to know the noise from a stage, then the correct method is to measure the noise output of the stage by itself. We did this for the 3rd-order allpass filter back in Chapter 13, and the actual output noise was −102.8 dBu, which shows you how dangerous it is to use the subtraction method inappropriately. If we add the internal noise of −102.8 dBu to the input noise of −95.2 dBu, in theory we get an output noise of −94.5 dBu, an increase of 0.7 dB (the small discrepancy is probably due to minor frequency response irregularities in the allpass filter caused by capacitor tolerances). We make a note that the noise contribution of the 3rd-order allpass filter is small but not negligible. More importantly, we can conclude that the noise performance of the HF path is dominated by the noise being fed to it from the input circuitry.

The final part of the HF path is the output attenuator and level trim network, which at its nominal setting has a 15.6 dB loss to undo the doubly-elevated level in the HF path. Since it is just a resistive attenuator, I expected no problems here. But . . . the noise level measured at the network output was −105.1 dBu, when it should have been −95.0 dBu −15.6 dBu =−110.6 dBu. A 5 dB discrepancy cannot be ignored; investigation showed that the problem was the output impedance of the network, which at 184 Ω is somewhat higher than usual. The Audio Precision SYS-2702 is designed to be fed from low-impedance sources, and one might conclude that its input amplifiers have relatively high current noise as a consequence of attaining very low voltage noise, but Bruce Hofer assures me this is not so. [1] Nevertheless, I found that driving the AP input via a 5532 unity-gain buffer and the usual 47 Ω cable-isolating resistor, to reduce the source impedance, gave a lower output noise reading of −109.4 dBu, despite the extra amplifier in the path. This minor puzzle remains unresolved at present.

In any event, I hope can all agree that is −109.4 dBu is pretty quiet.

## Noise Analysis: MID Path

The measured noise output from the first unity-gain 400 Hz highpass filter is −101.2 dBu. This is again less than the input noise because of the filtering action. The measured noise output from the second unity-gain 400 Hz highpass filter is −101.1 dBu, a tiny increase at the limit of measurement. We conclude that neither filter is making a significant contribution to the noise in the MID path.

After the two highpass filters come the two 3 kHz lowpass filters, fed with −101.1 dBu of noise from the second 400 Hz highpass filter. The noise output from the first lowpass filter is −108.6 dBu, and from the second lowpass filter −109.4 dBu. Both these figures are much lower than the input noise of −101.1 dBu as a consequence of the lowpass filtering, which cuts the bandwidth from 22 kHz to 3 kHz.

The final stage in the MID path, as in the HF path, is a 3rd-order allpass filter time-delay compensator, this time designed for a 400 usec delay. We saw when we looked at the HF allpass filter in isolation that it was relatively noisy, with a measured output noise of −102.8 dBu; in this case the different circuit values for 400 usec give us a slightly lower allpass-filter-only noise output of −104.1 dBu (the 2nd-order allpass filter that makes up the first section of the complete 3rd-order filter has a noise output of −104.9 dBu, so that is clearly where most of the noise comes from). When the 3rd-order allpass filter is placed in the MID path, the measured noise output is −102.9 dBu. Since this is more than 6 dB greater than the −109.4 dBu noise level going in, it is clear that by far the greater part of the noise is internally generated by the 3rd-order allpass filter. This is because the nominal level in the MID path is 6 dB lower than that in the HF path. We make a note that the noise contribution of the 3rd-order allpass filter is dominant and might repay some attention.

The final part of the MID path is the output attenuator and level trim network, which at its nominal setting has a 9.5 dB loss to undo the elevated level in the MID path. The noise level measured directly at its output was initially −107.9 dBu, but as we saw at the end of the HF path, that reading is exaggerated by the AP input current noise. Driving the AP input via a 5532 unity-gain buffer and 47 Ω cable-isolating resistor reduced the output noise reading to −110.4 dBu. That too, is rather quiet; the noise level is lower than that at the HF path output because of the presence of the 3 kHz lowpass filters.

## Noise Analysis: LF Path

This path consists only of the two 400 Hz lowpass filters and the fixed output attenuator. Given the 400 Hz cutoff frequency of the lowpass filters we expect a good noise performance, and we get it. The noise at the input of the LF path is the −101.0 dBu coming from the input circuitry. This is reduced to −111.0 dBu at the output of the first lowpass filter, and further to −112.2 dBu at the output of the second lowpass filter.

The final part of the LF path is the fixed output attenuator network, which has a 9.5 dB loss to undo the elevated level in the MID path. This in theory gives an output noise of −121.7 dBu. This is going to be hard to measure, as it is below the noise floor of the AP measuring equipment, which is, on my example, −119.6 dBu with the input short-circuited. Once more we have to drive the AP input via a 5532 unity-gain buffer and 47 Ω cable-isolating resistor to avoid misleadingly high readings due to current noise, and that 5532 will add its own voltage noise. The reading we get is −113.8 dBu. This is reduced to −115.1 dBu after we subtract the known AP noise floor. We then calculate the noise added by the 5532 buffer, as it is too low to measure accurately, using its typical input noise density of 5 nV/√Hz and a 22 kHz bandwidth (the effect of the 5532 current noise in the attenuator output impedance is negligible). The answer is that the 5532 buffer contributes −120.38 dBu. If we subtract that from our figure of −115.1 dBu, we get a stunningly low −116.6 dBu. This may not be the most accurate reading in the history of audio, but it is good evidence that everything is working properly, and we have a *very* low noise output from the LF path.

## Improving the Noise Performance: The MID Path

On our journey down the MID path, we noted that the 3rd-order allpass filter time-delay compensator was generating internally most of the noise that was measured at its output. It is the dominant noise generator in the MID path, so let's see if we can do something about that.

There is a very simple fix, which you may have already seen coming. The 3rd-order allpass filter is at the end of the MID path, so all its noise heads for the output unmolested. If, however, it is moved so it is after the 400 Hz highpass filters but before the 3 kHz lowpass filters, the latter will have a dramatic effect on the noise level. Making the change, the noise output of the 3rd-order allpass filter is now −99.6 dBu as it adds its own noise to the −101.1 dBu reaching it from the highpass filters upstream. The noise output of the first lowpass filter is pleasingly lower at −107.3 dBu, and the noise output of the second lowpass filter is even lower at −108.2 dBu.

After the output attenuator network, the noise at the output is reduced from −110.4 dBu to −113.3 dBu, an improvement of 2.9 dB that costs us nothing more than a moment's thought.

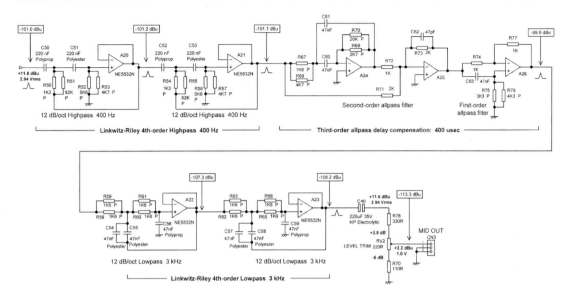

**Figure 23.8: Schematic diagram of the improved MID path, with the 3rd-order time-delay compensator moved to before the Linkwitz-Riley lowpass filter to reduce noise.**

The true noise output must be −108.2 −9.5 dBu = −117.7 dBu, because of the effect of the 9.6 dB output attenuator, but we have already described the difficulties of measuring noise at such low levels. The revised schematic is shown in Figure 23.8.

In case you're wondering why the MID path wasn't configured in this way from the start, the answer is that until I built the prototype circuit, I didn't know how the noise contributions of the stages was going to work out. It could have been predicted by a lot of calculation, but that gets complicated when you're dealing with stages like filters that have a non-flat-frequency response. Sometimes it's just quicker to get out the prototype board and start plugging bits in. You're going to be doing it sooner or later.

## Improving the Noise Performance: The Input Circuitry

The other major point we noted in our noise analysis of the crossover was that the noise performance of the HF path was dominated by the noise being fed to it from the input circuitry, the incoming level being −101.0 dBu. This is almost all coming from the balanced input amplifier, the contribution of the bandwidth definition filter being very small. Therefore the only way to improve things is to take a hard look at the balanced input amplifier; clearly our original design decision to make it a special low-noise type was sound, but what are the options for making it even quieter? Improvements to this stage are going to be value for money, as it will reduce the noise level being fed to all three signal paths.

Now, you might be questioning whether it is worth spending any more money at all on noise reduction, because the circuitry is already extremely quiet, and you might expect the equipment upstream, the preamplifier, mixing console, or whatever, to generate a higher noise level than the crossover. This to my mind is a matter of design philosophy rather than dogged pragmatism. We may suspect the source equipment may be noisier than our circuitry, but that is the responsibility of whoever designed it. If we make our device as good as we can without making obviously uneconomic decisions, then we not only get a virtuous warm glow, but we can be sure we are relatively future-proof in terms of improvements that may be brought about in the source equipment. There is also the point that we will get some excellent numbers for our specification, and they may impress prospective customers.

In Chapter 20 we looked at a series of improvements that could be made to unity-gain balanced input amplifiers to improve their noise performance, all of which, regrettably but perhaps inevitably, required more numerous or more costly opamps. Here we are dealing with a gain of +11.9 dB, so the optimal path of improvement may be somewhat different. The balanced input stage consists of two unity-gain buffers and a low-impedance balanced (differential) amplifier. From here on I shall just refer to them as "the buffers" and "the balanced amplifier". Given the difficulty of measuring the very low noise levels at the outputs of the attenuator network, the noise was measured just before the attenuator and the output noise calculated. In each case the input noise of the AP measuring system has been subtracted in an attempt to obtain accurate numbers.

We proceed as follows: Step 1: Replace the 5532 opamp A3 in the balanced amplifier with the expensive but quiet LM4562. This gives a rather disappointing reduction in the noise from the balanced stage of 1.1 dB, which feeds through to even smaller improvements in the path outputs

Step 2: We deduce that the input buffers A1, A2 must be generating more of the noise than we thought, so we put the 5532 back in the A3 position and replace both input buffers with a LM4562; since the LM4562 is a dual opamp, this makes layout simple. The noise from the balanced stage is now 1.9 dB better than the original design, which is more encouraging, and two of the paths have a 2.2 dB improvement. This sounds impossible but actually results from the three crossover paths dealing with different parts of the audio spectrum. At any rate there is no doubt that the modification of Step 2 is more effective than Step 1.

Step 3: We enhance the balanced amplifier by putting two identical 5532 stages in parallel and averaging their outputs by connecting them together with 10 Ω resistors. The LM4562 input buffers are retained, as their superior load-driving capability is useful for feeding two balanced stages in parallel. This drops the balanced input stage noise to 3.9 dB below the original design. The improvement at the output of the bandwidth definition filter is less at 3.1 dB. The filter makes its own noise contribution, and this is now more significant. Once more there are lesser but still useful reductions in the noise out from the three paths.

Step 4: LM4562s are not cheap, but we decide to splurge on two of them. We return to a single balanced amplifier using an LM4562, and the LM4562 input buffers are retained. This is very little better than Step 3 and significantly more expensive.

Step 5: It appears that abandoning the double balanced amplifier was not a good move, so we bring it back, this time with both of its sections employing an LM4562. This gives a very definite improvement, with the balanced input stage noise now 5.8 dB below the original design. The path output noise measurements also improve, though by a lesser amount, because, like the bandwidth definition filter, the noise generated by the path circuitry has become more significant as the noise from the balanced input stage has been reduced. We have only replaced two opamp packages with the LM4562 (which is about ten times more costly than the 5532), so the cost increase is not great.

The reduction in balanced input amplifier noise could be pursued much further by more extensive paralleling of opamps, as described in Chapter 20. However, is it worthwhile? This is questionable given the noise that is generated downstream of it.

These results are summarised in Table 23.4, together with their effect on the noise output (before the output attenuator) of each of the three crossover paths. You will note a few minor inconsistencies in the last decimal place of some of the figures; this is due partly to the fact that the three paths are handling different parts of the audio spectrum and to some extent due to the difficulties of measuring the very low noise levels accurately. Nonetheless, the overview it gives of the worth of the various modifications is correct.

Looking at the bottom two rows of the table which gives the final noise outputs and the signal/noise ratios for a 1 Vrms output (after the output attenuators), we can see that the HF path is the noisiest by a long way, despite its 6 dB higher internal level. This is because it contains a relatively noisy 3rd-order allpass filter but no lowpass filters to discriminate against noise. Clearly that 6 dB extra level is a very good idea. The MID path is quieter because it does have a lowpass filter which can be placed after its 3rd-order allpass filter. The LF path consists only of a lowpass filter and is consequently the quietest of the lot.

**Table 23.4: Improvements in noise performance.**

| Step | | Bal stage<br>Noise out<br>dB | BW defn<br>Noise out<br>dB | HF<br>Noise out<br>dB | MID<br>Noise out<br>dB | LF<br>Noise out<br>dB |
|---|---|---|---|---|---|---|
| 0 | Original design | 0 dB ref | 0 dB ref | 0 dB ref | 0 dB ref | 0 dB ref |
| 1 | Bal amp A3 = LM4562<br>Buffers A1, A2 = 5532 | 1.1 | 1.2 | 1.0 | 1.4 | 1.8 |
| 2 | Bal amp A3 = 5532<br>Buffers A1, A2 = LM4562 | 1.9 | 1.8 | 2.2 | 1.5 | 2.2 |
| 3 | 2 × Bal amp A3 = 5532<br>Buffers A1, A2 = LM4562 | 3.9 | 3.1 | 3.5 | 1.9 | 2.3 |
| 4 | Bal amp A3 = LM4562<br>Buffers A1, A2 = LM4562 | 4.0 | 3.3 | 2.3 | 2.0 | 2.5 |
| 5 | 2 × Bal amp A3 = LM4562<br>Buffers A1, A2 = LM4562 | 5.8 | 4.7 | 4.7 | 2.0 | 2.6 |
| | | | | dBu | dBu | dBu |
| 5 | Path output noise dBu | | | −115.3 | −120.0 | −125.2 |
| 5 | Path output signal/noise ratio dB | | | 117.5 | 122.2 | 127.4 |

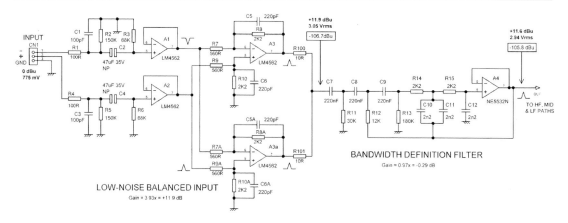

**Figure 23.9: Schematic diagram of the improved balanced input amplifier in its final state, using two paralleled balanced input amplifiers and two LM4562 opamps. The noise output from the balanced input stage has been reduced by 5.8 dB.**

The final version (Step 5) of the input circuitry is shown in Figure 23.9. Note that the bandwidth definition filter has not been modified in any way.

## The Noise Performance: Comparisons With Power Amplifier Noise

The noise performance of this active crossover, especially after the modifications, is, I hope you will agree, rather good. But how does it compare with the noise from a well-designed power amplifier? The EIN of a Blameless power amplifier is −120 dBu [2] if the input signal is applied directly to the power amplifier rather than through a balanced input stage. As we have seen, balanced input stages are relatively noisy and are much noisier than a power amplifier alone.

The figure of −120 dBu is therefore the most demanding case to compare the crossover noise against. What are we trying to achieve? If we accept that the noise from the loudspeakers can go up by 3 dB when we connect the crossover, its output noise must be no more than −120 dBu, at the same level as the power amplifier EIN. If however we are more demanding and will only stand for a noise increase of 1 dB, which is at the limit of audibility, then the crossover output noise must be reduced to −126 dB. We will take that figure as a target and see what can be done.

We will look at the noise output of the MID path, because the ear is most sensitive in this part of the audio spectrum (400 Hz–3 kHz). The noise output after the performance optimisation is −120.0 dBu. This is very quiet indeed for a piece of audio equipment but regrettably 6.0 dB higher than the ambitious target we have just adopted. You may be doubting if the crossover noise performance could be improved that much, and while I am sure it could be done, I am less sure that it could be done economically.

The resistance values in the crossover have already been reduced as much as possible without introducing extra distortion or demanding large and expensive capacitors, thus reducing the effects of opamp current noise and resistor Johnson noise, but this has no effect on opamp voltage noise. The effect of this can only be reduced by using more expensive opamps with a lower input voltage noise density or by paralleling opamp stages. Neither technique promises a radical reduction in noise when practically applied.

Nonetheless, let us conduct a thought experiment. We know that whatever amplifier technology you use, putting two amplifiers in parallel and averaging their outputs (usually by connecting them together with 10 Ω resistors) reduces the noise output by 3 dB; putting four in parallel reduces the noise by 6 dB; and so on. The same applies to putting two identical active crossovers in parallel, so . . . if we stacked up four of them, we could unquestionably get a 6 dB noise reduction and meet our target. This is perhaps not very sensible, but it does prove one thing—it is physically possible to perform the formidable task of meeting our demanding noise target, even if it is hardly economical to do so.

## Conclusion

The description of this active crossover has hopefully demonstrated the techniques and design principles described in this book. The use of elevated internal levels, plus other low-noise techniques, has allowed us to achieve a remarkable noise performance.

## References

[1] Hofer, Bruce Personal Communication, May 2015
[2] Self, Douglas "Audio Power Amplifier Design" Sixth Edn, Newnes, 2013, pp. 150–157 (power amp input noise), ISBN: 978-0-240-52613-3

# *Crossover Design References*

Not all these papers and articles are referenced in the main text of this book. They are of varying value and significance. These references are to crossover design itself; references relating to peripheral matters like filter design are not included. Papers relating solely to digital crossovers have also not been included. The majority of these papers come from the JAES, the *Journal of The Audio Engineering Society*. These can be purchased on-line at www.aes.org/. The cost is currently $5 per paper for members and $20 per paper for non-members.

| | |
|---|---|
| **Early Stuff** | |
| Hilliard & Kimball | "Dividing Networks for Loud Speaker Systems" Academy Research Council Technical Bulletin, March 1936 Reprinted *JAES* 26(11), November 1978, pp. 850–855 |
| **1960–69** | |
| Robert Ashley | "On the Transient Response of Ideal Crossover Networks" *JAES* 10(3), July 1962, pp. 241–244 |
| **1970–79** | |
| Richard H. Small | "Constant Voltage Crossover Network Design" *JAES*, 19(1), January 1971, pp. 12–19 |
| Richard H. Small | "Phase and Delay Distortion in Multiple-Driver Loudspeaker Systems" *JAES*, 19(1), January 1971, p. 56 |
| Richard H. Small | "Crossover Networks and Modulation Distortion" *JAES* 19(1), January 1971, pp. 55–56 |
| Ashley & Kaminsky | "Active and Passive Filters as Loudspeaker Crossover Networks" *JAES* 19(6), June 1971, pp. 494–502 |
| Don Keele | "What's So Sacred About Exponential Horns?" AES Preprint #1038, May 1975 |
| Albert Neville Thiele | "Optimum Passive Loudspeaker Dividing Networks", Proceedings of the IREE (Australia), July 1975, pp. 220–224 |
| Siegfried Linkwitz | "Active Crossover Networks for Non-Coincident Drivers" *JAES*, 24(1), January/February 1976, pp. 2–8 |
| Erik Baekgaard | "A Novel Approach to Linear Phase Loudspeakers Using Passive Crossover Networks" *JAES* 25,(5), 284 1977 (filler-driver) |
| Siegfried Linkwitz | "Passive Crossover Networks for Non-Coincident Drivers" *JAES* 26(3), March 1978, pp. 149–150 |

*(Continued)*

| 1980–89 | |
| --- | --- |
| W. Marshall Leach | "Loudspeaker Driver Phase Response: The Neglected Factor in Crossover Network Design" *JAES*28(6), June 1980, pp. 410–421 |
| P. Garde | "All-Pass Crossover Systems" *JAES*, 28(9), September 1980 |
| Dennis G. Fink | "Time Offset and Crossover Design" *JAES* 28, September 1980 pp. 601–611 |
| Robert Bullock | "Loudspeaker-Crossover Systems: An Optimal Crossover Choice" *JAES* 30(7/8), August 1982, pp. 486–495 |
| Adams & Roe | "Computer-Aided Design of Loudspeaker Crossover Networks" *JAES* 30(7/8), August 1982, pp. 496–503 |
| Richard Greiner | "Tone Burst Testing on Selected Electronic Crossover Networks" *JAES* 30(7/8), August 1982, pp. 522–527 |
| Lipshitz, Pocock & Vanderkooy | "Audibility of Midrange Phase Distortion in Audio Systems" *JAES* 30(9), September 1982, pp. 580–595 |
| Wieslaw R. Woszczyk | "Bessel Filters as Loudspeaker Crossovers" 72nd AES Convention, October 1982, Preprint 1949 |
| S.P. Lipshitz &J. Vanderkooy | "A Family of Linear Phase Crossover Networks of High Slope Derived by Time Delay" *JAES* 31(1/2) January/February 1983, pp. 2–20 |
| Greiner & Schoessow | "Electronic Equalisation of Closed-Box Loudspeakers" *JAES*, 31(3), March 1983, pp. 125–134 |
| Shanefield | Comment on "Audibility of Midrange Phase Distortion in Audio Systems" plus reply by Lipshitz, Pocock & Vanderkooy *JAES* 31(6), June 1983 p. 447 |
| Robert Bullock | "Satisfying Loudspeaker Crossover Constraints with Conventional Networks" *JAES*, 31(7), July/August 1983 p. 489–499 |
| James Moir | Comment on "Audibility of Midrange Phase Distortion in Audio Systems" *JAES* 31(12), December 1983 p. 939 |
| B. J. Sokol | "Practical Subwoofer Design" (crossover filter) *Electronics World* December 1983 p. 41 |
| Robert Bullock | "Passive 3-way Allpass Crossover Networks" *JAES* 32(9), September 1984 p. 626 |
| Joseph D'Appolito | "Active Realization of Multi-Way All-Pass Crossover Systems" 76th AES Convention, October 1984, Preprint 2125 |
| Lipshitz & Vanderkooy | "Phase Linearisation of Crossover Networks Possible by Time Offset & Equalisation" *JAES* 32(12), December 1984, p. 946 |
| S.P. Lipshitz &J. Vanderkooy | "Use of Frequency Overlap and Equalization to Produce High-Slope Linear Phase Loudspeaker Crossover Networks" *JAES* 33(3), March 1985, pp. 114–126 |

| | |
|---|---|
| Robert Bullock | Corrections to "Passive Three-Way All-Pass Crossover Networks" *JAES* 33(5), May 1985, p. 355 |
| Catrysse | "On the Design of Some Feedback Circuits for Loudspeakers" *JAES* 33(6), June 1985, p. 430 |
| L. R. Fincham | "Subjective Importance of Uniform Group Delay at Low Frequencies" *JAES* 33(6), June 1985, p. 436 |
| Recklinghausen | "Low-Frequency Range Extension of Loudspeakers" *JAES* 33(6), June 1985, p. 440 |
| Mitra & Damonte | "Tunable Active Crossover Networks" *JAES* 33(10), October 1985, p. 762 |
| Deer, Bloom & Preis | "Perception of Phase Distortion in All-Pass Filters" *JAES*, 33(10), October 1985, pp. 782–786 |
| de Wit, Kaizer, & Op de Beek | "Numerical Optimization of the Crossover Filters in a Multiway Loudspeaker System" *JAES* 34(3), pp. 115–123; March 1986 |
| Vanderkooy & Lipshitz | "Power Response of Loudspeakers with Non-coincident Drivers–The Influence of Crossover Design" *JAES* 34(4), April 1986, pp. 236–244 |
| Robert Bullock | "A New 3-Way Allpass Crossover Network Design" *JAES* 34(5), May 1986, p. 315 |
| Rudolf Chalupa | "A Subtractive Implementation of Linkwitz-Riley Crossover Design" *JAES* 34(7/8), July/August 1986, p. 556 |
| Lipshitz & Vanderkooy | "In-Phase Crossover Network Design" *JAES* 34(11), November 1986, p. 889 |
| Regalia, Fujii, Mitra, Sanjit, Neuvo | "Active RC Crossover Networks With Adjustable Characteristics" *JAES* 35(1/2), January/February 1987, pp. 24–30 |
| M. Hawksford | "A Family of Circuit Topologies for the Linkwitz-Riley (LR-4) Crossover Alignment C14 82nd AES Convention, March 1987, Preprint 2468 |
| J. D'Appolito | "Active Realization of Multiway All-Pass Crossover Systems" *JAES* 35(4), April 1987, pp. 239–245 |
| Harry Baggen | "Active Phase-Linear Crossover Network" Elektor, September 1987, p. 61 |
| Maffioli & Nicolao | "Another Approach to the Ideal Crossover: The Energy Filler" AES Convention 84 (March 1988) Paper #2642 |
| Ronald Aarts | "A New Method for the Design of Crossover Filters" *JAES* 37(6), June 1989, pp. 445–454 |
| **1990–99** | |
| M. Hawksford | "Asymmetric All-Pass Crossover Alignments" *JAES* 41(3), March 1993, pp. 123–134 |
| Neville Thiele | "Precise Passive Crossover Networks Incorporating Loudspeaker Driver Parameters" *JAES* 45(7/8), July 1997, pp. 585–594 |

*(Continued)*

| Bill Hardman | "Precise Active Crossover" *Electronics World*, August 1999, p. 652 |
| Alex Megann | www.soton.ac.uk/~apm3/diyaudio/Filler drivers. html December 1999 (2nd-and 4th-order crossovers with filler-drivers) |

**2000–09**

| Christhof Heinzerling | "Adaptable Active Speaker System" (subtractive) *Electronics World* February 2000 p. 105 |
| John Watkinson | "Speaker's Corner: Crossover Summation" *Electronics World* September 2000 p. 720 |
| Neville Thiele | "Loudspeaker Crossovers With Notched Responses" *JAES*48(9), September 2000 p784 |
| Neville Thiele | "Passive All-Pass Crossover System of Order 3 (Low Pass) + 5 (High Pass), Incorporating Driver Parameters" *JAES* 50(12), December 2002, pp. 1030–1038 |
| Cochenour, Chai, Rich, & David | "Sensitivity of High-Order Loudspeaker Crossover Networks With All-Pass Response" *JAES* 51(10), October 2003, pp. 898–911 |
| Neville Thiele | "An Active Biquadratic Filter for Equalising Overdamped Loudspeakers" AES Convention Paper 6153, 116th convention, May 2004 |
| Neville Thiele | "Crossover Filter System and Method" US Patent 6,854,005 February 2005 (Assigned to Techstream Pty Ltd, Victoria, AU) |
| R. Christensen | "Active All-Pass Crossover Networks With Equal Resistors and Equal Capacitors" *JAES* 54(1/2), January/February 2006 |
| Neville Thiele | "Implementing Asymmetrical Crossovers" *JAES* 55(10), October 2007, pp. 819–832 |

# US Crossover Patents

These US patents concentrate on both active and passive crossover design, but they also have much information on loudspeaker design in general. This list does not pretend to be complete, but following the references in each patent should give a comprehensive overview of the field. Dates are for publication, not filing.

| | | | |
|---|---|---|---|
| 2 771 518 | Sziklai | Frequency band separation amplifier | November 1956 |
| 2 802 054 | Corney | Sound reproducing apparatus | August 1957 |
| 3 457 370 | Boner | Impedance correcting networks | July 1969 |
| 3 657 480 | Cheng et al | Multi channel audio system with crossover network feeding separate amplifiers for each channel with direct coupling to low-frequency loudspeaker | April 1972 |
| 3 727 004 | Bose | Loudspeaker system | April 1973 |
| 3 814 857 | Thomasen | 2-way loudspeaker system with two tandem-connected high-range speakers | June 1974 |
| 3 838 215 | Haynes Jr | Speakers and crossover circuit | September 1974 |
| 4 015 089 | Ishii et al | Linear phase response multi-way speaker system | March 1977 |
| 4 031 321 | Bakgaard | Loudspeaker systems | June 1977 |
| 4 198 540 | Cizek | Compensated crossover network | April 1980 |
| 4 229 619 | Takahashi | Method and apparatus for driving multi way speakers | October 1980 |
| 4 243 840 | Kates | Loudspeaker system | January 1981 |
| 4 282 402 | Liontonia | Design of crossover network for hifi | August 1981 |
| 4 430 527 | Eberbach | Loudspeaker crossover delay equaliser | February 1984 |
| 4 475 233 | Watkins | Resistively damped loudspeaker system | October 1984 |
| 4 525 857 | Orban | Crossover network & clipper | June 1985 |
| 4 589 135 | Baker | Zero phase shift filtering | May 1986 |
| 4 769 848 | Eberbach | Electroacoustic network delay HF | Sep 1988 |
| 4 771 466 | Modafferi | Multidriver loudspeaker apparatus | Sep 1988 |
| 4 897 879 | Geluk (B&W) | Multi-way loudspeaker system | January 1990 |
| 4 991 221 | Rush | Active speaker system and components therefor | February 1991 |
| 5 129 006 | Hill | Electronic audio signal amplifier | July 1992 |
| 5 185 801 | Meyer | Correction circuit and method for 2-way speaker | February 1993 |
| 5 327 505 | Kim | Multiple output transformers network for sound reproducing system July 1994 | |
| 5 568 560 | Combest | Audio crossover circuit | October 1996 |
| 5 930 374 | Werrbach | Phase coherent crossover | July 1999 |

*(Continued)*

(Continued)

| | | | |
|---|---|---|---|
| 6 115 475 | Alexander | Capacitor-less crossover network for electro-acoustic loudspeakers | Sep 2000 |
| 6 381 334 | Alexander | Series-configured crossover network for electro-acoustic loudspeakers | April 2002 |
| 6 707 919 | Koval | Driver control circuit | March 2004 |
| 6 854 005 | Thiele | Crossover filter system and method | February 2005 |
| 7 085 389 | Modafferi | Infinite slope loudspeaker crossover | August 2006 |
| 7 876 920 | Harvey | Active crossover for use with multi-driver headphones | January 2011 |
| 8 391 535 | Harvey | Active crossover for use with multi-driver in-ear monitors | March 2013 |
| 9 258 646 | Hillman | Self-powered audio speaker having modular components | February 2016 |
| | | | |
| Patent Applications | | | |
| | | | |
| 20120213386 | Knight | Audio crossover system and method | August 2012 |
| 20140219480 | Collins | Phase-Unified Loudspeakers: Parallel Crossovers | August 2014 |

# Crossover and Loudspeaker Articles *in* Wireless World/Electronics World

These patents concentrate on both active and passive crossover design,

## Crossovers

| | | | |
|---|---|---|---|
| Anon | Motional Feedback Loudspeaker | September 1973 | p. 425 |
| D. C. Read | Active Filter Crossover Networks | December 1973 | p. 574 |
| H. D. Harwood | Motional Feedback in Loudspeakers | March 1974 | p. 51 |
| D. C. Read | Active Crossover Networks | November 1974 | p. 443 |
| David Read | More on Active Crossover Networks | April 1981 | p. 41 |
| Jeff Macaulay | 20 Hz Sub-woofer | August 1995 | p. 636 |
| Russel Breden | Roaring Subwoofer | February 1997 | p. 104 |
| Bill Hardman | Precise Active Crossover | August 1999 | p. 652 |
| Christof Heinzerling | Adaptable Active Speaker Design | February 2000 | p. 105 |
| Jeff Macaulay | Tying the Knot—MFB Loudspeaker: Part 1 | September 2002 | p. 26 |
| Jeff Macaulay | Tying the Knot—MFB Loudspeaker: Part 2 | October 2002 | p. 46 |

## Loudspeakers

| | | | |
|---|---|---|---|
| K. A. Exley | Bass Without Big Baffles | April 1951 | p. 132 |
| J Moir | Loudspeaker Diaphragm Control | July 1951 | p. 252 |
| Anon | Loudspeaker Without Diaphragm (Ionophone). | January 1952 | p. 1 |
| Anon | Distortion in Electrostatic Loudspeakers. | February 1956 | p. 54 |
| H D Harwood | Non-linearity of Air in Loudspeaker Cabinets | November 1974 | p. 459 |
| Siegfried Linkwitz | Loudspeaker System Design: Part 1 | May 1978 | p. 52 |
| Siegfried Linkwitz | Loudspeaker System Design: Part 2 | June 1978 | p. 67 |
| Jim Wilkinson | Bookshelf Loudspeaker Mk II | June 1979 | p. 49 |
| J H Buijs | Binaural Recordings & Loudspeakers | November 1982 | p. 39 |
| B. J. Sokol | Practical Subwoofer Design | December 1983 | p. 41 |
| Ian Hegglun | Speaker Feedback | May 1996 | p. 378 |
| Douglas Self | Loudspeaker undercurrents (Otala excess currents) | February 1998 | p. 98 |
| Russell Breeden | Loudspeaker Motional Feedback | November 1998 | p. 955 |
| John Watkinson | Speakers Corner: crossover summation | September 2000 | p. 720 |
| John Watkinson | Speaker's Corner: Voltage or Current Drive? | May 2001 | p. 354 |
| John Watkinson | Speaker's Corner: Motional Feedback | September 2001 | p. 698 |
| Joe Begin (AP) | Measuring Loudspeaker Impedance | May 2012 | p. 20 |

# *Loudspeaker Design References*

These references are only a small sample of the many that are applicable to loudspeaker design. These have been selected because they either relate closely to crossover design, for example in the equalising of diffraction effects, or are of special significance in themselves.

| | |
|---|---|
| **Early Stuff** | |
| Rice & Kellogg | "Notes on the Development of a New Type of Hornless Loudspeaker" AIEE Transactions, September 1925 (The first description of the modern moving-coil direct radiator cone loudspeaker) |
| Muller, Black & Davis | "The Diffraction Produced by Cylindrical and Cubical Obstacles and by Circular and Square Plates" *JAES* 10(1), July 1938, pp.6–13 |
| **1960–69** | |
| H. F. Olson | "Direct Radiator Loudspeaker Enclosures" *JAES* 17(1), January 1969, pp. 22–29 (Classic paper on frequency response anomalies due to diffraction) |
| **1970–79** | |
| Richard H. Small | Direct Radiator Loudspeaker System Analysis *JAES* 20(5), pp. 383–395; June 1972 |
| Richard H. Small | "Closed-Box Loudspeaker Systems Part 1: Analysis" *JAES* 20(10), December 1972, pp. 798–808 |
| Richard H. Small | "Closed-Box Loudspeaker Systems Part 2: Synthesis" *JAES* 21(1), February 1973, pp. 11–18 |
| Richard H. Small | Vented-Box Loudspeaker Systems—Part 1: Small-Signal Analysis *JAES* 21(5), June 1973, pp. 363–372 |
| Richard H. Small | Vented-Box Loudspeaker Systems-Part 2: Large-Signal Analysis *JAES* 21(6), August 1973, pp. 438–444 |
| Richard H. Small | Vented-Box Loudspeaker Systems-Part 3: Synthesis *JAES* 21(7), September 1973, pp. 549–554 |
| Richard H. Small | Vented-Box Loudspeaker Systems-Part 4: Appendices *JAES* 21(8), October 1973, pp. 635–639 |
| Richard H. Small | "Passive-Radiator Loudspeaker Systems Part 1: Analysis" *JAES* 22(8), October 1974, pp. 592–601 |
| Richard H. Small | "Passive-Radiator Loudspeaker Systems, Part 2: Synthesis" *JAES* 22(9) November 1974, pp. 683–689 |

*(Continued)*

(Continued)

| 1980–89 | |
|---|---|
| Bews & Hawksford | "Application of Geometric Theory of Diffraction (GTD) to Diffraction at the Edges of Loudspeaker Baffles" *JAES* 34(10), October 1986, pp. 771–779 |
| Juha Backman | "Computation of Diffraction for Loudspeaker Enclosures" *JAES* 37(5), May 1989, pp. 353–362 |
| James Porter and Earl Geddes | "Loudspeaker Cabinet Edge Diffraction" *JAES* 37(11), November 1989, pp. 908–918 |
| **1990–99** | |
| Ralph E. Gonzalez | "A Dual-Baffle Loudspeaker Enclosure for Balanced Reverberant Response" 91st AES Convention, October 1991, Preprint No. 3203 |
| John Vanderkooy | "A Simple Theory of Cabinet Edge Diffraction" *JAES* 39(12), December 1991, pp. 923–933 |
| Ian Hegglun | "Speaker Feedback" (motional feedback) *Electronics World* May 1996p. 378 |
| Russel Breden | "Roaring Subwoofer" (motional feedback) *Electronics World* February 1997p. 104 |
| J. R. Wright | "Fundamentals of Diffraction" *JAES* Vol. 45(5), May 1997, pp. 347–356 |
| **2000–2009** | |
| John Watkinson | "Speaker's Corner: Voltage or Current Drive?" *Electronics World* May 2001 p. 354 |
| John Watkinson | "Speaker's Corner: Motional Feedback" *Electronics World* September 2001 p. 698 |

# Component Series E3 to E96

| E3 | E6 | E12 | E24 | E96 | E96 | E96 | E96 |
|---|---|---|---|---|---|---|---|
| 100 | 100 | 100 | 100 | 100 | 178 | 316 | 562 |
|  |  |  | 110 | 102 | 182 | 324 | 576 |
|  |  | 120 | 120 | 105 | 187 | 332 | 590 |
|  |  |  | 130 | 107 | 191 | 340 | 604 |
|  | 150 | 150 | 150 | 110 | 196 | 348 | 619 |
|  |  |  | 160 | 113 | 200 | 357 | 634 |
|  |  | 180 | 180 | 115 | 205 | 365 | 649 |
|  |  |  | 200 | 118 | 210 | 374 | 665 |
| 220 | 220 | 220 | 220 | 121 | 215 | 383 | 681 |
|  |  |  | 240 | 124 | 221 | 392 | 698 |
|  |  | 270 | 270 | 127 | 226 | 402 | 715 |
|  |  |  | 300 | 130 | 232 | 412 | 732 |
|  | 330 | 330 | 330 | 133 | 237 | 422 | 750 |
|  |  |  | 360 | 137 | 243 | 432 | 768 |
|  |  | 390 | 390 | 140 | 249 | 442 | 787 |
|  |  |  | 430 | 143 | 255 | 453 | 806 |
| 470 | 470 | 470 | 470 | 147 | 261 | 464 | 825 |
|  |  |  | 510 | 150 | 267 | 475 | 845 |
|  |  | 560 | 560 | 154 | 274 | 487 | 866 |
|  |  |  | 620 | 158 | 280 | 499 | 887 |
|  | 680 | 680 | 680 | 162 | 287 | 511 | 909 |
|  |  |  | 750 | 165 | 294 | 523 | 931 |
|  |  | 820 | 820 | 169 | 301 | 536 | 953 |
|  |  |  | 910 | 174 | 309 | 549 | 976 |
| 1000 | 1000 | 1000 | 1000 |  |  |  | 1000 |

| E24: 6 pairs in 1:2 ratio | | E24: 4 pairs in 1:3 ratio | | E96: 16 pairs in 1:2 ratio | |
|---|---|---|---|---|---|
| 100 | 200 | 100 | 300 | 100 | 200 |
| 110 | 220 | 110 | 330 | 105 | 210 |
| 120 | 240 | 120 | 360 | 113 | 226 |
| 150 | 300 | 130 | 390 | 137 | 274 |
| 180 | 360 | | | 140 | 280 |
| 750 | 1500 | **E24: 2 pairs in 1:4 ratio** | | 147 | 294 |
| | | 300 | 1200 | 158 | 316 |
| | | 750 | 3000 | 162 | 324 |
| | | | | 174 | 348 |
| | | **E24: 6 pairs in 1:5 ratio** | | 187 | 374 |
| | | 150 | 750 | 196 | 392 |
| | | 200 | 1000 | 221 | 442 |
| | | 220 | 1100 | 232 | 464 |
| | | 240 | 1200 | 590 | 1180 |
| | | 300 | 1500 | 732 | 1464 |
| | | 360 | 1800 | 750 | 1500 |
| | | | | No pairs in 1:3 ratio | |
| | | | | No pairs in 1:4 ratio | |
| | | | | E96: 16 pairs in 1:5 ratio | |

# *Index*

Page numbers in *italics* indicate figures and in **bold** indicate tables on the corresponding pages.

1-bandpass notch filters 337
1.0 dB-Chebyshev crossovers *see* Chebyshev crossovers
1 dB-Chebyshev filters 162, *163–165*
3 dB-Chebyshev lowpass filters 162, *165–167*
3-way crossovers 70
6 dB/octave dipole equalisation 388
±15 V supply designing 616–619, *617–618,*
±17 V supply designing **619**, *619–620*
5532 low-noise unity-gain balanced input stage *597*
5532 opamps 623; common-mode distortion 474–477; distortion in shunt-feedback mode 478; internal circuitry 471–572; output loading in shunt-feedback mode 472–474; output stage biasing 477–482; series feedback 472–474, *476*, *575*; with shunt feedback 471–572, *472*; versions 482
5534 opamps 472, 483–484

*Acoustical Engineering* 35
active crossover design 584; balanced line input stage 627–628; balanced outputs 638; bandwidth definition filter 628; capacitor selection 627; crossover programming 638–639; design principles 625; gain structure 626–627; HF signal path 628–632, *629*, **630,**

**632,** *633*; level trim controls 636–638; output attenuators 636–638; resistor selection 627; specification 625–626, *626*
active crossover design, LF path: 400 Hz Linkwitz-Riley lowpass filter 635; no time-delay compensation 636
active crossover design noise analysis: HF path 640–641; input circuitry 639–640; LF path 642
active crossover design noise performance: input circuitry 643–646; power amplifier noise 646–647
active crossovers 5–6; advantages and disadvantages of 9–15; bi-amping and bi-wiring 6–8, *7*; multi-way connectors 22–23; as next step in hi-fi 16; popularising 21–22; system configurations 16–20, *16–20*
active filters 145–146, 425; distortion in 297–298; noise in 327–328
active gain controls 514–516
active servo technology (AST) 548
AD797 599
AD797 opamps 492–494; common-mode distortion 494; series-feedback THD *495*
AD797 voltage-follower **494**
adjustable peak/dip equalisers *398–404*, 398–405
air, sound absorption in 44–46, *45–46*

air-motion transformers 31
Akerberg-Mossberg highpass filter 279–280, *280*
Akerberg-Mossberg lowpass filter 276–279, *277–279*
allpass crossovers 62–63
allpass filters 147; 1st order 356–364, *357–363*; 2nd order 364–369, *365–368*; 3rd order 369–373, *369–373*; cascaded 1st order *363*, 363–364; distortion and noise in 1st order *362*, 362–363; distortion and noise in 2nd order *368*, 368–369; distortion and noise in 3rd order 370–373, *371–373*; in general 355–356, *356*; higher order 373–379, *374–379*, **376**; Tow-Thomas biquad notch and 285–286, *286*
all-pole crossovers 61; *see also* types, crossover
all-stop filters 147
amplifiers, multiple differential 596
amplitude/frequency distribution 507–510
amplitude peaking and Q in filters 150–151, *152*
arbitrary capacitance values capacitors 446–447
arbitrary resistance values resistors 438–439
asymmetric crossovers 62
Atlantic City Convention Hall 33
attenuation, lowpass filters with *236*, 236–237
attenuator 504

audio spectrum 518
automotive audio and active
crossovers 5
automotive audio subwoofers
540–541
aux-fed subwoofers 539–540
auxiliary bass radiators (ABR) 527,
528, 532; loudspeakers 28

Baekgaard, Erik 111
Baggen, Harry 132
Bainter notch filter 337–342, *340,
342*
balanced cables and interference
560–562
balanced circuit 593
balanced connectors 562–563
balanced inputs 559; amplifiers
556; stage, variations on 576
balanced interconnections *559,*
559–560,
balanced outputs *606,* 606–607
balanced signal levels 563
bandpass filters 146; 1-bandpass
notch filter 337; high-Q 334–335,
*335;* multiple-feedback 333–334,
*334;* Tow-Thomas 280–285,
*281–283,* **284–285**
bandpass subwoofers 528–529
bandwidth and noise 328–329,
**328–329**
bandwidth definition filters 628;
highpass 262–267, *262–267;*
lowpass 237
basic electronic balanced input
565–567, **566,** *566*
bass management facilities 535
bass response equalisation
388–389
Baxandall configuration 540
beaming and lobing 2–3, **3**
Bernoulli, Daniel 155
Bessel, Wilhelm 155
Bessel crossovers: 2nd order
79–80, *79–80,* 149; 3rd order
89, *90–91;* 4th order 99, *99;*
cutoff frequencies 148
Bessel filters 148, 155, 158–160,
*158–160;* bandwidth definition

*237–238,* 237–239, **238;** linear-
phase 182–183; with non-equal
resistors 224, *225;* for time
delays 382–383, *382–383;*
transitional 182; up to 8th order
173–174, **174,** *174*
bi-amping 6–8, *7*
bias current, opamps 463–464
Billam Test 298–299
bipolar-input opamps 463
bipolar junction transistor (BJT)
opamps 460, 467
bipolar opamps 598–599
biquad equalisers 407–413,
*408–409, 411–413;* capacitance
multiplication for *414,* 414–415
biquad filters 276; Akerberg-
Mossberg 276–280, *277–280;*
Tow-Thomas 280–288,
*281–283,* **284–285,** *286, 287*
bi-wiring 6–8, *7*
Boctor notch filters 343–344, *344*
Boltzmann's constant 441
brickwall filters 147
bridged-differentiator notch filter
342, *343*
bridged-T equalisers 406–407,
*406–407*
Butterworth crossovers: 2nd
order 71–78, *72–77,* 133–134,
*134–135,* 149; 3rd order 84–86,
*84–86,* 135, *136;* 4th order
93–95, *94–95,* 135, *136–137,*
137; cutoff frequencies 148;
subtractive 133–134, *134–135,*
135, *136, 136–137,* 137;
transitional 182; up to 8th order
*168,* 168–172, **169,** *170–171*
Butterworth filters 149, 151,
153, *153–155,* 207; 3rd order
**248,** *248,* 248–249; 4th order
215–219, **249–252,** *250,*
251–254, *252;* 5th order **229,**
*229,* 229–230, 257–258, **259,**
*259;* 6th order 232–234, *233,*
**233–234;** bandwidth and noise in
328–329, **328–329;** bandwidth
definition *237–238,* 237–239,
**238,** 262–267, *262–267;* defined

gains *245,* 245–246, **246;**
distortion aggravation factor
298–302, *299–300,* **300–302;**
equal resistors **244,** 244–245;
Legendre-Papoulis 187–189,
*188–189,* **189;** with non-equal
resistors 224, *225;* with other
filter characteristics 256, *257,*
**258;** single stage 217–219,
*218–220,* **220,** 232–234, *233,*
**233–234,** *250,* 252–253, **253,**
*253–254;* two stages 251–252,
**252,** *252,* 259, **260,** *260*

C2, DC-blocking capacitor 558
Cabasse La Sphère 39–40
cables, loudspeaker 8–9
Cambridge Audio 840 W power
amplifier 598
capacitance multiplication for
biquad equalisers *414,* 414–415
capacitor-coupled power amplifiers
522
capacitors: arbitrary capacitance
values 446–447; in audio
circuitry 448; distortion and
*320,* 320–321; electrolytic
capacitor non-linearity 454–457;
non-electrolytic capacitor non-
linearity 449–453; non-linearity
448; shortcomings 447–449;
values and tolerances 445–446
carbon composition resistors 436
carbon film resistors 426, 443
cardioid subwoofer arrays 539
cascaded 1st order allpass filters
*363,* 363–364
Cauer filters 179–181, *180–181*
CBT (constant beamwidth
transducer) line arrays 39
ceramic capacitors 449
characteristic frequencies 148
Chebyshev, Pafnuty Lvovich 160
Chebyshev crossovers: 2nd order
80–81, *81–82,* 149; 3rd order
89, *91–92,* 92, *93;* 4th order
99–101, *100–101*
Chebyshev filters 160–162,
*161–167;* 4th order 216–217;

inverse 177–178, *178–179*; linear-phase 183; with other filter characteristics 256, *257*, *258*; transitional 182; up to 8th order 175–176, **175–176**

chip surface-mount (SM) resistors 427

circuit noise and low-impedance design 504

clip-detect indicators 501

clipping 10

combined HF-boost and HF-cut equalisers 398

combined unbalanced and balanced inputs 576–577

common impedance coupling 562

common-mode distortion, opamps 467–468

common-mode input impedances 567

common-mode range, opamps 463

common-mode rejection ratio (CMRR) 507, 559, 563–565, *564*; amplifier component mismatches *571*, 571–573, *573*, *574*; opamp 570; opamp frequency response 569–570, **570**, *570*; opamp gain *568*, 569–570, **569**; theoretical balanced interconnection *564*

common-mode signals 567

component sensitivity in filter design 195–196, **196**, 202, **206**, 206–207

component series 459–460

component series tables E3 to E96 659–660

cone break-up 2

conjugate impedance compensation 12

constant-power crossovers 62–63

constant-voltage crossovers 63

control protection, system design 500–501

coupling 448

crossover design references 649–652, 655

crossover frequency control (lowpass filter) 534

crossover in/out switch 534

crossover output stages 604

crossover programming 638–639

crossovers: absolute phase 57–58; active (*see* active crossovers); beaming and lobing and 2–3, *3*; bi-amping and bi-wiring with 6–8, *7*; design references 649–652; function of 1; linear phase 56–57; matching loudspeakers and 20–21; minimum phase 57; necessity of 1–2; notch (*see* notch crossovers); passive 4, *4*, 9–10; phase perception 58–59; requirements (*see* requirements, crossover); subjectivism and 23; subtractive (*see* subtractive crossovers); target functions 59–60; topology 113–118, *114–118*; types (*see* types, crossover); US patents 653–654

damping factor 533

D'Appolito configuration 33–34

data-acquisition application 583

DC-blocking capacitors 558, 575

DC-offset protection circuitry 522

decoupling capacitors 448, *618*

defined gains: 2nd order Butterworth highpass *245*, 245–246, **246**; 2nd order Butterworth lowpass 207, **208**, *208*

delay lines for subtractive crossovers 379–381, *380–381*

delays 10

design, filter 271, **272**; 1st order highpass *241*, 241–242; 1st order lowpass 197; 2nd order highpass 242–247; 2nd order lowpass 198–209; 3rd order highpass **248**, *248*, 248–249; 3rd order lowpass 209–215; 4th order highpass 251–256, **252–253**, *252*; 4th order lowpass 214–224, *225*; 5th order highpass 252–253, 257–258, **259**, *259*; 5th order lowpass 225–232; Bainter notch

339–342, *340*, *342*; component sensitivity in 195–196, **196**; defined gains 207, **208**, *208*, 245, 245–246, **246**; equal capacitor *204*, 204–206, *205*; equal resistors *244*, 244–245; filter frequency scaling 202–204; multiple-feedback 272–275, *276*; non-equal capacitors 247; non-equal resistors 207, 209, 214–215, *215*; optimisation 209; real 195; unity gain *199*, 199–202, **201**, 243–244; variable-frequency *239*, 239–240, 267–270, *268–269*; *see also* filters; single stage filters; three stages highpass filters; three stages lowpass filters; two stage filters

Dickason, Vance 67–68, 93

dielectric absorption 447–448, **448**

diffraction 39–44, *40–44*

diffraction compensation equalisation 389–390, *390*

DIN connectors 22

dip equalisers with high Q *402–404*, 402–405

dipole subwoofers 529–530

discrete-opamp hybrid stage 594

discrete transistor technology 503

distortion: in active filters 297–298; aggravation factor 297, 298–302, *299–300*, **300–302**, **321**, 321–322; capacitors and *320*, 320–321; Doppler 48; drive unit 47–48; and looking for DAF 303–304, *303–305*; mixed capacitors in low-distortion 2nd order filters *310*, 310–311; modulation 46–47; in multiple-feedback filters **321**, 321–323, *322–323*; in 2nd order highpass filters 308–310, *309*; in 2nd order lowpass filters 305–307, *306–308*; negligible extra 55–56; and noise in 2nd order allpass filters *368*, 368–369; and noise in 3rd order allpass filters

370–373, *371–373*; and noise in 1st order allpass filters *362*, 362–363; in 3rd order lowpass single stage filters 311–312, *311–312*; simulations 318–320, *319*; in 4th order highpass single stage filters 317–318, *318*; in 4th order lowpass single stage filters 314–317, *315–317*; in Tow-Thomas filters 324–327, *325–327*

distortion aggravation factor (DAF) 297, 298–302, *299–300*, **300–302**; looking for 303–304, *303–305*; in multiple-feedback filters **321**, 321–322

distributed peak detection 520–522

domestic electronic crossover standard 506

Doppler distortion 48

drive unit distortion 47–48

drive unit equalisation 387–388

D-type connectors 22–23

dual opamp 599

dual-opamp 5532 460

dual-opamp TL072 460, *469–470*,

dual supplies from a single winding *622*, 622–623

Duddell, William 31

Duelund crossover 113

DuKane 31–32

E24 resistor *429*, *430*

E96 resistor *429*

electrical noise 432

electrolytic capacitors 446; distortion *454*, *455*; non-linearity 454–457

electrolytic coupling capacitors 604

electrolytic-sized decoupling capacitors 615–616

electromagnetic planar loudspeakers 30–31

electronic crossover 520; mode switching in 502

*Electronics World* 121, 132, 655–656

electronic vs transformer balanced inputs 563

electrostatic coupling 561–562

electrostatic interference 468

electrostatic loudspeakers 29–30

*Elektor* 132

elliptical filter crossovers 121–125, *122–124*

elliptical filters 179–181, *180–181*; using Bainter highpass notch 342

equal capacitor, Sallen & Key 2nd order lowpass *204*, 204–206, **205**

equalisation: by adjusting all filter parameters 422; bass response 388–389; with -3 dB/octave slopes 415–419, *416–418*; with -4.5 dB/octave slopes 420, *421*; with -3 dB/octave slopes over limited range 419–420, *419–420*; diffraction compensation 389–390, *390*; drive unit 387–388; by filter frequency offset 421–422; loudspeaker 387–393; need for 385; with non-standard slopes 415–421, *416–421*; outcomes of *386*, 386–387; room interaction correction 390–393, *392*; 6 dB/octave dipole 388

equalisation circuitry 393, 505; biquad 407–415, *408–409*, *411–414*; bridged-T 406–407, *406–407*; combined HF-boost and HF-cut 398; dip equalisers with high Q *402–404*, 402–405; fixed frequency and low Q *398–399*, 398–400; HF-boost and LF-cut 393–395, *393–395*; HF-cut and LF-boost 395–398, *396–397*; parametric *405*, 405–406; variable centre frequency and low Q 400–402, *401*

equal resistors, Sallen & Key 2nd order highpass *244*, 244–245

equivalent input noise (EIN) 584

equivalent series inductance (ESL) 447

equivalent series resistance (ESR) 447,

excess noise 441, 442, *442*

external noise 560

external signal levels 555

Fairchild 5532 data sheet 558

feedback resistor 472

FET-input opamps 464, 558,

fifteen-way D-type connectors 23

fifth order filters: single stage 260; three stages 225–230, *226*, **226–228**, 257–258, *259*, *259*; two stages 228–229, *229*

filler-driver crossovers *111–112*, 111–113

filter cutoff frequencies 148

filter order in signal path 516–518

filter performance 297–298; *see also* distortion; noise

filters 145; active (*see* active filters); 1st order 148, *241*, 241–242; 2nd order 149, 198–209, 242–249; 4th order 215–224, *225*; adding zeros to more complex 176–181, *178–181*; Akerberg-Mossberg highpass 279–280, *280*; Akerberg-Mossberg lowpass 276–279, *277–279*; allpass (*see* allpass filters); all-stop 147; amplitude peaking and Q 150–151, *152*; bandpass (*see* bandpass filters); bandwidth definition *237–238*, 237–239, **238**, 262–264, *262–264*; biquad 276; brickwall 147; characteristics 149–162, *150–167*; elliptical 179–181, *180–181*, 342; Gaussian 184–185, *186–187*; Halpern 193; higher order 149, 167–176, *168*, **169**, *170–171*, **172**, *173–174*, **174–176**; highpass (*see* highpass filters); Laguerre 190; Legendre-Papoulis 187–189, *188–189*, **189**; lesser-known characteristics of **182**, 182–193, *183*, **184**, *185–189*, **189**, *191–192*; linear-phase

**182**, 182–183, *183*, **184**; lowpass (*see* lowpass filters); multiple-feedback 272–275, *276*; notch (*see* notch filters); order of 147; performance 297–298 (*see also* distortion; noise); state-variable (*see* state-variable filters (SVF)); subsonic 263–264, *263–264*; synchronous 190–193, **191**, *191–192*; Tow-Thomas biquad lowpass and bandpass 280–285, *281–283*, **284–285**; transitional 182; ultraspherical 193; variable-frequency highpass 267–270, *268–269*; variable-frequency lowpass *239*, 239–240; *see also* Bessel filters; Butterworth filters; Chebyshev filters; design, filter; highpass filters; Linkwitz-Riley filters; lowpass filters
first order crossovers 63–69, *64–68*; Solen split crossovers 69–70, *69–70*; subtractive 132–133, *132–133*; 3-way 70
first order filters: allpass 356–364, *357–363*; distortion and noise in *362*, 362–363; highpass *241*, 241–242; lowpass 148, 197, *197*, 556–557,
fixed frequency and low Q equalisers 398–399, 398–400
fixed-gain amplifiers 514
flatness of summed amplitude/frequency response on-axis 51
fourth order crossovers 92–107, *107–108*; Bessel 99, *99*; Butterworth 93–95, *94–95*, 135, *136–137*, 137; Gaussian 103–106, *104–105*; Legendre 106–107, *107–108*; linear-phase 101–103, *102–103*; Linkwitz-Riley 95–97, *96–98*; 1.0 dB-Chebyshev 99–101, *100–101*; subtractive 135, *136–137*, 137
fourth order filters: highpass 251–257, **258**, 317–318, *318*; lowpass 215–224, *225*, 314–317,

*315–317*; noise in 331; state-variable 293–295, *294–295*
frequencies: characteristic 148; filter cutoff 148
frequency offsets, determination of 109–110, **110**
frequency response, negligible impairment of 56
frequency scaling, filter 202–204
function of crossovers 1

gain structures 510–513, **511**, 626–627
galvanic isolation 562
ganged potentiometers 581
Gauss, Johann Carl Friedrich 185
Gaussian 4th order crossover 103–106, *104–105*
Gaussian distribution **431**, 431–435, *432*, 437, *437*
Gaussian filters 184–185, *186–187*
Gaussian resistors **433**
ground-cancelling outputs 605–606, *606*
ground voltages 562
group delay **53**, 53–54

Halpern filter 193
Hardman, Bill 121
Hardman crossover 121–124
Heinzerling, Christhof 132
HF-boost and LF-cut equalisers 393–395, *393–395*
HF-cut and LF-boost equalisation 395–398, *396–397*
HF signal path 628–632, *629*, **630**, **632**, *633*
*Hifi News* 32
hi-fi subwoofers 530–531
hi-fi systems 15–16
higher order crossovers 108–109
higher order filters 149, 167–168; allpass 373–379, *374–379*, **376**; Bessel 173–174, **174**, *174*; Butterworth 168, 168–172, **169**, *170–171*; Chebyshev 175–176, **175–176**; Linkwitz-Riley **172**, 172–173, *173*
high-impedance balanced inputs 582

high-level inputs, subwoofers 532–533
high-level outputs, subwoofers 533
highpass filters 146–147, 499; 1st order *241*, 241–242; 2nd order 242–247; 3rd order **248**, *248*, 248–249, 312–314, *314*; 4th order 251–257, **258**; 5th order 257–258, **259**, *259*; 6th order 260–262; Akerberg-Mossberg 279–280, *280*; bandwidth definition 262–267, *262–267*; Butterworth single-stage 252–256, **253**, *253–255*, **256**; components 243; defined gains *245*, 245–246, **246**; distortion in 2nd order 308–310, *309*, 323, *323*; distortion in 3rd order 312–314, *313–314*; equal resistors *244*, 244–245; input impedance 262; Linkwitz-Riley single stage 254–256, *255*, **256**; multiple-feedback 274–275, *274–276*; noise in 329–332; non-equal capacitors 247; with other filter characteristics 256, *257*, **258**; single-stage 249; single-stage 5th order 260; subsonic filters 263–264, *263–264*; three stages 260–262, **261**, *261*; Tow-Thomas biquad 286–288, *287*, 326–327, *326–327*; two stages 3rd order **248**, *248*, 248–249; two stages 4th order 251, **252**, *252*; two stages 5th order 259, **260**, *260*; unity gain 243–244; variable-frequency 267–270, *268–269*
highpass subsonic filter 534
high-Q bandpass filters 334–335, *335*
Hill, Alan 32
home entertainment crossovers 535–536; fixed frequency 536; multiple variable 536; variable frequency 536
home entertainment subwoofers 531–532; crossover frequency

control (lowpass filter) 534; crossover in/out switch 534; high-level inputs 532–533; high-level outputs 533; highpass subsonic filter 534; level control 533; LFE input 533; low-level inputs (balanced) 532; low-level inputs (unbalanced) 532; mono summing 533; phase switch (normal/inverted) 534; power amplifiers for 536–537; signal activation out of standby 535; variable phase control 535
horn-loaded subwoofers 530
horn loudspeakers 29

impedance-balanced output 603
in-amps 583
IN+ and IN− 559
inductance 9, 11, 12
input amplifier functions 556
input buffers 581–582, 596
input impedance: highpass 262; lowpass 234
input level controls, system design 499
input overvoltage protection *592*, 592–593
instrumentation amplifier *583*, 583–584; applications 584–585; at unity gain 588–590, *589–590*; with 4x gain 585–588, ***585–588***
integration, subwoofer 537–538
intermodulation distortion 500
internal signal levels 555–556, **556**
Inverse Chebyshev filters 177–178, *178–179*
inverting ground-canceling output 605–606, *606*
Ionophone speakers 32
isobaric subwoofers 529

J-arrays 34, *35*, 37
JFET-input opamps 460
Johnson noise 441–442, 467, 469, 504, 515, 557, 566, 600

Key, E. L. 198
Klein, Siegfried 31
Klipsch, Paul 32

Laguerre filters 190
lapped screen 561
Legendre crossover 106–107, *107–108*
Legendre-Papoulis filters 187–189, *188–189*, **189**
level control, subwoofers 533
level trim controls 636–638
LFE input 533
LF non-linearity 611, *612*
LF path: 400 Hz Linkwitz-Riley lowpass filter 635; no time-delay compensation 636
linear phase crossovers 56–57; 4th order 101–103, *102–103*
linear-phase filters **182**, 182–183, *183*, **184**
line arrays: amplitude tapering 37, *38*; CBT (constant beamwidth transducer) 39; frequency tapering 37–39, *38*
line inputs and outputs: balanced cables and interference 560–562; balanced connectors 562–563; balanced input stage, variations on 576; balanced interconnections 559, *559*, 560; balanced signal levels 563; basic electronic balanced input 565–567, **566**, *566*; combined unbalanced and balanced inputs 576–577; common-mode rejection ratio (CMRR) 563–565, *564*; electronic vs transformer balanced inputs 563; external signal levels 555; high input impedance balanced inputs *582*, 582–583; input amplifier functions 556; input overvoltage protection *592*, 592–593; instrumentation amplifier *583*, 583–584; instrumentation amplifier applications 584–585; instrumentation amplifier at unity gain 588–590, ***589***, ***590***;

instrumentation amplifier with 4x gain 585–588, ***585–588***; internal signal levels 555–556, **556**; low-noise balanced inputs 593–598, *595*, *597*; noise and balanced inputs 593; practical balanced input 573–576, *575*, **576**; self variable-gain balanced input *581*, 581–582; Superbal Input 577–578, *578*; switched-gain balanced inputs *578*, 578–580; transformer balanced inputs 590–592, *591*; ultra-low-noise balanced inputs 598–600, **599**, **600**, **600–601**; unbalanced inputs 556–558, *557*; variable-gain balanced inputs *580*, 580–581
line isolating transformer 607
line or area arrays 539
line outputs: balanced outputs *606*, 606–607; ground-cancelling outputs 605–606, *606*; output transformer frequency response *608*, 608–609; transformer balanced outputs *607*, 607–608; transformer distortion 609–610, *610*; reducing 611–614; unbalanced outputs 603–604, *604*; zero-impedance outputs 604–605, *605*
Linkwitz, Siegfried 59
Linkwitz-Riley 4th-order crossover *626*, 629
Linkwitz-Riley crossovers: cutoff frequencies 148; 2nd order *78*, 78–79; 3rd order 86–87, *87–89*; 4th order 95–97, *96–98*
Linkwitz-Riley filters 148, 153, 155, *156–157*, 513, 516; 4th order highpass 254–256, *255*, **256**; loading effects 234–235, *235*; with other filter characteristics 256, *257*, **258**; single stage *221*, 221–224, **222**, *223*, **224**, 254–256, *255*, **256**; up to 8th order **172**, 172–173, *173*
LM4562 *491*, 598–599

LM4562 opamps 485–486; common-mode distortion 486–489

LME49990 opamps 489–491; common-mode distortion 491–492; in a voltage-follower circuit *493*

loading effects 234–235, *235*

lobing and beaming 2–3, **3**; MTM tweeter-mid configurations for 33–34

*Loudspeaker Design Cookbook* 68

loudspeakers: air-motion transformers 31; in automotive audio 5; auxiliary bass radiator (ABR) 28; basic arrangements 25, *25*; beaming and lobing 2–3, **3**; bi-amping and bi-wiring 6–8, *7*; cables for 8–9; CBT (constant beamwidth transducer) line arrays 39; design references 659–660; diffraction 39–44, *40–44*; disadvantages of active crossovers and 14–15; Doppler distortion 48; drive unit distortion 47–48; electromagnetic planar 30–31; electrostatic 29–30; equalisation of 387–393; horn 29; line array amplitude tapering 37, *38*; line array frequency tapering 37–39, *38*; matching crossovers and 20–21; modulation distortion 46–47; MTM tweeter-mid configurations 33–34; passive crossovers 4, *4*; plasma arc 31–32; reflex (ported) 27–28; ribbon 30; rotary woofer 32–33; sealed-box *26*, 26–27; sound absorption in air and 44–46, *45–46*; transmission line 28–29; vertical line arrays 34–39, *35–36, 38*

loudspeakers design references 457–458, 465–466, 655

low-frequency effects (LFE) channel 535

low-impedance passive output attenuator *519*

low-level inputs (balanced) 532

low-level inputs (unbalanced) 532

low-noise balanced inputs 593–598, **595**, *597*; card 598; in real life 598

low-noise preamplifier 582

lowpass filters 146, 198; 1st order 197; 2nd order 198–209, 305–307, *306–308*; 3rd order 209–215, 311–312, *311–312*; 4th order 215–224, *225*; 5th order 225–232; 6th order 230–234; Akerberg-Mossberg 276–279, *277–279*; with attenuation *236*, 236–237; bandwidth definition *237–238*, 237–239, **238**; components 199; component sensitivity 202, **206**, 206–207; defined gains 207, **208**, *208*; distortion in 2nd order 305–307, *306–308*, 322, *322–323*, 324–326, *325*; distortion in 3rd order 311–312, *311–312*; equal capacitor *204*, 204–206, **205**; filter components 199; input impedance 234; loading effects 234–235, *235*; multiple-feedback 273, *272*, 274–275, *274–276*; noise in 329–332; non-equal resistors 207, 209, 214–215, *215*, 224; optimisation 209; single stage 210–214, **211**, *212–213*, **213–214**; single stage Butterworth 217–219, *218–220*, **221**, 232–234, *233*, **233–234**; single stage Linkwitz-Riley *221*, 221–224, **222**, *223*, **224**; single stage 5th order **229**, *229*, 229–230; single stage 6th order 232–234, *233*, **233–234**; single stage with non-equal resistors 224, *225*; single stage with other filter characteristics 224; three stage 6th order 230–232, *231*, **231–232**; three stages 5th order 225–228, *226*, **226–228**; for time delays 382–383, *382–383*; Tow-Thomas biquad

280–285, *281–283*, **284–285**, 324–326, *325*; two stage 3rd order *209*, 209–210, **210**; two stages 5th order 228–229, *229*; two stage 4th order 215–217, *216*, **216–217**; unity gain *199*, 199–202, **201**; variable-frequency *239*, 239–240

low Q equalisers *398–399*, 398–400

magnetic coupling 562

magnetic-planar loudspeakers 30–31

metal electrode face-bonded (MELF) surface-mount resistors 427

metal film resistors 426

Metallica 510

metal oxide resistors 426

metering 501

MID/HF crossover frequency 502

MID/HLF/MID crossover frequency 502

Miller, Ray 155

Mini-DIN connectors 22

minimum phase crossovers 57

mixed capacitors in low-distortion 2nd order filters *310*, 310–311

mixed feedback 612

mode switching in electronic crossover 502

modulation distortion 46–47

Moir, James 59

monopole radiators 528

mono summing 533

Monte Carlo method 440–441

motional feedback 13

motional feedback (MFB) loudspeakers 543–544, *544*; acceleration 548–551; history 544; position 544–545; velocity 545–548

motional feedback of velocity *546*

MTM tweeter-mid configurations 33–34

multi-band distortion 6

multiple-feedback filters (MFB) 272–273; 2nd order highpass

274; 2nd order lowpass 273, *273*; 3rd order 274–275, *274–276*; bandpass 333–334, *334*; distortion in **321**, 321–323, *322–323*; noise in 331–332

multi-point bipolar peak detector *521*

multi-way connectors 22–23

mute switches 519–520

mutual shutdown circuitry *623*, 623–624

NE5532 opamps 471

necessity of crossovers 1–2

Neville Thiele Method™ (NTM) crossovers 125–127, *126–129*

New Order, the 210, 227, 248, 342, 343

nickel-chromium (Ni-Cr) film 427

nine-way D-type connectors 22–23

noise: in active filters 327–328; bandwidth and 328–329, **328–329**; and distortion in 1st order allpass filters *362*, 362–363; and distortion in 2nd order allpass filters *368*, 368–369; and distortion in 3rd order allpass filters 370–373, *371–373*; headroom, and internal levels 503–504; index 442; in 2nd order filters 329–330, 331–332; opamps 460–461; in 3rd order filters 330–331; in 4th order filters 331; in Tow-Thomas filters 332

noise, negligible extra 54–55

noise analysis: HF path 640–641; input circuitry 639–640; LF path 642

noise and balanced inputs 593

noise figure (NF) 593

noise gain 513–514, *514*

noise performance *512*; input circuitry 643–646; power amplifier noise 646–647

non-all-pole crossovers 61

non-electrolytic capacitor: non-linearity 449–453; tolerance 446

non-equal capacitors 247

non-equal resistors 207, 209, 214–215, *215*, 224, *225*

non-inverting ground-canceling output 605–606, *606*

non-linear junction capacitance 592

non-polarised DC-blocking capacitor 603

non-standard slopes, equalisation with 415–421, *416–421*

normal distribution *see* Gaussian distribution

notch crossovers: elliptical filter 121–125, *122–124*; Neville Thiele Method™ (NTM) 125–127, *126–129*

notch filters 146–147, 335–336, *336*; Bainter 337–342, *340*, *342*; Boctor 343–344, *344*; bridged-differentiator 342, *343*; 1-bandpass notch filter 337; other 345; simulating 345; Tow-Thomas biquad 285–286, *286*; Twin-T 336–337, *337*

ohmic resistance 607–608

Ohm's acoustic law 58

Ohm's law 427, 443

Olson tests 39–44

OP27 496

OP27 opamps 494–497

opamp: common-mode rejection ratio (CMRR) 570; frequency response 569–570, **569**, *570*; gain *568*, 568–569, **569**

opamps 459; 5532: common-mode distortion 474–477; output loading in shunt-feedback mode 472–474; output stage biasing 477–482; series feedback 472–474; with shunt feedback 471–572; versions 482; 5534 472, 483–484; AD797 492–494; LM4562 485–486; common-mode distortion 486–489; LME49990 489–491; common-mode distortion 491–492; NE5532 471; OP27 494–497; price per package *465*; properties: bias current

463–464; common-mode distortion 467–468; common-mode range 463; cost 464–465; distortion due to loading 466–467; input offset voltage 463; internal distortion 465–466; noise 460–461; slew rate 462; selection 497–498; surveyed 468; test circuits *468*; TL072 *469*, 469–470, *470*; types 460

opamp supply rail voltages 615–616, **616**

opamp voltage noise 513

order of filters 147

output attenuators 506–507, 636–638

output level controls 518–519; system design 500

output mute switches, system design 500

output phase-reverse switches, system design 500

output transformer frequency response *608*, 608–609

parametric equalisers *405*, 405–406

parasitic tweeters 2

passive components for active crossovers, capacitors: arbitrary capacitance values 446–447; electrolytic capacitor non-linearity 454–457; non-electrolytic capacitor non-linearity 449–453; shortcomings 447–449; values and tolerances 445–446

passive components for active crossovers, resistors 425–428; arbitrary resistance values 438–439; combinations 439–441; Gaussian distribution 431–435; Johnson and excess noise 441–442; non-linearity 443–445; surface-mount resistors 427–428; through-hole resistors 426–427; uniform distribution 437–438; value distributions 435–436; values and tolerances 428–430 passive components for active crossovers component values 425

passive crossovers 4, *4*, 9–10

passive preamplifiers 557

patents, crossover 653–654

peak-detect indicators 501

peak-to-peak ripple amplitude 617

perception, phase 58–59

performance, filter 297–298; *see also* distortion; noise

phase inversion 566

phase-invert switches 520

phase perception 58–59

phase response 53

phase switch (normal/inverted) 534

phono connectors 562

plasma arc loudspeakers 31–32

polar response 52–53

polyester capacitors 451

polyethylene terephthalate 449

polypropylene capacitors 447, 452

potentiometers 581

power amplifier 522, 536–537

power supply design: ±15 V supply designing 616–619, **617**, **618**; ±17 V supply designing **619**, 619–620; for discrete circuitry 624, *624*; dual supplies from a single winding *622*, 622–623; mutual shutdown circuitry *623*, 623–624; opamp supply rail voltages 615–616, **616**; ripple performance *621*, **621**, 621–622; variable-voltage regulators *620*, 620–621, **621**, *621*

power supply rejection ratios (PSRR) 615

practical balanced input 573–576, **575**, *576*

preamplifier-crossover-power amplifier chain *506*, *507*

PSRR 615

*Q*: dip equalisers with high *Q* *402–404*, 402–405; filter amplitude peaking and 150–151, *152*; high-*Q* bandpass filters 334–335, *335*; sealed-box

loudspeakers *26*, 26–27; variable centre frequency and low *Q* equalisers 400–402, *401*

Quad ESL-63 29–30

quasi-floating output 607

raised internal levels 504–506

recommended supply voltage 615

reflex (ported) loudspeakers 27–28

reflex (ported) subwoofers 527

relay output muting 501

reliability, negligible impairment of 56

requirements, crossover: acceptable group delay behaviour **53**, 53–54; acceptable phase response 53; acceptable polar response 52–53; adequate flatness of summed amplitude/frequency response on-axis 51; general 51–54; negligible extra distortion 55–56; negligible extra noise 54–55; negligible impairment of frequency response 56; negligible impairment of reliability 56; negligible impairment of system headroom 55; sufficiently steep roll-off slopes between the filter outputs 51–52

resistors 425–428, 594; arbitrary resistance values 438–439; chip surface-mount (SM) resistors 427; combinations *439*, 439–441; E24 resistor *429*, *430*; E96 resistor *429*; excess noise 441, 442, *442*; Gaussian distribution 431–435; Johnson and excess noise 441–442; MELF surface-mount resistors 427; non-linearity 443–445, *444*; surface-mount resistors 427–428; thick-film SM resistors 427; thin-film resistors 427; through-hole resistors 426–427; types **426**; uniform distribution 437–438; value distributions 435–436; values and tolerances 428–430; voltage coefficients **443**

reverse-biased regulators 618

ribbon loudspeakers 30

ripple performance *621*, **621**, 621–622

roll-off slopes 51–52

room interaction correction 390–393, *392*

rotary woofers 32–33

ruthenium oxide (RuO$_2$) 427

Sallen, R. P. 198

Sallen & Key highpass filters *see* highpass filters

Sallen & Key lowpass filters *see* lowpass filters

sealed-box loudspeakers *26*, 26–27

sealed-box (infinite baffle) subwoofers 526–527

second order crossovers 70–81, *82–83*; 1.0 dB-Chebyshev 80–81, *81–82*; Bessel 79–80, *79–80*; Butterworth 71–78, *72–77*, 149; Linkwitz-Riley *78*, 78–79;

second order filters: allpass 364–369, *365–368*; distortion in multiple-feedback **321**, 321–323, *322–323*; distortion in 2nd order 305–310, *306–309*; distortion in Tow-Thomas 324–326, *325*; highpass 242–247, 274, 308–310, *309*; lowpass 149, 198–209, 273, *273*; mixed capacitors in low-distortion *310*, 310–311; noise in 329–330, 331–332; state-variable 292–293, *293*

security cover 501

security screws 501

Self variable-gain balanced input *581*, 581–582

series resistance 9

servospeaker method *545*

shunt-feedback amplifier 472, 477

shunt-feedback circuit 515; 5532 distortion *473*, **474**

shunt-feedback mode 493

shunt protection diodes 618

signal activation out of standby 535

signal-present indicators 501
signal summation *352*, 352–353, **353**
signal-to-noise (S/N) ratio 499, 503, 508, 515, **516**
simulations: distortion 318–320, *319*; notch filter 345
single-amplifier biquad (SAB) filters 276
single-opamp 5534 460
single stage filters: 4th order 314–318, *315–318*; 5th order **229**, *229*, 229–230, 260; 6th order 232–234, *233*, **233–234**, 262; Butterworth 217–219, *218–220*, **220**, 232–234, *233*, **233–234**, 252–253, **253**, *253–254*; distortion in (*see* distortion); Linkwitz-Riley *221*, 221–224, **222**, *223*, **224**, 254–256, *255*, **256**; noise in 330–331; with non-equal resistors 224, *225*; with other filter characteristics 224, 256, *257*, **258**; 3rd order 249, 311–312, *311–312*
sixth order filters: single stage 232–234, *233*, **233–234**, 262; three stages 230–232, *231*, **231–232**, 260–262, **261**, *261*
slew rate 462; limiting distortion 466
Small, Dick 131
small-signal audio equipment 425
Solen split crossovers 69–70, *69–70*
sound absorption in air 44–46, *45–46*
sound pressure level (SPL) 36, 44–46, *45–46*; allpass crossovers and 62
sound-reinforcement crossovers 501
sound-reinforcement subwoofers 538–539; cardioid subwoofer arrays 539; line or area arrays 539
Sowter 3292 output transformer *608*, 609–610, *610*

speaker protection 500
spiral arrays 37
spiral-format film resistors 436
St. Paul's cathedral 36, 37, 38–39
state-variable filters (SVF) 288–292, *289*, **290**, *291–292*, 502; 2nd order, variable-frequency 292–293, *293*; 4th order, variable-frequency 293–295, *294–295*; other orders of 295
statistical noise 432
stereo hi-fi balanced interconnections 562
straight-line arrays 34, *35*, 37
subjectivism in crossover design 23
subsonic filters 263–264, *263–264*, 499–500; combined ultrasonic and 264–267, *265–267*
subtractive crossovers 131–132; 1st order 132–133, *132–133*; 2nd order Butterworth 133–134, *134–135*; 3rd order Butterworth 135, *136*; 4th order Butterworth 135, *136–137*, 137; delay lines for 379–381, *380–381*; performing the subtraction for 141–142, *142–143*; with time delays 137–141, *138–141*
subwoofer drive units 530
subwoofers: applications 525; automotive audio subwoofers 540–541; aux-fed subwoofers 539–540; auxiliary bass radiator (ABR) subwoofers 527; bandpass subwoofers 528–529; dipole subwoofers 529–530; enclosure *526*; hi-fi subwoofers 530–531; horn-loaded subwoofers 530; integration 537–538; isobaric subwoofers 529; reflex (ported) subwoofers 527; sealed-box (infinite baffle) subwoofers 526–527; subwoofer drive units 530; technologies 525–526, *526*; transmission line subwoofers 528
subwoofers, home entertainment 531–532; crossover frequency

control (lowpass filter) 534; crossover in/out switch 534; high-level inputs 532–533; high-level outputs 533; highpass subsonic filter 534; level control 533; LFE input 533; low-level inputs (balanced) 532; low-level inputs (unbalanced) 532; mono summing 533; phase switch (normal/inverted) 534; power amplifiers for 536–537; signal activation out of standby 535; variable phase control 535
subwoofers, home entertainment crossovers 535–536; fixed frequency 536; multiple variable 536; variable frequency 536
subwoofers, sound-reinforcement 538–539; cardioid subwoofer arrays 539; line or area arrays 539
sufficiently steep roll-off slopes between the filter outputs 51–52
Superbal Input 577–578, *578*
supply filtering 448
surface-mount resistors 427–428, *428*
switchable crossover modes 501–502
switched-gain balanced inputs *578*, 578–580
switched-gain inputs 579–580
Sydney Town Hall, Australia 33
symmetric crossovers 62
synchronous filters 190–193, **191**, *191–192*
system design, active crossover: active gain controls 514–516; amplitude/frequency distribution 507–510; circuit noise and low-impedance design 504; distributed peak detection 520–522; filter order in signal path 516–518; gain structures 510–513; mute switches 519–520; noise, headroom, and internal levels 503–504; noise gain 513–514; output attenuator 506–507; output

level controls 518–519; phase-invert switches 520; power amplifier considerations 522; raised internal levels 504–506; switchable crossover modes 501–502

system design, active crossover, features of 499; control protection 500–501; input level controls 499; output level trims 500; output mute switches 500; output phase-reverse switches 500; subsonic filters 499–500; ultrasonic filters 500; usually absent, metering 501; usually absent, relay output muting 501

system headroom, negligible impairment of 55

target functions 59–60

Thevenin-Norton transformation 442

thick-film resistors 426

thick-film SM resistors 427

Thiele, Neville 125

Thigpen, Bruce 32

thin-film resistors 427

thin-film surface-mount resistors *445*

third-harmonic distortion *450*

third order crossovers 83–92, *84–93*; 1.0 dB-Chebyshev 89, *91–92*, 92, *93*; Bessel 89, *90–91*; Butterworth 84–86, *84–86*, 135, *136*; Linkwitz-Riley 86–87, *87–89*

third order filters: allpass 369–373, *369–373*; distortion in 311–314, *311–314*; highpass **248**, *248*, 248–249, 274–275, *274–276*; lowpass 209–215, 274–275, *274–276*; multiple-feedback 274–275, *274–276*; noise in 330–331

Thomson, W. E. 155

three stages highpass filters: 5th order 257–258, **259**, *259*; 6th order 260–262, **261**, *261*

three stages lowpass filters: 5th order 225–228, *226*, **226–228**;

6th order 230–232, *231*, **231–232**

through-hole resistors 426–427, 430, 443

time-delay filters 631–632; 1st order allpass filters 356–361, *357–361*; 2nd order allpass filters 364–369, *365–368*; 3rd order allpass filters 369–373, *369–373*; allpass filters in general 355–356, *356*; cascaded 1st order allpass filters *363*, 363–364; filter technology 355; higher order allpass filters 373–379, *374–379*, **376**; physical methods of 353–354, *354*; requirement for delay compensation and 347–348, *348*; sample crossover and delay filter specification 355; signal summation *352*, 352–353, **353**

time delays: calculating 349–352, **350–351**, *351*; lowpass filters for 382–383, *382–383*; for subtractive crossovers 379–381, *380–381*; subtractive crossovers with 137–141, *138–141*; variable allpass 381–382

topology, crossover 113–118, *114–118*

Tow-Thomas biquad highpass filters 286–288, *287*; distortion in 326–327, *326–327*; noise in 332

Tow-Thomas biquad lowpass and bandpass filters 280–285, *281–283*, **284–285**; distortion in 324–326, *325*

Tow-Thomas biquad notch and allpass responses 285–286, *286*

transfer distortion 465

transformer 591–592, 594

transformer balanced inputs 590–592, *591*

transformer balanced outputs *607*, 607–608

transformer coupling 592

transformer distortion 609–610, *610*; reducing 611–614

transitional filters 182

transmission line loudspeakers 28–29

transmission line subwoofers 528

Trinity Church, Wall Street, New York 32–33

tri-wiring 6–7, 21

TRW-17 33

tweeters: 1st order crossovers and 63–64; active crossovers and 10; necessity of crossovers due to small size of 2; parasitic 2; ribbon loudspeakers and 30

Twin-T notch filter 336–337, *337*

two stage filters: 3rd order *209*, 209–210, **210**, **248**, *248*, 248–249; 4th order 215–217, *216*, **216–217**, 251, **252**, *252*; 5th order 228–229, *229*, 259, **260**, *260*

types, crossover: 1.0 dB-Chebyshev 80–81, *81–82*, 89, *91–92*, 92, *93*, 99–101, *100–101*; 1st order 63–70, *64–70*; 2nd order 70–81, *82–83*; 3rd order 83–92, *84–93*; 3 way 70; 4th order 92–107, *107–108*; allpass and constant-power 62–63; all-pole and non-all-pole 61; Bessel 79–80, *79–80*, 89, *90–91*, 99, *99*; Butterworth 71–78, *72–77*, 84–86, *84–86*, 93–95, *94–95*; constant-voltage 63; crossover topology and 113–118, *114–118*; determining frequency offsets for 109–110, **110**; Duelund 113; filler-driver *111–112*, 111–113; Gaussian 103–106, *104–105*; higher order 108–109; Legendre 106–107, *107–108*; linear-phase 56–57, 101–103, *102–103*; Linkwitz-Riley *78*, 78–79, 86–87, *87–89*, 95–97, *96–98*; Solen split 69–70, *69–70*; symmetric and asymmetric 62;

ultra-low-noise balanced inputs 598–600, **599**, **600–601**

ultrasonic and subsonic combined filters 264–267, *265–267*, 628
ultrasonic filters 500
ultraspherical filters 193
unbalanced inputs 556–558, *557*, 566; circuit resistances 574
unbalanced outputs 603–604, *604*
unity gain: 2nd order highpass 243–244; 2nd order lowpass *199*, 199–202, **201**
US Crossover Patents 653

variable allpass time delays 381–382
variable centre frequency and low Q 400–402, *401*

variable-frequency filters: highpass 267–270, *268–269*; lowpass *239*, 239–240; other orders of state-variable 295; state-variable 2nd order 292–293, *293*; state-variable 4th order 293–295, *294–295*
variable-gain balanced inputs *580*, 580–581
variable output voltage IC regulators 620, *620*
variable phase control 535
variable-voltage regulators *620*, 620–621, **621**, *621*
vertical line arrays 34–39, *35–36*, *38*
voice coil heating 11

voltage coefficient 426, **426**
voltage-doublers 622
voltage-follower configuration 491
voltage noise density 600
von Helmholtz, Hermann 58

white noise 441
*Wireless World* 36, *36*, 655–656
wirewound resistors 426, 443
woofers, rotary 32–33

XLR connectors 22, 560, 562

zero-impedance outputs 604–605, *605*, *607*, 611, *613*
zero source resistance trace 474
Zobel network 563